Discovering Computers 2001

BRIEF EDITION

Concepts for a Connected World

Web and CNN Turner Learning Enhanced

Web and CNN Turner Le@rning Enhanced

Gary B. Shelly
Thomas J. Cashman
Misty E. Vermaat
Tim J. Walker

Contributing Authors
Robert A. Safdie
Dale E. Torri-Safdie

COURSE TECHNOLOGY
ONE MAIN STREET
CAMBRIDGE MA 02142
Thomson Learning™

Australia • Canada • Denmark • Japan • Mexico • New Zealand • Philippines
Puerto Rico • Singapore • South Africa • Spain • United Kingdom • United States

Printed in the United States of America

For more information, contact:

Course Technology
One Main Street
Cambridge, Massachusetts 02142, USA

or find us on the World Wide Web at: www.course.com

Asia (excluding Japan)
Thomson Learning
60 Albert Street, #15-01
Albert Complex
Singapore 189969

Japan
Thomson Learning
Palaceside Building 5F
1-1-1 Hitotsubashi, Chiyoda-ku
Tokyo 100 0003 Japan

Australia/New Zealand
Nelson/Thomson Learning
102 Dodds Street
South Melbourne, Victoria 3205
Australia

Latin America
Thomson Learning
Seneca, 53
Colonia Polanco
11560 Mexico D.F. Mexico

South Africa
Thomson Learning
Zonnebloem Building,
Constantia Square
526 Sixteenth Road
P.O. Box 2459
Halfway House, 1685
South Africa

Canada
Nelson/Thomson Learning
1120 Birchmount Road
Scarborough, Ontario
Canada M1K 5G4

UK/Europe/Middle East
Thomson Learning
Berkshire House
168-173 High Holborn
London, WC1V 7AA United Kingdom

Spain
Thomson Learning
Calle Magallanes, 25
28015-MADRID
ESPANA

ISBN 0-7895-5938-2

1 2 3 4 5 6 7 8 9 10 BC 04 03 02 01 00

CONTENTS

DISCOVERING COMPUTERS 2001

Concepts for a Connected World, Web and CNN Enhanced
Brief Edition

PREFACE ix

CHAPTER 1
INTRODUCTION TO USING COMPUTERS

OBJECTIVES 1.1
COMPUTER LITERACY 1.2
WHAT IS A COMPUTER AND WHAT DOES IT DO? 1.3
THE COMPONENTS OF A COMPUTER 1.5
Input Devices 1.5
Output Devices 1.6
System Unit 1.6
Storage Devices 1.7
Communications Devices 1.8
WHY IS A COMPUTER SO POWERFUL? 1.8
Speed 1.8
Reliability 1.8
Accuracy 1.8
Storage 1.8
Communications 1.8
COMPUTER SOFTWARE 1.9
System Software 1.10
Application Software 1.12
Software Development 1.13
NETWORKS AND THE INTERNET 1.14
CATEGORIES OF COMPUTERS 1.18
PERSONAL COMPUTERS 1.19
Desktop Computers 1.20
Portable Computers 1.22
MINICOMPUTERS 1.24
MAINFRAME COMPUTERS 1.24
SUPERCOMPUTERS 1.24
EXAMPLES OF COMPUTER USAGE 1.25
Home User 1.26
Small Business User 1.28
Mobile User 1.30
Large Business User 1.32
Power User 1.34
COMPUTER USER AS A PROVIDER OF INFORMATION 1.36
TECHNOLOGY TRAILBLAZER — BILL GATES 1.38
COMPANY ON THE CUTTING EDGE — MICROSOFT 1.39
IN BRIEF 1.40
KEY TERMS 1.42
AT THE MOVIES 1.43
CHECKPOINT 1.44
AT ISSUE 1.45
MY WEB PAGE 1.46
HANDS ON 1.47
NET STUFF 1.48

SPECIAL FEATURE 1.50
Timeline 2001
Milestones in Computer History

CHAPTER 2
APPLICATION SOFTWARE AND THE WORLD WIDE WEB

OBJECTIVES 2.1
APPLICATION SOFTWARE 2.2
The Role of the Operating System 2.2
The Role of the User Interface 2.3
Starting a Software Application 2.4
PRODUCTIVITY SOFTWARE 2.7
Word Processing Software 2.8
Spreadsheet Software 2.12
Database Software 2.15
Presentation Graphics Software 2.18
Personal Information Managers 2.21
Software Suite 2.21
Project Management Software 2.22
Accounting Software 2.22
GRAPHICS AND MULTIMEDIA SOFTWARE 2.23
Computer-Aided Design 2.24
Desktop Publishing Software (Professional) 2.24
Paint/Image Editing Software (Professional) 2.25
Video and Audio Editing Software 2.26
Multimedia Authoring Software 2.26
Web Page Authoring Software 2.27
SOFTWARE FOR HOME, PERSONAL, AND EDUCATIONAL USE 2.27
Integrated Software 2.28
Personal Finance Software 2.28
Legal Software 2.29
Tax Preparation Software 2.29
Desktop Publishing (Personal) 2.30
Paint/Image Editing Software (Personal) 2.30
Clip Art/Image Gallery 2.31
Home Design/Landscaping Software 2.31
Educational/Reference/Personal Computer Entertainment Software 2.32
SOFTWARE FOR COMMUNICATIONS 2.32
Groupware 2.32
Electronic Mail Software 2.33
Web Browsers 2.34
BROWSING THE WORLD WIDE WEB 2.35
Connecting to the Web and Starting a Browser 2.36
Navigating Web Pages Using Links 2.37
Using the Browser Toolbars 2.39
Entering a URL 2.39
Searching for Information on the Web 2.40
APPLICATIONS ON THE WEB 2.42
Application Service Providers 2.43
LEARNING AIDS AND SUPPORT TOOLS 2.44
TECHNOLOGY TRAILBLAZER — MARC ANDREESSEN 2.46
COMPANY ON THE CUTTING EDGE — AMERICA ONLINE, INC. 2.47
IN BRIEF 2.48
KEY TERMS 2.51
AT THE MOVIES 2.52

CHECKPOINT 2.53
AT ISSUE 2.54
MY WEB PAGE 2.55
HANDS ON 2.56
NET STUFF 2.57

SPECIAL FEATURE 2.58

CHAPTER 3
THE COMPONENTS IN THE SYSTEM UNIT

OBJECTIVES 3.1
THE SYSTEM UNIT 3.2
- The Motherboard 3.3

CENTRAL PROCESSING UNIT 3.4
- The Control Unit 3.5
- The Arithmetic/Logic Unit 3.6
- Pipelining 3.7
- Registers 3.8
- The System Clock 3.8
- Microprocessor Comparison 3.8
- Processor Installation and Upgrades 3.10
- Heat Sinks and Heat Pipes 3.10
- Coprocessors 3.11
- Parallel Processing 3.11

DATA REPRESENTATION 3.11
MEMORY 3.14
- RAM 3.15
- Cache 3.17
- ROM 3.18
- CMOS 3.19
- Memory Access Times 3.19

EXPANSION SLOTS AND EXPANSION CARDS 3.20
- PC Cards 3.20

PORTS 3.21
- Serial Ports 3.23
- Parallel Ports 3.23
- Universal Serial Bus Port 3.24
- Special-Purpose Ports 3.24

BUSES 3.24
- Expansion Bus 3.26

BAYS 3.27
POWER SUPPLY 3.28
LAPTOP COMPUTERS 3.28
PUTTING IT ALL TOGETHER 3.29
TECHNOLOGY TRAILBLAZERS — ANDY GROVE AND GORDON MOORE 3.30
COMPANY ON THE CUTTING EDGE — INTEL 3.31
IN BRIEF 3.32
KEY TERMS 3.34
AT THE MOVIES 3.35
CHECKPOINT 3.36
AT ISSUE 3.37
MY WEB PAGE 3.38
HANDS ON 3.39
NET STUFF 3.40

SPECIAL FEATURE 3.41

CHAPTER 4
INPUT

OBJECTIVES 4.1
WHAT IS INPUT? 4.2
WHAT ARE INPUT DEVICES? 4.3
THE KEYBOARD 4.3
- Keyboard Types 4.6

POINTING DEVICES 4.7
- Mouse 4.7
- Trackball 4.10
- Touchpad 4.11
- Pointing Stick 4.11
- Joystick 4.12
- Touch Screen 4.12
- Pen Input 4.12

SCANNERS AND READING DEVICES 4.13
- Optical Scanner 4.14
- Optical Readers 4.16
- Magnetic Ink Character Recognition Reader 4.19
- Data Collection Devices 4.20

DIGITAL CAMERAS 4.20
AUDIO AND VIDEO INPUT 4.22
- Audio Input 4.22
- Speech Recognition 4.22
- Video Input 4.24
- Videoconferencing 4.25

INPUT DEVICES FOR PHYSICALLY CHALLENGED USERS 4.26
PUTTING IT ALL TOGETHER 4.27
TECHNOLOGY TRAILBLAZER — LOU GERSTNER 4.28
COMPANY ON THE CUTTING EDGE — IBM 4.29
IN BRIEF 4.30
KEY TERMS 4.32
AT THE MOVIES 4.33
CHECKPOINT 4.34
AT ISSUE 4.35
MY WEB PAGE 4.36
HANDS ON 4.37
NET STUFF 4.38

CHAPTER 5
OUTPUT

OBJECTIVES 5.1
WHAT IS OUTPUT? 5.2
WHAT ARE OUTPUT DEVICES? 5.3
DISPLAY DEVICES 5.3
- CRT Monitors 5.3
- Flat-Panel Displays 5.4
- Monitor Quality 5.6
- Video Cards 5.8
- Monitor Ergonomics 5.10
- High-Definition Television 5.11

PRINTERS 5.11
- Impact Printers 5.12
- Nonimpact Printers 5.14
- Portable Printers 5.18
- Plotters and Large-Format Printers 5.19
- Special-Purpose Printers 5.20

AUDIO OUTPUT 5.20

OTHER OUTPUT DEVICES 5.21
Data Projectors 5.21
Facsimile (Fax) Machine 5.22
Multifunction Devices 5.23
TERMINALS 5.23
OUTPUT DEVICES FOR PHYSICALLY CHALLENGED USERS 5.24
PUTTING IT ALL TOGETHER 5.25
TECHNOLOGY TRAILBLAZERS — WILLIAM HEWLETT AND DAVID PACKARD 5.26
COMPANY ON THE CUTTING EDGE — HEWLETT-PACKARD 5.27
IN BRIEF 5.28
KEY TERMS 5.30
AT THE MOVIES 5.31
CHECKPOINT 5.32
AT ISSUE 5.33
MY WEB PAGE 5.34
HANDS ON 5.35
NET STUFF 5.36

OBJECTIVES 6.1
MEMORY VERSUS STORAGE 6.2
Memory 6.2
Storage 6.2
FLOPPY DISKS 6.4
Characteristics of Magnetic Media 6.5
Characteristics of a Floppy Disk 6.6
Floppy Disk Drives 6.7
Care of Floppy Disks 6.9
High-Capacity Floppy Disks 6.9
HARD DISKS 6.10
Characteristics of a Hard Disk 6.11
How a Hard Disk Works 6.12
Removable Hard Disks 6.14
Hard Disk Controllers 6.14
RAID 6.15
Maintaining Data Stored on a Hard Disk 6.16
COMPACT DISCS 6.16
CD-ROMs 6.19
DVD-ROMs 6.22
TAPES 6.23
PC CARDS 6.24
OTHER TYPES OF STORAGE 6.25
Smart Cards 6.25
Microfilm and Microfiche 6.26
Enterprise Storage Systems and Data Warehouses 6.26
PUTTING IT ALL TOGETHER 6.27
TECHNOLOGY TRAILBLAZER — SCOTT McNEALY 6.28
COMPANY ON THE CUTTING EDGE — SILICON GRAPHICS 6.29
IN BRIEF 6.30
KEY TERMS 6.32
AT THE MOVIES 6.33
CHECKPOINT 6.34
AT ISSUE 6.35
MY WEB PAGE 6.36
HANDS ON 6.37
NET STUFF 6.38

CHAPTER 7
THE INTERNET

OBJECTIVES 7.1
THE INTERNET 7.2
HISTORY OF THE INTERNET 7.4
HOW THE INTERNET WORKS 7.6
Internet Service Providers and Online Services 7.6
Connecting to the Internet 7.7
How Data Travels the Internet 7.7
Internet Addresses 7.8
THE WORLD WIDE WEB 7.10
Search Engines 7.12
Multimedia on the Web 7.12
WEBCASTING 7.18
ELECTRONIC COMMERCE 7.19
WEB PUBLISHING 7.20
OTHER INTERNET SERVICES 7.23
E-mail 7.23
FTP 7.25
Telnet 7.26
Newsgroups and Message Boards 7.27
Mailing Lists 7.28
Chat Rooms 7.29
Instant Messaging 7.29
Portals 7.29
NETIQUETTE 7.30
USING THE INTERNET: COOKIES AND SECURITY 7.31
Cookies 7.31
Internet Security 7.32
NETWORK COMPUTERS 7.33
WEB APPLIANCES 7.33
WIRELESS WEB COMMUNICATIONS 7.34
TECHNOLOGY TRAILBLAZER — TIM BERNERS-LEE 7.36
COMPANY ON THE CUTTING EDGE — YAHOO! 7.37
IN BRIEF 7.38
KEY TERMS 7.40
AT THE MOVIES 7.41
CHECKPOINT 7.42
AT ISSUE 7.43
MY WEB PAGE 7.44
HANDS ON 7.45
NET STUFF 7.46

OPERATING SYSTEMS AND UTILITY PROGRAMS

OBJECTIVES **8.1**
SYSTEM SOFTWARE **8.2**
OPERATING SYSTEMS **8.2**
User Interfaces 8.3
Features of Operating Systems 8.4
Functions of an Operating System 8.6
POPULAR OPERATING SYSTEMS **8.10**
DOS 8.11
Windows 3.x 8.12
Windows 95 8.12
Windows NT 8.12
Windows 98 8.12
Windows 2000 8.12
Windows Millennium 8.14
Windows CE 8.14
Palm OS 8.14
Mac OS 8.14
OS/2 8.14
UNIX 8.16
Linux 8.16
NetWare 8.17
STARTING A COMPUTER **8.17**
Boot Disk 8.20
UTILITY PROGRAMS **8.21**
TECHNOLOGY TRAILBLAZERS — STEVE WOZNIAK AND STEVE JOBS **8.26**
COMPANIES ON THE CUTTING EDGE — APPLE VS. IBM COMPATIBLES — THE RIVALRY **8.27**
IN BRIEF **8.28**
KEY TERMS **8.30**
AT THE MOVIES **8.31**
CHECKPOINT **8.32**
AT ISSUE **8.33**
MY WEB PAGE **8.34**
HANDS ON **8.35**
NET STUFF **8.36**

SPECIAL FEATURE **8.37**
Buyer's Guide 2001
How to Purchase, Install, and Maintain a Personal Computer

APPENDIX

CODING SCHEMES AND NUMBER SYSTEMS

CODING SCHEMES **A.1**
ASCII and EBCDIC A.1
Unicode A.1
Parity A.2
NUMBER SYSTEMS **A.2**
The Decimal Number System A.3
The Binary Number System A.4
The Hexadecimal Number System A.4

INDEX I.1

Photo Credits I.20

PREFACE

The previous five editions of this textbook have been runaway best-sellers. Each of the these editions included new learning innovations, such as integration of the World Wide Web, CyberClass, Interactive Labs, and Teaching Tools that set it apart from its competitors. *Discovering Computers 2001: Concepts for a Connected World, Web and CNN Enhanced, Brief Edition* continues with the innovation, quality, timeliness, and reliability that you have come to expect from the Shelly Cashman Series®. The newest edition of *Discovering Computers* includes these enhancements:

- Currency, such as the latest hardware, software, and trends in the computer field
- A complete WebCT navigation system with Web-related activities customized specifically for this book; a *WebCT Student Guide* also is available
- An exercise section at the end of each chapter titled MY WEB PAGE that includes WebCT, CyberClass, and SCSITE Web-based exercises that offer a unique way for students to solidify, reinforce, and extend the concepts presented in the chapter
- A new 18-page Special Feature following Chapter 7 titled E-Commerce 2001: A Revolution in Merchandising that describes e-commerce basics, business-to-business and business-to-consumer e-commerce interactions, and how to set up an electronic storefront.
- Step-by-step illustrations that significantly simplify the complexity of the computer concepts presented
- Updated timeline and buyer's guide
- A Web site with dramatically improved functionality

OBJECTIVES OF THIS TEXTBOOK

Discovering Computers 2001: Concepts for a Connected World, Web and CNN Enhanced, Brief Edition is intended to be used in combination with an applications, Internet, or programming textbook in a one-quarter or one-semester introductory computer course. No experience with computers is assumed. The material presented provides an in-depth treatment of introductory computer subjects. Students will finish the course with a solid understanding of computers, how to use computers, and how to access information on the World Wide Web. The objectives of this book are as follows:

- Teach the fundamentals of computers and computer nomenclature, particularly with respect to personal computer hardware and software, and the World Wide Web
- Give students an in-depth understanding of why computers are essential components in business and society in general
- Present the material in a visually appealing and exciting manner that invites students to learn
- Make use of the World Wide Web as a repository of the latest information and as an integrated learning tool
- Offer alternative learning techniques with streaming audio and video on the Web, Learning Games, WebCT, and CyberClass
- Recognize the personal computer's position as the backbone of the computer industry and emphasize its use as a stand-alone and networked device
- Present strategies for purchasing, installing, and maintaining a personal computer
- Provide exercises and lab assignments that allow students to interact with a computer and actually learn by using the computer and the World Wide Web

DISTINGUISHING FEATURES

Discovering Computers 2001: Concepts for a Connected World, Web and CNN Enhanced, Brief Edition includes the following distinguishing features.

A Proven Book

More than three-and-one-half million students have learned about computers using Shelly and Cashman computer fundamentals textbooks. With the additional World Wide Web integration and interactivity, streaming up-to-date audio and video, extraordinary visual drawings and photographs, unprecedented currency, and the Shelly and Cashman touch, this book will make your computer concepts course exciting and dynamic, an experience your students will remember as a highlight of their educational careers.

World Wide Web and CNN Enhanced

Each of the Shelly Cashman Series computer fundamentals books has included significant educational innovations that have set them apart from all other textbooks in the field. This book sustains this tradition of innovation with its continued integration of the World Wide Web. The purpose of integrating the World Wide Web into the book is to (1) offer students additional information and currency on topics of importance; (2) make available alternative learning techniques with Web-based streaming audio and up-to-date, computer-related CNN videos; (3) underscore the relevance of the World Wide Web as a basic information tool that can be used in all facets of society; and (4) offer instructors the opportunity to organize and administer their traditional campus-based or distance-education-based courses on the Web using WebCT or CyberClass. The World Wide Web is integrated into the book in seven central ways:

- End-of-chapter pages and the special features in the book have been stored as Web pages on the World Wide Web. While working on an end-of-chapter page, students can display the corresponding Web page to obtain additional information on a term or exercise, and get an alternative point of view. See page xv for more information.
- Streaming audio on the Web in the end-of-chapter IN BRIEF sections.
- Streaming up-to-date, computer-related CNN videos on the Web in the end-of-chapter AT THE MOVIES sections.
- Interactive Labs on the Web in the end-of-chapter NET STUFF sections.
- Throughout the text, marginal annotations titled WEB INFO provide suggestions on how to obtain additional information via the Web on an important topic covered on the page.
- WebCT is a Web-based teaching and learning system. Included with this textbook is a customized navigation system and corresponding book-related content that can be uploaded to an institution's WebCT site or to WebCT's hosting server and used in a traditional classroom setting or in a distance education environment. The content includes a sample syllabus, practice tests, a bank of test questions, course overview, links to learning games, and a variety of Web-related exercises.
- CyberClass is a Web-based teaching and learning system that offers flash cards, practice tests, case scenarios, course administration, self-study games, and an online testing system.

This textbook, however, does not depend on Web access in order to be used successfully. The Web access adds to the already complete treatment of topics within the book.

Technology Trailblazer and Company on the Cutting Edge Features

Every student graduating from an institution of higher education should be aware of the leaders and major companies in the field of computers. Thus, each chapter ends with two full pages devoted to features titled Technology Trailblazer and Company on the Cutting Edge. The Technology Trailblazer feature presents people who have made a difference in the computer revolution, such as Bill Gates, Andy Grove, Steve Wozniak, Steve Jobs, Marc Andreessen, Tim Berners-Lee, William Hewlett, and David Packard. The Company on the Cutting Edge feature presents the major computer companies, such as Microsoft, Intel, Yahoo!, AOL, Hewlett-Packard, and IBM.

A Visually Appealing Book that Maintains Student Interest

Using the latest technology, pictures, drawings, and text have been artfully combined to produce a visually appealing and easy-to-understand book. Many of the figures show a step-by-step pedagogy, which simplifies the more complex computer concepts. Pictures and drawings reflect the latest trends in computer technology. Finally, the text was set in two columns, which research indicates is easier for students to read. This combination of pictures, step-by-step drawings, and text sets a new standard for computer textbook design.

Latest Computer Trends

The terms and technologies your students see in this book are those they will encounter when they start using computers. Only the latest application software packages are shown throughout the book. New topics and terms include Microsoft Windows 2000; Microsoft Office 2000; IrDA ports; smart phones, video cameras, color screens on handheld computers, Web appliance, TV out ports, TV tuner cards, AMD processors (Athlon, AMD-K6-III, and AMD-K6-2), Intel Itanium, integrated CPUs, SSE instructions, 3Dnow Technology, RDRAM, SDRAM II, fast infrared port, HDTV, photo printers, e-stamps, Internet button on keyboards, CD/DVD controls on keyboards, proposed OSHA ergonomic standards for employers, optical mouse (IntelliMouse with IntelliEye), pen scanners, flash cards, DVD-E, DVD-R, electronic mall, electronic storefront, hands-free internet service, Internet digital postage machine, Internet2, Next Generation Internet, photo community, StarOffice, open source, Linux, instant messaging, MP3, Webware, application service providers (ASP); streaming audio and video; Webcasting; electronic credit; Web publishing; and much more.

Shelly Cashman Series Interactive Labs

Eighteen unique, hands-on exercises, developed specifically for this book, allow students to use the computer to learn about computers. Students can step through each Lab exercise in about 15 minutes. Assessment is available. The Interactive Labs are described in detail on page xvi. These Labs are available free both on the Web (see page 1.48) or on the Teaching Tools CD-ROM, or on CD-ROM for an additional cost (ISBN 0-7895-5679-0).

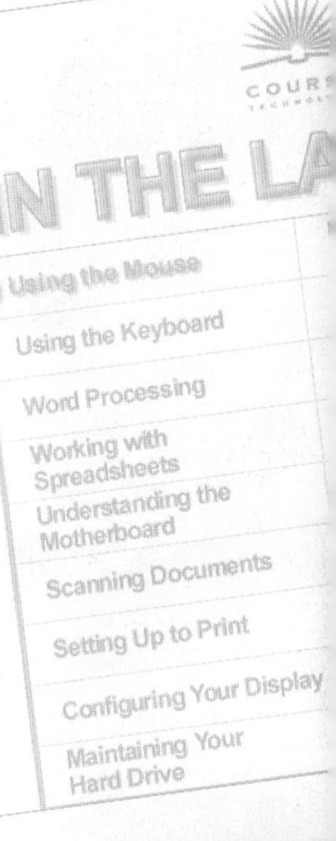

End-of-Chapter Exercises

Unlike other books on the subject of computer fundamentals, a major effort was undertaken in this book to offer exciting, rich, and thorough end-of-chapter material to reinforce the chapter objectives and assist you in making your course the finest ever offered. As indicated earlier, each and every one of the end-of-chapter pages is stored as a Web page on the World Wide Web to provide your students in-depth information and alternative methods of preparing for examinations. Each chapter ends with the following:

- **IN BRIEF** This section summarizes the chapter material in the form of questions and answers. Each question addresses a chapter objective, making this section invaluable in reviewing and preparing for examinations. Links on the Web page provide additional current information. With a single-click on the Web page, the review section is read to the student using streaming audio.
- **KEY TERMS** This list of the key terms found in the chapter together with the page number on which the terms are defined will aid students in mastering the chapter material. A complete summary of all key terms in the book, together with their definitions, appears in the Index at the end of the book. On the Web page, students can click terms to view a definition and a picture, and then click a link to visit a page that offers an alternative explanation.
- **AT THE MOVIES** In this section, students complete exercises that require them to click photographs on the Web page to view streaming up-to-date CNN videos. These videos, which present computer-related topics, reinforce the chapter or provide extended knowledge of important concepts.
- **CHECKPOINT** Matching and short-answer questions, together with a figure from the chapter that must be labeled, reinforce the material presented within the chapter. Students accessing the Web page answer the questions in an interactive forum.

- **AT ISSUE** The computer industry is not without its controversial issues. At the end of each chapter, several scenarios are presented that challenge students to examine critically their perspective of technology in society. The Web pages provide links to challenge students further.
- **MY WEB PAGE** These exercises have students connect through WebCT, the textbook Web site, or CyberClass where they complete tasks that include practice tests, visiting and evaluating Web sites, Web-based learning games, search activities, and flash cards.
- **HANDS ON** To complete their introduction to computers, students must interact with and use a computer. A series of Windows Lab exercises begins with the simplest exercises within Windows. Students then are led through additional activities that, by the end of the book, will enable them to be proficient in using Windows.
- **NET STUFF** In this section, students gain an appreciation for the World Wide Web by visiting interesting and exciting Web pages and completing suggested tasks. Also included in this section are exercises that have students complete the Shelly Cashman Series Interactive Labs. These Interactive Labs can be done directly from the World Wide Web. The last NET STUFF exercise sends students into a Chat room where they can discuss engaging issues and topics presented in the book with other students throughout the world.

Timeline 2001: Milestones in Computer History

A colorful, highly informative eleven-page timeline following Chapter 1 steps students through the major computer technology developments over the past 60 years, including the most recent advances.

Guide to World Wide Web Sites

More than 150 popular Web sites are listed and described in a new guide to Web sites that follows Chapter 2.

How Computer Chips Are Made

This feature following Chapter 3 steps through the intricate details of making a computer chip.

E-Commerce 2001: A Revolution in Merchandising

This eighteen-page feature following Chapter 7 describes e-commerce basics, business-to-business and business-to-consumer e-commerce interactions, and how to set up an electronic storefront.

Buyer's Guide 2001: How to Purchase, Install, and Maintain a Personal Computer

A twelve-page guide following Chapter 8 introduces students to purchasing, installing, and maintaining a desktop or laptop computer.

SHELLY CASHMAN SERIES TEACHING TOOLS

A comprehensive set of Teaching Tools accompanies this textbook in the form of two CD-ROM packages. The two packages titled Teaching Tools (ISBN 0-7895-5940-4) and Course Presenter (ISBN 0-7895-5939-0) are described in the following sections. Both packages are available free to adopters through your Course Technology representative or by calling one of the following telephone numbers: Colleges and Universities, 1-800-648-7450; High Schools, 1-800-824-5179; Career Colleges, 1-800-477-3692; Canada, 1-800-268-2222; and Corporations and Governments, 1-800-340-7450.

Teaching Tools

The Teaching Tools for this textbook include both teaching and testing aids. The contents of the Teaching Tools CD-ROM are listed below.

- **Instructor's Manual** The Instructor's Manual is made up of Microsoft Word files that include the following for each chapter: chapter objectives; chapter overview; detailed lesson plans with page number references; teacher notes and activities; answers to the exercises; test bank (100 true/false, 50 multiple-choice, and 70 fill-in-the-blank questions per chapter); and figure references. The figures are available in Figures in the Book. The test bank questions are numbered the same as in Course Test Manager. You can print a copy of the chapter test bank and use the printout to select your questions in Course Test Manager. Using your word processing software, you can generate quizzes and exams.
- **Figures in the Book** Illustrations for every picture, table, and screen in the textbook are available in electronic form. Use this ancillary to present a slide show in lecture or to print transparencies for use in lecture with an overhead projector. If you have a personal computer and LCD device, this ancillary can be an effective tool for presenting lectures.
- **Course Test Manager** Course Test Manager is a powerful testing and assessment package that enables instructors to create and print tests quickly from the large test bank. Instructors with access to a networked computer lab (LAN) can administer, grade, and track tests online. Students also can take online practice tests, which generate customized study guides that indicate where in the textbook students can find more information for each question.
- **Course Syllabus** Any instructor who has been assigned a course at the last minute knows how difficult it is to come up with a course syllabus. For this reason, a sample syllabus is included that can be customized easily to a course.
- **Student Files** A few of the exercises in the end-of-chapter HANDS ON section ask students to use these files. You can distribute the files on the Teaching Tools CD-ROM to your students over a network, or you can have them follow the instructions on the inside back cover of this book to obtain a copy of the *Discovering Computers 2001* Data Disk.
- **Interactive Labs** These are the non-audio versions of the eighteen hands-on Interactive Labs exercises. Students can step through each lab in about fifteen minutes to solidify and reinforce computer concepts. Assessment requires students to answer questions about the contents of the Interactive Labs.
- **Interactive Lab Solutions** This ancillary includes the solutions for the Interactive Labs assessment quizzes.
- **WebCT Content** This ancillary includes a comprehensive WebCT customized navigation system and corresponding book-related content that can be uploaded to your institution's WebCT site or to WebCT's hosting server and used in a traditional classroom setting or in a distance education environment. The content includes a sample syllabus, practice tests, a bank of test questions, a list of book-related links, course overview, links to learning games, and a variety of Web-related exercises described in the chapter-ending section titled MY WEB PAGE (see page 1.46).

Course Presenter with Figures, Animations, and CNN Video Clips

Course Presenter is a multimedia lecture presentation system that provides PowerPoint slides for every subject in each chapter. Use this presentation system to present well-organized lectures that are both interesting and knowledge-based. Fourteen presentation files are provided for the book, one for each chapter. Each file contains PowerPoint slides for every subject in each chapter together with optional choices to show any figure in the chapter as you introduce the material in class. More than 40 current, two- to three-minute up-to-date, computer-related CNN video clips and more than 35 animations that reinforce chapter material also are available for optional presentation. Course Presenter provides consistent coverage for multiple lecturers.

SUPPLEMENTS

Four supplements can be used in combination with *Discovering Computers 2001: Concepts for a Connected World, Web and CNN Enhanced, Brief Edition*.

Audio Chapter Review on CD-ROM

The Audio Chapter Review on CD-ROM (ISBN 0-7895-5941-2) vocalizes the end-of-chapter IN BRIEF pages (see page 1.40). Students can use this supplement with a CD player or PC to solidify their understanding of the concepts presented. It is a great tool for preparing for examinations. This same Audio Chapter Review also is available at no cost on the Web by clicking the Audio button on the IN BRIEF page at the end of any chapter.

Shelly Cashman Series Interactive Labs with Audio on CD-ROM

The Shelly Cashman Series Interactive Labs with Audio on CD-ROM (ISBN 0-7895-5679-0) may be used in combination with this textbook to augment your students' learning process. See page xvi for a description of each lab. These Interactive Labs also are available at no cost on the Web by clicking the appropriate button on the NET STUFF exercise pages (see page 1.48) and as a non-audio version on the Teaching Tools CD-ROM. A companion student guide for the Interactive Labs, titled *A Record of Discovery for Exploring Computers, Third Edition* (ISBN 0-7895-5840-8), enhances the Interactive Labs presentation, reinforces concepts, shows relationships, and provides additional facts.

Study Guide

This highly popular supplement (ISBN 0-7895-6081-X) includes a variety of activities that help students recall, review, and master introductory computer concepts. The *Study Guide* complements the end-of-chapter material with a guided chapter outline; a self-test consisting of true/false, multiple-choice, short answer, fill-in, and matching questions, an entertaining puzzle, and other challenging exercises.

WebCT Student Guide

The *WebCT Student Guide* (ISBN 0-7895-6105-0) was designed to introduce students to the WebCT course tools and the comprehensive Webcourselet for this textbook on the Teaching Tools CD-ROM that is available to adopters.

ACKNOWLEDGMENTS

The Shelly Cashman Series would not be the leading computer education series without the contributions of outstanding publishing professionals. First, and foremost, among them is Becky Herrington, director of production and designer. She is the heart and soul of the Shelly Cashman Series, and it is only through her leadership, dedication, and tireless efforts that superior products are made possible. Becky created and produced the award-winning Windows series of books.

Under Becky's direction, the following individuals made significant contributions to these books: Doug Cowley, production manager; Ginny Harvey, series specialist and developmental editor; Ken Russo, senior Web designer; Mike Bodnar, associate production manager; Stephanie Nance, graphic artist; Mark Norton, Web designer; Meena Mohtadi, production editor; Marlo Mitchem, Chris Schneider, Hector Arvizu, and Kenny Tran, graphic artists; Jeanne Black and Betty Hopkins, Quark experts; Nancy Lamm, copyeditor/proofreader; Cristina Haley, indexer; Sarah Evertson of Image Quest and Abby Reip, photo researchers; Susan Sebok and Ginny Harvey, contributing writers; Richard Keaveny, managing editor; Jim Quasney, series consulting editor; Lora Wade, product manager; Erin Bennett, associate product manager; Francis Schurgot, Web product manager; Marc Ouellette, associate Web product manager; Rajika Gupta, marketing manager; and Erin Runyon, editorial assistant.

Our sincere thanks go to Dennis Tani, who together with Becky Herrington, designed this book. In addition, Dennis designed the cover, performed all the initial layout, typography, and executed the magnificent drawings contained in this book.

Finally, thanks to Charles Aitkenhead, Virgil Brewer, Kay Delk, Jeffrey J. Quasney, and Harry Rosenblatt (Chapters 1 through 5) for reviewing the manuscript of the previous edition; William Vermaat for researching, reviewing the manuscript, and taking photographs; Darrell Ward of *Hyper*Graphics Corporation for the development of CyberClass; and Dolores Pusins for the development of the WebCT content for this textbook. We hope you find using this book an exciting and rewarding experience.

Gary B. Shelly
Thomas J. Cashman

Misty E. Vermaat
Tim J. Walker

NOTES TO THE STUDENT

If you have access to the World Wide Web, you can obtain current and additional information on topics covered in this book in the three ways listed below.

1. Throughout the book, marginal annotations called WEB INFO (Figure 1) specify subjects about which you can obtain additional current information. Enter the designated URL and then click the appropriate term on the Web page.
2. Each chapter ends with seven sections titled IN BRIEF, KEY TERMS, AT THE MOVIES, CHECKPOINT, AT ISSUE, MY WEB PAGE, HANDS ON, and NET STUFF. These sections in your book are stored as Web pages on the Web. You can visit them by starting your browser and entering the URL in the instructions at the top of the end-of-chapter pages. When the Web page displays, you can click links or buttons on the page to broaden your understanding of the topics and obtain current information about the topic.
3. Use WebCT, the textbook Web site, or CyberClass. WebCT and CyberClass are two different Web-based teaching and learning systems that adopters of this textbook can use in a traditional campus setting or distance learning setting. The textbook Web site includes thousands of links, learning games, animations to simplify complex concepts, audio, and video. See MY WEB PAGE at the end of each chapter (i.e., page 1.46) for examples of how these three Web-based learning systems can enhance your learning experience.

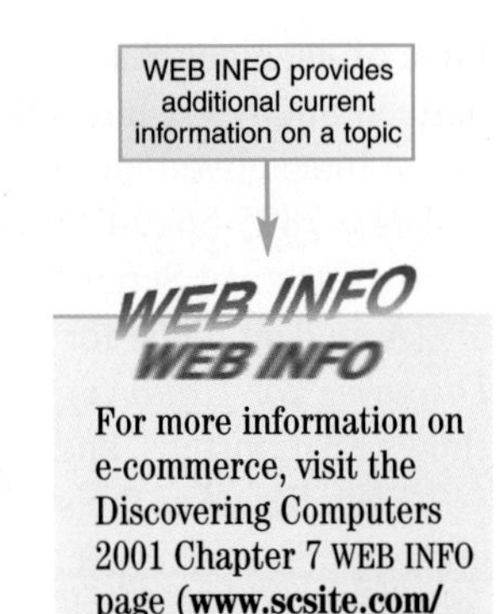

Figure 1

Each time you reference a Web page from *Discovering Computers 2001*, a sidebar displays on the left. To display one of the Student Exercises (Figure 2 on the next page), click the chapter number and then click the Student Exercise title in the sidebar. To display one of the Special Features, click the desired Special Feature title in the sidebar.

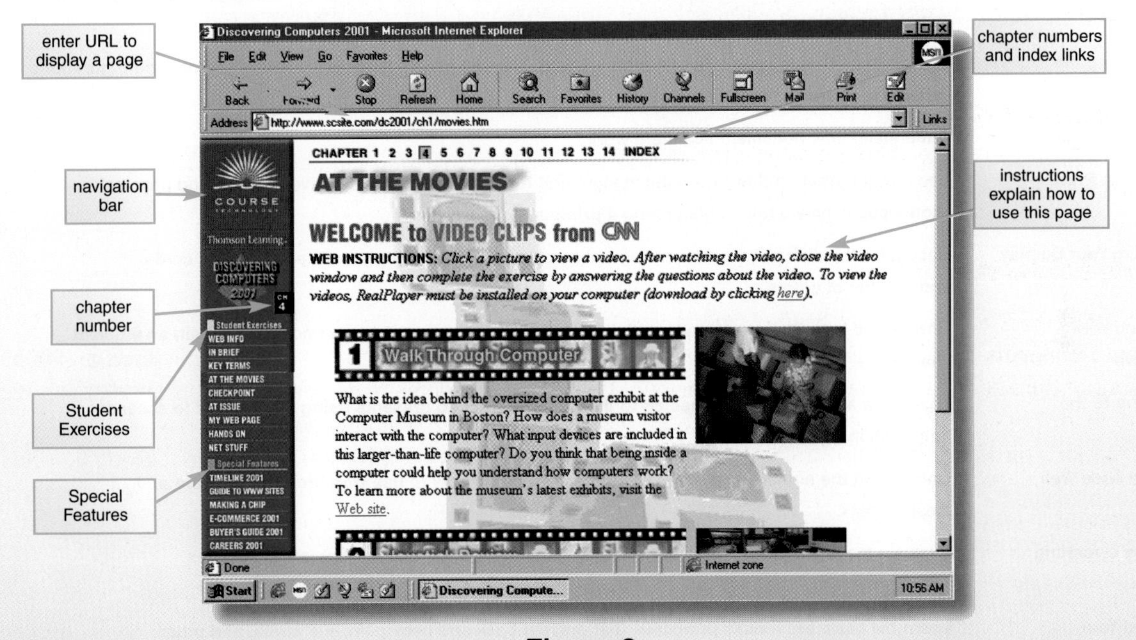

Figure 2

TO DOWNLOAD PLAYERS

For best viewing results of the Web pages referenced in this book, download the Shockwave and Flash Player. To play the audio in the IN BRIEF section and view the video in the AT THE MOVIES section at the end of each chapter, you must download RealPlayer. Follow the steps below:

Shockwave and Flash Player — (1) Launch your browser; (2) enter the URL, `www.macromedia.com`; (3) click downloads; (4) click shockwave and flash players; (5) click the AUTOINSTALL NOW! button in the Step 1 area; (6) follow the instructions in the dialog boxes to complete the installation.

RealPlayer — (1) Launch your browser; (2) enter the URL, `www.real.com`; (3) scroll down and click the RealPlayer 7 FREE button; (4) click the RealPlayer 7 Basic link; (5) step through and respond to the subsequent forms, requests, and dialog boxes; (6) when the File Download dialog box displays, click the Save this program to disk option button and then click the OK button; (7) save the file to a folder and remember the folder name; and (8) launch Windows Explorer and then double-click the downloaded file in step 7.

SHELLY CASHMAN SERIES INTERACTIVE LABS WITH AUDIO

Each of the eight chapters in this book includes the NET STUFF exercises, which utilize the World Wide Web. The thirteen Shelly Cashman Series Interactive Labs described below are included as exercises in the NET STUFF section. These Interactive Labs are available on the Web (see page 1.48) or on CD-ROM. The audio version on CD-ROM (ISBN 0-7895-5679-0) is available at an additional cost. A non-audio version also is available at no extra cost on the Shelly Cashman Series Teaching Tools CD-ROM that is available free to adopters.

A student guide for the Interactive Labs is available at an additional cost. The student guide is titled *A Record of Discovery for Exploring Computers, Third Edition* (ISBN 0-7895-5840-8), which reviews the Interactive Labs content, shows relationships, and provides additional facts.

Each lab takes the students approximately 15 minutes to complete using a personal computer and helps them gain a better understanding of a specific subject covered in the chapter.

Shelly Cashman Series Interactive Labs with Audio

Lab	Function	Page
Using the Mouse	Master how to use a mouse. The Lab includes exercises on pointing, clicking, double-clicking, and dragging.	1.48
Using the Keyboard	Learn how to use the keyboard. The Lab discusses different categories of keys, including the edit keys, function keys, ESC, CTRL, and ALT keys and how to press keys simultaneously.	1.49
Word Processing	Gain a basic understanding of word processing concepts, from creating a document to printing and saving the final result.	2.57
Working with Spreadsheets	Learn how to create and utilize spreadsheets, including entering formulas, creating graphs, and performing what-if analysis.	2.57
Understanding the Motherboard	Step through the components of a motherboard. The Lab shows how different motherboard configurations affect the overall speed of a computer.	3.40
Scanning Documents	Understand how document scanners work.	4.38
Setting Up to Print	See how information flows from the system unit to the printer and how drivers, fonts, and physical connections play a role in generating a printout.	5.36
Configuring Your Display	Recognize the different monitor configurations available, including screen size, display cards, and number of colors.	5.36
Maintaining Your Hard Drive	Understand how files are stored on disk, what causes fragmentation, and how to maintain an efficient hard drive.	6.38
Connecting to the Internet	Learn how a computer is connected to the Internet. The Lab presents using the Internet to access information.	7.46
The World Wide Web	Understand the significance of the World Wide Web and how to use Web browser software and search tools.	7.46
Evaluating Operating Systems	Evaluate the advantages and disadvantages of different categories of operating systems.	8.36
Working at Your Computer	Learn the basic ergonomic principles that prevent back and neck pain, eye strain, and other computer-related physical ailments.	8.36

INTRODUCTION TO USING COMPUTERS

OBJECTIVES

After completing this chapter, you will be able to:

- Explain why it is important to be computer literate
- Define the term computer
- Identify the components of a computer
- Explain why a computer is a powerful tool
- Differentiate among the various categories of software
- Explain the purpose of a network
- Discuss the uses of the Internet and the World Wide Web
- Describe the categories of computers and their uses

Whether you are attending classes, working in an office, or participating in recreational activities, you use computers every day. Even activities in your daily routine – typing a report, driving your car, paying for goods and services with a credit card, or using an ATM – can involve the use of computers. Computers have become the tools people use to access and provide information and communicate with others around the world. Today, computers are everywhere.

The purpose of this book is to present the knowledge you need to understand how computers work and how computers are used. Chapter 1 introduces you to basic computer concepts such as what a computer is, how it works, and what makes it a powerful tool. You will begin to learn about the many different categories of computers and their applications. In the process, you will gain an understanding of the vocabulary used to describe computers. While you are reading, remember that this chapter is an overview and that many of the terms and concepts that are introduced will be discussed further in later chapters.

COMPUTER LITERACY

The vocabulary of computing is all around you. Before the advent of computers, memory was the mental ability to recall previous experiences; storage was an area where you kept out-of-season clothing; and communication was the act of exchanging opinions and information through writing, speaking, or signs. In today's world, these words and countless others have taken on new meanings as part of the common terminology used to describe computers and their use.

When you hear the word computer, initially you may think of those found in the workplace – the computers used to create business letters, memos, and other correspondence; calculate payroll; track inventory; or generate invoices. In the course of a day or week, however, you encounter many other computers. Your home, for instance, may contain a myriad of electronic devices, such as cordless telephones, VCRs, handheld video games, cameras, and stereo systems, that include small computers.

Computers help you with your banking in the form of automatic teller machines (ATMs) used to deposit or withdraw funds. When you buy groceries, a computer tracks your purchases and calculates the amount of money you owe; and sometimes generates coupons customized to your buying patterns.

Even your car is equipped with computers that operate the electrical system, control the temperature, and run sophisticated antitheft devices. Figure 1-1 shows a variety of computers being used in everyday life.

Computers are valuable tools. As technology advances and computers extend into every facet of daily living, it is essential you gain some level of **computer literacy**. To be successful in today's world, you must have a knowledge and understanding of computers and their uses.

WHAT IS A COMPUTER AND WHAT DOES IT DO?

A **computer** is an electronic machine, operating under the control of instructions stored in its own memory, that can accept data (input), manipulate the data according to specified rules (process), produce results (output), and store the results for future use.

Data is a collection of unorganized facts, which can include words, numbers, images, and sounds. Computers manipulate and process data to create information.

WEB INFO

For more information on computer literacy, visit the Discovering Computers 2001 Chapter 1 WEB INFO page (**www.scsite.com/dc2001/ch1/webinfo.htm**) and click Computer Literacy.

Figure 1-1 *Computers are present in every aspect of daily living – in the workplace, at home, and in the classroom.*

Information is data that is organized, has meaning, and is useful. Examples are reports, newsletters, a receipt, a picture, an invoice, or a check. In Figure 1-2, data is processed and manipulated to create a check.

Data entered into a computer is called **input**. The processed results are called **output**. Thus, a computer processes input to create output. A computer also can hold data and information for future use in an area called **storage**. This cycle of input, process, output, and storage is called the **information processing cycle**.

A person that communicates with a computer or uses the information it generates is called a **user**.

The electric, electronic, and mechanical equipment that makes up a computer is called **hardware**. **Software** is the series of instruction that tells the hardware how to perform tasks. Without software, hardware is useless; hardware needs the instructions provided by software to process data into information.

The next section discusses various hardware components. Later in the chapter, categories of software are discussed.

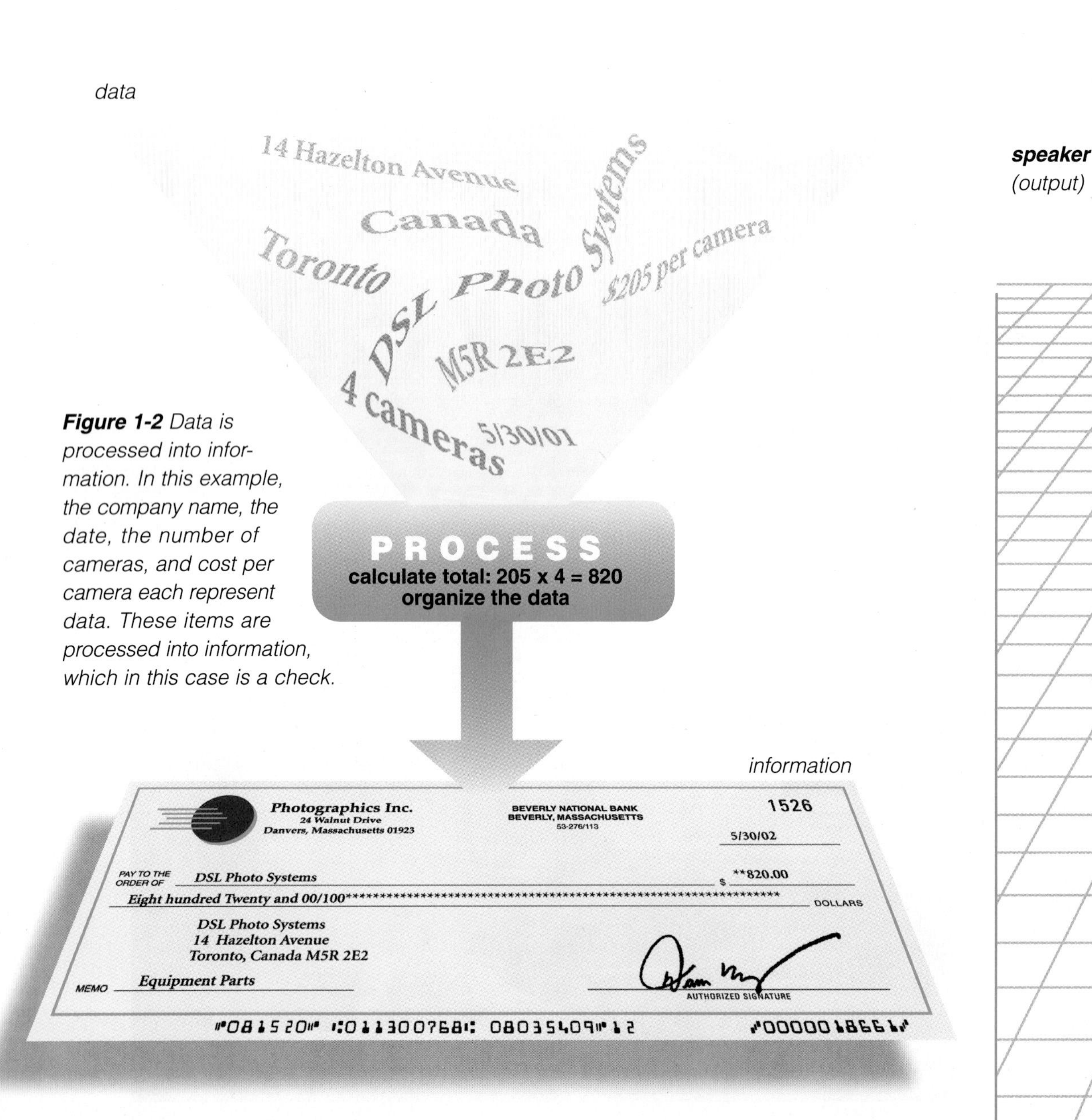

Figure 1-2 Data is processed into information. In this example, the company name, the date, the number of cameras, and cost per camera each represent data. These items are processed into information, which in this case is a check.

THE COMPONENTS OF A COMPUTER

A computer consists of a variety of hardware components that work together with software to perform calculations, organize data, and communicate with other computers.

These hardware components include input devices, output devices, a system unit, storage devices, and communications devices. Figure 1-3 shows some common computer hardware components.

Input Devices

An **input device** allows a user to enter data and commands into the memory of a computer. Four commonly used input devices are the keyboard, the mouse, a microphone, and a PC camera.

A computer keyboard contains keys that allow you to type letters of the alphabet, numbers, spaces, punctuation marks, and other symbols. A computer keyboard also contains special keys that allow you to perform specific functions on the computer.

WEB INFO

For more information on input devices, visit the Discovering Computers 2001 Chapter 1 WEB INFO page (**www.scsite.com/dc2001/ch1/webinfo.htm**) and click Input Devices.

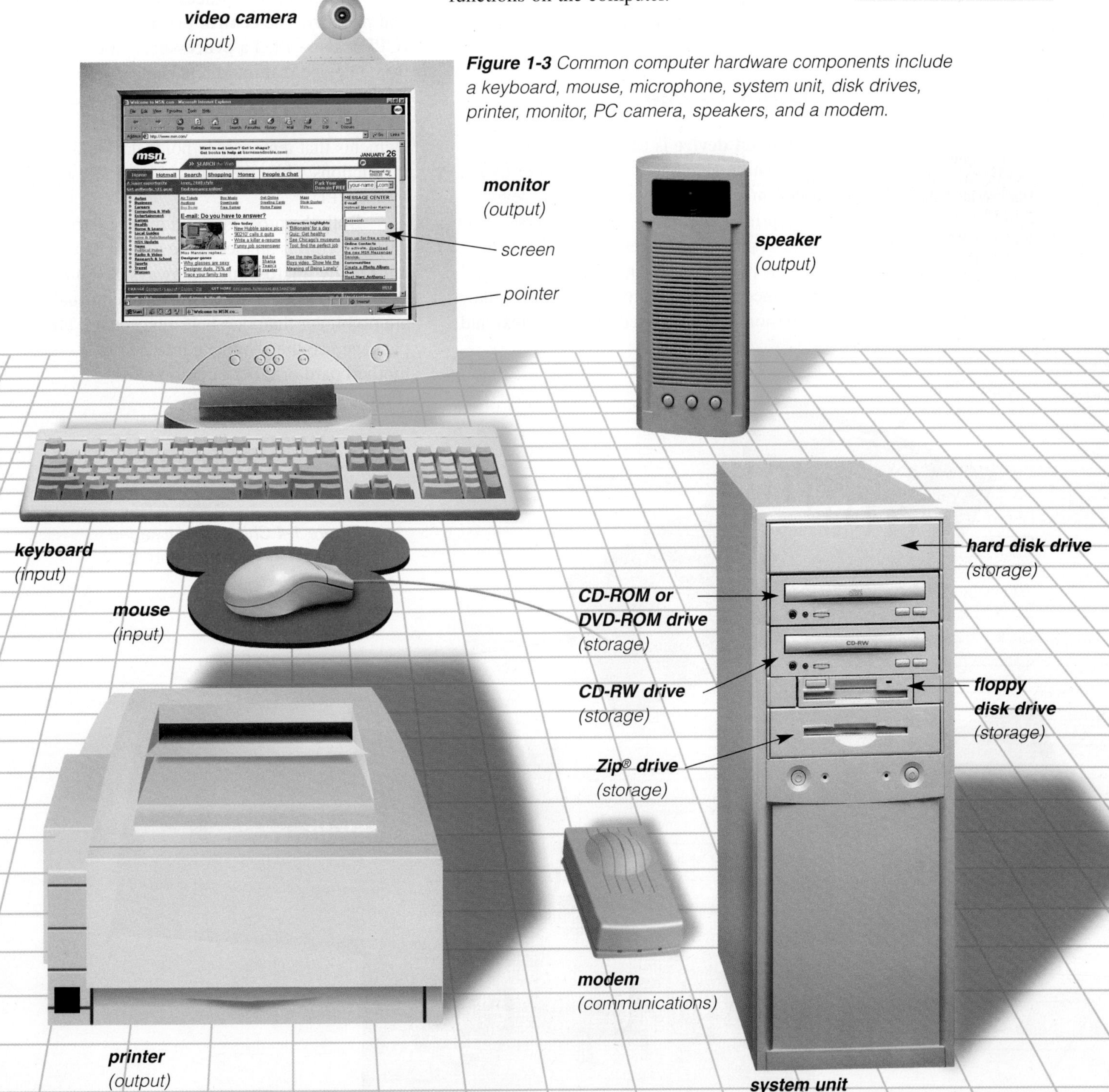

Figure 1-3 *Common computer hardware components include a keyboard, mouse, microphone, system unit, disk drives, printer, monitor, PC camera, speakers, and a modem.*

A mouse is a small handheld device that contains at least one button. The mouse controls the movement of a symbol on the screen called a pointer. For example, moving the mouse across a flat surface allows you to move the pointer on the screen. You also can make choices and initiate processing on the computer by using a mouse.

A microphone allows you to speak to the computer in order to enter data and control the actions of the computer. A PC camera allows others to see you while communicating with you, as well as allowing you to edit videos, create a movie, and take digital photographs.

WEB INFO

For more information on output devices, visit the Discovering Computers 2001 Chapter 1 WEB INFO page (**www.scsite.com/dc2001/ch1/webinfo.htm**) and click Output Devices.

Output Devices

An **output device** is used to convey the information generated by a computer to a user. Three commonly used output devices are a printer, a monitor, and speakers.

A printer produces text and graphics, such as photographs, on paper or other hard-copy medium. A monitor, which looks like a television screen, is used to display text and graphics. Speakers allow you to hear music, voice, and other sounds generated by the computer.

WEB INFO

For more information on CPUs, visit the Discovering Computers 2001 Chapter 1 WEB INFO page (**www.scsite.com/dc2001/ch1/webinfo.htm**) and click CPUs.

System Unit

The **system unit** is a box-like case made from metal or plastic that houses the computer electronic circuitry. The circuitry in the system unit usually is part of or is connected to a circuit board called the motherboard.

Two main components on the motherboard are the central processing unit (CPU) and memory. The **central processing unit (CPU)**, also called a **processor**, is the electronic device that interprets and carries out the instructions that operate the computer.

Memory is a series of electronic elements that temporarily holds data and instructions while they are being processed by the CPU.

Both the processor and memory are chips. A chip is an electronic device that contains many microscopic pathways designed to carry electrical current. Chips, which usually are no bigger than one-half inch square, are packaged so they can be connected to a motherboard or other circuit boards (Figure 1-4).

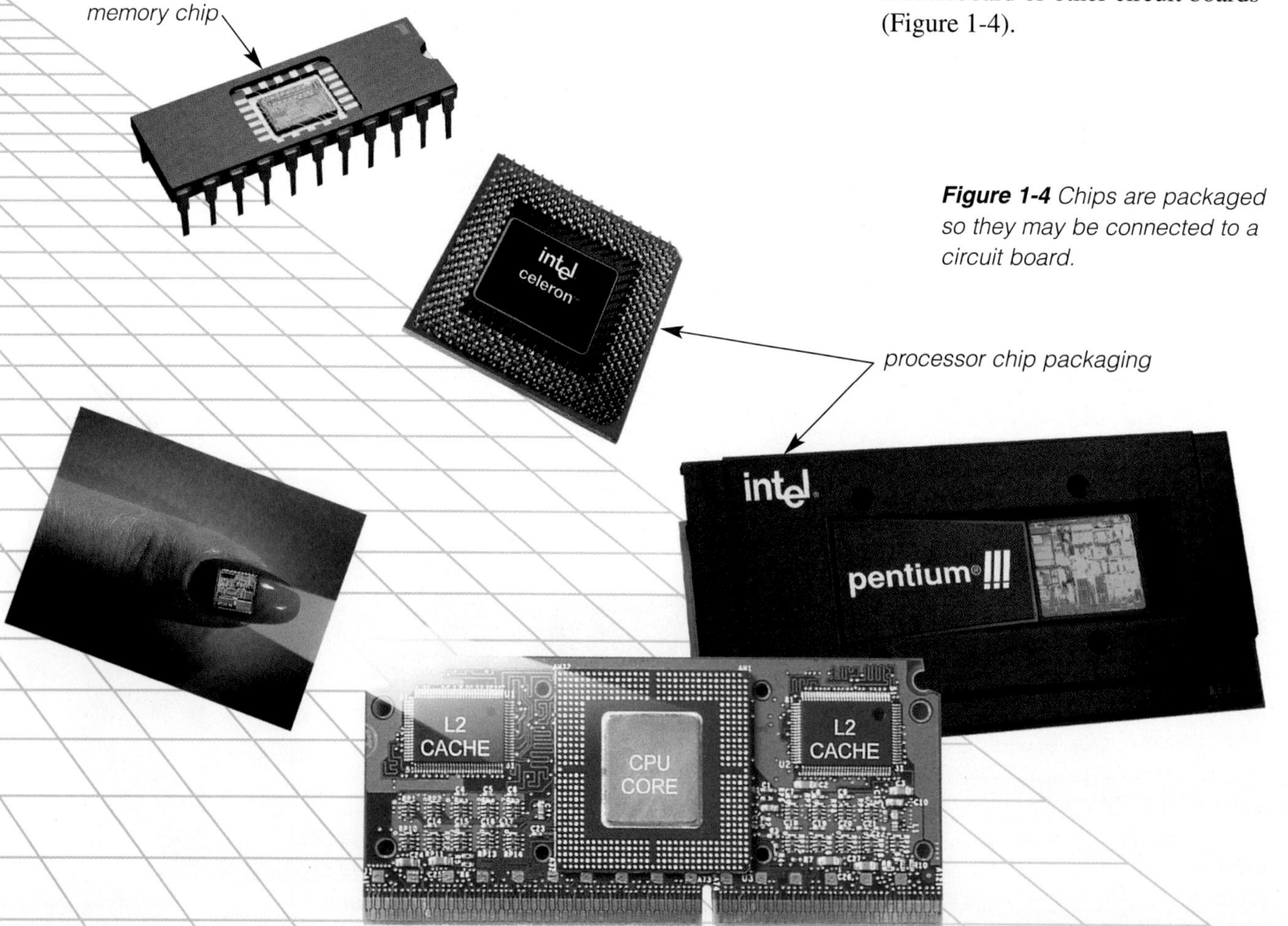

Figure 1-4 *Chips are packaged so they may be connected to a circuit board.*

Some computer components, such as the processor and memory, reside inside the system unit; that is, they are internal. Other components, such as the keyboard, mouse, microphone, monitor, PC camera, and printer, often are located outside the system unit. These devices are considered external. Any external device that attaches to the system unit is called a **peripheral** device.

Storage Devices

Storage holds data, instructions, and information for future use. Storage differs from memory, in that it can hold these items permanently, whereas memory holds these items only temporarily while they are being processed. A **storage medium** (media is the plural) is the physical material on which data, instructions, and information are stored. One commonly used storage medium is a disk, which is a round, flat piece of plastic or metal on which items can be encoded, or written.

Figure 1-5 A floppy disk is inserted and removed from a floppy disk drive.

For more information on storage devices, visit the Discovering Computers 2001 Chapter 1 WEB INFO page (**www.scsite.com/dc2001/ch1/webinfo.htm**) and click Storage Devices.

A **storage device** is used to record and retrieve data, instructions, and information to and from a storage medium. Storage devices often function as a source of input because they transfer items from storage into memory. Four common storage devices are a floppy disk drive, a hard disk drive, a CD-ROM drive, and a DVD-ROM drive. A disk drive is a device that reads from and may write onto a disk.

A floppy disk consists of a thin, circular, flexible disk enclosed in a plastic shell. A floppy disk stores data, instructions, and information using magnetic patterns and can be inserted into and removed from a floppy disk drive (Figure 1-5). A Zip® disk is a higher capacity floppy disk that can store the equivalent of about 70 standard floppy disks.

A hard disk provides much greater storage capacity than a floppy disk. A hard disk usually consists of several circular disks on which data, instructions, and information are stored magnetically. These disks are enclosed in an airtight, sealed case, which often is housed inside the system unit (Figure 1-6). Some hard disks are removable, which means they can be inserted and removed from a hard disk drive, much like a floppy disk. Removable disks are enclosed in plastic or metal cartridges so that they can be removed from the drive. The advantage of

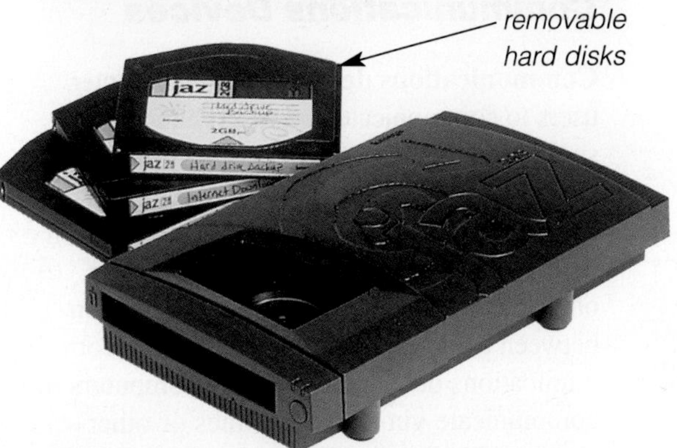

Figure 1-6 Some hard disks are self-contained devices housed inside the system unit (top picture). Removable hard disks, in contrast, are inserted and removed from a drive (bottom picture).

removable media such as a floppy disk and removable hard disk is it can be taken out of the computer and transported or secured.

Another type of disk used to store data is the compact disc (Figure 1-7). A compact disc stores data using microscopic pits, which are created by a laser light. One type of compact disc is a *CD-ROM,* which is accessed or played using a CD-ROM drive. A variation of the standard CD-ROM is the rewriteable CD, also called a CD-RW. Whereas you only can access data on a CD-ROM, you also can erase and store data on a CD-RW. A newer type of compact disc is a *DVD-ROM*, which has tremendous storage capacities – enough for a full-length movie. To use a DVD-ROM, you need a DVD-ROM drive.

Figure 1-7 *A compact disc (CD) is a round, flat piece of metal with a protective plastic coating. Three types of compact discs are CD-ROMs, CD-RWs, and DVD-ROMs.*

Communications Devices

Communications devices enable computer users to communicate and to exchange items such as data, instructions, and information with another computer. Communications devices transmit these items over transmission media, such as cables, telephone lines, or other means, used to establish a connection between two computers. A modem is a communications device that enables computers to communicate via telephone lines or other means. Although modems are available as both external and internal devices, most are internal; that is, contained within the system unit.

For more information on communications devices, visit the Discovering Computers 2001 Chapter 1 WEB INFO page (**www.scsite.com/dc2001/ch1/webinfo.htm**) and click Communications Devices.

WHY IS A COMPUTER SO POWERFUL?

A computer's power is derived from its capability of performing the information processing cycle operations (input, process, output, and storage) with amazing speed, reliability, and accuracy; its capacity to store huge amounts of data and information; and its ability to communicate with other computers.

Speed

Inside the system unit, operations occur through electronic circuits. When data, instructions, and information flow along these circuits, they travel at close to the speed of light. This allows billions of operations to be carried out in a single second.

Reliability

The electronic components in modern computers are dependable because they have a low failure rate. The high reliability of the components enables the computer to produce consistent results.

Accuracy

Computers can process large amounts of data and generate error-free results, provided the data is entered correctly. If inaccurate data is entered, the resulting output will be incorrect. This computing principle – known as garbage in, garbage out (GIGO) – points out that the accuracy of a computer's output depends on the accuracy of the input.

Storage

Many computers can store enormous amounts of data and make this data available for processing any time it is needed. Using current storage devices, the data can be transferred quickly from storage to memory, processed, and then stored again for future use.

Communications

Most computers today have the capability of communicating with other computers. Computers with this capability can share any of the four information processing cycle operations – input, process, output, and storage – with another computer. For example, two computers connected by a communications device such as a modem can share stored data, instructions, and information. When two or more computers are connected together via communications media and devices, they comprise a network. The most widely known network is the Internet, a worldwide collection of networks that links together millions of businesses, government installations, educational institutions, and individuals (Figure 1-8).

COMPUTER SOFTWARE

Software, also called a **computer program** or simply a **program**, is a series of instructions that tells the hardware of a computer what to do. For example, some instructions direct the computer to allow you to input data from the keyboard and store it in memory. Other instructions cause data stored in memory to be used in calculations such as adding a series of numbers to obtain a total. Some instructions compare two values stored in memory and direct the computer to perform alternative operations based on the results of the comparison; and some instructions direct the computer to print a report, display information on the monitor, draw a color graph on the monitor, or store information on a disk.

Before a computer can perform, or **execute**, a program, the instructions in the program must be placed, or loaded, into the memory of the computer. Usually, they are loaded into memory from storage. For example, a program might be loaded from the hard disk of a computer into memory for execution.

When you purchase a program, such as one that contains legal documents, you will receive one or more floppy disks, one or more CD-ROMs, or a single DVD-ROM on which the software is stored (Figure 1-9). To use this software, you often must **install** the software on the computer's hard disk.

WEB INFO

For more information on computer programs, visit the Discovering Computers 2001 Chapter 1 WEB INFO page (**www.scsite.com/dc2001/ch1/webinfo.htm**) and click Computer Programs.

Figure 1-9 *When you buy software, you receive media such as floppy disks, CD-ROMs, or a DVD-ROM that contains the software program.*

Figure 1-8 *The Internet is a worldwide collection of networks that links together millions of businesses, the government, educational institutions, and individuals.*

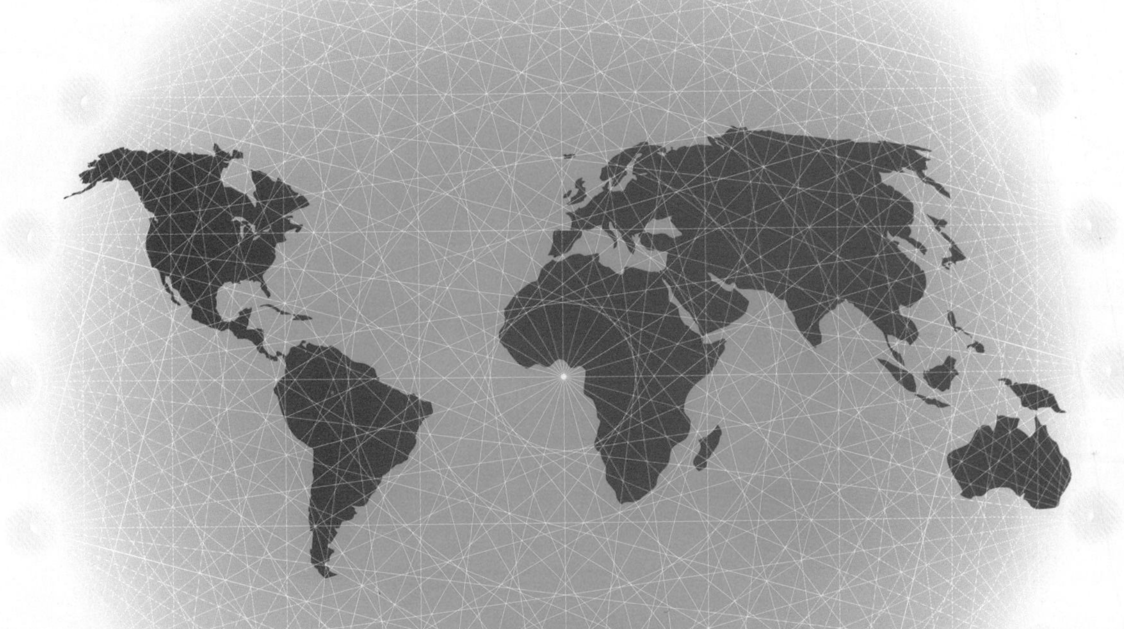

Sometimes, a program can be loaded in memory directly from a floppy disk, CD-ROM, or DVD-ROM so you do not have to install it on a hard disk first.

When you buy a computer, it usually has some software already installed on its hard disk. Thus, you can use the computer as soon as you receive it.

Figure 1-10 illustrates the steps for running a computer program that simulates flying an airplane.

Software is the key to productive use of computers. With the correct software, a computer can become a valuable tool.

Software can be categorized into two types: system software and application software. The following sections describe these categories of software.

WEB INFO

For more information on operating systems, visit the Discovering Computers 2001 Chapter 1 WEB INFO page (**www.scsite.com/dc2001/ch1/webinfo.htm**) and click Operating Systems.

System Software

System software, which consists of programs that control the operations of the computer and its devices, serves as the interface between a user and the computer's hardware. Two types of system software are the operating system and utility programs.

OPERATING SYSTEM The **operating system** contains instructions that coordinate all of the activities of hardware devices. The operating system also contains instructions that allow you to run application software. Microsoft Windows is the name of a popular operating system that is used on many of today's computers.

When you start a computer, the operating system is loaded, or copied, into memory from the computer's hard disk. It remains in memory while the computer is running and allows you to communicate with the computer and other software.

Figure 1-10 *Running a program.*

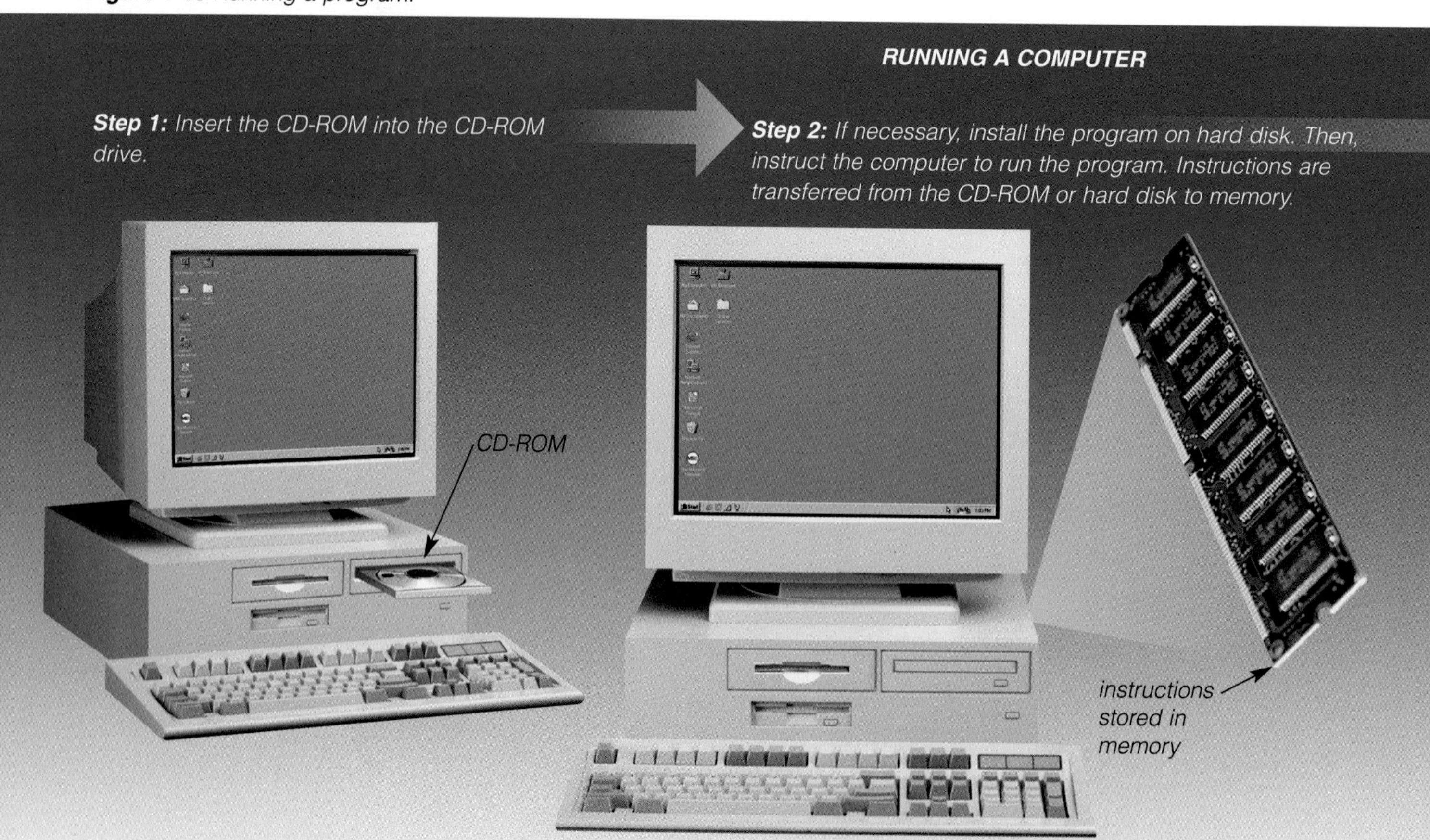
RUNNING A COMPUTER
Step 1: Insert the CD-ROM into the CD-ROM drive.
Step 2: If necessary, install the program on hard disk. Then, instruct the computer to run the program. Instructions are transferred from the CD-ROM or hard disk to memory.
CD-ROM
instructions stored in memory

UTILITY PROGRAMS A utility program is a type of system software that performs a specific task, usually related to managing a computer, its devices, or its programs. An example of a utility program is an uninstaller, which removes a program that has been installed on a computer. Most operating systems include several utility programs for managing disk drives, printers, and other devices. You also can buy stand-alone utility programs to perform additional computer management functions.

USER INTERFACE All software has a **user interface** that is the part of the software with which you interact. The user interface controls how data and instructions are entered and how information is presented on the screen. Many of today's software programs have a **graphical user interface**, or **GUI** (pronounced gooey), which allows you to interact with the software using visual images such as icons. An **icon** is a small image that represents a program, an instruction, or some other object. Figure 1-11 shows a widely used operating system, Microsoft Windows, which has a graphical user interface.

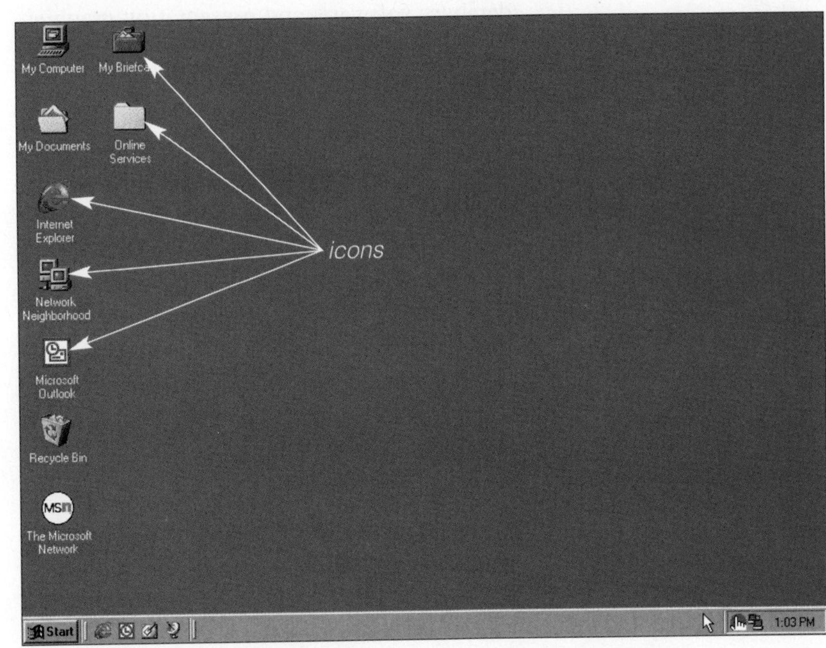

Figure 1-11 *Microsoft Windows is an operating system with a graphical user interface.*

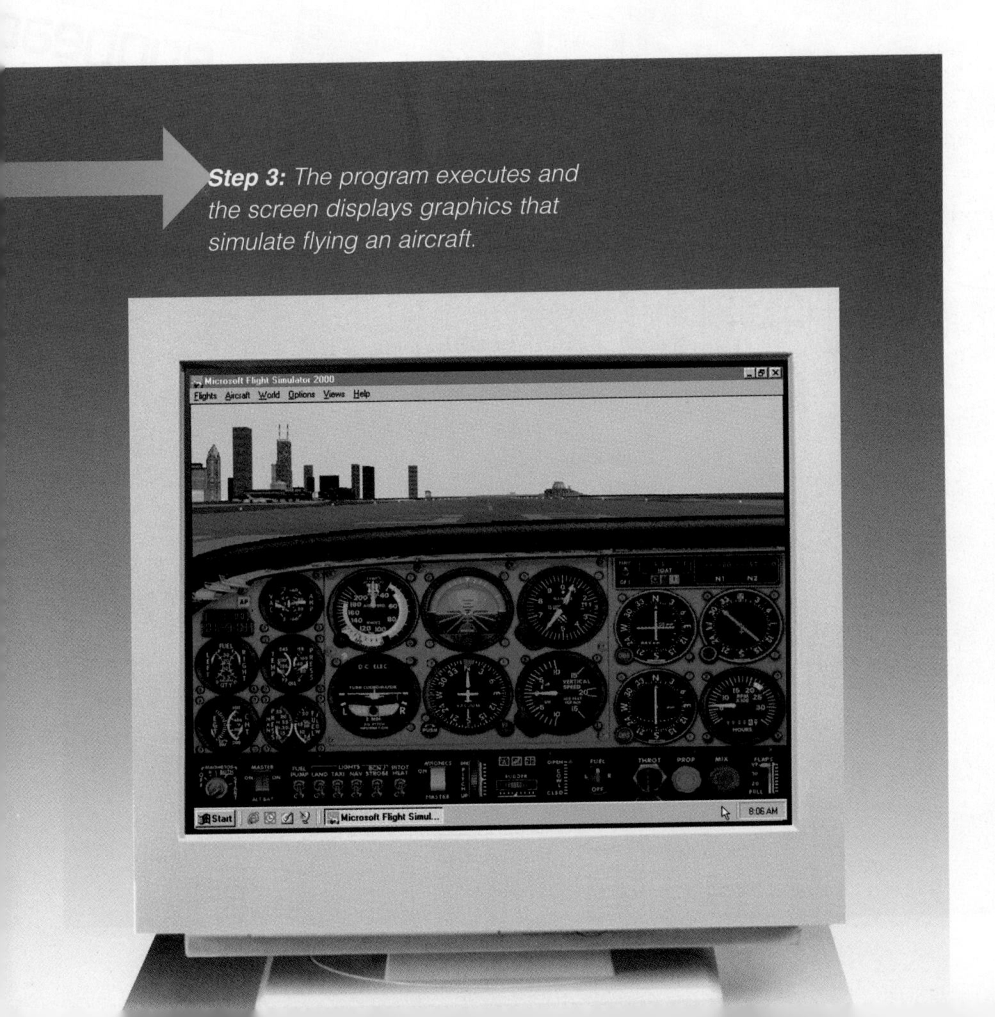

Application Software

Application software consists of programs designed to perform specific tasks for users. Popular application software includes word processing software, spreadsheet software, database software, and presentation graphics software. Word processing software allows you to create documents such as letters and memos. Spreadsheet software allows you to calculate numbers arranged in rows and columns and often is used for budgeting, forecasting, and other financial tasks. Database software is used to store data in an organized fashion, as well as to retrieve, manipulate, and display that data in a meaningful form. Presentation graphics software allows you to create documents called slides that are used in making presentations. These four applications often are sold together as a single unit, called a suite, in which individual applications are packaged in the same box and sold for a price that is significantly less than buying the applications individually.

Many other types of application software exist, thus enabling users to perform a variety of tasks. Some widely used software applications include: reference, education, and entertainment; desktop publishing; photo and video editing; multimedia authoring; network, communications, electronic mail, and Web browsers; accounting; project management; and personal information management. Each of these applications is discussed in depth in Chapter 2.

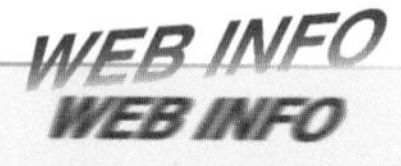

For more information on application software, visit the Discovering Computers 2001 Chapter 1 WEB INFO page (**www.scsite.com/dc2001/ch1/webinfo.htm**) and click Application Software.

Figure 1-12a

Application software is available as packaged software, custom software, shareware, freeware, and public-domain software.

PACKAGED SOFTWARE **Packaged software** is designed to meet the needs of a wide variety of users, not just a single user or company. Packaged software sometimes is called commercial off-the-shelf software because you can purchase these programs off the shelf from software vendors or stores that sell computer products (Figure 1-12a). You also can purchase packaged software on the Internet (Figure 1-12b). Some companies today offer products on the Internet; that is, instead of installing the software onto your computer, you run the programs on the Internet.

Figure 1-12b

Figure 1-12 *Packaged software programs can be purchased from computer stores, office equipment suppliers, retailers, and software vendors.*

CUSTOM SOFTWARE Sometimes a user or organization with unique software requirements cannot find packaged software that meets all of its needs. In this case, the user or organization can use **custom software**, which is a program or programs developed at a user's request to perform specific functions.

SHAREWARE **Shareware** is software that is distributed free for a trial period. If you want to use a shareware program beyond that period of time, you are expected to send a payment to the person or company that developed the program. Upon sending this small fee, the developer registers you to receive service assistance and updates.

FREEWARE AND PUBLIC-DOMAIN SOFTWARE **Freeware** is software that is provided at no cost to a user by an individual or company. Although free, freeware is copyrighted, meaning you cannot resell it as your own. **Public-domain software** is free software that has been donated for public use and has no copyright restrictions.

Examples of shareware, freeware, and public-domain software include utility programs, graphics programs, and games. Thousands of these programs are available on the Internet; you also can obtain copies of the program from the developer, a coworker, or a friend.

Software Development

People who write software programs are called **computer programmers** or **programmers**. Programmers write the instructions necessary to direct the computer to process data into information. The instructions must be placed in the correct sequence so the desired results occur. Complex programs can require hundreds of thousands of program instructions.

When writing complex programs for large businesses, programmers often follow a plan developed by a systems analyst. A **systems analyst** manages the development of a program, working with both the user and the programmer to determine and design the desired output of the program.

Programmers use a programming language to write computer programs. Some programming languages, such as Microsoft Visual Basic, have a graphical user interface that makes it easier to include the correct instructions in a program. Figure 1-13 illustrates the steps involved to create a program using the Microsoft Visual Basic programming language.

NETWORKS AND THE INTERNET

A **network** is a collection of computers and devices connected together via communications media and devices such as cables, telephone lines, modems, or other means. Sometimes a network is wireless; that is, uses no physical lines or wires. When your computer is connected to a network, you are said to be online.

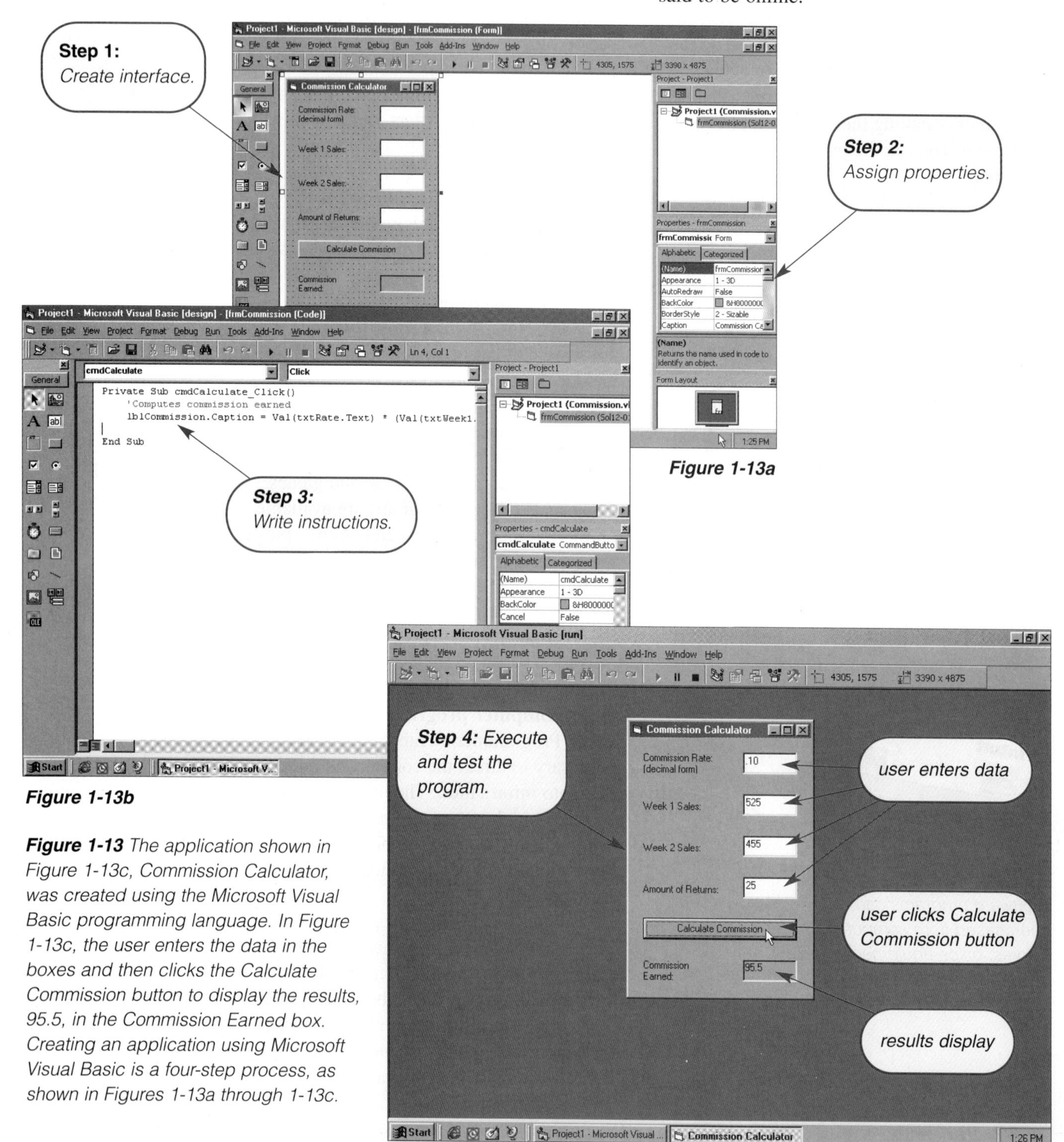

Figure 1-13a

Figure 1-13b

Figure 1-13c

***Figure 1-13** The application shown in Figure 1-13c, Commission Calculator, was created using the Microsoft Visual Basic programming language. In Figure 1-13c, the user enters the data in the boxes and then clicks the Calculate Commission button to display the results, 95.5, in the Commission Earned box. Creating an application using Microsoft Visual Basic is a four-step process, as shown in Figures 1-13a through 1-13c.*

Computers are networked together so users can share **resources**, such as hardware devices, software programs, data, and information. Sharing resources saves time and money. For example, instead of purchasing one printer for every computer in a company or in a home, you can connect a single printer and all computers via a network (Figure 1-14); the network enables all of the computers to access the same printer.

Most business computers are networked together. These networks can be relatively small or quite extensive. A network that connects computers in a limited geographic area, such as a school computer laboratory, office, or group of buildings, is called a local area network (LAN). A network that covers a large geographical area, such as one that connects the district offices of a national corporation, is called a wide area network (WAN) (Figure 1-15).

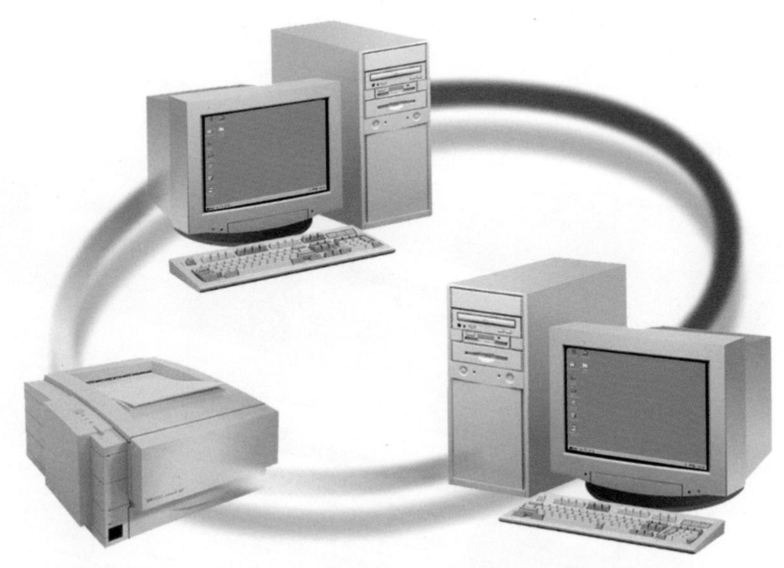

Figure 1-14 *This local area network (LAN) enables two separate computers to share the same printer.*

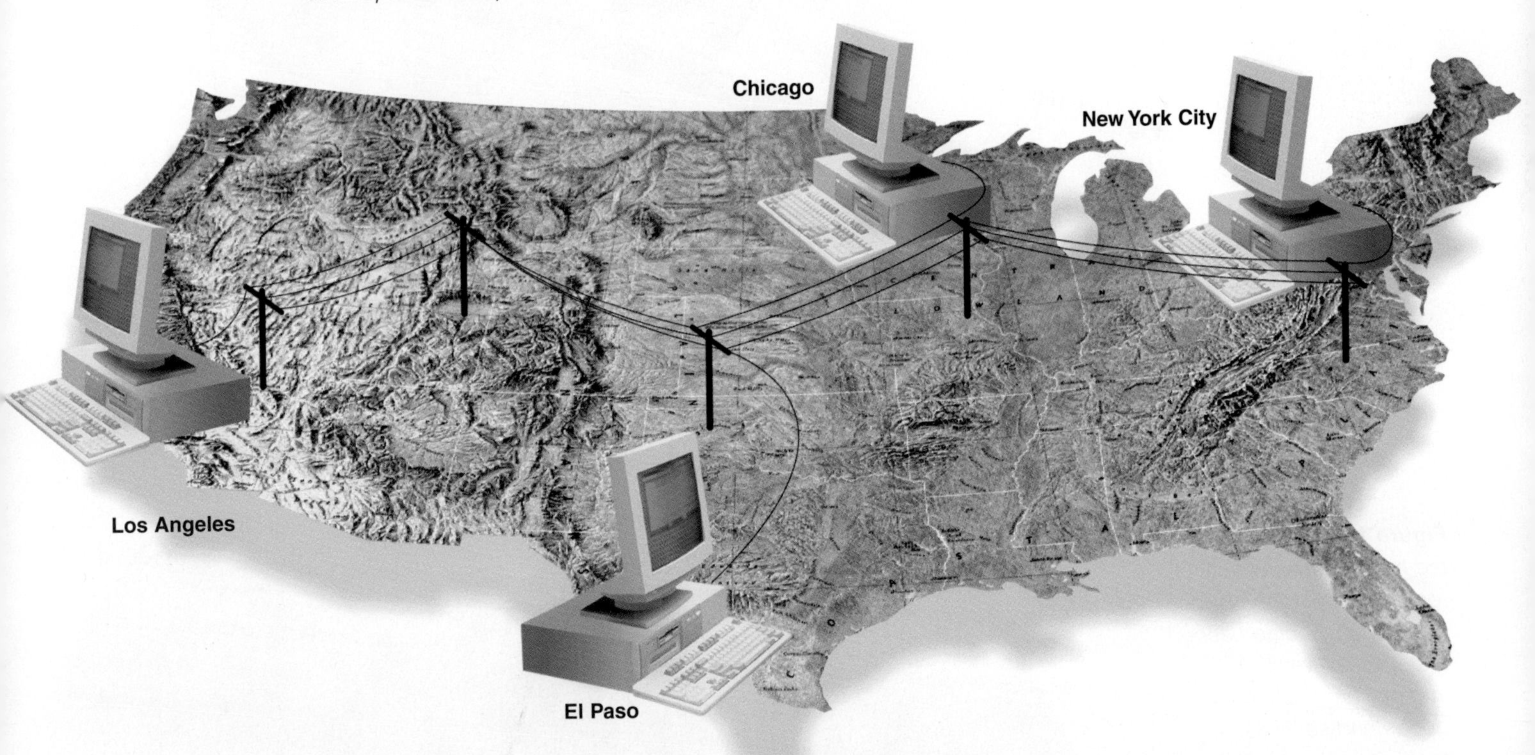

Figure 1-15 *A network can be quite large and complex, connecting users in district offices around the country (WAN).*

The world's largest network is the **Internet**, which is a worldwide collection of networks that links together millions of computers by means of modems, telephone lines, wireless technology, and other communications devices and media. With an abundance of resources and data accessible via the Internet, more than 125 million users around the world are making use of the Internet for a variety of reasons, some of which include the following (Figure 1-16):

- Sending messages to other connected users (e-mail)
- Accessing a wealth of information, such as news, maps, airline schedules, and stock market data
- Shopping for goods and services
- Meeting or conversing with people around the world
- Accessing sources of entertainment and leisure, such as online games, magazines, and vacation planning guides

WEB INFO

For more information on the Internet, visit the Discovering Computers 2001 Chapter 1 WEB INFO page (**www.scsite.com/dc2001/ch1/webinfo.htm**) and click Internet.

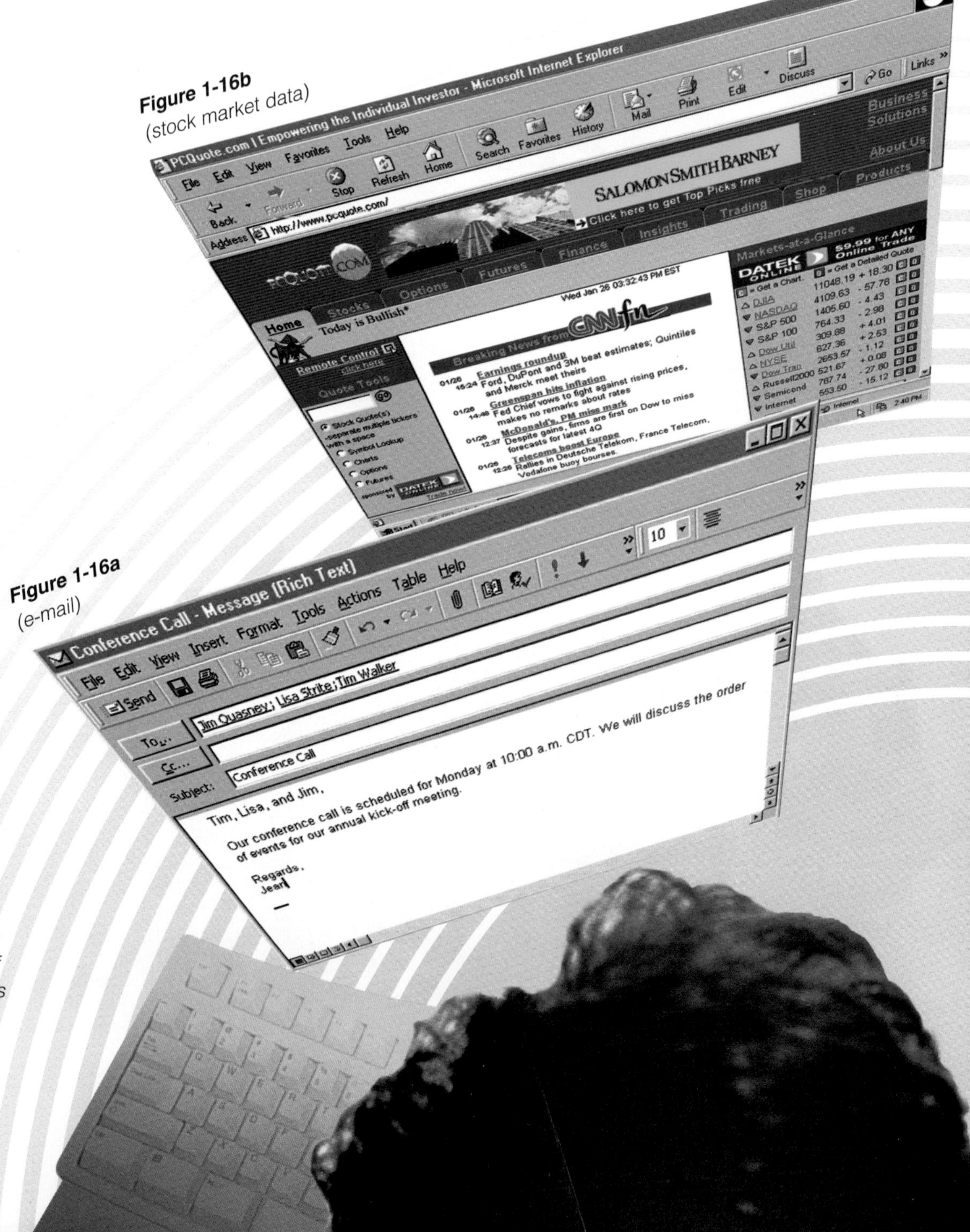

Figure 1-16 Users access the Internet for a variety of reasons: to send messages to other connected users, to access a wealth of information, to shop for goods and services, to meet or converse with people around the world, and for entertainment.

Most users connect to the Internet in one of two ways: through an Internet service provider or through an online service. An Internet service provider (ISP) is an organization that supplies connections to the Internet for a monthly fee. Like an ISP, an online service provides access to the Internet, but it also provides a variety of other specialized content and services such as financial data, hardware and software guides, news, weather, legal information, and other similar commodities. For this reason, the fees for using an online service usually are slightly higher than fees for using an ISP. Two popular online services are America Online and The Microsoft Network.

Figure 1-16c (shopping)

Figure 1-16d (meeting people)

Figure 1-16e (vacation planning)

One of the more popular segments of the Internet is the World Wide Web, also called the Web, which contains billions of documents called Web pages. A Web page is a document that contains text, graphics, sound, or video, and has built-in connections, or links, to other Web documents. Web pages are stored on computers throughout the world. A Web site is a related collection of Web pages. You access and view Web pages using a software program called a Web browser. The two most popular Web browsers are Microsoft Internet Explorer and Netscape Navigator. Figure 1-17 illustrates one method of connecting to the Web and displaying a Web page.

CATEGORIES OF COMPUTERS

The four major categories of computers are personal computers, minicomputers, mainframe computers, and supercomputers. These categories are based on the differences in the size, speed, processing capabilities, and price of computers. Due to rapidly changing

Figure 1-17 *One method of connecting to the Web and displaying a Web page.*

Step 1: *Use your computer and modem to make a local telephone call to an online service, such as The Microsoft Network.*

CONNECTING TO THE WEB AND DISPLAYING A WEB PAGE

Step 2: *A Web browser such as Internet Explorer displays on your screen.*

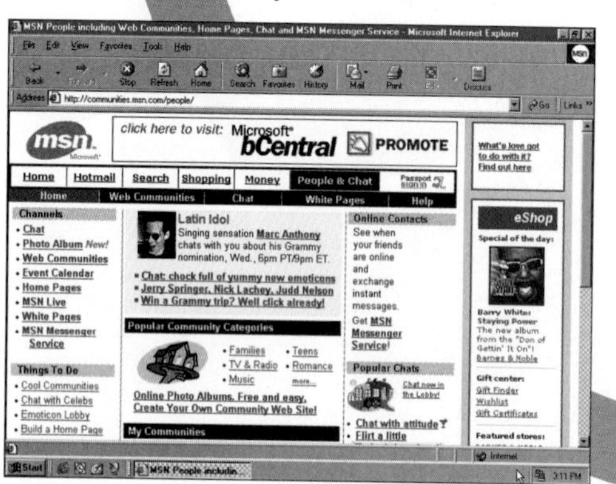

Web address

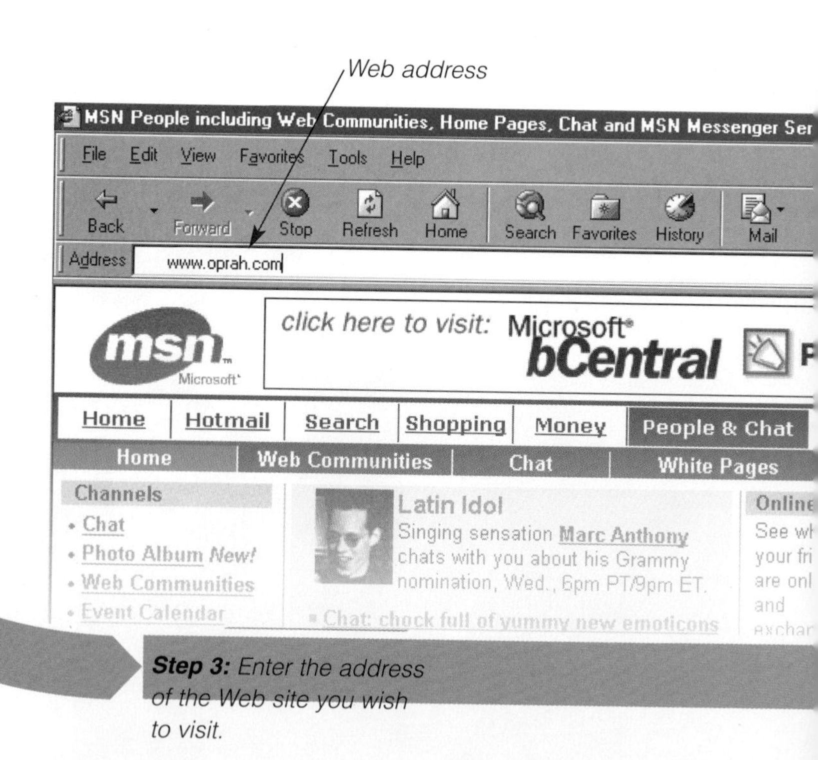

Step 3: *Enter the address of the Web site you wish to visit.*

technology, the categories cannot be defined precisely. For example, the speed used to define a mainframe today may be used to define a minicomputer next year. Some characteristics may overlap categories. Still, they frequently are used and should be understood. Figure 1-18 summarizes the four categories of computers, which are discussed in the following pages.

PERSONAL COMPUTERS

A **personal computer** (**PC**) is a computer that can perform all of its input, processing, output, and storage activities by itself; that is, it contains at least one input device, one output device, one storage device, memory, and a processor. The processor, sometimes called a microprocessor, is a central processing unit (CPU) on a single chip and is the basic building block of a PC.

WEB INFO

For more information on personal computers, visit the Discovering Computers 2001 Chapter 1 WEB INFO page (**www.scsite.com/dc2001/ch1/webinfo.htm**) and click Personal Computers.

CATEGORIES OF COMPUTERS

Category	Physical size	Number of instructions executed per second	Number of simultaneously connected users	General price range
Personal computer	Fits in your hand or on a desk	Up to 400 million	One stand-alone or many networked	Several thousand dollars or less
Minicomputer	Small cabinet	Thousands to millions	Two to 4,000	$5,000 to $150,000
Mainframe	Partial room to a full room of equipment	Millions	Hundreds to thousands	$300,000 to several million dollars
Supercomputer	Full room of equipment	Millions to billions	Hundreds to thousands	Several million dollars and up

Figure 1-18 *This table summarizes some of the differences among the categories of computers. Because of rapid changes in technology, these should be considered general guidelines only.*

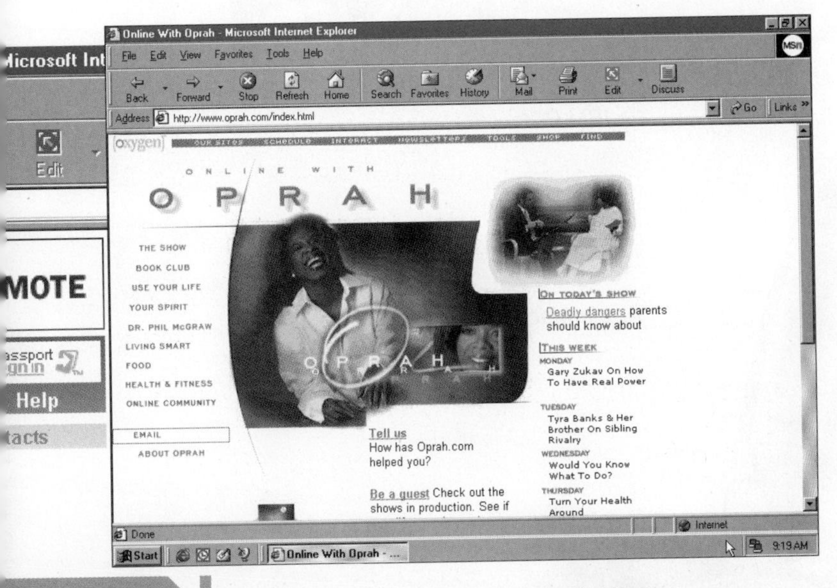

Step 4: *The Web browser locates the Web site for the entered address and displays a Web page on your screen.*

Two popular series of personal computers are the PC (Figure 1-19) and the Apple Macintosh (Figure 1-20). These two types of computers have different processors and use different operating systems. The PC and compatibles use the Windows operating system, whereas the Apple Macintosh uses the Macintosh operating system. Today, the terms PC and compatible are used to refer to any personal computer that is based on specifications of the original IBM PC computer. Companies such as Gateway, Compaq, Dell, and Toshiba all sell PC-compatible computers.

Two major categories of personal computers are desktop computers and portable computers. These types of personal computers are discussed in the next two sections.

Desktop Computers

A **desktop computer** is designed so the system unit, input devices, output devices, and any other devices fit entirely on or under a desk or table (Figure 1-21). In some desktop models, the system unit is placed horizontally on top of a desk along with the other devices. A **tower model**, in contrast, has a tall and narrow system unit that is designed to be placed on the floor vertically. Tower model desktop computers are available in a variety of heights: a full tower is at least 21 inches tall, a mid-tower is about 16 inches tall, and a mini-tower is usually 13 inches tall. The model of desktop computer you use often is determined by the design of your workspace.

A less expensive desktop computer that combines the monitor and system unit into a single device is called an **all-in-one computer** (Figure 1-22). These compact computers are ideal for the casual home user.

Figure 1-19 The PC and compatibles use the Windows operating system.

Figure 1-20 The Apple Macintosh uses the Macintosh operating system.

A more expensive and powerful desktop computer, called a **workstation**, is designed for work that requires intense calculations and graphics capabilities. Users in fields such as engineering, desktop publishing, and graphic art use workstations. For example, a workstation would be used to view and create maps or create computer animated special effects for Hollywood movies.

A computer that is not connected to a network and has the capability of performing the information processing cycle operations (input, process, output, and storage) by itself is called a **stand-alone** computer. Most desktop computers today, however, have networking capabilities. Another use of the term workstation refers to any computer connected to a network, regardless of its category.

A desktop computer also can function as a server on a network. A **server** is a computer that manages the resources on a network. Servers control access to the software,

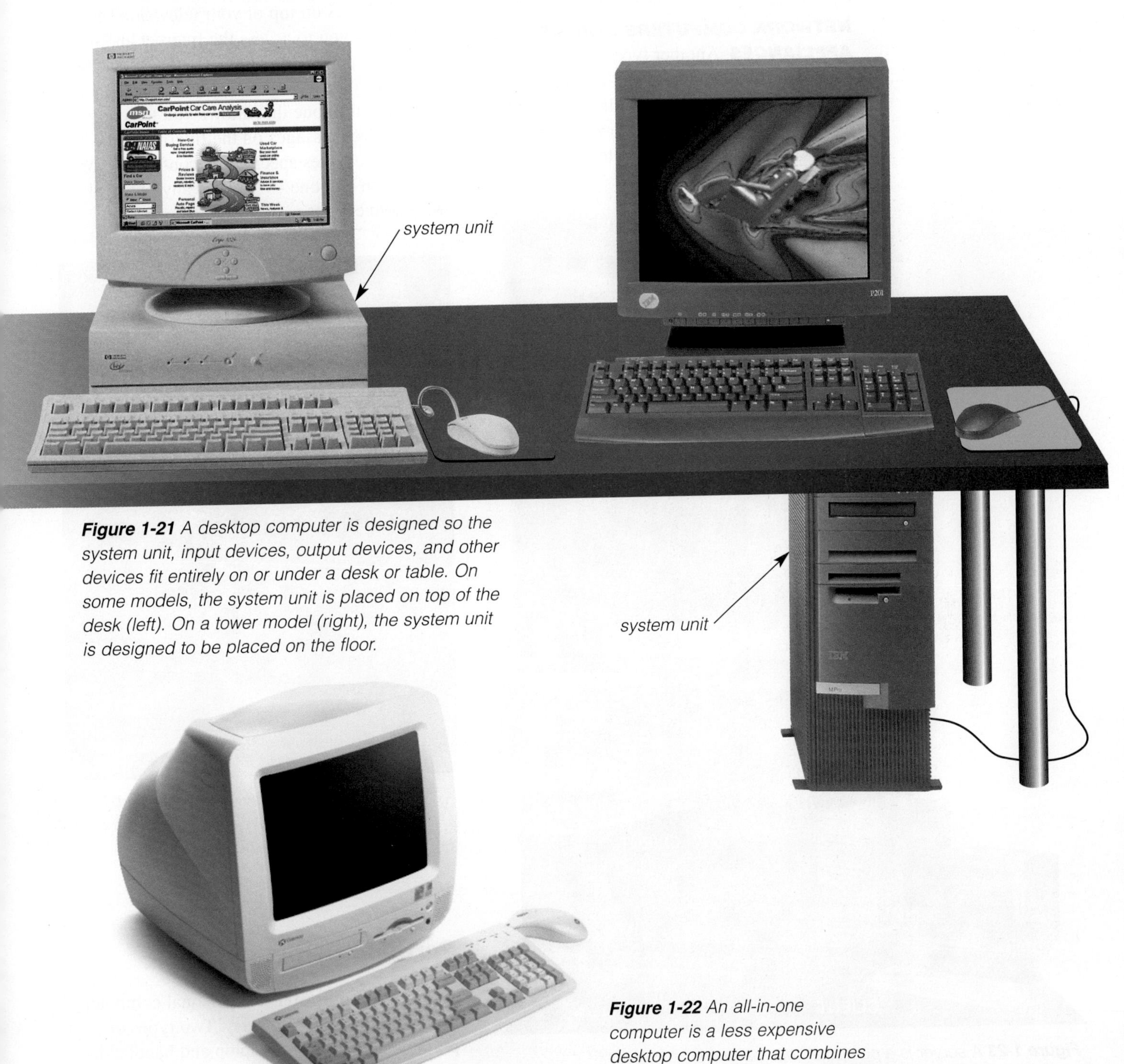

Figure 1-21 *A desktop computer is designed so the system unit, input devices, output devices, and other devices fit entirely on or under a desk or table. On some models, the system unit is placed on top of the desk (left). On a tower model (right), the system unit is designed to be placed on the floor.*

Figure 1-22 *An all-in-one computer is a less expensive desktop computer that combines the monitor and system unit into a single device.*

software, printers, and other devices on the network as well as provide a centralized storage area for software programs and data. The other computers on the network, called clients, can access the contents of the storage area on the servers (Figure 1-23). In a network, one or more computers usually are designated as the server(s). The major difference between the server and client computers is that the server ordinarily is faster and has more storage space.

NETWORK COMPUTERS AND WEB APPLIANCES Another type of personal computer is called a **network computer (NC)**. It is designed specifically to connect to a network, especially the Internet. Most network computers cannot operate as stand-alone computers; that is, they have to be connected to a network to be functional. Because a network computer relies on the network for storage, it typically does not have a hard disk or CD-ROM drive.

Many consumers today use set-top boxes, smart phones, and other Web-enabled devices. Called **Web appliances** or **Internet appliances**, these devices are designed specifically to connect to the Internet at home or on the road. A set-top box, such as WebTV™, sits on top of your television set and allows you to access the Internet and navigate Web pages using a device that looks like a remote control (Figure 1-24). A smart phone is a cellular phone that allows you to send and receive messages on the Internet and browse Web sites specifically configured for display on a phone — without any physical wire connections.

Figure 1-23 *A server is a computer that manages resources on a network. Other computers on the network are called clients.*

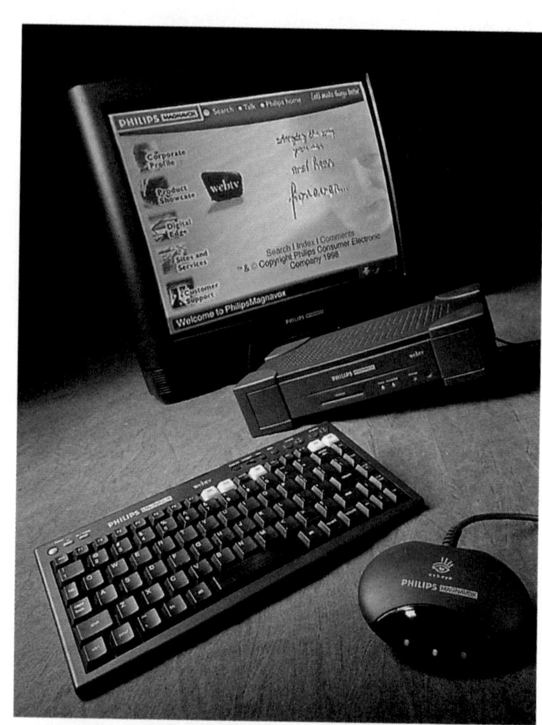

Figure 1-24 *With a device such as WebTV™, you can access the Web from the comfort of your family room or a hotel room.*

Portable Computers

A **portable computer** is a personal computer that is small enough to carry. Two types of portable computers are laptop and handheld. Each of these types of portable computers is discussed next.

LAPTOP COMPUTERS Designed for mobility, a **laptop computer**, also called a **notebook computer**, is a personal computer small enough to fit on your lap. Today's laptop computers are thin, lightweight, and can be just as powerful as the average desktop computer. Laptop computers generally are more expensive than their desktop counterparts.

On a typical laptop computer, the keyboard is located on top of the system unit, the monitor attaches to the system unit with a hinge, and the drives are built into the system unit (Figure 1-25). Weighing on average between four and ten pounds, these computers can be transported easily from place to place. Most laptop computers can run either on batteries or using a standard power supply. Users with mobile computing needs, such as business travelers, often use a laptop.

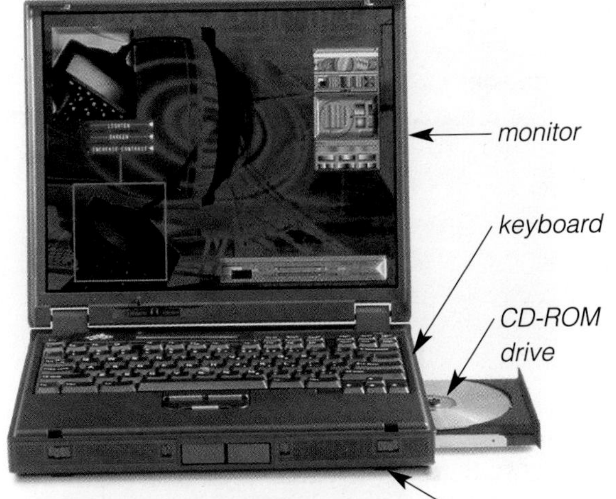

Figure 1-25 *On a typical laptop computer, the keyboard is located on top of the system unit, the monitor attaches to the system unit with a hinge, and the drives are built into the system unit.*

WEB INFO

For more information on laptop computers, visit the Discovering Computers 2001 Chapter 1 WEB INFO page (**www.scsite.com/dc2001/ch1/webinfo.htm**) and click Laptop Computers.

HANDHELD COMPUTERS A **handheld computer** or **palmtop computer** is a small personal computer designed to fit in your hand (Figure 1-26). Because of their reduced size, the keyboards and screens on handheld computers are quite small. Computers in the handheld category usually do not have disk drives; instead, programs and data are stored on chips inside the system unit. Many handheld computers can be connected to a larger computer for the purpose of exchanging information. A business traveler or other mobile user might use a handheld computer if a laptop computer is too large.

Figure 1-26 *A handheld computer is a small personal computer designed to fit in your hand.*

A popular type of handheld computer is the **Personal Digital Assistant (PDA)**. A PDA often supports personal information management (PIM) applications such as a calendar, appointment book, calculator, memo pad, and even telephone services and Internet access. Today, a wide variety of people use PDAs instead of writing in a pocket-sized appointment book. Some PDAs are Web-enabled, allowing you to access the Internet wirelessly.

WEB INFO

For more information on PDAs, visit the Discovering Computers 2001 Chapter 1 WEB INFO page (**www.scsite.com/dc2001/ch1/webinfo.htm**) and click PDAs.

Because handheld computers have such small keyboards, many of these computers use pen input, which allows you to write on the screen instead of typing on a keyboard (Figure 1-27). These computers, called **pen computers**, contain special software that permits the computer to recognize handwritten characters and other symbols. Many PDAs

Figure 1-27 *Pen computers, such as this PDA, allow you to write directly on the screen of the handheld computer.*

Figure 1-28 *Minicomputers are more powerful than a workstation but less powerful than a mainframe.*

use pen input and can be considered pen computers. Pen computers also are used by parcel delivery individuals, meter readers, and other people whose jobs require them to move from place to place.

MINICOMPUTERS

A **minicomputer**, such as the one shown in Figure 1-28, is more powerful and larger than a workstation computer. Minicomputers often can support up to 4,000 connected users at the same time. Users often access a minicomputer via a **terminal**, which is a device with a monitor and keyboard. Such terminals – sometimes called **dumb terminals** because they have no processing power – cannot act as stand-alone computers and must be connected to the minicomputer to operate.

A minicomputer also can act as a server in a network environment. In this case, personal computers access the minicomputer.

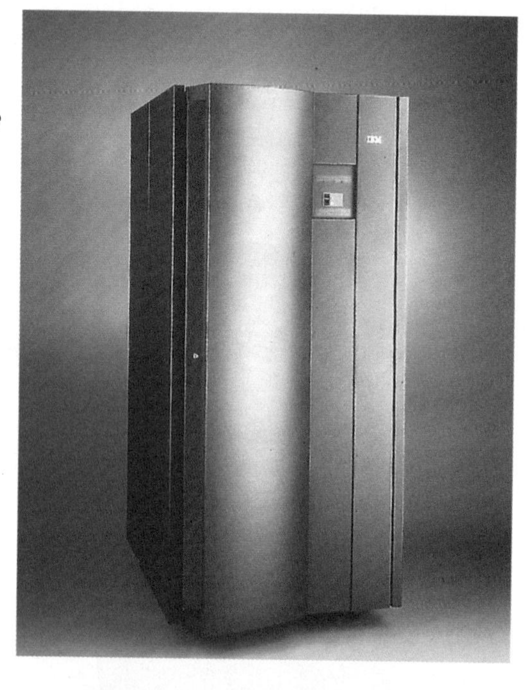

Figure 1-29 *Mainframe computers are large, expensive, powerful machines that can handle thousands of connected users simultaneously and process up to millions of instructions per second.*

MAINFRAME COMPUTERS

A **mainframe** is a large, expensive, very powerful computer that can handle hundreds or thousands of connected users simultaneously (Figure 1-29). Like minicomputers, mainframes also can act as a server in a network environment. Mainframes can store tremendous amounts of data, instructions, and information, which users can access with terminals or personal computers.

SUPERCOMPUTERS

A **supercomputer**, shown in Figure 1-30, is the fastest, most powerful computer – and the most expensive. Capable of processing more than 64 billion instructions in a single second, supercomputers are used for applications requiring complex, sophisticated mathematical calculations. For example, a supercomputer would be used for weather forecasting, nuclear energy research, and petroleum exploration.

Figure 1-30 *Supercomputers are the fastest, most powerful, and most expensive computers.*

EXAMPLES OF COMPUTER USAGE

Every day, numerous users rely on different types of computers for a variety of applications. Whether used to run complex application software, connect to a network, or perform countless other functions, computers are powerful tools at home, at work, and at play. To illustrate the variety of uses for computers, this section takes you on a visual and narrative tour of five categories of users: a home user, a small business user, a mobile user, a large business user, and a power user (Figure 1-31). Examples of hardware and software listed in the table below are presented on the following pages.

USER	HARDWARE/NETWORK	SOFTWARE
Home User	• Desktop computer; set-top box or other Web appliance; handheld computer • Internet	• Reference (e.g., encyclopedias, medical dictionaries, road atlas) • Entertainment (e.g., games, music composition, greeting card publishing) • Educational (e.g., foreign language tutorials, children's math and reading software) • Computer-based training • Productivity (e.g., word processor, spreadsheet) • Personal finance, online banking • Communications and Web browser • E-mail
SMALL BUSINESS USER	• Desktop computer; handheld computer (PDA); shared network printer • Local area network • Internet	• Productivity (e.g., word processor, spreadsheet, database) • Company specific (e.g., accounting, legal reference) • Communications and Web browser • May use network versions of some software packages • E-mail
MOBILE USER	• Laptop computer equipped with a modem; laptop carrying case; video projector • Web-enabled PDA • Internet • Local area network	• Productivity (e.g., word processor, spreadsheet, presentation graphics) • Personal information management • Communications and Web browser • E-mail
LARGE BUSINESS USER	• Minicomputer or mainframe computer • Desktop or laptop computer; handheld computer (PDA); kiosk • Local area network or wide area network, depending on the size of the company • Internet	• Productivity (e.g., word processor, spreadsheet, database, presentation graphics) • Personal information management • Desktop publishing • Accounting • Network management • Communications and Web browser • May use network versions of some software packages • E-mail
POWER USER	• Workstation or other powerful computer with multimedia capabilities • Local area network • Internet	• Desktop publishing • Multimedia authoring • Photo, sound, and video editing • Communications and Web browser • Computer-aided design

Figure 1-31 *Today, computers are used in millions of businesses and homes to support work tasks and leisure activities. Depending on their intended usage, different computer users require different kinds of hardware and software to meet their needs effectively. The types of users are listed here together with the hardware, software, and network types used most commonly by each.*

Home User

An increasingly common item in the home is a desktop computer, which can be used for different purposes by various family members, including entertainment; research and education; budgeting and personal financial management; home business management; personal and business communications; Web access; and shopping.

More than likely, the **home user** has purchased a variety of software applications (Figure 1-32). Reference software, such as encyclopedias, medical dictionaries, or a road atlas, provides valuable and thorough information for everyone in the family. Software also provides hours of entertainment. For example, you can play games such as solitaire, chess, and Monopoly™; compose music; make a family tree; or create a greeting card. Educational software helps adults learn to speak a foreign language and youngsters to read, write, count, and spell. To make computers easier for children to use, many companies design special hardware just for kids (Figure 1-33).

Figure 1-32b (greeting card)

Figure 1-32a (personal finance)

Figure 1-32 A variety of software products are used in a typical household.

Figure 1-32c (word processing)

Figure 1-33 To assist young children at the computer, special keyboards are designed for their small hands.

For other users in the house, a variety of productivity software exists. Most computers today are sold with word processing and spreadsheet software already installed. Personal finance software helps to prepare taxes, balance a checkbook, and manage investments and family budgets, as well as allowing you to connect to your bank via the Internet to pay bills online. Other software assists in organizing names and addresses, setting up maintenance schedules, preparing legal documents, and purchasing stocks online.

Today, more than 50 million home users also access the Web (Figure 1-34). Once online, users can retrieve a tremendous amount of information, shop for products and services, communicate with others around the world using e-mail or chat rooms, or even take college classes. Some home users access the Web through the desktop computer, while others use Web appliances.

Handheld computers also find a place in the home. For example, a handheld computer commonly is used to maintain daily schedules and address lists or is used for a more specialized purpose such as managing and monitoring the health condition of a family member.

Already, half of today's homes have one or more computers. As computers continue to drop in price, they will become a more common component of the family household.

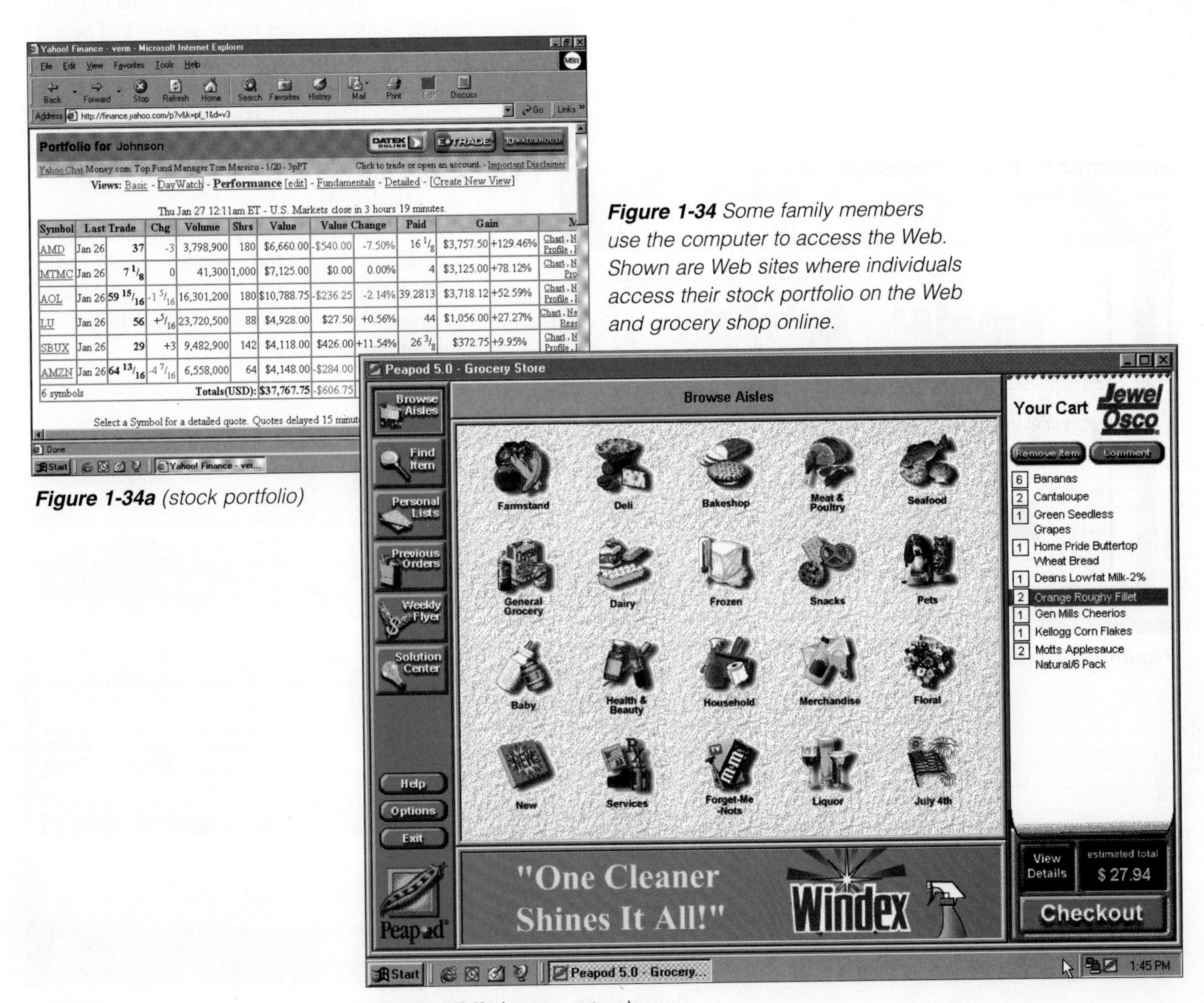

Figure 1-34 *Some family members use the computer to access the Web. Shown are Web sites where individuals access their stock portfolio on the Web and grocery shop online.*

Figure 1-34a *(stock portfolio)*

Figure 1-34b *(grocery store)*

Figure 1-35 Small businesses often provide a desktop personal computer for some or all of their employees.

Small Business User

Computers also play an important role in helping **small business users** manage their resources effectively. A small business, sometimes called a **small office/home office (SOHO)**, such as a local law practice, accounting firm, travel agency, or florist, usually has fewer than fifty employees. Small businesses often provide a desktop personal computer for some or all of their employees (Figure 1-35). Many individuals also will have handheld computers, specifically Web-enabled PDAs, to manage appointments and contact information.

Small businesses often have a local area network to connect the computers in the company. Networking the computers saves money on both hardware and software. For example, a company can avoid the expense of buying multiple printers by connecting a single shared printer to the network. The company also can purchase a network version of a software package, which usually costs less than purchasing a separate software license for each individual desktop computer. Employees can access the software on a server as needed.

For business document preparation, finances, and tracking, a small business owner usually purchases basic productivity software such as word processing and spreadsheet software (Figure 1-36).

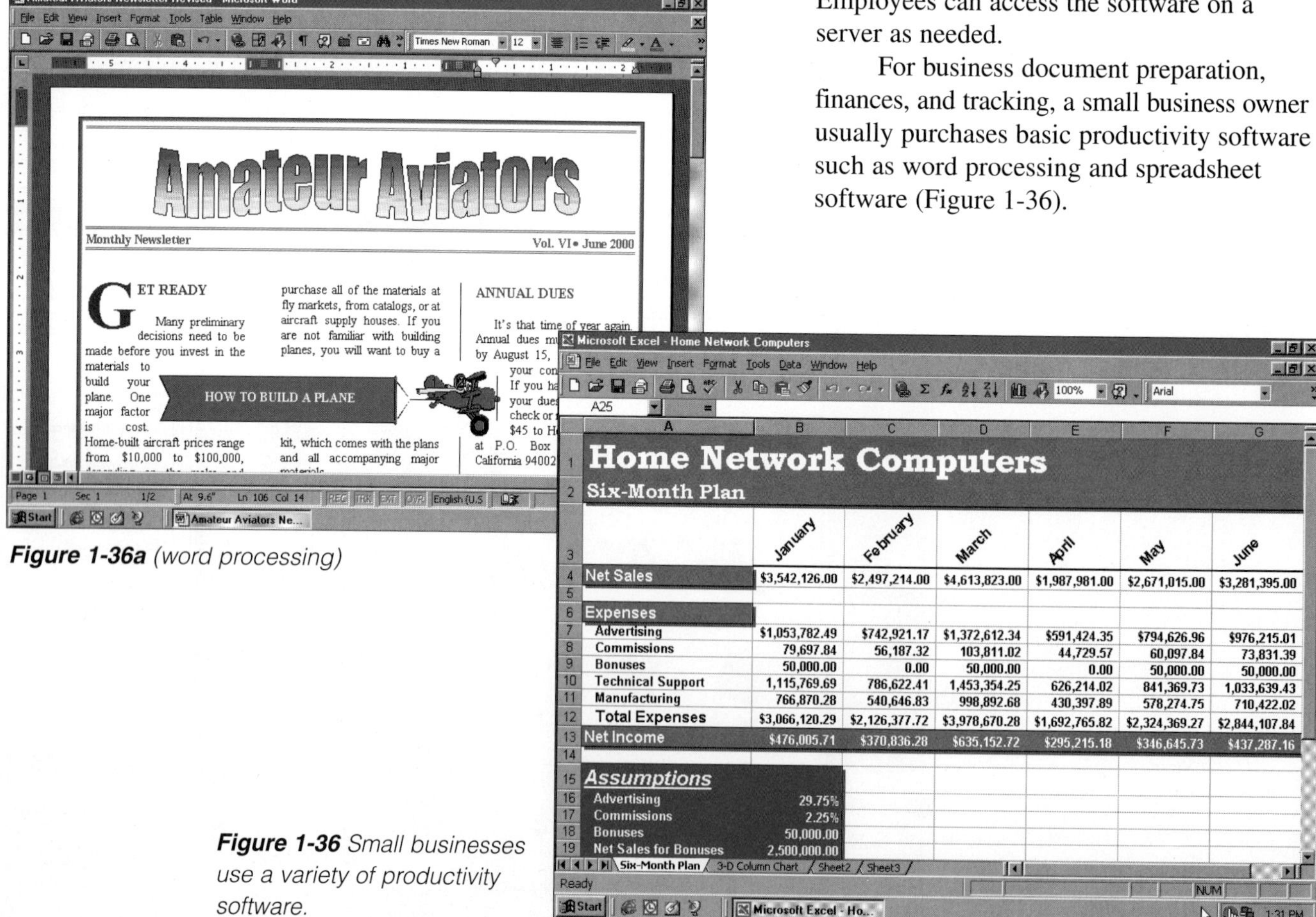

Figure 1-36a (word processing)

Figure 1-36 Small businesses use a variety of productivity software.

Figure 1-36b (spreadsheet)

Businesses may use other types of software, specific to the industry or company. An accounting firm, for example, will have accounting software to prepare journals, ledgers, income statements, balance sheets, and other documents.

The employees in a small business usually have access to the Web, so they can connect to online references such as the Yellow Pages, travel information (Figure 1-37), package shipping information (Figure 1-38), and postal rates, as well as legal information relevant to small businesses. Today, many small businesses are building their own Web sites to advertise their products and services and take orders and requests from customers (Figure 1-39).

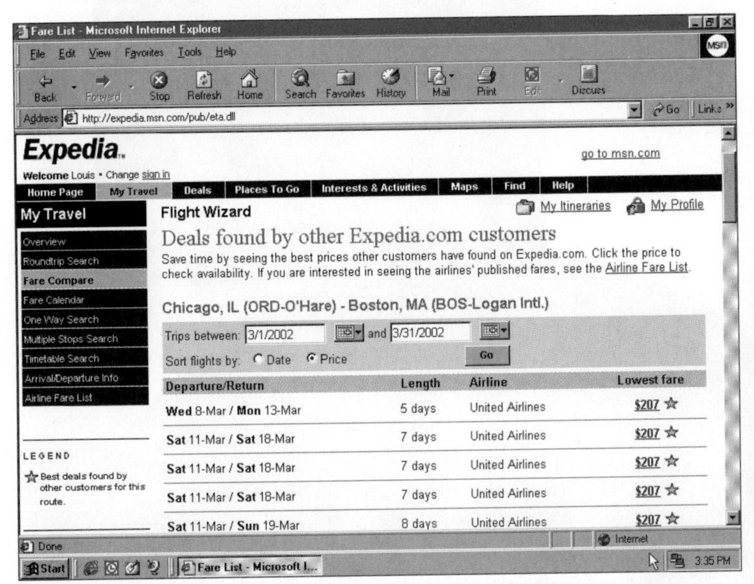

Figure 1-37 *Making travel arrangements on the Web saves time and money.*

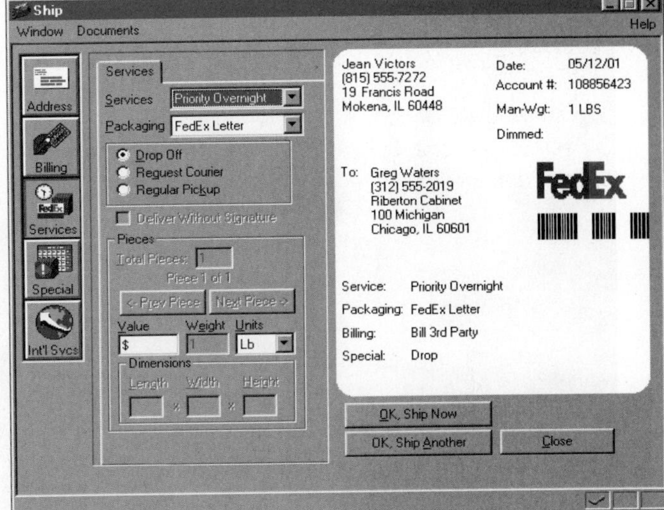

Figure 1-38 *With access to the Web, small businesses can send or track packages.*

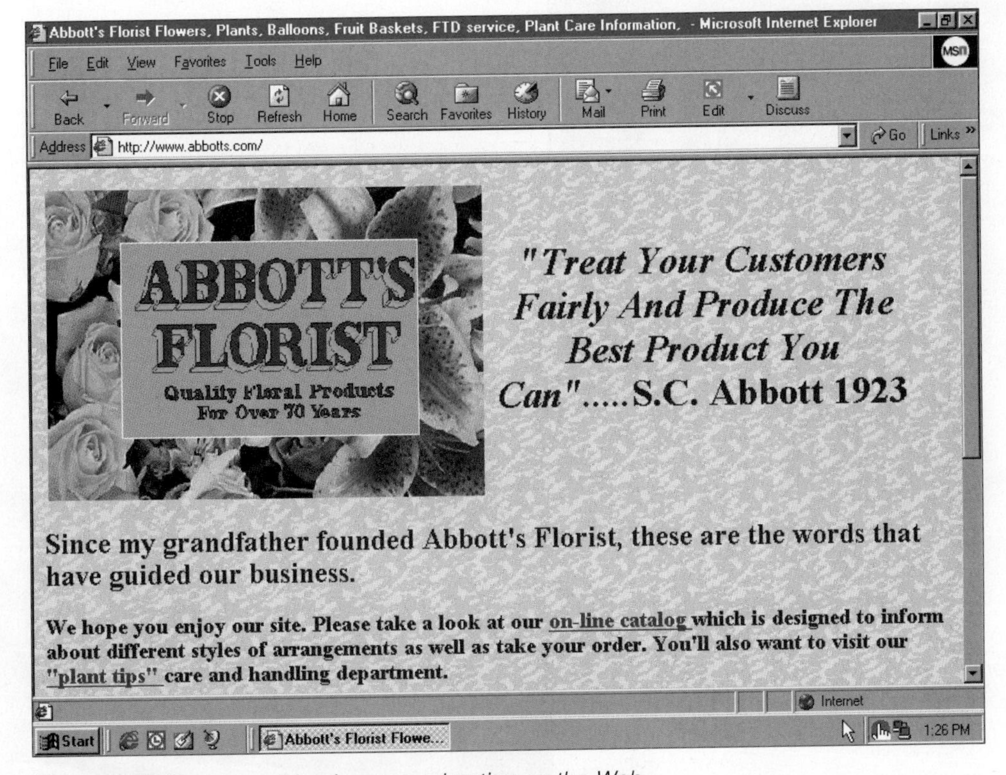

Figure 1-39 *Some small businesses advertise on the Web.*

Mobile User

As businesses expand to serve customers across the country and around the world, more and more people find themselves traveling to and from a main office to conduct business. Such users – who need to use a computer while on the road – are examples of those with mobile computing needs (Figure 1-40). **Mobile users** include a range of people, such as outside sales representatives,

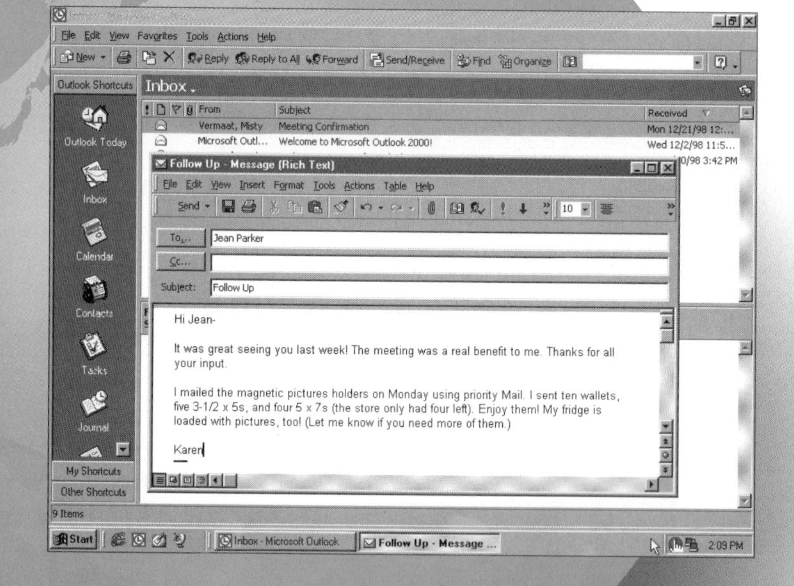

Figure 1-40 *Mobile users have laptop computers, handheld computers, smart phones, and other Web-enabled devices so they can work while on the road.*

marketing managers, real estate and insurance agents, and consultants.

Mobile users often have a laptop computer equipped with a modem, which enables them to transfer information between their computer and another computer, such as one at the main office. They also use smart phones, Web-enabled PDAs, and other handheld devices.

The software utilized by mobile users includes basic productivity software such as word processing and spreadsheet software, as well as presentation graphics software that allows the creation and delivery of presentations.

Figure 1-41 *Mobile users often connect their laptops to multimedia video projectors for presentations.*

To deliver such presentations to a large audience, the user connects the laptop to a video projector that displays the presentation on a full screen (Figure 1-41). Other types of software, such as communications software and a Web browser, allow a mobile user to access the company network and the Internet.

While transporting the laptop computer, a mobile user needs a durable, well-insulated carrying case to protect the computer if it is dropped or bumped. Back in their main office or residence, many laptop owners have a **docking station**, which is a platform into which you place a laptop. The docking station contains connections to peripherals such as a keyboard, monitor, printer, and other devices (Figure 1-42). When a laptop is in the docking station, it essentially functions as a desktop computer. With a docking station, a mobile user can enjoy the features of a desktop computer, such as a full-sized keyboard and monitor, while accessing the software and data stored on the laptop computer.

Figure 1-42 *When you insert a laptop into a docking station, you can use peripherals such as a full-sized keyboard, a full-sized monitor, and a printer with your laptop.*

Large Business User

A large business can have hundreds or thousands of employees in offices across a region, the country, or the world. The company may have an equally large number of computers, all of which are connected in a network. This network – a local area network or a wide area network depending on the size of the company – enables communications between employees at all locations.

Throughout a large business, computers help employees perform a variety of tasks related to their job. For example, **large business users** in a typical large company use an automated telephone system to route calls to the appropriate department or person. The inside sales representatives enter orders into desktop computers while on the telephone with a customer (Figure 1-43). Outside sales representatives – the mobile users in the firm – use laptop computers as described in the previous section. The marketing department uses desktop publishing software to prepare marketing literature such as newsletters, product brochures, and advertising material. The accounting department uses software to pay invoices, bill customers, and process payroll. The employees in the information systems department have a huge responsibility: to keep the computers and the network running and determine when and if new hardware or software is required.

Figure 1-43 *Computers are used throughout a large business for tasks such as entering orders.*

Figure 1-44a *(PDA being used at a meeting)*

Figure 1-44 *Employees in a large business often use PDAs. Many models allow you to transfer data from a desktop computer to the PDA so you have important information at your fingertips while you attend meetings.*

In addition to word processing, spreadsheet, database, and presentation graphics software, employees in a large firm also may use calendar programs to post their schedules on the network and PDAs to maintain personal or company information (Figure 1-44). Electronic mail and Web browsers also enable communication between employees and others around the world.

Most large organizations have their own Web site to showcase products, services, and selected company information (Figure 1-45). Customers, vendors, and any other interested parties can access the information on the Web without having to speak to a company employee.

Some large businesses also use a kiosk to provide information to the public. A **kiosk** is a freestanding computer, usually with a touch screen that serves as an input device (Figure 1-46). More advanced kiosks allow customers to place orders, make payments, or even access the Web.

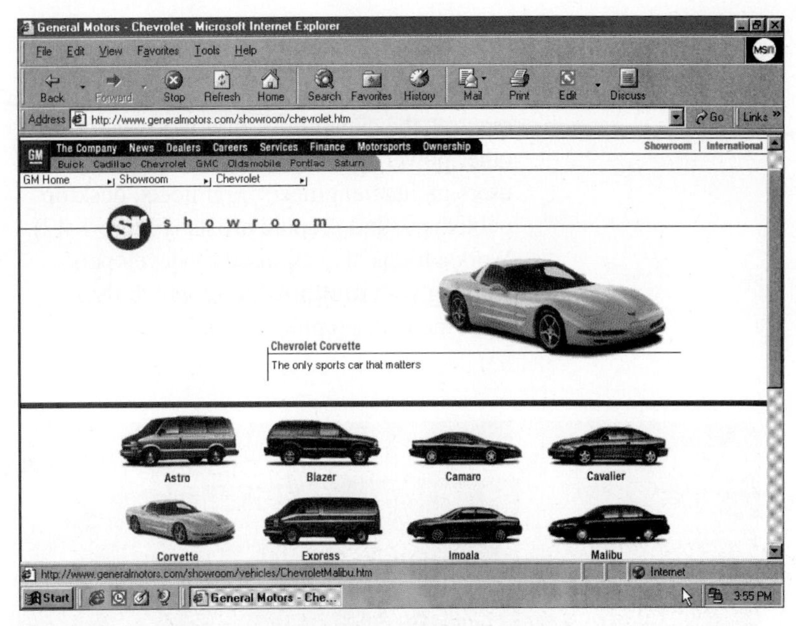

Figure 1-45 *Large businesses such as General Motors, often have their own Web site to showcase products, services, and company information.*

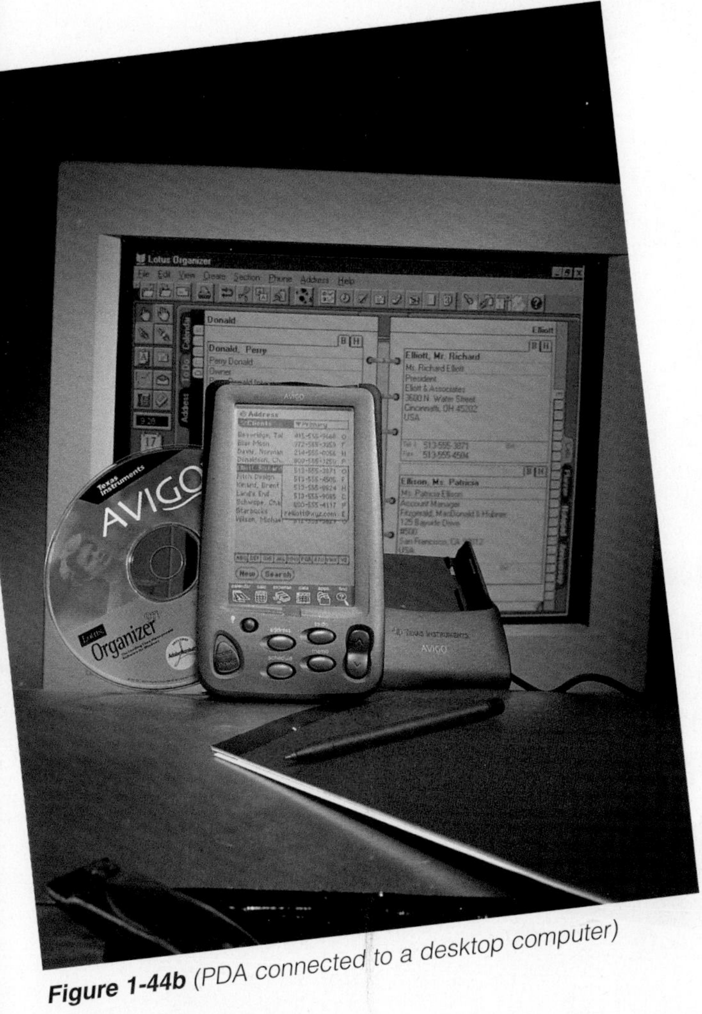

Figure 1-44b *(PDA connected to a desktop computer)*

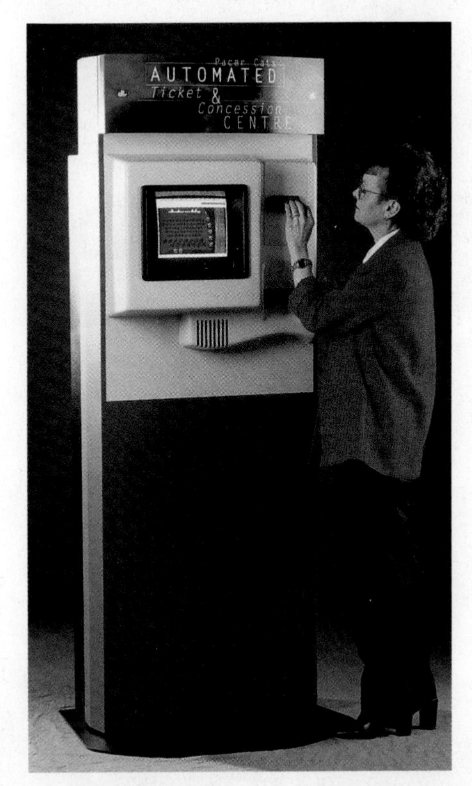

Figure 1-46 *A kiosk is a freestanding computer, usually with multimedia capabilities and a touch screen. This kiosk allows customers to place orders.*

Power User

Another category of user, called a **power user**, requires the capabilities of a workstation or other powerful computer. Examples of power users include engineers, architects, desktop publishers, and graphic artists (Figure 1-47). Workstations also are used by developers working with **multimedia**, in which they combine text, graphics, sound, video and other media elements into one application. Because of the nature of their work, all of these users need computers with extremely fast processors that have multimedia capabilities.

In addition to powerful hardware, a workstation contains software specific to the needs of the power user (Figure 1-48). For example, engineers and architects use software to draft and design items such as floor plans, mechanical assemblies, and computer chips. The desktop publisher uses specialized software to prepare marketing literature such as newsletters, brochures, and annual reports. A multimedia developer uses multimedia authoring software to create presentations containing text, graphics, video, sound, and animation. Animation is the appearance of motion. Many of these users also have software that enables them to create and edit drawings, photographs, audio, and video. Because of its specialized design, this software usually is quite expensive.

Power users are found in all types of businesses, both large and small; some also work at home. Depending on where he or she works, a power user might fit into one of the previously discussed categories, as well. Thus, in addition to their specific needs, these users often have additional hardware and software requirements such as network capabilities and Internet access.

Figure 1-47 *Examples of power users are engineers, architects, desktop publishers, graphic artists, and multimedia authors.*

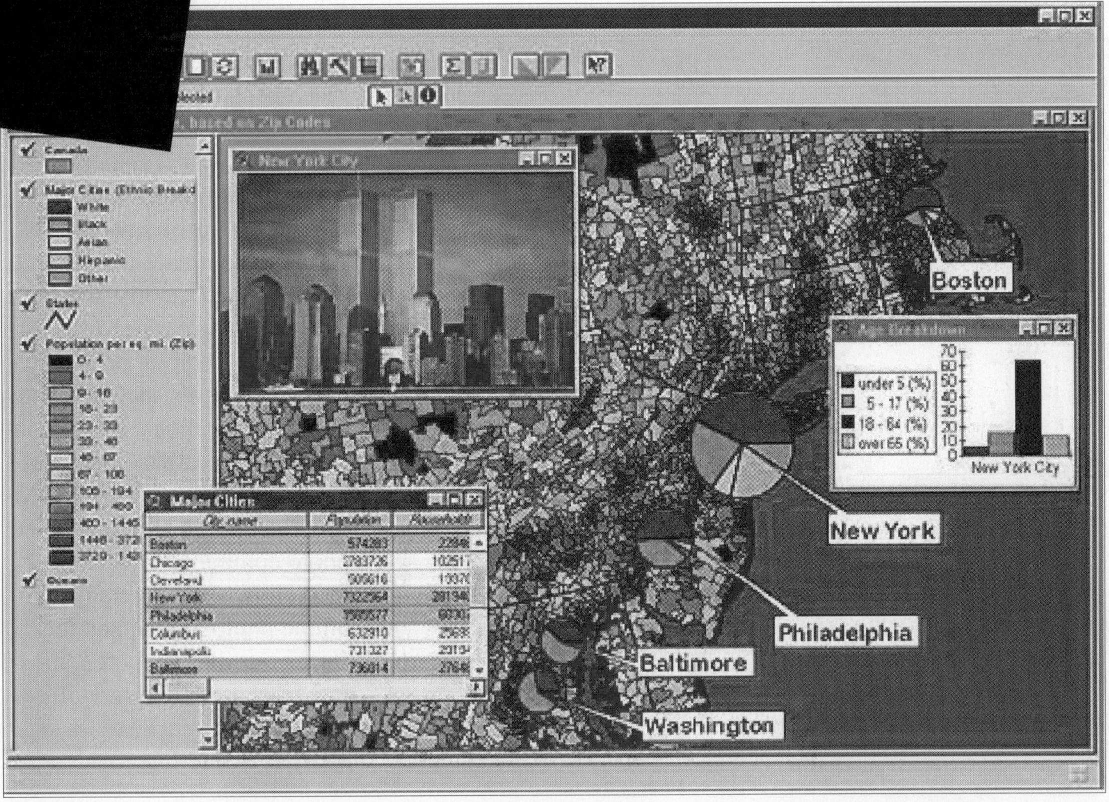

Figure 1-48 *One application for a power user is to create maps.*

Graphic image supplied courtesy of Environmental Systems Research Institute, Inc.

COMPUTER USER AS A PROVIDER OF INFORMATION

You have learned that individuals in each of the five categories of users (home user, small business user, mobile user, large business user, and power user) use the Internet to access a wealth of information and shop for goods and services. In addition to being a passive user of information, however, each of these types of users has the ability also to provide information to other connected users around the world. Users today have embraced this growing service of the Internet, where they can be active participants that provide personal and business information, photographs, items for sale, and even live conversation.

One type of application software that is fairly easy to learn is called Web page authoring software, which helps you create Web pages. Once you have created a Web page, you publish it, which is the process of making the Web page available on the Internet. You have learned that you connect to the Internet through an Internet service provider (ISP) or an online service. Many ISPs and online services store personal Web pages for their subscribers free of charge. Through your application software, you can copy Web pages from your computer to the ISPs – making your Web pages available to the world.

All types of users create Web pages for a variety of reasons:

- Home users publish Web pages that provide information about their family
- Small business users publish Web pages that provide information about their business
- Job seekers often publish Web pages that resemble a resume (Figure 1-49)
- Some colleges and high schools post student term papers
- Educators publish online courses, which are called distance-learning courses

Users access and view these published Web pages using a software program called a Web browser. When you publish a Web page, you can instruct browsers to display your Web pages when a user searches for a certain type of information. For example, if you are seeking a position as a sales representative of personal computers and related hardware and software products and have published a resume-type Web page, your Web page should display when a potential employer searches for the phrase, computer sales representative.

Home and small business users also display photographs, videos, artwork, and other images on personal Web pages or as advertisements on other's Web pages. To display photographs, you use a scanner, a digital camera, or a PC camera. Operating similar to a copy machine, a scanner converts a picture to a form the computer can use. A digital camera allows you to take pictures and store the photographed images in a form the computer understands. To create and modify graphical images, many easy-to-use paint/image editing programs exist; some even include photo-editing capabilities so that you can touch-up digital photographs such as removing red-eye. Some Web sites are called photo communities because they allow you to create an online photo album, and they store your digital photographs free of charge (Figure 1-50).

If you are a small business and you would like to advertise

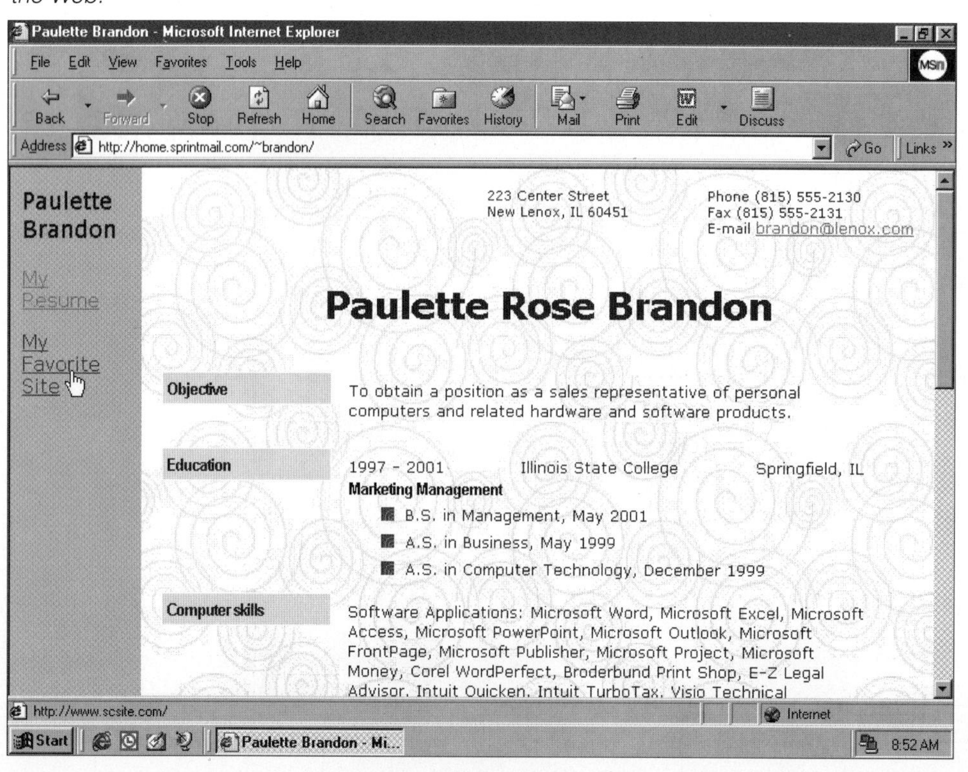

Figure 1-49 *Job seekers often publish Web pages so that potential employers easily can locate their resume on the Web.*

and take orders on the Web, you can sell your products at an electronic storefront. Some Web sites provide a means for you to create a storefront directly at their site, from which users around the world can view and make purchases (Figure 1-51a). If you have only a single item for sale, you instead might consider putting it for sale on an online auction (Figure 1-51b).

If you simply want to communicate with others on the Web, you can use e-mail, chat rooms, instant messaging, or online PDA-type applications. A chat is an interactive typed conversation that takes place on a computer. You chat with others in chat rooms, which typically are organized by topic (Figure 1-52). Instant messaging (IM) is a service that notifies you when one or more people are online and then allows you to exchange messages with them or join a private chat room with them. Many Web sites offer PDA-type applications such as calendars and address books so that you can share your appointments and contacts with others.

Today, millions of people use the Internet to provide personal and business information, photographs, items for sale, and even live conversation. For a more detailed discussion of the Web and its services, see Chapter 7.

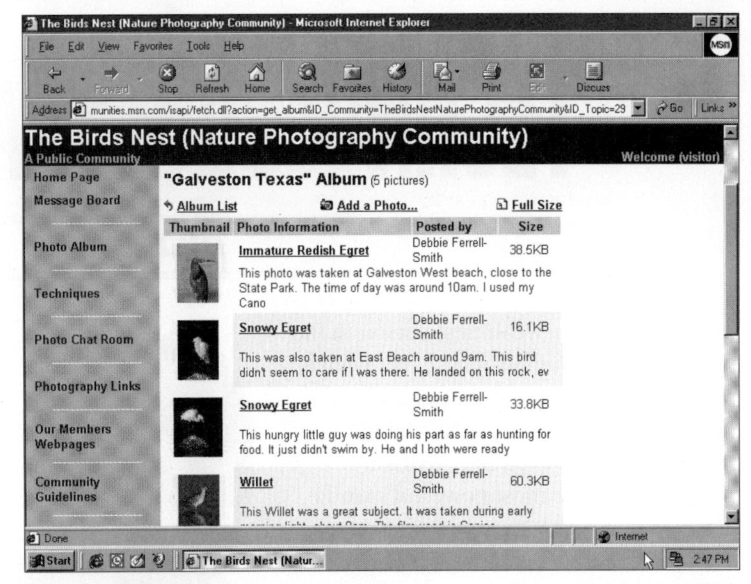

Figure 1-50 This photo community has a collection of nature photography.

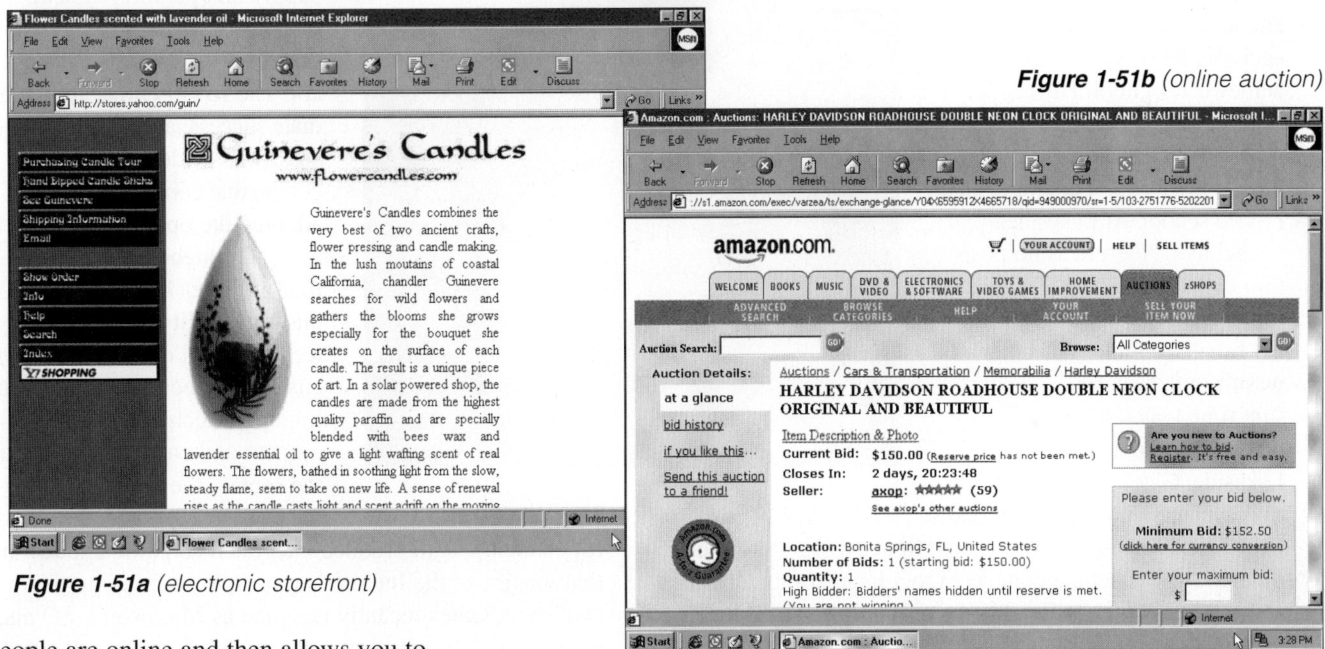

Figure 1-51b (online auction)

Figure 1-51a (electronic storefront)

Figure 1-51 Users wishing to sell items can create an electronic storefront for their home business or sell individual items through online auctions.

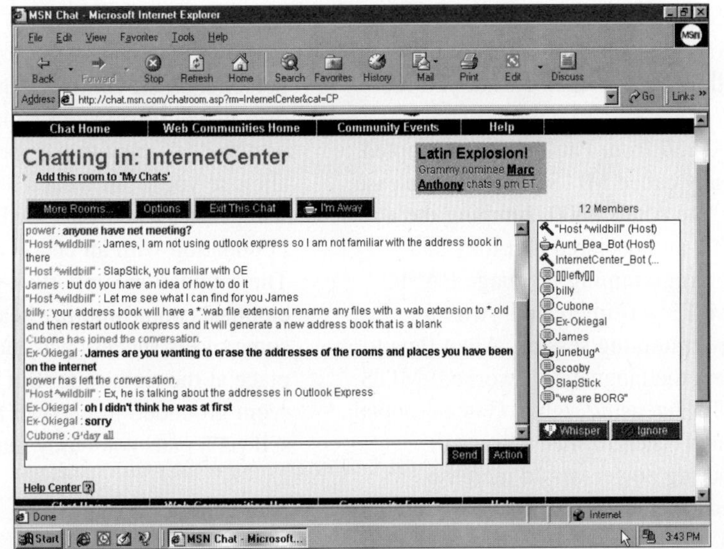

Figure 1-52 This chat room is discussing Internet-related topics, such as how to use e-mail programs.

TECHNOLOGY TRAILBLAZER

BILL GATES

Bill Gates is the wealthiest person in the world. Admirers call him insightful, innovative, and dedicated, while critics claim he is haughty, opportunistic, and lucky. Admirers and critics agree, however, that William Henry Gates III is the most powerful man in the computer industry.

Despite his larger-than-life presence today, Gates's early years were inauspicious. Even with parental encouragement (25 cents for each A), he was an underachiever in school. His reputation as "class clown" changed when his parents transferred him to Lakeside, a distinguished private school. At Lakeside, Gates says, "there was no position called the clown. I applied for it, but either they didn't like my brand of humor or humor wasn't in that season." For the first time Gates studied – and he earned straight As. At Lakeside, Gates was introduced to computers and lifelong friend, Paul Allen. Gates and Allen spent hours writing programs on a teletype machine connected to a large computer and soon were writing programs for local companies. After graduation, Gates enrolled at Harvard, but he spent more time playing games and in the computer laboratory than in class. The best classes, Gates feels, dealt in fundamentals. "The goal [of school] should be to learn how the world works, the underlying principles. You must study specifics in order to do that, but the specifics aren't what you're learning."

In 1975, an excited call from Paul Allen interrupted Gates's studies. A company called MITS planned to release the first microcomputer, the Altair 8800. Jumping the gun slightly, they called MITS and declared that they had developed a form of the programming language BASIC for the new computer. MITS was interested. After a marathon eight-week programming session, Allen flew to MITS headquarters to test the language. It worked! MITS agreed to pay a royalty for licensing rights; Gates dropped out of Harvard; and he and Allen formed Microsoft.

Certain that computing power would increase steadily, Gates convinced Allen they should concentrate on software. "What is it that limits being able to get value out of infinite computing power? Software." Microsoft's big break came in 1980, when IBM sought an operating system for its first personal computer. IBM's conservative, middle-aged, business-suited executives met with Microsoft's twenty-five year old, tousle-haired headman in his smudged glasses and wrinkled chinos. Gates agreed to supply an operating system, provided he kept the rights to market it to other manufacturers. The IBM PC was an immediate success, and Gates's foresight was evident when IBM-compatible computers also adopted the operating system.

Throughout his administration, Gates has shown an uncanny ability to anticipate technological developments and customer needs. Today, the most important technological development and customer need are the same – the Internet. Gates believes that Microsoft's greatest challenge is to produce integrated software applications that maximize the Internet's power. In response to this challenge, Gates recently resigned as Microsoft CEO and enthusiastically assumed a new position as chairman and chief software architect. Despite the change in title, Bill Gates's vision remains unchanged: to do "better software than anyone else."

It is a profitable vision, to be sure. Yet, even with personal assets estimated at more than $100 billion (much of it in Microsoft stock), Gates is wary of the trappings of wealth. "It's easy to get spoiled," he says, "by things that alienate you from what's important." Together with his wife, Gates has established the Bill and Melinda Gates Foundation with an endowment of more than $20 billion. The foundation supports a variety of charities, with an emphasis on promoting health and learning. Gates admits some of his success was luck – being born in the right place at the right time. But in the end, his achievements seem grounded in his own advice: "To succeed, put yourself on a path you enjoy that leads somewhere important."

COMPANY ON THE CUTTING EDGE

COMPANY ON THE CUTTING EDGE

MICROSOFT

In 1975, Microsoft had three programmers, one product, and revenues of $16,000. Less than twenty-five years later, the software giant employs 25,000 people, offers scores of software titles, and has earnings of more than $11 billion. Microsoft's rapid ascent is an achievement almost unparalleled in the history of American business.

Bill Gates and Paul Allen, former high-school classmates, founded Microsoft in 1975. The company's name is a contraction of "microcomputer software;" an advantage of being the first in a field, Gates points out, is the ability to choose an obvious name. The founders had no business plan, no capital, and no financial backing, but they did have a product – a form of the BASIC programming language tailored for the first microcomputer – that they licensed for about $30 a copy. Within five years, Microsoft employed a staff of forty and had developed and licensed versions of other programming languages, opened an international office, and marketed some computer games.

In 1980 IBM, a $30 billion behemoth, asked Microsoft to provide an operating system for its new IBM personal computer. The deadline? Three months. Luckily, Gates was able to find and purchase the core of a suitable operating system, dubbed Q-DOS ("Quick and Dirty Operating System"), from another company. Microsoft's version, called MS-DOS, would become the international standard for IBM and IBM-compatible personal computers. With 80 percent of the PC market, Microsoft's sales rocketed to $97 million in 1984.

Microsoft did not rest. While the company continued to update MS-DOS, it also worked on the operating system's successor, Windows. Like Apple's Macintosh operating system, Windows had a graphical user interface. (Apple Corporation later sued unsuccessfully for copyright infringement.) Acceptance was slow when Windows was released in 1985. Critics wondered if graphical user interfaces were as inevitable as Microsoft believed. Yet, while Windows struggled, Microsoft's application software packages enjoyed great success. In 1988, Microsoft surpassed Lotus Development Corporation to become the world's top software vendor.

Microsoft had not given up on Windows. A new version, Windows 3.0, was released in 1990. Aided by more powerful processors, a greater variety of compatible software, and a $10 million promotion, four million copies were shipped in just one year. With the introduction of a networking version, called Windows NT, Windows became the most popular graphical operating system in the world with more than 25 million users. "We bet the company on Windows," Gates later said, "and we deserve the benefit."

In 1995, Microsoft reinvented itself in response to the growing Internet. The new Windows 95 operating system incorporated Internet Explorer, a Web browser, and elements that made it easy to access Microsoft's new online service. Windows 95 sold one million copies in its first four days. The next year Microsoft debuted a cable television news network and a corresponding Web site. In less than two years, Microsoft had moved from the periphery to the vanguard of the information revolution.

Almost ninety percent of personal computers sold use a Microsoft operating system. Ironically, this dominance resulted in what may be Microsoft's greatest challenge to date – an antitrust suit. Sparked by manufacturers of rival Web browsers, the United States Department of Justice claimed that, with actions such as including Internet Explorer in Windows 98, Microsoft stifled competition. Microsoft denied the charge, countering that restrictions on these innovations punish creativity, harming consumers. In 1999, U.S. District Judge Thomas Penfield Jackson ruled against the software giant.

Despite the lawsuit, Microsoft's huge customer acceptance, sound financial base, state-of-the-art facilities, and talented staff point to a bright future. In the ever-changing arena of computer software, however, even a colossus may be vulnerable. "Microsoft won't be immortal," Gates warns. "All companies fail...My goal is to keep my company vital as long as possible."

IN BRIEF

www.scsite.com/dc2001/ch1/brief.htm

WEB INSTRUCTIONS: *To display this page from the Web, launch your browser and enter the URL, www.scsite.com/dc2001/ch1/brief.htm. Click the links for current and additional information. To listen to an audio version of this IN BRIEF, click the Audio button to the right of the title, IN BRIEF, at the top of the page. To play the audio, RealPlayer must be installed on your computer (download by clicking here).*

1. Why Is Computer Literacy Important?

To be successful in today's world, it is crucial to have knowledge and understanding of computers and their uses. This knowledge, called computer literacy, is essential as technology advances and computers extend into every facet of daily living.

2. What Is a Computer?

A computer is an electronic machine, operating under the control of instructions stored in its own memory, that can accept data (input), manipulate the data according to specified rules (process), produce results (output), and store the results for future use (**storage**). **Data** is a collection of unorganized facts, which can include words, numbers, images, and sounds. Computers manipulate data to create information. **Information** is data that is organized, has meaning, and is useful. Examples are reports, newsletters, a receipt, or a check. Data entered into a computer is called **input**. The processed results are called **output**. The cycle of input, process, output, and storage is called the **information processing cycle**.

3. What Are the Components of a Computer?

Hardware is the electric, electronic, and mechanical equipment that makes up a computer. An **input device** allows a user to enter data and commands into the memory of a computer. Three commonly used input devices are the keyboard, the mouse, and a microphone. An **output device** is used to convey information generated by a computer to the user. Three commonly used output devices are a printer, a monitor, and speakers. The **system unit** is a box-like case made from metal or plastic that houses the computer circuitry. The system unit contains the central processing unit (CPU), which interprets and carries out the instructions that operate a computer, including computations; and **memory**, which is a series of electronic elements that temporarily holds the data and instructions while the CPU is processing them. **Storage devices** are mechanisms used to record and retrieve data, information, and instructions to and from a storage medium. Common storage devices are a floppy disk drive, hard disk drive, CD-ROM drive, and DVD-ROM drive. **Communications devices** enable a computer to exchange items such as data, instructions, and information with another computer.

4. Why Is a Computer a Powerful Tool?

A computer's power is derived from its capability of performing the information processing cycle operations with speed, reliability, and accuracy; its capacity to store huge amounts of data, instructions, and information; and its ability to communicate with other computers.

5. What Are the Categories of Computer Software?

Software is the series of instructions that tells the hardware of a computer what to do. Software can be categorized into two types: system software and application software. **System software** controls the operation of the computer and its devices and serves as the interface between a user

 www.scsite.com/dc2001/ch1/brief.htm

Student Exercises
WEB INFO
IN BRIEF
KEY TERMS
AT THE MOVIES
CHECKPOINT
AT ISSUE
MY WEB PAGE
HANDS ON
NET STUFF
Special Features
TIMELINE 2001
GUIDE TO WWW SITES
MAKING A CHIP
E-COMMERCE 2001
BUYER'S GUIDE 2001
CAREERS 2001
TRENDS 2001
LEARNING GAMES
CHAT
INTERACTIVE LABS
NEWS
HOME

and the computer's hardware. Two types of system software are the **operating system**, which contains instructions that coordinate the activities of hardware devices; and utility programs, which perform specific tasks usually related to managing a computer. **Application software** performs specific tasks for users, such as creating documents, spreadsheets, databases, or presentation graphics. **Computer programmers** write software programs, often following a plan developed by a **systems analyst**.

6. What Is the Purpose of a Network?

A network is a collection of computers and devices connected together via communications media. Computers are networked so users can share resources such as hardware devices, software programs, data, and information.

7. How Is the Internet Used?

The world's largest network is the **Internet**, which is a worldwide collection of networks that links together millions of computers. The Internet is used to send messages to other users, obtain information, shop for goods and services, meet or converse with people around the world, and access sources of entertainment and leisure. The World Wide Web, which contains billions of Web pages with text, graphics, sound, and built-in connections to other Web pages, is one of the more popular segments of the Internet.

8. What Are the Categories of Computers?

The four major categories of computers are personal computers, minicomputers, mainframe computers, and supercomputers. These categories are based on differences in size, speed, processing capabilities, and price. A **personal computer** (**PC**) can perform all of its input, processing, output, and storage by itself. Two types of personal computers are **desktop computers**, which are designed to fit entirely on or under a desk or table, and **portable computers**, which are small enough to carry. **Minicomputers** are larger and more powerful than personal computers and often can support up to 4,000 connected users. A **mainframe** is a large, expensive, very powerful computer that can handle hundreds or thousands of connected users simultaneously. **Supercomputers** – the fastest, most powerful, and most expensive computers – are capable of processing more than 64 billion instructions in a single second.

9. How Are Computers Used?

Every day, people depend on different types of computers for a variety of applications. **Home users** rely on their computers for entertainment; communications; research and education; Web access; shopping; personal finance; and productivity applications such as word processing and spreadsheets. Small business users utilize productivity software as well as communications software, Web browsers, e-mail, and specialized software. **Mobile users** have laptop computers so they can work on the road. They often use presentation software. **Large business users** use computers to run their businesses by using productivity software, communications software, automated systems for most departments in the company, and large networks. **Power users** require the capabilities of workstations or other powerful computers to design plans, produce publications, create graphic art, and work with multimedia that includes text, graphics, sound, video, and other media elements. Through the Internet, each category of users can access a wealth of information and has the ability to provide information to other connected users around the world.

Student Exercises
WEB INFO
IN BRIEF
KEY TERMS
AT THE MOVIES
CHECKPOINT
AT ISSUE
MY WEB PAGE
HANDS ON
NET STUFF
Special Features
TIMELINE 2001
GUIDE TO WWW SITES
MAKING A CHIP
E-COMMERCE 2001
BUYER'S GUIDE 2001
CAREERS 2001
TRENDS 2001
LEARNING GAMES
CHAT
INTERACTIVE LABS
NEWS
HOME

CHAPTER 1 2 3 4 5 6 7 8 9 10 11 12 13 14 INDEX

KEY TERMS

www.scsite.com/dc2001/ch1/terms.htm

WEB INSTRUCTIONS: *To display this page from the Web, launch your browser and enter the URL, www.scsite.com/dc2001/ch1/terms.htm. Scroll through the list of terms. Click a term to display its definition and a picture. Click KEY TERMS on the left to redisplay the KEY TERMS page. Click the TO WEB button for current and additional information about the term from the Web. To see animations, Shockwave and Flash Player must be installed on your computer (download by clicking here).*

all-in-one computer (1.20)
application software (1.12)
central processing unit (CPU) (1.6)
communications devices (1.8)
computer (1.3)
computer literacy (1.3)
computer program (1.9)
computer programmers (1.13)
custom software (1.13)
data (1.3)
desktop computer (1.20)
docking station (1.31)
dumb terminals (1.24)
execute (1.10)
freeware (1.13)
graphical user interface (GUI) (1.11)
handheld computer (1.23)
hardware (1.4)
home users (1.26)
icon (1.11)
information (1.4)
information processing cycle (1.4)
input (1.4)
input device (1.5)
install (1.10)
Internet (1.16)
Internet appliances (1.22)
kiosk (1.33)
laptop computer (1.23)
large business users (1.32)

APPLICATION SOFTWARE: Programs designed to perform specific tasks for users; include word processing software, spreadsheet software, database software, presentation graphics software, and many other types of software for a variety of tasks. (1.12)

mainframe (1.24)
memory (1.6)
minicomputer (1.24)
mobile users (1.30)
multimedia (1.34)
network (1.14)
network computer (NC) (1.22)
notebook computer (1.23)
operating system (1.10)
output (1.4)
output device (1.6)
packaged software (1.13)
palmtop computer (1.23)
pen computers (1.23)
personal computer (PC) (1.19)
Personal Digital Assistant (PDA) (1.23)
portable computer (1.22)
power user (1.34)
processor (1.6)
program (1.9)
programmers (1.13)
public-domain software (1.13)
resources (1.15)
server (1.21)
shareware (1.13)
small business users (1.28)
small office/home office (SOHO) (1.28)
software (1.4)
stand-alone (1.21)
storage (1.4)
storage device (1.7)
storage medium (1.7)
supercomputer (1.24)
system software (1.10)
system unit (1.6)
systems analyst (1.13)
terminal (1.24)
tower model (1.20)
user (1.4)
user interface (1.11)
Web appliances (1.22)
workstation (1.21)

AT THE MOVIES

www.scsite.com/dc2001/ch1/movies.htm

WELCOME to VIDEO CLIPS from CNN

WEB INSTRUCTIONS: *To display this page from the Web, launch your browser and enter the URL, www.scsite.com/dc2001/ch1/movies.htm. Click a picture to view a video. After watching the video, close the video window and then complete the exercise by answering the questions about the video. To view the videos, RealPlayer must be installed on your computer (download by clicking here).*

1 Personal Computers for Students

When purchasing a personal computer, should you choose your cheapest option? The video states that a good college computer should have 32 MB of memory, 4 GB of hard disk storage, a 233 or 266 MHz CPU, and a 56K modem. What is your opinion of this recommendation? If you had a choice, would you buy a desktop or a laptop computer? Why? In addition to the computer, what accessories might you purchase? Does your school require students to own their own computer? Does it provide Internet access?

2 Handheld Computers

A handheld computer is a small personal computer designed to fit in your hand. What are some other terms used to describe handheld computers? Many of these small computers run basic operating systems such as Windows CE. What does CE stand for? What other productivity or Internet software can you run on a handheld computer? Who might use one of these com-puters and why? Would you purchase a handheld computer instead of desktop or laptop? Why?

3 Microsoft Inner Workings

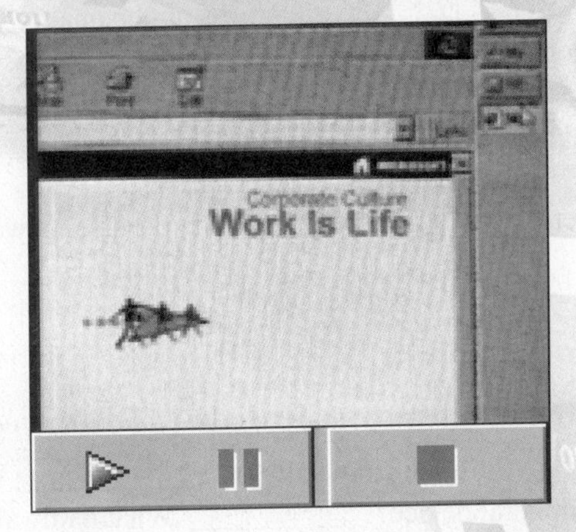

Have you ever wondered what it would be like to work for the world's largest software company? At Microsoft Corporation, a $200 billion company that holds a 90 percent market share in key PC markets, the only rule is to win. How would you describe the corporate culture at Microsoft? How do you think Microsoft's culture differs from the culture at other large corporations? What do you think the advantages and disadvantages would be of working for a company whose motto is, work is life?

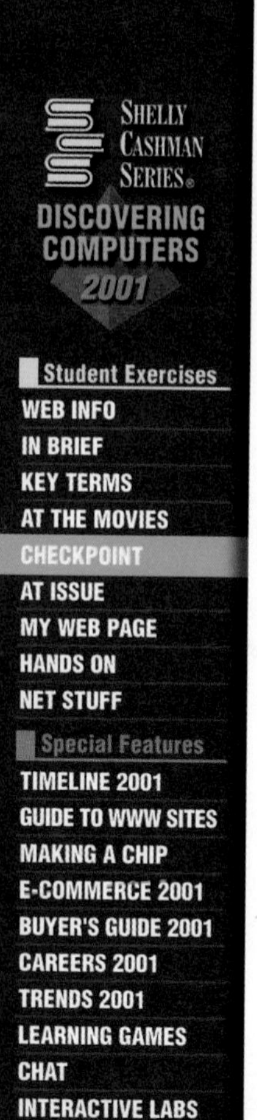

CHECKPOINT

www.scsite.com/dc2001/ch1/check.htm

WEB INSTRUCTIONS: *To display this page from the Web, launch your browser and enter the URL, www.scsite.com/dc2001/ch1/check.htm. Click the links for current and additional information. To experience the animation and interactivity, Shockwave and Flash Player must be installed on your computer (download by clicking here).*

Label the Figure

Instructions: *Categorize these common computer hardware components. Write the letter next to each component on the right in an appropriate blue box. Then write the words from the list on the left in the appropriate yellow boxes to identify the hardware components.*

monitor
speakers
keyboard
mouse
printer
system unit
hard disk drive
CD-ROM drive
floppy disk drive
Zip drive
modem
microphone

INPUT | OUTPUT | STORAGE | COMMUNICATIONS | PROCESSING

a b c d e f g h i j k l

Matching

Instructions: *Match each term from the column on the left with the best description from the column on the right.*

_______ 1. network computer (NC)
_______ 2. workstation
_______ 3. Personal Digital Assistant (PDA)
_______ 4. laptop computer
_______ 5. server

a. PC with a tall and narrow system unit designed to be placed vertically on the floor.
b. Powerful desktop computer designed for work that requires intense calculation and graphics capabilities.
c. A type of personal computer designed specifically to connect to a network.
d. Portable computer designed for mobility, small enough to fit on a lap.
e. Desktop computer that manages the resources on a network.
f. Popular handheld computer that often supports personal information management applications.
g. Large, expensive, powerful computers that can handle thousands of connected users simultaneously.

Short Answer

Instructions: *Write a brief answer to each of the following questions.*

1. How is hardware different from software? __________ Why is hardware useless without software? __________
2. What is a peripheral device? __________ What hardware components are considered peripheral devices? __________
3. What are four common storage devices? __________ How are they different? __________
4. How is packaged software different from custom software? __________ How is shareware different from freeware and public-domain software? __________
5. Why do people use the Internet? __________ How do most users connect to the Internet? __________

CHAPTER 1 2 3 4 5 6 7 8 9 10 11 12 13 14 INDEX

AT ISSUE www.scsite.com/dc2001/ch1/issue.htm

WEB INSTRUCTIONS: *To display this page from the Web, launch your browser and enter the URL, www.scsite.com/dc2001/ch1/issue.htm. Click the links for current and additional information.*

"To err is human, but to really foul things up requires a computer." This anonymous quote, from a 1982 BBC radio broadcast, reflects the way many people once felt about computers. Today, computers are much more prevalent than they were two decades ago. It is estimated that almost 50 percent of American households have computers, and more than 25 percent are connected to the Internet. As hardware and software prices continue to fall, these percentages are expected to rise. Even people who do not own computers often have access to them through schools, libraries, and community centers. Has the increasingly widespread availability and use of computers changed people's attitudes? Are people today more optimistic about the impact of computers than people were a generation ago? Why? Has greater familiarity with computers bred greater affection or greater contempt? Why?

TWO

A survey of senior executives found 75 percent could not explain the purpose of a modem, 65 percent believed the Internet was privately controlled, and 50 percent thought an Arch Deluxe was a PC part. Almost 100 percent of sixth graders questioned knew that modems let computers communicate over telephone lines, the Internet is not owned by anyone, and an Arch Deluxe is a McDonald's sandwich. In the race for computer literacy, many fear that some people have an unfair head start because of their youth, education, or economic status. Is this fear justified? Why or why not? What, if anything, can be done to level the playing field?

THREE

In one county, the number of computers purchased by local school districts increased by 85 percent, while the supply of library books declined by almost ten percent. School officials claim computers extend learning opportunities and develop the computer literacy needed in today's technological world. Yet, some parents complain that computer purchases represent frivolous, status-seeking spending. How should a school district's money be spent? If you were an administrator, what percentage of your budget would you spend on computers? On library books and textbooks? Why? What factors would influence your decision?

Popular theory says that if an auditorium full of monkeys each was given a typewriter, eventually they would produce a known classic. Monkeys may or may not be able to emulate Shakespeare, but today, computers can perform assignments once thought exclusively human. Computers have been programmed to read written work and answer questions demonstrating comprehension. One computer has been programmed to paint, producing pictures that have sold for more than $20,000. Computer-controlled robots have been programmed to play soccer. What types of activities or problems, if any, still do not embrace computer involvement? Why? Will computers someday be capable of handling these activities or problems? Why or why not?

FIVE

In 1997, the Graduate Management Admissions Test (GMAT) became the first computerized standardized test. Other tests – including the GRE and even the SAT – intend to follow suit. Computer Adaptive Tests show one question at a time, and a student's response dictates the difficulty of the next question. Grades are based on the difficulty of questions and the number of correct answers. Advocates claim the tests are multidimensional and better reflect individual abilities. Critics argue the tests favor those with computer experience, demand unfamiliar test-taking skills, and evaluate computer literacy as much as knowledge of the subject matter. Are computerized standardized tests a good idea? Why or why not? Would you be comfortable taking one? How would you prepare differently for a computerized standardized test?

SHELLY CASHMAN SERIES®
DISCOVERING COMPUTERS 2001

Student Exercises
WEB INFO
IN BRIEF
KEY TERMS
AT THE MOVIES
CHECKPOINT
AT ISSUE
MY WEB PAGE
HANDS ON
NET STUFF

Special Features
TIMELINE 2001
GUIDE TO WWW SITES
MAKING A CHIP
E-COMMERCE 2001
BUYER'S GUIDE 2001
CAREERS 2001
TRENDS 2001
LEARNING GAMES
CHAT
INTERACTIVE LABS
NEWS
HOME

Student Exercises
WEB INFO
IN BRIEF
KEY TERMS
AT THE MOVIES
CHECKPOINT
AT ISSUE
MY WEB PAGE
HANDS ON
NET STUFF
Special Features
TIMELINE 2001
GUIDE TO WWW SITES
MAKING A CHIP
E-COMMERCE 2001
BUYER'S GUIDE 2001
CAREERS 2001
TRENDS 2001
LEARNING GAMES
CHAT
INTERACTIVE LABS
NEWS
HOME

CHAPTER 1 2 3 4 5 6 7 8 9 10 11 12 13 14 INDEX

MY WEB PAGE www.scsite.com/dc2001/ch1/myweb.htm

WEB INSTRUCTIONS: *The icons to the left of the numbered activities on this page indicate the learning system availability. If you are using WebCT, follow the instructions in the exercises below. If you are using the textbook Web site or CyberClass, launch your browser, enter the URL www.scsite.com/dc2001/ch1/myweb.htm, and then click the corresponding icon to the left of the exercise.*

1. Practice Test: Click My Web Page and then click Chapter 1. Click Practice Test. Answer each question and then click the Save Answer button. When completed, click the Finish button. Click the OK button to submit the quiz for grading. Click the View Results button, and then make a note of any missed answers.

2. Web Guide: Click My Web Page and then click Chapter 1. Click Web Guide to display the Guide to WWW Sites page. Click Reference and then click AskEric Virtual Library. Click AskEric InfoGuides and search for Computer Literacy. Click a search results link of your choice. Use your word processing program to prepare a brief report on what you learned and submit your assignment to your instructor.

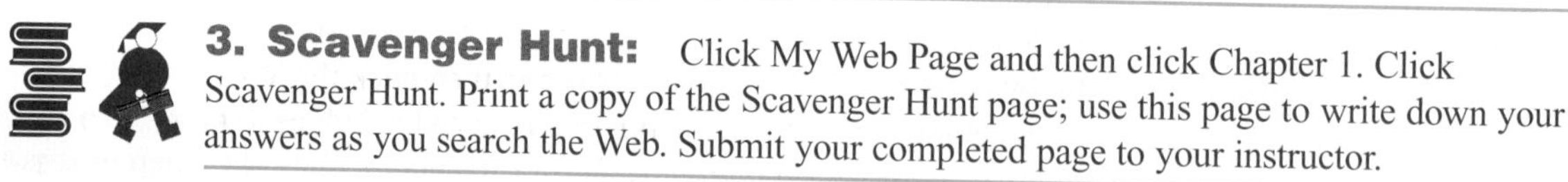

3. Scavenger Hunt: Click My Web Page and then click Chapter 1. Click Scavenger Hunt. Print a copy of the Scavenger Hunt page; use this page to write down your answers as you search the Web. Submit your completed page to your instructor.

4. Who Wants to Be a Computer Genius?: Click My Web Page and then click Chapter 1. Click Computer Genius to find out if you are a computer genius. For directions, click the How to Play button. When you are ready to play, click the Game button. Submit your score to your instructor.

5. Wheel of Terms: Click My Web Page and then click Chapter 1. Click Wheel of Terms to reinforce important terms you learned in this chapter by playing the Shelly Cashman Series version of this popular game. For directions, click the How to Play button. When you are ready to play, click the Game button. Submit your score to your instructor.

6. Career Corner: Click My Web Page and then click Chapter 1. Click Career Corner to display the Hire-Ed page. Search for jobs in your state. Write a brief report on the jobs you found. Submit the report to your instructor.

7. Search Sleuth: Click My Web Page and then click Chapter 1. Click Search Sleuth to learn search techniques that will help make you a research expert. Submit the completed assignment to your instructor.

8. Crossword Puzzle Challenge: Click My Web Page and then click Chapter 1. Click Crossword Puzzle Challenge. Complete the puzzle to reinforce skills you learned in this chapter. For directions, click the How to Play button. When you are ready to play, click the Game button. Submit the completed puzzle to your instructor.

9. Flash Cards: This exercise uses CyberClass. Follow the instructions at the top of this page and then click the CyberClass icon to the left; or ask your instructor for logon instructions. Click Flash Cards on the Main Menu. Click the plus sign. Answer all the questions in any two subjects of your choice. Click the Close button in the upper-right corner to close all windows. Notice the many other exercises at this site. Complete all other exercises as assigned by your instructor.

HANDS ON

www.scsite.com/dc2001/ch1/hands.htm

WEB INSTRUCTIONS: *To display this page from the Web, launch your browser and enter the URL,* www.scsite.com/dc2001/ch1/hands.htm. *Click the links for current and additional information.*

One

Using Windows Help

This exercise uses Windows 98 or Windows 2000 procedures. In the past, when you purchased computer software, you also received large printed manuals that attempted to answer any questions you might have. Today, Help usually is offered directly on the computer. To make it easy to find exactly the Help you need, Windows Help is arranged on three sheets: Contents, Index, and Search. Click the Start button on the taskbar and then click Help on the Start menu. Click the Contents tab. What do you see? When would you use the Contents sheet to find Help? Click the Index tab. What do you see? When would you use the Index sheet to find Help? Click the Search tab. What do you see? When would you use the Search sheet to find Help?

Two

What's New in Microsoft Windows?

This exercise uses Windows 98 or Windows 2000 procedures. Click the Start button on the taskbar and then click Help on the Start menu. Click the Contents tab in the Windows Help window. Click the Introducing Windows 98 or Windows 2000 book, and then click the What's New in Windows 98 or Windows 2000 book. Click a topic in which you are interested. Click each topic in the right pane. How is this version of Windows better than previous versions of Windows? Will the improvement make your work more efficient? Why or why not? What improvement, if any, would you still like to see? Close the Windows Help window.

Three

Improving Mouse Skills

This exercise uses Windows 98 or Windows 2000 procedures. Click the Start button on the taskbar. Point to Programs, then point to Accessories on the Programs submenu. Point to Games on the Accessories submenu, and then click Solitaire on the Games submenu. When the Solitaire window displays, click the Maximize button. Click Help on the Solitaire menu bar, and then click Help Topics. Click the Contents tab. Click the object of Solitaire topic and read the information. Click the Playing Solitaire topic. Read and print the information by clicking the Solitaire Help window's Option button, clicking Print, and then clicking the OK button. Click the Close button in the Solitaire Help window. Play the game of Solitaire. Close the Solitare window.

Four

Learning About Your System

You can learn some important information about your computer system by studying the System Properties. Click the Start button. Point to Settings and then click Control Panel on the Settings submenu. Double-click System in the Control Panel window. Click the General tab in the System dialog box. Use the General sheet to find out the answers to these questions:

- What operating system does your computer use?
- To whom is your system registered?
- What type of processor does your computer have?
- How much memory (RAM) does your computer have?

Close the System Properties dialog box.

SHELLY CASHMAN SERIES®
DISCOVERING COMPUTERS 2001

Student Exercises
WEB INFO
IN BRIEF
KEY TERMS
AT THE MOVIES
CHECKPOINT
AT ISSUE
MY WEB PAGE
HANDS ON
NET STUFF

Special Features
TIMELINE 2001
GUIDE TO WWW SITES
MAKING A CHIP
E-COMMERCE 2001
BUYER'S GUIDE 2001
CAREERS 2001
TRENDS 2001
LEARNING GAMES
CHAT
INTERACTIVE LABS
NEWS
HOME

CHAPTER 1 2 3 4 5 6 7 8 9 10 11 12 13 14 INDEX

NET STUFF

www.scsite.com/dc2001/ch1/net.htm

WEB INSTRUCTIONS: *To display this page from the Web, launch your browser and enter the URL, www.scsite.com/dc2001/ch1/net.htm. To use the Mouse lab or the Keyboard lab from the Web, Shockwave and Flash Player must be installed on your computer (download by clicking here).*

MOUSE LAB

1. Shelly Cashman Series Mouse Lab

a) To start the Shelly Cashman Series Mouse Lab, complete the step that applies to you.
   (1) **Running from the World Wide Web:** Enter the URL, www.scsite.com/sclabs/menu.htm; or display the NET STUFF page (see instructions at the top of this page) and then click the MOUSE LAB button.
   (2) **Running from a CD:** Insert the Shelly Cashman Series Labs with Audio CD in your CD-ROM drive.
   (3) **Running the No-Audio Version from a Hard Disk or Network:** Click the Start button on the taskbar, point to Shelly Cashman Series Labs on the Programs submenu, and click Interactive Labs.

b) When the Shelly Cashman Series IN THE LAB screen displays (Figure 1-49), if necessary maximize the window, and then follow the instructions on the screen to start the Using the Mouse lab.

c) When the Using the Mouse screen displays, if necessary maximize the window, and then read the objectives.

d) If assigned, follow the instructions on the screen to print the questions associated with the lab.

e) Follow the instructions on the screen to continue in the lab.

f) When completed, follow the instructions on the screen to terminate the lab.

g) If assigned, hand in your answers for the printed questions to your instructor.

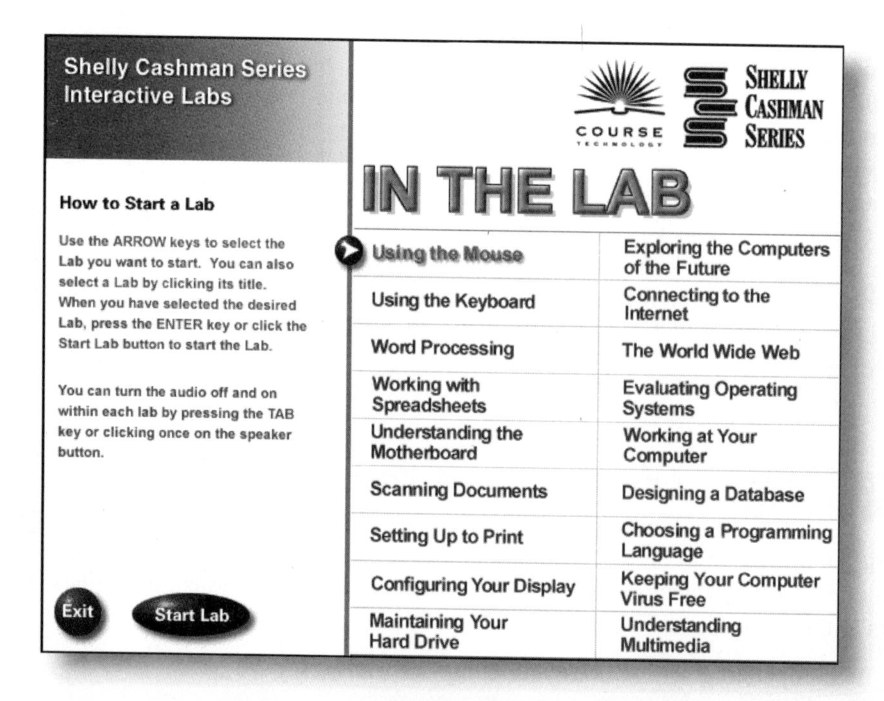

Figure 1-49

CHAPTER 1 2 3 4 5 6 7 8 9 10 11 12 13 14 INDEX

NET STUFF

www.scsite.com/dc2001/ch1/net.htm

KEYBOARD LAB

2. Shelly Cashman Series Keyboard Lab

Follow the appropriate instructions in NET STUFF Exercise 1 on the previous page to start and use the Using the Keyboard lab. If you are running from the Web, enter the URL, www.scsite.com/sclabs/menu.htm; or display the NET STUFF page (see instructions at the top of the previous page) and then click the KEYBOARD LAB button.

LEARN THE NET

3. Learn the Net

No matter how much computer experience you have, navigating the Net for the first time can be intimidating. How do you get started? Click the LEARN THE NET button and complete this exercise to discover how you can find out everything you wanted to know about the Internet.

IN THE NEWS

4. In the News

Companies can have problems with their products, or problems with perceptions of their products. In the computer industry, a flaw in a software program or a failure in a hardware device would be a product problem. An accusation of unfair business practices or a rumor of a program malfunction would be a perception problem. Click the IN THE NEWS button and read a news article about a problem experienced by a company in the computer industry. What is the problem? Is the problem affecting the company's profits? If so, why? How is the company addressing the problem? Do you think the company's efforts will be successful? Why or why not?

WEB CHAT

5. Web Chat

Everyone who works with computers has experienced moments of enormous frustration – incomprehensible error messages, software glitches that produce unanticipated results, or even the entire system freezing up. Many people feel reactions to computer problems tend to be more extreme than reactions to problems with other tools they use. If you are viewing this page on the Web, start the video on the right to see how one individual handled a problem with his computer. Do computer problems make people angrier than problems with other tools? Why? How can individuals reduce their frustration when dealing with computer failures? Click the WEB CHAT button to enter a Web Chat discussion related to this topic.

Student Exercises
WEB INFO
IN BRIEF
KEY TERMS
AT THE MOVIES
CHECKPOINT
AT ISSUE
MY WEB PAGE
HANDS ON
NET STUFF
Special Features
TIMELINE 2001
GUIDE TO WWW SITES
MAKING A CHIP
E-COMMERCE 2001
BUYER'S GUIDE 2001
CAREERS 2001
TRENDS 2001
LEARNING GAMES
CHAT
INTERACTIVE LABS
NEWS
HOME

www.scsite.com/dc2001/ch1/timeline1.htm

WEB INSTRUCTIONS: *To gain World Wide Web access to additional and up-to-date information regarding this special feature, launch your browser and enter the URL shown at the top of the page you want to view.*

Milestones in Computer History

Timeline 2001

1937

Dr. John V. Atanasoff and Clifford Berry design and build the first electronic digital computer. Their machine, the Atanasoff-Berry-Computer, or ABC, provides the foundation for advances in electronic digital computers.

1946

Dr. John W. Mauchly and J. Presper Eckert, Jr. complete work on the first large-scale electronic, general-purpose digital computer. The ENIAC (Electronic Numerical Integrator And Computer) weighs thirty tons, contains 18,000 vacuum tubes, occupies a thirty-by-fifty-foot space, and consumes 160 kilowatts of power. The first time it is turned on, lights dim in an entire section of Philadelphia.

1943

During World War II, British scientist Alan Turing designs the Colossus, an electronic computer created for the military to break German codes. The computer's existence is kept secret until the 1970s.

1945

Dr. John von Neumann writes a brilliant paper describing the stored program concept. His breakthrough idea, where memory holds both data and stored programs, lays the foundation for all digital computers that have since been built.

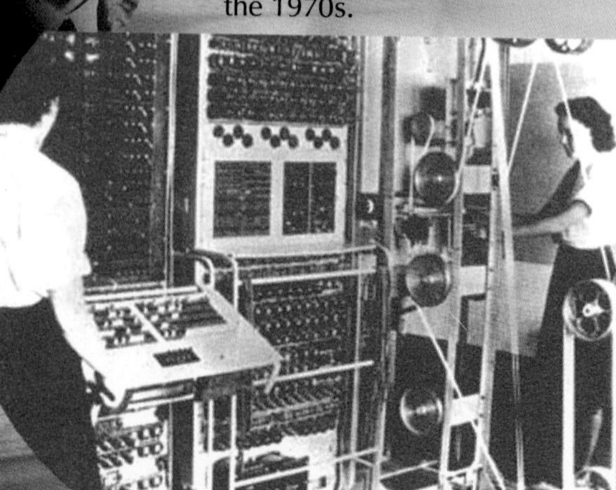

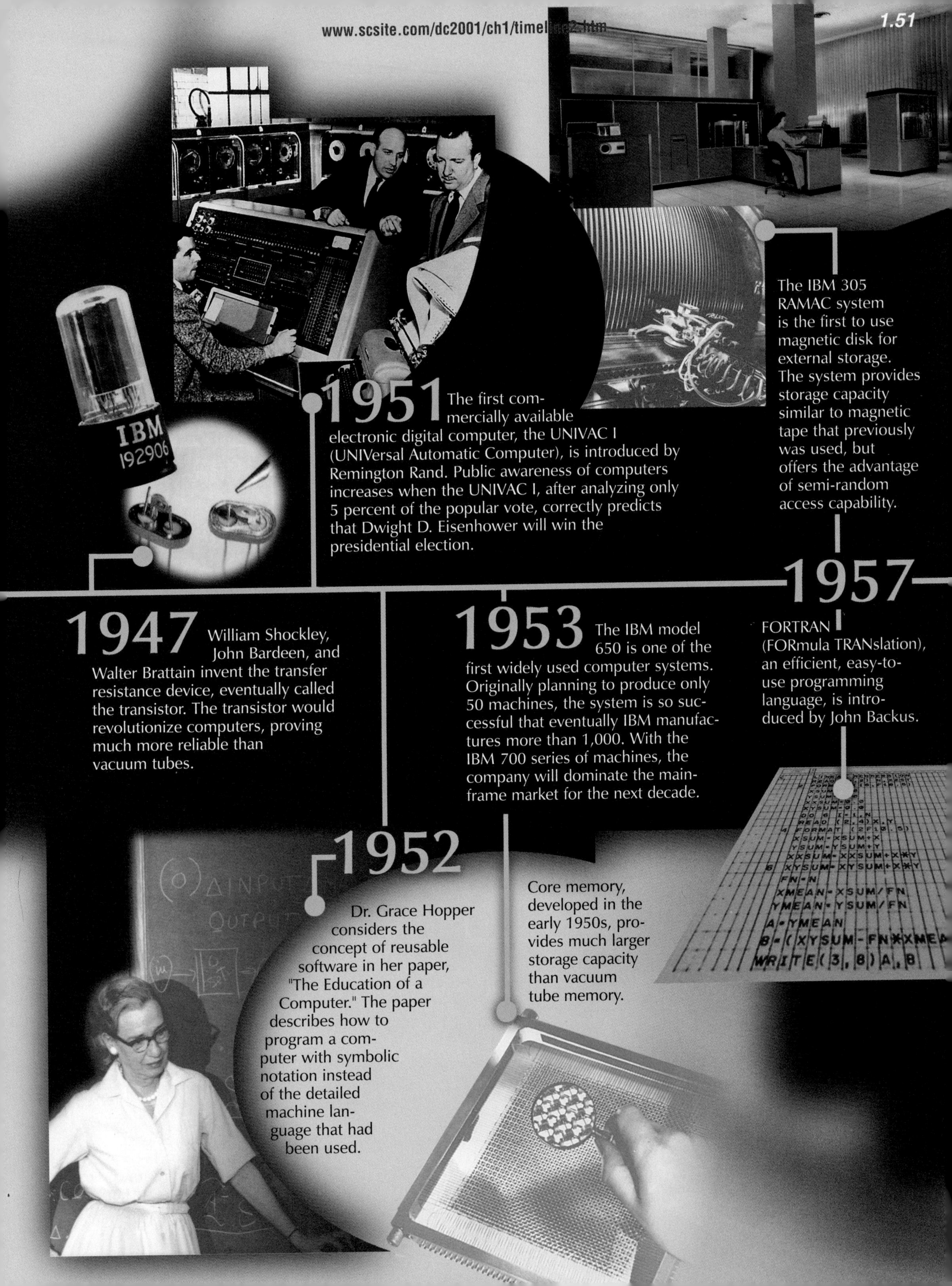

1951 The first commercially available electronic digital computer, the UNIVAC I (UNIVersal Automatic Computer), is introduced by Remington Rand. Public awareness of computers increases when the UNIVAC I, after analyzing only 5 percent of the popular vote, correctly predicts that Dwight D. Eisenhower will win the presidential election.

The IBM 305 RAMAC system is the first to use magnetic disk for external storage. The system provides storage capacity similar to magnetic tape that previously was used, but offers the advantage of semi-random access capability.

1947 William Shockley, John Bardeen, and Walter Brattain invent the transfer resistance device, eventually called the transistor. The transistor would revolutionize computers, proving much more reliable than vacuum tubes.

1953 The IBM model 650 is one of the first widely used computer systems. Originally planning to produce only 50 machines, the system is so successful that eventually IBM manufactures more than 1,000. With the IBM 700 series of machines, the company will dominate the mainframe market for the next decade.

1957 FORTRAN (FORmula TRANslation), an efficient, easy-to-use programming language, is introduced by John Backus.

1952 Dr. Grace Hopper considers the concept of reusable software in her paper, "The Education of a Computer." The paper describes how to program a computer with symbolic notation instead of the detailed machine language that had been used.

Core memory, developed in the early 1950s, provides much larger storage capacity than vacuum tube memory.

www.scsite.com/dc2001/ch1/timeline3.htm

1958

Computers built with transistors mark the beginning of the second generation of computer hardware.

1960

COBOL, a high-level business application language, is developed by a committee headed by Dr. Grace Hopper. COBOL uses English-like phrases and runs on most business computers, making it one of the more widely used programming languages.

1965

Dr. John Kemeny of Dartmouth leads the development of the BASIC programming language. BASIC will be widely used on personal computers.

Digital Equipment Corporation (DEC) introduces the first minicomputer, the PDP-8. The machine is used extensively as an interface for time-sharing systems.

1959

More than 200 programming languages have been created.

IBM introduces two smaller, desk-sized computers: the IBM 1401 for business and the IBM 1602 for scientists. The IBM 1602 initially is called the CADET, but IBM drops the name when campus wags claim it is an acronym for, Can't Add, Doesn't Even Try.

1964

The number of computers has grown to 18,000.

Third-generation computers, with their controlling circuitry stored on chips, are introduced. The IBM System/360 computer is the first family of compatible machines, merging science and business lines.

1968

Alan Shugart at IBM demonstrates the first regular use of an 8-inch floppy (magnetic storage) disk.

In a letter to the editor titled, "GO TO Statements Considered Harmful," Dr. Edsger Dijsktra introduces the concept of structured programming, developing standards for constructing computer programs.

Computer Science Corporation becomes the first software company listed on the New York Stock Exchange.

IBM

Under pressure from the industry, IBM announces that some of its software will be priced separately from the computer hardware. This unbundling allows software firms to emerge in the industry.

The ARPANET network, a predecessor of the Internet, is established.

1969

Dr. Ted Hoff of Intel Corporation develops a microprocessor, or microprogrammable computer chip, the Intel 4004.

1971

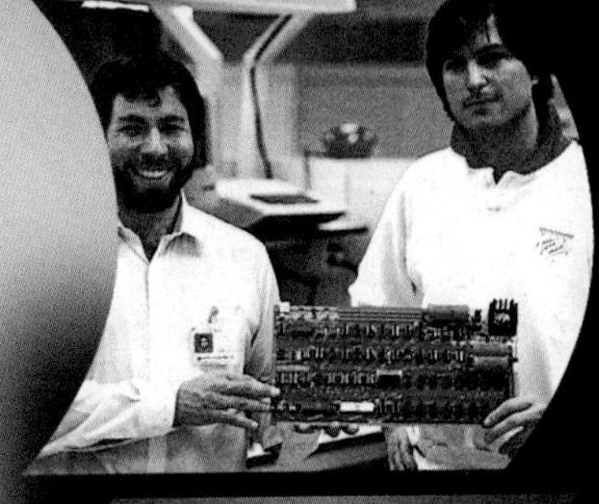

1975

MITS, Inc. advertises one of the first microcomputers, the Altair. Named for the destination in an episode of *Star Trek*, the Altair is sold in kits for less than $400. Although initially it has no keyboard, no monitor, no permanent memory, and no software, 4,000 orders are taken within the first three months.

1976

Steve Wozniak and Steve Jobs build the first Apple computer. A subsequent version, the Apple II, is an immediate success. Adopted by elementary schools, high schools, and colleges, for many students the Apple II is their first contact with the world of computers.

1970

Fourth-generation computers, built with chips that use LSI (large-scale integration) arrive. While the chips used in 1965 contained as many as 1,000 circuits, the LSI chip contains as many as 15,000.

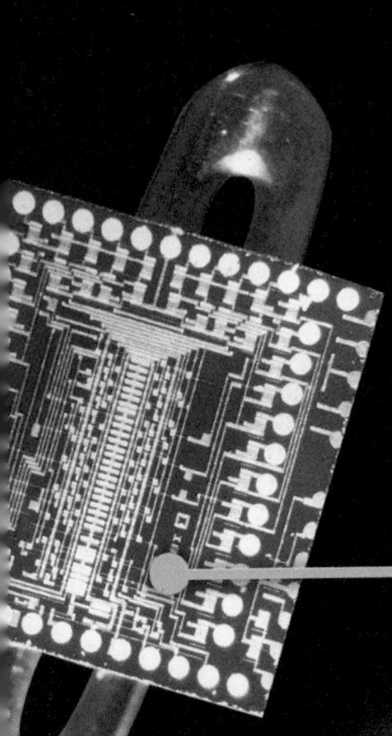

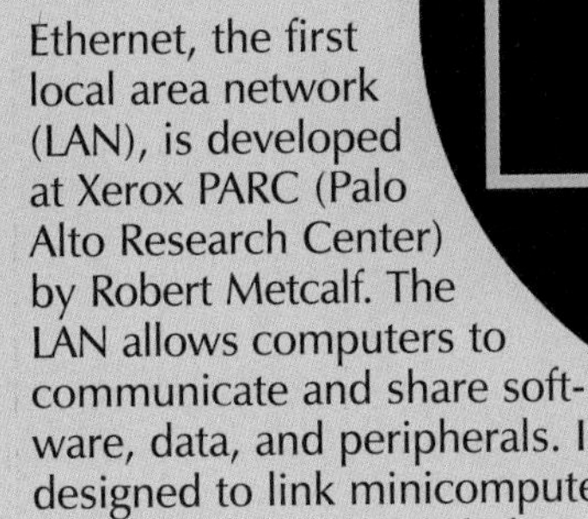

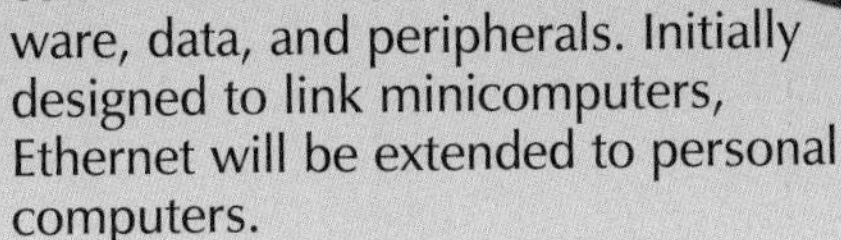

Ethernet, the first local area network (LAN), is developed at Xerox PARC (Palo Alto Research Center) by Robert Metcalf. The LAN allows computers to communicate and share software, data, and peripherals. Initially designed to link minicomputers, Ethernet will be extended to personal computers.

www.scsite.com/dc2001/ch1/timeline5.htm

1979

VisiCalc, a spreadsheet program written by Bob Frankston and Dan Bricklin, is introduced. Originally written to run on Apple II computers, VisiCalc will be seen as the most important reason for the acceptance of personal computers in the business world.

The first public online information services, CompuServe and the Source, are founded.

1980

Alan Shugart presents the Winchester hard drive, revolutionizing storage for personal computers.

IBM offers Microsoft Corporation co-founder, Bill Gates, the opportunity to develop the operating system for the soon-to-be announced IBM personal computer. With the development of MS-DOS, Microsoft achieves tremendous growth and success.

1981

The IBM PC is introduced, signaling IBM's entrance into the personal computer marketplace. The IBM PC quickly garners the largest share of the personal computer market and becomes the personal computer of choice in business.

1982

3,275,000 personal computers are sold, almost 3,000,000 more than in 1981.

3,275,000

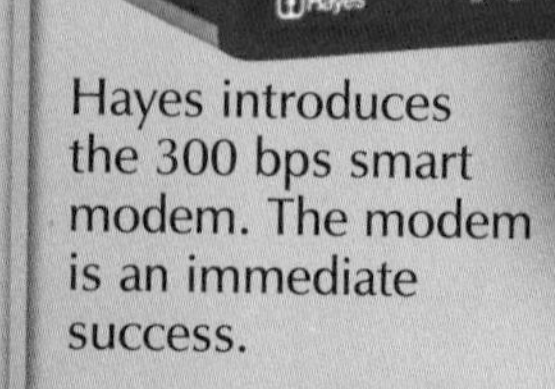

Hayes introduces the 300 bps smart modem. The modem is an immediate success.

COMPAQ

Compaq, Inc. is founded to develop and market IBM-compatible PCs.

1983

Lotus Development Corporation is founded. Its spreadsheet software, Lotus 1-2-3, which combines spreadsheet, graphics, and database programs in one package, becomes the best-selling program for IBM personal computers.

Instead of choosing a person for its annual award, *TIME* magazine names the computer Machine of the Year for 1982, acknowledging the impact of computers on society.

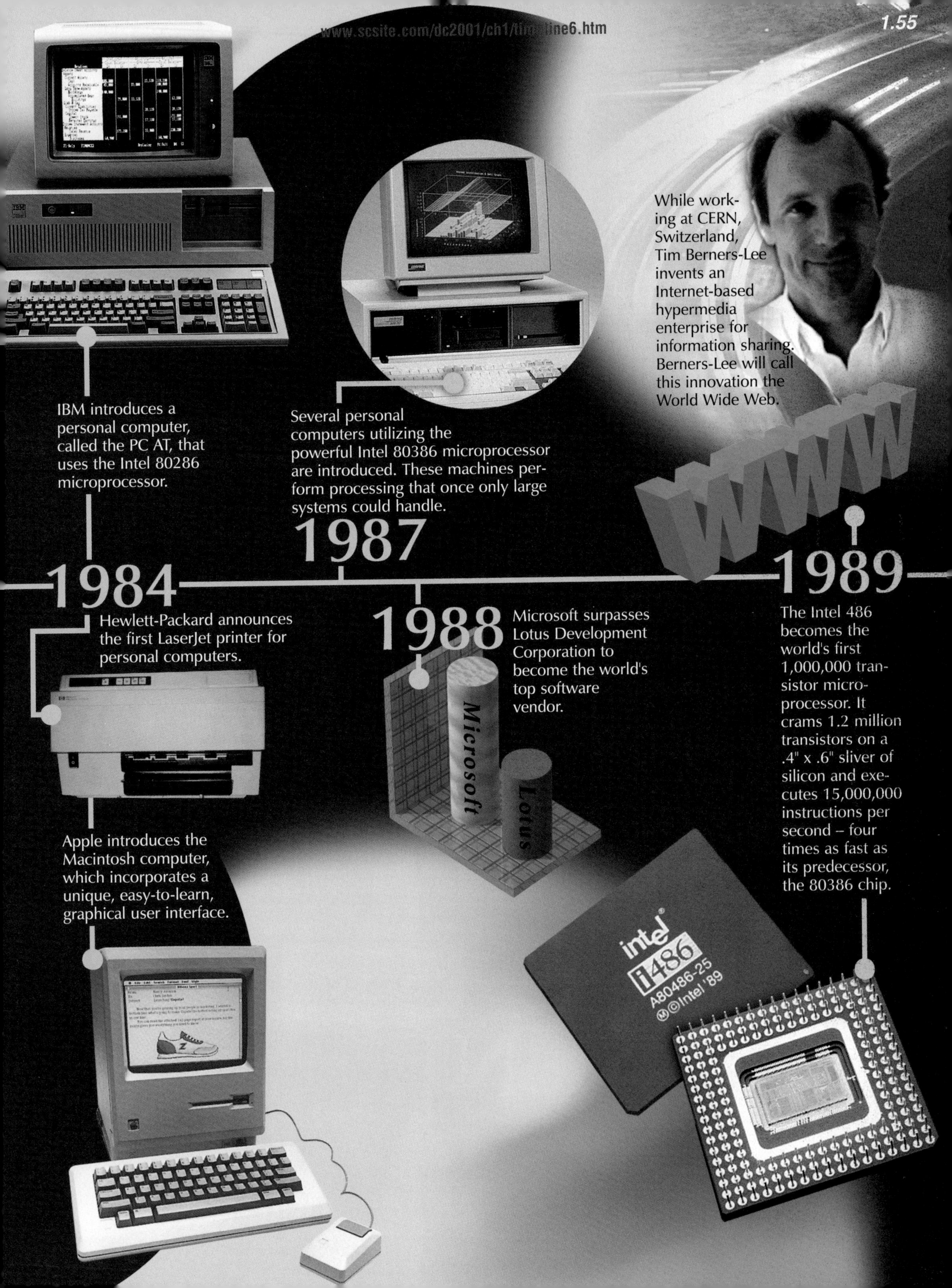
1984
IBM introduces a personal computer, called the PC AT, that uses the Intel 80286 microprocessor.
Hewlett-Packard announces the first LaserJet printer for personal computers.
Apple introduces the Macintosh computer, which incorporates a unique, easy-to-learn, graphical user interface.
1987
Several personal computers utilizing the powerful Intel 80386 microprocessor are introduced. These machines perform processing that once only large systems could handle.
1988
Microsoft surpasses Lotus Development Corporation to become the world's top software vendor.
Microsoft
Lotus
1989
While working at CERN, Switzerland, Tim Berners-Lee invents an Internet-based hypermedia enterprise for information sharing. Berners-Lee will call this innovation the World Wide Web.
WWW
The Intel 486 becomes the world's first 1,000,000 transistor microprocessor. It crams 1.2 million transistors on a .4" x .6" sliver of silicon and executes 15,000,000 instructions per second – four times as fast as its predecessor, the 80386 chip.
intel
i486
A80486-25
©Intel '89

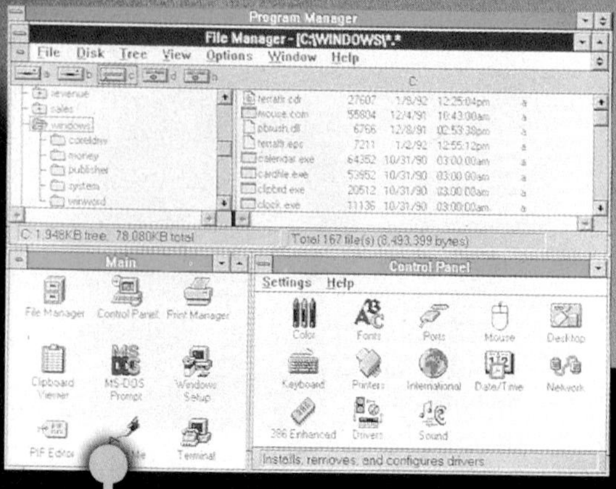

Microsoft releases Windows 3.1, the latest version of its Windows operating system. Windows 3.1 offers improvements such as TrueType fonts, multimedia capability, and object linking and embedding (OLE). In two months, 3,000,000 copies of Windows 3.1 are sold.

1992

Several companies introduce computer systems using the Pentium® microprocessor from Intel. The Pentium® chip is the successor to the Intel 486 processor. It contains 3.1 million transistors and is capable of performing 112,000,000 instructions per second.

Jim Clark and Marc Andreessen found Netscape and launch Netscape Navigator 1.0, a browser for the World Wide Web.

1991

World Wide Web Consortium releases standards that describe a framework for linking documents on different computers.

1993

The White House launches its Web page. The site includes an interactive citizens' handbook and White House history and tours.

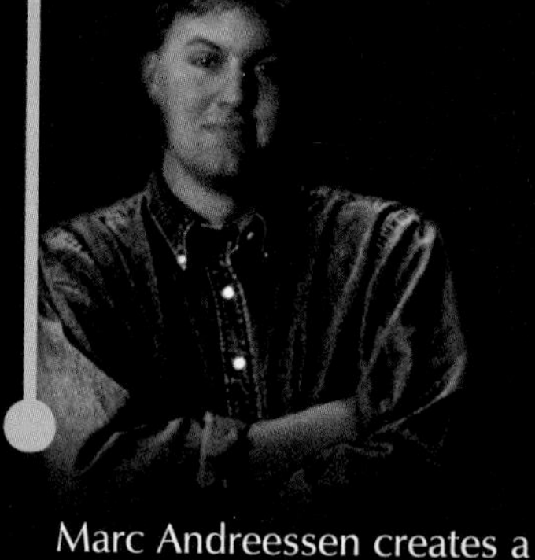

Marc Andreessen creates a graphical Web browser called Mosaic. This success leads to the organization of Netscape Communications Corporation.

1994

Linus Torvalds creates the Linux kernal, a UNIX-like operating system that he releases free across the Internet for further enhancement by other programmers.

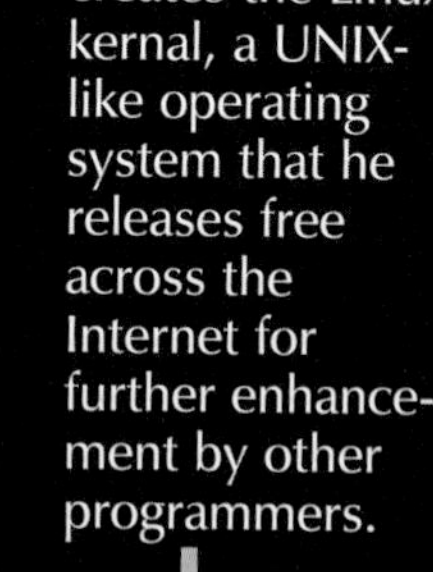

Microsoft releases Windows 95, a major upgrade to its Windows operating system. Windows 95 consists of more than 10,000,000 lines of computer instructions developed by 300 person-years of effort. More than 50,000 individuals and companies test the software before it is released.

U.S. Robotics introduces PalmPilot, a handheld personal organizer. The PalmPilot's user friendliness and low price make it a standout next to more expensive personal digital assistants (PDAs).

The Summer Olympics in Atlanta makes extensive use of computer technology, using an IBM network of 7,000 personal computers, 2,000 pagers and wireless devices, and 90 industrial-strength computers to share information with more than 150,000 athletes, coaches, journalists, and Olympics staff members, and millions of Web users.

1995

Sun Microsystems launches Java, an object-oriented programming language that allows users to write one application for a variety of computer platforms. Java becomes one of the hottest Internet technologies.

1996

Two out of three employees in the United States have access to a PC, and one out of every three homes has a PC. Fifty million personal computers are sold worldwide and more than 250,000,000 are in use.

Microsoft releases Windows NT 4.0, an operating system for client-server networks. Windows NT's management tools and Wizards make it easier for developers to build and deploy business applications.

An innovative technology called webtv combines television and the Internet by providing viewers with tools to navigate the Web.

www.scsite.com/dc2001/ch1/timeline9.htm

Intel introduces the Pentium® II processor with 7.5 million transistors. The new processor, which incorporates MMX™ technology, processes video, audio, and graphics data more efficiently and supports applications such as movie-editing, gaming, and more.

Microsoft releases Internet Explorer 4.0 and seizes a key place in the Internet arena. This new Web browser is greeted with tremendous customer demand.

DVD (Digital Video Disc), the next generation of optical disc storage technology, is introduced. DVD can store computer, audio, and video data in a single format, with the capability of producing near-studio quality. By year's end, 500,000 DVD players are shipped worldwide.

Microsoft ships Windows 98, an upgrade to Windows 95. Windows 98 offers improved Internet access, better system performance, and support for a new generation of hardware and software. In six months, more than 10,000,000 copies of Windows 98 are sold worldwide.

1997

1998

Deep Blue, an IBM supercomputer, defeats world chess champion Gary Kasparov in a six-game chess competition. Millions of people follow the 9-day long rematch on IBM's Web site.

Apple and Microsoft sign a joint technology development agreement. Microsoft buys $150,000,000 of Apple stock.

E-commerce, or electronic commerce – the marketing of goods and services over the Internet – booms. Companies such as Dell, E*TRADE, and Amazon.com spur online shopping, allowing buyers to obtain everything from hardware and software to financial and travel services, insurance, automobiles, books, and more.

Fifty million users are connected to the Internet and World Wide Web.

More than 10,000,000 people take up telecommuting– the capability of working at home and communicating with an office via computer. More and more firms embrace telecommuting to help increase productivity, reduce absenteeism, and provide greater job satisfaction.

Apple Computer introduces the iMac, the latest version of its popular Macintosh computer. The iMac abandons such conventional features as a floppy disk drive but wins customers with its futuristic design, see-through case, and easy setup. Consumer demand outstrips Apple's production capabilities, and some vendors are forced to begin waiting lists.

Intel releases its Pentium® III processor, which provides enhanced multimedia capabilities.

U.S. District Judge Thomas Penfield Jackson rules in the antitrust lawsuit brought by the Department of Justice and 19 states that Microsoft used its monopoly power to stifle competition.

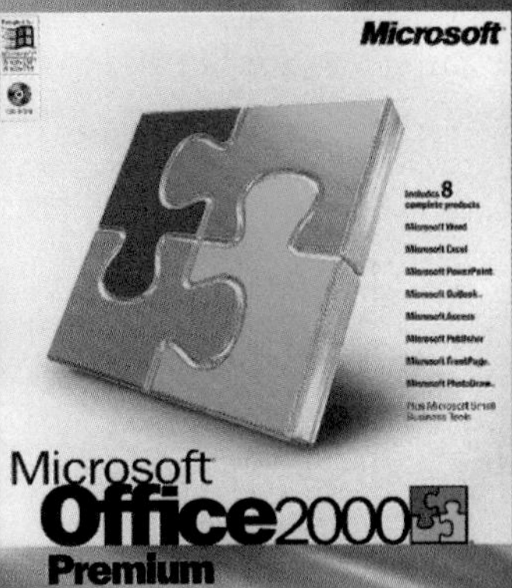

Microsoft introduces Office 2000, its premier productivity suite, offering new tools for users to create content and save it directly to a Web site without any file conversion or special steps.

1998

Compaq Computer, the United States' leading personal computer manufacturer, buys Digital Equipment Corporation in the biggest takeover in the history of the computer industry. Compaq becomes the world's second largest computer firm, behind IBM.

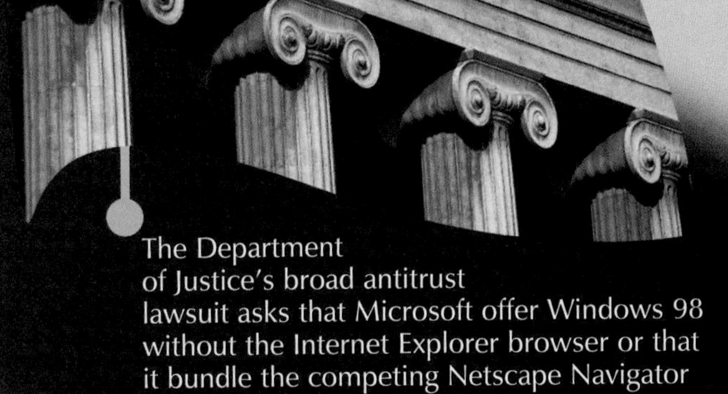

The Department of Justice's broad antitrust lawsuit asks that Microsoft offer Windows 98 without the Internet Explorer browser or that it bundle the competing Netscape Navigator browser with the operating system.

1999

Open Source Code software, such as the Linux operating system and the Apache Web server created by unpaid volunteers, begin to gain wide acceptance among computer users.

Governments and businesses frantically work to make their computer systems Y2K (Year 2000) compliant, spending more than $500 billion worldwide. Y2K non-compliant computers cannot distinguish if 01/01/00 refers to 1900 or 2000, and thus may operate using a wrong date. This Y2K bug can affect any application that relies on computer chips, such as ATMs, airplanes, energy companies, and the telephone system.

Microsoft ships Windows 2000. Windows 2000 offers improved behind-the-scene security and reliability over its predecessor Windows NT. Its new Active Directory stores information about network devices and users and makes this information available to administrators, users, and applications.

Internet providers offer computers free to those who sign up for their service, while hardware manufacturers offer application software free with their computers.

According to the U.S. Commerce Department, Internet traffic is doubling every 100 days, resulting in an annual growth rate of more than 700 percent. It has taken radio and television 30 years and 15 years to reach 60 million people, respectively. The Internet has achieved the same audience base in 3 years.

2000

Wireless technology achieves significant market penetration. Prices drop, usage increases, and wireless carriers scramble for new services, particularly for a mobile workforce.

e-commerce achieves mainstream acceptance. Annual e-commerce sales exceed $100 billion and Internet advertising expenditures reach more than $5 billion.

Application service providers offer a return to a centralized computing environment, in which large megaservers warehouse your data, information, and software, so it is accessible using a variety of devices from any location.

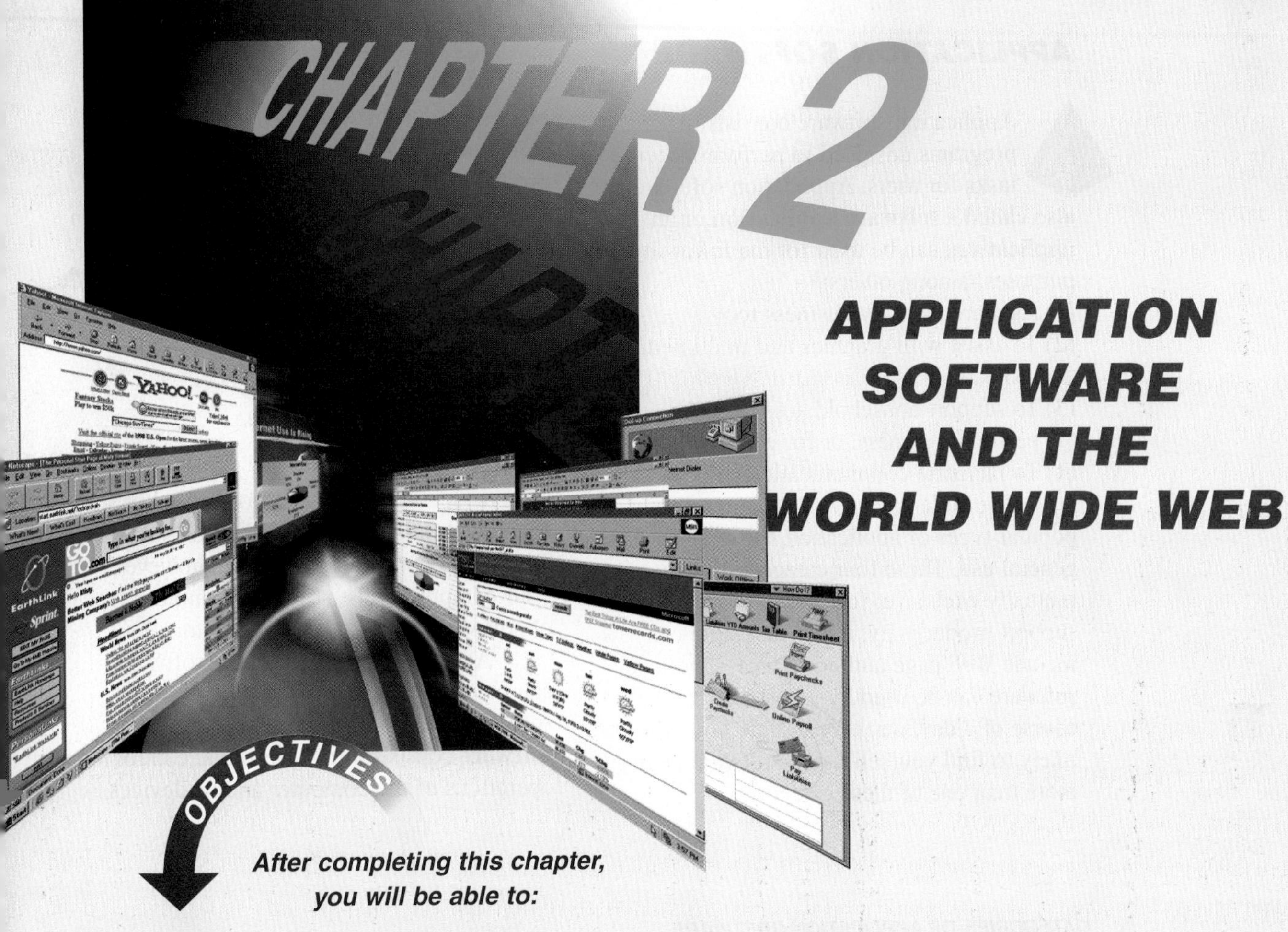

CHAPTER 2

APPLICATION SOFTWARE AND THE WORLD WIDE WEB

OBJECTIVES

After completing this chapter, you will be able to:

- Define application software
- Explain how to start a software application
- Explain the key features of widely used software applications
- Provide examples illustrating the importance of the World Wide Web
- Describe how to use a Web browser
- Explain how to search for information on the Web
- Describe the learning aids available with many software applications

A key aspect of building computer literacy is learning about software, which is the series of instructions that tells computer hardware how to perform tasks. Having a solid understanding of software — especially application software — will help you use your computer to be more productive, organized, and well informed.

Application software, such as word processing, spreadsheets, and e-mail, can help you perform tasks such as creating documents, analyzing finances, and sending messages. One type of application software, called a Web browser, allows you to view pages on the World Wide Web, thus giving you access to a vast resource of information. Another allows you to create Web pages so you can share information with the world.

Understanding application software also can help advance your personal and professional goals. In fact, many employers today consider an understanding of software to be a hiring requirement. Because application software concepts are so important, this book discusses them early so you can refer back to them as you learn more about computers, how they are used today, and how they can help you in your future.

APPLICATION SOFTWARE

Application software consists of programs designed to perform specific tasks for users. Application software, also called a **software application** or an **application**, can be used for the following purposes, among others:

(1) As a productivity/business tool
(2) To assist with graphics and multimedia projects
(3) To support household activities, for personal business, or for education
(4) To facilitate communications

The table in Figure 2-1 categorizes popular types of application software by their general use. These four categories are not mutually exclusive; for example, e-mail can support productivity, a software suite can include Web page authoring tools, and legal software can be used by a business. In the course of a day, week, or month, you are likely to find yourself using software from more than one of these categories.

A variety of application software is available as packaged software that you can purchase from software vendors in retail stores or on the Web. A specific software product, such as Microsoft Word, often is called a **software package**. Many application software packages also are available as shareware, freeware, and public-domain software; these packages, however, usually have fewer capabilities than retail software packages.

The Role of the Operating System

Like most computer users, you probably are somewhat familiar with application software. To run any application software, however, your computer must be running another type of software — an operating system.

As described in Chapter 1, software can be categorized into two types: system software and application software. **System software** consists of programs that control the operations of the computer and its devices.

CATEGORIES OF APPLICATION SOFTWARE

Productivity/ Business	Graphic Design/ Multimedia	Home/Personal/ Educational	Communications
• Word Processing	• Computer-Aided Design	• Integrated Software	• Groupware
• Spreadsheet	• Desktop Publishing (Professional)	• Personal Finance	• E-mail
• Database	• Paint/Image Editing (Professional)	• Legal	• Web Browser
• Presentation Graphics	• Video and Audio Editing	• Tax Preparation	• Chat Rooms
• Personal Information Management	• Multimedia Authoring	• Desktop Publishing (Personal)	• Newsgroups
• Software Suite	• Web Page Authoring	• Paint/Image Editing (Personal)	
• Accounting		• Home Design/ Landscaping	
• Project Management		• Educational	
• Webware		• Reference	
		• Entertainment	

Figure 2-1 *The four major categories of popular application software. You likely will use software from more than one of these categories.*

As shown in Figure 2-2, system software serves as the interface between you (the user), your application software, and your computer's hardware. One type of system software, the **operating system**, contains instructions that coordinate all of the activities of the hardware devices in a computer. The operating system also contains instructions that allow you to run application software. The other type of system software is a utility program, or utility, which is discussed in Chapter 8.

Before a computer can run any application software, the operating system must be loaded from the hard disk into the computer's memory. Each time you start your computer, the operating system is loaded, or copied, into memory from the computer's hard disk. Once the operating system is loaded, it tells the computer how to perform functions such as processing program instructions and transferring data among input and output devices and memory. The operating system, which remains in memory while the computer is running, allows you to communicate with the computer and other software, such as application software. The operating system continues to run until the computer is turned off.

The Role of the User Interface

All software, including the operating system, communicates with the user in a certain way, through a portion of the program called a user interface. A **user interface** controls how you enter data or instructions and how information and processing options are presented to you.

One of the more common user interfaces is a graphical user interface (GUI). A **graphical user interface**, or **GUI**, combines text, graphics, and other visual cues to make software easier to use.

In 1984, Apple Computer introduced the Macintosh operating system, which used a graphical user interface. Recognizing the value of this easy-to-use interface, many software companies followed suit, developing their own GUI software. Since then, Apple has developed several new versions (modifications) of their original GUI operating system for their Macintosh computers. Today's most widely used personal computer operating system and graphical user interface, however, is Microsoft Windows, which often is referred to simply as **Windows**.

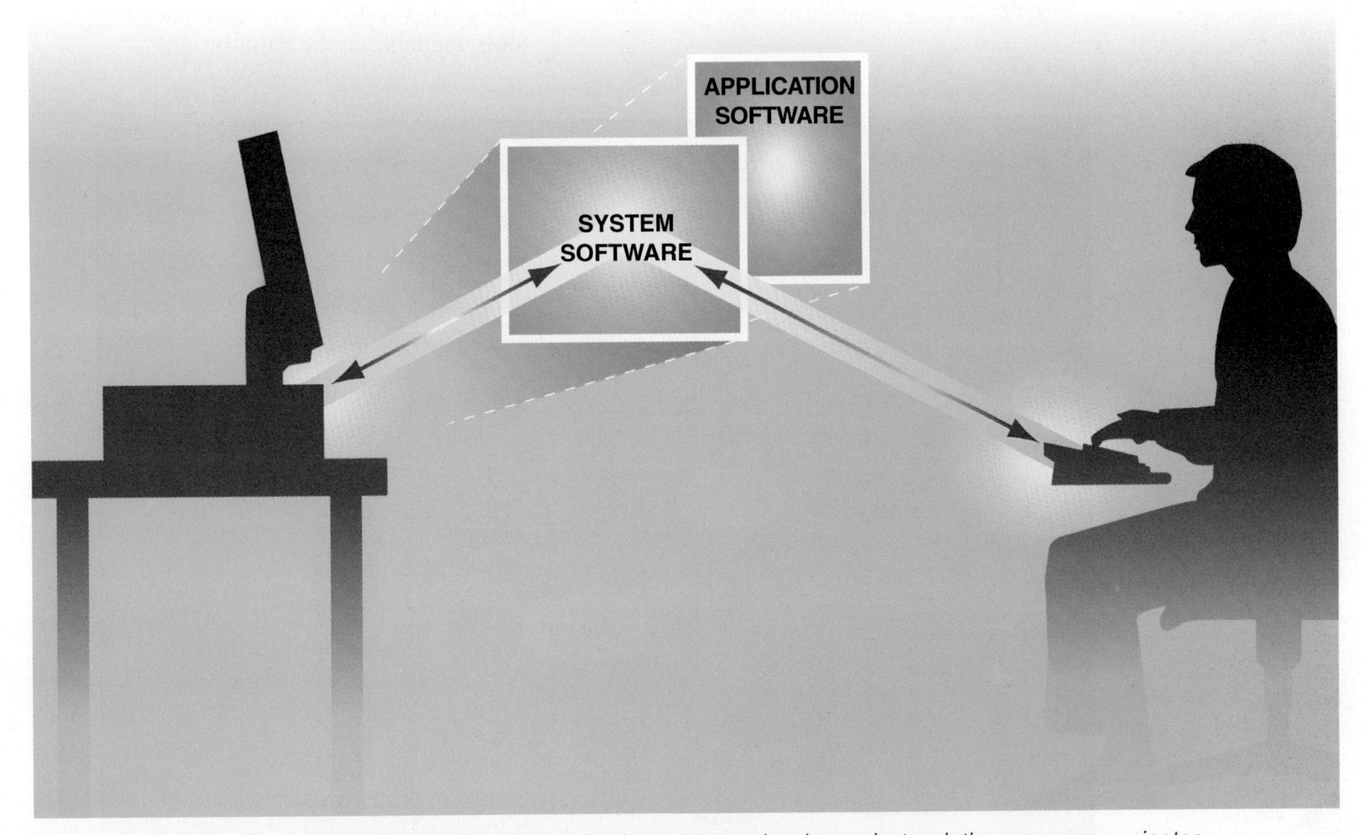

Figure 2-2 *A user does not communicate directly with the computer hardware; instead, the user communicates with the system software or with the application software to control the hardware.*

Starting a Software Application

Both the Apple Macintosh and Microsoft Windows operating systems use the concept of a desktop to make the computer easier to use. The **desktop** is an onscreen work area that uses common graphical elements such as icons, buttons, windows, menus, links, and dialog boxes, all of which can display on the desktop. The Windows desktop shown in Figure 2-3 contains many icons and buttons.

An **icon** is a small image that displays on the screen to represent a program, a document, or some other object. A **button** is a graphical element (usually a rectangular or circular shape) that you can click to cause a specific action to take place. To click a button, you typically point to it with a mouse and then press and release a mouse button.

The Windows desktop contains a Start button in its lower-left corner, which can be used to start an application. When you click the Start button, the Start menu displays on the desktop. A **menu** contains commands you can select. **Commands** are instructions that cause a computer program to perform a specific action.

Some menus have a **submenu**, which displays when you point to a command on a previous menu. For example, as shown in Figure 2-4, when you click the Start button and point to the Programs command on the Start menu, the Programs submenu displays. Pointing to the Accessories command on the Programs submenu displays the Accessories submenu. As shown on the Accessories submenu, Windows includes several applications such as Calculator, Paint, and WordPad.

You can start an application by clicking its program name on a menu or submenu. Doing so instructs the operating system to start the application by transferring the program's instructions from a storage medium into memory. For example, if you click Paint on the Accessories submenu, Windows transfers the Paint program instructions from the computer's hard disk into memory.

Once started, an application displays in a window on the desktop. A **window** is a rectangular area of the screen that is used to display a program, data, and/or information. The top of a window has a **title bar**, which is a horizontal space that contains the window's name. Figure 2-5 shows the Paint window. This window contains a photographic image that has been converted from a paper document to an electronic document using special hardware and software. The words, Our Hawaiian Vacation, have been added using tools available in the Paint program.

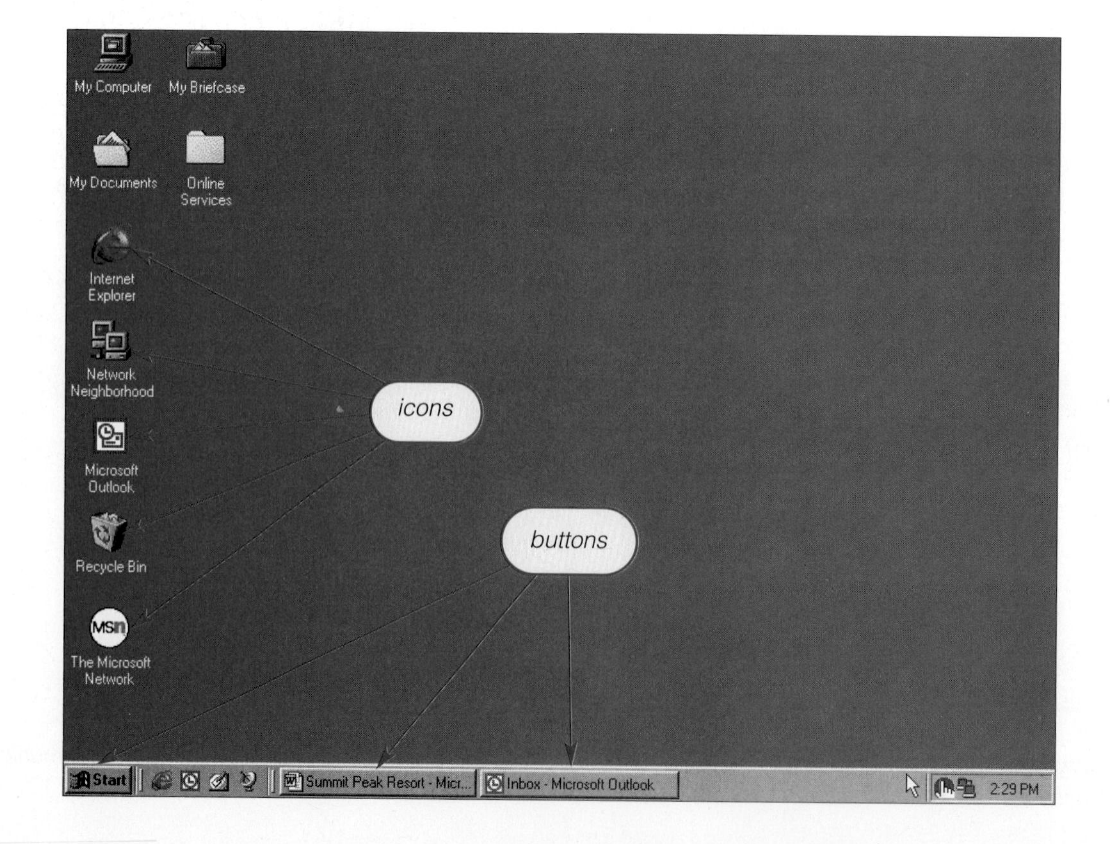

Figure 2-3 *This Windows desktop shows a variety of icons and buttons.*

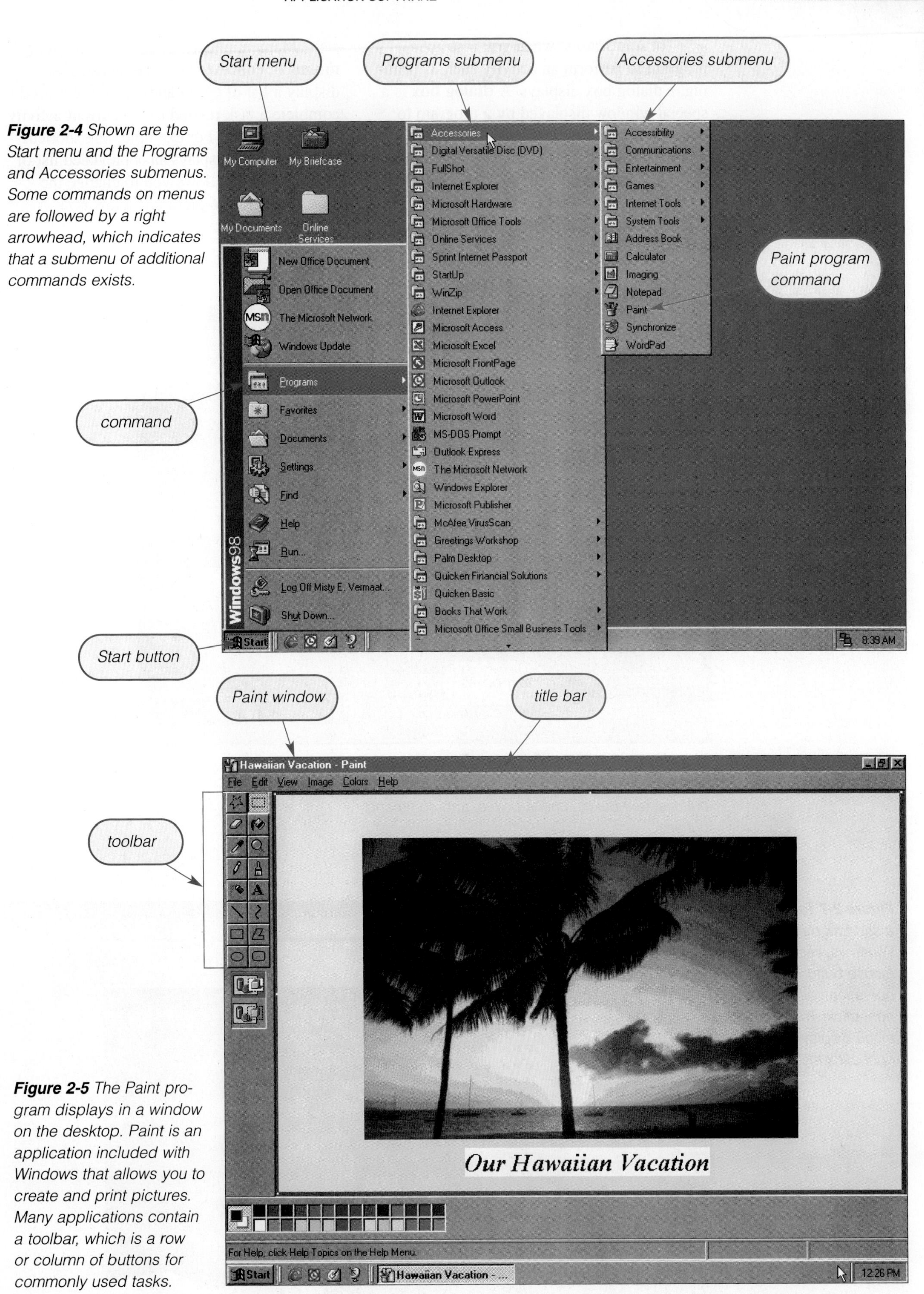

Figure 2-4 *Shown are the Start menu and the Programs and Accessories submenus. Some commands on menus are followed by a right arrowhead, which indicates that a submenu of additional commands exists.*

Figure 2-5 *The Paint program displays in a window on the desktop. Paint is an application included with Windows that allows you to create and print pictures. Many applications contain a toolbar, which is a row or column of buttons for commonly used tasks.*

In some cases, when you instruct a program to perform an activity such as printing, a dialog box displays. A **dialog box** is a special window displayed by a program to provide information, present available options, or request a response (Figure 2-6). A Print dialog box, for example, gives you many printing options, such as printing multiple copies, using different printers, or printing all or part of a document.

Many applications also use **shortcut menus** or **context-sensitive menus** that display a list of commands commonly used to complete a task related to the current activity or selected item. For example, one shortcut menu for Paint displays an abbreviated list of commands that allows you to change the appearance of the picture (Figure 2-7).

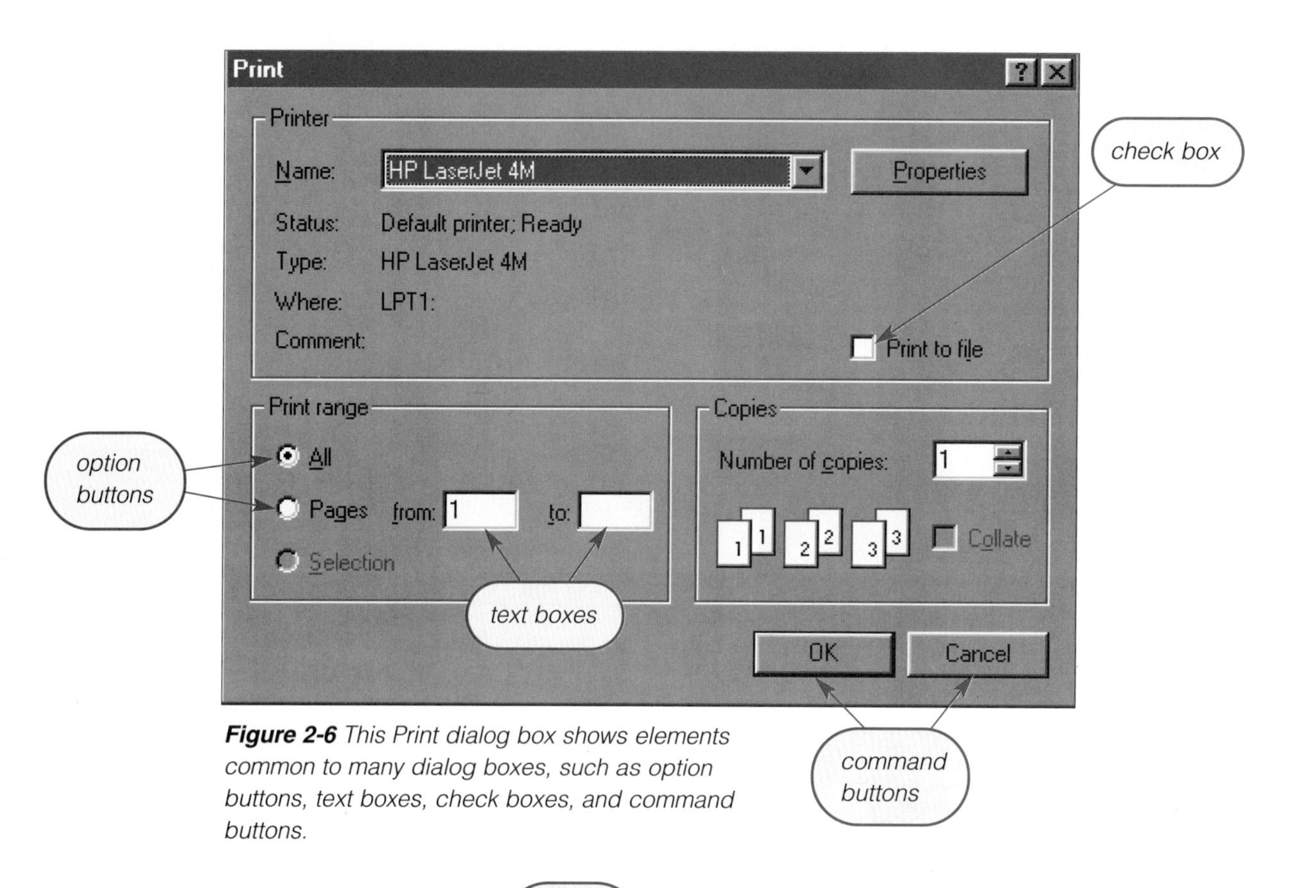

Figure 2-6 This Print dialog box shows elements common to many dialog boxes, such as option buttons, text boxes, check boxes, and command buttons.

Figure 2-7 *To display a shortcut menu in Windows, click the right mouse button, a mouse operation called a right-click. This shortcut menu displays when you right-click the picture.*

Many elements shown on the previous pages, such as icons, buttons, and menus are part of the graphical user interface that allows you to communicate with software. One of the major advantages of a graphical user interface is that these elements usually are common across most applications. Once you learn the purpose and functionality of these elements, you can apply that knowledge to other software applications.

PRODUCTIVITY SOFTWARE

Productivity software is designed to make people more effective and efficient while performing daily activities. Productivity software includes applications such as word processing, spreadsheet, database, presentation graphics, personal information management, accounting, project management, and other related types of software. Figure 2-8 lists popular software packages for each of these applications. The features and functions of each of these applications are discussed in the following sections.

POPULAR PRODUCTIVITY SOFTWARE PACKAGES

Software Application	Popular Packages/Companies
Word Processing	• Corel WordPerfect • Lotus Word Pro • Microsoft Word 2000
Spreadsheet	• Corel Quattro Pro • Lotus 1-2-3 • Microsoft Excel 2000
Database	• Corel Paradox • Lotus Approach • Microsoft Access 2000 • Microsoft Visual FoxPro
Presentation Graphics	• Corel Presentations • Lotus Freelance Graphics • Microsoft PowerPoint 2000
Personal Information Manager	• 3Com Palm Desktop • CorelCENTRAL • Lotus Organizer • Microsoft Outlook 2000
Software Suite	• Corel WordPerfect Suite • Lotus SmartSuite • Microsoft Office 2000
Accounting	• Intuit QuickBooks • Peachtree Complete Accounting
Project Management	• Microsoft Project • Primavera SureTrak Project Manager
Webware	• Intuit • Microsoft • Sun Microsystems

Figure 2-8 *Popular productivity software products.*

WEB INFO

For more information on word processing software, visit the Discovering Computers 2001 Chapter 2 WEB INFO page **(www.scsite.com/dc2001/ch2/webinfo.htm)** and click Word Processing Software.

Word Processing Software

A widely used application software package is **word processing software**, which is used to create, edit, and format textual documents (Figure 2-9). Millions of people use word processing software every day to create documents such as letters, memos, reports, fax cover sheets, mailing labels, and newsletters.

DEVELOPING A DOCUMENT While using many software applications, you have the ability to create, edit, format, print, and save documents. During the process of developing a document, you likely will switch back and forth among all of these activities.

Creating involves developing the document by entering text or numbers, inserting graphical images, and performing other tasks using an input device such as a keyboard or mouse. If you design an announcement in Microsoft Word, for example, you are creating a document.

Editing is the process of making changes to the document's existing content. Common editing features include inserting, deleting, cutting, copying, and pasting items into a document. For example, using Microsoft Word, you can insert, or add, text to an announcement, such as the cost of various vacation packages. When you delete, you remove text or other content. To cut involves removing a portion of the document and electronically storing it in a temporary storage location called the **Clipboard**. When you copy, a portion of the document is duplicated and stored on the Clipboard. To place whatever is stored on the Clipboard into the document, you paste it into the document.

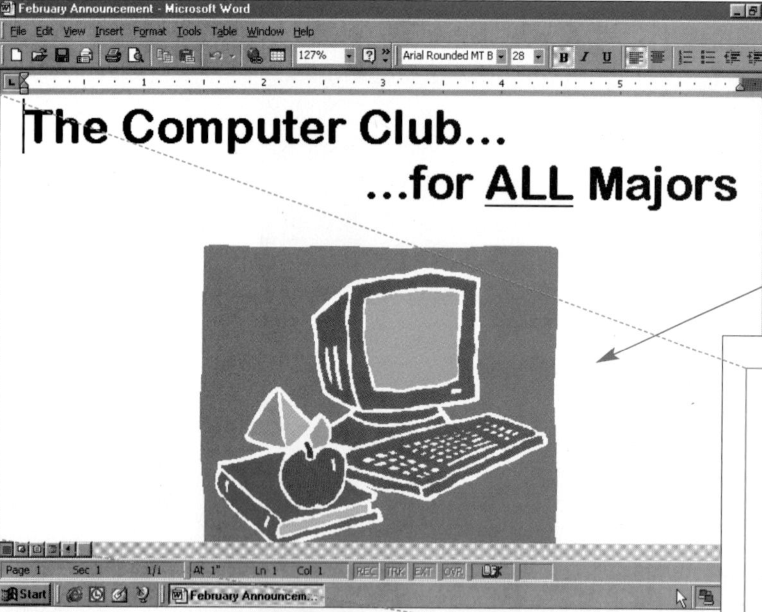

Figure 2-9 *With word processing software, you can create documents that are professional and visually appealing.*

The Computer Club...
...for ALL Majors

MONTHLY MEETING ANNOUNCEMENT

- This month's meeting of The Computer Club is scheduled for **Monday, February 19**, from 11:30 a.m. to 2:00 p.m. in Conference Center 102. Bring a sack lunch.
- The roundtable discussion is on Internet service providers and begins at 12:00 noon.
- Anne Vance, director of office automation at our school, is our guest speaker. Anne will demonstrate how to use *Word 2000* to create a newsletter. Her presentation begins at 1:00 p.m.

Questions? Call Mark at 555-5587.

Formatting involves changing the appearance of a document. Formatting is important because the overall look of a document can significantly affect its ability to communicate effectively.

Examples of formatting tasks are changing the font, font size, or font style of text (Figure 2-10). A **font** is a name assigned to a specific design of characters. Times New Roman and Arial are examples of fonts. The **font size** specifies the size of the characters in a particular font. Font size is gauged by a measurement system called points. A single **point** is about 1/72 of an inch in height. The text you are reading in this book is 10 point. Thus, each character is about 10/72 of an inch in height. A **font style** adds emphasis to a font. Examples of font styles are **bold**, *italic*, and underline.

While you are creating, editing, and formatting a document, it is held temporarily in memory. Once you have completed these steps, you may want to save your document for future use. **Saving** is the process of copying a document from memory to a storage medium such as a floppy disk or hard disk. You also should save the document frequently while working with it, to ensure your work is not lost. Many applications have an optional **AutoSave** feature that automatically saves open documents at specified time periods.

12-Point

Times New Roman

22-Point

Times New Roman
Bold *Italic* Underlined

48-Point

Times New Roman
Bold *Italic* Underlined

12-Point

Arial

22-Point

Arial
Bold *Italic* Underlined

48-Point

Arial
Bold *Italic* Underlined

Figure 2-10 *The Times New Roman and Arial fonts are shown in three font sizes and a variety of font styles.*

Any document that you are working with or have saved exists as a file. A **file** is a named collection of data, instructions, or information, such as a document that you create. To distinguish among various files, each file has a **file name**, which is a unique set of letters of the alphabet, numbers, and other characters that identifies the file.

Once you have created a document, you can print it many times, with each copy looking just like the first. **Printing** is the process of sending a file to a printer to generate output on a medium such as paper. You also can send the document to others electronically, if your computer is connected in a network.

BASIC WORD PROCESSING FEATURES

Word processing software has many formatting features to make documents look professional and visually appealing. For example, you can change the font and font size of headlines and headings, change the color of characters, or organize text into newspaper-style columns. Any colors used for characters or other formatting will print as black or gray unless you have a color printer.

Most word processing software allows you to incorporate graphical images of many types into your document. For example, you can enhance a document by adding a **border**, which is a decorative line or pattern along one or more edges of a page or around a graphical image. One type of graphical image commonly included with word processing software is **clip art**, which is a collection of drawings, diagrams, and photographs that can be inserted in other documents. Figure 2-9 on page 2.8 includes a clip art image of a skier. Clip art collections, which can contain several hundred to several thousand images, usually are grouped by type, such as buildings, nature, or people.

While some clip art is included in your word processing package, you can create clip art and other graphics using Paint or another application and **import** (bring in) the clip art into the word processing document. Once you insert or import a clip art image or other graphical image into a document, you can move it, resize it, rotate it, crop it, and adjust its color.

All word processing software provides at least some basic capabilities to help you create, edit, and format documents. For example, you can define the size of the paper on which to print, as well as the **margins** — that is, the portion of the page outside the main body of text, on the top, bottom, and sides of the paper. The word processing software automatically readjusts any text so it fits within the new definitions.

With **wordwrap**, if you type text that extends beyond the right page margin, the word processing software automatically positions text at the beginning of the next line. Wordwrap allows you to type words in a paragraph continually without pressing the ENTER key at the end of each line.

As you type more lines of text than can display on the screen, the top portion of the document moves upward, or scrolls, off the screen. Because of the size of the screen, you can view only a portion of a document on the screen at a time. **Scrolling** is the process of moving different portions of the document on the screen into view.

A major advantage of using word processing software is that you easily can change what you have written. You can insert, delete, or rearrange words, sentences, paragraphs, or entire sections. You can use the **find** or **search** feature to locate all occurrences of a particular character, word, or phrase. This feature can be used in combination with the **replace** feature to substitute existing characters or words with new ones. For example, you can instruct the word processing software to locate the word, vacation, and replace it with the word, holiday.

Current word processing packages even have a feature that automatically corrects errors and makes word substitutions as you type text. For instance, you can type an abbreviation such as asap and the word processing software will replace this abbreviation with the phrase, as soon as possible.

To review the spelling of individual words, sections of a document, or the entire document, you can use a **spelling checker** (sometimes called a spell checker). The spelling checker compares the words in the document to an electronic dictionary that is part of the word processing software. You can customize the electronic dictionary by adding words such as companies, streets, cities, and personal names, so the software can check the spelling of those words as well. Many word processing software packages allow you to check the spelling of a whole document at one time, or to check the spelling of individual words as you type them.

You also can insert headers and footers into a word processing document. A **header** is text you want at the top of each page; a **footer** is text you want at the bottom of each page. Page numbers, as well as company names, report titles, or dates are examples of items frequently included in headers and footers.

In addition to these basic features, most current word processing packages provide numerous additional features. The table in Figure 2-11 lists these additional features.

POPULAR WORD PROCESSING FEATURES

AutoCorrect	As you type words, the AutoCorrect feature corrects common spelling errors. For example, if you type the word, adn, the word processing software automatically changes it to the correct word, and. AutoCorrect also corrects errors in capitalization. For example, it capitalizes names of days and the first letter in a sentence.
AutoFormat	As you type, the AutoFormat feature automatically applies formatting to your text. For example, it automatically can number a list or convert a Web address to a hyperlink. AutoFormat also automatically creates symbols, fractions, and ordinal numbers. For example, when you type :), it changes to a smiling face symbol ☺; the fraction 1/2 is created when you type 1/2; and the ordinal 2nd is created when you type 2nd.
Columns	Most word processing software can arrange text in two or more columns like a newspaper or magazine. The text from the bottom of one column automatically flows to the top of the next column.
Grammar Checker	You can use the grammar checker to proofread documents for grammar, writing style, and sentence structure errors in your document. You can check the grammar of a document all at one time, or instruct the word processing software to check grammar as you enter text.
Tables	Tables are a way of organizing information into rows and columns. With tables, you easily can rearrange rows and columns, change column widths, sort rows and columns, sum the contents of rows and columns, or format the contents of a table. Instead of evenly spaced rows and columns, some word processing packages allow you to draw the tables of any size or shape directly into the document.
Templates	A template is a document that contains the formatting necessary for a specific document type. For example, a letter template would contain the proper spacing and indicate the position of elements common to a business letter such as a date, inside address, salutation, body, closing, and signature block. Templates usually exist for documents such as memos, fax cover sheets, and letters.
Thesaurus	With a thesaurus, you can look up synonyms (words with the same meaning) for words in a document while you are using your word processing software.
Tracking Changes/ Comments	If multiple users work with a document, you can instruct the word processing software to highlight or color-code the changes made by various users. This way, you can see easily what changes have been made to the document. You also can add comments to a document, without changing the text itself. These comments allow you to communicate with the other users working on the document.
Voice Recognition	With some of the newer word processing packages, you can speak into your computer's microphone and watch the spoken words display on your screen as you talk. With these packages, you also can edit and format the document by speaking or spelling an instruction.
Web Page Development	Most word processing software supports Internet connectivity, allowing you to create, edit, and format documents for the World Wide Web. You automatically can convert an existing word processing document into the standard document format for the World Wide Web. You also can view and browse Web pages directly from your word processing software.

Figure 2-11 *Additional features included with many word processing software packages.*

WEB INFO

For more information on spreadsheet software, visit the Discovering Computers 2001 Chapter 2 WEB INFO page (**www.scsite.com/dc2001/ch2/webinfo.htm**) and click Spreadsheet Software.

Spreadsheet Software

Another widely used software application is **spreadsheet software**, which allows you to organize data in rows and columns. These rows and columns collectively are called a **worksheet**. For years, people used manual methods, such as those performed with pencil and paper, to organize data in rows and columns. The data in an electronic worksheet is organized in the same manner as it is in a manual worksheet (Figure 2-12).

As with word processing software, most spreadsheet software has basic features to help you create, edit, and format worksheets. These features, as included in several popular spreadsheet software packages, are described in the following sections.

SPREADSHEET ORGANIZATION A spreadsheet file is like a notebook with up to 255 individual worksheets. On each worksheet, data is organized vertically in columns and horizontally in rows. Each worksheet typically has 256 columns and 65,536 rows. A letter identifies each column, and a number identifies each row. The column letters begin with A and end with IV; row numbers begin with 1 and end with 65,536. Only a small fraction of these columns and rows displays on the screen at one time. To view a different part of a worksheet, you can scroll to display it on your screen.

The intersection of a column and row is called a **cell**. Each worksheet has more than 16 million (256 x 65,536) cells into which you can enter data. Cells are identified by the column and row in which they are located. For example, the intersection of column C and row 9 is referred to as cell C9. In Figure 2-12, cell C9 contains the number, 15,235.00, which represents the Rent expense for January through March.

Cells may contain three types of data: labels (text), values (numbers), and formulas. The text, or **label**, entered in a cell identifies the data and helps organize the worksheet. Using descriptive labels, such as Community College Revenue, Total Income, and Classroom Rental, helps make a worksheet more meaningful.

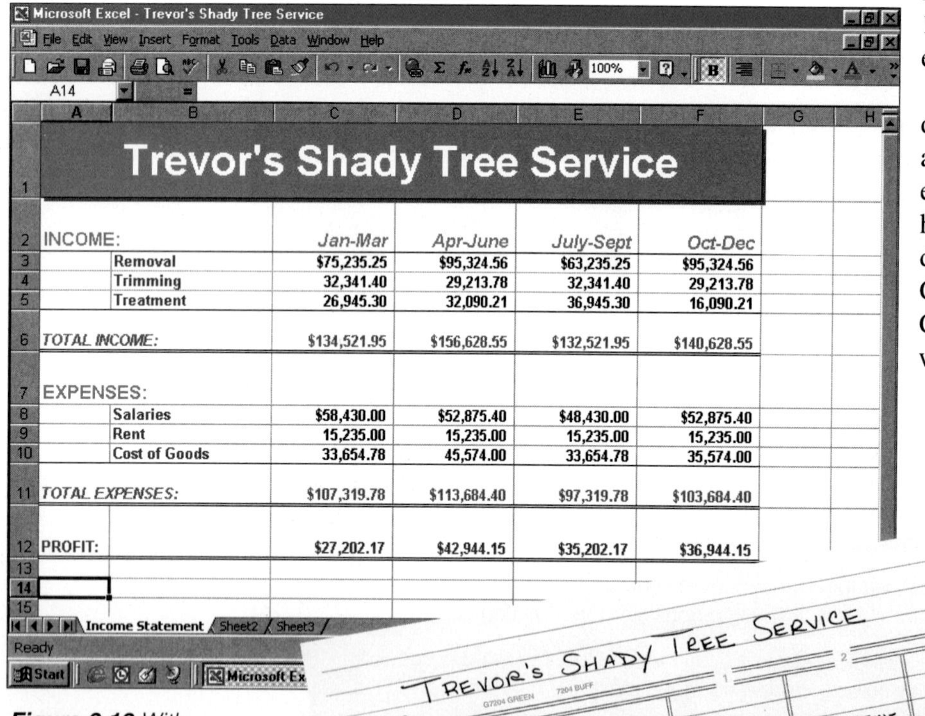

Figure 2-12 *With spreadsheet software, you create worksheets that contain numeric data arranged in rows and columns.*

CALCULATIONS Many of the worksheet cells shown in Figure 2-12 contain a number, or a **value**. Other cells, however, contain formulas that generate values. A **formula** performs calculations on the data in the worksheet and displays the resulting value in a cell, usually the cell containing the formula. In Figure 2-12, for example, cell C6 could contain the formula =C3+C4+C5 to calculate the total income for January through March.

When creating a worksheet, you can enter your own formulas or, in some cases, you can use a function that is included with the spreadsheet software. A **function** is a predefined formula that performs common calculations such as adding the values in a group of cells or generating a value such as the time or date. For example, instead of using the formula =C3+C4+C5 to calculate the total income for January through March, you could use the function =SUM(C3:C5), which adds, or sums, the contents of cells C3, C4, and C5. Figure 2-13 is a list of functions commonly included in spreadsheet software packages.

MACROS Spreadsheet software and other programs often include a timesaving feature called a macro. A **macro** is a sequence of keystrokes and instructions that are recorded and saved. When you run the macro, the macro performs the sequence of keystrokes and instructions. Creating a macro can help you save time by allowing you to enter a single character or word to perform frequently used tasks. For example, you can create a macro to format cells automatically or print a portion of a worksheet.

SPREADSHEET FUNCTIONS

FINANCIAL	
FV (rate, number of periods, payment)	Calculates the future value of an investment
NPV (rate, range)	Calculates the net present value of an investment
PMT (rate, number of periods, present value)	Calculates the periodic payment for an annuity
PV (rate, number of periods, payment)	Calculates the present value of an investment
RATE (number of periods, payment, present value)	Calculates the periodic interest rate of an annuity
DAY & TIME	
DATE	Returns the current date
NOW	Returns the current date and time
TIME	Returns the current time
MATHEMATICAL	
ABS (number)	Returns the absolute value of a number
INT (number)	Rounds a number down to the nearest integer
LN (number)	Calculates the natural logarithm of a number
LOG (number, base)	Calculates the logarithm of a number to a specified base
ROUND (number, number of digits)	Rounds a number to a specified number of digits
SQRT (number)	Calculates the square root of a number
SUM (range)	Calculates the total of a range of numbers
STATISTICAL	
AVERAGE (range)	Calculates the average value of a range of numbers
COUNT (range)	Counts how many cells in the range have entries
MAX (range)	Returns the maximum value in a range
MIN (range)	Returns the minimum value in a range
STDEV (range)	Calculates the standard deviation of a range of numbers
LOGICAL	
IF (logical test, value if true, value if false)	Performs a test and returns one value if the result of the test is true and another value if the result is false

Figure 2-13 *Functions included with many spreadsheet software packages.*

RECALCULATION One of the more powerful features of spreadsheet software is its capability of recalculating the rest of the worksheet when data in a worksheet changes. To appreciate this capability, consider that each time you change a value in a manual worksheet, you must erase the old value, write in a new value, erase any totals that contain calculations referring to the changed value, and then recalculate these totals and enter the new results. When working with a manual worksheet, accurately making changes and updating the affected values can be time consuming and result in new errors.

Making changes in an electronic worksheet is much easier and faster. When you enter a new value to change data in a cell, any value that is affected by the change is updated automatically and instantaneously. In Figure 2-12 on page 2.12 for example, if you change the Rent Expenses for April through June from 15,235.00 to 16,235.00, the total in cell D11 automatically changes to $114,684.40 and the profit in cell D12 changes to $41,944.15.

Spreadsheet software's capability of recalculating data also makes it a valuable tool for decision making using what-if analysis. **What-if analysis** is a process in which certain values in a spreadsheet are changed in order to reveal the effects of those changes.

CHARTING **Charting**, which is another standard feature of spreadsheet software, allows you to display data in a chart that shows the relationship of data in graphical, rather than numerical, form. A visual representation of data through charts often makes it easier to analyze and interpret information. Three popular chart types are line charts, column charts, and pie charts (Figure 2-14).

Line charts are effective for showing a trend over a period of time, as indicated by a rising or falling line. A line chart for an Income Statement, for example, could show total expenses and total income over a period of time. **Column charts**, also called **bar charts**, display bars of various lengths to show the relationship of data. The bars can be horizontal, vertical, or stacked on top of one another. A column chart for an Income Statement might show the income breakdown by

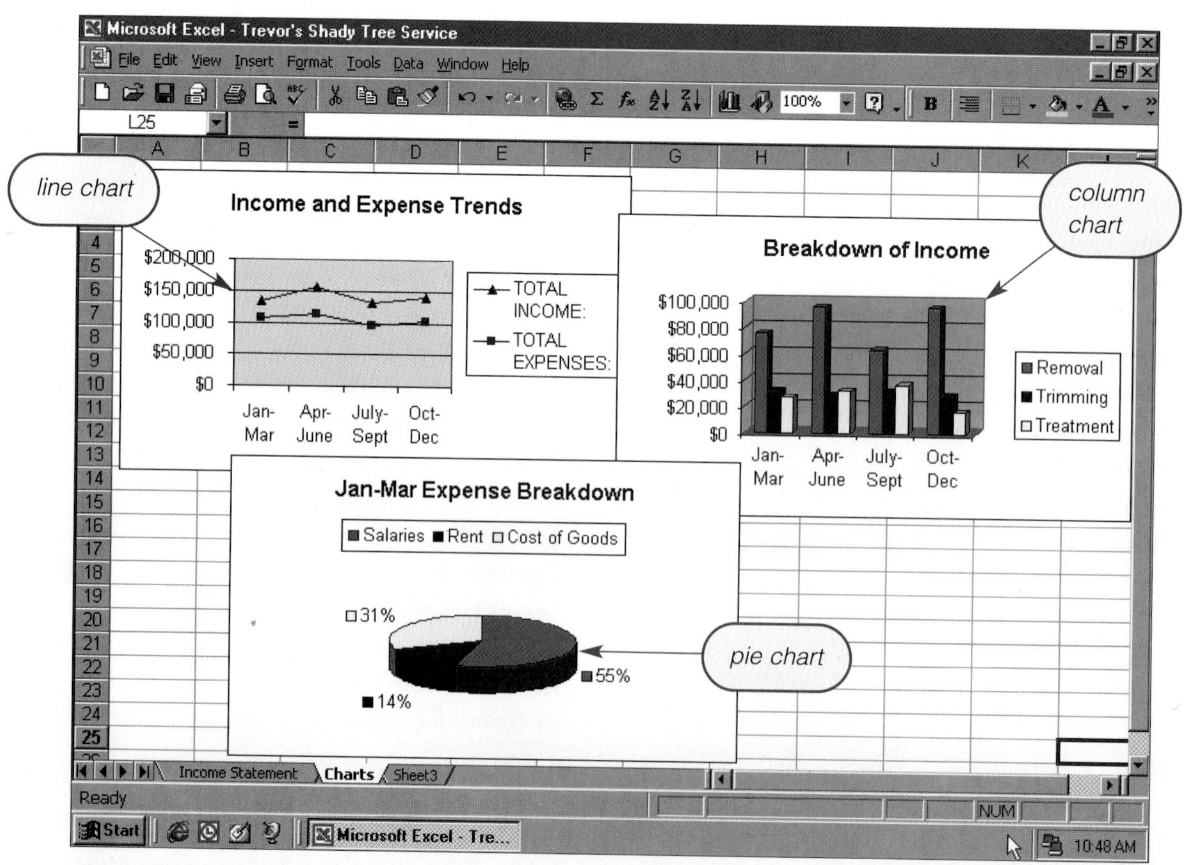

Figure 2-14 *Three basic types of charts provided with spreadsheet software are line charts, column charts, and pie charts. The line chart, column chart, and pie chart shown were created from the data in the worksheet in Figure 2-12 on page 2.12.*

category, with each bar representing a different category. **Pie charts**, which have the shape of round pies cut into pieces or slices, are used to show the relationship of parts to a whole. You might use a pie chart to show what percentage (part) each expense category contributed to the total expense (whole) for a time period.

Spreadsheet software also incorporates many of the features found in word processing software such as a spelling checker, changing fonts and font sizes, adding colors, tracking changes, and the capability of converting an existing spreadsheet document into the standard document format for the World Wide Web.

Database Software

A **database** is a collection of data organized in a manner that allows access, retrieval, and use of that data. In a manual database, data might be recorded on paper and stored in a filing cabinet. In a computerized database, such as the one shown in Figure 2-15, data is stored in an electronic format on a storage medium. **Database software**, also called a **database management system (DBMS)**, allows you to create a computerized database; add, change, and delete data; sort and retrieve data from the database; and create forms and reports using the data in the database.

With most popular personal computer database software packages, a database consists of a collection of **tables**, organized in rows and columns. A row in a table is called a **record**, and contains information about a given person, product, or event. A column in a table is called a **field**, and contains a specific piece of information within a record.

The Museum Gift Shop database shown in Figure 2-15 consists of two tables: a Product table and a Vendor table. The Product table contains ten records (rows), each of which contains data about one product. The product data is listed in the table's six fields (columns): product identification number, description, quantity on hand, cost of product, selling price, and vendor code. The description field, for instance, contains a name of a particular product.

WEB INFO

For more information on database software, visit the Discovering Computers 2001 Chapter 2 WEB INFO page (**www.scsite.com/dc2001/ch2/webinfo.htm**) and click Database Software.

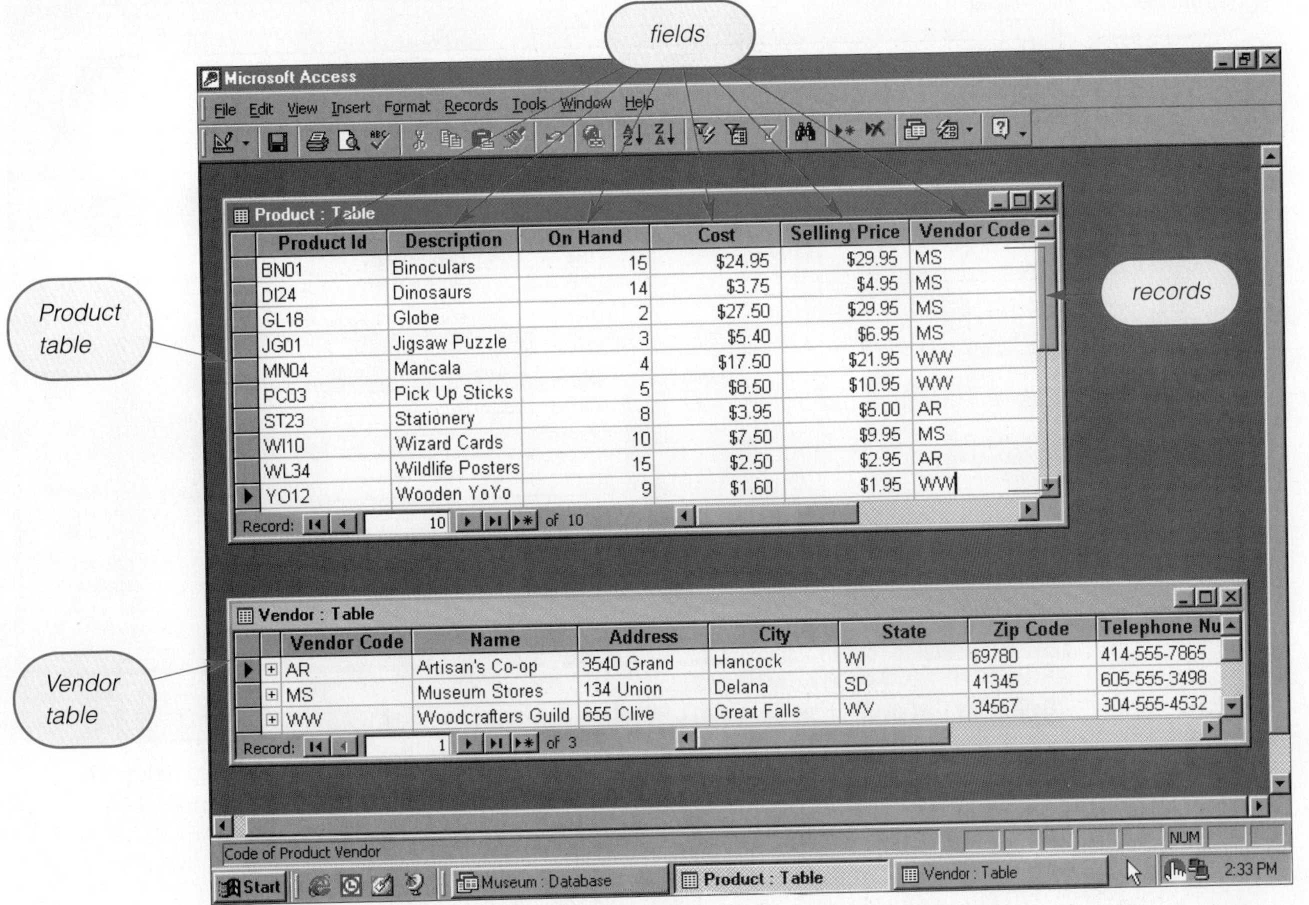

Product Id	Description	On Hand	Cost	Selling Price	Vendor Code
BN01	Binoculars	15	$24.95	$29.95	MS
DI24	Dinosaurs	14	$3.75	$4.95	MS
GL18	Globe	2	$27.50	$29.95	MS
JG01	Jigsaw Puzzle	3	$5.40	$6.95	MS
MN04	Mancala	4	$17.50	$21.95	WW
PC03	Pick Up Sticks	5	$8.50	$10.95	WW
ST23	Stationery	8	$3.95	$5.00	AR
WI10	Wizard Cards	10	$7.50	$9.95	MS
WL34	Wildlife Posters	15	$2.50	$2.95	AR
YO12	Wooden YoYo	9	$1.60	$1.95	WW

Vendor Code	Name	Address	City	State	Zip Code	Telephone Nu
AR	Artisan's Co-op	3540 Grand	Hancock	WI	69780	414-555-7865
MS	Museum Stores	134 Union	Delana	SD	41345	605-555-3498
WW	Woodcrafters Guild	655 Clive	Great Falls	WV	34567	304-555-4532

Figure 2-15 *This database contains two tables: one for products and one for vendors. The product table has ten records and six fields; the vendor table has three records and seven fields.*

DATABASE ORGANIZATION Before you begin creating a database, you should perform some preliminary tasks. Make a list of the data items you want to organize; each of these will become a field in the database. To identify the different fields, assign each field a unique name that is short, yet descriptive. For example, the field name for a product identification number could be Product Id, and the field name for the quantity on hand could be On Hand.

Once you have determined the fields and field names, you also must decide the field lengths and data types. The **field length** is the maximum number of characters to be stored for data in a particular field. The Description field, for instance, may be defined as 25 characters in length. The **data type** specifies the type of data that the field can contain. Common data types include:

- **Text**: letters, numbers, or special characters
- **Numeric**: numbers only
- **Currency**: dollar and cents amounts
- **Date**: month, day, and year information
- **Memo**: freeform text of any type or length
- **Hyperlink**, or **link**: Web address that links to Web page

Completing these steps provides a general description of the records and fields in a table, including the number of fields, field names, field lengths, and data types. These items collectively are referred to as the table **structure** (Figure 2-16).

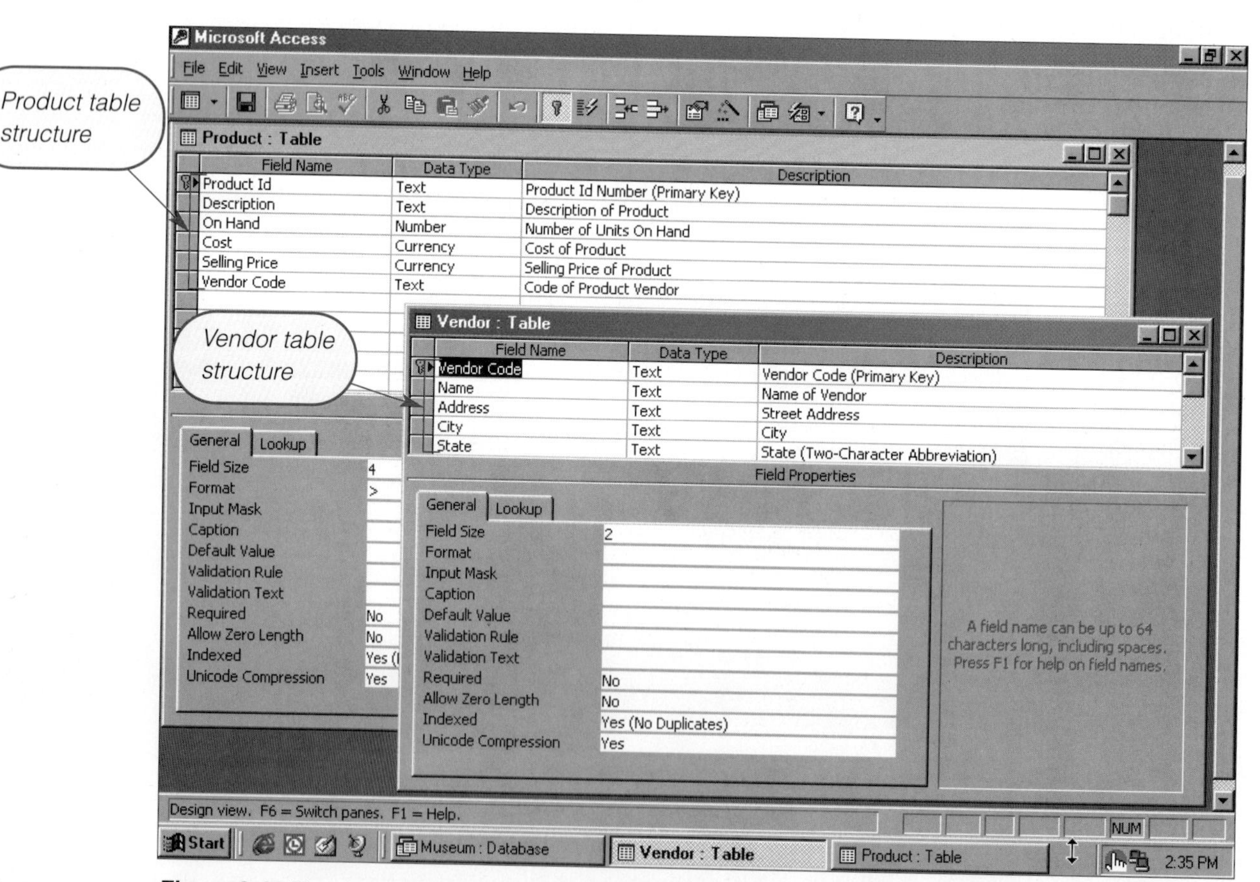

Figure 2-16 *The structure of a table includes the field names, field lengths, and data types. This Microsoft Access 2000 screen illustrates the structures for the Product and the Vendor tables.*

ENTERING DATA After a table structure is created, you can enter individual records into a table. Records usually are entered one at a time using the keyboard, often through a data entry form (Figure 2-17). As you are entering the data, the database software checks, or validates, the data. **Validation** is the process of comparing the data to a set of defined rules or values to determine if the data is acceptable. For example, designating a field as a numeric data type allows a user to enter only numbers into the field. Validation is important because it helps to ensure that data entered into the database is error free.

Another way to enter data into a database is to import data from an existing file. For example, you can import data saved in a spreadsheet file into a database.

MANIPULATING DATA Once the records are entered, you can use the database software to manipulate the data to generate information. For example, you can **sort**, or organize, a set of records in a particular order, such as alphabetical or by entry date. You also can retrieve information from the database by running a query. A **query** is a specific set of instructions for retrieving data from the database. You can specify which data the query retrieves by specifying **criteria**, or restrictions that the data must meet. For example, suppose you wanted to generate a list of all products that cost less than $15.00. You could set up a query to list the Product Id, Description, On Hand, and Cost for all records that meet this criteria. The list could be sorted to display the most expensive products first (Figure 2-18).

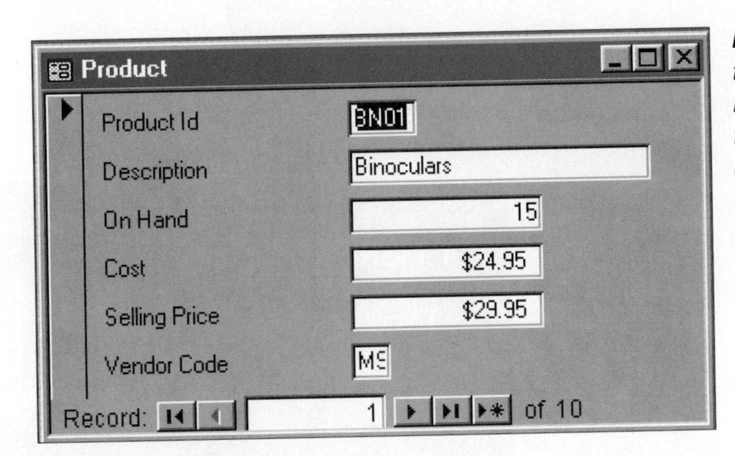

Figure 2-17 *After you define the table structure, you can enter data into the database using a data entry form. Most database software will create a data entry form automatically, based on the way fields are defined. In this data entry form, you enter data into the Product table.*

Product Query : Select Query

Product Id	Description	On Hand	Cost
PC03	Pick Up Sticks	5	$8.50
WI10	Wizard Cards	10	$7.50
JG01	Jigsaw Puzzle	3	$5.40
ST23	Stationery	8	$3.95
DI24	Dinosaurs	14	$3.75
WL34	Wildlife Posters	15	$2.50
YO12	Wooden YoYo	9	$1.60

Record: 7 of 7 (Filtered)

Figure 2-18 *Database software can produce reports based on criteria specified by a user. For example, this screen shows the result of a request, called a query, to list the Product Id, Description, On Hand, and Cost fields for all records that have a cost less than $15.00. The results of the query can be displayed or printed.*

Presentation Graphics Software

WEB INFO

For more information on presentation graphics software, visit the Discovering Computers 2001 Chapter 2 WEB INFO page (**www.scsite.com/dc2001/ch2/webinfo.htm**) and click Presentation Graphics Software.

Presentation graphics software allows you to create documents called presentations, which are used to communicate ideas, messages, and other information to a group. The presentations can be viewed as slides that display on a large monitor or on a projection screen (Figure 2-19).

Presentation graphics software typically provides an array of predefined presentation formats that define complementary colors for backgrounds, text, and other items on the slides. Presentation graphics software also provides a variety of layouts for each individual slide such as a title slide, a two-column slide, and a slide with clip art. Any text, charts, and graphical images used in a slide can be enhanced with 3-D and other special effects such as shading, shadows, and textures.

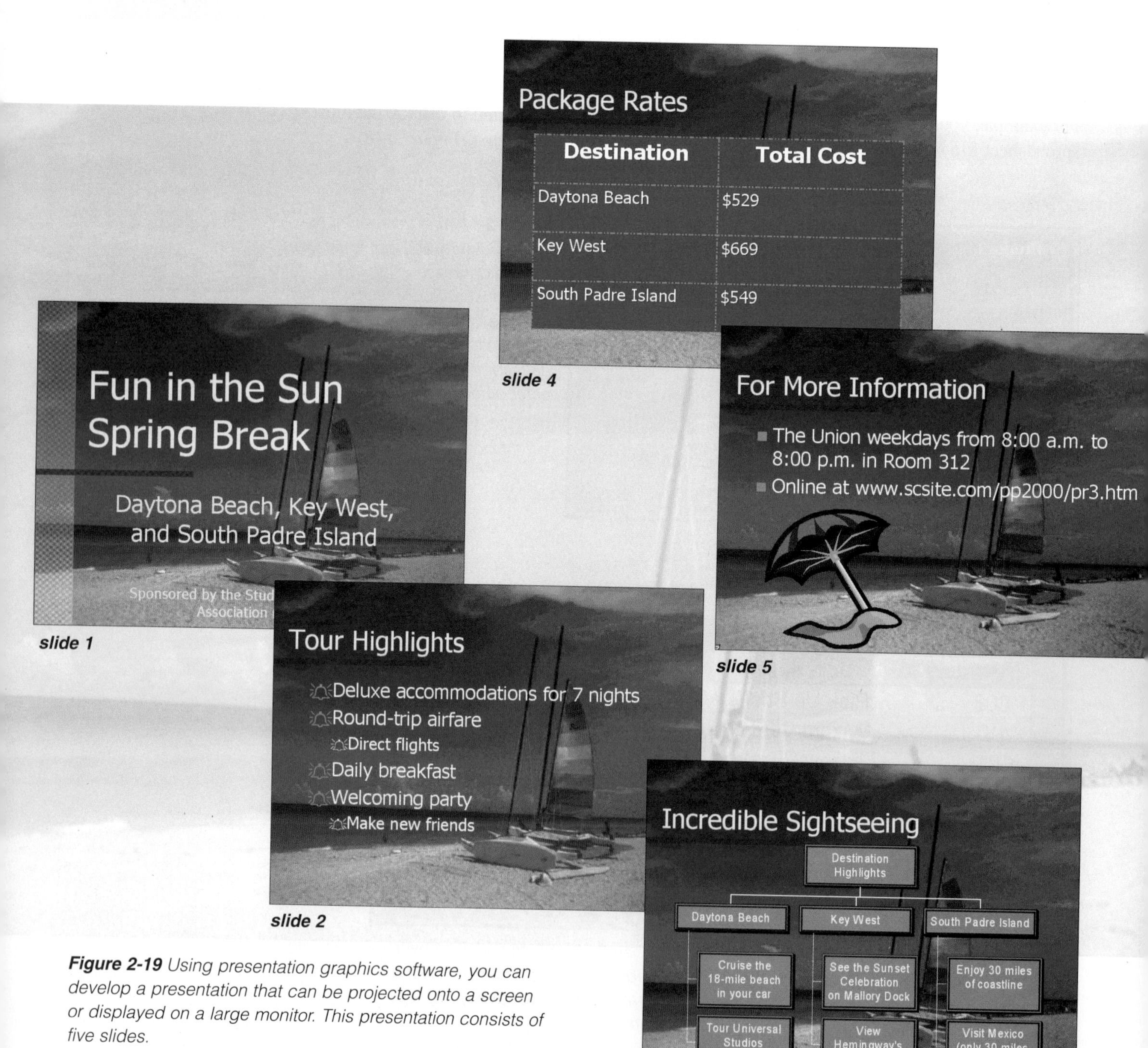

Figure 2-19 *Using presentation graphics software, you can develop a presentation that can be projected onto a screen or displayed on a large monitor. This presentation consists of five slides.*

With presentation graphics software, you can incorporate objects from the clip gallery into your slides to create multimedia presentations. A **clip gallery** includes clip art images, pictures, video clips, and audio clips. A clip gallery can be stored on your hard disk, a CD-ROM, or a DVD-ROM; in other cases, you access the clip gallery on the Web. As with clip art collections, a clip gallery typically is organized by categories that can include academic, business, entertainment, transportation, and so on. For example, the transportation category may contain a clip art image of a bicycle, a photograph of a locomotive, a video clip of an airplane in flight, and an audio clip of a Model T car horn.

When building a presentation, you also can set the slide timing, so that the presentation automatically displays the next slide after a predetermined delay. Special effects can be applied to the transition between each slide. One slide, for example, might slowly dissolve as the next slide comes into view.

To help organize the presentation, you can view small versions of all the slides in slide sorter view (Figure 2-20). Slide sorter view presents a screen view similar to how 35mm slides would look on a photographer's light table. The slide sorter allows you to arrange the slides in any order.

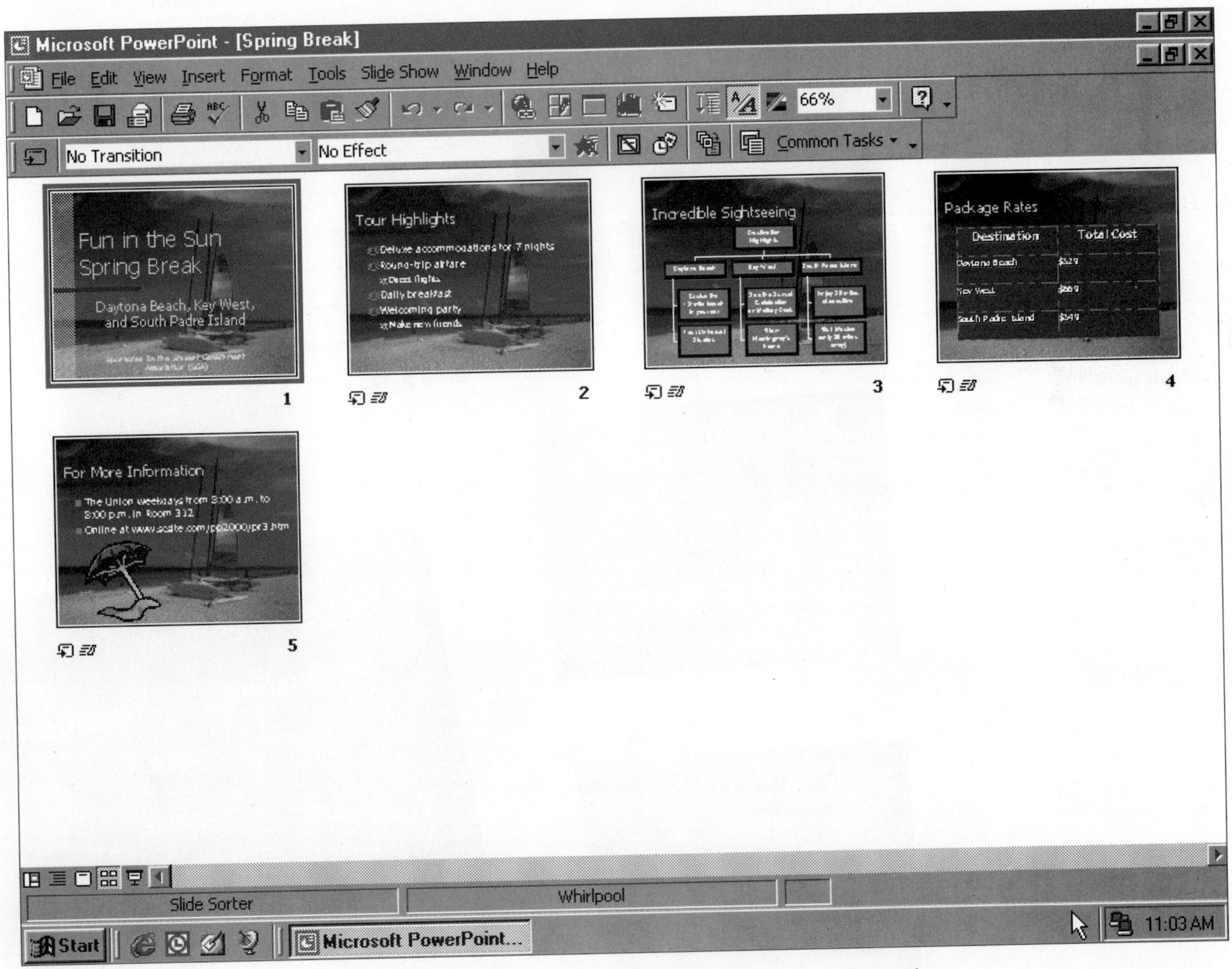

Figure 2-20 *Slide sorter view shows a small version of each slide. Using a pointing device or the keyboard, you can rearrange the slides to change the order of the presentation.*

Once you have created a presentation, you can view or print the presentation as slides or in several other formats. An outline includes only the text from each slide, such as the slide title and the key points (Figure 2-21a). Audience handouts include images of two or more slides on a page that you can distribute to presentation attendees (Figure 2-21b). You also may wish to print a notes page to help you deliver the presentation; a notes page shows a picture of the slide along with any notes you want to see while discussing a topic or slide (Figure 2-21c).

Presentation graphics software also incorporates some of the features found in word processing software such as a spelling checker, font formatting capabilities, and the capability of converting an existing slide show into the standard document format for the World Wide Web.

Figure 2-21 *In addition to viewing the presentation as slides, presentation graphics packages allow you to view or print the presentation as an outline, as audience handouts, or as notes pages for the speaker.*

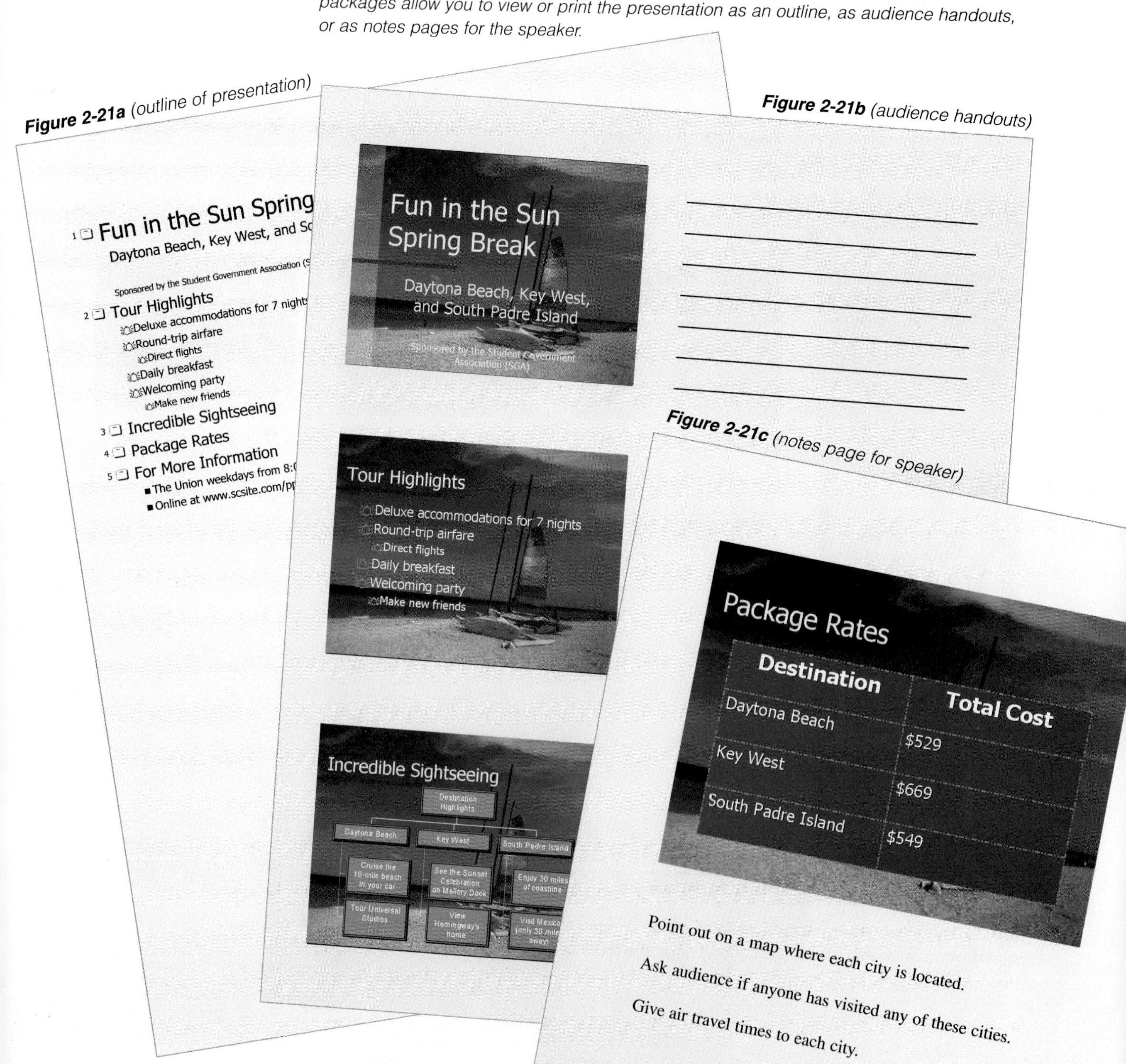

Figure 2-21a *(outline of presentation)*

Figure 2-21b *(audience handouts)*

Figure 2-21c *(notes page for speaker)*

Personal Information Managers

A **personal information manager** (**PIM**) is a software application that includes an appointment calendar, address book, and notepad to help you organize personal information such as appointments and task lists (Figure 2-22). A PIM allows you to take information that you previously tracked in a weekly or daily calendar, and organize and store it on your computer. PIMs can manage many different types of information such as telephone messages, project notes, reminders, task and address lists, and important dates and appointments.

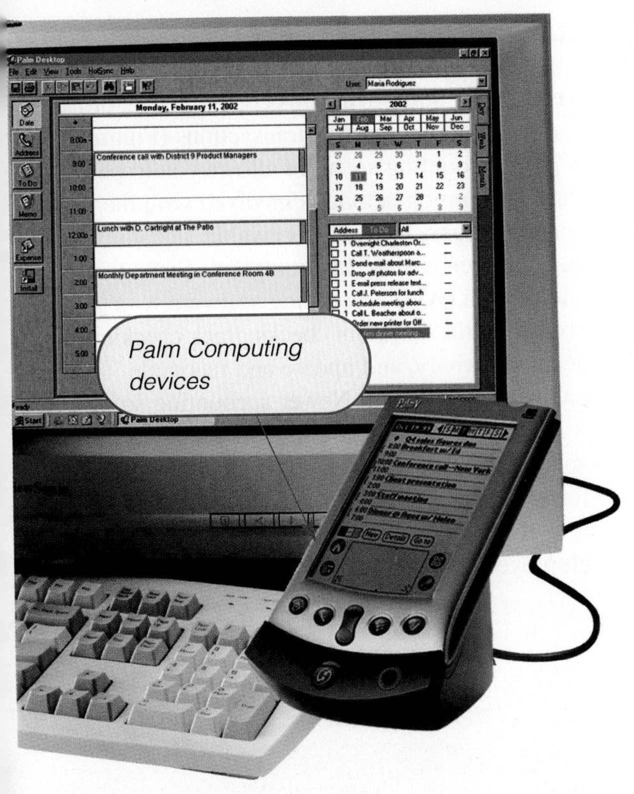

Figure 2-22 Some handheld computers such as the Palm Computing devices run PIM software. With these computers, you can transfer information from the handheld computer to your desktop computer so appointments, address lists, and other important information always are available.

Precisely defining a PIM is difficult because personal information managers offer a range of capabilities. As noted, however, most include at least an appointment calendar, address book, and notepad. An **appointment calendar** allows you to schedule activities for a particular day and time. With an **address book**, you can enter and maintain names, addresses, and telephone numbers of customers, co-workers, family members, and friends. Instead of writing notes on a piece of paper, you can use a **notepad** to record ideas, reminders, and other important information. Many PIMs also include a calculator or a simple spreadsheet application; some also include e-mail capabilities.

Software Suite

A software **suite** is a collection of individual application software packages sold as a single package (Figure 2-23). When you install the suite, you install the entire collection of applications at once instead of installing each application individually. At a minimum, suites typically include the following software applications: word processing, spreadsheet, database, and presentation graphics.

Software suites offer two major advantages: lower cost and ease of use. Typically, buying a collection of software packages in a suite costs significantly less than purchasing each of the application packages separately. Software suites provide ease of use because the applications within a suite normally use a similar interface and have some common features. Thus, once you learn how to use one application in the suite, you are familiar with the interface in the other applications in the suite. For example, once you learn how to print using the suite's word processing package, you can apply the same skill to the spreadsheet, database, and presentation graphics software in the suite.

Figure 2-23 Two popular software suites are Lotus SmartSuite and Microsoft Office 2000. Lotus SmartSuite includes WordPro, 1-2-3, Freelance, Approach, Organizer, FastSite, and ScreenCam. Microsoft Office 2000 Premium contains Word, Excel, PowerPoint, Access, Outlook, Publisher, FrontPage, and PhotoDraw.

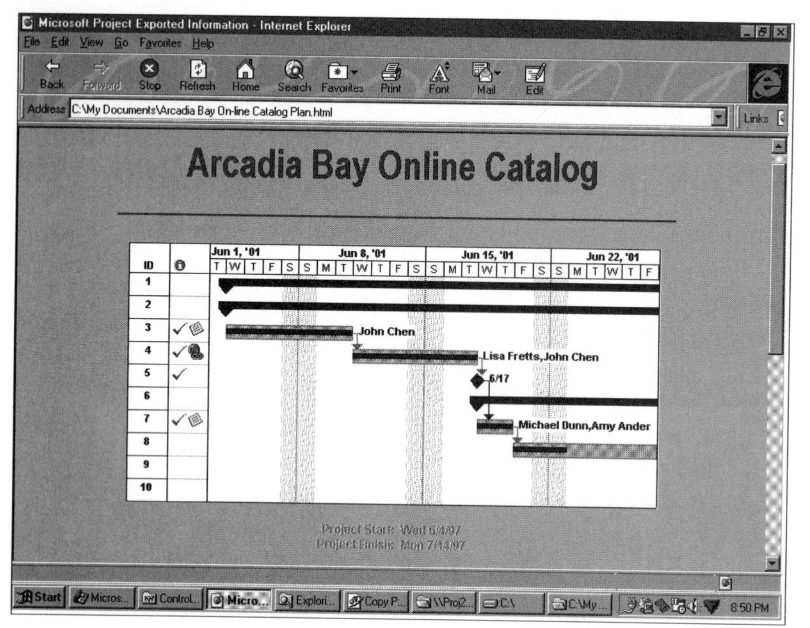

Figure 2-24 Project management software allows you to plan, schedule, track, and analyze the events, resources, and costs of a project.

Project Management Software

Project management software allows you to plan, schedule, track, and analyze the events, resources, and costs of a project (Figure 2-24). A general contractor, for example, might use project management software to manage a home-remodeling schedule, or a publisher might use it to coordinate the process of producing a textbook. The value of project management software is that it helps managers track, control, and manage project variables, thereby allowing them to complete a project on time and within budget. Project management is discussed in more detail in Chapter 11.

Accounting Software

Accounting software helps companies record and report their financial transactions (Figure 2-25). Accounting software allows you to perform accounting activities involved with the general ledger, accounts receivable, accounts payable, purchasing, invoicing, job costing, and payroll. With accounting software, you also can write and print checks, track checking account activity, and update and reconcile balances on demand. Newer accounting software packages support online direct deposit and payroll services, which makes it possible for a company to deposit paychecks directly into employee's checking accounts and pay employee taxes electronically.

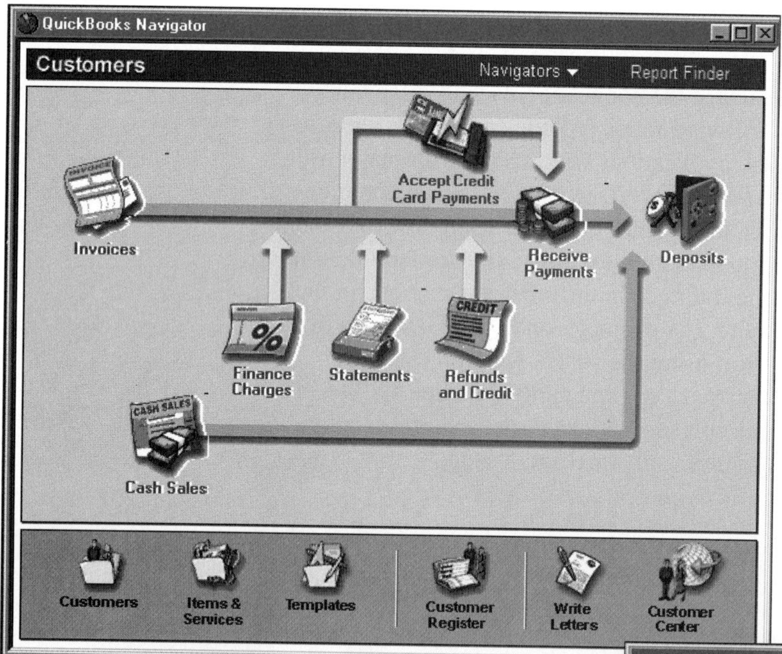

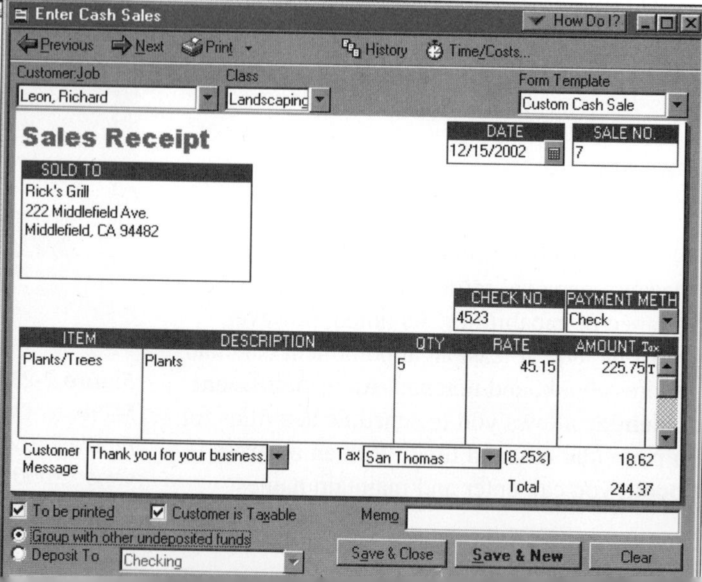

Figure 2-25 Accounting software helps companies record and report their financial transactions.

Some accounting software offers more sophisticated features such as multiple company reporting, foreign currency reporting, and forecasting the amount of raw materials needed for products. The cost of accounting software for small businesses ranges from less than one hundred to several thousand dollars. Accounting software for large businesses can cost several hundred thousand dollars.

GRAPHICS AND MULTIMEDIA SOFTWARE

In addition to productivity software, many individuals also work with software designed specifically for their fields of work. Power users such as engineers, architects, desktop publishers, and graphic artists, for example, often use powerful software that allows them to work with graphics and multimedia. Types of graphics and multimedia software include computer-aided design, desktop publishing, paint/image editing, video and audio editing, multimedia authoring, and Web page authoring. Figure 2-26 lists the more popular products for each of these applications. The features and functions of each of these applications are discussed in the following sections.

POPULAR GRAPHICS AND MULTIMEDIA SOFTWARE PACKAGES

Software Application	Popular Packages
Computer-Aided Design (CAD)	• Autodesk AutoCAD • Visio Technical
Desktop Publishing (Professional)	• Adobe InDesign • Adobe PageMaker • QuarkXPress
Paint/Image Editing (Professional)	• Adobe Illustrator • Adobe Photoshop • CorelDRAW • Macromedia FreeHand • MetaCreations Painter 3D
Video and Audio Editing	• Adobe Premiere • click2learn.com Digital Video Producer
Multimedia Authoring	• click2learn.com ToolBook • Macromedia Authorware • Macromedia Director
Web Page Authoring	• Adobe GoLive • Adobe PageMill • Corel WEB.SiteBuilder • Lotus FastSite • Macromedia Dreamweaver • Macromedia Flash • Microsoft FrontPage 2000

Figure 2-26 *Popular graphics and multimedia software products.*

Computer-Aided Design

Computer-aided design (CAD) software is a sophisticated type of application software that assists a user in creating engineering, architectural, and scientific designs (Figure 2-27). For example, using CAD, engineers can create design plans for airplanes and security systems; architects can design building structures and floor plans; and scientists can design drawings of molecular structures.

CAD software eliminates the laborious manual drafting that design processes can require. With CAD, designers can make changes to a drawing or design and view the results. Three-dimensional CAD programs allow designers to rotate designs of 3-D objects to view them from any angle. Some CAD software even can generate material lists for building designs.

Desktop Publishing Software (Professional)

Desktop publishing (DTP) software allows you to design and produce sophisticated documents that contain text, graphics, and brilliant colors (Figure 2-28). Although many word processing packages have some of the capabilities of DTP software, professional designers and graphic artists use DTP software because it is designed specifically to support **page layout**, which is the process of arranging text and graphics in a document. DTP software thus is ideal for the production of high-quality color documents such as newsletters, marketing literature, catalogs, and annual reports. In the past, documents of this type were created by slower, more expensive traditional publishing methods such as typesetting. Today's DTP software also allows you to convert a color document into a format for use on the World Wide Web.

WEB INFO

For more information on desktop publishing software, visit the Discovering Computers 2001 Chapter 2 WEB INFO page (**www.scsite.com/dc2001/ch2/webinfo.htm**) and click Desktop Publishing Software.

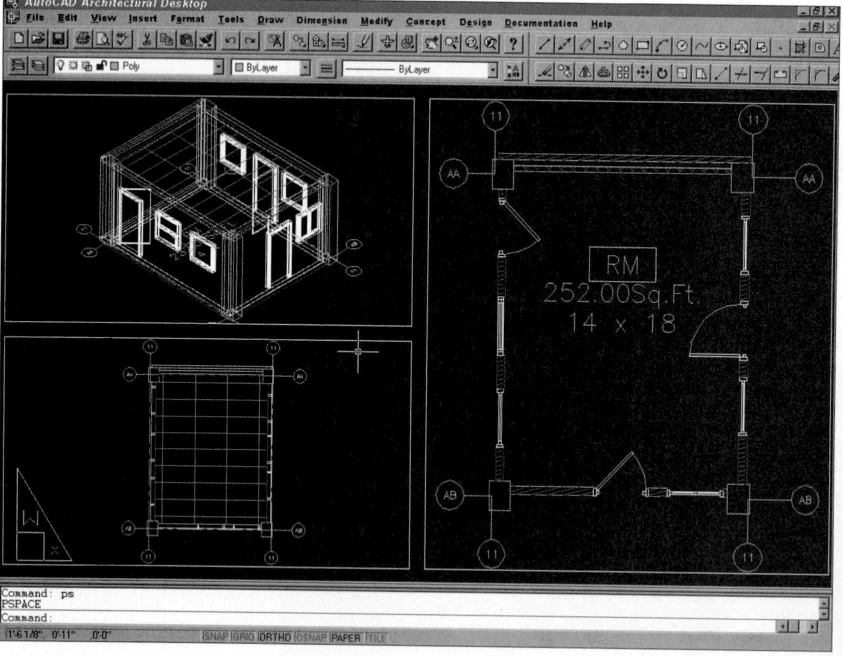

Figure 2-27 *CAD software is sophisticated software that assists engineers, architects, and scientists in creating designs.*

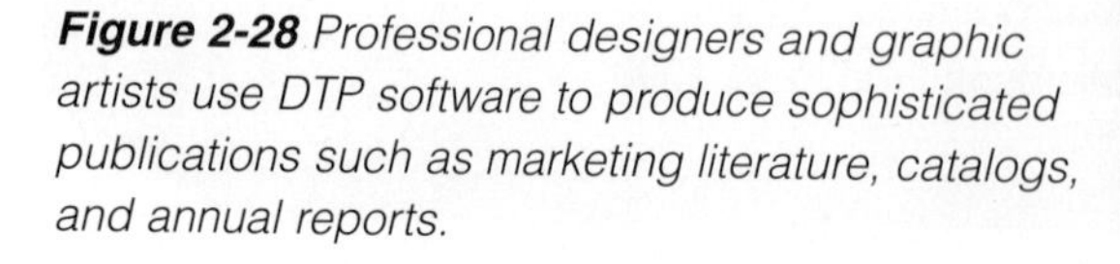

Figure 2-28 *Professional designers and graphic artists use DTP software to produce sophisticated publications such as marketing literature, catalogs, and annual reports.*

When creating a document using DTP software, you can add text and graphical images directly into the document, or you can import existing text and graphics from other files. For example, you can import text from a word processing file into a desktop publishing document. Graphics files such as illustrations and photographs also can be imported into a DTP document. One type of input device, called a scanner, can be used to convert printed graphics such as photographs and drawings into files that DTP software can use.

Once you have created or inserted a graphical image into a document, the DTP software can crop, sharpen, and change the colors in the image by adding tints or percentages of colors. To help you select a color for a graphical image or text, DTP software packages include color libraries. A **color library** is a standard set of colors used by designers and printers to ensure that colors will print exactly as specified. Using a color library, you can choose standard colors or specialty colors such as metallic or fluorescent colors.

Paint/Image Editing Software (Professional)

Graphic artists, multimedia professionals, technical illustrators, and desktop publishers use paint software and image editing software to create and modify graphical images such as those used in DTP documents and Web pages (Figure 2-29). **Paint software**, also called **illustration software**, allows you to draw pictures, shapes, and other graphical images using various tools on the screen such as a pen, brush, eyedropper, and paint bucket. **Image editing software** provides the capabilities of paint software as well as the capability of modifying existing images. For example, you can retouch photographs; adjust or enhance image colors; and add special effects such as shadows and glows.

WEB INFO

For more information on paint/image editing software, visit the Discovering Computers 2001 Chapter 2 WEB INFO page (**www.scsite.com/dc2001/ch2/webinfo.htm**) and click Paint/Image Editing Software.

Figure 2-29 With image editing software, artists can create and modify a variety of graphical images.

Video and Audio Editing Software

Video consists of images that are played back at speeds that provide the appearance of full motion. With **video editing software** (Figure 2-30), you can modify a segment of a video, called a clip. For example, you can reduce the length of a video clip, reorder a series of clips, or add special effects such as words that move horizontally across the screen. Video editing software typically includes audio editing capabilities.

Audio is any music, speech, or other sound that is stored and produced by the computer. With **audio editing software**, you can modify audio clips. Audio editing software usually includes filters, which are designed to enhance audio quality. A filter, for example, might remove a distracting background noise from the audio clip.

Multimedia Authoring Software

Multimedia authoring software is used to create electronic interactive presentations that can include text, images, video, audio, and animation (Figure 2-31). The software helps you create presentations by allowing you to control the placement of text and images and the duration of sounds, video, and animation. Once created, such multimedia presentations often take the form of interactive computer-based presentations designed to facilitate learning and elicit direct student participation. Multimedia presentations usually are stored and delivered via a CD-ROM or DVD-ROM, over a local area network, or via the Internet.

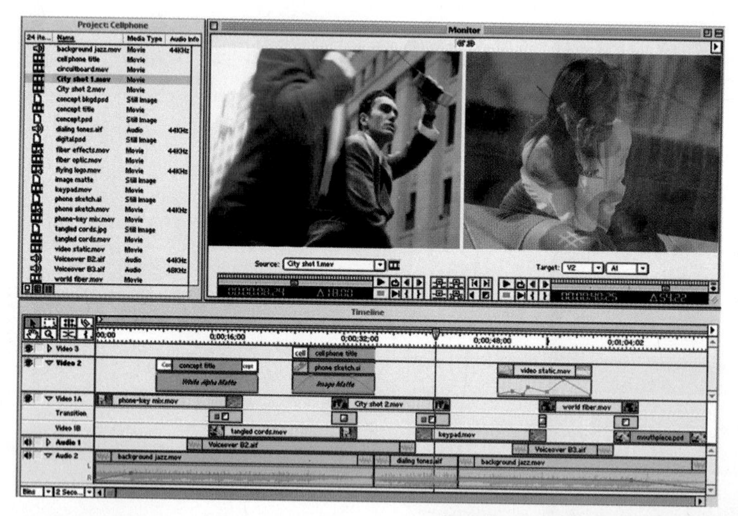

Figure 2-30 *With video editing software, you can modify video images and accompanying audio.*

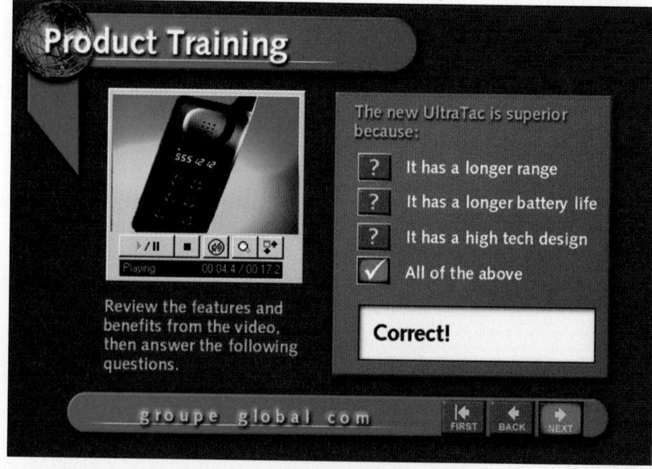

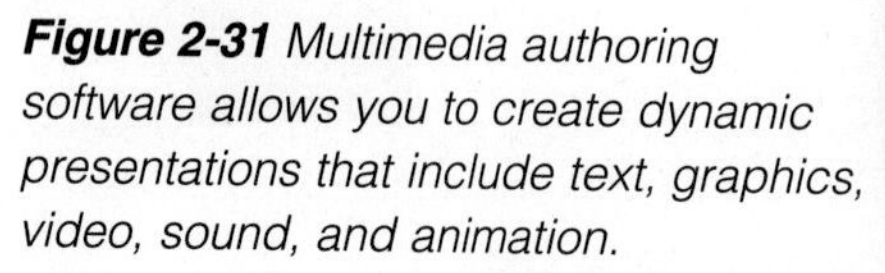

Figure 2-31 *Multimedia authoring software allows you to create dynamic presentations that include text, graphics, video, sound, and animation.*

Web Page Authoring Software

Web page authoring software is software specifically designed to help you create Web pages, in addition to organizing, managing, and maintaining Web sites. As noted in previous sections, many application software packages include Web page authoring features you can use to create basic Web pages that contain text and graphical images. Web page authoring software, however, includes features that also allow you to create sophisticated multimedia Web pages that include graphical images, video, audio, animation, and other special effects. With Web page authoring software, both new and experienced users can create fascinating Web sites. Web page authoring software is presented in more detail in Chapter 7.

SOFTWARE FOR HOME, PERSONAL, AND EDUCATIONAL USE

Many software applications are designed specifically for use at home or for personal or educational use. Examples of such software packages are integrated software that includes word processing, spreadsheet, database, and other software in a single package; personal finance; legal; tax preparation; desktop publishing; paint image/editing; clip art/image gallery; home design/landscaping; educational; reference; and entertainment. Most of the products in this category are relatively inexpensive, often priced at less than $100. Figure 2-32 lists popular software packages for many of these applications. The features and functions of each of these applications are discussed in the following sections.

WEB INFO

For more information on Web page authoring software, visit the Discovering Computers 2001 Chapter 2 WEB INFO page (**www.scsite.com/dc2001/ch2/webinfo.htm**) and click Web Page Authoring Software.

POPULAR SOFTWARE PACKAGES FOR HOME/PERSONAL/EDUCATIONAL USE

Software Application	Popular Packages
Integrated Software	• Microsoft Works
Personal Finance	• Intuit Quicken • Microsoft Money
Legal	• E-Z Legal Advisor • Kiplinger's Home Legal Advisor • WillMaker
Tax Preparation	• Intuit TurboTax • Kiplinger TaxCut
Desktop Publishing (Personal)	• Broderbund Print Shop Pro Publisher • Microsoft Publisher 2000
Paint/Image Editing (Personal)	• Adobe PhotoDeluxe • Corel PHOTO-PAINT • Microsoft PhotoDraw 2000 • Paint Shop Pro
Clip Art/Image Gallery	• Corel GALLERY • Nova Development Art Explosion
Home Design/Landscaping	• Autodesk Planix Complete Home Suite • Broderbund 3D Home Design Suite
Reference	• Microsoft Encarta • Mosby's Medical Encyclopedia • Rand McNally StreetFinder • Rand McNally TripMaker • Websters Gold

Figure 2-32 *Popular software products for home, personal, and educational use.*

WEB INFO

For more information on personal finance software, visit the Discovering Computers 2001 Chapter 2 WEB INFO page (**www.scsite.com/dc2001/ch2/webinfo.htm**) and click Personal Finance Software.

Integrated Software

Integrated software is software that combines applications such as word processing, spreadsheet, and database into a single, easy-to-use package. Like a software suite, the applications within the integrated software package use a similar interface and share some common features. Once you learn how to use one application in the integrated software package, you are familiar with the interface in the other applications.

Unlike a software suite, however, you cannot purchase the applications in the integrated software package individually. Each application in an integrated software package is designed specifically to work as part of a larger set of applications (thus the name integrated).

The applications within the integrated software package typically do not have all the capabilities of stand-alone productivity software applications such as word processing and spreadsheets. Integrated software thus is less expensive than a more powerful software suite. For many home and personal users, however, the capabilities of an integrated software package more than meet their needs.

Personal Finance Software

Personal finance software is a simplified accounting program that helps you pay bills; balance your checkbook; track your personal income and expenses, such as credit-card bills; track investments; and evaluate financial plans (Figure 2-33).

Using personal finance software can help you determine where, and for what purpose, you are spending money so that you can manage your finances. Reports can summarize transactions by category (such as dining), by payee (such as the electric company), or by time period (such as the last two months). Bill-paying features include the capability of printing checks on your printer or having an outside service print your checks.

Personal finance software packages usually offer a variety of online services which require access to the Web. For example, you can track your investments online; compare insurance rates from leading insurance companies; and even do your banking transactions online. With online banking, you can transfer money electronically from your checking or credit card accounts to payees' accounts. To obtain current credit card statements, bank statements, and account balances, you can download monthly transaction information

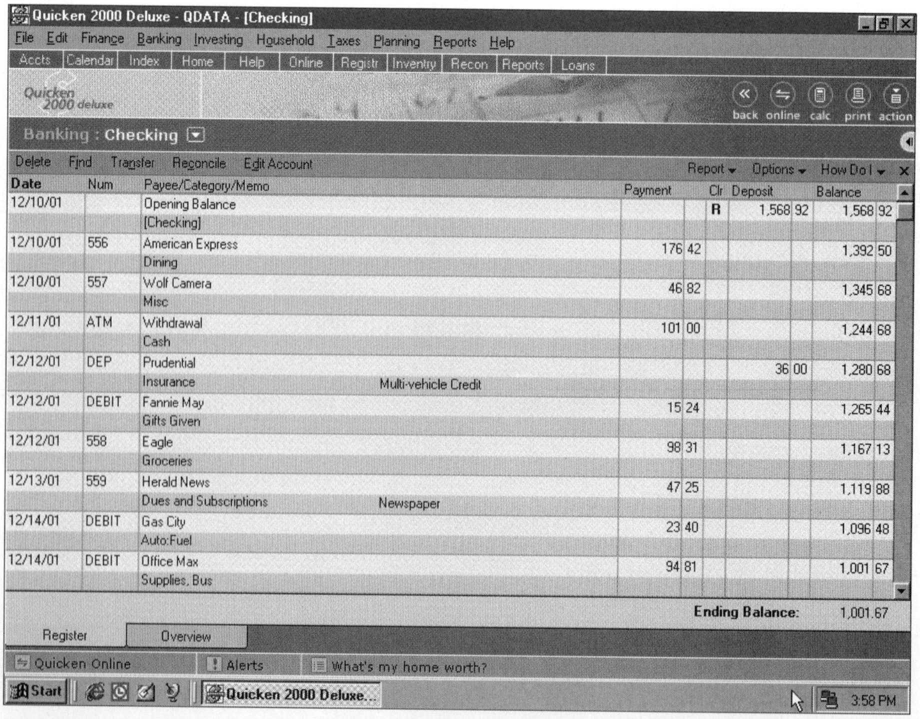

Figure 2-33 *Personal finance software assists you with paying bills; balancing your checkbook; tracking credit card activity, personal income and expenses, and investments; and evaluating financial plans.*

from the Web or copy it from a monthly transaction disk.

Financial planning features include analyzing home and personal loans, preparing income taxes, and managing retirement savings. Other features found in many personal finance packages include home inventory, budgeting, and tax-related transactions.

Legal Software

Legal software assists in the preparation of legal documents and provides legal advice to individuals, families, and small businesses (Figure 2-34). Legal software provides standard contracts and documents associated with buying, selling, and renting property; estate planning; and preparing a will. By answering a series of questions or completing a form, the legal software tailors the legal document to your needs.

Once the legal document is created, you can file the paperwork with the appropriate agency, court, or office; or you can take the document to your attorney for his or her review and signature.

Tax Preparation Software

Tax preparation software guides individuals, families, or small businesses through the process of filing federal taxes (Figure 2-35). These software packages also offer money saving tax tips, designed to lower your tax bill. After you answer a series of questions and complete basic forms, the tax preparation software creates and analyzes your tax forms to search for missed potential errors and deduction opportunities.

Once the forms are complete, you can print any necessary paperwork, completed and ready for you to file. Some tax preparation software packages even allow you to file your tax forms electronically.

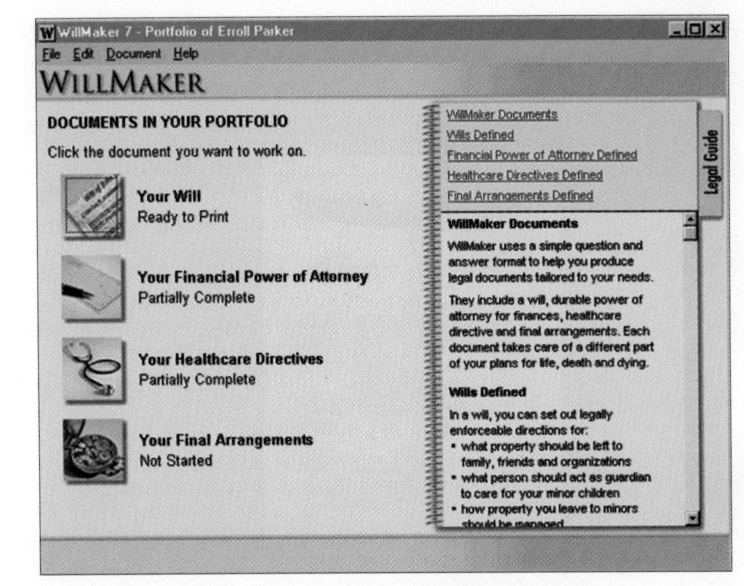

Figure 2-34 *Legal software provides legal advice to individuals, families, and small businesses and assists in the preparation of legal documents.*

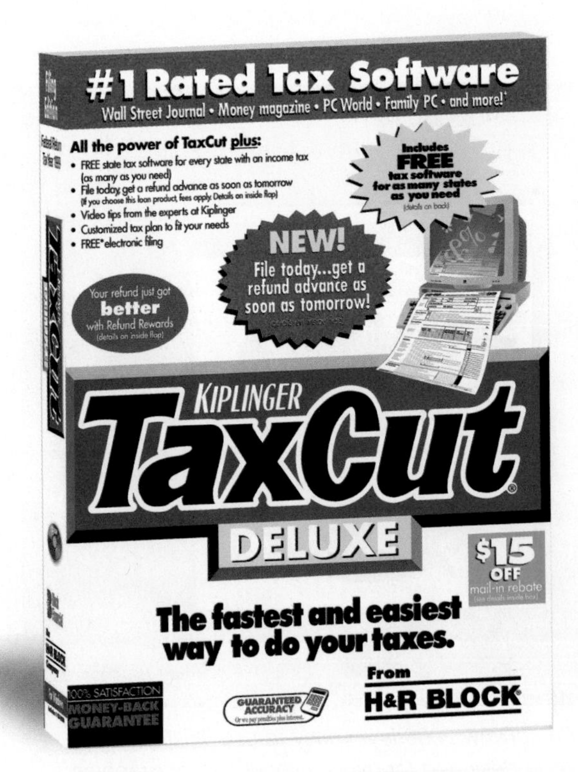

Figure 2-35 *Tax preparation software guides individuals, families, or small businesses through the process of filing federal taxes.*

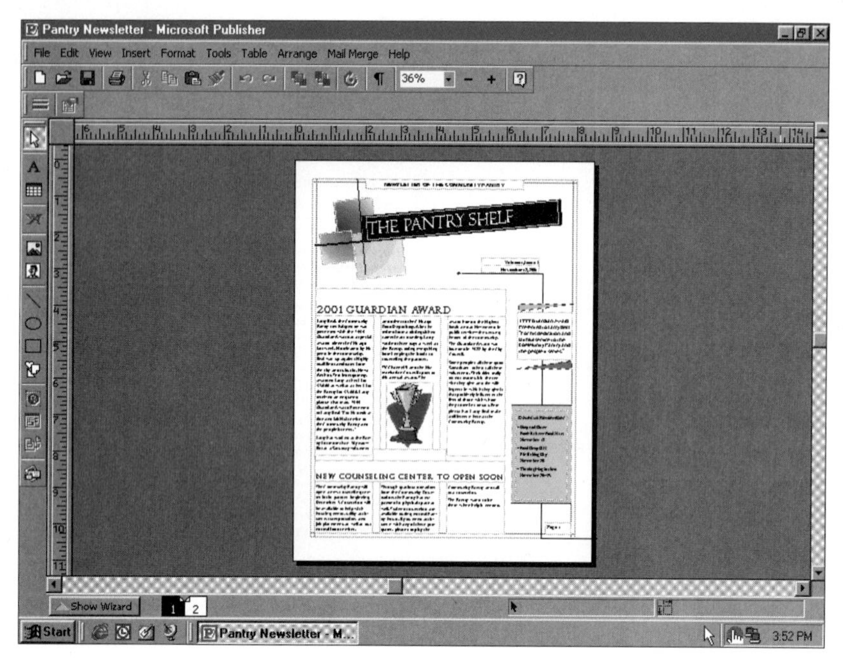

Figure 2-36 *With Microsoft Publisher 2000, home and small business users can create professional looking publications, such as this newsletter.*

Desktop Publishing (Personal)

Instead of using professional DTP software (as discussed earlier in this chapter), many home and small business users utilize much simpler, easy-to-understand DTP software designed for smaller-scale desktop publishing projects (Figure 2-36). Using **personal DTP software**, you can create newsletters, brochures, and advertisements; postcards and greeting cards; letterhead and business cards; banners, calendars, and logos. Personal DTP software guides you through the development of these documents by asking a series of questions, offering numerous predefined layouts, and providing standard text you can add to documents. In some packages, as you enter text, the personal DTP software checks your spelling. You can print your finished publications on a color printer or place them on the Web.

Paint/Image Editing Software (Personal)

Personal paint/image editing software provides an easy-to-use interface, usually with more simplified capabilities than its professional counterpart, including functions tailored to meet the needs of the home and small business user (Figure 2-37).

Like the professional versions, personal paint software includes various simplified tools that allow you to draw pictures, shapes, and other images.

Personal image editing software provides the capabilities of paint software and the capability of modifying existing graphics. One popular type of image editing software, called **photo-editing software**, allows you to edit digital photographs by removing red-eye, adding special effects, or creating electronic photo albums. When the photograph is complete, you can print it on labels, calendars, business cards, and banners; or place it on a Web page.

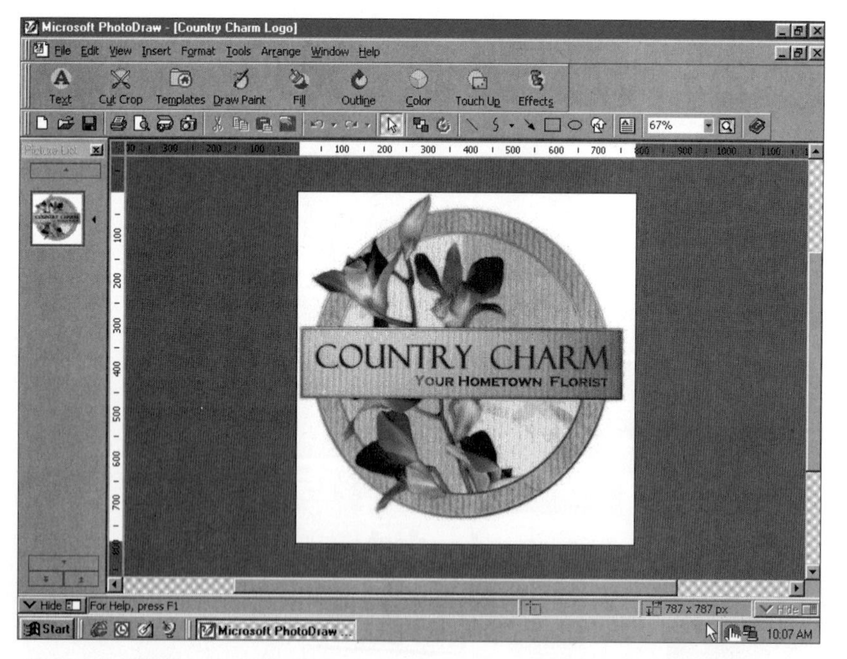

Figure 2-37 *Microsoft PhotoDraw 2000 is an easy-to-use illustration and photo-editing package designed for users without formal graphics or design training.*

Clip Art/Image Gallery

Many applications include a **clip art/image gallery**, which is a collection of clip art and photographs (Figure 2-38). You also can purchase clip art/image galleries if you need a wider selection of images. In addition to clip art, many clip art/image galleries provide fonts, animations, sounds, video clips, and audio clips. You can use the images, fonts, and other items from the clip art/image gallery in all types of documents, including word processing, desktop publishing, spreadsheets, and presentation graphics.

Home Design/Landscaping Software

Homeowners or potential homeowners can use **home design/landscaping software** to assist with the design or remodeling of a home, deck, or landscape (Figure 2-39). Home design/landscaping software includes hundreds of predrawn plans which you can customize to meet your needs. Once designed, many home design/landscaping packages will print a material list outlining costs and quantities for the entire project.

WEB INFO

For more information on clip art/image galleries, visit the Discovering Computers 2001 Chapter 2 WEB INFO page (**www.scsite.com/dc2001/ch2/webinfo.htm**) and click Clip Art/Image Gallery.

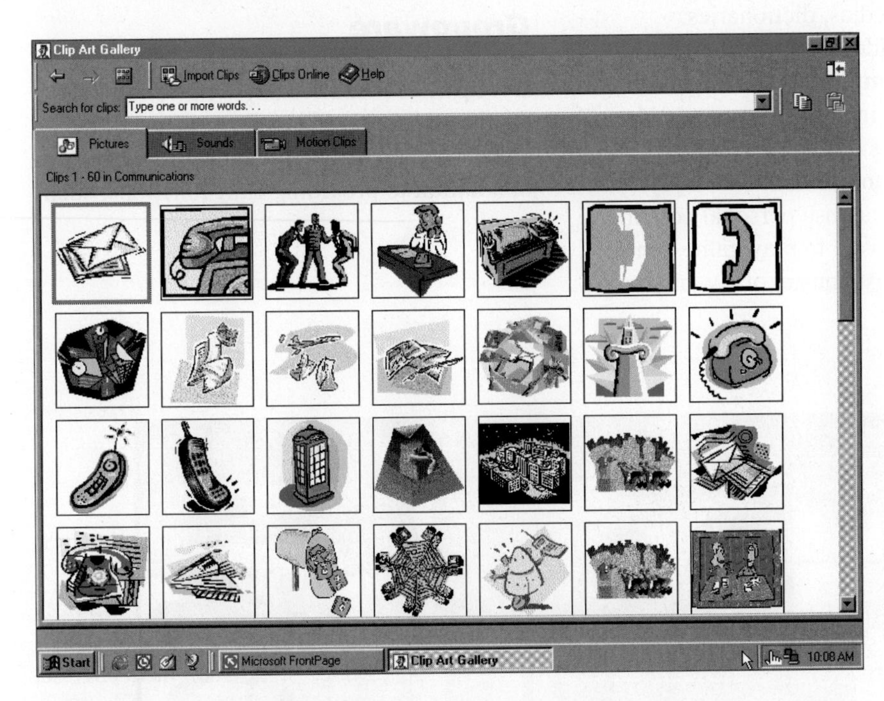

Figure 2-38 Clip art/image galleries are included with many applications, such as Microsoft FrontPage 2000 shown here.

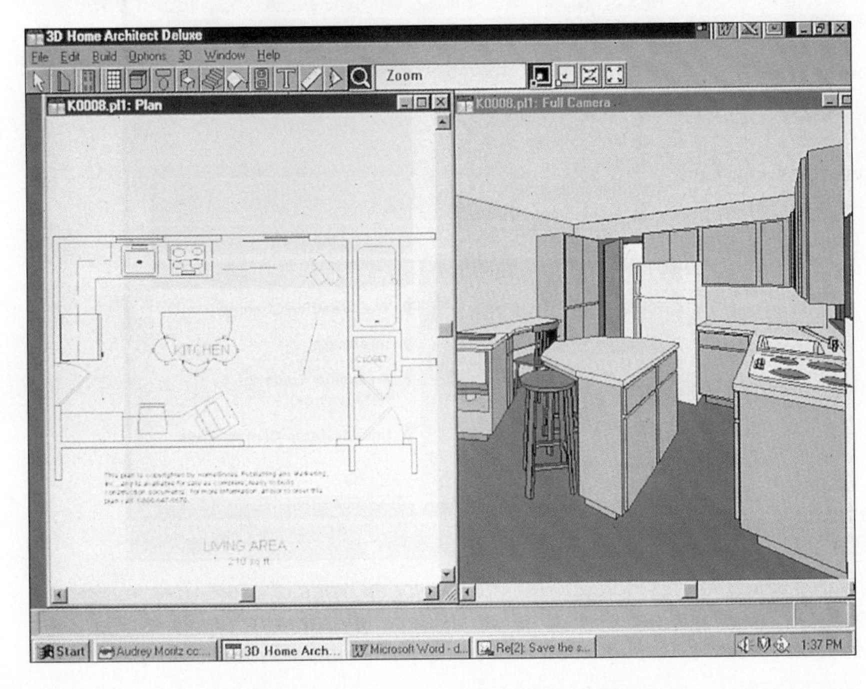

Figure 2-39 Home design/landscaping software can help you design or remodel a home, deck, or landscape.

Educational/Reference/ Personal Computer Entertainment Software

Educational software is software designed to teach a particular skill. Educational software exists for just about any subject, from learning a foreign language to learning how to cook. Pre-school to high school learners can use educational software to assist them with subjects such as reading and math, or to prepare them for class or college entry exams.

Reference software provides valuable and thorough information for all individuals (Figure 2-40). Popular reference software includes encyclopedias, dictionaries, health/medical guides, and travel directories.

Personal computer entertainment software includes interactive games, videos, and other programs designed to support a hobby or provide amusement and enjoyment. For example, you can use personal computer entertainment software to play games, make a family tree, compose music, or fly an aircraft.

For more information on reference software, visit the Discovering Computers 2001 Chapter 2 WEB INFO page (**www.scsite.com/dc2001/ch2/webinfo.htm**) and click Reference Software.

SOFTWARE FOR COMMUNICATIONS

One of the more valuable aspects of software is its capability of supporting communications. Certain applications specifically are designed to facilitate communications, thus allowing you to share information with others. Communications software, which is a utility program that allows you to dial a modem, is discussed in a later chapter. Software for communications discussed in the following sections includes groupware, e-mail, and Web browsers.

Groupware

Groupware identifies any type of software that helps groups of people on a network collaborate on projects and share information, in addition to providing PIM functions, such

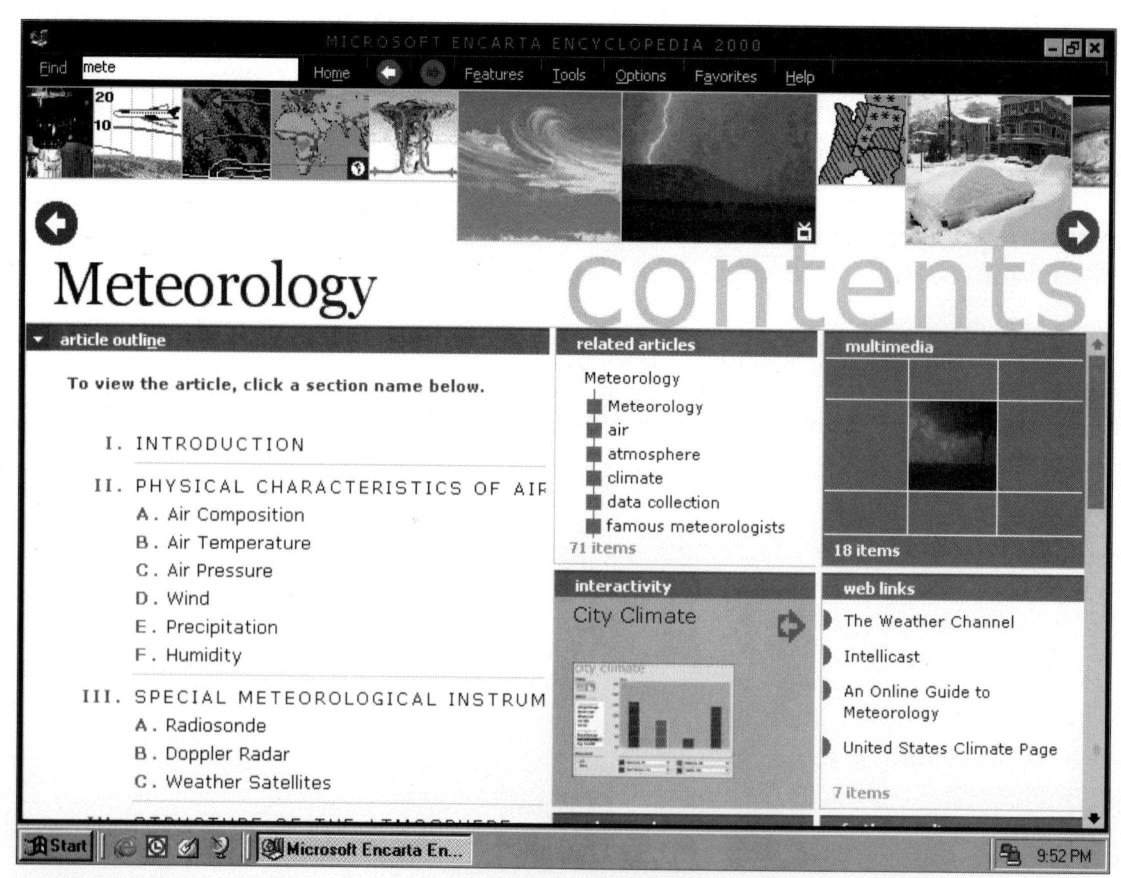

Figure 2-40 *Reference software provides valuable and thorough information for all types of users. This reference package includes text you can read about meteorology, a variety of pictures, maps, charts, and illustrations you can view of meteorology, and a video you can watch about thunderstorms.*

as an address book (Figure 2-41) and appointment calendar. A major feature of groupware is group scheduling, in which a group calendar tracks the schedules of multiple users and helps coordinate appointments and meeting times.

Electronic Mail Software

E-mail (**electronic mail**) is the transmission of messages via a computer network such as a local area network or the Internet. The message can be simple text or can include an attachment such as a word processing document, a graphical image, or an audio or video clip. Using **electronic mail software**, you can create, send, receive, forward, store, print, and delete e-mail messages (Figure 2-42). Most e-mail software has a mail notification alert that informs you via a message or sound that you have received new mail, even if you are working in another application.

When you receive an e-mail message, the message is placed in your **mailbox**, which is a storage location usually residing on the computer that connects you to the local area network or the Internet, such as the server operated by your Internet service provider (ISP). The server that contains the mailboxes often is called a **mail server**. Most ISPs and online services provide an Internet e-mail program and a mailbox on a mail server as a standard part of their Internet access services.

To make the sending of messages more efficient, e-mail software allows you to send a single message to a distribution list consisting of two or more individuals. The e-mail software copies the message and sends it to each person on the distribution list. For example, a message addressed to the Accounting Department distribution list would be sent to each of the employees in the accounting department.

Just as you address a letter when using the postal system, you must address an e-mail message with the e-mail address of your intended recipient. Likewise, when someone sends you a message, they must have your e-mail address. An Internet **e-mail address**, which is a combination of a user name and a domain name, identifies a user so he or she can receive Internet e-mail (Figure 2-43).

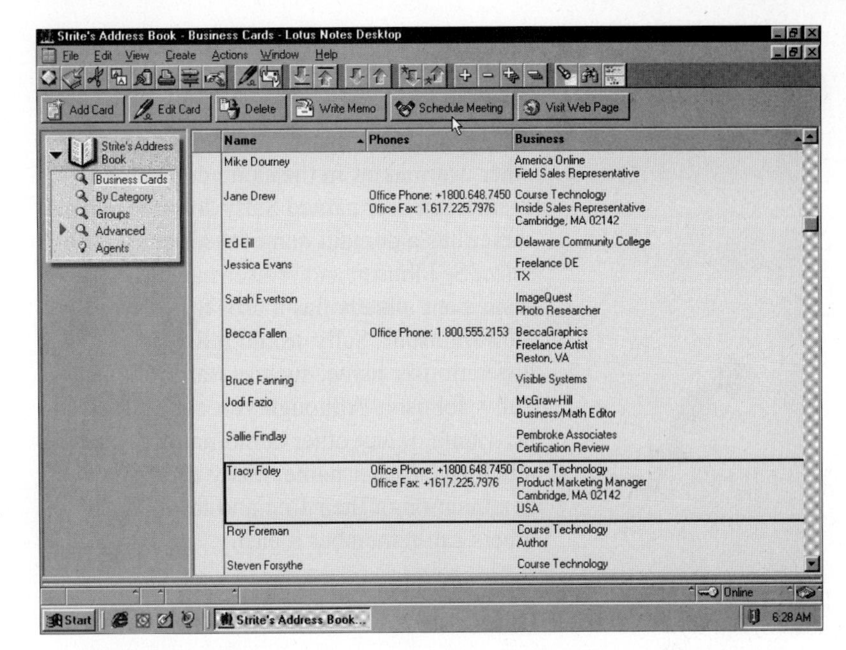

Figure 2-41 Groupware often provides services for personal information management, such as the business cards shown in this example.

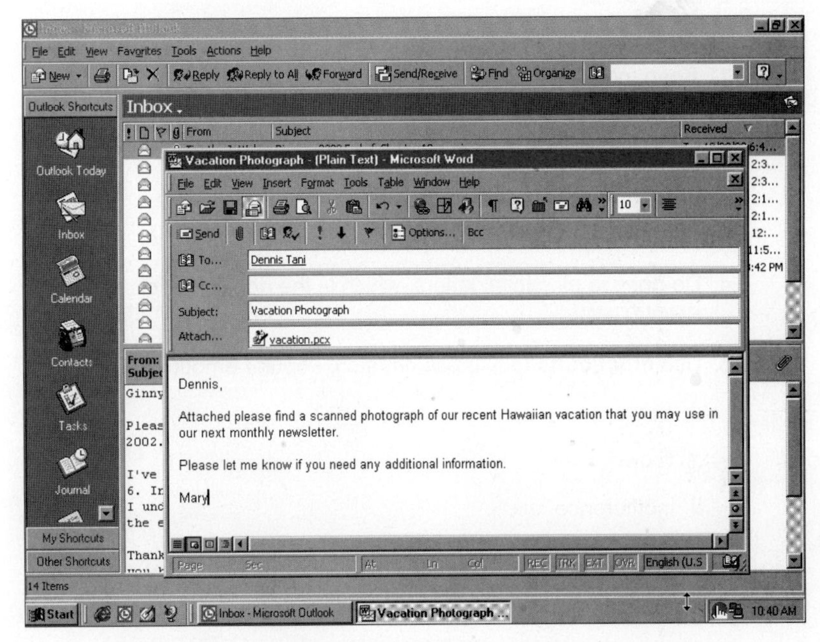

Figure 2-42 Using e-mail software, you can create, send, receive, forward, store, print, and delete e-mail messages.

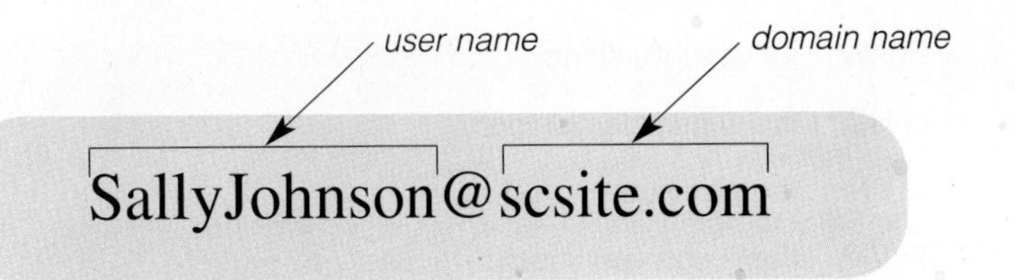

Figure 2-43 An Internet e-mail address is a combination of a user name and a domain name.

A **user name**, or **user-ID**, is a unique combination of characters, such as letters of the alphabet or numbers, that identifies you. Your user name must be different from the other user names in the same domain. For example, a user named Sally Johnson whose server has a domain name of scsite.com might select S_Johnson as her user name. If scsite.com already has a user S_Johnson (for Sam Johnson), Sally would have to select a different user name, such as SallyJohnson or Sally_Johnson. Although you can select a nickname or any other combination of characters for your user name, many users select a combination of their first and last names so others can remember it easily.

In an Internet e-mail address, an at symbol (@) separates the user name from the domain name. Using the example in Figure 2-43 on the previous page, a possible e-mail address would be SallyJohnson@scsite.com, which would be read as follows: Sally Johnson at s c site dot com. Most e-mail programs allow you to create an **address book**, which contains a list of names and e-mail addresses.

Although no complete listing of Internet e-mail addresses exists, several Internet sites list addresses collected from public sources. These sites also allow you to list your e-mail address voluntarily so others can find it. The site also might ask for other information, such as your high school or college, so others can determine if you are the person they want to reach.

When sending messages, you can append additional information to the message. Called a **signature**, this information usually is located at the bottom of the message and is attached automatically to each outgoing message. Your signature, for example, could include your name, company affiliation, and a favorite quotation.

Today, e-mail quickly is becoming a primary communication method for both personal and business use. Mobile users can send and receive e-mail messages at any time of the day. As e-mail has gained in popularity, several informal rules for using e-mail have developed; some of these are outlined in Figure 2-44.

E-MAIL RULES

1. Keep messages brief using proper grammar and spelling.
2. Be careful when using sarcasm and humor as it might be misinterpreted.
3. Be polite, be diplomatic, and avoid offensive language.
4. Do not use all capital letters, which is the equivalent of SHOUTING!
5. Use **emoticons** to express emotion. Popular emoticons include

   :) Smile

   :(Frown

   :l Indifference

   :\ Undecided

   :o Surprised
6. Use abbreviations and acronyms for phrases such as

   BTW by the way

   FYI for your information

   FWIW for what it's worth

   IMHO in my humble opinion

   TTFN ta ta for now

   TYVM thank you very much

Figure 2-44 *E-mail rules.*

Web Browsers

A software application called a **Web browser**, or **browser**, allows you to access and view Web pages. Today's browsers have graphical user interfaces and are quite easy to learn and use. The two more popular browsers are Netscape Navigator and Microsoft Internet Explorer (Figure 2-45). Browsers have many special features including buttons and navigation to help guide you through Web sites. In addition to displaying Web pages, most browsers allow you to use other Internet services such as electronic mail (e-mail). Using a Web browser to navigate the World Wide Web is discussed in depth in the next section.

BROWSING THE WORLD WIDE WEB

Today's computer users of all types must have a Web browser to access the vast resources available on the World Wide Web. In fact, one of the major reasons that business, home, and other users purchase computers is for Internet access. It is apparent in advertising that many companies and organizations assume you are familiar with the Internet. Web addresses appear on television, in radio broadcasts, and in printed newspapers and magazines. Companies encourage you to browse their Web-based catalogs and buy their products online. Many colleges give tours of their campuses on the Web, accept applications online, and offer classes on the Internet. To be successful today, you must have an understanding of the Internet — particularly the World Wide Web. Without it, you are missing a tremendous resource for products, services, and information. This chapter discusses the World Wide Web; other services on the Internet are discussed in Chapter 7.

Although many people use the terms World Wide Web and Internet interchangeably, the World Wide Web is just one of the many services available on the Internet. The **World Wide Web** (**WWW**), or **Web**, consists of a worldwide collection of electronic documents. Each of these electronic documents on the Web is called a Web page. A **Web page** can contain text, graphical images, sound, and video, as well as connections to other documents. A collection of related Web pages that you can access electronically is called a **Web site**. The following sections explain how to navigate Web pages on the World Wide Web.

WEB INFO

For more information on Web browsers, visit the Discovering Computers 2001 Chapter 2 WEB INFO page (**www.scsite.com/dc2001/ch2/webinfo.htm**) and click Web Browsers.

Figure 2-45a *(Netscape Communicator)*

Figure 2-45b *(Microsoft Internet Explorer)*

Figure 2-45 *The two more popular Web browsers are Netscape Communicator and Microsoft Internet Explorer.*

Connecting to the Web and Starting a Browser

As noted, to access and view Web pages you need browser software and a computer that is connected to the Internet. You can connect to the Internet in one of several ways. Some users connect through an Internet service provider or through an online service, using a modem to establish a connection by dialing a specific telephone number. Organizations such as schools and businesses often provide Internet access for students and employees. These users connect to the Internet through the business or school network, which is connected to an Internet service provider.

An **Internet service provider** (**ISP**) is an organization that has a permanent connection to the Internet and provides temporary connections to individuals and companies for a fee, usually about $20 per month. Like an ISP, an **online service** provides access to the Internet, but online services also have members-only features that offer a variety of special content and services such as news, weather, legal information, financial data, hardware and software guides, games, and travel guides. Two popular online services are The Microsoft Network (MSN) and America Online (AOL). The specifics of connecting to the Internet and using ISPs and online services are discussed in more depth in Chapter 7.

To establish the connection, you typically click an icon on your desktop to start the Web browser (Figure 2-46). Your modem will dial the telephone number to the ISP or online service. Once a telephone connection to the

Figure 2-46

ONE METHOD OF CONNECTING TO THE INTERNET

Step 1: *Click an icon on the desktop to start your browser.*

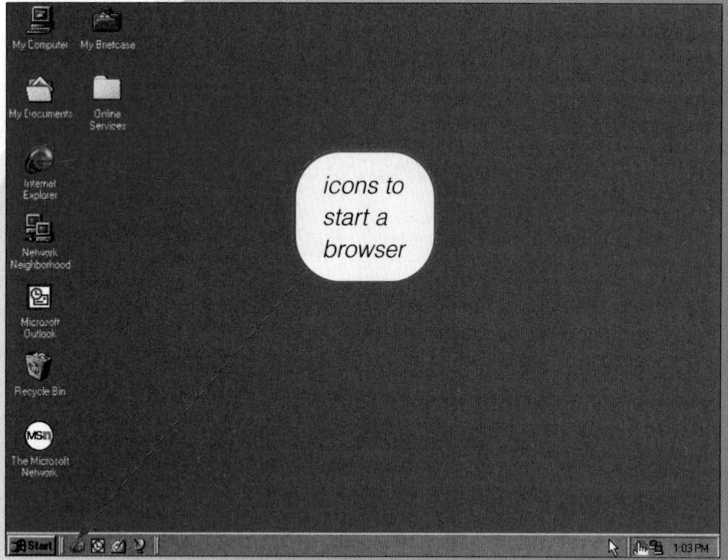

Step 4: *Connection to the Internet occurs as shown in Figure 2-47.*

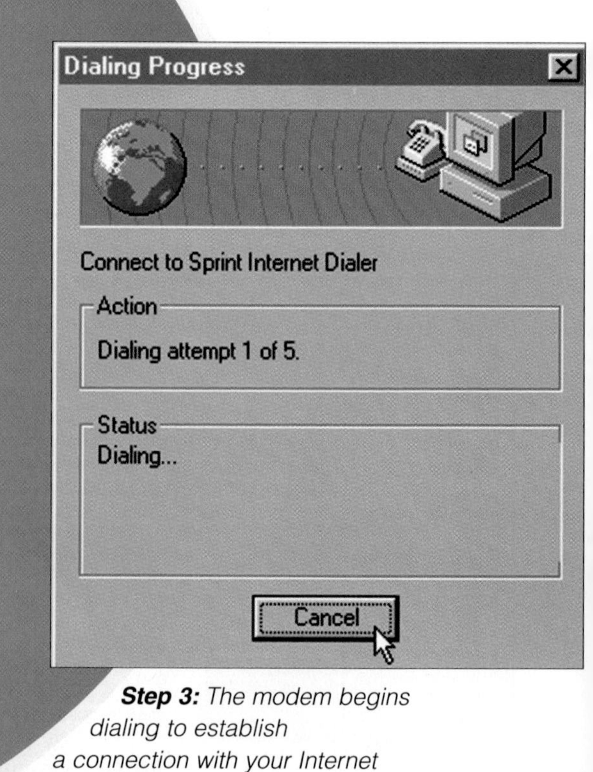

Step 3: *The modem begins dialing to establish a connection with your Internet service provider or online service.*

Dial-up Connection
Connect to Sprint Internet Dialer
User name:
ELN/02G5H
Password:
Save password
Connect automatically
Settings... Connect Work Offline

Step 2: *If you are not connected to the Internet already, you may be asked if you want to connect.*

Internet is established, the browser retrieves and displays a home page (Figure 2-47). A starting page for a Web site, called a **home page** or sometimes a portal, is similar to a book cover or a table of contents for the site and provides information about the site's purpose and content. Most browsers use the manufacturer's Web page as their initial home page, but you can change your browser's home page at any time. Many sites allow you to personalize the home page so it displays areas of interest to you.

The process of receiving information, such as a Web page, onto your computer from a server on the Internet is called **downloading**. While your browser is downloading a page, it typically displays an animated logo or icon in the top-right corner of the browser window; when the download is complete, the animation stops. Downloading a Web page to your screen can take from a few seconds to several minutes, depending on the speed of your connection and the amount of graphics on the Web page. To speed up the display of pages, most Web browsers let you turn off the graphics and display only text.

Navigating Web Pages Using Links

Most Web pages contain hyperlinks. A **hyperlink**, also called a **link**, is a built-in connection to another related Web page.

Links allow you to obtain information in a nonlinear way; that is, to make associations between topics instead of moving sequentially through the topics. Reading a book from cover to cover is a linear way of learning. Branching off and investigating related topics as you encounter them is a nonlinear way of learning. For example, while reading an article on nutrition, you might want to learn more about counting calories. Having linked to and read information on counting calories, you might want to find several low-fat, low-calorie recipes. Reading these might inspire you to learn about a chef that specializes in healthy but tasty food preparation. The capability of branching from one related topic to another in

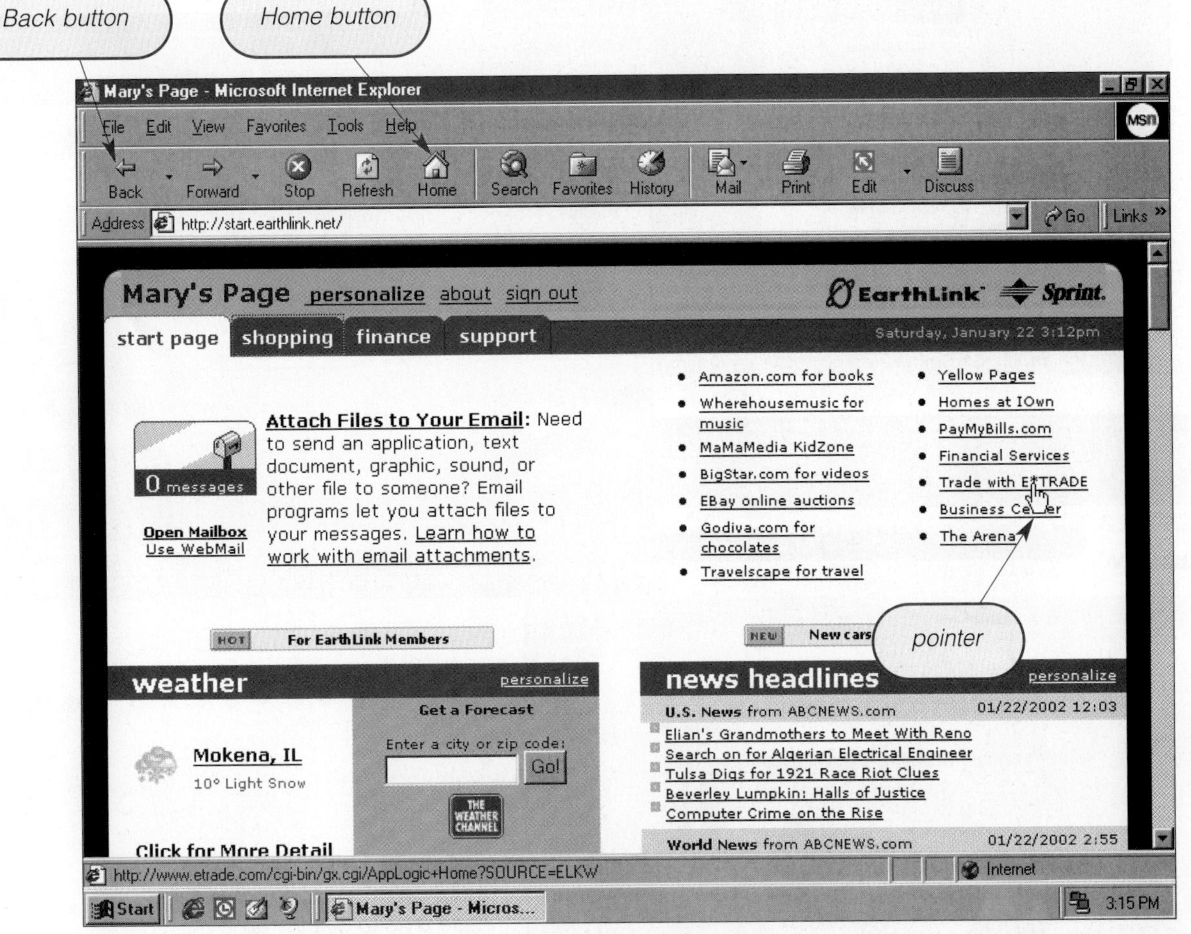

Figure 2-47 *When you start Microsoft Internet Explorer, a popular Web browser, a home page displays. You can change, and often personalize, the home page that displays when you start a browser.*

a nonlinear fashion is what makes links so powerful and the World Wide Web such an interesting place to explore.

On the Web, links can be a word, phrase, or image; you can identify them because they are underlined, are a different color from the rest of the document, or are highlighted graphical images. In many cases, when you point to a link, the pointer shape changes to a small hand with a pointing index finger. For example, in Figure 2-47 on the previous page, the pointer is on the Trade with E*TRADE link.

To activate a link, you click it, which then causes the Web page or the location within a Web page associated with the link to display on the screen. The link can point to the same Web page, a different Web page at the same Web site, or a separate Web page at a different Web site in another city or country. In essence, when you navigate using links, you are jumping from Web page to Web page. Displaying pages from one Web site after another is called **surfing the Web**. The steps in Figure 2-48 illustrate navigating using a variety of types of links.

To remind you visually that you have visited a location or document, some browsers change the color of a text link after you click it.

Figure 2-48
NAVIGATING USING A VARIETY OF LINKS

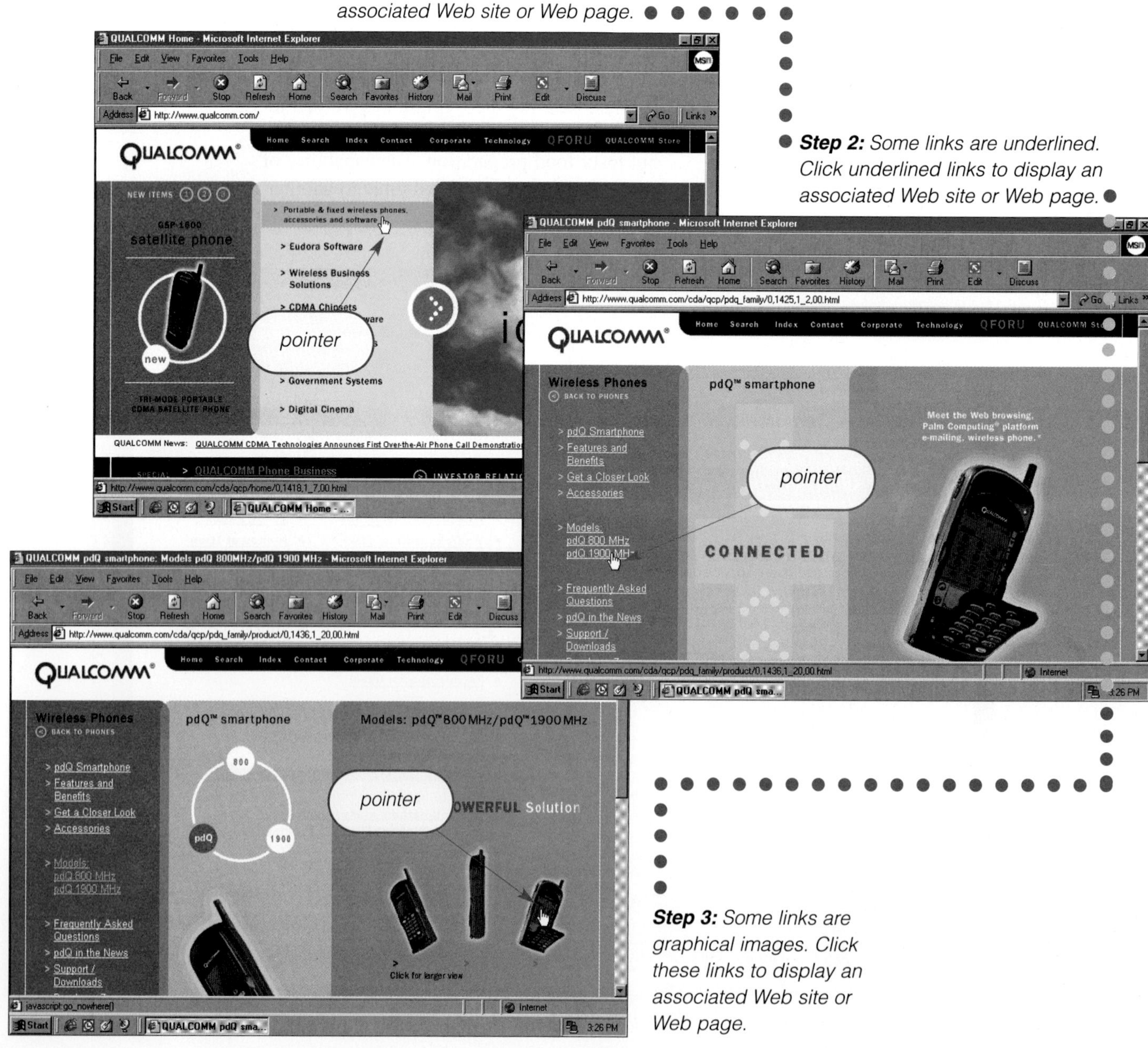

Step 1: Some links are a different color. Click these links to display an associated Web site or Web page.
Step 2: Some links are underlined. Click underlined links to display an associated Web site or Web page.
Step 3: Some links are graphical images. Click these links to display an associated Web site or Web page.
pointer
pointer
pointer

Using the Browser Toolbars

In addition to navigating using links, browsers have buttons, text boxes, and other features to help you navigate and work with Web sites. The Back button, for example, displays the previously displayed Web page. The Home button redisplays the home page (see Figure 2-47 on page 2.37). The table in Figure 2-49 identifies the functions of buttons on the Microsoft Internet Explorer toolbar.

Entering a URL

A Web page has a unique address, called a **Uniform Resource Locator** (**URL**). A browser retrieves a Web page by using its URL, which tells the browser where the document is located. URLs make it possible for you to navigate using links because a link is associated with a URL. When you click a link, you are issuing a request to display the Web site or the document specified by the URL.

MICROSOFT INTERNET EXPLORER TOOLBAR BUTTONS

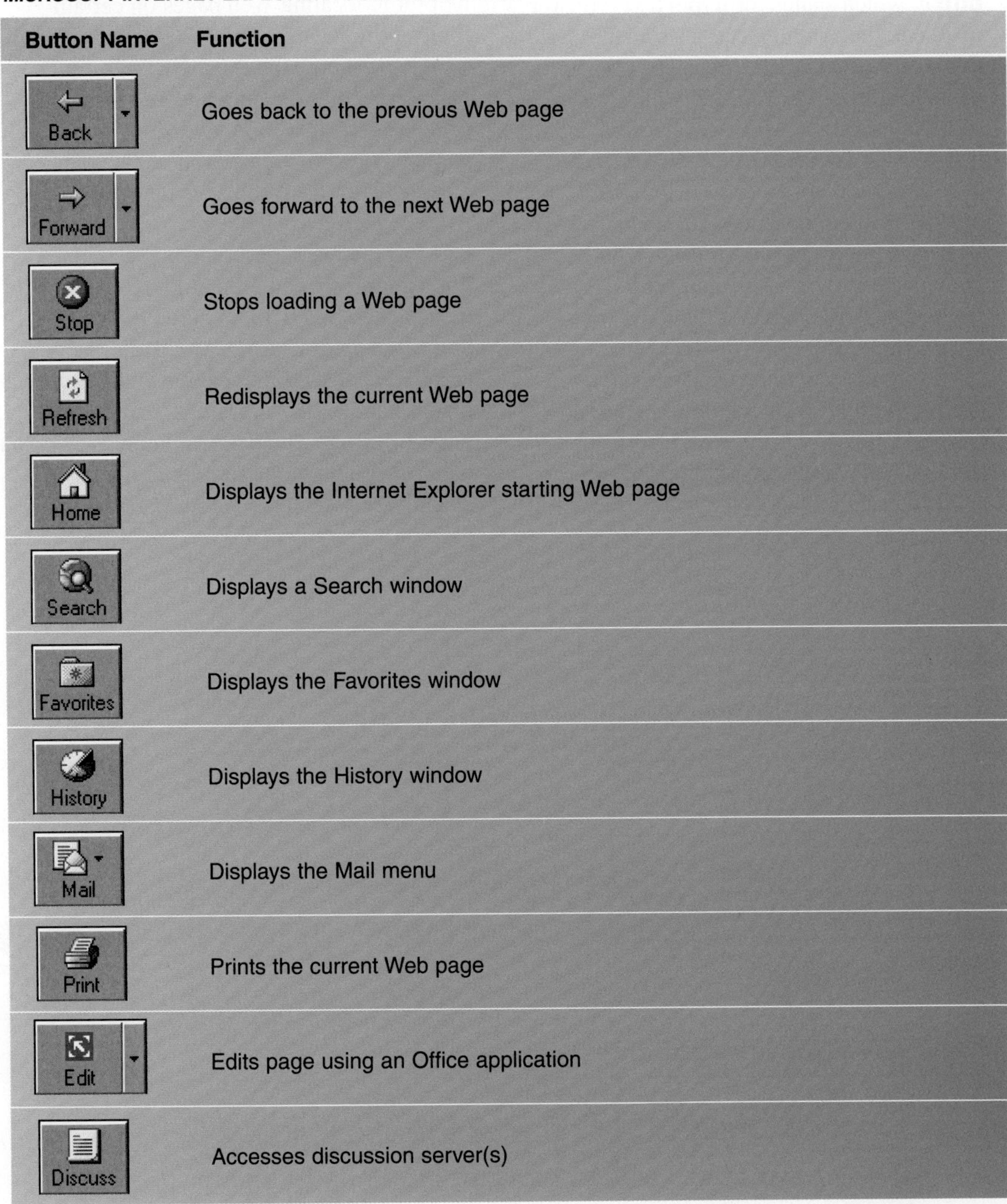

Button Name	Function
Back	Goes back to the previous Web page
Forward	Goes forward to the next Web page
Stop	Stops loading a Web page
Refresh	Redisplays the current Web page
Home	Displays the Internet Explorer starting Web page
Search	Displays a Search window
Favorites	Displays the Favorites window
History	Displays the History window
Mail	Displays the Mail menu
Print	Prints the current Web page
Edit	Edits page using an Office application
Discuss	Accesses discussion server(s)

Figure 2-49 *Functions of the buttons on the Microsoft Internet Explorer toolbar.*

WEB INFO

For more information on search engines, visit the Discovering Computers 2001 Chapter 2 WEB INFO page (**www.scsite.com/dc2001/ch2/webinfo.htm**) and click Search Engines.

If you know the URL of a Web page, you can type it into a text box at the top of the browser window. For example, if you type the URL of http://www.suntimes.com/index/business.html in the Address text box and then press the ENTER key, the browser will download and display the Business Section of the Chicago Sun-Times Online Newspaper (Figure 2-50).

As shown in Figure 2-50, a URL consists of a protocol, domain name, and sometimes the path to a specific Web page or location in a Web page. Most Web page URLs begin with **http**://, which stands for **hypertext transfer protocol**, the communications standard used to transfer pages on the Web. The domain name identifies the Web site, which is stored on a Web server. A **Web server** is a computer that delivers (serves) requested Web pages.

If you do not enter a URL exactly, your browser will not be able to locate the site or Web page you want to visit (view). To help minimize errors, most current browsers allow you to type URLs without the http:// portion. For example, you can type the text www.suntimes.com/index/business.html instead of http:// www.suntimes.com/index/business.html. If you enter an incorrect address, some browsers search for similar addresses and provide a list from which you can select.

Searching for Information on the Web

No single organization controls additions, deletions, and changes to Web sites, which means no central menu or catalog of Web site content and addresses exists. Several companies, however, maintain organized directories of Web sites to help you find information on specific topics.

A **search engine** is a software program you can use to find Web sites, Web pages, and Internet files. Search engines are particularly helpful in locating Web pages on certain topics or in locating specific pages for which you do not know the exact URL. To find a page or pages, you enter a word or phrase, called **search text** or **keywords**, in the search engine's text box; the search engine then displays a list of all Web pages that contain the word or phrase you entered. Any Web page listed as the result of a search is called a **hit**. For example, if you did not know the URL of the Chicago Sun-Times Online newspaper, you could enter Chicago Sun-Times as your search text. The search engine would return a list of hits, or Web pages, that contain the phrase Chicago Sun-Times (Figure 2-51). You then click an appropriate link in the list to display the associated Web site or Web page.

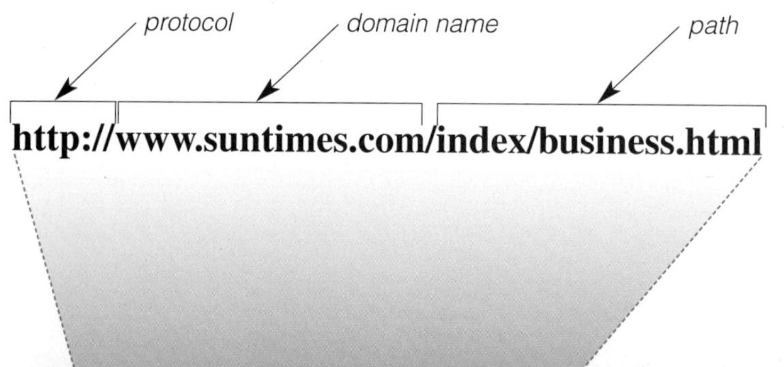

Figure 2-50 The URL for the Chicago Sun-Times Online Business section is http://www.suntimes.com/index/business.html. When you enter this URL in the Address text box, the associated Web page displays.

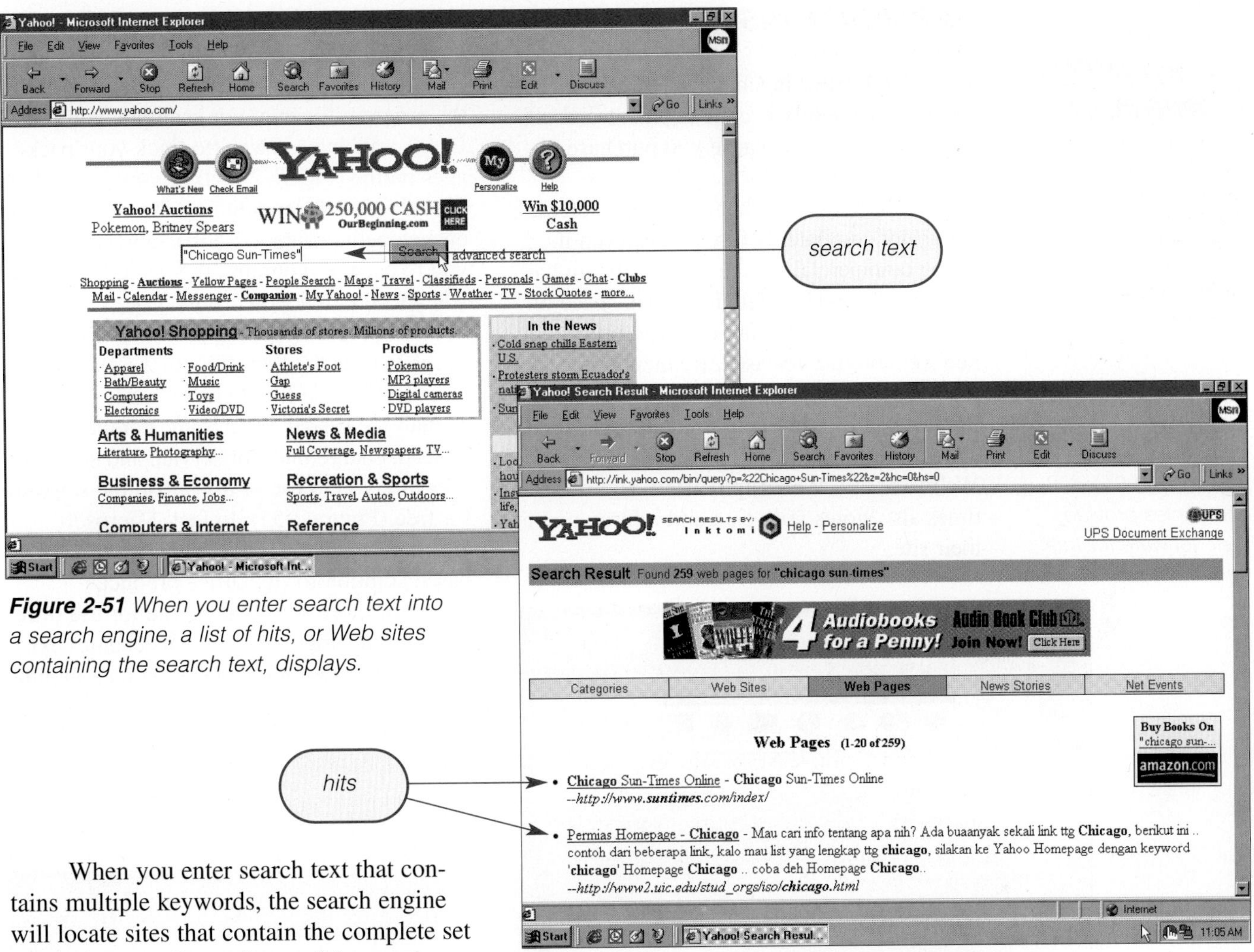

Figure 2-51 When you enter search text into a search engine, a list of hits, or Web sites containing the search text, displays.

When you enter search text that contains multiple keywords, the search engine will locate sites that contain the complete set of words, as well as subsets of the words. For example, a search with the keywords, Chicago Sun-Times, results in 645,573 hits, or Web pages, that contain the word Chicago or the word Sun-Times. To reduce the number of hits, you should search for Web pages that contain both words. In one search engine, for example, surrounding your keywords with quotation marks instructs the search engine to look for Web pages containing that exact phrase. The search text, "Chicago Sun-Times," for example, reduces the number of hits to 259.

Search engines actually do not search the entire Internet; such a search would take an extremely long time. Instead, they search an index of Internet sites and Web pages that constantly is updated by the company that provides the search engine.

The table in Figure 2-52 lists the Web site addresses of several Internet search engines. Most of these sites also provide directories of Web sites organized in categories such as sports, entertainment, or business.

Widely Used Search Engines	
AltaVista	www.altavista.com
Excite	www.excite.com
GoTo.com	www.goto.com
HotBot	www.hotbot.com
Infoseek	www.infoseek.com
Lycos	www.lycos.com
Yahoo!	www.yahoo.com

Figure 2-52 Widely used search engines and their URLs.

APPLICATIONS ON THE WEB

WEB INFO

For more information on Web applications, visit the Discovering Computers 2001 Chapter 2 WEB INFO page (**www.scsite.com/dc2001/ch2/webinfo.htm**) and click Web Applications.

You learned in Chapter 1 that packaged software frequently is called commercial off-the-shelf software because you purchase it from a software vendor, retail store, or Internet business and then install the software onto your computer before you can run it. Using commercial off-the-shelf software has the disadvantages of requiring disk space on your computer and being costly to upgrade as new versions are released. Realizing these disadvantages, some companies today offer products and services on the Web. That is, a Web site stores the program, often called **Webware** or a **Web application**, and sometimes also stores your data and information at their site.

To access these programs on the Web, you simply visit the Web site that offers the program. Some Web sites provide access to the program for free. For example, you use UPS's tracking software to track your packages online (Figure 2-53a) or you can look up a wealth of information in Britannica's online encyclopedia (Figure 2-53b). Others allow you to use the program for free and pay a fee when a certain action occurs. For example, you can prepare your tax return for free using TurboTax for the Web (Figure 2-54), but if you elect to also file it electronically, you pay a small fee (under $10). Some companies, instead, charge only for service and support – allowing you to use or download the software for free (Figure 2-55). In fact, Microsoft has indicated it will offer some type of Web-based solution for its Office product.

For those sites that charge for use of the program, a variety of payment schemes exist. Some rent use of the application on a monthly basis, some charge based on the number of user accesses, and others charge a one-time fee.

Figure 2-53 *Some companies provide their software for free on the Web.*

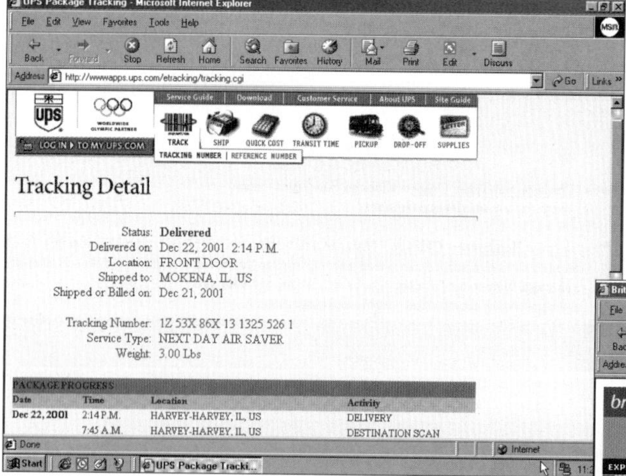

Figure 2-53a *(package tracking on the Web)*

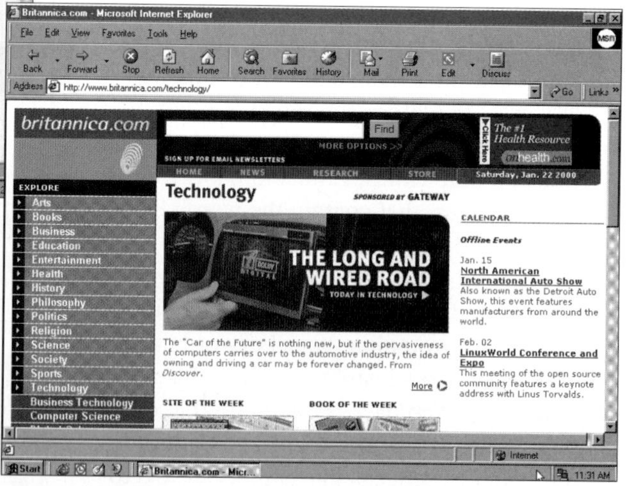

Figure 2-53b *(encyclopedia on the Web)*

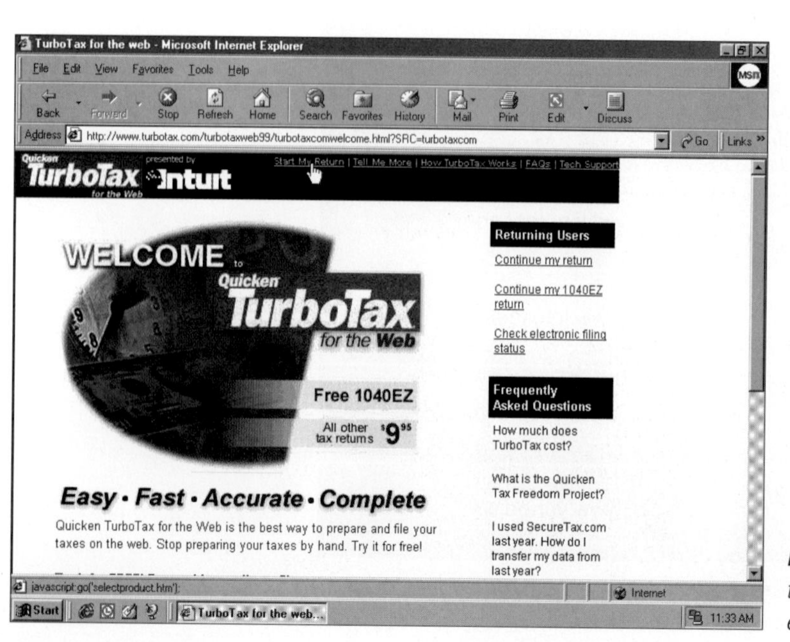

Figure 2-54 *With TurboTax on the Web, you prepare your taxes for free but pay a small fee if you elect to file them electronically.*

Application Service Providers

Storing and maintaining programs can be a costly investment for businesses. Thus, some have elected to outsource to an application service provider one or more facets of their information technology needs. An **application service provider** (**ASP**) is a third-party organization that manages and distributes software and services on the Web. For example, click2learn.com is an ASP that creates, delivers, and manages training products on the Web (Figure 2-56).

Five categories of ASPs have emerged:

- Enterprise ASP: customizes and delivers high-end business applications, such as finance and database
- Local/Regional ASP: offers a variety of software applications to a specific geographic region
- Specialist ASP: delivers applications to meet a specific business need, such as preparing taxes
- Vertical Market ASP: provides applications for a particular industry, such as construction or healthcare
- Volume Business ASP: supplies prepackaged applications, such as accounting, to businesses

Despite the advantages, some companies will wait to outsource to an ASP until they have faster Internet connections.

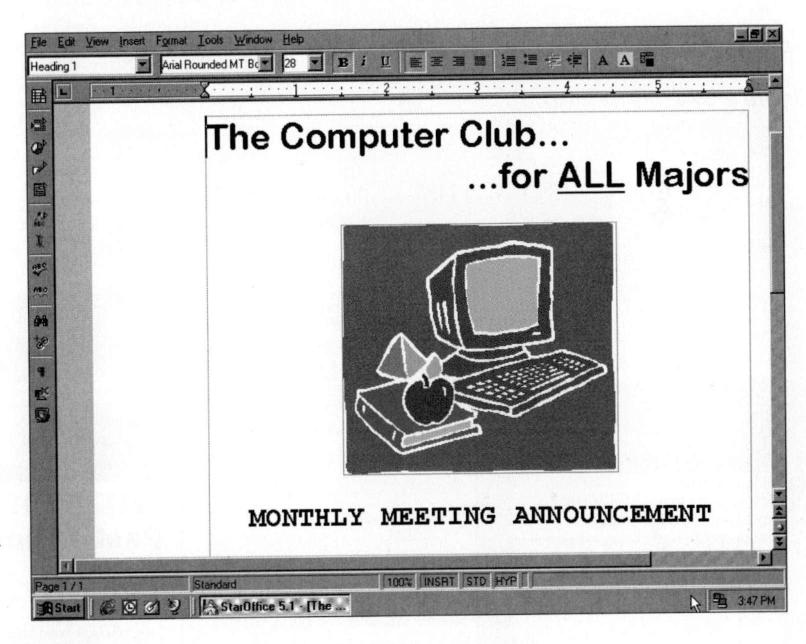

Figure 2-55 *Sun Microsystems only charges for service and support on its StarOffice product, which is an integrated word processing, spreadsheet, presentation graphics, database, photo-editing, personal information manager, and communications suite. Shown here is a Microsoft Word document opened in the word processing window of StarOffice.*

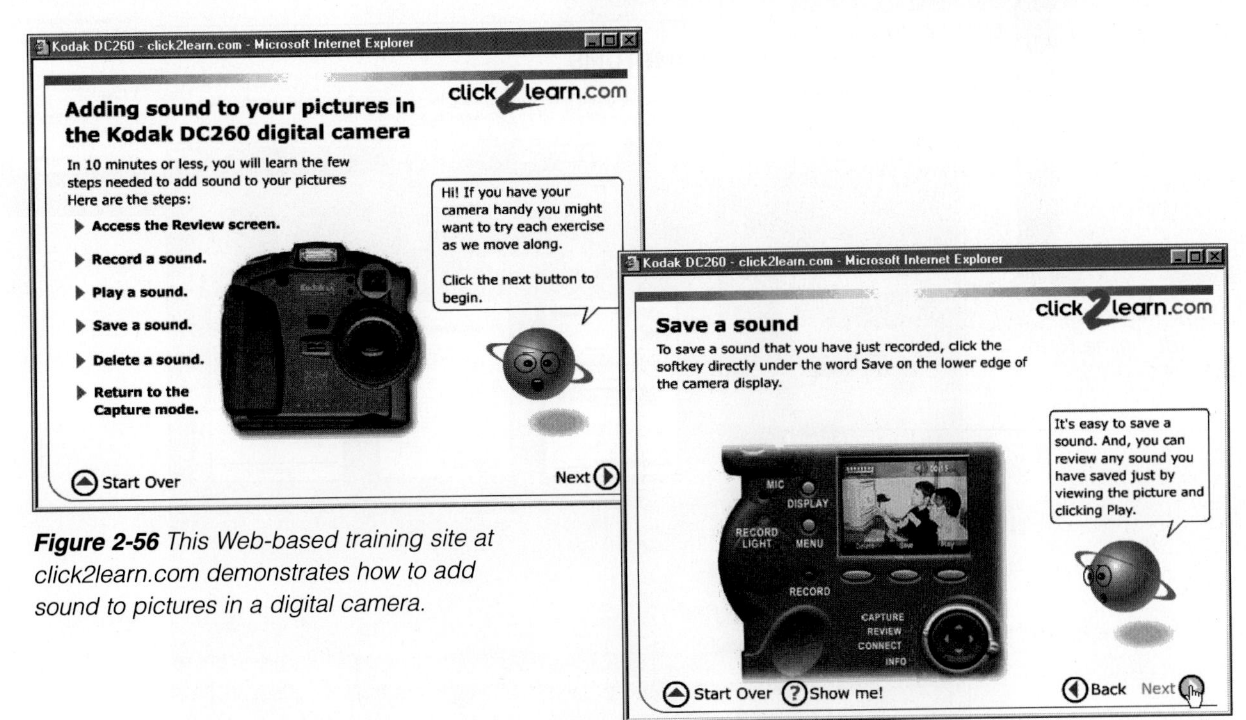

Figure 2-56 *This Web-based training site at click2learn.com demonstrates how to add sound to pictures in a digital camera.*

LEARNING AIDS AND SUPPORT TOOLS

Learning how to use an application software package or the Web effectively involves time and practice. To aid you in that learning process, many software applications and Web sites provide online Help, FAQs, tutorials, and wizards (Figure 2-57).

Online Help is the electronic equivalent of a user manual; it usually is integrated into an application software package. Many packages also have links to their Web site, which provides additional help and support. Online Help provides assistance that can increase your productivity and reduce your frustrations by minimizing the time you spend learning how to use an application software package.

In most packages, a function key or a button on the screen starts the Help feature. When you are using an application and have a question, you can use the Help feature to ask a question or access the Help topics in subject or alphabetical order. Often the Help is **context-sensitive**, meaning that the Help information is related to the current task being attempted. Most online Help also points you to Web sites that provide updates and more comprehensive resources to answer your software questions.

Figure 2-57a *(online Help)*

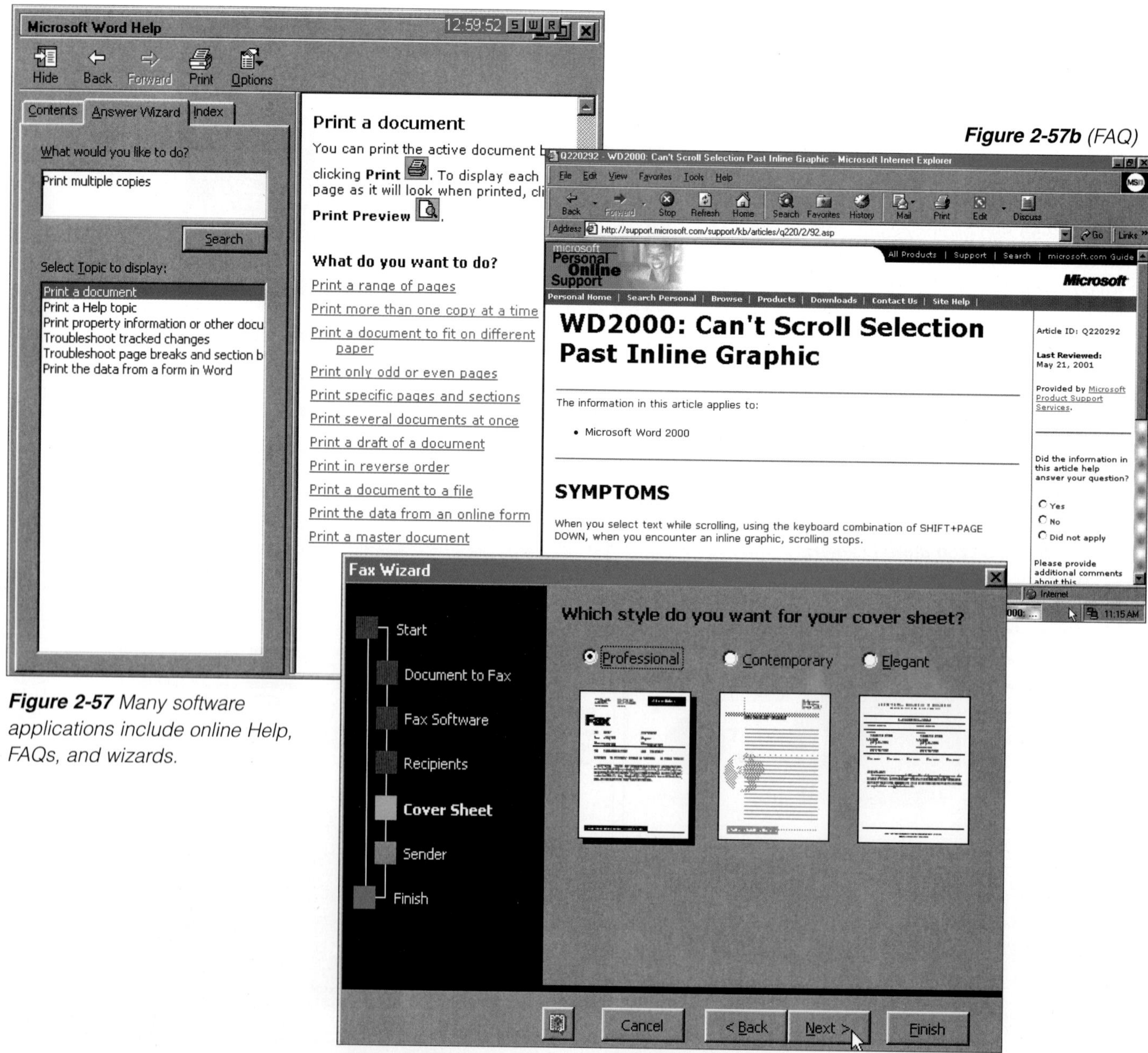

Figure 2-57b *(FAQ)*

Figure 2-57 *Many software applications include online Help, FAQs, and wizards.*

Figure 2-57c *(wizard)*

In many cases, online Help has replaced the user manual altogether, and software developers no longer include user's manuals with the software. If you want to learn more about the software package from a printed manual, however, many books are available to help you learn to use the features of personal computer application packages (Figure 2-58). These books typically are available in bookstores and software stores.

You also can find Web pages with an application software **FAQ** (Frequently Asked Questions) to help you find answers to common questions.

Tutorials are step-by-step instructions using real examples that show you how to use an application. Some tutorials are available in printed manuals; others are software-based or Internet-based, thus allowing you to use your computer to learn about a package.

A **wizard** is an automated assistant that helps you complete a task by asking you questions and then automatically performing actions based on your answers. Many software applications include wizards. For example, word processing software uses wizards to help you create memorandums, meeting agendas, fax cover sheets, and letters; spreadsheet software includes chart and function wizards; and database software has form and report wizards.

Many colleges and schools provide training on several of the applications discussed in this chapter. If you would like more direction than is provided in online Help, FAQs, tutorials, wizards, and trade books, contact your local school for a list of classes they offer.

Figure 2-58 Many bookstores sell trade books to help you learn to use the features of personal computer application packages.

TECHNOLOGY TRAILBLAZER

MARC ANDREESSEN

In 1982, Jim Clark founded Silicon Graphics, a billion-dollar manufacturer of visual computing systems. A dozen years later, Clark was eager for a fresh venture. Seeking a horse on which to hitch his new wagon, Clark asked an engineer for promising names in the computer industry. He wanted more than a technician. "I believe in athletes," he insists, "not first basemen." With Clark's requirements in mind, the engineer offered only one name — Marc Andreessen.

Andreessen grew up in Wisconsin but earned his reputation at the University of Illinois. Working part-time, for about $7.00 an hour, at the school's National Center for Supercomputing Applications (NCSA), he developed UNIX programs for supercomputers. Here, Andreessen became familiar with the Internet, a primarily text-based resource used almost exclusively by scientists and academics. Andreessen envisioned a program that would make the Internet and its subset, the World Wide Web, accessible to the general population. The program itself, Andreessen admits, "wasn't a breakthrough. It was just that nobody had developed a tool that would allow the next wave to come along and be able to take advantage of the Internet."

Andreessen convinced a friend, Eric Bina, to help with the project. The pair worked eighteen hours a day, writing code and debating other issues – from current events ("I'm a news junkie," Andreessen says) to the best snack food (Andreessen liked Pepperidge Farm pastries, while Bina preferred Skittles candy). In three months, they completed the program, which they called Mosaic.

On the face of it, Mosaic did not look like much. Only 9,000 lines long – the Windows operating system is almost a thousand times longer – the program was fairly slow and unstable. Mosaic's impact, however, was phenomenal. Mosaic translated the hypertext markup language (HTML) formatting code in which Web documents are written into colorful, graphical pages connected by hyperlinks. For the first time people had a user-friendly, graphical tool they could use to *browse* the Web, jumping from hyperlinks to other Web pages with a click of the mouse. NCSA engineers developed versions of the program for Windows and Macintosh computers. People could download Mosaic free on the Internet, and its popularity exploded. "We introduced Mosaic to twelve people at the beginning of 1993," Andreessen recalls, "and by the end, there were one million people using it. It grew like a virus."

After graduating from the University of Illinois, Andreessen accepted a job as a programmer at a small company in California's Silicon Valley. Here, Jim Clark contacted him, and the two met to consider ideas for a new company. Eventually, the talk came around to Mosaic. Several people who helped develop the browser were about to leave NCSA, and this gave Andreessen an idea. "We can always create a Mosaic killer — do the program right," he suggested. Andreessen recruited his NCSA friends, who were happy to rejoin him. Later, in a *Newsweek* article, an NCSA coworker would claim he and the other programmers were plumbers, but Andreessen was the architect who "showed us where all the bathrooms went." Clark invested four million dollars in the new company, which was called Netscape Communications Corporation. *Mozilla* (Mosaic killer) became the company's virtual mascot. The Netscape browser was its product.

The company was an immediate success, and stock options made the Andreessen wealthy. Not content to rest on his laurels, the Web wizard recently co-founded a new company, called Loudcloud, to provide technology and software to Internet startups. "What always amazes me is that people…find all sorts of creative uses for [our] product that I would never have thought of, and finding out what they're doing is really interesting to me. I try to figure out what we should do next."

COMPANY ON THE CUTTING EDGE

COMPANY ON THE CUTTING EDGE

AMERICA ONLINE, INC.

Each day, more than 15 million people in the world hear the trademark voice intoning the familiar, "you've got mail." Millions more are likely to be introduced to it if recent events play out for America Online, Inc.

America Online, Inc. (AOL) is the world leader in online services, including electronic mail, software, computer support services, and Internet access. Headquartered in Dulles, Virginia, AOL employs more than 12,000 people in 45 locations worldwide. It was founded in 1985 as Quantum Computer Services Corporation, but the idea really started earlier than that. In 1982, Stephen M. Case, a marketer for PepsiCo, Inc., became interested in electronic communications. Then, the Internet still was a place where only scholars and specialists felt comfortable. Case believed that online communications could be used and enjoyed by more people if using the Internet could be simplified.

In 1985, he founded Quantum and partnered with Commodore International, Ltd., a personal computer manufacturer. Case created an online service strictly for use by owners of Commodore computers. Dubbed Q-link, it was an archaic precursor to today's online offerings. The idea came at the right time.

Having sown the seeds of his idea, Case began a series of partnerships with strategic businesses. Early alliances included Tandy Corporation, a maker of IBM-clones, and Apple Computer. By 1989, Quantum introduced its America Online network, and by 1991 had changed the company name to America Online. AOL then began focusing on growth, concentrating mainly on services for IBM compatibles and Apple computers.

AOL has been called a niche-based product, and its early alliances with specific groups explains why. One of AOL's first joint ventures was with The Tribune Company, owner of the *Chicago Tribune* newspaper. Another was with SeniorNet, a senior citizens organization formed primarily to get senior citizens to use computers. Explained Case in 1992, "We see ourselves as a series of specialized magazines catering to specific interests."

This narrower view did not last long. Between 1991 and 1992, AOL's subscriber base grew by 50 percent. Between 1992 and 1993, its annual revenues had increased by 50 percent, and its subscriber base by nearly 80 percent. As AOL continued to expand, it began offering Internet access. Although the Internet still was considered the private playground of techies, AOL hoped to market software to make using it easier. The company aggressively began to make its presence known through mailing campaigns, membership kits, and magazine inserts, easing the way for new subscribers to get online. Additionally, IBM, Apple, Tandy, and Compaq, among others, began incorporating AOL software into their products. It worked; revenues and users alike doubled that year and growth continued.

In 1998, AOL acquired Netscape Communications Corporation, forming a particularly vital alliance. AOL's Internet service has been criticized for being too simplistic and lacking in sophistication. As one media pundit put it, "AOL is the gateway through which the unwashed masses have flooded the Internet." AOL's marketing skills and popularity, coupled with Netscape's cutting-edge technology, made a formidable competitor for Microsoft, with ideas for information appliances (i.e., Internet set-top boxes) that compete with Microsoft's WebTV.

AOL's growth continues today. Partnerships with news media, manufacturers (including Dell Computer Corporation and U.S. Robotics), television networks, and others ensure AOL's place on top of the online services heap. In addition, in 2000 America Online jump-started the new century by proposing the world's largest corporate merger – a $172 *billion* deal joining AOL, the world's largest Internet company, with Time Warner, one of the world's largest media conglomerates.

With all its success and growth, perhaps AOL's ultimate nod of recognition as a household catchword comes from Hollywood. In December 1998, a movie starring Tom Hanks and Meg Ryan was released. The title? *You've Got Mail*, what else?

IN BRIEF

 www.scsite.com/dc2001/ch2/brief.htm

WEB INSTRUCTIONS: *To display this page from the Web, launch your browser and enter the URL, www.scsite.com/dc2001/ch2/brief.htm. Click the links for current and additional information. To listen to an audio version of this IN BRIEF, click the Audio button to the right of the title, IN BRIEF, at the top of the page. To play the audio, RealPlayer must be installed on your computer (download by clicking here).*

1. What Is Application Software?

Application software consists of programs designed to perform specific tasks for users. Application software can be grouped into four major categories: productivity software; graphics and multimedia software; home, personal, and educational software; and communications software.

2. How Is a Software Application Started?

Both the Microsoft Windows and the Apple Macintosh operating systems use the concept of a desktop. The **desktop** is an onscreen work area with common graphical elements such as **icons**, **buttons**, menus, links, windows, and **dialog boxes**. A software application can be started by clicking its program name on a **menu**, or list of **commands**. Clicking the program name instructs the operating system to transfer the program's instructions from a storage medium into memory. Once started, the application displays in a window on the desktop. A **window** is a rectangular area of the screen that is used to show the program, data, and/or information.

3. What Are Key Features of Productivity Software?

Productivity software makes people more effective and efficient while performing daily activities. **Word processing software** is used to **create**, **edit**, and **format** documents that consist primarily of text. **Spreadsheet software** organizes numeric data in a **worksheet** made up of rows and columns. **Database software** is used to create a **database**, an organized collection of data that can be accessed, retrieved, and used. **Presentation graphics software** creates documents called **presentations** that communicate ideas, messages, and other information to a group. A **personal information manager (PIM)** is software that includes an **appointment calendar**, **address book**, and **notepad** to help organize personal information. **Project management software** is used to plan, schedule, track, and analyze the progress of a project. **Accounting software** helps companies record and report their financial transactions.

4. What Are Key Features of Graphics and Multimedia Software?

Power users often use software that allows them to work with graphics and multimedia. **Computer-aided design (CAD) software** assists in creating engineering, architectural, and scientific designs. **Desktop publishing (DTP)** software is used in designing and producing sophisticated documents. **Paint software** is used to draw graphical images with various tools, while **image editing software** provides the capability of modifying existing images. **Video editing software** and **audio editing software** modify **video** and **audio** segments called *clips*. **Multimedia authoring software** creates electronic interactive presentations that can include text, images, video, audio, and animation. **Web page authoring software** is designed to create Web pages and to organize, manage, and maintain Web sites.

 www.scsite.com/dc2001/ch2/brief.htm

5. What Are Key Features of Home, Personal, and Educational Software?

Many applications are designed for use at home, or for personal or educational use. **Integrated software** combines several productivity software applications into a single package. **Personal finance software** is an accounting program that helps pay bills, balance a checkbook, track income and expenses, follow investments, and evaluate financial plans. **Legal software** assists in the creation of legal documents and provides legal advice. **Tax preparation software** guides users through the process of filing federal taxes. **Personal DTP software** helps develop conventional documents by asking questions, offering predefined layouts, and providing standard text. **Photo-editing software** is used to edit digital photographs. **Clip art/image galleries** are collections of clip art and photographs. **Home design/landscaping software** assists with design or remodeling. **Educational software** teaches a particular skill, **reference software** provides information, and **personal computer entertainment software** is designed to support a hobby or provide amusement.

6. What Are Key Features of Communications Software?

One of the more valuable aspects of software is its capability of supporting communications. **Groupware** identifies any type of software that helps groups of people on a network collaborate on projects and share information. **Electronic mail software** is used to create, send, receive, forward, store, print, and delete **e-mail (electronic mail)**. A **Web browser** is a software application used to access and view Web pages.

7. Why Is the World Wide Web Important?

The **World Wide Web (WWW)** is a tremendous resource for products, services, and information. The **Web** consists of a worldwide collection of electronic documents called Web pages. A **Web page** can contain text, graphical images, sound, video, and connections to other documents. A **hyperlink**, or **link**, is a built-in connection on a Web page. Links can be used to obtain information in a *nonlinear* way; that is, to make associations between topics instead of moving sequentially through topics. When a link is clicked, the location associated with it displays on the screen. Navigating using links is called **surfing the Web**.

8. How Is a Web Browser Used?

When a Web browser is started, a connection to the Internet typically is established through an **Internet service provider (ISP)** or an **online service**. The browser retrieves and displays a **home page**, or starting page. Other Web pages can be viewed in the browser window by clicking a link on the home page, typing the **Uniform Resource Locator (URL)** for a Web page, or using buttons on the browser toolbar.

IN BRIEF

www.scsite.com/dc2001/ch2/brief.htm

Student Exercises
WEB INFO
IN BRIEF
KEY TERMS
AT THE MOVIES
CHECKPOINT
AT ISSUE
MY WEB PAGE
HANDS ON
NET STUFF
Special Features
TIMELINE 2001
GUIDE TO WWW SITES
MAKING A CHIP
E-COMMERCE 2001
BUYER'S GUIDE 2001
CAREERS 2001
TRENDS 2001
LEARNING GAMES
CHAT
INTERACTIVE LABS
NEWS
HOME

9. How Do You Search for Information on the Web?

A **search engine** is a software program that finds Web sites, Web pages, and Internet files. To look up the URL of a page or pages, a word or phrase called **search text** or **keywords** is entered in the search engine text box. The search engine then displays a list of Web pages that contain the word or phrase entered.

10. What Learning Aids Are Available with Software Applications?

Many software applications and Web sites provide learning aids such as online Help, FAQs, tutorials, and wizards. **Online Help** is the electronic equivalent of a user manual. **FAQs** (Frequently Asked Questions) provide answers to common queries. **Tutorials** are step-by-step instructions that show how to use an application. A **wizard** is an automated assistant that helps complete a task by asking questions and then performing actions based on the answers.

KEY TERMS

www.scsite.com/dc2001/ch2/terms.htm

WEB INSTRUCTIONS: *To display this page from the Web, launch your browser and enter the URL, www.scsite.com/dc2001/ch2/terms.htm. Scroll through the list of terms. Click a term to display its definition and a picture. Click KEY TERMS on the left to redisplay the KEY TERMS page. Click the TO WEB button for current and additional information about the term from the Web. To see animations, Shockwave and Flash Player must be installed on your computer (download by clicking here).*

accounting software (2.22)
address book (2.21, 2.34)
application (2.2)
application service provider (ASP) (2.43)
appointment calendar (2.21)
audio (2.26)
audio editing software (2.26)
AutoSave (2.9)
bar charts (2.14)
bold (2.9)
border (2.10)
browser (2.34)
button (2.4)
cell (2.12)
charting (2.14)
clip art (2.10)
clip art/image gallery (2.31)
Clipboard (2.8)
clip gallery (2.19)
color library (2.25)
column charts (2.14)
commands (2.4)
computer-aided design (CAD) software (2.24)
context-sensitive (2.44)
context-sensitive menus (2.6)
creating (2.8)
criteria (2.17)
currency (2.16)
data type (2.16)
database (2.15)
database management system (DBMS) (2.15)
database software (2.15)
date (2.16)
desktop (2.4)
desktop publishing (DTP) software (2.24)
dialog box (2.6)
downloading (2.37)
editing (2.8)
educational software (2.32)
electronic mail software (2.33)
e-mail (electronic mail) (2.33)
e-mail address (2.33)
FAQ (2.45)
field (2.15)
field length (2.16)
file (2.10)
file name (2.10)
find (2.10)
font (2.9)
font size (2.9)
font style (2.9)
footer (2.11)
formatting (2.9)
formula (2.13)
function (2.13)
graphical user interface (GUI) (2.3)
groupware (2.32)

SUITE: Collection of individual application software packages sold as a single package. Two popular software suites are Microsoft Office 2000 and Lotus SmartSuite. (2.21)

header (2.11)
hit (2.40)
home design/landscaping software (2.31)
home page (2.37)
http:// (2.40)
hyperlink (2.16, 2.37)
hypertext transfer protocol (2.40)
icon (2.4)
illustration software (2.25)
image editing software (2.25)
import (2.10)
integrated software (2.28)
Internet service provider (ISP) (2.36)
keywords (2.40)
label (2.12)
legal software (2.29)
line charts (2.14)
link (2.16, 2.37)
macro (2.13)
mailbox (2.33)
mail server (2.33)
margins (2.10)
memo (2.16)
menu (2.4)
multimedia authoring software (2.26)
notepad (2.21)
numeric (2.16)
online Help (2.44)
online service (2.36)
operating system (2.3)
page layout (2.24)
paint software (2.25)
personal computer entertainment software (2.32)
personal DTP software (2.30)
personal finance software (2.28)
personal information manager (PIM) (2.21)
photo-editing software (2.30)
pie charts (2.15)
point (2.9)
presentation graphics software (2.18)
printing (2.10)
productivity software (2.7)
project management software (2.22)
query (2.17)
record (2.15)
reference software (2.32)
replace (2.10)
saving (2.9)
scrolling (2.10)
search (2.10)
search engine (2.40)
search text (2.40)
shortcut menus (2.6)
signature (2.34)
software application (2.2)
software package (2.2)
sort (2.17)
spelling checker (2.10)
spreadsheet software (2.12)
structure (2.16)
submenu (2.4)
suite (2.21)
surfing the Web (2.38)
system software (2.2)
tables (2.15)
tax preparation software (2.29)
text (2.16)
title bar (2.4)
tutorials (2.45)
Uniform Resource Locator (URL) (2.39)
user-ID (2.34)
user interface (2.3)
user name (2.34)
validation (2.17)
value (2.13)
video (2.26)
video editing software (2.26)
Web (2.35)
Webware (2.42)
Web application (2.42)
Web browser (2.34)
Web page (2.35)
Web page authoring software (2.27)
Web server (2.40)
Web site (2.35)
what-if analysis (2.14)
window (2.4)
Windows (2.3)
wizard (2.45)
word processing software (2.8)
wordwrap (2.10)
worksheet (2.12)
World Wide Web (WWW) (2.35)

Student Exercises
WEB INFO
IN BRIEF
KEY TERMS
AT THE MOVIES
CHECKPOINT
AT ISSUE
MY WEB PAGE
HANDS ON
NET STUFF

Special Features
TIMELINE 2001
GUIDE TO WWW SITES
MAKING A CHIP
E-COMMERCE 2001
BUYER'S GUIDE 2001
CAREERS 2001
TRENDS 2001
LEARNING GAMES
CHAT
INTERACTIVE LABS
NEWS
HOME

CHAPTER 1 2 3 4 5 6 7 8 9 10 11 12 13 14 INDEX

AT THE MOVIES

www.scsite.com/dc2001/ch2/movies.htm

WELCOME to VIDEO CLIPS from CNN

WEB INSTRUCTIONS: *To display this page from the Web, launch your browser and enter the URL, www.scsite.com/dc2001/ch2/movies.htm. Click a picture to view a video. After watching the video, close the video window and then complete the exercise by answering the questions about the video. To view the videos, RealPlayer must be installed on your computer (download by clicking here).*

1 Internet Spam

Business advertisements take many forms. One form of advertisement and Internet communication that has been difficult for the Federal Trade Commission to regulate and Congress to legislate against is spam. What is spam? Have you ever received spam? Why would a company prefer e-mail advertisements to hard copy? If one form of advertisement is regulated, should all forms be regulated? What is the difference between unsolicited e-mail and junk mail? How might a U.S. Constitutional amendment protect the right to spam? What recourse do you have against spam?

2 History Software

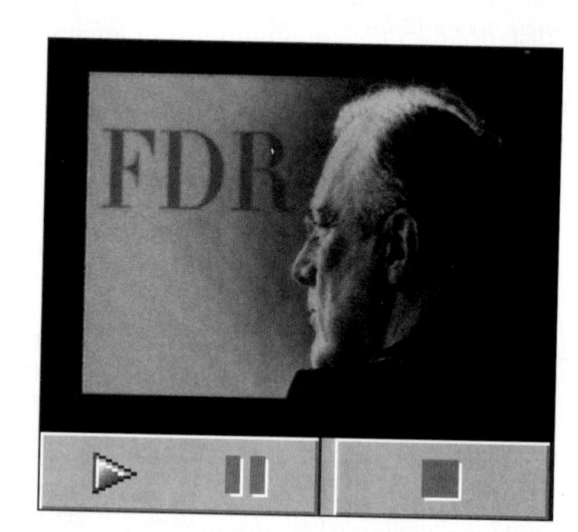

Have any of your history classes used software such as shown in the video clip? Do you think seeing film of speeches makes the person and event you are studying more real? How much interactivity do you think is necessary in software that teaches a subject? Would you rather just view the software or do you want to interact with it? How can you use software such as described in the video to research subject matter and prepare a report? What would be the best way to use the software shown in the video for your history assignments? In what ways could you use integrated application software suites, such as Microsoft Office or Corel PerfectSuite, with history software to enhance your reports? How would you reference the multimedia CD in a footnote or endnote?

3 WebTV

What is WebTV? Is it expensive? How can users interact online by using WebTV? Is it compatible with Windows 98? What is the relationship between the Net Channel and WebTV? Explain how WebTV expands Microsoft's presence into the consumer electronics arena? Would you want to use WebTV to access the Internet?

CHECKPOINT

www.scsite.com/dc2001/ch2/check.htm

WEB INSTRUCTIONS: *To display this page from the Web, launch your browser and enter the URL, www.scsite.com/dc2001/ch2/check.htm. Click the links for current and additional information. To experience the animation and interactivity, Shockwave and Flash Player must be installed on your computer (download by clicking here).*

Label the Figure

Instructions: *Identify the indicated elements in the Windows 98 graphical user interface.*

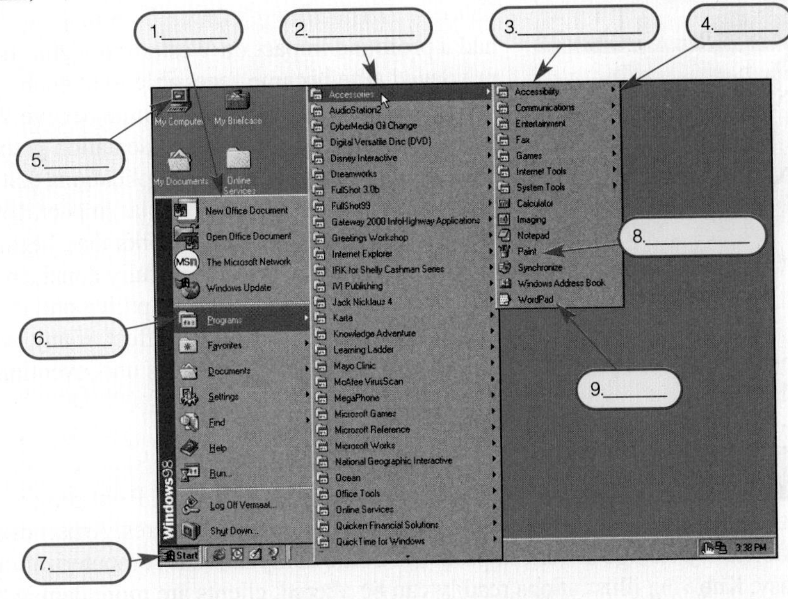

Matching

Instructions: *Match each application software feature from the column on the left with the best description from the column on the right.*

_______ 1. wordwrap

_______ 2. what-if analysis

_______ 3. validation

_______ 4. clip gallery

_______ 5. query

a. Database feature that compares data to a set of defined rules or values to determine if it is acceptable.

b. Presentation graphics feature consisting of images, pictures, and clips that can be incorporated into slides.

c. Word processing feature used to locate all occurrences of a particular character, word, or phrase.

d. Spreadsheet feature that displays data relationships in a graphical, rather than numerical, form.

e. Word processing feature that allows typing continually without pressing the ENTER key at the end of each line.

f. Database feature that is a specific set of instructions for retrieving data.

g. Spreadsheet feature in which certain values are altered to reveal the effects of those changes.

Short Answer

Instructions: *Write a brief answer to each of the following questions.*

1. How are creating, editing, and formatting a word processing document different? __________ What is the Clipboard and during which activity is it usually used? __________
2. What are the major advantages of a software suite? __________ How is a software suite different from integrated software? __________
3. Why do professional designers and graphic artists use DTP software instead of word processing packages? __________ What is a color library? __________
4. What is an Internet e-mail address? __________ What two parts of an e-mail address are separated by the at (@) sign? __________
5. What is online Help? __________ How do FAQs, wizards, and tutorials help software users? __________

AT ISSUE

www.scsite.com/dc2001/ch2/issue.htm

WEB INSTRUCTIONS: *To display this page from the Web, launch your browser and enter the URL, www.scsite.com/dc2001/ch2/issue.htm. Click the links for current and additional information.*

ONE

Johann Guttenberg's invention of printing from moveable type had a profound impact on Western thought. Books once available only to a privileged elite became accessible to a much wider audience, thereby broadening the distribution of ideas. Some believe Web page authoring software, desktop publishing software, presentation graphics software, multimedia authoring software, and other applications that help people communicate more effectively will have a similar impact. Unpublished authors can use these applications to produce works that, because of their professional-looking appearance, are thoughtfully considered and extensively circulated. Will these applications really help give previously unheard speakers a louder voice? Why or why not? What effect, if any, will these applications have on the delivery, and possible acceptance, of material that reflects unconventional, or not generally accepted, ideas?

TWO

Today, many commercial artists, creators of cartoons, book covers, and billboards use paint software. Paint programs authentically mimic art produced by hand, right down to brush strokes and surface textures. Artists can import graphic files and effortlessly change a work in progress. Some complain that computerization has made their work too easy. Knowing illustrations readily can be altered, clients are more demanding and less forgiving. Even worse, digital art has been denounced as having a bland quality that reflects little effort, feeling, or imagination. What is the future of paint software? Will it ever be widely accepted? Why or why not? How might paint software change commercial art? Will it ever be used by noncommercial artists? Why or why not?

THREE

Trained individuals have used polygraphs, or lie detectors, for years. Now, an application called Truster turns a PC into a polygraph. When a subject speaks into a microphone, a Truster algorithm measures vocal stress and reports relative veracity (Truth, Inaccurate, Slightly Inaccurate, Not Sure, False) on the computer screen. Sales have been brisk, much to some people's dismay. Polygraph use legally is restricted, but anyone can use Truster, even an instructor trying to find out if the dog really ate your homework. When would you be comfortable, and uncomfortable, with the use of lie-detecting software? Why? Truster claims an 85 percent accuracy rate. As accuracy rates improve, will you be more or less accepting of lie-detecting software? Why?

FOUR

Industry experts claim more than 80 percent of entertainment and educational software is purchased by males. As computer literacy becomes increasingly important to career advancement, this figure indicates an added obstacle for females entering the workplace. Analysts insist much of the disparity in computer use is a result of the nature of entertainment and educational software. Men like the shoot-'em-up, win-lose character of most entertainment software. Women, on the other hand, prefer exploratory, less competitive software. Unless software adjusts, analysts predict female interest in using computers for entertainment and education will continue to wane. Are the analysts right? Do males and females favor different types of software? If so, should software developers adapt their products? Why or why not? Is it important to modify educational/entertainment software to meet the interests of different groups? Why?

FIVE

Software developed for elementary school children, kindergartners, and even preschoolers has won the praise of educators and child psychologists. Yet, controversy has erupted about "how young is too young" over Knowledge Adventure's® JumpStart Baby™ program, which is aimed at children nine- to twenty-four-months old. According to developers, JumpStart Baby™ makes even young children comfortable with computers. The software is tailored to tots and, supporters insist, certainly is more beneficial than an equal amount of time spent watching television. Knowledge Adventure® advocates that this software is designed as "lap-ware," meaning it is intended to be used by baby and parent together, which can serve as a springboard to stimulating activities rich in communication and social interaction during the critical developmental years. Critics feel, however, that digital blocks are no substitute for the real thing. Children need to experience the real world, not a cyber representation. When should children be introduced to computers? Why? How can parents ensure that a child's computer experience is worthwhile?

Student Exercises
WEB INFO
IN BRIEF
KEY TERMS
AT THE MOVIES
CHECKPOINT
AT ISSUE
MY WEB PAGE
HANDS ON
NET STUFF
Special Features
TIMELINE 2001
GUIDE TO WWW SITES
MAKING A CHIP
E-COMMERCE 2001
BUYER'S GUIDE 2001
CAREERS 2001
TRENDS 2001
LEARNING GAMES
CHAT
INTERACTIVE LABS
NEWS
HOME

MY WEB PAGE

www.scsite.com/dc2001/ch2/myweb.htm

WEB INSTRUCTIONS: *The icons to the left of the numbered activities on this page indicate the learning system availability. If you are using WebCT, follow the instructions in the exercises below. If you are using the textbook Web site or CyberClass, launch your browser, enter the URL www.scsite.com/dc2001/ch2/myweb.htm, and then click the corresponding icon to the left of the exercise.*

1. Practice Test: Click My Web Page and then click Chapter 2. Click Practice Test. Answer each question and then click the Save Answer button. When completed, click the Finish button. Click the OK button to submit the quiz for grading. Click the View Results button, and then make a note of any missed answers.

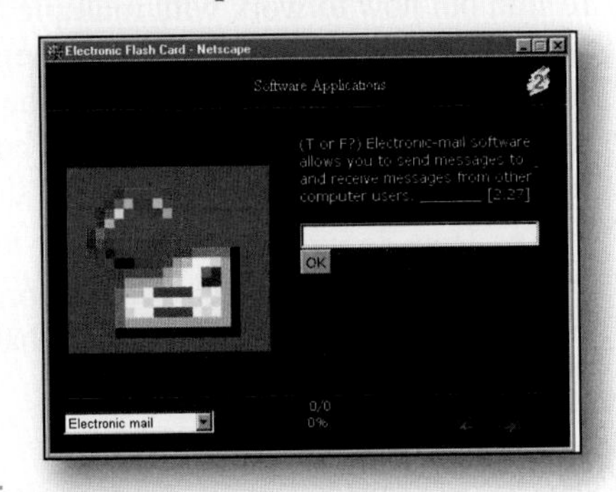

2. Web Guide: Click My Web Page and then click Chapter 2. Click Web Guide to display the Guide to WWW Sites page. Click Shopping and then click eBay. Search for Software. Use your word processing program to prepare a brief report on the software programs you found. Submit your assignment to your instructor.

3. Scavenger Hunt: Click My Web Page and then click Chapter 2. Click Scavenger Hunt. Print a copy of the Scavenger Hunt page; use this page to write down your answers as you search the Web. Submit your completed page to your instructor.

4. Who Wants to Be a Computer Genius?: Click My Web Page and then click Chapter 2. Click Computer Genius to find out if you are a computer genius. For directions, click the How to Play button. When you are ready to play, click the Game button. Submit your score to your instructor.

5. Wheel of Terms: Click My Web Page and then click Chapter 2. Click Wheel of Terms to reinforce important terms you learned in this chapter by playing the Shelly Cashman Series version of this popular game. For directions, click the How to Play button. When you are ready to play, click the Game button. Submit your score to your instructor.

6. Career Corner: Click My Web Page and then click Chapter 2. Click Career Corner to display the Penn State's Career Services page. Click a link of your choice. Write a brief report on the information you found. Submit the report to your instructor.

7. Search Sleuth: Click My Web Page and then click Chapter 2. Click Search Sleuth to learn search techniques that will help make you a research expert. Submit the completed assignment to your instructor.

8. Crossword Puzzle Challenge: Click My Web Page and then click Chapter 2. Click Crossword Puzzle Challenge. Complete the puzzle to reinforce skills you learned in this chapter. For directions, click the How to Play button. When you are ready to play, click the Game button. Submit the completed puzzle to your instructor.

9. Flash Cards: This exercise uses CyberClass. Follow the instructions at the top of this page and then click the CyberClass icon to the left; or ask your instructor for logon instructions. Click Flash Cards on the Main Menu. Click the plus sign. Answer all the questions in any two subjects of your choice. Click the Close button in the upper-right corner to close all windows. Notice the many other exercises at this site. Complete all other exercises as assigned by your instructor.

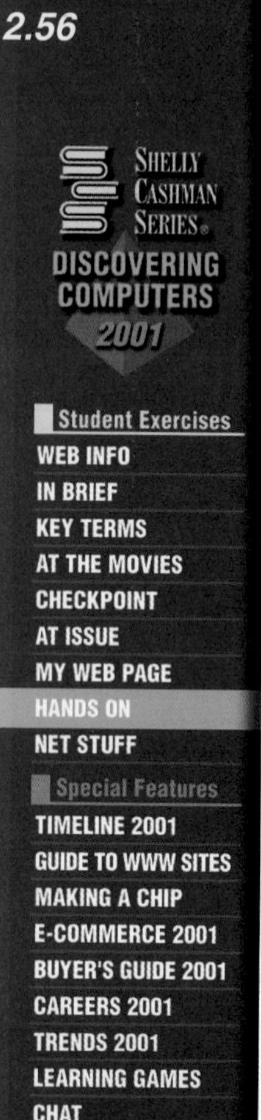

HANDS ON

www.scsite.com/dc2001/ch2/hands.htm

WEB INSTRUCTIONS: *To display this page from the Web, launch your browser and enter the URL, www.scsite.com/dc2001/ch2/hands.htm. Click the links for current and additional information.*

One

Working with Application Programs

This exercise uses Windows 98 procedures. Windows is a multitasking operating system, meaning you can work on two or more applications that reside in memory at the same time. To find out how to work with multiple application programs, click the Start button on the taskbar, and then click Help on the Start menu. Click the Contents tab. Click the Exploring Your Computer book. Click the Work with Programs book. Click an appropriate topic to answer each of the following questions:

- How do you start a program?
- How do you switch between programs?
- How do you quit a program that is not responding?
- How do you quit a program?

Close the Windows Help window.

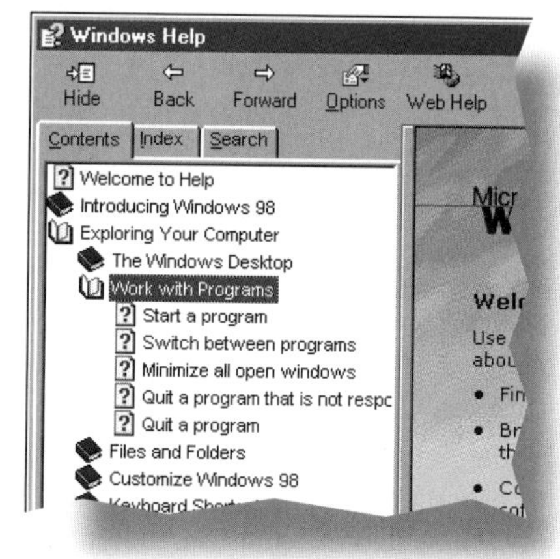

Two

Creating a Word Processing Document

WordPad is a simple word processor included with the Windows operating system. To create a document with WordPad, click the Start button on the taskbar, point to Programs on the Start menu, point to Accessories on the Programs submenu, and then click WordPad on the Accessories submenu. If necessary, when the WordPad window displays, click its Maximize button. Click View on the menu bar. If a check mark does not display before Toolbar, click it. Type a complete answer to one of the AT ISSUE questions posed in this chapter or in Chapter 1. Your answer should be at least two paragraphs long. Press the TAB key to indent the first line of each paragraph and the ENTER key to begin a new paragraph. To correct errors, press the BACKSPACE key to erase to the left of the insertion point and press the DELETE key to erase to the right. To insert text, move the I-beam mouse pointer to where the text should be inserted, click, and then begin typing. At the end of your document, press the ENTER key twice and then type your name. When your document is complete, save it by inserting a floppy disk into drive A, clicking the Save button on the toolbar, typing `a:\h2-2` in the File name text box in the Save As dialog box, and then clicking the Save button. Click the Print button on the toolbar to print your document. Close WordPad.

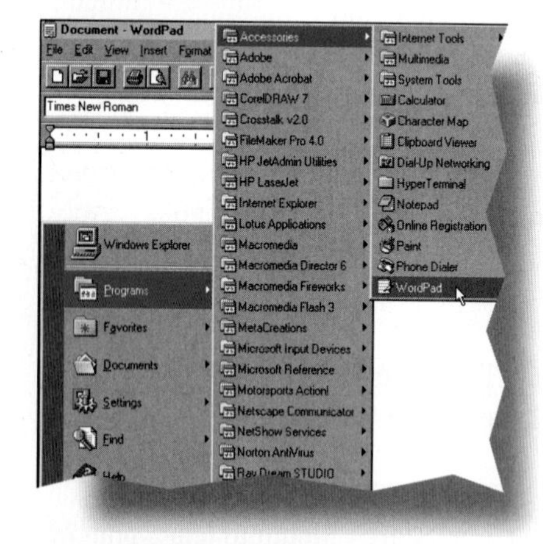

Three

Using WordPad's Help

This exercise uses Windows 98 procedures. Open WordPad as described in HANDS ON 2 above. Click Help on WordPad's menu bar and then click Help Topics. When the Help Topics, WordPad Help window displays, click the Index tab. Type `saving documents` in the text box and then press the ENTER key. Click To save changes to a document in the Topics Found window and then click the Display button. How can you save changes to a document? How can you save an existing document with a new name? Close the Help window and WordPad.

Four

Productivity Software Products

What productivity software packages are on your computer? Click the Start button on the taskbar and point to Programs on the Start menu. Scan the Programs submenu (if necessary, point to the arrow at the top or bottom of the submenu to move the submenu up or down) for the names of popular productivity packages. Write down the package name and the type of software application (see the chart on page 2.7 for help). When you are finished, click an empty area of the desktop.

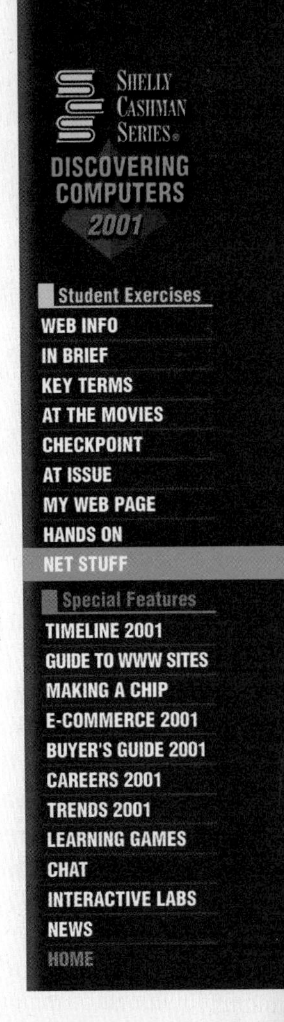

CHAPTER 1 2 3 4 5 6 7 8 9 10 11 12 13 14 INDEX

NET STUFF

www.scsite.com/dc2001/ch2/net.htm

WEB INSTRUCTIONS: *To display this page from the Web, launch your browser and enter the URL, www.scsite.com/dc2001/ch2/net.htm. To use the Word Processing lab or the Spreadsheet lab from the Web, Shockwave and Flash Player must be installed on your computer (download by clicking here).*

WORD PROCESSING LAB

1. Shelly Cashman Series Word Processing Lab

Follow the instructions in NET STUFF 1 on page 1.48 to start and use the Word Processing lab. If you are running from the Web, enter the URL, www.scsite.com/sclabs/menu.htm; or display this NET STUFF page (see instructions at the top of this page) and then click the WORD PROCESSING LAB button.

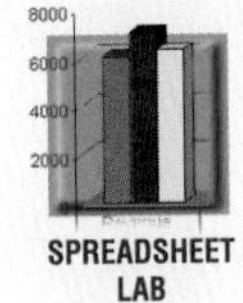
SPREADSHEET LAB

2. Shelly Cashman Series Spreadsheet Lab

Follow the instructions in NET STUFF 1 on page 1.46 to start and use the Working with Spreadsheets lab. If you are running from the Web, enter the URL, www.scsite.com/sclabs/menu.htm; or display the NET STUFF page (see instructions at the top of this page) and then click the SPREADSHEET LAB button.

SET UP E-MAIL

3. Setting Up an E-Mail Account

The fastest growing software application may be electronic mail (e-mail). One free e-mail service reports 30 million current subscribers with an additional 80,000 joining every day. To set up a free e-mail account, click the SET UP E-MAIL button. Follow the procedures to establish an e-mail account. When you are finished, send yourself an e-mail.

IN THE NEWS

4. In the News

It is a computer user's nightmare – a button is clicked accidentally or a key is pressed unintentionally and an important message, document, or presentation is deleted. Happily, software called Save Butt can restore a sound night's sleep. Save Butt continuously copies open files to the hard drive. Not only are files kept safe, but you always can return to earlier versions of a project. Click the IN THE NEWS button and read a news article about a new software program. Who is introducing the program? What is the program called? What does it do? Who will benefit from using this software? Why? Where can the software be obtained? Would you be interested in this software? Why or why not?

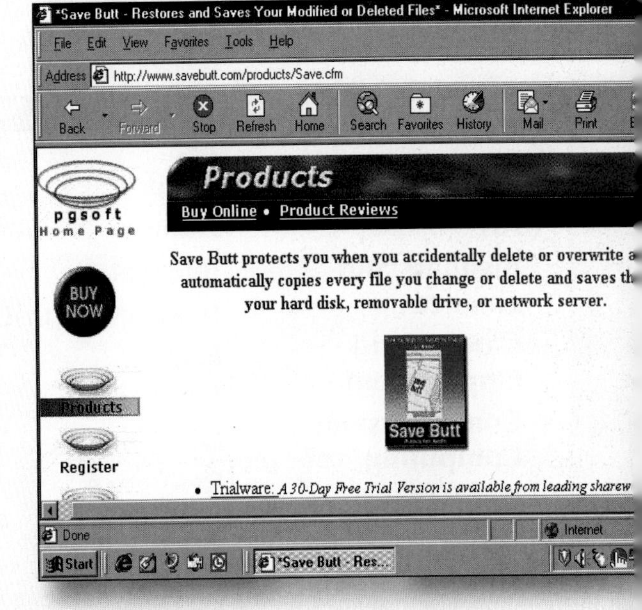

WEB CHAT

5. Web Chat

A noted university psychology professor is advocating an Intelligent Essay Assessor, a software program that grades written essays. To establish a foundation, the program is given examples of good and bad essays that have been manually graded, and then is supplied with selections from experts. With this background, the Intelligent Essay Assessor understands student essays by comparing word usage to patterns in the samples. An essay that relates the same knowledge as that in the admirable examples receives a high score. In what subject areas, or at what levels, could an Intelligent Essay Assessor be used effectively? Why? What might be some drawbacks of the software? How could these drawbacks be addressed? Would you be willing to have your work evaluated by a software program? Why or why not? Click the WEB CHAT button to enter a Web Chat discussion related to this topic.

GUIDE TO WORLD WIDE WEB SITES

The World Wide Web is an exciting and highly dynamic medium. Every day, new Web sites are added, existing ones are changed, and still others cease to exist. Because of this, you may find that a URL listed here has changed or no longer is valid. In order to offer the most up-to-date information available, the *Discovering Computers 2001* Web site is continually checked and updated.

WEB INSTRUCTIONS: *To gain World Wide Web access to the most current version of the sites included in this special feature, launch your browser and enter the URL, www.scsite.com/dc2001/ch2/websites.htm.*

Categories

- Animation
- Art
- Business and Finance
- Careers and Employment
- Computers and Computing
- Digital Music
- Education
- Entertainment
- Environment
- Fitness
- Government and Politics
- Health and Medicine
- History
- Humor
- Internet
- Internet Security
- Law
- Museums
- News Sources
- Reference
- Robotics
- Science
- Shopping
- Sports
- Travel
- Unclassified
- Weather
- Zines

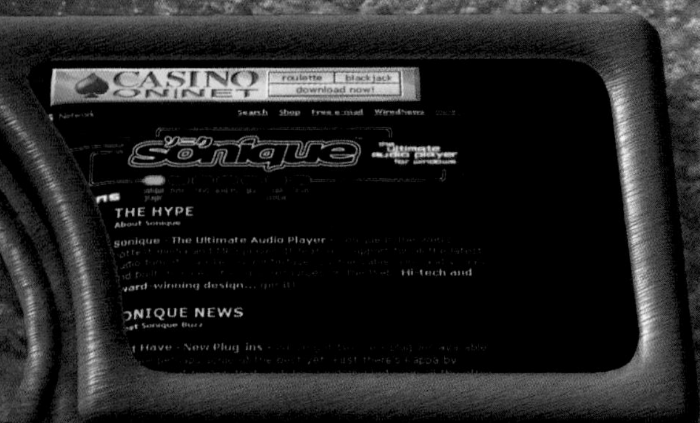

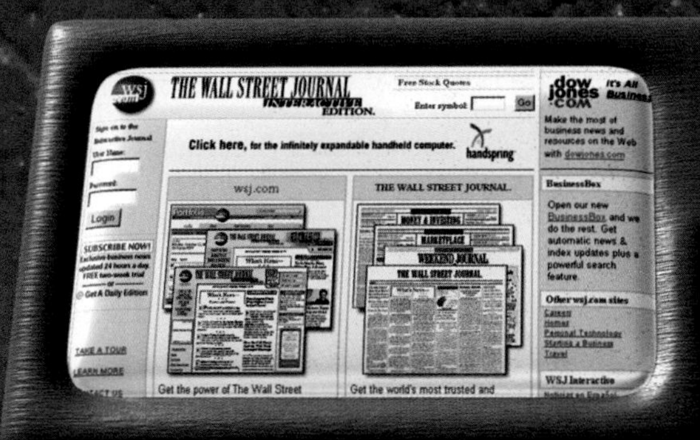

CATEGORY/SITE NAME	LOCATION	COMMENT
Animation		
Animation Express	www.animationexpress.com	More cool animations
rgb gallery	hotwired.lycos.com/rgb	Art animations
Shockwave	www.shockwave.com	Cool shockwave animations
Art		
Fine Art Forum	www.msstate.edu/Fineart_Online/home.html	Art and technology news
Leonardo da Vinci	metalab.unc.edu/wm/paint/auth/vinci	Works of the famous Italian artist and thinker
The Andy Warhol Museum	www.clpgh.org/warhol	Famous American pop artist
The WebMuseum	www.louvre.fr/louvrea.htm	Web version of Louvre Museum, Paris
World Art Treasures	sgwww.epfl.ch/BERGER	100,000 slides organized by civilization
World Wide Arts Resources	wwar.com	Links to many art sites
Business and Finance		
FinanCenter, Inc.	www.financenter.com	Personal finance information
FINWEB All Business Network	www.all-biz.com	Links to Web business information
PC Quote	www.pcquote.com	Free delayed stock quotes
Raging Bull	www.ragingbull.com	Real-time stock quotes
Stock Research Group	www.smallcapcenter.com	Investment information
The Wall Street Journal	interactive.wsj.com	Financial news page
Yahoo! Finance	quote.yahoo.com	Free delayed stock quotes
Careers and Employment		
Career Magazine	www.careermag.com	Career articles and information
CareerMosaic®	www.careermosaic.com	Jobs from around the world
CareerPath	new.careerpath.com	Job listings from U.S. newspapers
E-Span Job Options	www.joboptions.com	Searchable job database
Monster.com	www.monster.com	Jobs finder
Computers and Computing		
Computer companies	Insert name or initials of most computer companies between www. and .com to find their Web site. Examples: www.ibm.com, www.microsoft.com, www.dell.com.	
Internet Guide	www.internet.com	E-Business and Technology Network
MIT Media Lab	www.media.mit.edu	Information on computer trends
PC Guide	www.pcguide.com	PC Reference Information
The Computer Museum	www.computerhistory.org	Exhibits and history of computing
Virtual Computer Library	www.utexas.edu/computer/vcl	Information on computers and computing
Virtual Museum of Computing	www.museums.reading.ac.uk/vmoc	History of computing and online computer based exhibits
Digital Music		
Live Concerts	www.liveconcerts.com	RealMedia streamed concerts
MP3.com	www.mp3.com	Music files
Sonique	www.sonique.com	MP3 player made by aliens!!
This American Life	www.thislife.org	Public Radio program

For an updated list: www.scsite.com/dc2001/ch2/websites.htm

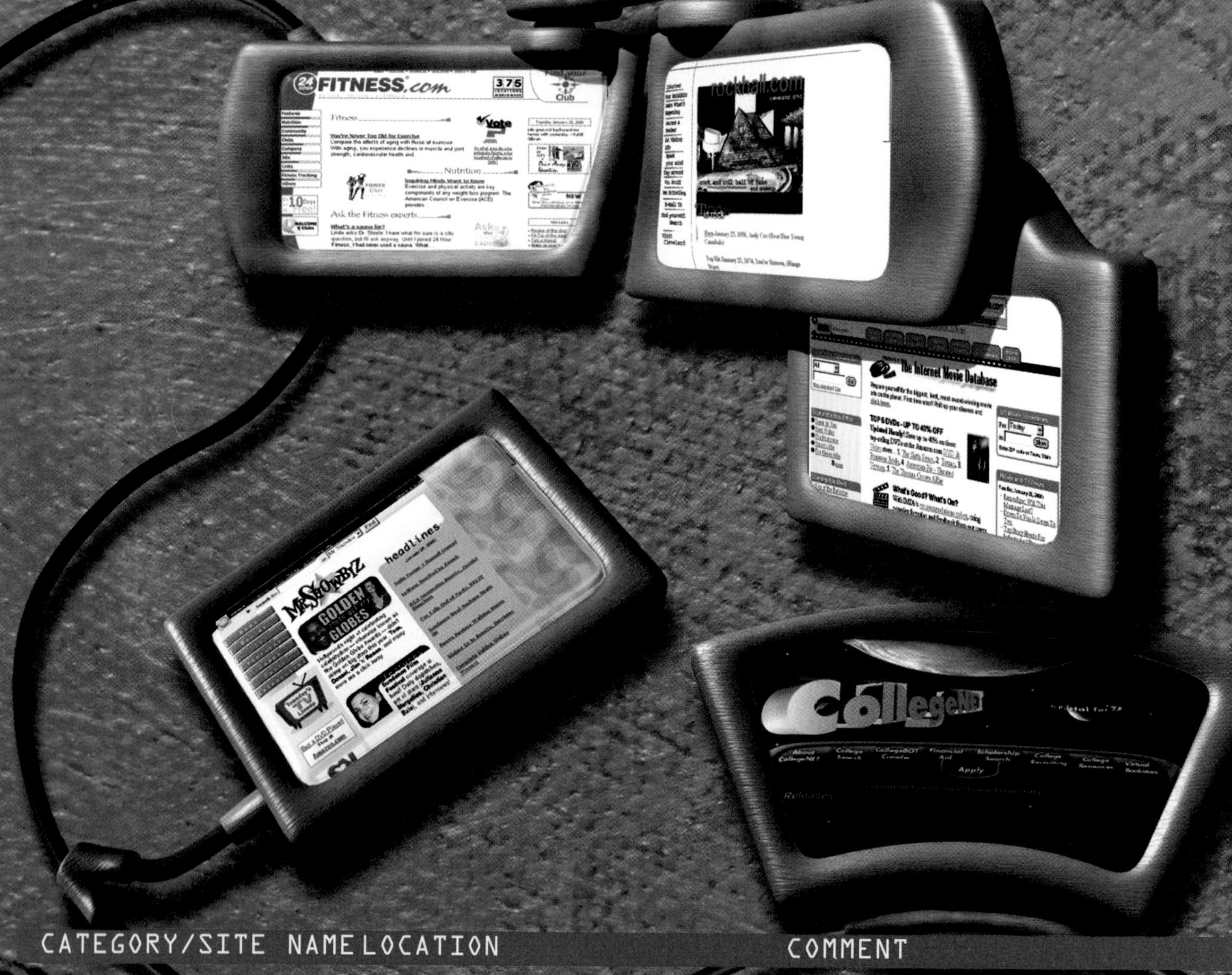

CATEGORY/SITE NAME	LOCATION	COMMENT
Education		
CollegeNET	www.collegenet.com	Searchable database of more than 2,000 colleges and universities
EdLinks	webpages.marshall.edu/~jmullens/edlinks.html	Links to many educational sites
The Open University	www.open.ac.uk	Independent study courses from the U.K.
Entertainment		
Classics World	www.bmgclassics.com/classics/index.shtml	Classical music information
Internet Movie Database	www.imdb.com	Movies
Internet Underground Music Archive	www.iuma.com	Underground music database
Mr. Showbiz	www.mrshowbiz.com	Information on latest films
Music Boulevard	www.musicblvd.com	Search for and buy all types of music
Playbill On-Line	www.playbill.com	Theater news
Rock & Roll Hall of Fame	www.rockhall.com	Cleveland Museum Site
Environment		
EnviroLink Network	www.envirolink.org	Environmental information
Greenpeace	www.greenpeace.org	Environmental activism
U.S. Environmental Protection Agency (EPA)	www.epa.gov	U.S. government environmental news
Fitness		
24 Hour Fitness	www.24hourfitness.com	A Health and Fitness Community
Global Fitness.com	www.globalfitness.com	Health and Fitness

For an updated list: www.scsite.com/dc2001/ch2/websites.htm

CATEGORY/SITE NAME	LOCATION	COMMENT
Government and Politics		
Canada Info	www.clo.com/~canadainfo/canada.html	List of Canadian Web Sites
CIA	www.odci.gov	Political and economic information on countries
Democratic National Committee	www.democrats.org/index.html	Democratic Party News
FedWorld	www.fedworld.gov	Links to U.S. government sites
PoliSci.com	www.polisci.com	Politics on the web
Republican National Committee	www.rnc.org	GOP Party News
The Library of Congress	www.loc.gov	Variety of U.S. government information
The White House	www.whitehouse.gov	Take tour and learn about occupants
U.S. Census Bureau	www.census.gov	Population and other statistics
United Nations	www.un.org	Latest UN projects and information
Health and Medicine		
Centers for Disease Control and Prevention (CDC)	www.cdc.gov	How to prevent and control disease
CODI	codi.buffalo.edu	Resource for disability products and services
The Interactive Patient	medicus.marshall.edu/medicus.htm	Simulates visit to doctor
Women's Medical Health Page	www.cbull.com/health.htm	Articles and links to other sites
History		
American Memory	rs6.loc.gov/amhome.html	American History
Historical Text Archive	www.geocities.com/Athens/Forum/9061/USA/usa.html	U.S. documents, photos, and databases
Virtual Library History	www.ukans.edu/history/VL	Organized links to history sites
World History Archives	www.hartford-hwp.com/archives/index.html	Links to history sites

For an updated list: www.scsite.com/dc2001/ch2/websites.htm

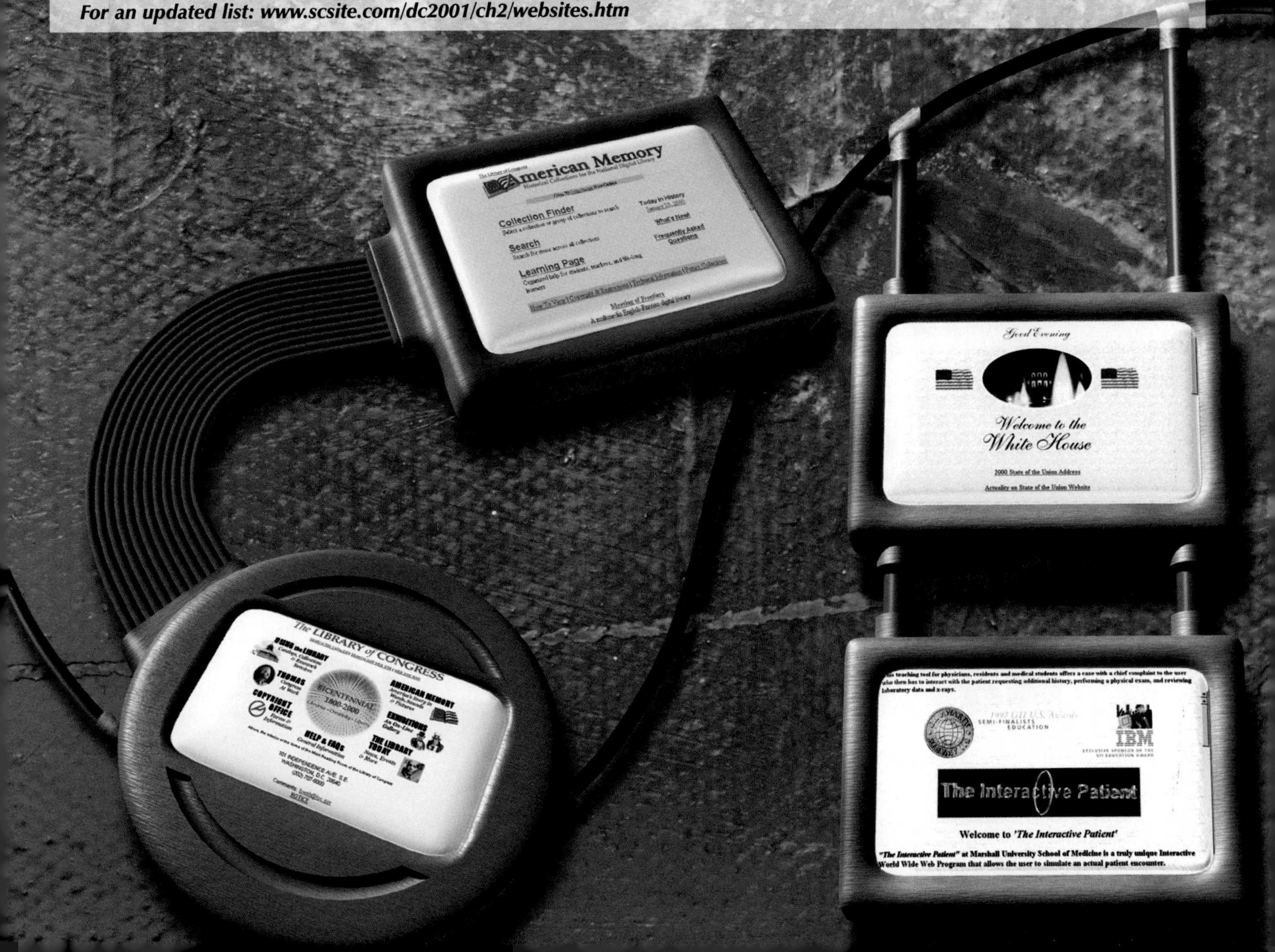

CATEGORY/SITE NAME	LOCATION	COMMENT
Humor		
Calvin & Hobbes Gallery	www.calvinandhobbes.com	Comic strip gallery
Comedy Central Online	www.comcentral.com	From comedy TV network
Late Show Top 10 Archive	marketing.cbs.com/lateshow/topten	David Letterman Top 10 lists
The Dilbert Zone	www.unitedmedia.com/comics/dilbert	Humorous insights about working
Internet		
Beginners Central Internet	northernwebs.com/bc	Beginner's guide to the Internet
Internet Glossary	www.matisse.net/files/glossary.html	Definitions of Internet terms
WWW Frequently Asked Questions	www.boutell.com/faq/oldfaq/index.html	Common Web questions and answers
Internet Security		
Computer Incident Advisory Capability (CIAC)	ciac.llnl.gov/ciac/CIACVirusDatabase.html	Virus database
F-secure Hoax warnings	www.datafellows.com/news/hoax.htm	Industry standard information source for new virus hoaxes and false alerts
Secure Solutions Experts (SSE)	www.sse.ie/securitynews.html	Over 60 of the best information security news sites, many of which are updated daily
Law		
APB News.com	www.apbnews.com	Crime, Justice & Safety News
Copyright Website	www.benedict.com	Provides copyright information
FindLaw	www.findlaw.com	Law resource portal
KuesterLaw	www.kuesterlaw.com	The Technology Law Resource
Privacy Rights Clearinghouse	www.privacyrights.org	Informational privacy issues
Museums		
The Smithsonian	www.si.edu	Information and links to Smithsonian museums
U.S. Holocaust Memorial Museum	www.ushmm.org	Dedicated to World War II victims
University of California Museum of Paleontology	www.ucmp.berkeley.edu	Great information on dinosaurs and other exhibits

For an updated list: www.scsite.com/dc2001/ch2/websites.htm

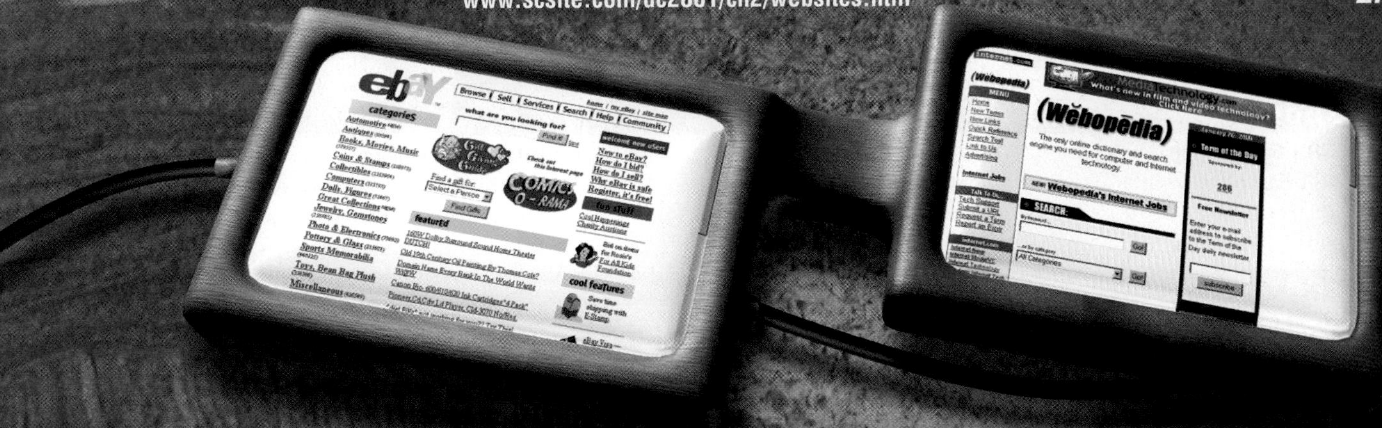

CATEGORY/SITE NAME	LOCATION	COMMENT
News Sources		
C/NET	www.cnet.com	Technology news
CNN Interactive	www.cnn.com	CNN all-news network
Pathfinder	www.pathfinder.com	Excerpts from Time-Warner magazines
The Electronic Newsstand	www.enews.com	Articles from worldwide publications
USA TODAY	www.usatoday.com	Latest U.S. and international news
Wired News	www.wired.com	Wired magazine online and HotWired Network
Reference		
About.com	www.about.com	Search engine and portal
AskEric Virtual Library	www.askeric.org/Virtual	Educational resources
AskJeeves	www.askjeeves.com	Search engine
Bartlett's Quotations	www.columbia.edu/acis/bartleby/bartlett	Organized, searchable database of famous quotes
Dictionary Library	www-math.uni-paderborn.de/dictionaries/ Dictionaries.html	Links to many types of dictionaries
Internet Public Library	www.ipl.org	Literature and reference works
The New York Public Library	www.nypl.org	Extensive reference and research material
Webopedia	www.webopedia.com	Online dictionary and search engine
Robotics		
KhepOnTheWeEb	khepontheweb.epfl.ch	Control a Robot with your Netscape Web Browser!
Robotics and Intelligent Machines Laboratory	robotics.eecs.berkeley.edu	Robotics
Science		
American Institute of Physics	www.aip.org	Physics research information
Exploratorium	www.exploratorium.edu	Interactive science exhibits
Internet Chemistry Index	www.chemie.de	List of chemistry information sites
National Institute for Discovery Science (NIDS)	www.accessnv.com/nids	Research of anomalous phenomena
Solar System Simulator	space.jpl.nasa.gov	JPL's spyglass on the cosmos
The NASA Homepage	www.nasa.gov	Information on U.S. space program
The Nine Planets	seds.lpl.arizona.edu/nineplanets/ nineplanets	Tour the solar system
Shopping		
Amazon.com	www.amazon.com	Books and gifts
Bartleby	www.bartleby.com	Great Books online
BizRate	www.bizrate.com	Rates e-commerce sites
BizWeb	www.bizweb.com	Search for products from more than 1,000 companies
CommerceNet	www.commerce.net	Index of products and services
Consumer World	www.consumerworld.org	Consumer Information
Ebay	www.ebay.com	Online auctions
Internet Bookshop	www.bookshop.co.uk	780,000 titles on more than 2,000 subjects
Internet Shopping Network	www.internet.net	Specialty stores, hot deals, computer products
PriceLine.com	www.priceline.com	Lots of stuff from groceries to airfare; name your price
The Internet Mall™	www.internet-mall.com	Comprehensive list of Web businesses

For an updated list: www.scsite.com/dc2001/ch2/websites.htm

CATEGORY/SITE NAME	LOCATION	COMMENT
Sports		
ESPN SportsZone	espn.go.com	Latest sports news
NBA Basketball	www.nba.com	Information and links to team sites
NFL Football	www.nfl.com	Information and links to team sites
Sports Illustrated	www.cnnsi.com	Leading sports magazine
Travel		
Excite City.Net	www.excite.com/travel	Guide to world cities
InfoHub WWW Travel Guide	www.infohub.com	Worldwide travel information
Lonely Planet Travel Guides	www.lonelyplanet.com	Budget travel guides and stories
Microsoft Expedia	expedia.msn.com	Complete travel resource
Travelocity℠	www.travelocity.com	Online travel agency
TravelWeb℠	www.travelweb.com	Places to stay
Unclassified		
Cool Site of the Day	cool.infi.net	Different site each day
Cupid's Network™	www.cupidnet.com	Links to dating resources
Taxi Cam	www.ny-taxi.com	NYC from a taxi web-cam
Where's George	www.wheresgeorge.com	Dollar Bill Locator
Weather		
INTELLICAST Guides	www.intellicast.com	International weather and skiing information
The Weather Channel	www.weather.com	National and local forecasts
Weather Underground	www.wunderground.com	Cool weather maps
Zines		
AFU & Urban Legends Archive	www.urbanlegends.com	Urban Legends
Breakup Girl	www.breakupgirl.com	Saving Love Lives all over the world
Rock School	www.rockschool.com	Everything you need to know about being in a rock band
The Smoking Gun	www.thesmokinggun.com	Confidential documents

For an updated list: www.scsite.com/dc2001/ch2/websites.htm

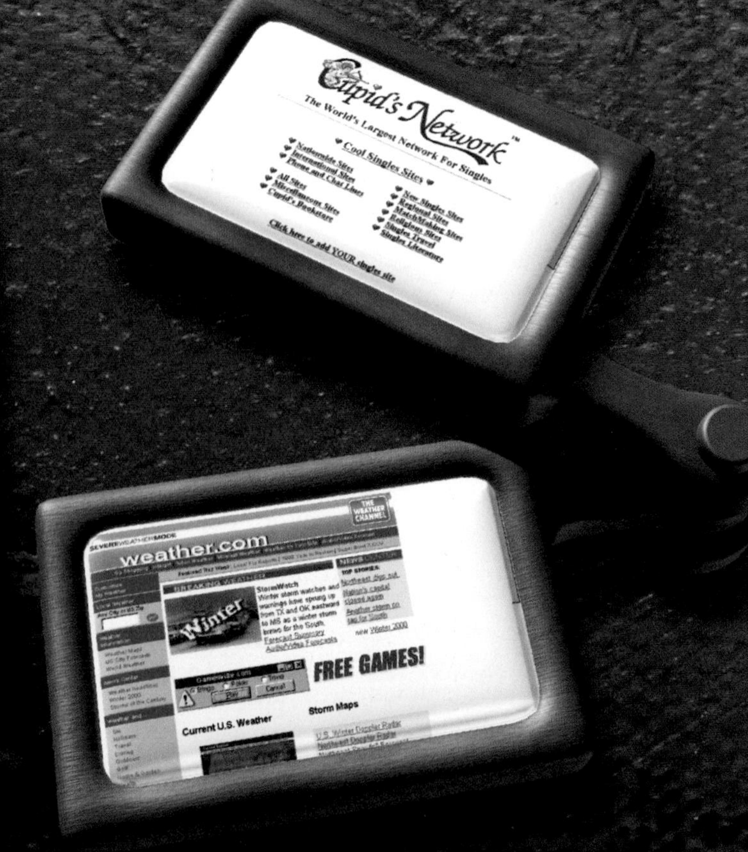

CHAPTER 3

THE COMPONENTS IN THE SYSTEM UNIT

OBJECTIVES

After completing this chapter, you will be able to:

- Identify the components in the system unit and explain their functions
- Explain how the CPU uses the four steps of a machine cycle to process data
- Compare and contrast various microprocessors on the market today
- Define a bit and describe how a series of bits is used to represent data
- Differentiate between the various types of memory
- Describe the types of expansion slots and expansion cards in the system unit
- Explain the difference between a serial and a parallel port
- Describe how buses contribute to a computer's processing speed

At some point during your professional career or personal endeavors, you probably will be involved in the decision to purchase a new computer or upgrade an existing computer. Thus, it is important that you understand the purpose of the components of a computer. As discussed in Chapter 1, a computer includes devices used for input, processing, output, storage, and communications. Many of these components are housed together in the system unit.

Chapter 3 presents the components in the system unit, describes how memory stores data, instructions, and information, and discusses the sequence of operations that occur when a computer executes an instruction. The chapter also includes a comparison of various microprocessors on the market today.

THE SYSTEM UNIT

The **system unit** is a box-like case that houses the electronic components of the computer that are used to process data. The system unit is made of metal or plastic and is designed to protect the electronic components from damage. On a desktop personal computer, the electronic components and most storage devices reside inside the system unit. Other devices, such as a keyboard, mouse, monitor, and printer, normally are located outside the system unit. A laptop computer houses almost all of its electronic components in the system unit (Figure 3-1).

At some point, you might find it necessary to open the system unit on a personal computer and replace or install a new component. For this reason, it is important that you have some familiarity with the inside of the system unit.

Figure 3-2 identifies some of the components inside a system unit, including the processor, memory module, several expansion cards, ports, and connectors. The processor, short for central processing unit, is the device that interprets and carries out the basic instructions that operate a computer. A memory module is a package that houses the memory that temporarily holds data and instructions while they are being processed by the CPU.

An expansion card is a circuit board that adds devices or capabilities to the computer. Three types of expansion cards found in most of today's personal computers are a sound card, a modem card, and a video card.

Finally, devices outside the system unit, such as a keyboard, monitor, printer, mouse,

Figure 3-1 *On desktop personal computers, the system unit usually is separated from the monitor and keyboard. Some models, called tower models, are positioned vertically on the floor. Other system units are designed to be placed on top of a desk. On laptop computers, the keyboard is built into the system unit.*

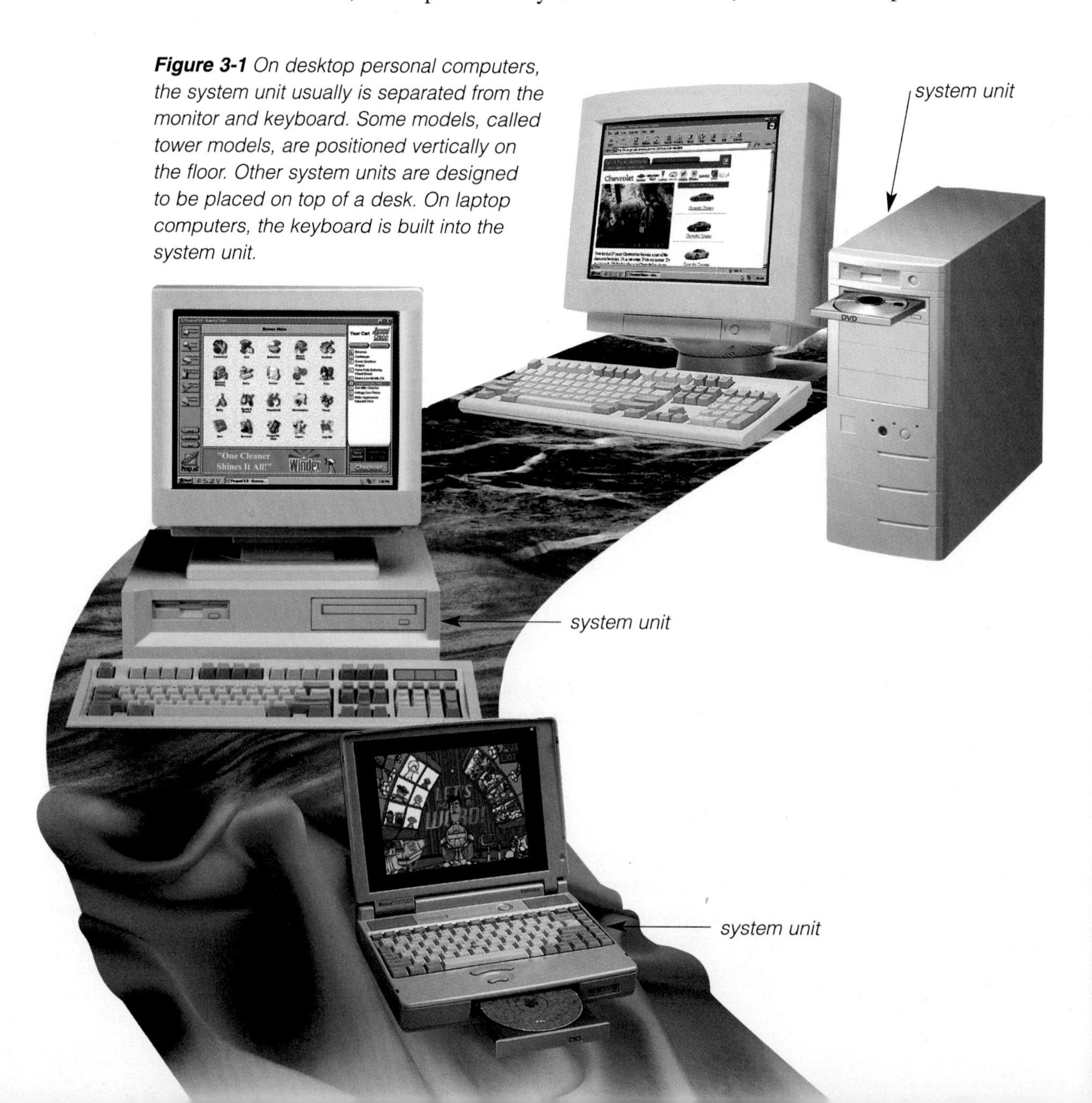

or microphone, are attached by a cable to a port on the system unit. These and other electronic components in the system unit are discussed in the following sections.

The Motherboard

Many of the electronic components in the system unit reside on a circuit board called the **motherboard** or **system board**. Figure 3-3 on the next page shows a photograph of a personal computer motherboard and identifies some of its components, including several different types of chips.

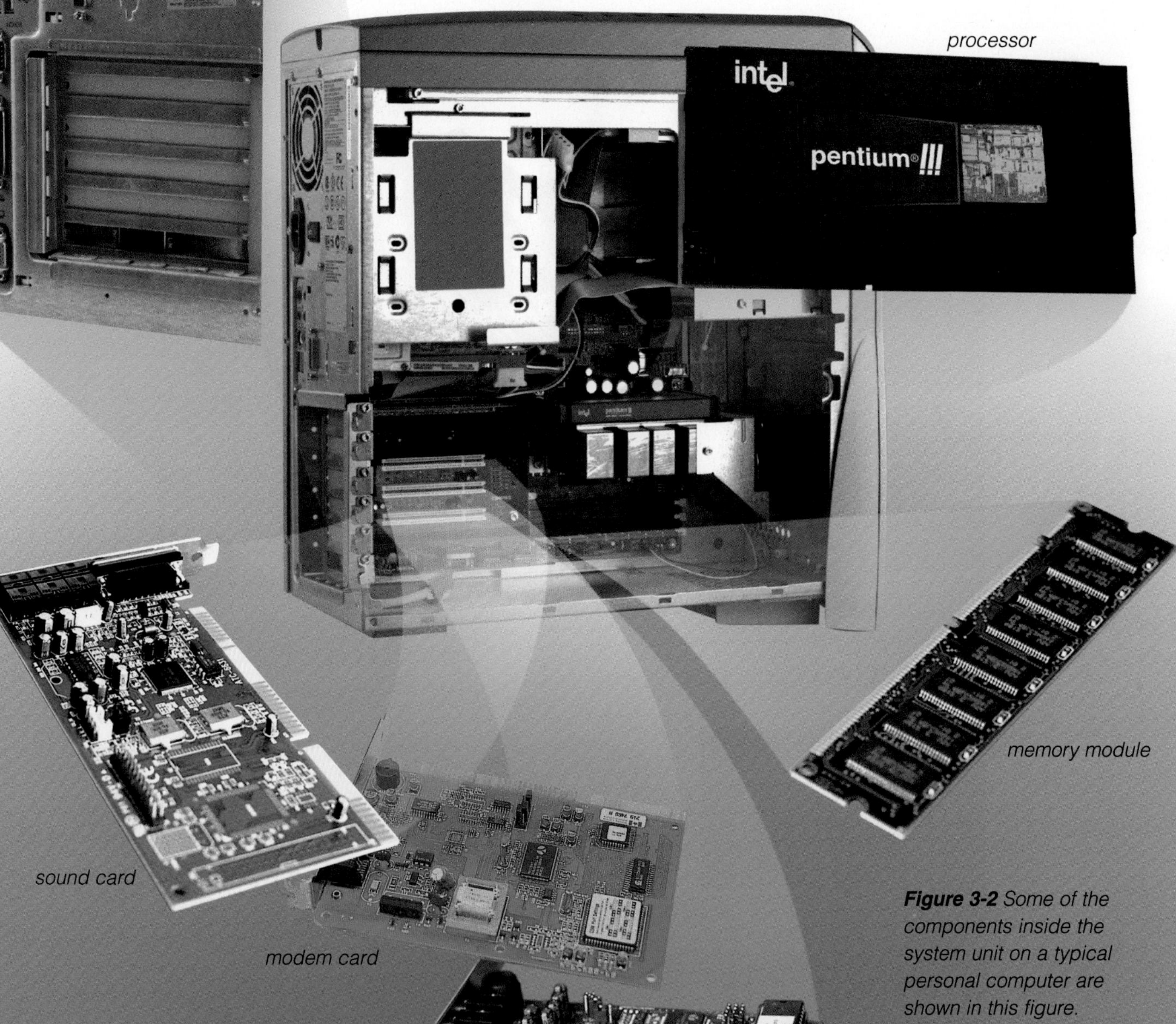

Figure 3-2 *Some of the components inside the system unit on a typical personal computer are shown in this figure.*

CHIPS A **chip** is a small piece of semi-conducting material, usually no bigger than one-half-inch square, on which one or more integrated circuits are etched (Figure 3-4). An **integrated circuit** (**IC**) is a microscopic pathway capable of carrying electrical current. Each integrated circuit can contain millions of elements such as **transistors**, which act as electronic switches, or gates, that open or close the circuit for electronic signals. The motherboard in the system unit contains many different types of chips. Of these, one of the more important is the central processing unit (CPU).

WEB INFO

For more information on motherboards, visit the Discovering Computers 2001 Chapter 3 WEB INFO page (**www.scsite.com/dc2001/ch3/webinfo.htm**) and click Motherboards.

CENTRAL PROCESSING UNIT

The **central processing unit** (**CPU**) interprets and carries out the basic instructions that operate a computer. The CPU, also referred to as the **processor**, significantly impacts overall computing power and manages most of a computer's operations. That is, most of the devices connected to the computer communicate with the CPU in order to carry out a task (Figure 3-5). The CPU contains the control unit and the arithmetic/logic unit. These two components work together to perform the processing operations.

On larger computers, such as mainframes and supercomputers, the various

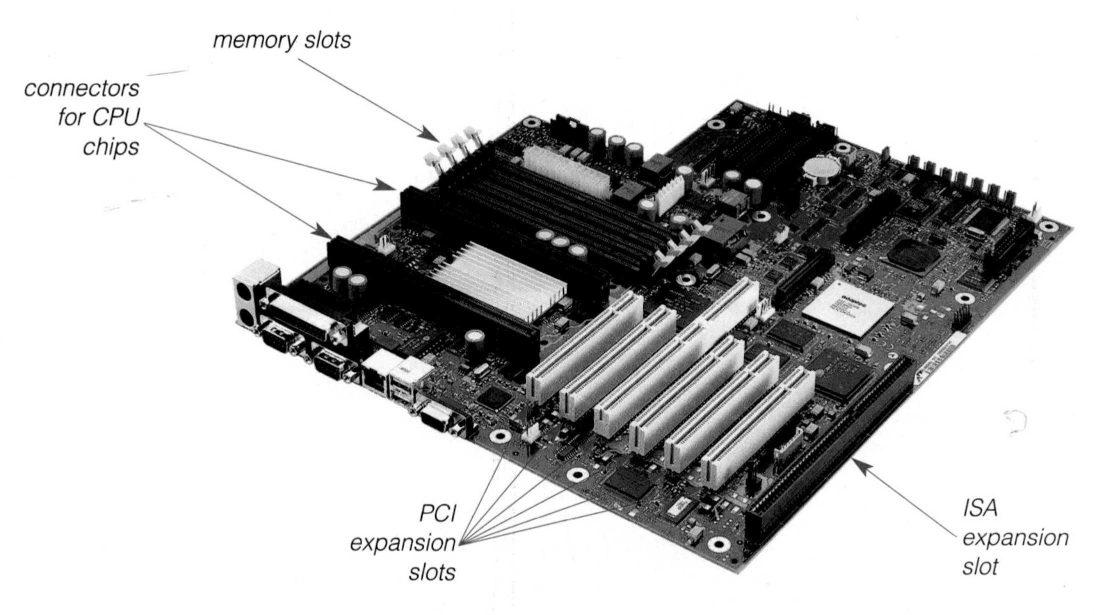

Figure 3-3 The motherboard in a personal computer contains many chips and other electronic components.

single edge contact (SEC) cartridge

Figure 3-4 Chips are packaged so they can be connected to a circuit board. DIP and PGA packages contain thin metal feet, called pins, which attach the package to the circuit board. A dual inline package (DIP) consists of two parallel rows of downward-pointing pins. A pin grid array (PGA) package holds a larger number of pins because the pins are mounted on the bottom surface of the package. A single edge contact (SEC) cartridge, does not use pins; instead the cartridge connects to the motherboard on one of its edges.

functions performed by the CPU are spread across many separate chips and sometimes multiple circuit boards. On a personal computer, the CPU usually is contained on a single chip and sometimes is called a **microprocessor** (Figure 3-6). In addition to the control unit and the arithmetic/logic unit, a microprocessor usually contains the registers and system clock. Each of these microprocessor components is discussed in the following sections.

The Control Unit

As you know, a program or set of instructions must be stored in memory for a computer to process data. The CPU uses its control unit to execute these instructions. The **control unit**, one component of the CPU, directs and coordinates most of the operations in the computer. The control unit has a role much like a traffic cop: it interprets each instruction issued by a program and then initiates the appropriate action to carry out the instruction. For every instruction, the control unit repeats a set of four basic operations: (1) fetching an instruction, (2) decoding the instruction, (3) executing the instruction, and, if necessary, (4) storing the result. Together, these four operations comprise the **machine cycle** or **instruction cycle** (Figure 3-7). **Fetching** is the process of obtaining a program instruction

WEB INFO

For more information on microprocessors, visit the Discovering Computers 2001 Chapter 3 WEB INFO page (**www.scsite.com/dc2001/ch3/webinfo.htm**) and click Microprocessors.

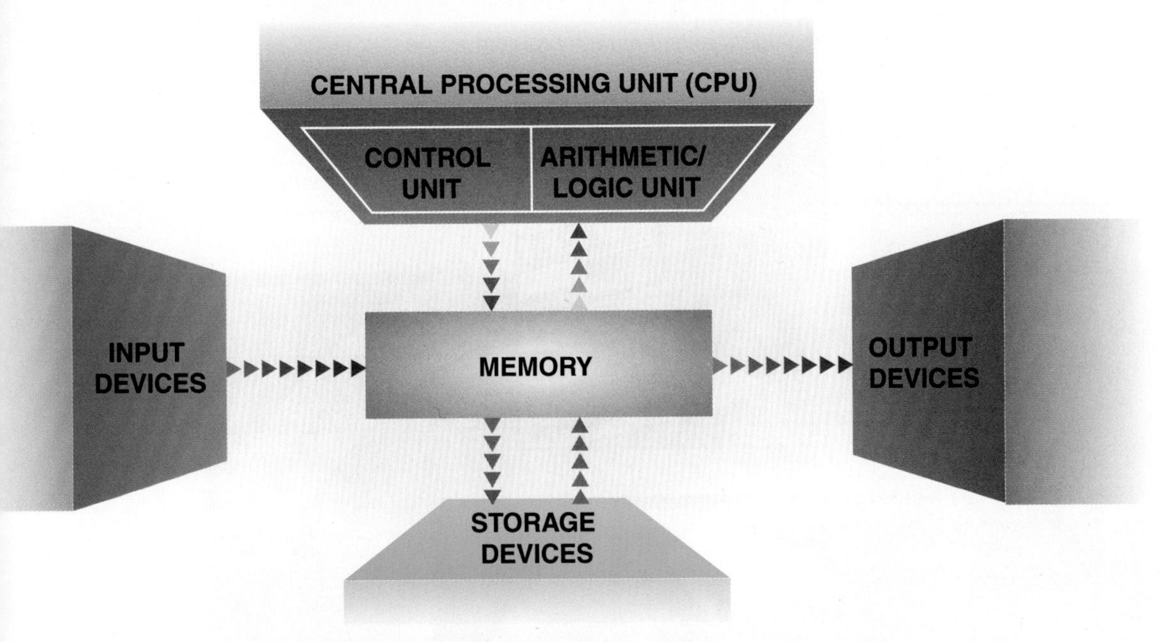

Figure 3-5 *Most of the devices connected to the computer communicate with the CPU in order to carry out a task. The arrows in this figure represent the flow of data, instructions, and information.*

Figure 3-6 *The Pentium® and Athlon™ processors are designed for higher-performance PCs, while the Celeron™ processor is geared toward lower costing PCs. The Xeon™ processor is geared toward servers and workstations.*

or data item from memory. **Decoding** is the process of translating the instruction into commands the computer understands. **Executing** is the process of carrying out the commands. **Storing** is the process of writing the result to memory. The time it takes to fetch is called **instruction time**, or **I-time**. The time it takes to decode and execute is called **execution time** or **E-time**. If you add together the I-time and E-time, you have the total time required for a machine cycle.

Some computer professionals measure a computer's speed according to the number of instructions it can process in one second. Sometimes, this speed is measured in **MIPS**, which stands for **m**illion **i**nstructions **p**er second. Current personal computers, for example, can process more than 300 MIPS. Because different instructions require different amounts of processing time, however, no real standard for measuring MIPS exists. In addition, MIPS refers only to the CPU speed, whereas applications generally are limited by other factors such as input and output speed.

The Arithmetic/Logic Unit

The **arithmetic/logic unit** (ALU), another component of the CPU, performs the execution part of a machine cycle as shown in Figure 3-7. Specifically, the ALU performs the arithmetic, comparison, and logical operations.

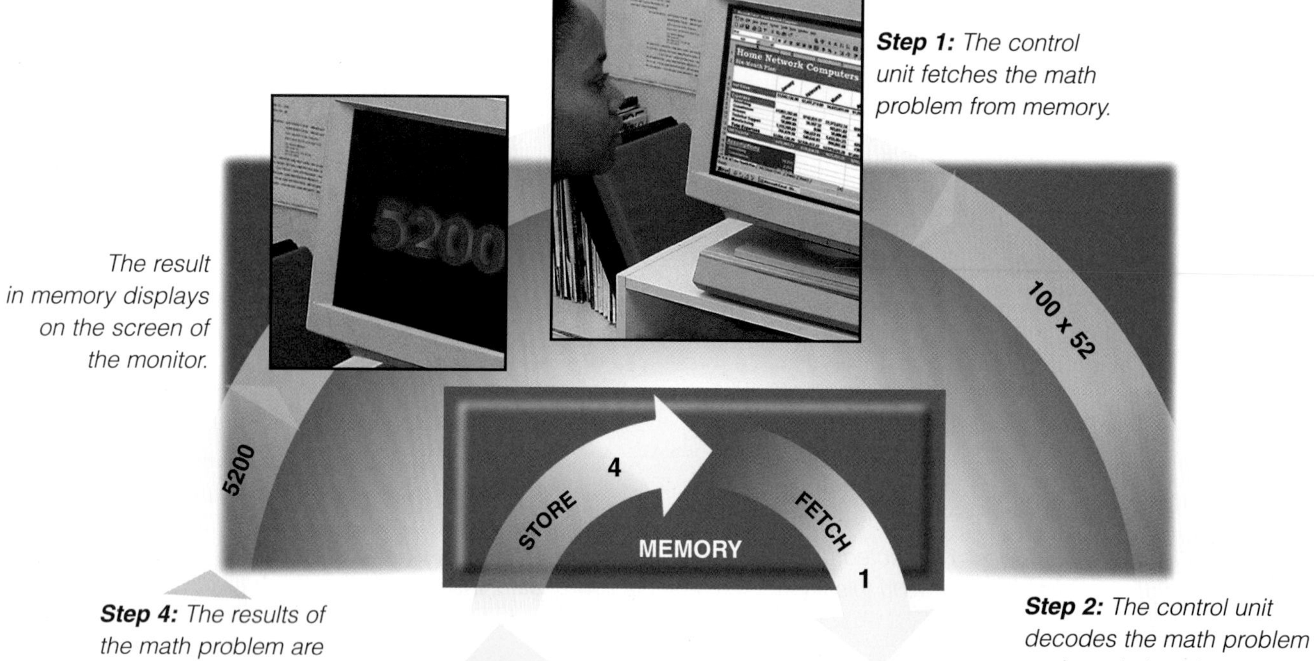

A student enters a math problem into memory of the computer.
Step 1: The control unit fetches the math problem from memory.
The result in memory displays on the screen of the monitor.
5200
100 x 52
4
STORE
FETCH
MEMORY
1
5200

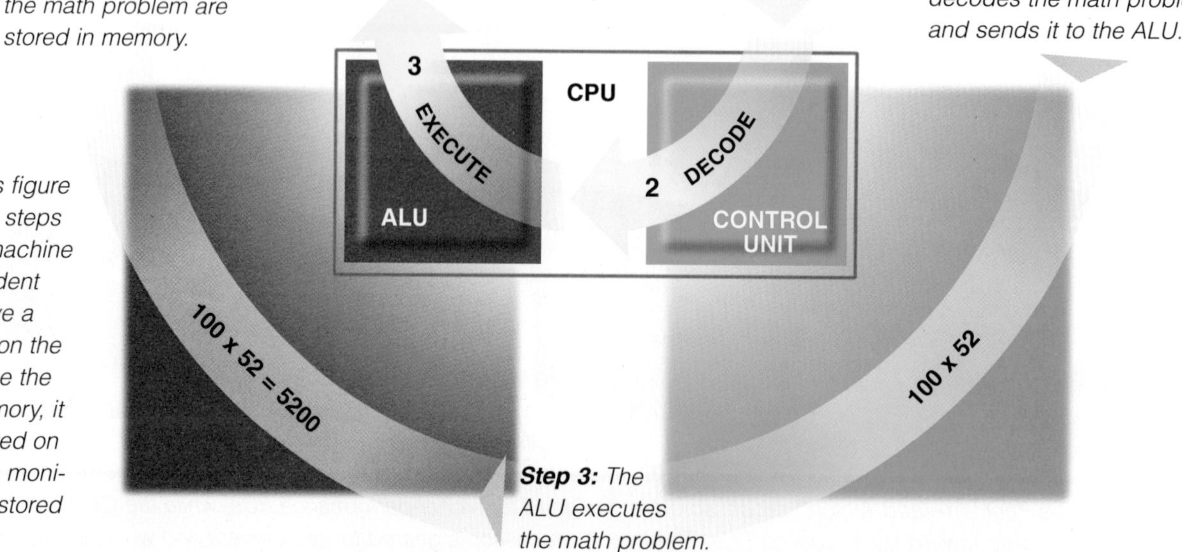

Figure 3-7 *This figure shows the four steps involved in a machine cycle for a student wanting to solve a math problem on the computer. Once the result is in memory, it can be displayed on the screen of a monitor, printed, or stored on a disk.*

Arithmetic operations include addition, subtraction, multiplication, and division. **Comparison operations** involve comparing one data item to another to determine if the first item is greater than, equal to, or less than the other item. Depending on the result of the comparison, different actions may occur. For example, to determine if an employee should receive overtime pay, the total hours the employee worked during the week have to be compared to straight-time hours allowed (40 hours, for instance). If the total hours worked is greater than 40, then an overtime wage is calculated; if total hours worked is not greater than 40, no overtime wage is calculated. **Logical operations** work with conditions and logical operators such as AND, OR, and NOT. For example, if you wanted to search a job database for part-time work in the admissions office, you would search for any jobs classified as part-time *AND* listed under admissions.

Pipelining

In some computers, the CPU processes only a single instruction at a time. That is, the CPU waits until an instruction completes all four stages of the machine cycle (fetch, decode, execute, and store) before beginning work on the next instruction. With **pipelining**, the CPU begins executing a second instruction before the first instruction is completed. Pipelining results in faster processing, because the CPU does not have to wait for one instruction to complete the machine cycle before fetching the next. For example, by the time the first instruction is in the last stage of the machine cycle, three other instructions could have been fetched and started through the machine cycle (Figure 3-8).

Although formerly used only in high-performance processors, pipelining now is common in processors used in today's personal computers. For instance, most newer processor chips can pipeline up to four instructions. Superscalar CPUs have two or more pipelines that can process instructions simultaneously.

Registers

The CPU uses temporary storage locations, called **registers**, to hold data and instructions. A microprocessor contains many different types of registers, each with a specific function. These functions include storing the location from where an instruction was fetched, storing an instruction while it is being decoded, storing data while the ALU processes it, and storing the results of a calculation.

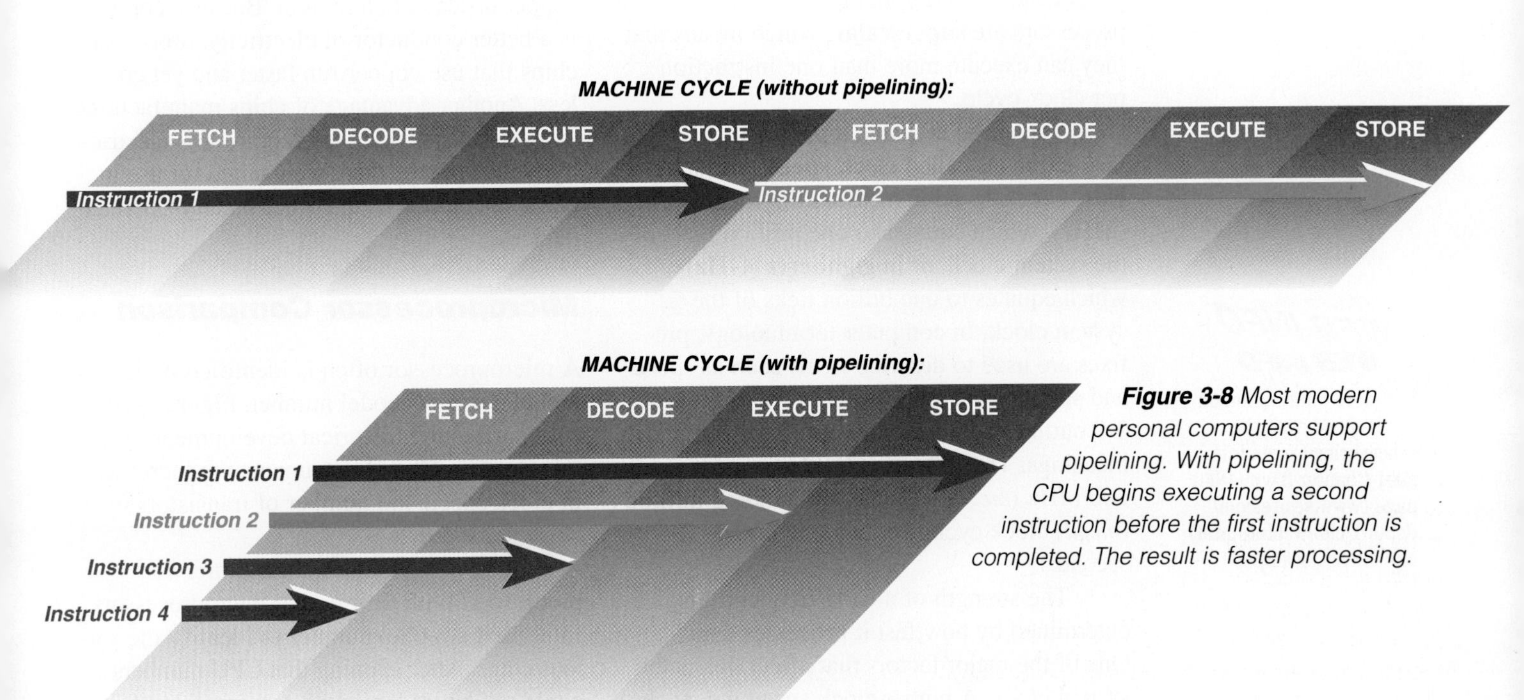

Figure 3-8 *Most modern personal computers support pipelining. With pipelining, the CPU begins executing a second instruction before the first instruction is completed. The result is faster processing.*

COMMON PREFIXES AND THEIR MEANINGS

Prefixes for Small Amounts	Meaning	Decimal Notation
MILLI	One thousandth of	.001
MICRO	One millionth of	.000001
NANO	One billionth of	.000000001
PICO	One trillionth of	.000000000001

Prefixes for Large Amounts	Meaning	Decimal Notation
KILO	One thousand	1,000
MEGA	One million	1,000,000
GIGA	One billion	1,000,000,000
TERA	One trillion	1,000,000,000,000

Figure 3-9 *This table outlines common prefixes and their meanings.*

The System Clock

The control unit relies on a small chip called the **system clock** to synchronize, or control the timing of, all computer operations. Just as your heart beats at a regular rate to keep your body functioning, the system clock generates regular electronic pulses, or ticks, that set the operating pace of components in the system unit. Each tick is called a **clock cycle**. A CPU requires a fixed number of clock cycles to execute each instruction. The faster the clock, the more instructions the CPU can execute per second. In addition, most of today's processors are **superscalar**, which means that they can execute more than one instruction per clock cycle.

The speed at which a processor executes instructions is called **clock speed** or **clock rate**. Clock speed is measured in **megahertz (MHz)**, which equates to one million ticks of the system clock, or in **gigahertz (GHz)**, which equates to one billion ticks of the system clock. In computer terminology, prefixes are used to describe items such as speed and storage capabilities. The table in Figure 3-9 outlines some common prefixes and their meanings. Thus, a computer that operates at 600 MHz (*mega*hertz) has six hundred million (*mega*) clock cycles, or ticks, in one second.

For more information on clock speed, visit the Discovering Computers 2001 Chapter 3 WEB INFO page (**www.scsite.com/dc2001/ch3/webinfo.htm**) and click Clock Speed.

The strength of a CPU frequently is determined by how fast it processes data. One of the major factors that affect this is the system clock. A higher clock speed means the CPU can process more instructions per second than the same CPU with a lower clock speed. For example, a 700 MHz CPU is faster than the same CPU operating at 500 MHz. The speed of the system clock affects only the CPU; it has no effect on peripherals such as a printer or disk drive.

The speed of the system clock varies among processors. Due to a technological breakthrough by IBM, most processors today operate at speeds in excess of 400 MHz. For nearly 30 years, aluminum was used to create the electronic circuitry on a single chip of silicon crystal. Now, a process exists that uses copper instead of aluminum. Because copper is a better conductor of electricity, processor chips that use copper run faster and yet cost less. Another advantage of chips manufactured using copper is that they require less electricity, which makes them well-suited for use in portable computers and other battery-operated devices.

Microprocessor Comparison

A microprocessor often is identified by its model name or model number. Figure 3-10 summarizes the historical development of the microprocessor and documents the increases in clock speed and number of transistors in chips since 1982.

Intel is a leading manufacturer of processors. With their earlier microprocessors, Intel used a model number to identify the various chips. After learning that CPU numbers could not be trademarked and protected from use by competitors, Intel decided to identify their microprocessors with names, not numbers – thus emerged their series of processors known as **Pentium® processors**. A second brand of Intel processor called the **Celeron**™ is designed for less expensive PCs. Two more brands, called the **Xeon**™ and **Itanium**™ processors, are geared toward workstations and servers.

Other companies such as Cyrix and AMD currently make **Intel-compatible microprocessors**. These microprocessors have the same internal design or architecture as Intel processors and perform the same functions, but often are less expensive. Intel and Intel-compatible processors are used in PCs.

An alternative to the Intel-style microprocessor is the **Motorola microprocessor**, which is found in Apple Macintosh and Power Macintosh systems. The processor used in Apple's PowerPC introduced a new architecture that increased the speed of the processor.

The **Alpha microprocessor**, which originally was developed by Digital Equipment Corporation, is used primarily in workstations and high-end servers. Current models of the Alpha chip run at clock speeds from 300 to 700 MHz.

A new type of microprocessor, called an **integrated CPU**, combines functions of a CPU, memory, and a graphics card on a single chip. These chips are designed for lower-costing personal computers and smaller-sized computers such as a set-top box.

Determining which processor to obtain when you purchase a computer depends on the type of computer you buy and how you plan to use the computer. If you purchase a PC (IBM-compatible), you will have the choice of an Intel processor or an Intel-compatible processor. Apple Macintosh and Power Macintosh users will want to choose a PowerPC processor.

Your intended use also will determine the clock speed of the processor you choose. The selection of the speed of the processor is an important consideration. A home user surfing the Web, for example, will need a less powerful processor than an artist working with graphics or applications requiring multimedia capabilities such as full-motion video.

Most of today's processors are equipped with **MMX**™ **technology**, in which a set of instructions is built into the processor so it can manipulate and process multimedia data more efficiently. To further improve

COMPARISON OF WIDELY USED MICROPROCESSORS

NAME	DATE INTRODUCED	MANUFACTURER	CLOCK SPEED (MHz)	NUMBER OF TRANSISTORS
Itanium™	2000	Intel	800 and up	25.4-60 million
Pentium® III Xeon™	1999	Intel	500-1000 (1 GHz)	9.5-28 million
Pentium® III	1999	Intel	400-1000 (1 GHz)	9.5-28 million
Celeron™	1998	Intel	266-633	19 million
Pentium® II Xeon™	1998	Intel	400-450	7.5-27.4 million
Athlon™	1999	AMD	500-1100 (1.1 GHz)	22 million
AMD-K6®-III	1999	AMD	400-450	21.3 million
AMD-K6®-2	1998	AMD	366-533	9.3 million
AMD-K6®	1998	AMD	300	8.8 million
Pentium® II	1997	Intel	233-450	7.5 million
Pentium® with MMX™ technology	1997	Intel	166-233	4.5 million
Pentium® Pro	1995	Intel	150-200	5.5 million
Pentium®	1993	Intel	75-200	3.3 million
80486DX	1989	Intel	25-100	1.2 million
80386DX	1985	Intel	16-33	275,000
80286	1982	Intel	6-12	134,000
PowerPC	1994	Motorola	50-450	Up to 50 million
68040	1989	Motorola	25-40	1.2 million
68030	1987	Motorola	16-50	270,000
68020	1984	Motorola	16-33	190,000
Alpha	1993	Digital; Compaq	150-700	Up to 100 million

Figure 3-10 *A comparison of some of the more widely used microprocessors. The greater the number of transistors, the more complex and powerful the chip.*

performance of multimedia, the Web, and 3D graphics, Intel's latest processors also include **SSE instructions** and AMD's latest processors also have **3DNow!™ technology**.

Figure 3-11 describes guidelines for selecting an Intel processor. Remember, the higher the clock speed, the faster the processor – but also the more expensive the computer.

Processor Installation and Upgrades

A processor chip is inserted into an opening, or socket, on the motherboard. Most of today's computers are equipped with a **zero-insertion force (ZIF) socket**, which has a small lever or screw designed to facilitate the installation and removal of processor chips.

Instead of buying an entirely new computer, some processors can be upgraded to increase their performance. Processor upgrades take one of three forms: chip for chip, piggyback, or daughterboard. With a **chip for chip upgrade**, the existing processor chip is replaced with a new one. Because a ZIF socket requires no force to remove and install a chip, users easily can upgrade the processor on computers equipped with this type of socket. Some motherboards also have a second ZIF socket, which is designed to hold an upgrade chip. In this case, the existing processor chip remains on the motherboard, and the upgrade chip is installed into the second ZIF socket. With a **piggyback upgrade**, the new processor chip is stacked on top of the old one. With a **daughterboard upgrade**, the new processor chip is located on a daughterboard. A **daughterboard** is a small circuit board that plugs into the motherboard, often to add additional capabilities to the motherboard.

Heat Sinks and Heat Pipes

Newer processor chips generate a lot of heat, which could cause the chip to burn up. Often, the computer's main fan generates enough airflow to cool the processor. Sometimes, however, a heat sink is required – especially

Figure 3-11
Determining which processor to obtain when you purchase a computer depends on your computer usage.

GUIDELINES FOR SELECTING AN INTEL PROCESSOR

INTEL PROCESSOR	CLOCK SPEED	USE
Xeon™	600-1000 MHz	Power users with workstations; servers on a network
Pentium® family	Above 800	Users that design professional graphics and drawings, produce and edit videos, record and edit music, participate in video conference calls, create professional Web sites, play graphic-intensive multi-player Internet games
	700-800	Users that design professional documents containing graphics such as newsletters or number intensive spreadsheets; produce multimedia presentations; use the Internet as a intensive research tool; edit photographs; send documents and graphics via the Web; watch video; play graphic-intensive games on CD or DVD; create personal Web sites
	Below 700	Home users that manage personal finances; create basic documents with word processing and spreadsheet software; communicate with others on the Web via e-mail, chats, and discussions; shop on the Web; create basic Web pages
Celeron™	400-633	Home users that manage personal finances, create basic documents with word processing and spreadsheet software; edit photographs; make greeting cards and calendars; use educational or entertainment CD-ROMs, communicate with others on the Web via e-mail, chats, and discussions

when upgrading to a more powerful processor. A **heat sink** is a small ceramic or metal component with fins on its surface that is designed to absorb and ventilate heat produced by electrical components. Some heat sinks are packaged as part of the processor chip, while others must be installed on top of the chip. Because a heat sink consumes a lot of room, a smaller device called a **heat pipe** is used to cool laptop computers.

Coprocessors

Another way to increase the performance of a computer is through the use of a **coprocessor**, which is a special processor chip or circuit board designed to assist the processor in performing specific tasks. Users running engineering, scientific, or graphics applications, for instance, will notice a dramatic increase in speed with a **floating-point coprocessor**, provided the application is designed to take advantage of the coprocessor. Floating-point coprocessors also are called math or numeric coprocessors. Most of today's computers are equipped with a floating-point coprocessor upon purchase; for others, the coprocessor chip or card is installed later.

Parallel Processing

Some computers use more than one processor to speed processing times. Known as **parallel processing**, this method uses multiple processors simultaneously to execute a program (Figure 3-12). That is, parallel processing divides up a problem so that multiple processors work on their assigned portion of the problem at the same time. As you might expect, parallel processors require special software designed to recognize how to divide up the problem and then bring the results back together again. Supercomputers use parallel processing for applications such as weather forecasting.

DATA REPRESENTATION

To fully understand the way a computer processes data, it is important to understand how data is represented in a computer. People communicate using words that are combined

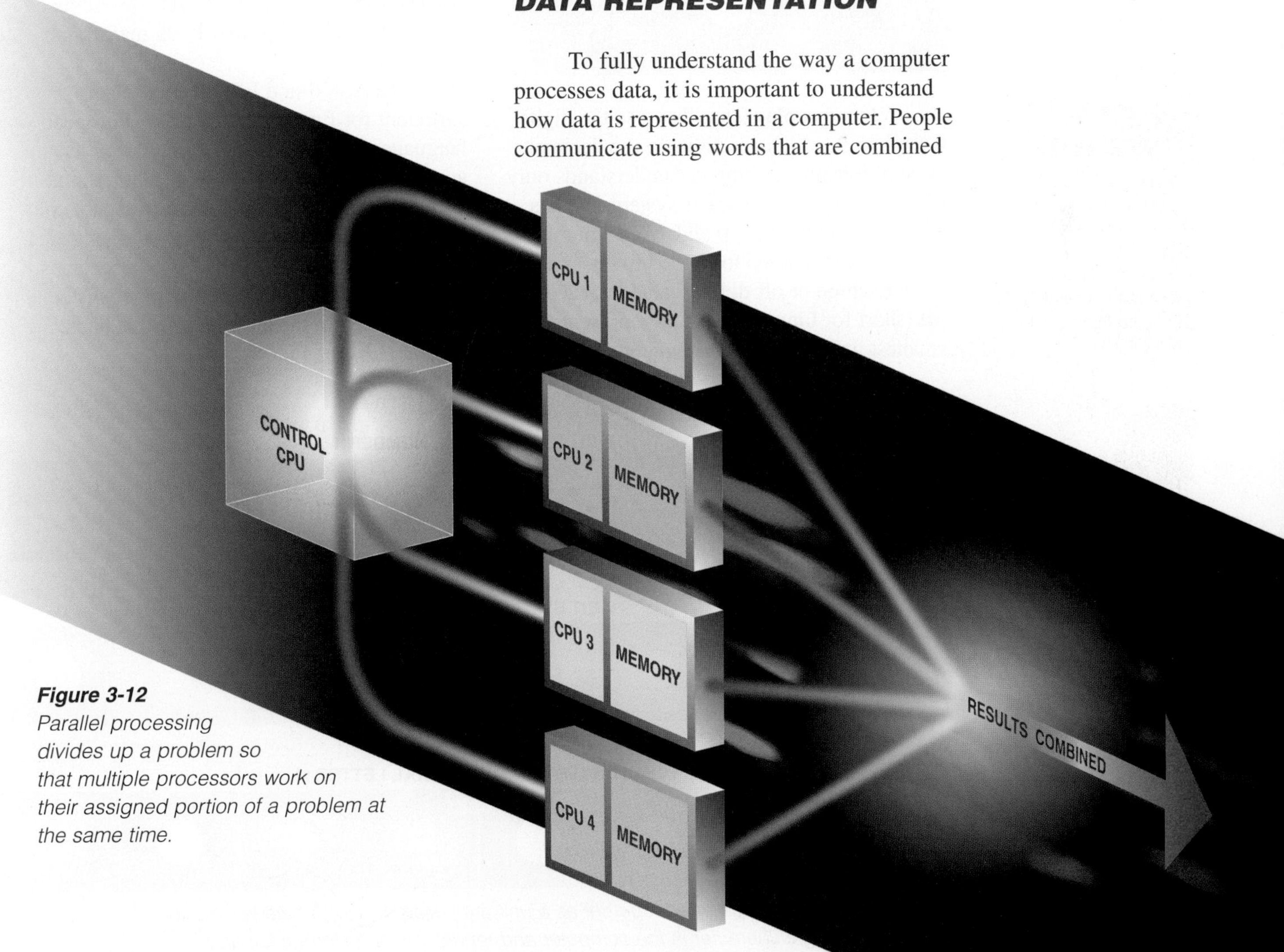

Figure 3-12
Parallel processing divides up a problem so that multiple processors work on their assigned portion of a problem at the same time.

into sentences in different ways each time we speak. Human speech is **analog**, meaning that it uses continuous signals to represent data and information. Most computers, by contrast, are **digital**, meaning that they understand only two discrete states: on and off. This is because computers are electronic devices powered by electricity, which has only two states: (1) on or (2) off.

These two states are represented easily by using two digits; 0 is used to represent the electronic state of off (absence of an electronic charge) and 1 is used to represent the electronic state of on (presence of an electronic charge) (Figure 3-13).

BINARY DIGIT (BIT)	ELECTRONIC CHARGE	ELECTRONIC STATE
1		ON
0		OFF

Figure 3-13 *A computer circuit represents the binary digits 0 or 1 electronically by the presence or absence of an electronic charge.*

When people count, they use the digits 0 through 9, which are digits in the decimal system. Because a computer understands only two states, it uses a number system that has just two unique digits, 0 and 1. This numbering system is referred to as the **binary** system.

Each on or off digital value is called a **bit** (short for **bi**nary digi**t**) and represents the smallest unit of data the computer can handle.

WEB INFO

For more information on the binary system, visit the Discovering Computers 2001 Chapter 3 WEB INFO page (**www.scsite.com/dc2001/ch3/webinfo.htm**) and click Binary.

By itself, a bit is not very informative. When eight bits are grouped together as a unit, they are called a **byte**. A byte is informative because it provides enough different combination of 0s and 1s to represent 256 individual characters including numbers, uppercase and lowercase letters of the alphabet, punctuation marks, and other characters such as the letters of the Greek alphabet.

The combinations of 0s and 1s used to represent characters are defined by patterns called a coding scheme. Using one type of coding scheme, the number 1 is represented as 00110001, the number 2 as 00110010, and the capital letter M as 01001101 (Figure 3-14). Two popular coding schemes are ASCII and EBCDIC (Figure 3-15). The **American Standard Code for Information Interchange**, called **ASCII** (pronounced *ASK-ee*), is the most widely used coding system to represent data. ASCII is used on many personal computers and minicomputers. The **Extended Binary Coded Decimal Interchange Code**, or **EBCDIC** (pronounced *EB-see-dic*) is used primarily on mainframe computers.

The ASCII and EBCDIC codes are sufficient for English and Western European languages but are not large enough for Asian and other languages that use different alphabets. **Unicode** is a coding scheme capable of representing all the world's current languages. The ASCII, EBCDIC, and Unicode schemes, along with the parity bit and number systems, are discussed in the appendix of this book.

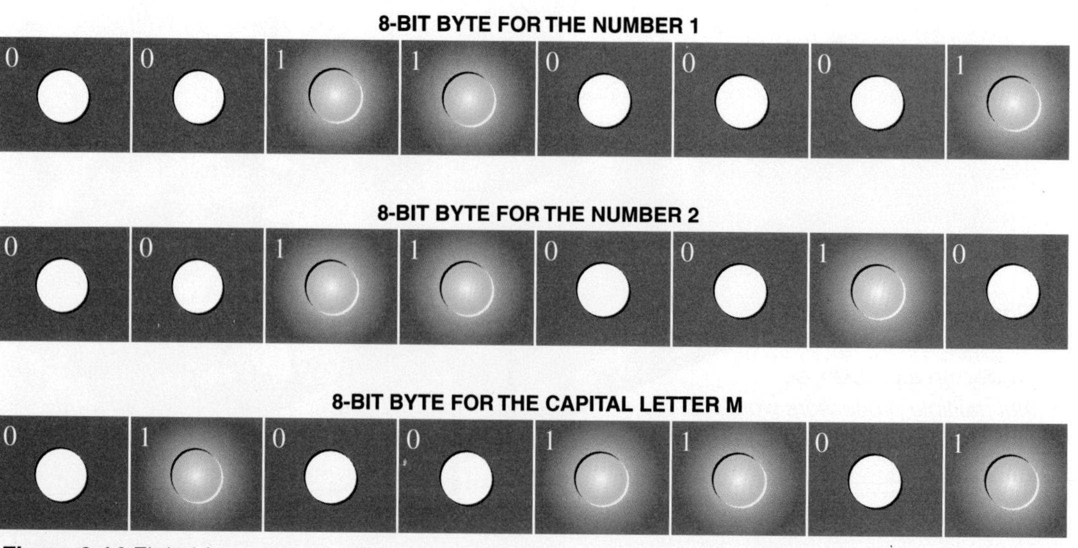

Figure 3-14 *Eight bits grouped together as a unit are called a byte. A byte is used to represent a single character in the computer and represents one storage location.*

ASCII	SYMBOL	EBCDIC
00110000	0	11110000
00110001	1	11110001
00110010	2	11110010
00110011	3	11110011
00110100	4	11110100
00110101	5	11110101
00110110	6	11110110
00110111	7	11110111
00111000	8	11111000
00111001	9	11111001
01000001	A	11000001
01000010	B	11000010
01000011	C	11000011
01000100	D	11000100
01000101	E	11000101
01000110	F	11000110
01000111	G	11000111
01001000	H	11001000
01001001	I	11001001
01001010	J	11010001
01001011	K	11010010
01001100	L	11010011
01001101	M	11010100
01001110	N	11010101
01001111	O	11010110
01010000	P	11010111
01010001	Q	11011000
01010010	R	11011001
01010011	S	11100010
01010100	T	11100011
01010101	U	11100100
01010110	V	11100101
01010111	W	11100110
01011000	X	11100111
01011001	Y	11101000
01011010	Z	11101001
00100001	!	01011010
00100010	"	01111111
00100011	#	01111011
00100100	$	01011011
00100101	%	01101100
00100110	&	01010000
00101000	(	01001101
00101001	)	01011101
00101010	*	01011100
00101011	+	01001110

Figure 3-15 *Two popular coding schemes are ASCII and EBCDIC.*

Coding schemes such as ASCII make it possible for humans to interact with a digital computer that recognizes only bits. When you press a key on a keyboard, the electronic signal is converted into a binary form the computer understands and is stored in memory. That is, every character is converted to its corresponding byte. The computer then processes that data in terms of bytes, which actually is a series of on/off electrical states. When processing is finished, the bytes are converted back into numbers, letters of the alphabet, or special characters to be displayed on a screen or be printed (Figure 3-16). All of these conversions take place so quickly that you do not realize they are occurring.

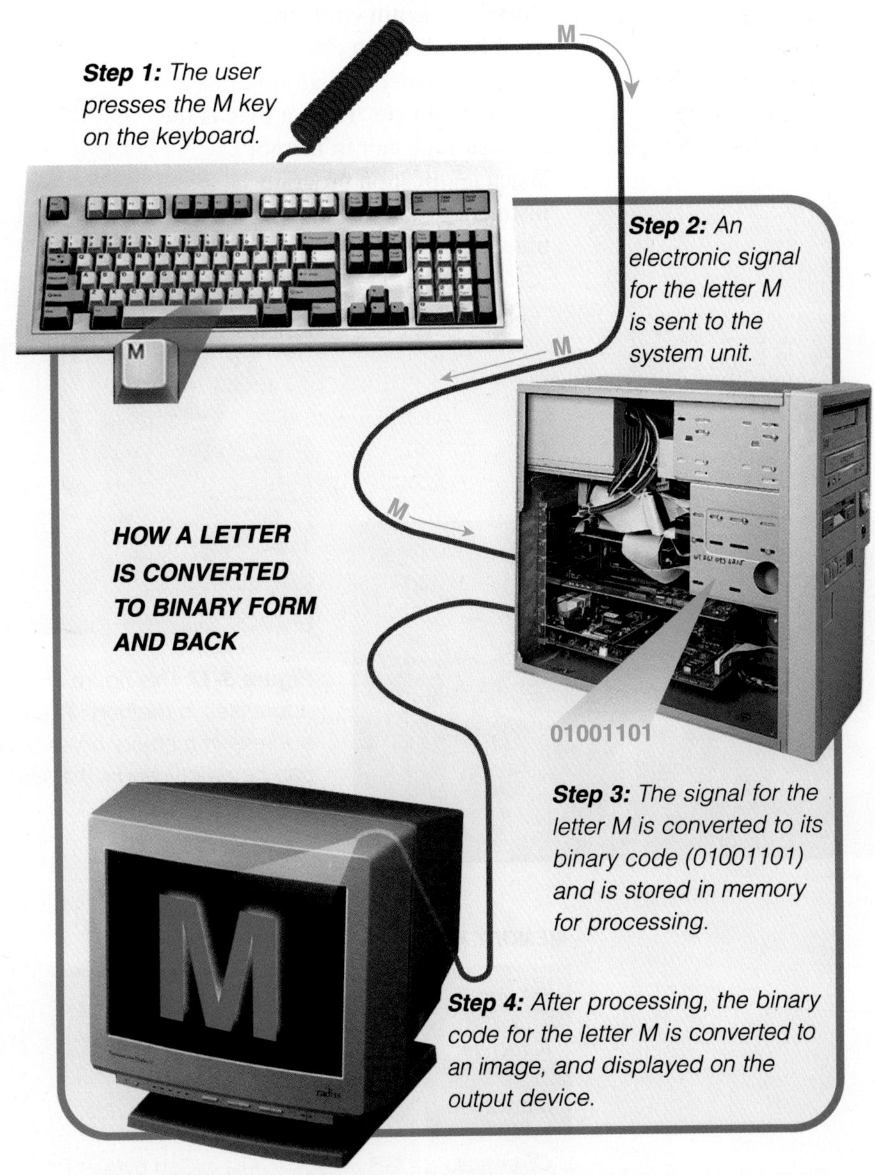

Step 1: The user presses the M key on the keyboard.
M
Step 2: An electronic signal for the letter M is sent to the system unit.
M
M
HOW A LETTER IS CONVERTED TO BINARY FORM AND BACK
01001101
Step 3: The signal for the letter M is converted to its binary code (01001101) and is stored in memory for processing.
M
Step 4: After processing, the binary code for the letter M is converted to an image, and displayed on the output device.

Figure 3-16 *Converting a letter to binary form and back.*

MEMORY

While performing a processing operation, a processor needs a place to temporarily store instructions to be executed and the data to be used with those instructions. A computer's **memory** in the system unit is used to store data, instructions, and information. The memory chips on the circuit boards in the system unit perform this function. Memory stores three basic items: (1) the operating system and other system software that control the usage of the computer equipment; (2) application programs designed to carry out a specific task such as word processing; and (3) the data being processed by the application programs. This role of memory to store both data and programs is known as the **stored program concept**.

Recall that a character is stored in the computer as a group of 0s and 1s, called a byte. Thus, a byte is the basic storage unit in memory. When application program instructions and data are transferred into memory from storage devices, they are stored as bytes, each of which is placed in a precise location in memory, called an **address**. This address is simply a unique number identifying the location of the byte in memory. The illustration in Figure 3-17 shows how seats in a stadium are similar to addresses in memory: (1) a seat holds one person at a time and an address in memory holds a single byte, (2) both a seat and an address can be empty, and (3) a seat has a unique identifying number and so does a memory address. Thus, to access data or instructions in memory, the computer references the addresses that contain bytes of data.

The size of memory is measured by the number of bytes available for use (Figure 3-18). A **kilobyte** of memory, abbreviated **KB** or **K**, is equal to exactly 1,024 bytes

Figure 3-17 *This figure shows how seats in a stadium are similar to addresses in memory: (1) a seat holds one person at a time and an address in memory holds a single byte, (2) both a seat and an address can be empty, and (3) a seat has a unique identifying number and so does an address.*

MEMORY AND STORAGE SIZES

Term	Abbreviation	Approximate Memory Size	Exact Memory Amount	Approximate Number of Pages of Text
Kilobyte	KB or K	1 thousand bytes	1,024 bytes	1/2
Megabyte	MB	1 million bytes	1,048,576 bytes	500
Gigabyte	GB	1 billion bytes	1,073,741,824 bytes	500,000
Terabyte	TB	1 trillion bytes	1,099,511,627,776 bytes	500,000,000

Figure 3-18 *This table outlines terms used to define storage size.*

(see Figure 3-18). To make storage definitions easier to identify, computer users often round a kilobyte down to 1,000 bytes. For example, if a memory chip can store 100 KB, it is said to hold 100,000 bytes (characters). A **megabyte**, abbreviated **MB**, is equal to approximately one million bytes.

The system unit contains two types of memory: volatile and nonvolatile. The contents of **volatile memory** are lost (erased) when the computer power is turned off. The contents of **nonvolatile memory**, on the other hand, are not lost when power is removed from the computer. RAM is an example of volatile memory. ROM, flash memory, and CMOS all are examples of nonvolatile memory. The following sections discuss each of these types of memory.

RAM

The memory chips in the system unit are called **RAM (random access memory)**. When the computer is powered on, certain operating system files (such as the files that determine how your Windows desktop displays) are loaded from a storage device such as a hard disk into RAM. These files remain in RAM as long as the computer is running. As additional programs and data are requested, they also are read from storage into RAM. The processor acts upon the data while it is in RAM. During this time, the contents of RAM may change as the data is processed (Figure 3-19). Multiple programs can be loaded into RAM simultaneously, provided you have enough RAM to accommodate all the programs. The program with which you are working currently displays on the screen.

WEB INFO

For more information on RAM, visit the Discovering Computers 2001 Chapter 3 WEB INFO page (**www.scsite.com/dc2001/ch3/webinfo.htm**) and click RAM.

Figure 3-19 *How application programs transfer in and out of RAM.*

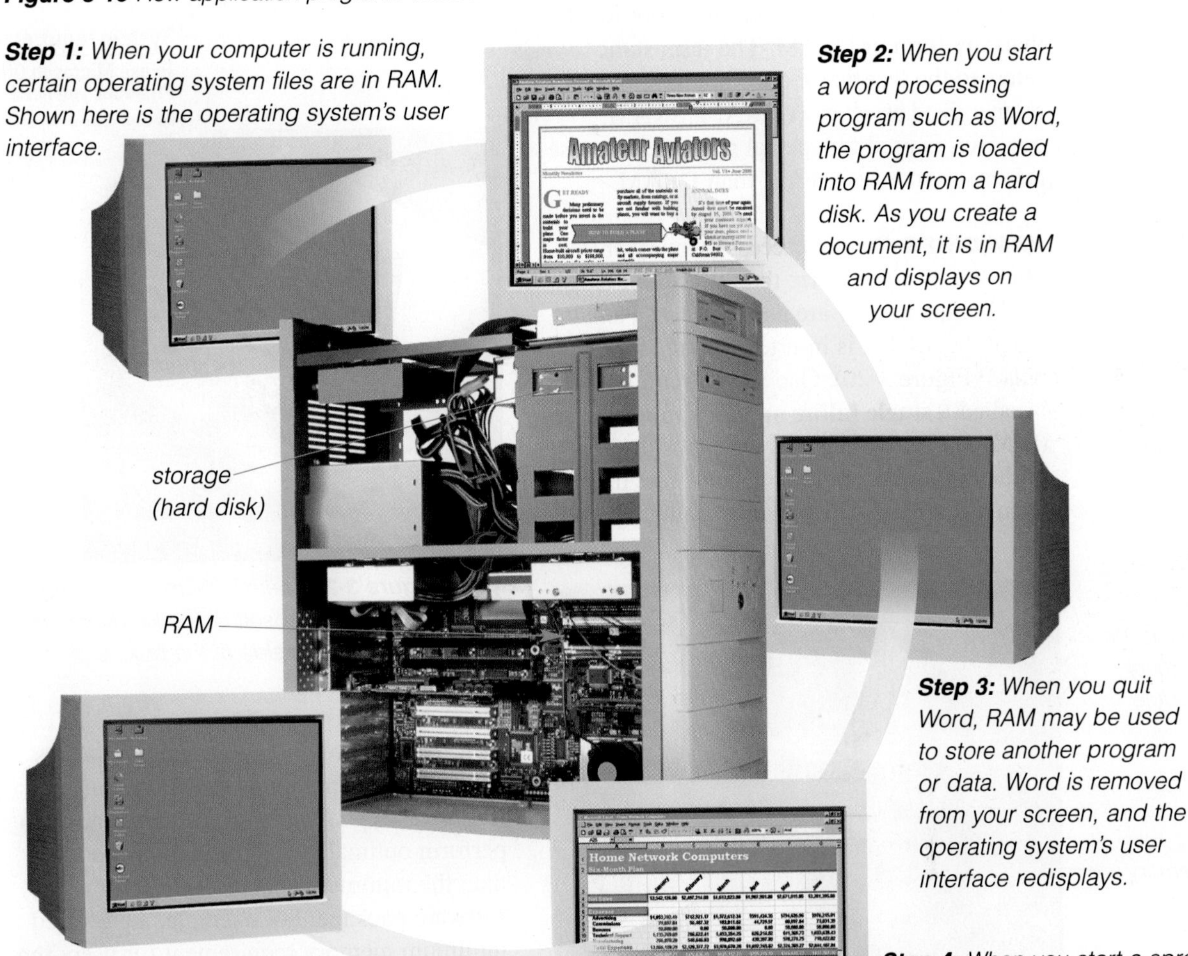

Step 1: When your computer is running, certain operating system files are in RAM. Shown here is the operating system's user interface.
Step 2: When you start a word processing program such as Word, the program is loaded into RAM from a hard disk. As you create a document, it is in RAM and displays on your screen.
Amateur Aviators
storage (hard disk)
RAM
Step 3: When you quit Word, RAM may be used to store another program or data. Word is removed from your screen, and the operating system's user interface redisplays.
Home Network Computers
Step 4: When you start a spreadsheet program such as Excel, the program is loaded into RAM from a hard disk. As you create a spreadsheet, it is in RAM and displays on your screen.
Step 5: When you quit Excel, RAM may be used to store another program or data. Excel is removed from your screen and the operating system's user interface redisplays.

RAM is volatile, which means items stored in RAM are lost when the power to the computer is turned off. For this reason, any items needed for future use must be **saved**, or copied from RAM to a storage device such as a hard disk, before the power to the computer is turned off.

Two basic types of RAM exist: dynamic RAM and static RAM. When discussing RAM, users normally are referring to **dynamic RAM**, also called **DRAM** (pronounced DEE-ram), a type of memory that must be re-energized constantly or it loses its contents. One type of DRAM, called **Synchronous DRAM (SDRAM)**, is much faster than DRAM because it is synchronized to the system clock. **RDRAM® (Rambus® DRAM)** is a newer type of DRAM that is even faster than SDRAM. Double-data rate (DDR) SDRAM, also called **SDRAM II**, also is faster than SDRAM. Most computers today use some type of SDRAM or RDRAM.

Static RAM, also called **SRAM** (pronounced ESS-ram), is faster and more reliable than any form of DRAM. The term static refers to the fact that it does not have to be re-energized as often as DRAM. SRAM, however, is used for special purposes because it is much more expensive than DRAM.

RAM CHIPS Random access memory chips often are smaller in size than processor chips. RAM chips usually are packaged on a small circuit board that is inserted into the motherboard (Figure 3-20). One such circuit board is called a **single inline memory module (SIMM)** because the pins on opposite sides of the circuit board are connected together to form a single set of contacts. Another type of circuit board is called a **dual inline memory module (DIMM)** because the pins on opposite sides of the circuit board are not connected and thus form two sets of contacts. The RAM chips commonly used in SIMMs and DIMMs are SDRAM chips. RDRAM chips, by contrast, are packaged on a circuit board called a **Rambus® inline memory module (RIMM)**.

Figure 3-20 This photograph shows a dual inline memory module (DIMM).

CONFIGURING RAM The amount of RAM a computer requires often depends on the types of applications to be used on the computer. Remember that a computer only can manipulate data that is in memory. RAM is similar to the workspace you have on the top of your desk. Just as a desktop needs a certain amount of space to hold papers, pens, your computer, and so on, a computer needs a certain amount of memory to be able to store an applicaton program and files. The more RAM a computer has, the more programs and files it can work on at once.

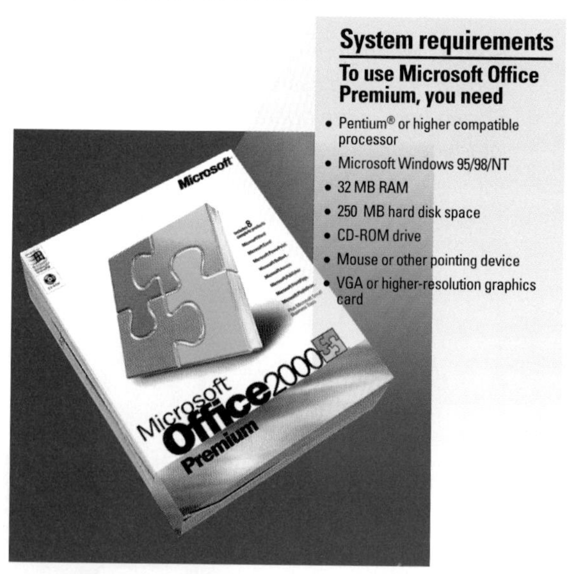

Figure 3-21 The minimum system requirements for Microsoft Office Premium are printed on the side of the box.

A software package usually indicates the minimum amount of RAM it requires (Figure 3-21). If you want the application to perform optimally, typically you need more than the minimum specifications on the software package. For example, a preferred minimum memory requirement for users running Microsoft Office Premium is 64 MB of RAM. In general, home users running Windows and using standard application software such as word processing should have at least 32 MB of RAM. Most business computers should be equipped with a minimum of 64 MB of RAM, so users can run accounting, financial, or spreadsheet programs, and programs requiring multimedia capabilities. Users composing multimedia presentations or using graphics-intensive applications may want even more RAM. The table in Figure 3-22a provides guidelines for the amount of RAM you need on your computer.

The necessary amount of RAM varies according to the type of work you do and the type of software applications you are using. Remember, however, that the amount of RAM on your computer determines how many programs and how much data a computer can handle at one time and thus affects overall performance. As shown in Figure 3-22b, advertisements for computers normally contain the type of processor, the speed of the processor measured in MHz, and the amount of RAM installed.

Cache

Most of today's computers improve their processing times by using **cache** (pronounced cash). **Memory cache**, also called a cache store or RAM cache, helps speed the processes of the computer by storing frequently used instructions and data.

The rationale is that the processor is likely to request these items over and over again. When the processor needs an instruction or data, it first searches cache.

Most modern computers have two types, or layers, of cache: Level 1 and Level 2. **Level 1 (L1) cache,** also called **primary cache** or **internal cache**, is built directly into the processor chip. L1 cache usually has a very small capacity, ranging from 8 KB to 64 KB. For example, Pentium®, Pentium® Pro, and Pentium® II processors all have 16 KB of L1 cache.

Level 2 (L2) cache, or **external cache**, is not part of the processor chip; instead L2 cache consists of high-speed SRAM chips. L2 cache is slightly slower than L1 cache, but has a larger capacity. When discussing cache, most users are referring to L2 cache, which ranges in size from 64 KB to 2 MB.

As noted, cache speeds processing time by storing frequently used instructions and data. When the processor needs an instruction or data, it searches memory in this order: L1 cache, then L2 cache, then RAM – with a greater delay in processing for each level of memory it must search. If the instruction or data is not found in memory, then a slower speed storage device such as a hard disk or CD-ROM must be searched.

WEB INFO

For more information on cache, visit the Discovering Computers 2001 Chapter 3 WEB INFO page (**www.scsite.com/dc2001/ch3/webinfo.htm**) and click Cache.

Figure 3-22a
(RAM guidelines)

RAM (in MB)	**32 to 64 MB** (minimum)	**64 to 128 MB** (minimum)	**256 MB and up**
Use	Home users managing personal finances, using standard application software such as word processing; using educational or entertainment CD-ROMs, communicating with others on the Web	Users requiring more advanced multimedia capabilities; running number-intensive accounting, financial, or spreadsheet programs; working with videos and music; creating Web sites; participating in video conferences; playing Internet games	Power users creating professional Web sites; running sophisticated CAD, 3-D design, or other graphics-intensive software

Figure 3-22b
(Computers for sale)

Model	C500Z	A533Q	A800M	P850G	P933R	A1000D
Processor	500MHz Celeron™ processor	533Hz AMD-K6®-2 processor	800MHz Athlon™ processor	850MHz Pentium® III processor	933MHz Pentium® III processor	1000MHz Athlon™ processor
Memory	32MB SDRAM Memory	64MB SDRAM Memory	96MB SDRAM Memory	128MB SDRAM Memory	512MB RDRAM Memory	384MB SDRAM Memory

Figure 3-22 *Determining how much RAM you need depends on the applications you intend to run on your computer. Today, most computers running Windows and Windows applications include at least 32 MB of RAM. Advertisements for computers normally contain the type of processor, the speed of the processor measured in MHz, as well as the amount of RAM installed.*

WEB INFO

For more information on ROM, visit the Discovering Computers 2001 Chapter 3 WEB INFO page (**www.scsite.com/dc2001/ch3/webinfo.htm**) and click ROM.

CONFIGURING CACHE A computer system with L2 cache usually performs at speeds that are 10 to 40 percent faster than those without any cache. To realize the largest increase in performance, the system should have from 256 K to 512 K of L2 cache (above that, the increases in performance are not significant). As shown in the advertisement in Figure 3-23, most current systems are equipped with 256 K or 512 K of L2 cache.

GP7-850

- Intel® Pentium® III Processor 850MHz with 512K Cache
- 128MB SDRAM
- VX900 19" Monitor (18" viewable)
- 16MB AGP Graphics Card
- 20GB 7200 RPM Ultra ATA Hard Drive
- MS Windows 98
- MS Office

L2 cache

9200XL

- 15" XGA TFT Color Display
- Intel® Pentium® III Processor 500MHz with 256K Cache
- 128MB SDRAM (expandable to 512MB)
- 4MB SGRAM 3D Graphics Card
- Removable Combo DVD-ROM & 3.5" Floppy Disk Drive
- 18GB Ultra ATA Hard Drive
- Two (2) Lithium Ion Batteries & AC Pack
- 56K Capable PC Card Modem
- MS Office

Figure 3-23 *As shown in these advertisements, most current systems are equipped with 256 K or 512 K of L2 cache.*

ROM

Read-only memory (**ROM**) is the name given to memory chips storing data that only can be read. That is, the data stored in ROM chips cannot be modified – hence, the name read only. While RAM is volatile, ROM is nonvolatile; its contents are not lost when power to the computer is turned off. ROM chips thus contain data, instructions, or information that is recorded permanently. For example, ROM contains the sequence of instructions the computer follows to load the operating system and other files when you first turn the computer on.

The data, instructions, or information stored on ROM chips often are recorded when the chip is manufactured. ROM chips that contain permanently written data, instructions, or information are called **firmware**.

Another type of ROM chip, called a **programmable read-only memory** (**PROM**) chip, is a blank ROM chip on which you can permanently place items. The instructions used to program a PROM chip are called **microcode**. Once the microcode is programmed into the PROM chip, it functions like a regular ROM chip and cannot be erased or changed.

FLASH MEMORY Another type of nonvolatile memory is called **flash memory** or **flash ROM**. Unlike a PROM chip that can be programmed only once, flash memory can be erased electronically and reprogrammed. Flash memory is used to store programs on personal computers, as well as cellular telephones, printers, digital cameras, pagers, and personal digital assistants (Figure 3-24). Flash memory is available in sizes ranging from 1 to 40 MB.

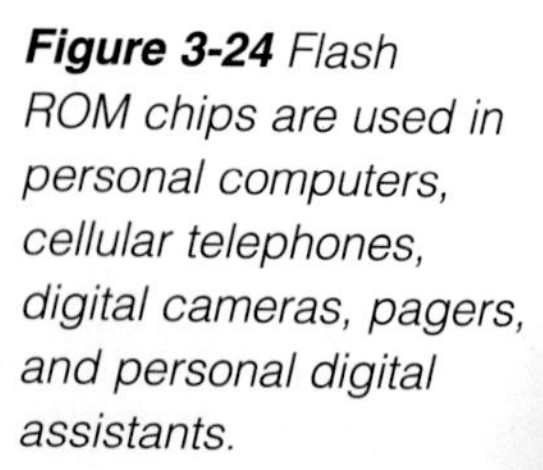

Figure 3-24 *Flash ROM chips are used in personal computers, cellular telephones, digital cameras, pagers, and personal digital assistants.*

CMOS

Another type of memory chip in the system unit is complementary metal-oxide semiconductor memory. **Complementary metal-oxide semiconductor** memory, abbreviated **CMOS** (pronounced SEE-moss), is used to store configuration information about the computer, such as the type of disk drives, keyboard, and monitor; the current date and time; and other startup information needed when the computer is turned on. CMOS chips use battery power to retain information even when the power to the computer is turned off. Battery-backed CMOS memory thus keeps the calendar, date, and time current even when the computer is off. Unlike ROM, information stored in CMOS memory can be changed, such as when you change from standard time to daylight savings time.

Memory Access Times

The speed at which the processor can access data from memory directly affects how fast the computer processes data. This speed often is defined as **access time**. Access time is measured in fractions of a second (Figure 3-25). For memory, access times are measured in terms of a **nanosecond** (abbreviated **ns**), which is one billionth of a second. A nanosecond is extremely fast (Figure 3-26). In fact, electricity travels about one foot in a nanosecond.

The access time (speed) of memory contributes to the overall performance of the computer. DRAM chips normally have access times ranging from 50 to 70 ns, while SRAM chips commonly range in speed from 7 to 20 ns. ROM's access times range from 55 to 250 ns. For comparison purposes, accessing data on a fast hard disk takes between 8 and 15 milliseconds. A **millisecond**, abbreviated **ms**, is one thousandth of a second. This means that accessing data in memory with a 70 ns access time is over 200,000 times faster than accessing data on a hard disk with a 15 ms access time.

While access times of memory greatly affect overall computer performance, manufacturers and retailers usually list a

ACCESS TIMES

TERM	ABBREVIATION	SPEED
Millisecond	ms	One-thousandth of a second
Microsecond	µs	One-millionth of a second
Nanosecond	ns	One-billionth of a second
Picosecond	psec	One-trillionth of a second

Figure 3-25 *Access times are measured in fractions of a second. This table outlines terms used to define access times.*

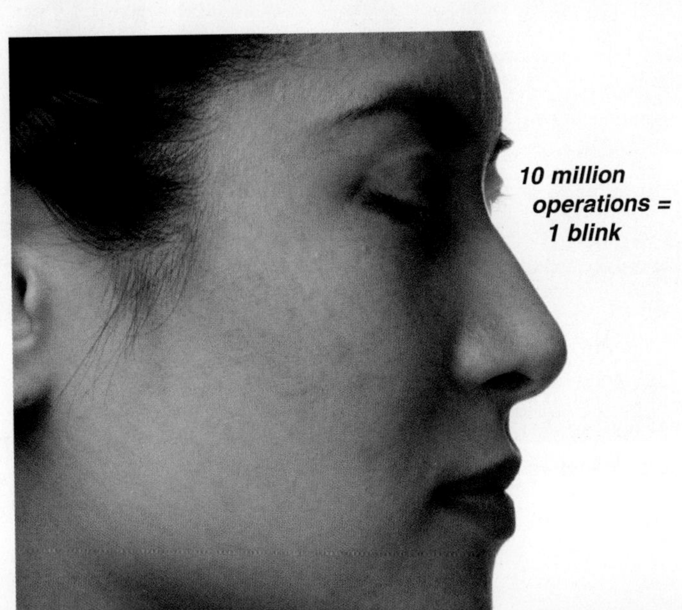

Figure 3-26 *It takes about one-tenth of a second to blink your eye, which is the equivalent of 100 million nanoseconds. A computer can perform some operations in as little as 10 nanoseconds. Thus, in the time it takes to blink your eye, a computer can perform some operations 10 million times.*

computer's memory in terms of its size, not its access time. Thus, an advertisement might describe a computer as having 512 K of L2 cache or 16 MB of SDRAM expandable to 256 MB.

Memory capacity can be expanded in a number of ways, such as installing additional memory in an expansion slot.

EXPANSION SLOTS AND EXPANSION CARDS

An **expansion slot** is an opening, or socket, where a circuit board can be inserted into the motherboard. These circuit boards add new devices or capabilities to the computer such as more memory, higher-quality sound devices, a modem, or graphics capabilities (Figure 3-27). Many terms are used to refer to this type of circuit board: **expansion card**, **expansion board**, **adapter card**, **interface card**, **card**, **add-in**, and **add-on**. Sometimes a device or feature is built into the expansion card; other times a cable is used to connect the expansion card to a device such as a scanner outside the system unit. Figure 3-28 shows an expansion card being plugged into an expansion slot on a personal computer motherboard.

WEB INFO

For more information on expansion cards, visit the Discovering Computers 2001 Chapter 3 WEB INFO page (**www.scsite.com/dc2001/ch3/webinfo.htm**) and click Expansion Cards.

Three types of expansion cards found in most of today's computers are a video card, a sound card, and an internal modem. A **video card**, also called a **video adapter** or **graphics card**, converts computer output into a video signal that is sent through a cable to the monitor, which displays an image on the screen. A **sound card** is used to enhance the sound-generating capabilities of a personal computer by allowing sound to be input through a microphone and output through speakers. An **internal modem** is a communications device that enables computers to communicate via telephone lines or other means.

In the past, installing an expansion card required setting switches and other elements on the motherboard. Many of today's computers support **Plug and Play**, which refers to the computer's capability to automatically configure expansion cards and other devices as they are installed. Having Plug and Play support means a user can plug in a device, turn on the computer, and then use, or play, the device without having to configure the system manually.

TYPES OF EXPANSION CARDS

Expansion Card	Function
Accelerator	To increase the speed of the CPU
Controller	To connect disk drives; being phased out because newer motherboards support these connections
Game	To connect a joystick
I/O	To connect input and output devices such as a printer or mouse; being phased out because newer motherboards support these connections
Interface	To connect other peripherals such as mouse devices, CD-ROMs, and scanners
Memory	To add more memory to the computer
Modem	To connect to other computers through telephone lines
Network	To connect to other computers and peripherals via a network
PC-to-TV Converter	To connect to a television
Sound	To connect speakers or a microphone
TV Tuner	To view television channels on your monitor
Video	To connect a monitor
Video Capture	To connect a video camera

Figure 3-27 *This table lists some of the types of expansion cards and their functions.*

PC Cards

Laptop and other portable computers have a special type of expansion slot used for

Figure 3-28 *This expansion card is being inserted into an expansion slot on the motherboard of a personal computer.*

installing PC Cards. A **PC Card** is a thin credit card-sized device that is used to add memory, disk drives, sound, and fax/modem capabilities to a laptop computer (Figure 3-29).

All PC Cards conform to standards developed by the **P**ersonal **C**omputer **M**emory **C**ard **I**nternational **A**ssociation (these cards originally were called **PCMCIA cards**), which help ensure that PC Cards can be interchanged between laptop computers. PC Cards thus are designed with the same length and width so they fit in a standard PC Card slot.

The height or thickness of PC Cards varies among three types, which are named Type I, Type II, and Type III. The thinnest **Type I cards** are used to add memory capabilities to the computer. **Type II cards** contain communications devices such as modems. The thickest **Type III cards** are used to house devices such as hard disks.

A PC Card slot usually is located on the side of a laptop computer. Unlike other expansion cards that require you to open the system unit and install the card onto the motherboard, a PC Card can be changed as needed without having to open the system unit or restart the computer. For example, if you need to send a fax, you can just insert the fax/modem card in the PC Card slot while the computer is running. The operating system automatically recognizes the new card and allows you to send the fax. The ability to add and remove devices while a computer is running is called **hot plugging** or **hot swapping**. Because of their small size and versatility, PC Cards also are used with consumer electronics products such as cable TV, automobiles, and digital cameras.

Figure 3-29 This picture shows a cellular telephone plugged into a modem card that is being inserted into a PC Card slot on a laptop computer.

WEB INFO

For more information on PC Cards, visit the Discovering Computers 2001 Chapter 3 WEB INFO page (**www.scsite.com/dc2001/ch3/webinfo.htm**) and click PC Cards.

PORTS

External devices such as a keyboard, monitor, printer, mouse, and microphone, often are attached by a cable to the system unit. The interface, or point of attachment, to the system unit is called a **port**. Most of the time, ports are located on the back of the system unit (Figure 3-30), but they also can be placed on the front.

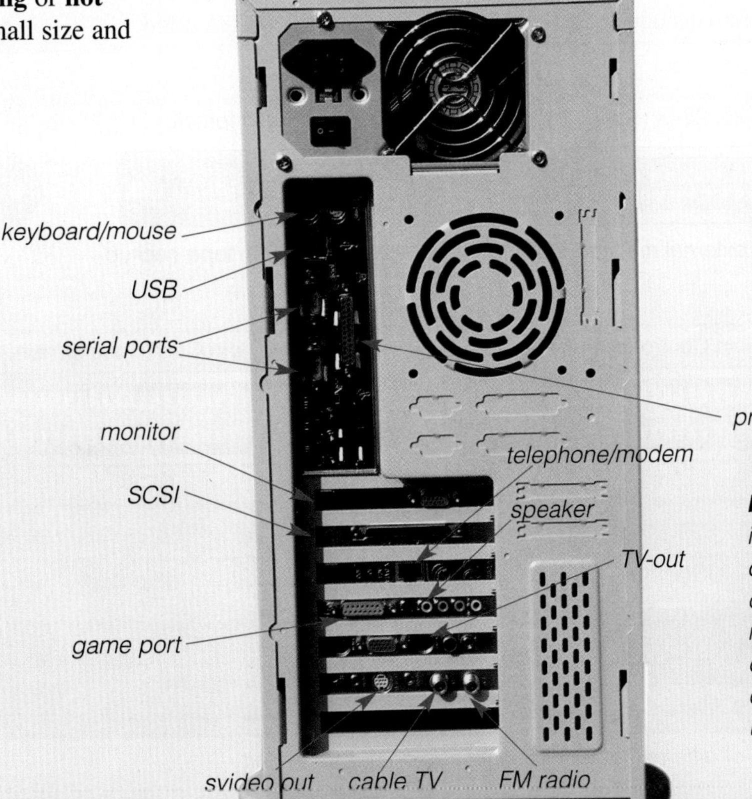

Figure 3-30 A port is an interface that allows you to connect a peripheral device such as a printer, mouse, or keyboard to the computer. Usually, ports are on the back of the system unit and often are labeled.

Ports have different types of connectors. A **connector** is used to join a cable to a device (Figure 3-31). One end of a cable is attached to the connector on the system unit and the other end of the cable is attached to the peripheral device.

Most connectors are available in one of two genders: male or female. **Male connectors** have one or more exposed pins, like the end of an electrical cord you plug into the wall. **Female connectors** have matching holes to accept the pins on a male connector, like an electrical wall outlet.

Figure 3-32 shows the different types of connectors on a system unit. Some of these connectors are equipped with the computer when you buy it. Other connectors are added by inserting expansion cards into the computer. That is, the expansion card has a port, which enables you to attach a device to the expansion card. Understanding the differences among connector types is important, because the cables you purchase to connect peripheral devices to your computer often are identified by their types. For example, types of printer ports include a 25-pin female, 36-pin female, 36-pin Centronics female, and USB.

Sometimes a new peripheral device cannot be attached to the computer because the connector on the system unit is the same gender as the connector on the cable. Using a **gender changer**, which is a device used to join two connectors that are either both female or both male, can solve this problem.

Most computers are equipped with two types of ports: serial and parallel. The next section discusses each of these ports.

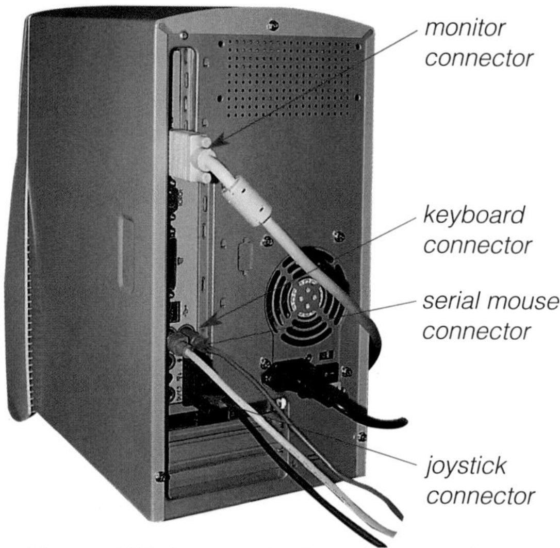

Figure 3-31 *A connector is used to attach an external device to the system unit.*

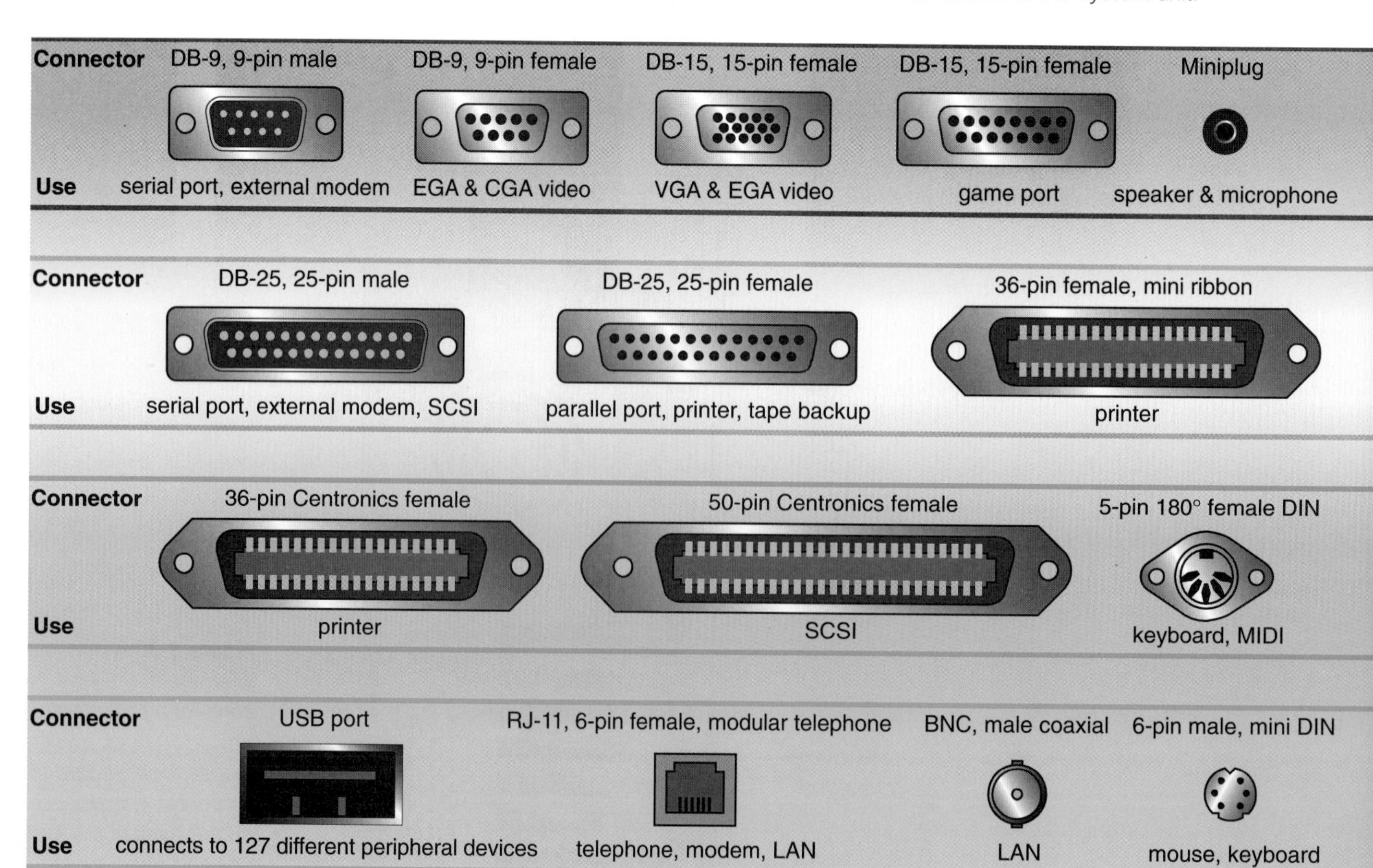

Figure 3-32 *Examples of different types of connectors on a system unit.*

Serial Ports

A **serial port** is one type of interface used to connect a device to the system unit. Because a serial port transmits only one bit of data at a time, it usually is used to connect devices that do not require fast data transmission rates, such as a mouse, keyboard, or modem (Figure 3-33). A modem, which connects the system unit to a telephone line, uses a serial port because the telephone line expects the data in a serial form. Serial ports conform to either the RS-232 or RS-422 standard, which specifies the number of pins used on the port's connector. Two common connectors for serial ports are a male 25-pin connector or a male 9-pin connector.

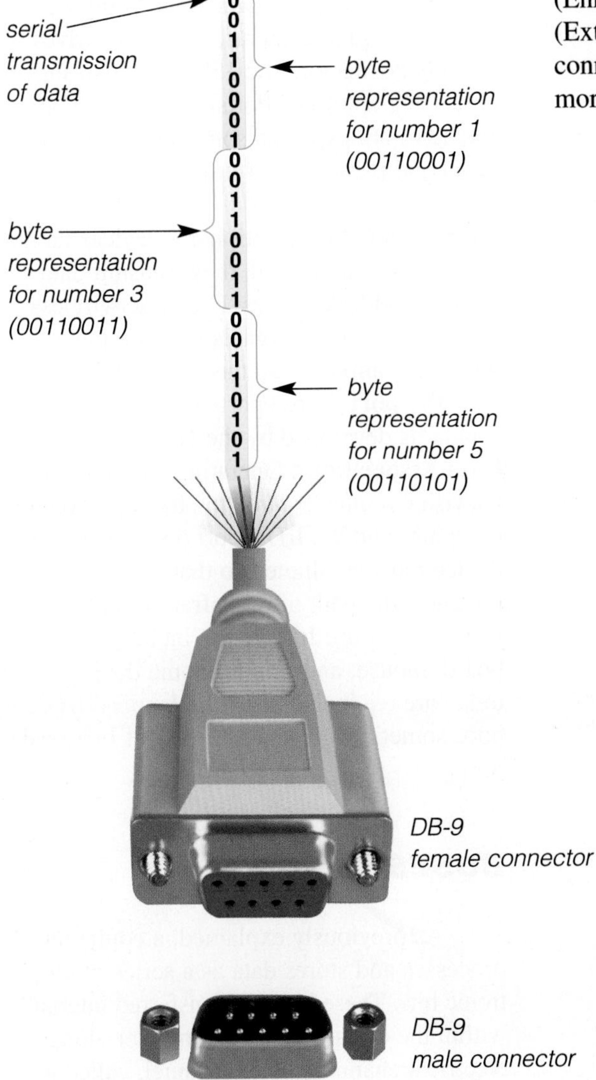

Figure 3-33 A serial port transmits data one bit at a time. One wire is used to send data; another is used to receive data; and the remaining wires are used for other communications operations.

Parallel Ports

Unlike a serial port, a **parallel port** is an interface used to connect devices that are capable of transferring more than one bit at a time. Parallel ports originally were developed as an alternative to the slower speed serial ports.

Many printers connect to the system unit using a parallel port with a 25-pin female connector. This parallel port can transfer eight bits of data (one byte) simultaneously through eight separate lines in a single cable (Figure 3-34). A parallel port sometimes is called a Centronics interface, after the company that first defined the standard for communication between the system unit and a printer. Two newer types of parallel ports, the EPP (Enhanced Parallel Port) and the ECP (Extended Capabilities Port), use the same connectors as the Centronics port, but are more than ten times faster.

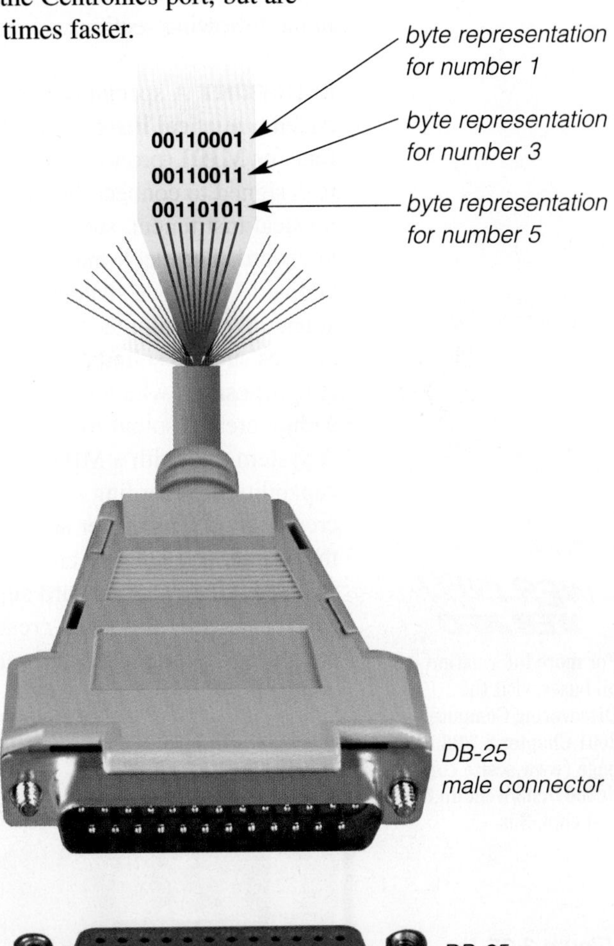

Figure 3-34 A parallel port is capable of transmitting more than one bit at a time. The port shown in this figure has eight wires that transmit data; the remaining wires are used for other communications operations.

Universal Serial Bus Port

A **universal serial bus (USB) port** can connect up to 127 different peripheral devices with a single connector (Figure 3-35). Using this port, devices are daisy chained together outside the system unit.

A USB port also connects to newer peripherals such as digital cameras and joysticks. Having a standard port and connector greatly simplifies the process of attaching devices to a personal computer. The USB also supports hot plugging and Plug and Play, which means you can install peripherals while the computer is running.

Special-Purpose Ports

Four special-purpose ports used on many of today's computers are MIDI, SCSI, 1394, and IrDA ports. Each of these ports is discussed in the following sections.

MIDI PORT A special type of serial port, called a **musical instrument digital interface,** or **MIDI** (pronounced MID-dee) **port**, is designed to connect the system unit to a musical instrument, such as an electronic keyboard. The electronic music industry has adopted MIDI as a standard to define how sounds are represented electronically by devices such as sound cards and synthesizers. A **synthesizer**, which can be a peripheral or a chip, creates sound from digital instructions. A system unit with a MIDI port has the capability of recording sounds that have been created by a synthesizer and then processing the sounds (the data) to create new sounds. Just about every sound card supports the MIDI standard, so sounds created using one computer can be played and manipulated by another.

WEB INFO

For more information on buses, visit the Discovering Computers 2001 Chapter 3 WEB INFO page (**www.scsite.com/dc2001/ch3/webinfo.htm**) and click Bus.

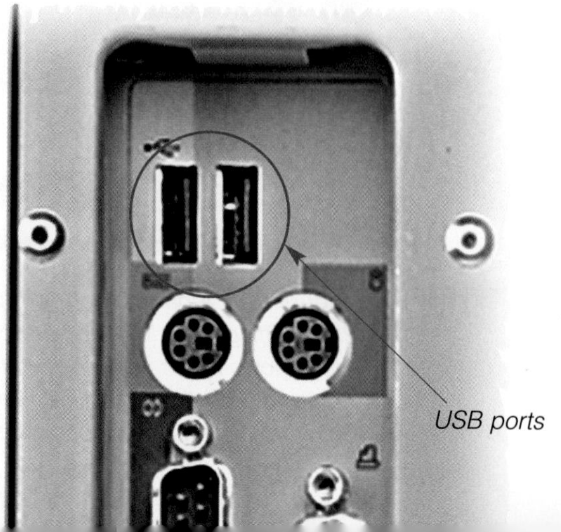

Figure 3-35 *The universal serial bus (USB) port can connect up to 127 peripheral devices with a single connector.*

USB ports

SCSI PORT A special high-speed parallel port used to attach peripheral devices such as disk drives and printers is called a **SCSI port**. A SCSI port can transmit up to 32 bits at a time. Pronounced scuzzy, SCSI stands for **small computer system interface**. A total of seven SCSI devices can be daisy chained together, which means the first SCSI device connects to the computer, the second SCSI device connects to the first SCSI device, and so on. Some new computers are equipped with a SCSI port, while others have a slot that supports a SCSI expansion card.

1394 PORT Similarly to the USB port, the **1394 port**, also called **FireWire**, can connect multiple types of devices requiring faster data transmission speeds such as digital video camcorders, digital VCRs, color printers, scanners, digital cameras, and DVD drives to a single connector. The 1394 port also supports Plug and Play. Ports such as the USB and 1394 are expected someday to replace serial and parallel ports completely.

IrDA PORT Some peripheral devices do not use any cables; instead, they transmit data via infrared light waves. For these wireless devices to transmit signals to a computer, both the computer and the device must have an **IrDA port**. These ports must conform to standards developed by the **IrDA (Infrared Data Association)**. Operating similar to a television remote control, the IrDA port on the computer and the IrDA port on the peripheral device must be aligned so that nothing obstructs the path of the infrared light wave. Devices that use IrDA ports include the keyboard, mouse, and printer. Some devices today are equipped with a high-speed IrDA port, sometimes called a **FIR (fast infrared)** port.

BUSES

As previously explained, a computer processes and stores data as a series of electronic bits. These bits are transferred internally within the circuitry of the computer along electrical channels. Each channel, called a **bus**, allows the various devices inside and attached to the system unit to communicate with each other. Just as vehicles travel on a highway to move from one destination to another, bits travel on a bus (Figure 3-36).

Buses are used to transfer bits from input devices to memory, from memory to the CPU, from the CPU to memory, and from memory to output or storage devices. All buses consist of two parts: a data bus and an address bus. The data bus transfers actual data and the address bus transfers information about where the data should go in memory.

A bus is measured by its size. The size of a bus, called the **bus width**, determines the number of bits that can be transmitted at one time. For example, a 32-bit bus can transmit 32 bits (four bytes) at a time. On a 64-bit bus, bits are transmitted from one location to another 64 bits (eight bytes) at a time. The larger the number of bits handled by the bus, the faster the computer transfers data.

Using the highway analogy again, assume that one lane on a highway can carry one bit. A 32-bit bus, then, is like a 32-lane highway; and a 64-bit bus is like an 64-lane highway.

If a number in memory occupies eight bytes, or 64 bits, it must be transmitted in two separate steps when using a 32-bit bus: once for the first 32 bits and once for the second 32 bits. Using a 64-bit bus, however, the number can to be transmitted in a single step, transferring all 64 bits at once. The wider the bus, the fewer number of transfer steps required and the faster the transfer of data. Figure 3-37 summarizes some of the microprocessors currently in use and their bus widths.

In conjunction with the bus width, many computer professionals discuss a computer's word size. **Word size** is the number of bits

COMPARISON OF BUS WIDTHS

Name	Bus Width
Itanium™	64
Pentium® III Xeon™	64
Pentium® III	64
Celeron™	64
Pentium® II Xeon™	64
Pentium® II	64
Pentium® with MMX™ technology	64
Pentium® Pro	64
Pentium®	64
80486DX	32
80386DX	32
80286	16
PowerPC	64
68040	32
68030	32
68020	32
Alpha	64

Figure 3-37 *A comparison of bus widths on some of the more widely used microprocessors.*

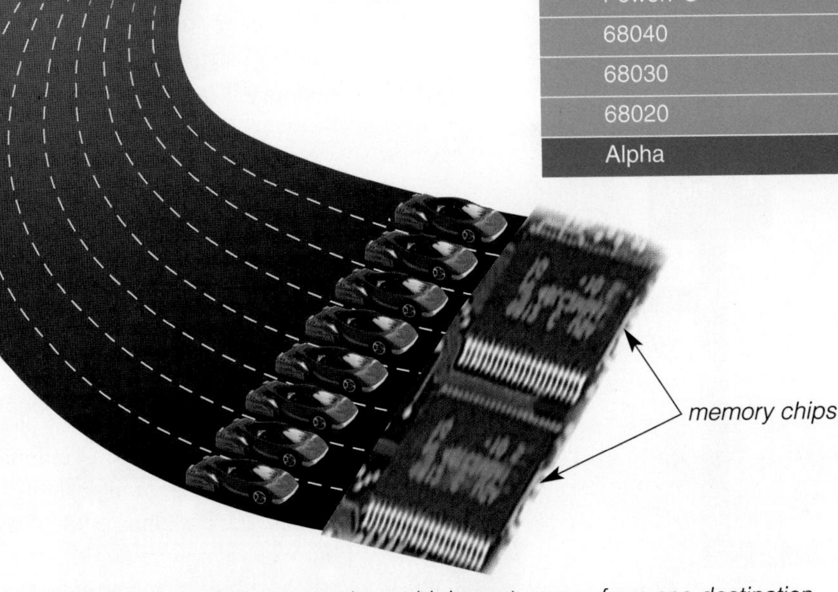

Figure 3-36 *Just as vehicles travel on a highway to move from one destination to another, bits travel on a bus. Buses are used to transfer bits from input devices to memory, from memory to the CPU, from the CPU to memory, and from memory to output or storage devices.*

the CPU can process at a given time. That is, a 64-bit processor can manipulate 64 bits at a time. Computers with a larger word size can process more data in the same amount of time than computers with a smaller word size. In most computers, the word size is the same as the bus width.

Every bus also has a clock speed. Just like the processor, the clock speed for a bus is measured in megahertz. Recall that one megahertz (MHz) is equal to one million ticks per second. Most of today's processors have a bus speed of either 100 MHz or 133 MHz. The higher the bus clock speed, the faster the transmission of data, which results in applications running faster.

Two basic types of buses are found in a computer: a system bus and an expansion bus. A **system bus** connects the CPU to main memory. An **expansion bus** allows the CPU to communicate with peripheral devices. When computer professionals use the term bus by itself, they usually are referring to the system bus.

Expansion Bus

Recall that a device outside the system unit is connected to a port on an expansion card, and an expansion card is inserted into an expansion slot. This expansion slot connects to the expansion bus, which allows the CPU to communicate with the peripheral device attached to the expansion card. Data transmitted to memory or the CPU travels from the expansion bus via the expansion bus and the system bus. The types of expansion buses on a motherboard determine the types of expansion cards you can add (Figure 3-38). For this reason, you should understand the following types of expansion buses: ISA bus, PCI bus, AGP bus, USB, 1394 bus, and PC Card bus.

- The most common and slowest expansion bus is the **ISA** (**Industry Standard Architecture**) **bus**. A mouse, modem card, sound card, and low-speed network card are examples of devices that connect to the ISA bus directly or through an ISA bus expansion slot.

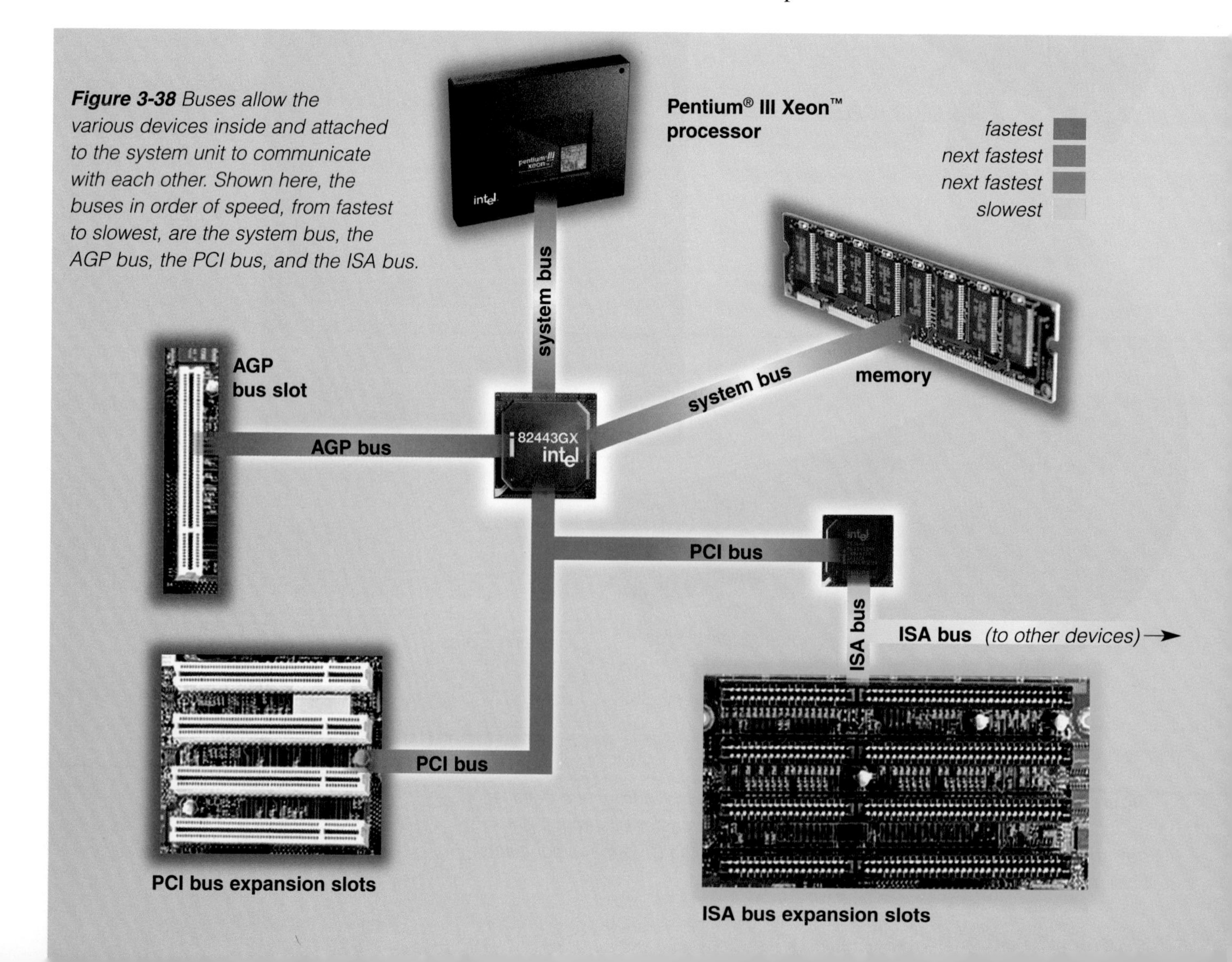

Figure 3-38 *Buses allow the various devices inside and attached to the system unit to communicate with each other. Shown here, the buses in order of speed, from fastest to slowest, are the system bus, the AGP bus, the PCI bus, and the ISA bus.*

- A **local bus** is a high-speed expansion bus used to connect higher speed devices such as hard disks. The first standard local bus was the **VESA local bus**, which was used primarily for video cards. The current local bus standard, however, is the **PCI (Peripheral Component Interconnect) bus** because it is more versatile than the VESA local bus. Types of cards inserted into a PCI bus expansion slot include video cards, SCSI cards, and high-speed network cards. The PCI bus transfers data about four times faster than the ISA bus. Most current personal computers have a PCI bus as well as an ISA bus.
- The **Accelerated Graphics Port (AGP)** is actually a bus designed by Intel to improve the speed with which 3-D graphics and video are transmitted. When an AGP video card is inserted in an AGP bus slot, the AGP bus provides a faster, dedicated interface between the video card and memory. Newer processors, such as the Pentium® II and Xeon™, support AGP technology.
- The **universal serial bus (USB)** and **1394 bus** are buses that eliminate the need to install expansion cards into expansion slots. In a computer equipped with a USB, for example, USB devices are connected to each other outside the system unit and then a single cable attaches to the USB port. The USB port then connects to the USB, which connects to the PCI bus on the motherboard. The 1394 bus works in a similar fashion. With these buses, you need not be concerned with running out of expansion slots.
- The expansion bus for a PC Card is the **PC Card bus**. With a PC Card inserted into a PC Card slot, data travels on the PC Card bus to the PCI bus.

BAYS

After you purchase a computer, you may want to install an additional device such as a disk drive to add storage capabilities to the system unit. A **bay** is an open area inside the system unit used to install additional equipment. Note that a bay is different from a slot, which is used for the installation of expansion cards. Because bays most often are used for disk drives, these spaces commonly are called **drive bays**.

Two types of drive bays exist: internal and external. An **external drive bay** or **exposed drive bay** allows access to the drive from outside the system unit. Floppy disk drives, CD-ROM drives, DVD-ROM drives, Zip® drives, and tape drives are examples of devices installed in external drive bays (Figure 3-39). An **internal drive bay** or **hidden drive bay** is concealed entirely within the system unit. Hard disk drives are installed in internal bays.

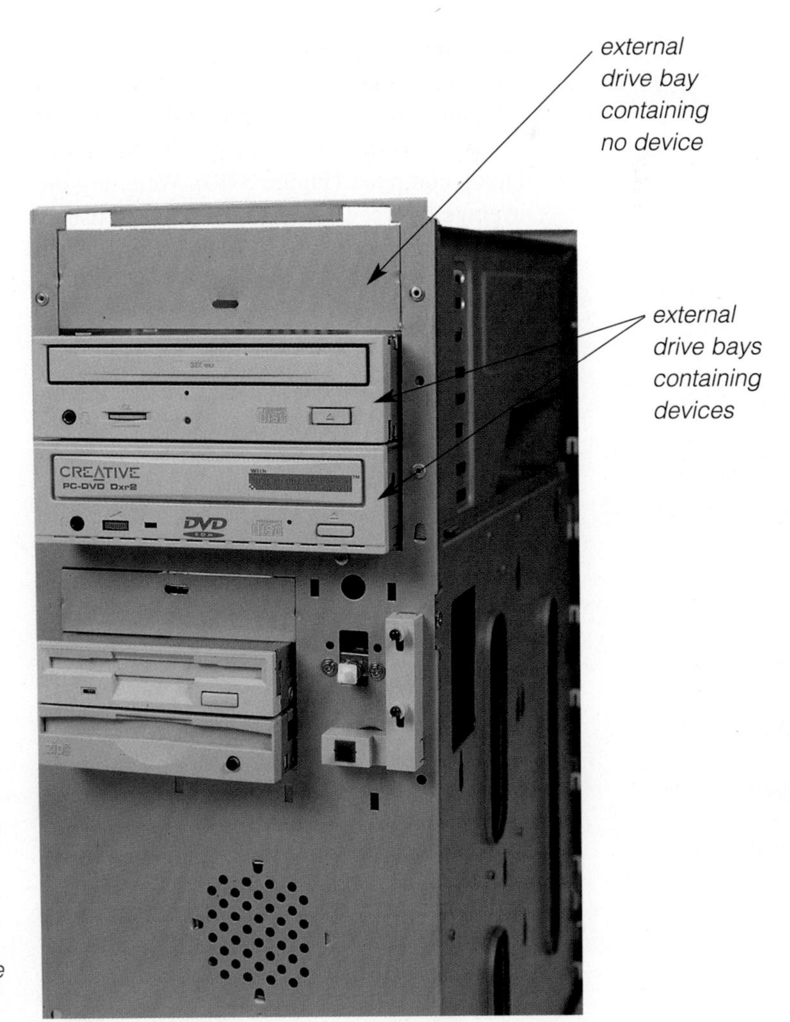

Figure 3-39 *Bays, also called drive bays, usually are located beside or on top of one another. Each bay is about 6 inches wide and 1.75 inches high.*

POWER SUPPLY

Many personal computers are plugged into standard wall outlets, which supply an alternating current (AC) of 115 to 120 volts. This type of power is unsuitable for use with a computer, which requires a direct current (DC) ranging from 5 to 12 volts. The **power supply** is the component in the system unit that converts the wall outlet AC power into DC power.

Some external peripheral devices such as an external modem or tape drive have an **AC adapter**, which is an external power supply. One end of the AC adapter plugs into the wall outlet and the other end attaches to the peripheral device. The AC adapter converts the AC power into DC power that the device requires.

LAPTOP COMPUTERS

As businesses expand to serve customers across the country and around the world, more and more people need to use a computer while traveling to and from a main office to conduct business. As noted in Chapter 1, users with such mobile computing needs – known as mobile users – often have a laptop computer (Figure 3-40). Weighing on average between four and ten pounds, usually these computers can run either using batteries or using a standard power supply.

Like their desktop counterparts, laptop computers have a system unit that contains electronic components used to process data (Figure 3-41). The difference is that many other devices also are built into the system unit. In addition to the motherboard, processor, memory, sound card, PC Card slot, and drive bay, the system unit also houses devices such as the keyboard, pointing device, and speakers. A laptop computer usually is more expensive than a desktop computer with the same capabilities.

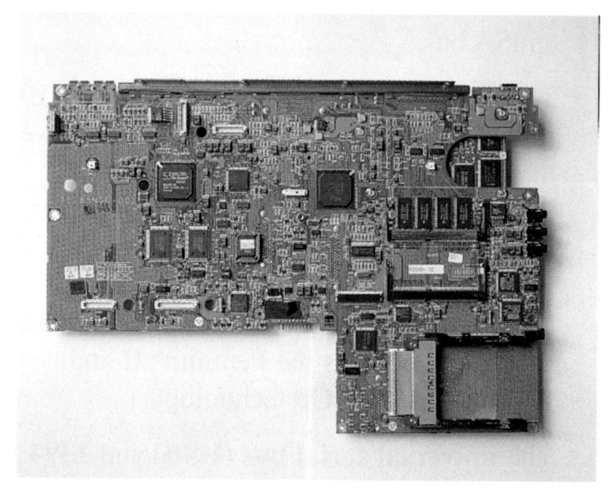

Figure 3-41 *Laptop computers have a system unit that contains electronic components used to process data.*

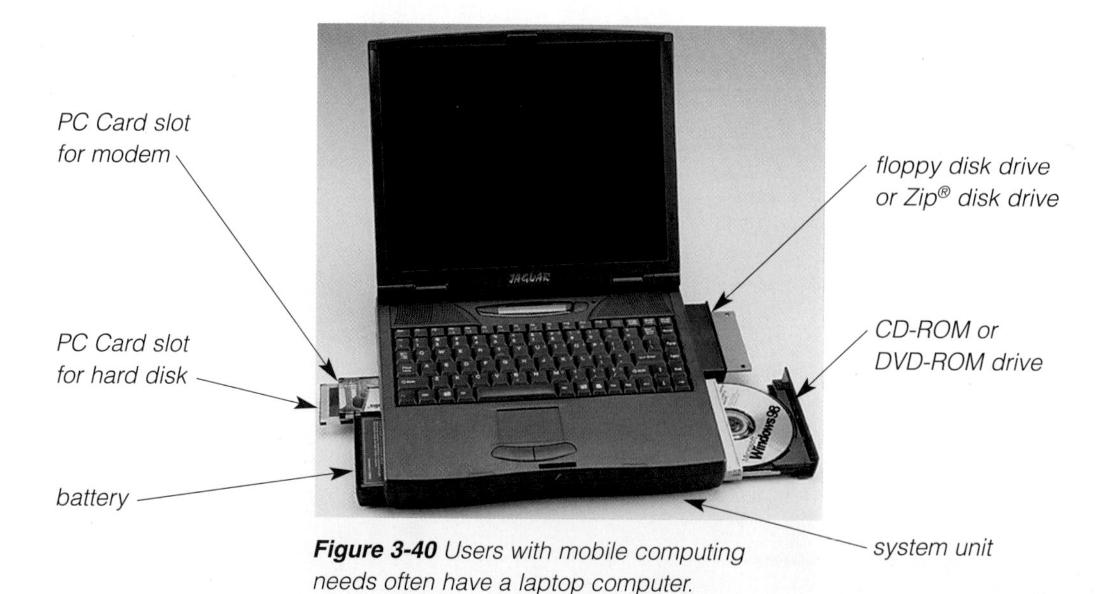

Figure 3-40 *Users with mobile computing needs often have a laptop computer.*

The typical laptop computer often is equipped with serial, parallel, mouse, USB, video, IrDA, and docking station ports (Figure 3-42). Recall that a docking station is a device into which you place a laptop. The docking station contains connections to peripherals such as a keyboard, monitor, printer, and other devices. Some laptops also have a **port replicator**, which is a device that allows you to connect many peripheral devices (such as a printer, modem, and mouse) into it; the port replicator then is connected to the laptop computer.

PUTTING IT ALL TOGETHER

When you purchase a computer, you should have an understanding of the components in the system unit. Many factors inside the system unit influence the speed and power of a computer. The type of computer configuration you require depends on your intended use. The table in Figure 3-43 lists the suggested processor, clock speed, and RAM requirements based on the needs of various types of computer users.

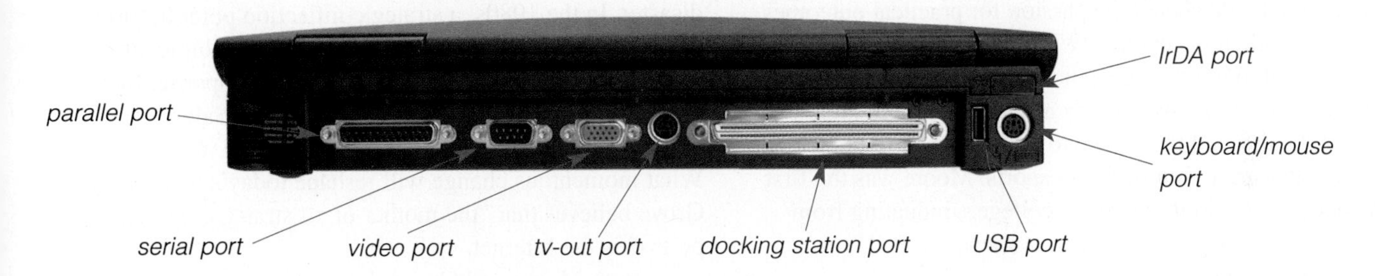

Figure 3-42 *A laptop computer often is equipped with serial, parallel, mouse, USB, video, IrDA, and docking station ports.*

SUGGESTED CONFIGURATIONS BY USER

USER	PROCESSOR AND CLOCK SPEED	MINIMUM RAM
Home User	Pentium® III or Athlon™ – 600 MHz or higher; or Celeron™ or AMD-K6®-2 – 533 MHz or higher	32 MB
SMALL BUSINESS USER	Pentium® III or Athlon™ – 650 MHz or higher	64 MB
MOBILE USER	Pentium® III or AMD-K6®-2-P 500 MHz or higher	64 MB
LARGE BUSINESS USER	Pentium® III or Athlon™ – 700 MHz or higher	128 MB
POWER USER	Pentium® III Xeon™ or Itanium™ or Athlon™ – 800 MHz or higher	256 MB

Figure 3-43 *This table recommends suggested processor, clock speed, and RAM configurations.*

TECHNOLOGY TRAILBLAZERS

ANDY GROVE AND GORDON MOORE

The name Intel has become synonymous with microprocessors. From its inception, Gordon Moore and Andy Grove have been synonymous with Intel.

Gordon Moore's life-long interest in technology was kindled at an early age by a neighbor's chemistry set. Even then, he displayed the passion for practical outcomes that has typified his career. "With the chemistry set," he says, "I had to get a good explosion at the end or I wasn't happy." Yet, Moore was hardly a *science geek*. In high school, he devoted more time to athletics than to homework, lettering in four different sports. Moore was the first member of his family to attend college, graduating from the California Institute of Technology with a Ph.D. in chemistry and physics.

Moore worked with Bill Shockley, inventor of the transistor, at Shockley Semiconductor. There, he met Robert Noyce, and eventually the two left to join Fairchild Semiconductor. At Fairchild in 1965, Moore made a startling prediction. The power of silicon chips, he claimed, would *double* every eighteen months. This bold forecast, now known as Moore's Law, would prove amazingly accurate. Convinced of the future of silicon chips, and frustrated by the company's response to their work, in 1968 Moore and Noyce quit Fairchild to start Intel. "We saw a new way of storing information for computers," Moore notes. "A product area where... the existing semiconductor companies were not active."

Moore and Noyce were Intel's inspiration, but Andy Grove was the key to Intel's execution. Born in Budapest, Hungary, Grove survived the reigns of Hitler and Stalin, escaping to the United States in 1957. He completed his undergraduate degree in three years, while learning English and working as a waiter. After earning a Ph.D. in chemical engineering from the University of California at Berkeley, Grove worked as Moore's assistant at Fairchild. He was one of the first people recruited by Intel, where he was put in charge of production. According to Grove, by modern standards Intel's early assembly lines "looked like Willy Wonka's chocolate factory, with hoses and wires and contraptions chugging along." Grove fueled production with his firm, demanding style, emphasizing turnout over mere motion. "Stressing output is the key to productivity, while looking to increase activity can result in just the opposite." Grove adopted a team-based approach, encouraging managers and employees to meet one-on-one to impart data and foster a feeling of a shared effort. "People in the trenches," he points out, "are usually in touch with impending changes early."

Grove insists that recognizing and responding to changes is a crucial element in business success. In his book, *Only the Paranoid Survive*, Grove writes of "strategic inflection points" – changes that alter the fundamentals of a business. These changes can lead to opportunity or disaster. In the 1980s, a strategic inflection point occurred in the computer industry when a production glut resulted in an overabundance of memory chips. In response, Intel concentrated its efforts on microprocessors. Sales soared $789 million in 1981 to more than $3 billion in 1990. What momentous change will reshape today's businesses? Grove believes that "the mother of all strategic inflection points" is the Internet.

Both Moore and Grove share a key characteristic: the willingness to make a commitment even when results are unknown. This forward-looking approach is a hallmark of Intel's corporate philosophy. Grove writes, "The best thing is to make the right decision. Making a wrong decision is okay, too. The worst thing to do is hedge. To hedge is to fail." Moore concurs, noting that although they involve some risk, decisions should not be feared. After all, Moore says, "If everything you try works, then you are not trying hard enough."

Andrew S. Grove

Gordon E. Moore

COMPANY ON THE CUTTING EDGE

COMPANY ON THE CUTTING EDGE

INTEL

On an August day in 1968, Robert Noyce was mowing his lawn when Gordon Moore, a coworker at Fairchild Semiconductor, stopped by to talk. The two shared gripes, frustrated by the company's apparent disinterest in their work. Colleagues, who some called "Fairchildren," had left to form new companies. In a moment of inspiration, the pair decided to start their own company and manufacture a new product – semiconductors to replace the magnetic cores that comprised computer memory. Noyce typed a one-page business plan, an investment banker raised $2.5 million, the partners each put up $250,000, and a company called Moore Noyce emerged. The name, unfortunately, sounded a lot like "More Noise" – hardly suitable for an electronics firm – so they incorporated as NM Electronics. Later, after purchasing rights from a motel chain, a new name was adopted – Intel (for integrated electronics). From such humble beginnings came a company that today has more than $25 billion in sales.

Even if semiconductors could be used for memory storage, critics maintained it would cost much more to manufacture semiconductors than magnetic cores. Undismayed, Noyce and Moore, together with Andy Grove, another Fairchild expatriate, struggled to reduce production costs while packing more transistors on a chip. Moore had predicted that the power of silicon chips would continue to grow. If he was right, eventually memory chips would be less expensive and more popular than magnetic cores. As Intel perfected increasingly more powerful memory chips, sales climbed steadily.

Intel's most revolutionary product, however, was developed in answer to a customer request. A calculator manufacturer asked Intel to produce twelve custom chips. Instead of twelve separate chips, Ted Hoff, an Intel engineer, suggested designing a single chip that could function as twelve. The entire central processing unit would be placed on one, general-purpose programmable chip, saving money and time. This chip, introduced in 1971, was the first microprocessor. Dubbed the Intel 4004, the chip measured only 1/8 inch by 1/6 inch but had 2,300 transistors, could perform 60,000 operations per second, and packed as much power as ENIAC, the room-sized, vacuum-tube computer. The 4004 was one of the more important inventions in the history of technology.

Manufacturing and design rights were purchased from the calculator maker, and Intel began promoting its general-purpose programmable chip in the engineering community. Ever more powerful chips drove earnings to more than $660 million in the late 1970s. When IBM chose the Intel 8088 chip for its new personal computer in 1980, Intel chips became the standard for all IBM-compatible personal computers. Yet, competition was fierce as other chip manufacturers developed and marketed less-expensive clones. To meet the challenge, Intel introduced increasingly capable microprocessors, including the 386 and 486, the first chip to speed computations with a math coprocessor.

Intel's climb was not without missteps. The most jarring stumble occurred in 1994 when a design flaw was discovered in Intel's new Pentium® chip. For the average user, the defect resulted in a mistake only once in every 27,000 years, so Intel ignored the error. Faced with an unanticipated public outcry, however, the company reconsidered and offered replacement chips, at a cost of almost $500 million. An increased awareness of public perceptions caused Intel to launch an extensive new advertising campaign, immortalizing the slogan, "Intel Inside." By 1997, Intel controlled ninety percent of the microprocessor market.

Today, Intel employs more than 60,000 workers at sites around the world. The company supports a number of charities and encourages employees to aid schools and nonprofit organizations by volunteering via "Intel Involved." Through these efforts, Intel hopes to demonstrate its commitment not only to the development of new products, but also to the communities that use them.

CHAPTER 1 2 3 4 5 6 7 8 9 10 11 12 13 14 INDEX

IN BRIEF

www.scsite.com/dc2001/ch3/brief.htm

WEB INSTRUCTIONS: *To display this page from the Web, launch your browser and enter the URL, www.scsite.com/dc2001/ch3/brief.htm. Click the links for current and additional information. To listen to an audio version of this IN BRIEF, click the Audio button to the right of the title, IN BRIEF, at the top of the page. To play the audio, RealPlayer must be installed on your computer (download by clicking here).*

1. What Are the Components of the System Unit?

The **system unit** is a box-like case housing the electronic components of a computer that are used to process data. System unit components include the processor, memory module, expansion cards, ports, and connectors. Many components reside on a circuit board called the **motherboard** or **system board.** The motherboard contains different types of **chips**, or small pieces of semiconducting material on which one or more **integrated circuits** (**IC**) are etched. One of the more important chips is the central processing unit.

2. How Does the CPU Process Data?

The **central processing unit** (**CPU**), sometimes referred to as the **processor**, interprets and carries out the basic instructions that operate a computer. The **control unit**, one component of the CPU, directs and coordinates most of the operations in the computer. For every instruction, the control unit repeats a set of four basic operations called the machine cycle: (1) **fetching** the instruction or data item from memory, (2) **decoding** the instruction into commands the computer understands, (3) **executing** the commands, and, if necessary, (4) **storing**, or writing, the result to memory. The **arithmetic/logic unit (ALU)**, another component of the CPU, performs the execution part of the machine cycle.

3. How Do Pipelining and the System Clock Affect Processing Speed?

With **pipelining**, the CPU begins executing a second instruction before the first instruction is completed. Pipelining results in faster processing because the CPU does not have to wait for one instruction to complete the machine cycle. The **system clock** is a small chip that the control unit relies on to synchronize computer operations. The faster the clock, the more instructions the CPU can execute per second. The speed at which a processor executes instructions is called **clock speed.** Clock speed is measured in **megahertz (MHz)**, which equates to one million ticks of the system clock.

4. What Are Some Microprocessors Available Today?

A personal computer's CPU usually is contained on a single chip called a **microprocessor**. Intel, a leading manufacturer of microprocessors, produces **Pentium® processors** for high-end personal computers, the **Celeron™** processor for less expensive PCs, and the **Xeon™** and **Itanium™** processors for workstations and servers. **Intel-compatible microprocessors** have the same internal design as Intel processors and perform the same functions, but are made by other companies and often are less expensive. The **Motorola microprocessor** is an alternative to the Intel-style microprocessor and is found in Apple Macintosh and Power Macintosh systems. The **Alpha microprocessor**, originally from Digital Equipment Corporation, is used primarily in workstations and high-end servers. A new type of microprocessor, called an **integrated CPU**, combines functions of a CPU, memory, and a graphics card on a single chip.

IN BRIEF

 www.scsite.com/dc2001/ch3/brief.htm

5. How Do Series of Bits Represent Data?

Most computers are **digital**, meaning they understand only two discrete states: on and off. These states are represented using two digits, 0 (off) and 1 (on). Each on or off value is called a **bit** (short for **bi**nary digi**t**), which is the smallest unit of data a computer can handle. Eight bits grouped together as a unit are called a **byte**. A byte can represent 256 individual characters including numbers, letters of the alphabet, punctuation marks, and other characters. Combinations of 0s and 1s used to represent data are defined by patterns called coding schemes. Popular coding schemes are **ASCII**, **EBCDIC**, and **Unicode**.

6. What Are Different Types of Memory?

In the system unit, a computer's **memory** stores data, instructions, and information. The number of bytes it can store measures memory size – a **kilobyte (KB)** is approximately one thousand bytes, and a **megabyte (MB)** is approximately one million bytes. **RAM (random access memory)** is a memory chip that the processor can read from and write to. Two types of RAM exist; **dynamic RAM (DRAM)**, which must be re-energized constantly, and **static RAM (SRAM)**, which must be reenergized less often but is more expensive. Most computers improve processing times by using **memory cache** to store frequently used instructions. **ROM (read-only memory)** is a memory chip that only can be read and used; that is, it cannot be modified. **Flash memory**, or **flash ROM**, is nonvolatile memory that can be erased electronically and reprogrammed. **CMOS** memory is nonvolatile memory used to store configuration information about the computer.

Rom Permanent
can't change
A. Flash
B. Cmos

RAM - Temporary
Lost if not saved
D Rom
S Rom.

7. What Are Expansion Slots and Expansion Boards?

An **expansion slot** is an opening, or socket, where a circuit board can be inserted into the motherboard. These circuit boards, sometimes referred to as **expansion boards** or **expansion cards**, add new devices or capabilities to the computer, such as a modem or more memory. **Plug and Play** refers to a computer's capability to configure expansion boards automatically and other devices as they are installed.

8. How Is a Serial Port Different from a Parallel Port?

A cable often attaches external devices to the system unit. The interface, or point of attachment, to the system unit is called a **port**. Ports have different types of **connectors** used to join a cable to a device. A **serial port** is an interface that transmits only one bit of data at a time. Serial ports usually are used to connect devices that do not require fast data transmission rates, such as a mouse, keyboard, or modem. A **parallel port** is an interface used to connect devices that are capable of transferring more than one bit at a time. Many printers connect to the system unit using a parallel port.

9. How Do Buses Contribute to a Computer's Processing Speed?

Bits are transferred internally within the circuitry of the computer along electrical channels. Each channel, called a bus, allows various devices inside and attached to the system unit to communicate with each other. The **bus width**, or size of the bus, determines the number of bits that can be transferred at one time. The larger the bus width, the faster the computer transfers data.

DISCOVERING COMPUTERS 2001

Student Exercises
WEB INFO
IN BRIEF
KEY TERMS
AT THE MOVIES
CHECKPOINT
AT ISSUE
MY WEB PAGE
HANDS ON
NET STUFF

Special Features
TIMELINE 2001
GUIDE TO WWW SITES
MAKING A CHIP
E-COMMERCE 2001
BUYER'S GUIDE 2001
CAREERS 2001
TRENDS 2001
LEARNING GAMES
CHAT
INTERACTIVE LABS
NEWS
HOME

CHAPTER 1 2 3 4 5 6 7 8 9 10 11 12 13 14 INDEX

KEY TERMS

www.scsite.com/dc2001/ch3/terms.htm

WEB INSTRUCTIONS: *To display this page from the Web, launch your browser and enter the URL, www.scsite.com/dc2001/ch3/terms.htm. Scroll through the list of terms. Click a term to display its definition and a picture. Click KEY TERMS on the left to redisplay the KEY TERMS page. Click the TO WEB button for current and additional information about the term from the Web. To see animations, Shockwave and Flash Player must be installed on your computer (download by clicking here).*

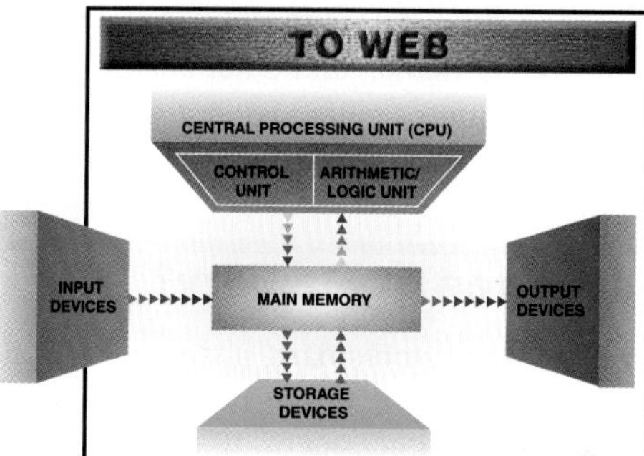

CENTRAL PROCESSING UNIT (CPU): Interprets and carries out the basic instructions that operate a computer. The CPU contains the control unit and the arithmetic/logic unit that work together to perform the processing operations. Also called processor, and on the personal computer it is called a microprocessor because it usually is contained on a single chip. (3.5)

1394 bus (3.27)
1394 port (3.24)
3DNow!™ technology (3.10)
AC adapter (3.28)
Accelerated Graphics Port (AGP) (3.27)
access time (3.19)
adapter card (3.20)
add-in (3.20)
add-on (3.20)
address (3.14)
Alpha microprocessor (3.9)
American Standard Code for Information Interchange (ASCII) (3.12)
analog (3.12)
arithmetic operations (3.7)
arithmetic/logic unit (ALU) (3.7)
Athlon™ (3.9)
bay (3.27)
binary (3.12)
bit (3.12)
bus (3.24)
bus width (3.25)
byte (3.12)
cache (3.17)
card (3.20)
Celeron™ (3.9)
central processing unit (CPU) (3.5)
chip (3.4)
chip for chip upgrade (3.10)
clock rate (3.8)
clock speed (3.8)
comparison operations (3.7)
complementary metal-oxide semiconductor (CMOS) (3.19)
connector (3.22)
control unit (3.6)
coprocessor (3.11)
daughterboard (3.10)
daughterboard upgrade (3.10)
decoding (3.6)
digital (3.12)
drive bays (3.27)
dual inline memory module (DIMM) (3.16)
dynamic RAM (DRAM) (3.16)
E-time (3.6)
executing (3.6)
execution time (3.6)
external cache (3.17)
expansion board (3.20)
expansion bus (3.26)
expansion card (3.20)
expansion slot (3.20)
exposed drive bay (3.27)
Extended Binary Coded Decimal Interchange Code (EBCDIC) (3.12)
external drive bay (3.27)
female connectors (3.22)
FIR (fast infrared) (3.24)
FireWire (3.24)
fetching (3.6)
firmware (3.18)
flash memory (3.18)
flash ROM (3.18)
floating-point coprocessor (3.11)
gender changer (3.22)
gigahertz (GHz) (3.8)
graphics card (3.20)
heat pipe (3.11)
heat sink (3.11)
hidden drive bay (3.27)
hot plugging (3.21)
hot swapping (3.21)
ISA (Industry Standard Architecture) bus (3.26)
instruction cycle (3.6)
instruction time (3.6)
integrated circuit (IC) (3.4)
integrated CPU (3.9)
Intel-compatible microprocessors (3.9)
interface card (3.20)
internal cache (3.17)
internal drive bay (3.27)
internal modem (3.20)
IrDA (Infrared Data Association) (3.34)
IrDA port (3.24)
Itanium™ (3.9)
I-time (3.6)
kilobyte (KB or K) (3.14)
Level 1 (L1) cache (3.17)
Level 2 (L2) cache (3.17)
local bus (3.27)
logical operations (3.7)
machine cycle (3.6)
male connectors (3.22)
megabyte (MB) (3.15)
megahertz (MHz) (3.8)
memory (3.14)
memory cache (3.17)
microcode (3.18)
microprocessor (3.5)
millisecond (ms) (3.20)
MIPS (3.7)
MMX™ technology (3.10)
motherboard (3.4)
Motorola microprocessor (3.9)
musical instrument digital interface (MIDI) port (3.24)
nanosecond (ns) (3.19)
nonvolatile memory (3.15)
parallel port (3.23)
parallel processing (3.11)
PC Card (3.21)
PC Card bus (3.27)
PCI (Peripheral Component Interconnect) bus (3.27)
PCMCIA cards (3.21)
Pentium® processors (3.9)
piggyback upgrade (3.10)
pipelining (3.7)
Plug and Play (3.20)
port (3.21)
port replicator (3.29)
power supply (3.28)
processor (3.5)
primary cache (3.17)
programmable read-only memory (PROM) (3.18)
Rambus® DRAM (RDRAM®) (3.16)
Rambus® inline memory module (RIMM) (3.16)
random access memory (RAM) (3.15)
read-only memory (ROM) (3.18)
registers (3.8)
saved (3.16)
SCSI port (3.24)
SDRAM II (3.16)
serial port (3.23)
single inline memory module (SIMM) (3.16)
small computer system interface (SCSI) port (3.24)
sound card (3.20)
SSE instructions (3.10)
static RAM (SRAM) (3.16)
storing (3.6)
stored program concept (3.14)
superscaler (3.8)
synchronous DRAM (SDRAM) (3.16)
synthesizer (3.24)
system board (3.4)
system bus (3.26)
system clock (3.8)
system unit (3.2)
transistors (3.4)
Type I cards (3.21)
Type II cards (3.21)
Type III cards (3.21)
Unicode (3.12)
universal serial bus (USB) (3.24, 3.27)
VESA local bus (3.27)
video adapter (3.20)
video card (3.20)
volatile memory (3.15)
word size (3.25)
Xeon™ processor (3.9)
zero-insertion force (ZIF) socket (3.10)

CHAPTER 1 2 3 4 5 6 7 8 9 10 11 12 13 14 INDEX

AT THE MOVIES

www.scsite.com/dc2001/ch3/movies.htm

Student Exercises
WEB INFO
IN BRIEF
KEY TERMS
AT THE MOVIES
CHECKPOINT
AT ISSUE
MY WEB PAGE
HANDS ON
NET STUFF
Special Features
TIMELINE 2001
GUIDE TO WWW SITES
MAKING A CHIP
E-COMMERCE 2001
BUYER'S GUIDE 2001
CAREERS 2001
TRENDS 2001
LEARNING GAMES
CHAT
INTERACTIVE LABS
NEWS
HOME

WELCOME to VIDEO CLIPS from CNN

WEB INSTRUCTIONS: *To display this page from the Web, launch your browser and enter the URL, www.scsite.com/dc2001/ch3/movies.htm. Click a picture to view a video. After watching the video, close the video window and then complete the exercise by answering the questions about the video. To view the videos, RealPlayer must be installed on your computer (download by clicking here).*

1 Grove Profile

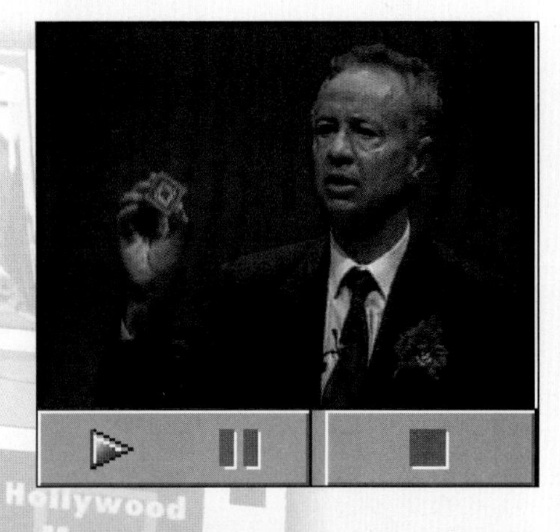

Intel is the leading manufacturer of microprocessors, including the Pentium, Celeron, and Xeon processors. Intel grew to its present size with more than 60,000 employees as a result of the outstanding leadership of Andrew Grove. Based on the personal information you learned about Mr. Grove, describe how his early life struggles, his strong work ethic, and his vision of capitalizing on business trends have been the foundation for Intel's worldwide microprocessor empire. How might Andrew Grove's forward-growth visions for Intel continue to drive the company toward even greater success in the future?

2 IBM

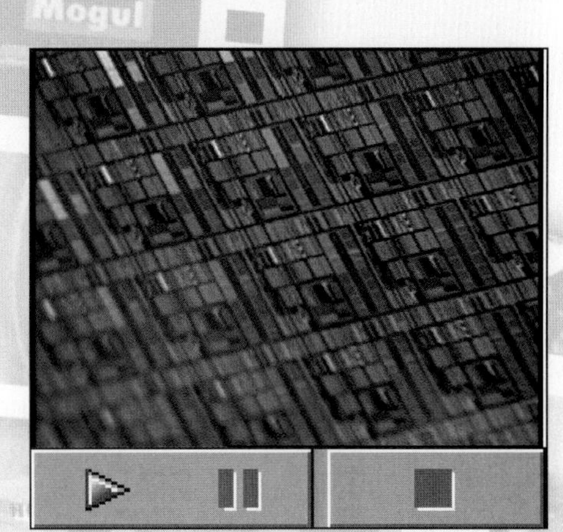

IBM is keeping pace with the phenomenal rate of change in the computer technology industry by manufacturing microchips in a more desirable way than in the past. They have developed a breakthrough in semiconductor production that promises to affect IBM's business favorably. Confidence in the new process has been reflected by an increase in the value of the company's stock. Explain the new technology IBM has developed. How does it differ from the way chips have been made in the past? What are the benefits of the new technology to consumers and businesses?

3 Clone Buster

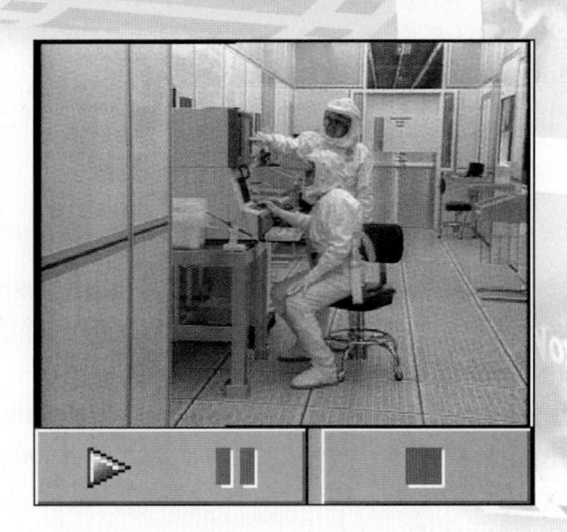

Although some felt the name, Pentium, would never take hold, the name has withstood the test of time. Why did Intel decide to move away from using numbers to name its chips? Why did Intel choose the name Pentium? Would you have chosen this name? If not, why?

CHECKPOINT

www.scsite.com/dc2001/ch3/check.htm

WEB INSTRUCTIONS: *To display this page from the Web, launch your browser and enter the URL, www.scsite.com/dc2001/ch3/check.htm. Click the links for current and additional information. To experience the animation and interactivity, Shockwave and Flash Player must be installed on your computer (download by clicking here).*

Label the Figure

Instructions: *Identify these components of the motherboard.*

Matching

Instructions: *Match each term from the column on the left with the best description from the column on the right.*

_______ 1. DRAM (dynamic RAM)

_______ 2. SRAM (static RAM)

_______ 3. SIMM (single inline memory module)

_______ 4. DIMM (dual inline memory module)

_______ 5. RAM cache (or memory cache)

a. Circuit board with RAM chips on only one side.
b. Type of memory that must be re-energized constantly or it loses its contents.
c. Nonvolatile memory that can be read and used but not modified.
d. Speeds processing time by storing frequently used instructions and data.
e. Circuit board with RAM chips on both sides.
f. Blank, nonvolatile memory chip on which items can be permanently placed.
g. More expensive type of memory that can be re-energized less often; used for special purposes.

Short Answer

Instructions: *Write a brief answer to each of the following questions.*

1. What is the purpose of the CPU? ___________ What are registers? ___________
2. How is instruction time, or I-time, different from execution time, or E-time? ___________ In what unit is a computer's speed measured? ___________
3. How are arithmetic operations, comparison operations, and logical operations different? ___________
4. What are some special-purpose ports used in many of today's computers? ___________ For what purpose is each port used? ___________
5. How is a system bus different from an expansion bus? ___________ How is an internal drive bay different from an external drive bay? ___________

Student Exercises
WEB INFO
IN BRIEF
KEY TERMS
AT THE MOVIES
CHECKPOINT
AT ISSUE
MY WEB PAGE
HANDS ON
NET STUFF
Special Features
TIMELINE 2001
GUIDE TO WWW SITES
MAKING A CHIP
E-COMMERCE 2001
BUYER'S GUIDE 2001
CAREERS 2001
TRENDS 2001
LEARNING GAMES
CHAT
INTERACTIVE LABS
NEWS
HOME

AT ISSUE

www.scsite.com/dc2001/ch3/issue.htm

WEB INSTRUCTIONS: *To display this page from the Web, launch your browser and enter the URL, www.scsite.com/dc2001/ch3/issue.htm. Click the links for current and additional information.*

ONE

As a rule, higher quality means higher prices. Computers, however, seem to challenge this rule. Today, personal computers purchased for less than $1,000 are more powerful than PCs that cost almost twice as much just a few years ago. Part of the plunge in prices is lower cost components, but another factor is a growing demand for cheaper machines. Many contemporary consumers have less income than previous purchasers. These bargain-conscious buyers want inexpensive PCs that are adequate for basic tasks, such as word processing and Internet access. As one industry analyst asks, "Why buy a Porsche when you're only going to drive 55 miles per hour?" How might an increased demand for low-cost computers affect PC manufacturers? How might a continued decline in PC prices impact consumers? How might a greater availability of low-cost PCs change the way computers are used in schools and businesses?

TWO

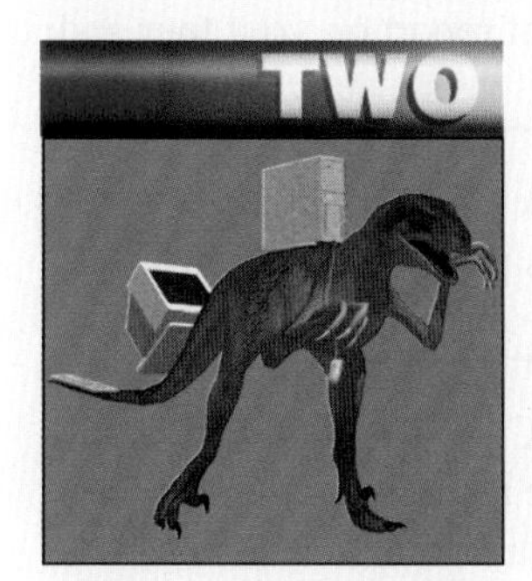

From 1989 to 1999, the maximum clock speed of Intel's microprocessors increased from 25 MHz to more than 700 MHz. New micron production processes and copper-based circuits continue the acceleration. Faster and less expensive microprocessors are anticipated, with clock speeds approaching 1GHz (1,000 MHz). In the midst of this progress, whispers of planned obsolescence are heard among some naysayers. Faster processor speed and new software that demands it, they argue, will make a PC bought today obsolete in two years. Besides, they continue, is the increased speed really necessary? What do you think? Is greater processor speed a benefit for manufacturers and consumers alike, or a boon for builders and a burden for buyers? Why?

THREE

At a recent auto show, Citroen unveiled the first car equipped with Intel's Connected Car PC technology. The car PC lets drivers access e-mail (which is read by a text-to-speech converter), receive weather and traffic data, and obtain navigation information. Passengers can use the PC to watch movies on DVD. Developing and producing car PCs is expensive. Because of their special environment, they must be more durable and offer a simpler interface than desktop PCs. Many automakers wonder if car PCs offer buyers enough to justify their cost. Would you be interested in a car PC? Why or why not? How much would you be willing to add to the price of a new car for the convenience of a car PC?

FOUR

Gordon Moore, cofounder of Intel, fears that society is becoming a two-class society, "those who are wired and those who are not." A recent Commerce Department report echoes Moore's concerns. The report asserts that age, income, and education all are factors in computer use. Do those who "are wired" have an advantage over those who "are not wired?" In what way? If you do not have a computer and cannot access the Internet, can you still compete for good grades in school? What can be done to make computing available to everyone? Do the computers in public and school libraries solve the problem? Why or why not? If this situation continues, what will be the effects on society as a whole?

In 1994, a design flaw in Intel's Pentium micro-processor chip caused a rounding error once in nine billion division operations. For most users, this would result in a mistake only once in every 27,000 years, so Intel initially ignored the problem. After an unexpected public outcry, however, eventually Intel supplied replacements to anyone who wanted one at a cost of almost $500 million. Did people overreact? Did the demand for perfection divert funds that could have been better spent elsewhere? (Intel's costs were equivalent to half a year's research and development budget.) How much perfection do consumers have a right to expect?

MY WEB PAGE

www.scsite.com/dc2001/ch3/myweb.htm

WEB INSTRUCTIONS: *The icons to the left of the numbered activities on this page indicate the learning system availability. If you are using WebCT, follow the instructions in the exercises below. If you are using the textbook Web site or CyberClass, launch your browser, enter the URL www.scsite.com/dc2001/ch3/myweb.htm, and then click the corresponding icon to the left of the exercise.*

1. Practice Test: Click My Web Page and then click Chapter 3. Click Practice Test. Answer each question and then click the Save Answer button. When completed, click the Finish button. Click the OK button to submit the quiz for grading. Click the View Results button, and then make a note of any missed answers.

2. Web Guide: Click My Web Page and then click Chapter 3. Click Web Guide to display the Guide to WWW Sites page. Click Computers and Computing and then click Virtual Museum of Computing. Scroll down the page and locate History of Computers. Click a link of your choice. Use your word processing program to prepare a brief report on your tour and submit your assignment to your instructor.

3. Scavenger Hunt: Click My Web Page and then click Chapter 3. Click Scavenger Hunt. Print a copy of the Scavenger Hunt page; use this page to write down your answers as you search the Web. Submit your completed page to your instructor.

4. Who Wants to Be a Computer Genius?: Click My Web Page and then click Chapter 3. Click Computer Genius to find out if you are a computer genius. For directions, click the How to Play button. When you are ready to play, click the Game button. Submit your score to your instructor.

5. Wheel of Terms: Click My Web Page and then click Chapter 3. Click Wheel of Terms to reinforce important terms you learned in this chapter by playing the Shelly Cashman Series version of this popular game. For directions, click the How to Play button. When you are ready to play, click the Game button. Submit your score to your instructor.

6. Career Corner: Click My Web Page and then click Chapter 3. Click Career Corner to display the Making College Count page. Click Picking a Major. Review this page, clicking the links below College Major/Career Choices. Write a brief report on one or two information technology careers. Submit the report to your instructor.

7. Send Sleuth: Click My Web Page and then click Chapter 3. Click Search Sleuth to learn search techniques that will help make you a research expert. Submit the completed assignment to your instructor.

8. Crossword Puzzle Challenge: Click My Web Page and then click Chapter 3. Click Crossword Puzzle Challenge. Complete the puzzle to reinforce skills you learned in this chapter. For directions, click the How to Play button. When you are ready to play, click the Game button. Submit the completed puzzle to your instructor.

9. Flash Cards: This exercise uses CyberClass. Follow the instructions at the top of this page and then click the CyberClass icon to the left; or ask your instructor for logon instructions. Click Flash Cards on the Main Menu. Click the plus sign. Answer all the questions in any two subjects of your choice. Click the Close button in the upper-right corner to close all windows. Notice the many other exercises at this site. Complete all other exercises as assigned by your instructor.

Student Exercises
WEB INFO
IN BRIEF
KEY TERMS
AT THE MOVIES
CHECKPOINT
AT ISSUE
MY WEB PAGE
HANDS ON
NET STUFF
Special Features
TIMELINE 2001
GUIDE TO WWW SITES
MAKING A CHIP
E-COMMERCE 2001
BUYER'S GUIDE 2001
CAREERS 2001
TRENDS 2001
LEARNING GAMES
CHAT
INTERACTIVE LABS
NEWS
HOME

HANDS ON

www.scsite.com/dc2001/ch3/hands.htm

WEB INSTRUCTIONS: *To display this page from the Web, launch your browser and enter the URL, www.scsite.com/dc2001/ch3/hands.htm. Click the links for current and additional information.*

One

Installing New Hardware

This exercise uses Windows 98 procedures. Plug and Play technology is a key feature of the Windows operating system. Plug and Play technology allows users to install new devices without having to reconfigure the system manually. To find out how to install a new device with Plug and Play technology, click the Start button on the taskbar, and then click Help on the Start menu. Click the Contents tab. Click the Managing Hardware and Software book and then click the Installing New Hardware and Software book. Click Install a Plug and Play device. What are the three steps in installing a Plug and Play device? When would Windows not detect a Plug and Play device? How is a device that is not Plug and Play installed?

Two

Setting the System Clock

Double-click the time on the taskbar. In the Date/Time Properties dialog box, click the question mark button on its title bar and then click the picture of the calendar. Read the information in the pop-up window and then click the pop-up window to close it. Repeat this process for other areas of the dialog box and then answer these questions:

- What is the purpose of the calendar?
- How do you change the time zone?
- What is the difference between the OK and the Apply buttons?

Close the Date/Time Properties dialog box.

Three

Using Calculator to Perform Number System Conversion

Instead of the decimal (base 10) number system that people use, computers use the binary (base 2) or hexadecimal (base 16) number systems. It is not necessary to understand these number systems to use a computer, but it is interesting to see how decimal numbers look when in binary or hexadecimal form. Click the Start button on the taskbar, point to Programs on the Start menu, point to Accessories on the Programs submenu, and then click Calculator on the Accessories submenu. Click View on the menu bar and then click Scientific to display the scientific calculator. Perform the following tasks:

- Click Dec to select decimal. Enter 35 by clicking the numeric buttons or using the numeric keypad. Click Bin to select binary. What number displays? Click Hex to select hexadecimal. What number displays? Click the C (Clear) button.
- Convert the following decimal numbers to binary and hexadecimal: 7, 256, and 3,421.
- What decimal number is equal to 10010 in the binary system? What decimal number is equal to 2DA9 in the hexadecimal system?

Close Calculator.

Four

Power Management

This exercise uses Windows 98 or Windows 2000 procedures. Environmental and financial considerations make it important to manage the amount of power a computer uses. Click the Start button on the taskbar, point to Settings on the Start menu, and then click Control Panel on the Settings submenu. Double-click the Power Management icon in the Control Panel dialog box. In the Power Management Properties dialog box, click the Power Schemes tab. What is a power scheme? What power scheme currently is being used on your computer? After how many minutes of inactivity is the monitor turned off? After how many minutes of inactivity are the hard disks turned off? Close the Power Management dialog box and the Control Panel dialog box. How can the Power Management dialog box be used to make a computer more energy efficient?

NET STUFF

www.scsite.com/dc2001/ch3/net.htm

WEB INSTRUCTIONS: *To display this page from the Web, launch your browser and enter the URL, www.scsite.com/dc2001/ch3/net.htm. To use the Motherboard lab from the Web, Shockwave and Flash Player must be installed on your computer (download by clicking here).*

MOTHERBOARD LAB

1. Shelly Cashman Series Motherboard Lab

Follow the appropriate instructions in NET STUFF 1 on page 1.48 to start and use the Understanding the Motherboard lab. If you are running from the Web, enter the URL, www.scsite.com/sclabs/menu.htm, or display the NET STUFF page (see instructions at the top of this page) and then click the MOTHERBOARD LAB button.

MICRO-PROCESSOR

2. How a Microprocessor Works

After reading about what a microprocessor does and the way it interacts with other system unit components, it still can be difficult to understand how a microprocessor performs even a simple task such as adding two plus three. To find the answer, click the MICROPROCESSOR button and complete this exercise to learn what a microprocessor does to find the answer.

NEWSGROUPS

3. Newsgroups

Would you like more information about a special interest? Perhaps you would like to share opinions and advice with people who have the same interest. If so, you might be interested in newsgroups, also called discussion groups or forums. A newsgroup offers the opportunity to read articles on a specific subject, respond to the articles, and even post your own articles. Click the NEWSGROUPS button to find out more about newsgroups. What is lurking? What is Usenet? Click the Searching Newsgroups link at the bottom of the page. Read and print the Searching Newsgroups page. How can you locate a newsgroup on a particular topic?

IN THE NEWS

4. In the News

The ENIAC (Electronic Numerical Integrator and Computer) often is considered the first modern computer. Invented in 1946, the ENIAC weighed thirty tons and filled a thirty-by-fifty-foot room, yet its capabilities are dwarfed by current laptop computers. The ENIAC performed fewer than one thousand calculations per minute; today, PCs can process more than 300 million instructions per second. The rapid development of computing power and capabilities is astonishing, and that development is accelerating. Click the IN THE NEWS button and read a news article about the introduction of a new or improved computer component. What is the component? Who is introducing it? Will the component change the way people use computers? If so, how?

WEB CHAT

5. Web Chat

Winnie the Pooh may be less rumbley in the tumbley. Toy makers have put something in the stomach of stuffed versions of the bear – microprocessors. The stuffed Pooh chatters through twenty minutes of talk and song and can be programmed to use a child's name, discuss favorite foods and activities, and play games both at and away from a computer. Yet, some parents are not impressed with the bear's accomplishments. They maintain that a simple stuffed toy develops creativity through imaginary conversations and fanciful play, but the processor-enriched bear promotes little more than passive observation. Stuffed toys with microprocessors are available for about $100. Would you buy one for your child? Why or why not? Click the WEB CHAT button to enter a Web Chat discussion related to this topic.

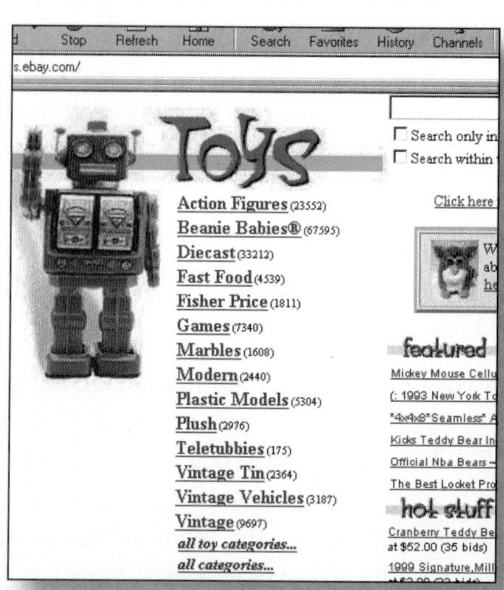

"The human tendency to regard little things as important has produced very many great things."

– *G. C. Lichtenberg*

WEB INSTRUCTIONS: *To gain World Wide Web access to additional and up-to-date information regarding this special feature, launch your browser and enter the URL shown at the top of this page.*

How Computer Chips Are Made

Computer chips are made by placing and removing thin layers of insulating, conducting, and semiconducting materials in hundreds of distinct steps. The chips are incredibly small — tinier than a human fingernail — and are becoming smaller. Intel recently released two chips using a 0.18-micron production process (a micron, or micrometer, is 0.000001 of a meter), instead of the 0.25-micron process used to produce Pentium and Pentium II processors. Smaller production processes make smaller chips and smaller circuitry possible. Because electricity travels faster over smaller circuits, these smaller chips are faster than ever. The smaller chips also are less expensive because manufacturers can create more chips from the same amount of raw materials.

Every computer chip consists of many layers of circuits and microscopic electronic components such as transistors, diodes, capacitors, and resistors. Connected together on a chip, these components are referred to as an **integrated circuit (IC)**. Most chips have at least four to six layers, but some have more than fifteen.

[FIGURE 1]

Computer chips, glass, artificial sweeteners, sandpaper, and even bathroom cleansers all share at least one common raw material: silicon. **Silicon**, which is the second most common element on earth, is found in sand, clay, bauxite, and quartz. Although some companies use other materials to make chips (a German company, for example, is experimenting with plastic), most computer chips are made from silicon crystals refined from quartz rocks. At 99.999999% pure, the refined silicon is the purest material produced commercially in large quantities.

The first step in the manufacturing process involves melting the silicon crystals. Next, a seed crystal is dipped into the melted silicon and then slowly drawn out to form a cylindrical ingot that is five to ten inches in diameter and several feet long. After being smoothed, a diamond saw blade slices the silicon ingot into wafers that are four to eight inches in diameter and 4/1000 of an inch thick (about as thick as a credit card). The creation of these silicon wafers is an essential phase in the chip manufacturing process and usually takes from ten to thirty days. Each wafer forms the foundation for hundreds of chips.

[FIGURE 3]

Much of the chip manufacturing process is performed in special laboratories called **clean rooms**. Because chip components are so small — sometimes less than one-hundredth the diameter of a human hair — even the smallest dust particle can ruin a chip. Clean rooms are one thousand times cleaner than a hospital operating room, with less than one particle per cubic foot of air. People who work in these facilities wear special protective clothing called **bunny suits**. Before entering the manufacturing area, the workers use an air shower to remove any dust from their suits.

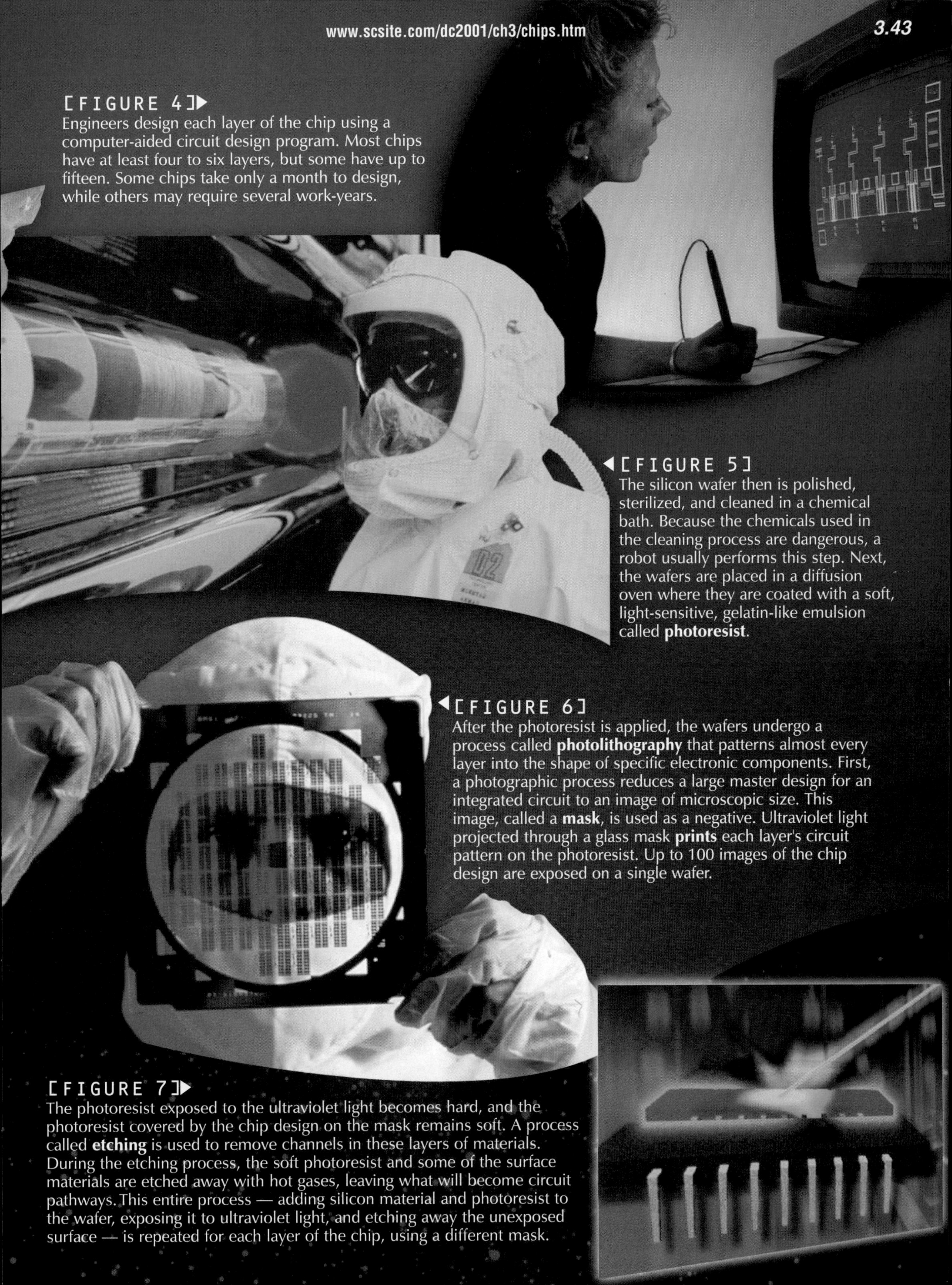

[FIGURE 4]

Engineers design each layer of the chip using a computer-aided circuit design program. Most chips have at least four to six layers, but some have up to fifteen. Some chips take only a month to design, while others may require several work-years.

[FIGURE 5]

The silicon wafer then is polished, sterilized, and cleaned in a chemical bath. Because the chemicals used in the cleaning process are dangerous, a robot usually performs this step. Next, the wafers are placed in a diffusion oven where they are coated with a soft, light-sensitive, gelatin-like emulsion called **photoresist**.

[FIGURE 6]

After the photoresist is applied, the wafers undergo a process called **photolithography** that patterns almost every layer into the shape of specific electronic components. First, a photographic process reduces a large master design for an integrated circuit to an image of microscopic size. This image, called a **mask**, is used as a negative. Ultraviolet light projected through a glass mask **prints** each layer's circuit pattern on the photoresist. Up to 100 images of the chip design are exposed on a single wafer.

[FIGURE 7]

The photoresist exposed to the ultraviolet light becomes hard, and the photoresist covered by the chip design on the mask remains soft. A process called **etching** is used to remove channels in these layers of materials. During the etching process, the soft photoresist and some of the surface materials are etched away with hot gases, leaving what will become circuit pathways. This entire process — adding silicon material and photoresist to the wafer, exposing it to ultraviolet light, and etching away the unexposed surface — is repeated for each layer of the chip, using a different mask.

www.scsite.com/dc2001/ch3/chips.htm

[FIGURE 8]

Silicon is a natural **semiconductor**, meaning that it can be either a conductor or an insulator. In pure form, silicon does not conduct electricity at room temperature. To alter the way chips conduct electricity, chip manufacturers **dope**, or treat, the silicon with impurities such as boron and phosphorous. These materials, called **dopants**, are added to the surface of the wafer in a process called **ion implementation**. The dopants create areas that will conduct electricity.

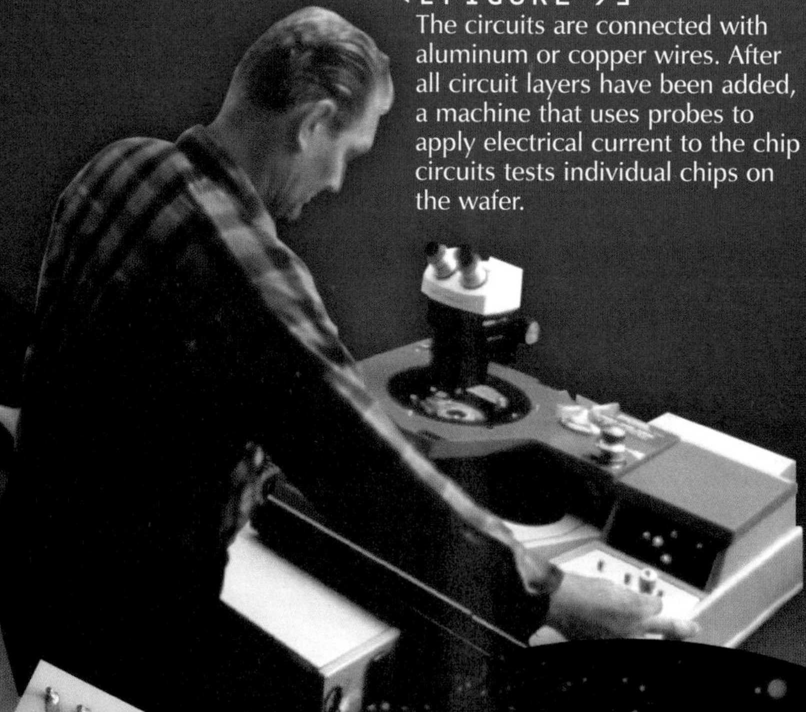

[FIGURE 9]

The circuits are connected with aluminum or copper wires. After all circuit layers have been added, a machine that uses probes to apply electrical current to the chip circuits tests individual chips on the wafer.

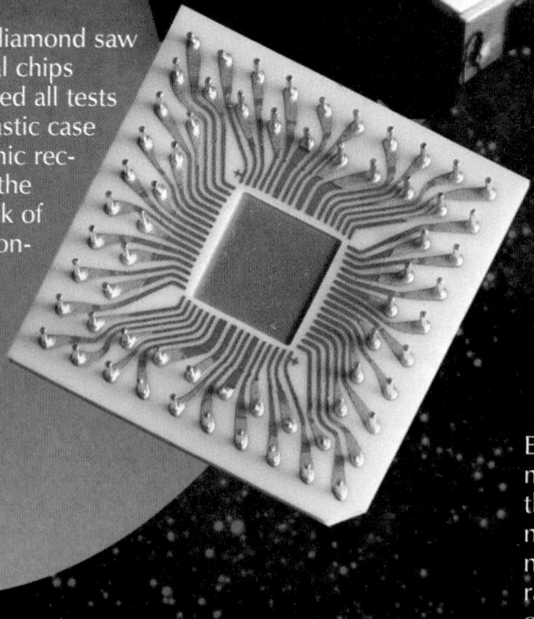

[FIGURE 10]

In a process called dicing, a diamond saw cuts the wafers into individual chips called **die**. Die that have passed all tests are placed in a ceramic or plastic case called a package — the ceramic rectangles with rows of pins on the bottom that most people think of as microprocessors. Highly conductive, noncorrosive gold wires connect the circuits on the chip to the pins on the package. The pins will connect the chip to a socket on a circuit board.

Each packaged chip is tested one more time, marking the last step in the chip-making process. The chips now are ready to be sent to companies that will include them in a wide range of items — from locomotives, cars, and traffic lights to electric guitars, coffee makers, and computers.

CHAPTER 4

INPUT

OBJECTIVES

After completing this chapter, you will be able to:

- Describe the four types of input
- List the characteristics of a keyboard
- Identify various types of keyboards
- Identify various types of pointing devices
- Explain how a mouse works
- Describe different mouse types
- Explain how scanners and other reading devices work
- Identify the purpose of a digital camera
- Describe the various techniques used for audio and video input
- Identify alternative input devices for physically challenged users

During the information processing cycle, a computer executes instructions and processes data (input) into information (output) and stores the information for future use. Input devices are used to enter instructions and data into the computer. This chapter describes the devices used for input and the various methods of entering input. Devices that can be used for both input and storage, such as disk drives, are covered in the discussion of storage in Chapter 6.

WHAT IS INPUT?

Input is any data or instructions you enter into the memory of a computer. Once input is in memory, the CPU can access it and process the input into output. Four types of input are data, programs, commands, and user responses (Figure 4-1):

- **Data** is a collection of unorganized facts that can include words, numbers, pictures, sounds, and videos. A computer manipulates and processes data into information, which is useful. Although technically speaking, a single item of data should be called a datum, the term data commonly is used and accepted as both the singular and plural form of the word.
- A **program** is a series of instructions that tells a computer how to perform the tasks necessary to process data into information. Programs are kept on storage media such as a floppy disk, hard disk, CD-ROM, or DVD-ROM. Programs respond to commands issued by a user.
- A **command** is an instruction given to a computer program. Commands can be issued by typing keywords or pressing special keys on the keyboard. A **keyword** is a specific word, phrase, or code that a program understands as an instruction. Some keyboards include keys that send a command to a program when you press them.

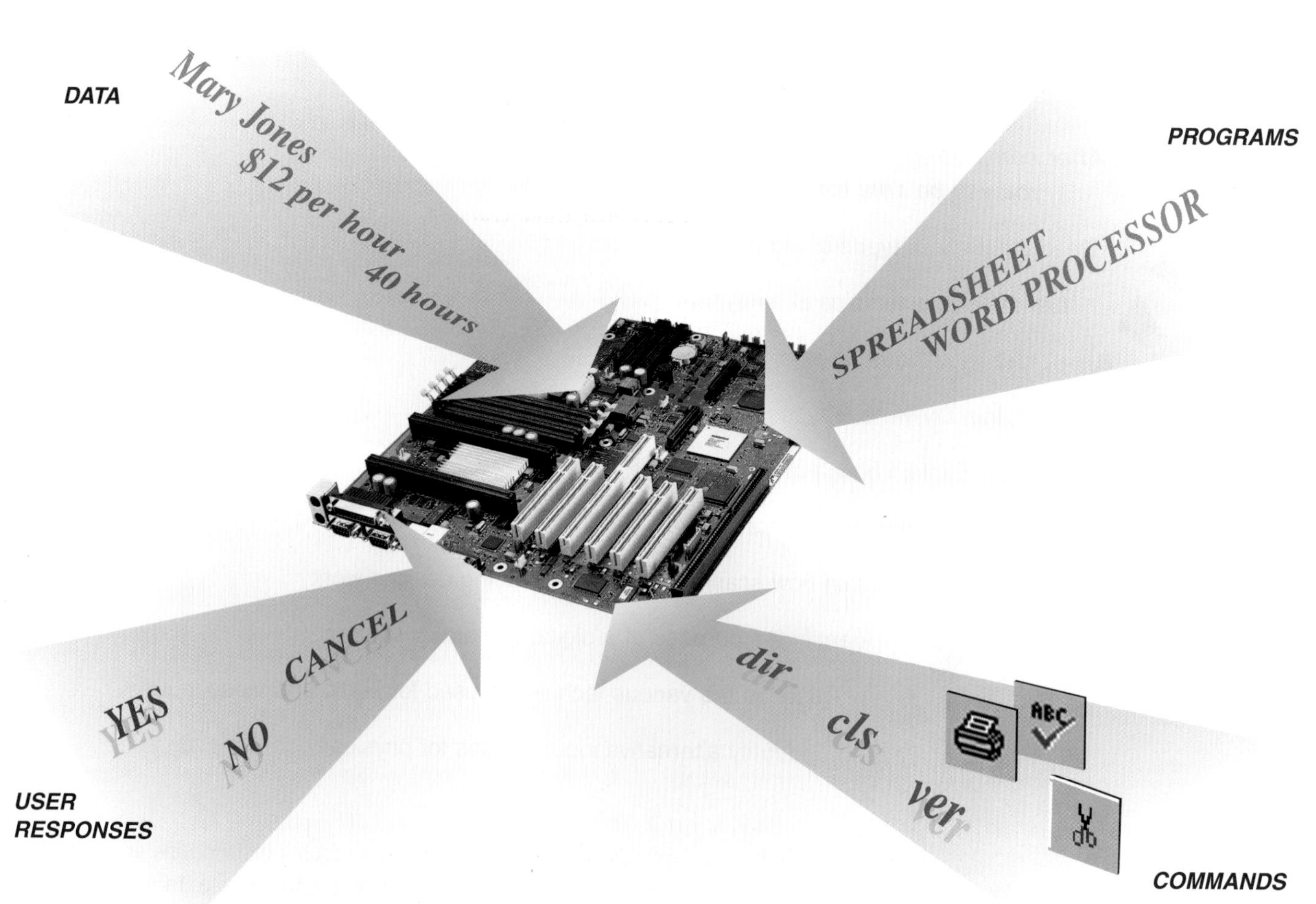

Figure 4-1 *Four types of input are data, programs, commands, and user responses.*

Instead of requiring you to remember keywords or special keys, many programs allow you to issue commands by selecting menu choices or graphical objects. For example, programs that are **menu-driven** provide menus as a means of entering commands. Today, most programs have a **graphical user interface** that use icons, buttons, and other graphical objects to issue commands. Of all of these methods, a graphical user interface is the most user-friendly way to issue commands.

- A **user response** is an instruction you issue to the computer by replying to a question posed by a computer program, such as *Do you want to save the changes you made?* Based on your response, the program performs certain actions. For example, if you answer, Yes, to this question, the program saves your changed file on a storage device.

WHAT ARE INPUT DEVICES?

An **input device** is any hardware component that allows you to enter data, programs, commands, and user responses into a computer. Input devices include the keyboard, pointing devices, scanners and reading devices, digital cameras, audio and video input devices, and input devices for physically challenged users. Each of these input devices is discussed in the following pages.

THE KEYBOARD

One of the primary input devices on a computer is the **keyboard** (Figure 4-2). You enter data into a computer by pressing the keys on the keyboard.

Desktop computer keyboards usually have from 101 to 105 keys, while keyboards for smaller computers such as laptops contain fewer keys. A computer keyboard includes keys that allow you to type letters of the alphabet, numbers, spaces, punctuation marks, and other symbols such as the dollar sign ($) and asterisk (*). A keyboard also contains special keys that allow you to enter data and

For more information on keyboards, visit the Discovering Computers 2001 Chapter 4 WEB INFO page (**www.scsite.com/dc2001/ch4/webinfo.htm**) and click Keyboard.

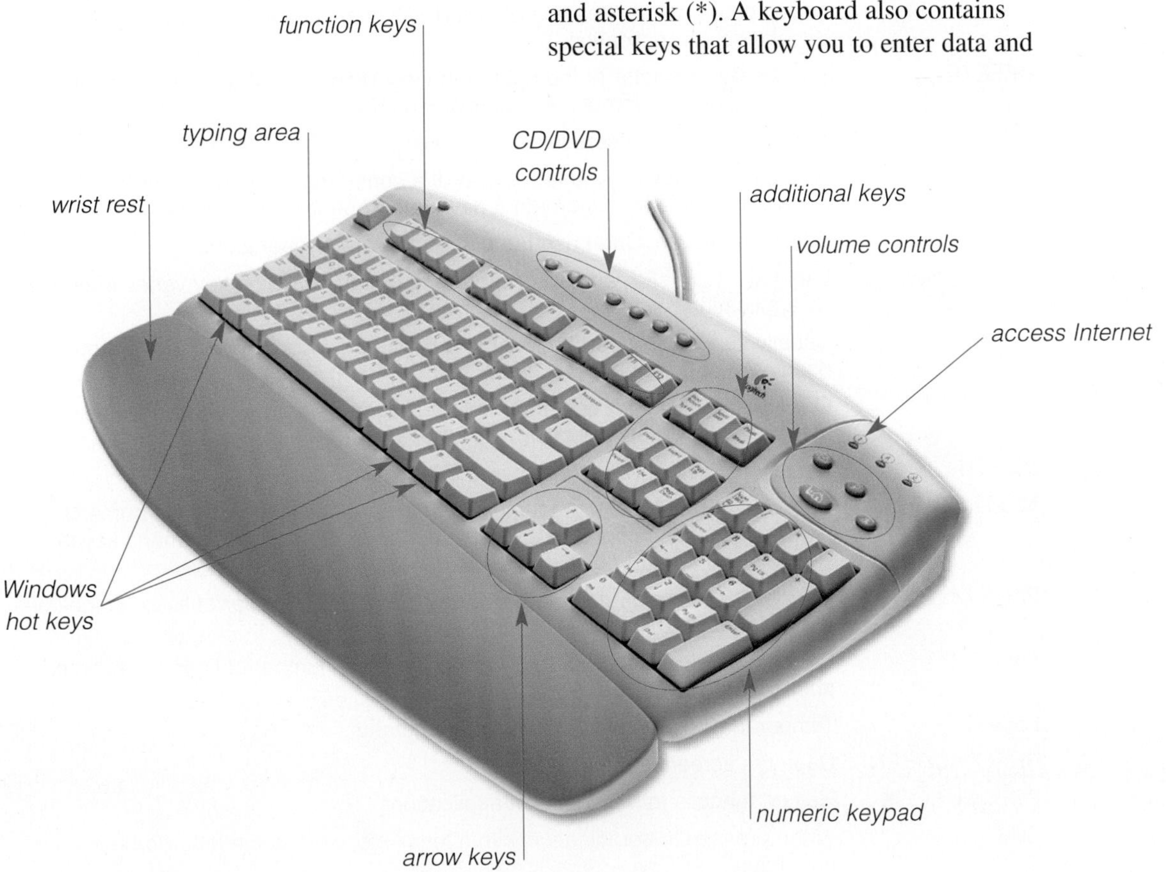

Figure 4-2 *A typical desktop computer keyboard. You can type using keys in the typing area and on the numeric keypad.*

instructions into the computer. The table in Figure 4-3 summarizes the purpose of special keys found on personal computer keyboards.

All computer keyboards have a typing area that includes the letters of the alphabet, numbers, punctuation marks, and other basic keys. Many desktop computer keyboards also have a numeric keypad located on the right side of the keyboard. A **numeric keypad** is a calculator-style arrangement of keys representing numbers, a decimal point, and some basic mathematical operators (see Figure 4-2 on the previous page). The numeric keypad is designed to make it easier to enter numbers.

Across the top, most keyboards contain function keys, which are labeled with the letter F followed by a number (see Figure 4-2). **Function keys** are special keys programmed to issue commands and accomplish certain tasks. The command associated with a function key depends on the program you are using. For example, in many programs, pressing the function key F1 displays a Help window. When instructed to press a function key such as F1, do not press the letter F followed by the number 1; instead press the key labeled

PC KEYBOARD SUMMARY

Key Name	Key Purpose
ALT	Short for Alternate. When pressed in combination with another key(s), usually issues a command. Meaning varies depending on application.
Arrow keys	Moves the insertion point in the direction of the arrow. For example, the UP ARROW key moves the insertion point up one position.
BACKSPACE	Erases the character to the left of the insertion point.
Break	Often pressed in combination with another key to stop or suspend execution of a program.
CAPS LOCK	A toggle key that when activated, shifts alphabetic letters to uppercase. It does not affect keys with numbers, punctuation marks, or other symbols.
CTRL	Short for Control. When pressed in combination with another key(s), usually issues a command. Meaning of CONTROL key combinations varies depending on application.
DELETE	Erases the character to the right of the insertion point. Also used to erase selected objects. Sometimes abbreviated DEL.
END	Usually moves the insertion point to an ending position, such as the end of a line.
ENTER	Also called Return key. Used at end of a command to direct computer to process command. Also used to create a new paragraph in a word processing application.
ESC	Short for Escape. Often used to quit a program or operation.
Function keys	Labeled F1, F2, F3, and so on. Meaning of each function key varies depending on application.
HOME	Usually sends the insertion point to a beginning location, such as the top of a document or beginning of a line.
HOT KEYS	Quickly and easily access programs or program features such as Windows menus, the Internet, or e-mail.
INSERT	Usually toggles between insert and overwrite modes.
NUM LOCK	Short for Numeric Lock. A toggle key that when activated, causes numeric keypad keys to function like a calculator. When deactivated, numeric keypad keys move insertion point.
PAGE DOWN	Causes the insertion point to move down a certain number of lines. Sometimes abbreviated PGDN.
PAGE UP	Causes the insertion point to move up a certain number of lines. Sometimes abbreviated PGUP.
Pause	Temporarily suspends a program or command.
PRINT SCREEN	Captures screen images.
SCROLL LOCK	Has no function in most current applications.
SHIFT	When pressed in combination with a letter key, causes the letter to be in uppercase.
TAB	Moves insertion point from place to place. Also used to insert tab characters into a word processing document.

Figure 4-3 *Summary of keys found on personal computer keyboards.*

F1. Function keys often are used in combination with other special keys (SHIFT, CTRL, ALT, and others) to issue commands. Many programs let you use a shortcut menu, a button, a menu, or a function key to obtain the same result (Figure 4-4).

Keyboards also contain keys that can be used to position the insertion point on the screen. The **insertion point** is a symbol that indicates where on the screen the next character you type will display. Depending on the program, the symbol may be a vertical bar, a rectangle, or an underline (Figure 4-5). **Arrow keys** allow you to move the insertion point left, right, up, or down. Most keyboards also contain keys such as HOME, END, PAGE UP, and PAGE DOWN, that you can press to move the insertion point to the beginning or end of a line, page, or document.

Most keyboards also include **toggle keys**, which can be switched between two different states. The NUM LOCK key, for example, is a toggle key. When you press it once, it locks the numeric keypad so you can use it to type numbers. When you press the NUM LOCK key again, the numeric keypad is unlocked so the same keys serve as arrow keys that move the insertion point. Many keyboards have status lights in the upper-right corner that light up to indicate that a toggle key is activated.

COMMAND SUMMARY

Command	Function Key(s)	Menu	Button
Close Word	ALT+F4	File|Exit	
Copy	SHIFT+F2	Edit|Copy	
Microsoft Word Help	F1	Help|Microsoft Word Help	
Open	CTRL+F12	File|Open	
Print	CTRL+SHIFT +F12	File|Print	
Print Preview	CTRL+F2	File|Print Preview	
Save	SHIFT+F12	File|Save	
Spelling and Grammar	F7	Tools|Spelling and Grammar	

Figure 4-4 *Many programs allow you to use a button, a menu, or a function key to obtain the same result, as shown by these examples from Microsoft Word.*

Many of the newer keyboards include buttons that allow you to access your CD/DVD drive and adjust speaker volume.

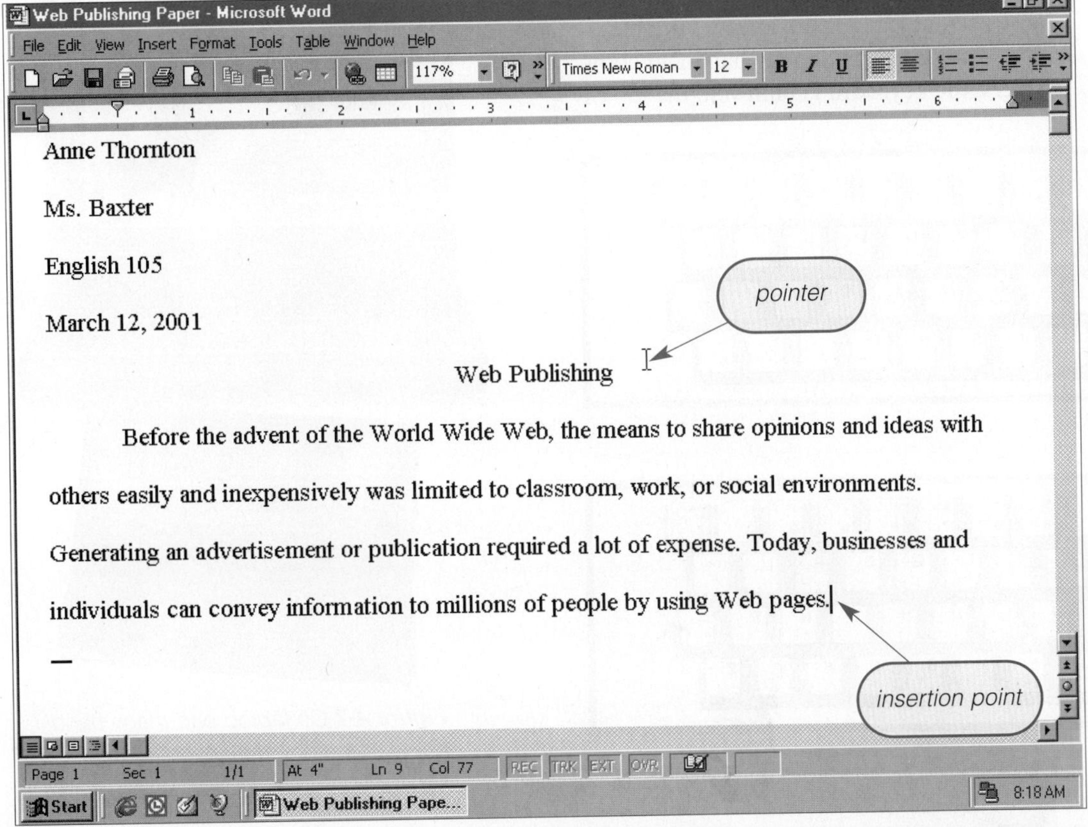

Figure 4-5 *In most Windows programs, such as Word, the insertion point is a blinking vertical bar. You can use the keyboard or the mouse to move the insertion point. The pointer, another symbol that displays on the screen, is controlled by using a mouse.*

Keyboard Types

A standard computer keyboard sometimes is called a **QWERTY keyboard** because of the layout of its typing area (Figure 4-6a). Pronounced KWER-tee, this keyboard layout is named after the first six leftmost letters on the top alphabetic line of the keyboard. Because of the way the keys are organized, a QWERTY keyboard might limit your typing speed.

A keyboard with an alternative layout was designed to improve typing speed. Called the **Dvorak keyboard** (pronounced de-VOR-zhak), this type of keyboard places the most frequently typed letters in the middle of the typing area (Figure 4-6b). Despite the more logical design of the Dvorak keyboard, the QWERTY keyboard is more widely used.

Most of today's desktop computer keyboards are **enhanced keyboards**, which means they have twelve function keys along the top, two CTRL keys, two ALT keys, and a set of arrow and additional keys between the typing area and the numeric keypad (see Figure 4-2 on page 4.3).

Although most keyboards attach to a serial port on the system unit via a cable, some keyboards – called **wireless keyboards** – transmit data via infrared light waves. For a wireless keyboard to transmit signals to a computer, both the computer and the wireless keyboard must have an IrDA port. These IrDA ports must be aligned so that nothing obstructs the path of the infrared light wave.

On laptops and many handheld computers, the keyboard is built into the top of the system unit (Figure 4-7). To fit in these smaller computers, the keyboards usually are smaller and have fewer keys. A typical laptop computer keyboard, for example, has only 85 keys, compared to the 105 keys on most desktop computer keyboards. To provide all of the functionality of a desktop computer keyboard, manufacturers design many of the keys to serve two or three different purposes.

Regardless of size, many keyboards have a rectangular shape with the keys aligned in rows. Users who spend a significant amount of time typing on these keyboards sometimes experience repetitive strain injuries of their wrists. For this reason, some manufacturers have redesigned their keyboards to minimize the chance of these types of workplace injuries (Figure 4-8). Keyboards such as these are called ergonomic keyboards. The goal of **ergonomics** is to incorporate comfort, efficiency, and safety into the design of items in the workplace. Because employees can be injured or develop disorders of the muscles, nerves, tendons, ligaments, and joints from working in a nonergonomically designed area, OSHA (Occupational Safety and Health Administration) has proposed standards whereby employers must establish programs that prevent these types of injuries or disorders.

Figure 4-6 *Comparison of the QWERTY and Dvorak keyboard layouts.*

Figure 4-6a *(QWERTY)*

Figure 4-6b *(Dvorak)*

Figure 4-7 *On laptop and many handheld computers, the keyboard is built into the top of the system unit.*

Figure 4-8 *The Microsoft Natural Keyboard Pro is an ergonomic keyboard designed to minimize strain on your hands and wrists.*

POINTING DEVICES

A **pointing device** is an input device that allows you to control a pointer on the screen. In a graphical user interface, a **pointer** is a small symbol on the display screen (see Figure 4-5 on page 4.5). A pointer often takes the shape of a block arrow ( ), an I-beam ( ), or a pointing hand ( ). Using a pointing device, you can position the pointer to move or select items on the screen. For example, you can use a pointing device to move the insertion point; select text, graphics, and other objects; and click buttons, icons, links, and menu commands.

Common pointing devices include the mouse, trackball, touchpad, pointing stick, joystick, touch screen, light pen, and graphics tablet. Each of these devices is discussed in the following sections.

Mouse

The mouse is the most widely used pointing device on desktop computers because it takes full advantage of a graphical user interface. Designed to fit comfortably under the palm of your hand, a **mouse** is an input device that is used to control the movement of the pointer, often called a **mouse pointer**, on the screen and to make selections from the screen. The top of the mouse has one to four buttons; some also have a small wheel. The bottom of a mouse is flat and contains a multi-directional mechanism, usually either a small ball or an optical sensor, which detects movement of the mouse (Figure 4-9). Mouse devices that contain a small ball often rest on a **mouse pad**, which usually is a rectangular rubber or foam pad that provides better traction for the mouse than the top of a desk. The mouse pad also protects the ball mechanism from a build up of dust and dirt, which could cause it to malfunction.

WEB INFO

For more information on a mouse, visit the Discovering Computers 2001 Chapter 4 WEB INFO page (**www.scsite.com/dc2001/ch4/webinfo.htm**) and click Mouse.

Figure 4-9 *A mouse is used to control the movement of a pointer on the screen and make selections from the screen.*

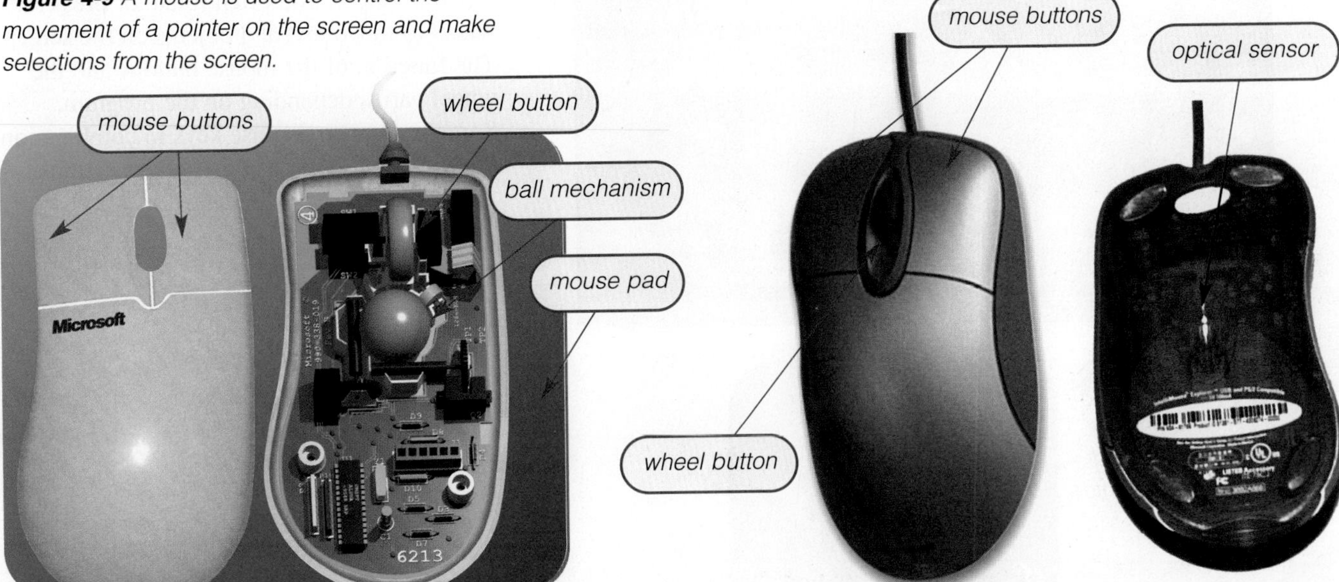

Figure 4-9a *(mechanical mouse contains a small ball)*

Figure 4-9b *(optical mouse contains an optical sensor)*

Figure 4-10
MOVING THE MOUSE POINTER

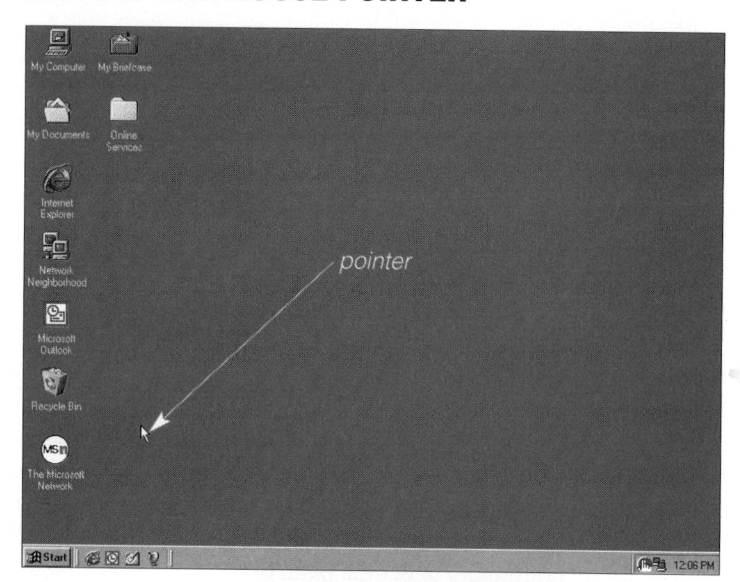

Step 1: *Position the mouse on the lower-left edge of the mouse pad.*

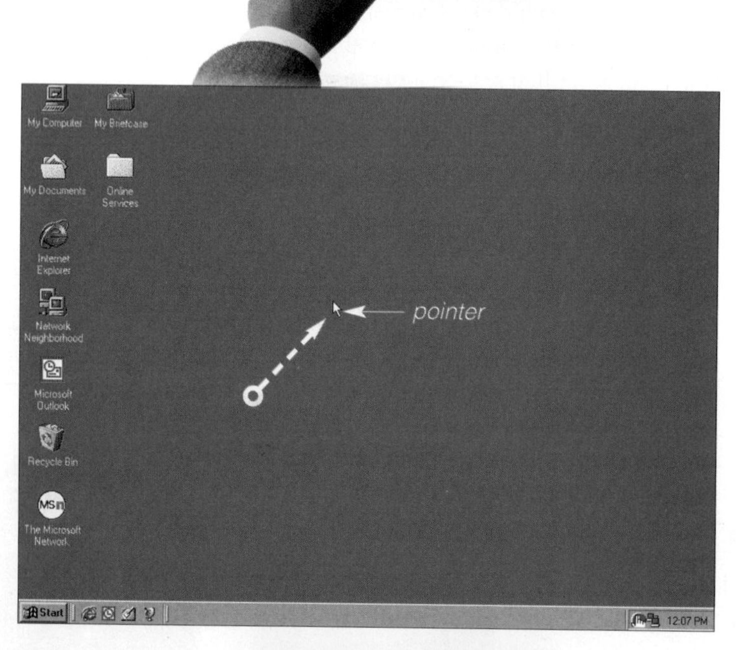

Step 2: *Move the mouse diagonally toward the middle of the mouse pad to move the pointer on the screen.*

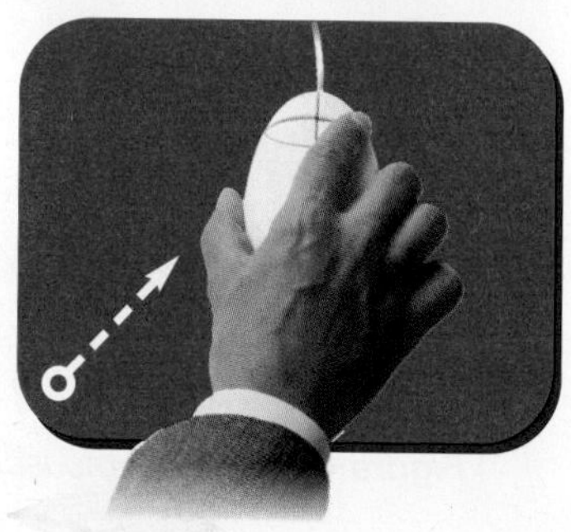

USING A MOUSE As you move the mouse across a flat surface such as a desktop, the pointer on the screen also moves. For example, when you move the mouse to the right, the pointer moves right on the screen (Figure 4-10). When you move the mouse to the left, the pointer moves left on the screen, and so on. If you have never worked with a mouse, you might find it a little awkward at first; with a little practice, however, you will discover that a mouse is quite easy to use.

Generally, you use the mouse to move the pointer on the screen to an object such as a button, a menu, an icon, a link, or text and then press one of the mouse buttons to perform a certain action on that object. In Windows 98, for example, if you point to the Start button on the taskbar and then press, or **click**, the primary mouse button, the Start menu displays on the screen (Figure 4-11).

For a right-handed user, the primary mouse button typically is the left button, and the secondary mouse button typically is the right mouse button. The function of these buttons, however, can be reversed to accommodate left-handed people.

You can perform other operations using the mouse in addition to clicking. These operations include point, right-click, double-click, drag, and right-drag. The table in Figure 4-12 explains how to perform each of these mouse operations and the general function of each operation.

Some mouse devices also have a wheel located between two buttons that can be used with certain programs (see Figure 4-9 on the previous page). You often rotate or press the wheel to move text and objects on the screen. The function of the mouse buttons and the wheel varies depending on the program. Some programs also use keys in combination with the mouse to perform certain actions.

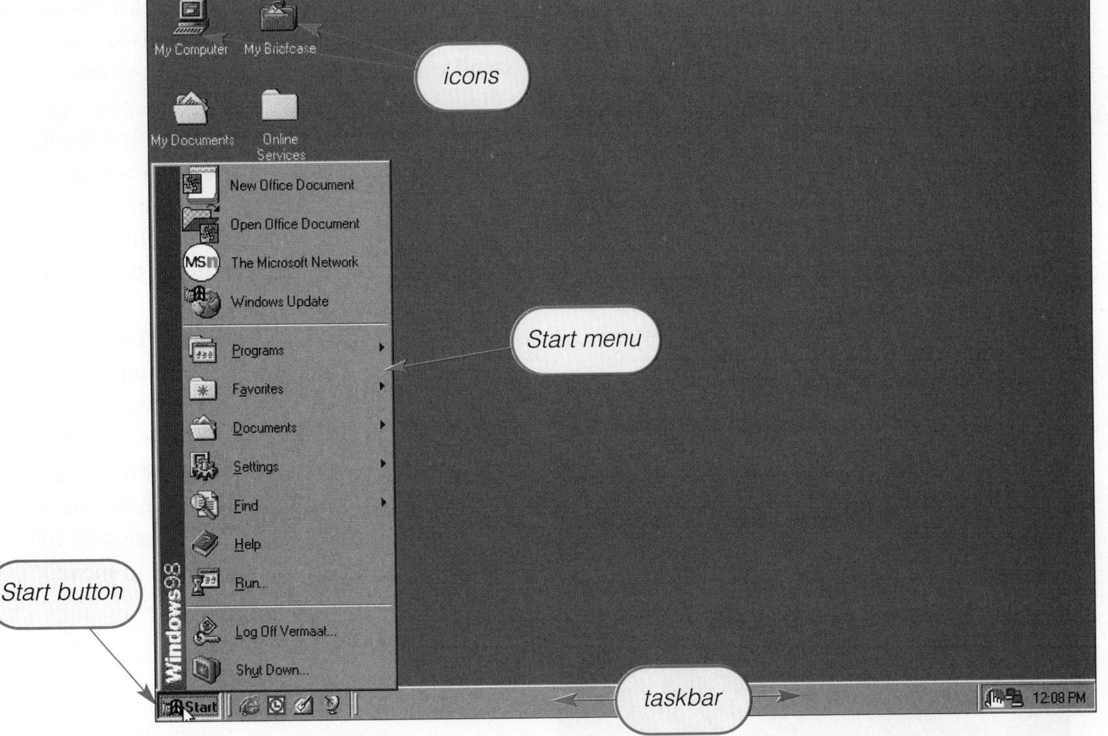

Figure 4-11 In Windows 98, if you point to the Start button on the taskbar and then click the primary mouse button, the Start menu displays on the screen.

MOUSE TYPES A mouse that has a rubber or metal ball on its underside is called a **mechanical mouse**. When the ball rolls in a certain direction, electronic circuits in the mouse translate the movement of the mouse into signals that are sent to the computer. Another type of mouse, called an **optical mouse**, has no moving mechanical parts inside; instead it uses devices that emit and sense light to detect the mouse's movement. The newer optical mouse devices can be used on nearly all types of surfaces, eliminating the need for a mouse pad. An optical mouse is more precise than a mechanical mouse and does not require cleaning like a mechanical mouse, but it also is more expensive.

MOUSE OPERATIONS

Operation	Mouse Action	Example
Point	Move the mouse across a flat surface until the pointer rests on the item of choice on the desktop.	Position the pointer on the screen.
Click	Press and release the primary mouse button, which usually is the left mouse button.	Select or deselect items on the screen or start a program or program feature.
Right-click	Press and release the secondary mouse button, which usually is the right mouse button.	Display a shortcut menu.
Double-click	Quickly press and release the primary mouse button twice without moving the mouse.	Start a program or program feature.
Drag	Point to an item, hold down the left mouse button, move the item to the desired location on the screen, and then release the left mouse button.	Move an object from one location to another or draw pictures.
Right-drag	Point to an item, hold down the right mouse button, move the item to the desired location on the screen, and then release the right mouse button.	Display a shortcut menu after moving an object from one location to another.
Rotate wheel	Roll the wheel forward or backward.	Scroll up or down a few lines.
Press wheel button	Press the wheel button while moving the mouse on the desktop.	Scroll continuously.

Figure 4-12 The more common mouse operations.

A mouse connects to your computer in one of two ways. Most connect using a cable that attaches to an RS-232C serial port, which usually is located on the back of the system unit. Some mouse devices are cordless, relying on battery power. Operating similarly to a television remote control, a **cordless mouse** or **wireless mouse**, uses infrared or radio waves to communicate with a receiver (Figure 4-13). A cordless mouse frees up desk space and eliminates the clutter of a cord.

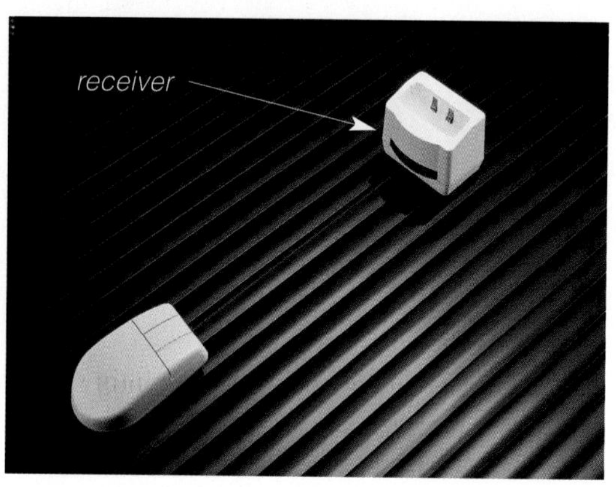

Figure 4-13 *This cordless mouse uses infrared remote transmission to communicate with the computer.*

Trackball

WEB INFO

For more information on trackballs, visit the Discovering Computers 2001 Chapter 4 WEB INFO page (**www.scsite.com/dc2001/ch4/webinfo.htm**) and click Trackballs.

Some users opt for alternative pointing devices other than a mouse, such as a trackball. Whereas a mechanical mouse has a ball mechanism on the bottom, a **trackball** is a stationary pointing device with a ball mechanism on its top (Figure 4-14). The ball mechanism in a larger trackball is about the size of a Ping-Pong ball; some mouse devices also have a small trackball about the size of a marble (Figure 4-15).

Figure 4-14 *A trackball is like an upside-down mouse. You rotate the ball mechanism with your thumb, fingers, or the palm of your hand to move the pointer.*

To move the pointer using a trackball, you rotate the ball mechanism with your thumb, fingers, or the palm of your hand. Around the ball mechanism, usually a trackball also has one or more buttons that work just like mouse buttons.

Although it shares characteristics with a mouse, a trackball is not as accurate as a mouse. A trackball's ball mechanism also requires frequent cleaning because it picks up oils from your fingers and dust from the environment. If you have limited desk space, however, a trackball is a good alternative to a mouse because you do not have to move the entire device.

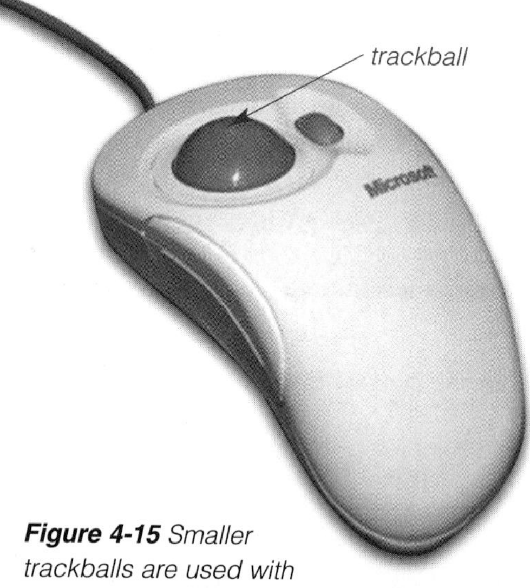

Figure 4-15 *Smaller trackballs are used with some mouse devices.*

Touchpad

A **touchpad** or **trackpad** is a small, flat, rectangular pointing device that is sensitive to pressure and motion (Figure 4-16). To move the pointer using a touchpad, you slide your fingertip across the surface of the pad. Some touchpads have one or more buttons around the edge of the pad that work like mouse buttons; on others, you tap the pad's surface to simulate mouse operations such as clicking.

Although you can attach a stand-alone touchpad to any personal computer, touchpads are found more often on laptop computers.

Figure 4-16 Laptop computers sometimes have a touchpad to control the movement of the pointer.

WEB INFO

For more information on touchpads, visit the Discovering Computers 2001 Chapter 4 WEB INFO page (**www.scsite.com/dc2001/ch4/webinfo.htm**) and click Touchpads.

Pointing Stick

A **pointing stick** is a pressure-sensitive pointing device shaped like a pencil eraser that was first developed by IBM for its laptop computers. Because of its small size, the pointing stick is positioned between keys on the keyboard (Figure 4-17). To move the pointer using a pointing stick, you push the pointing stick with your finger. The pointer on the screen moves in the direction that you push the pointing stick.

One advantage of using a pointing stick is that it does not require any additional desk space. Another advantage is that it does not require cleaning like a mechanical mouse or trackball. Whether you select a laptop that has a trackball, touchpad, or pointing stick is a matter of personal preference.

WEB INFO

For more information on pointing sticks, visit the Discovering Computers 2001 Chapter 4 WEB INFO page (**www.scsite.com/dc2001/ch4/webinfo.htm**) and click Pointing Sticks.

Figure 4-17 Some laptop computers use a pointing stick to control the movement of the pointer.

WEB INFO

For more information on joysticks, visit the Discovering Computers 2001 Chapter 4 WEB INFO page (**www.scsite.com/dc2001/ch4/webinfo.htm**) and click Joysticks.

Joystick

Users running game software such as a driving or flight simulator may prefer to use a joystick as their pointing device. A **joystick** is a vertical lever mounted on a base (Figure 4-18). You move the lever in different directions to control the actions of a vehicle or player. The lever usually includes buttons called *triggers* that you can press to activate certain events. Some joysticks also have additional buttons that you can set to perform other actions.

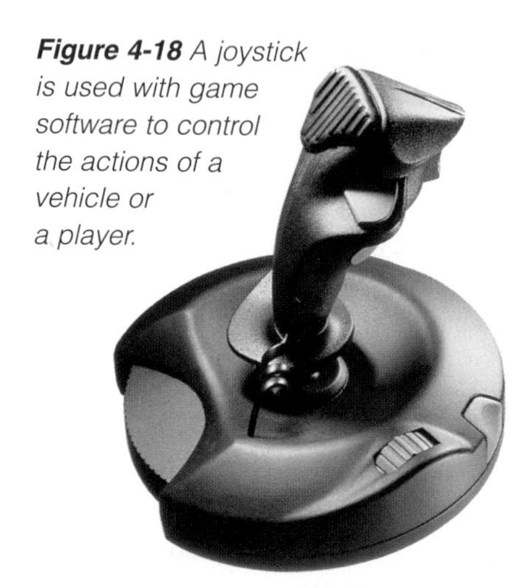

Figure 4-18 *A joystick is used with game software to control the actions of a vehicle or a player.*

Touch Screen

A monitor that has a touch-sensitive panel on the screen is called a **touch screen.** You interact with the computer by touching areas of the screen with your finger, which acts as an input device. Because they require a lot of arm movements, touch screens are not used to enter large amounts of data. Instead you touch words, pictures, numbers, or locations identified on the screen.

Touch screens often are used in kiosks located in stores, hotels, airports, and museums. Customers at Hallmark stores, for example, can use a kiosk to create personalized greeting cards (Figure 4-19). Some laptop computers even have a touch screen.

Figure 4-19 *This kiosk allows you to create personalized Hallmark greeting cards.*

Pen Input

Many input devices use an electronic pen instead of a keyboard or mouse for input. Some of these devices require you to point to onscreen objects with the pen; others allow you to input data using drawings, handwriting, and other symbols that are written with the pen on a surface.

LIGHT PEN A **light pen** is a handheld input device that contains a light source or can detect light. Some light pens require a specially designed monitor, while others work with a standard monitor (Figure 4-20). Instead of touching the screen with your finger to interact with the computer, you press the light pen against the surface of the screen or point the light pen at the screen and then press a button on the pen. Light pens are used in applications where desktop space is limited such as in the health-care field or when a wide variety of people use the application, such as electronic voting.

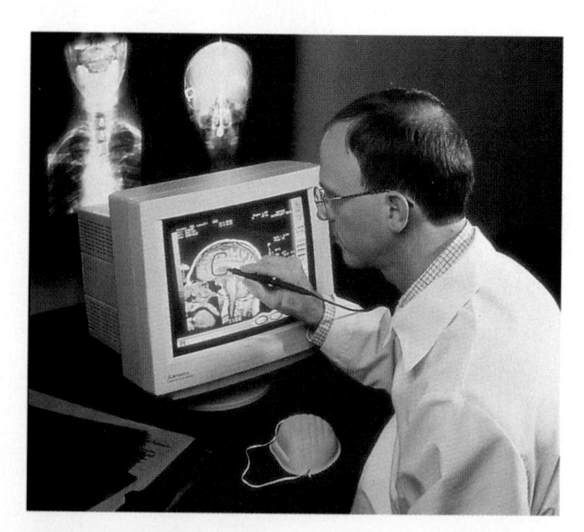

Figure 4-20 *To make selections with a light pen, you touch the pen against the surface of the screen or point the pen at the screen and then press a button on the pen.*

PEN COMPUTING Many handheld computers also allow you to input data using an electronic pen (Figure 4-21). The **pen** (also called a **stylus**) looks like a ballpoint pen but uses an electronic head instead of ink. Pen computers use **handwriting recognition software** that translates the letters and symbols used in handwriting into character data that the computer can use. Although most handwriting recognition software recognizes printed letters and can be trained to distinguish writing styles, pen-computing technology continues to be refined.

Figure 4-21 Many handheld computers support handwriting input through a pen.

GRAPHICS TABLET A **graphics tablet**, also called a **digitizer** or **digitizing tablet**, consists of a flat, rectangular, electronic plastic board used to input drawings, sketches, or other graphical data (Figure 4-22). Each location on the graphics tablet corresponds to a specific location on the screen. When you draw on the tablet with either an electronic pen or a puck, the tablet detects and converts the movements into digital signals that are sent into the computer. A **puck** is a device that looks similar to a mouse, except that is has a window with cross hairs so the user can see through to the tablet. Users with precise pointing requirements such as mapmakers and architects use a puck.

SCANNERS AND READING DEVICES

Some devices make the input process more efficient by eliminating the manual entry of data. Instead of a person entering data using a keyboard or pointing device, these devices capture data from a **source document**, which is the original form of the data. When using a keyboard or pointing device to enter data, the source document might be a timecard, order blank, invoice, or any other document that contains data to be processed.

Devices that capture data directly from source documents include optical scanners, optical character recognition devices, optical mark recognition devices, bar code scanners, and magnetic-ink character recognition readers. Examples of source documents used with these devices include advertisements, brochures, photographs, inventory tags, or checks. Each of these devices is discussed in the following pages.

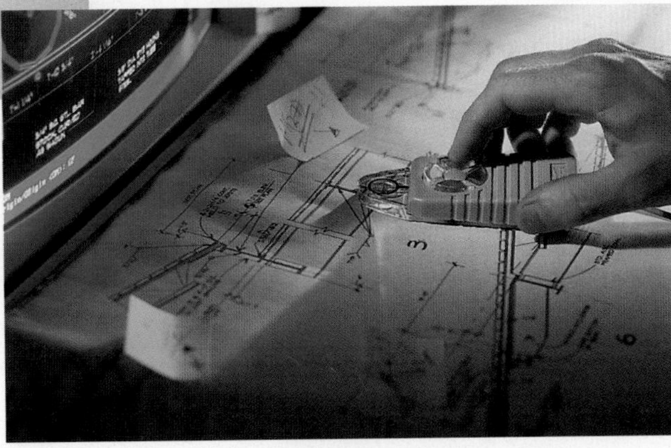

Figure 4-22 Some graphics tablets use a pen as their input device. Architects and engineers, however, often use a puck with a graphics tablet to input drawings, sketches, or other graphical data.

WEB INFO

For more information on scanners, visit the Discovering Computers 2001 Chapter 4 WEB INFO page (**www.scsite.com/dc2001/ch4/webinfo.htm**) and click Scanners.

Optical Scanner

An **optical scanner**, usually simply called a **scanner**, is a light-sensing input device that reads printed text and graphics and then translates the results into a form the computer can use (Figure 4-23). A scanner is similar to a copy machine except that it creates a file of the document instead of a paper copy. The file that contains the scanned object then can be stored on a disk, displayed on the screen, printed, faxed, sent via electronic mail, or included in another document. For example, you can scan a picture and then incorporate the picture into a brochure using a desktop publishing program.

When a document is scanned, the results are stored in rows and columns of dots called a **bitmap** (Figure 4-24). Each dot on a bitmap consists of one or more bits of data. The more bits used to represent a dot, the more colors and shades of gray that can be represented. For instance, one bit per dot is enough to represent simple one-color images, but for colors and shades of gray, each dot requires more than one bit of data.

Today's scanners range from 30 bit to 48 bit, with the latter being a higher quality, but more expensive.

Figure 4-23
HOW AN OPTICAL SCANNER WORKS

Step 1: *The document to be scanned is placed face down on the glass window.*

Step 2: *A bright light moves underneath the scanned document.*

Step 3: *An image of the document is reflected into a series of mirrors.*

Step 4: *The light is converted to an analog electrical current by a chip called a charge-coupled device (CCD).*

Step 5: *The analog signal is converted to a digital signal by a device called an analog-to-digital converter (ADC).*

Step 6: *The digital information is sent to software in the computer to be used by illustration, or desktop publishing, or other software.*

Step 7: *The scanned image displays on the screen.*

The density of the dots, known as the **resolution**, determines sharpness and clearness of the resulting image (Figure 4-25). Resolution typically is measured in **dots per inch (dpi)**, and is stated as the number of columns and rows of dots. For example, a 600 x 1200 (pronounced 600 by 1200) dpi scanner has 600 columns and 1200 rows of dots. If just one number is stated, such as 1200 dpi, that number refers to both the number of rows and the number of columns. The more dots, the better the resolution, and the resulting image is of higher quality.

Some manufacturers refer to the actual scanned resolution as the *optical resolution*, differentiating it from *enhanced* or *interpolated resolution*. The enhanced resolution usually is higher because it uses a special formula to add dots between those generated by the optical resolution.

Most of today's affordable color desktop scanners for the home or small business user have an optical resolution ranging from 600 to 2000 dpi. Commercial scanners designed for power users range from 4000 to 8000 dpi. The table in Figure 4-26 summarizes the four basic types of scanners.

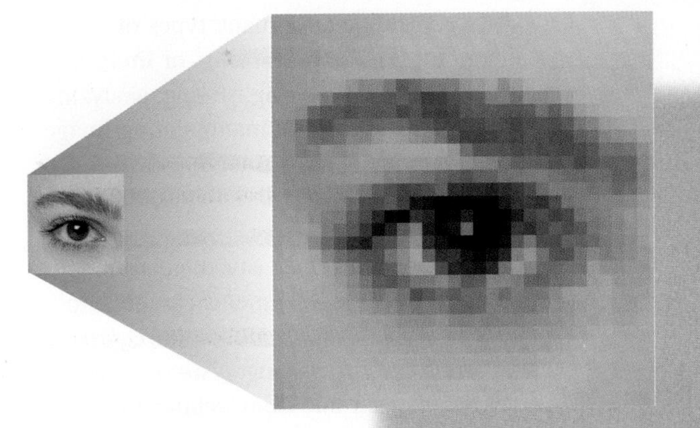

Figure 4-24 *Each dot on a bitmap consists of one or more bits of data. The more bits used to represent a dot, the more colors and shades of gray that can be represented. This is a bitmap for a woman's eye.*

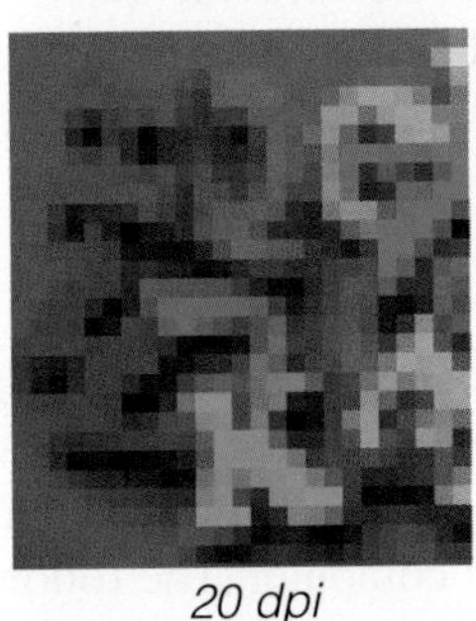

20 dpi

72 dpi

600 dpi

Figure 4-25 *The higher the resolution or dots per inch (dpi), the sharper and clearer the resulting image. Today's scanners have a resolution of at least 600 dpi.*

TYPES OF SCANNERS

Scanner Photo	Method of Scanning/ Use	Scannable Items
Pen Scanners	• Move pen over text to be scanned, then transfer data to computer • Ideal for mobile users, students, researchers • Some include PDA functionality	• Any printed text
Flatbed	• Similar to a copy machine. • Scanning mechanism passes under the item to be scanned, which is placed on a glass surface.	• Single-sheet documents • Bound material • Photographs • Slides (with an adapter)
Sheet-fed	• Item to be scanned is pulled into a stationary scanning mechanism. • Smaller and less expensive than a flatbed.	• Single-sheet documents • Photographs • Slides (with an adapter)
Drum	• Item to be scanned rotates around a stationary scanning mechanism. • Very large and expensive. • Used in publishing industry.	• Single-sheet documents • Photographs • Slides • Negatives

Figure 4-26 *This table lists the various types of scanners.*

Organizations use many types of scanners for **image processing**, or **imaging**, which consists of capturing, storing, analyzing, displaying, printing, and manipulating images (bitmaps). Image processing enables organizations to convert paper documents such as reports, memos, and procedure manuals into an electronic form. Once saved electronically, the routing of these documents can be automated. They also can be stored and indexed using an **image processing system**, which serves as an electronic filing cabinet that provides access to exact reproductions of the original documents. The government, for example, uses an image processing system to store property deeds and titles to provide quick access to the public, lawyers, and loan officers.

Optical Readers

An **optical reader** is a device that uses a light source to read characters, marks, and codes and then converts them into digital data that can be processed by a computer. The following sections discuss three types of optical readers: optical character recognition, optical mark recognition, and bar code.

OPTICAL CHARACTER RECOGNITION

Optical character recognition (OCR) is a technology that involves reading typewritten, computer-printed, or handwritten characters from ordinary documents and translating the images into a form that the computer can understand. Most **OCR devices** include a small optical scanner for reading characters and sophisticated software for analyzing what is read.

OCR devices range from large machines that can read thousands of documents per minute to handheld wands that read one document at a time. OCR devices are used to read characters printed using an OCR font. Although others exist, the standard OCR font is called OCR-A (Figure 4-27). During the scan of a document, an OCR device determines the shapes of characters by detecting patterns of light and dark. **Optical character recognition (OCR) software** then compares these shapes with predefined shapes stored in memory and converts the shapes into characters the computer can understand.

OCR software also is used with optical scanners such as flatbed, sheet-fed, and pen scanners. For example, suppose you need to modify a business report, but do not have the original word processing file. You could use a flatbed scanner to scan the document, but you still would not be able to edit the report. The scanner, which does not differentiate between text and graphics, will save the report as a bitmap image, which cannot be edited directly in a word processing program. To convert it into an editable text file that can be edited, you must have optical character recognition (OCR) software that works with the scanner. The resulting output can be stored in a variety of file formats, including those recognized by word processing software.

Current OCR software has a very high success rate and usually can identify more than 99 percent of scanned material. OCR software also will mark text it could not read, allowing you to make corrections easily.

Companies use OCR devices to increase the speed and accuracy of data entry. OCR is very useful when a significant amount of data must be entered into a computer and only the printed pages are available.

OCR also is used frequently for **turn-around documents**, which are documents designed to be returned (turned around) to the organization that created and sent them. For

WEB INFO

WEB INFO

For more information on OCR, visit the Discovering Computers 2001 Chapter 4 WEB INFO page (**www.scsite.com/dc2001/ch4/webinfo.htm**) and click OCR.

ABCDEFGHIJKLMNOPQRSTUVWXYZ

1234567890-=■;',./

Figure 4-27 *A portion of the characters in the OCR-A font. Notice how characters such as the number 0 and the letter O are shaped differently so the reading device easily can distinguish between them.*

example, when you receive a gas bill, you tear off a portion of the bill and send it back to the gas company with your payment (Figure 4-28). The portion of the bill you return usually has your account number, payment amount, and other information printed in optical characters.

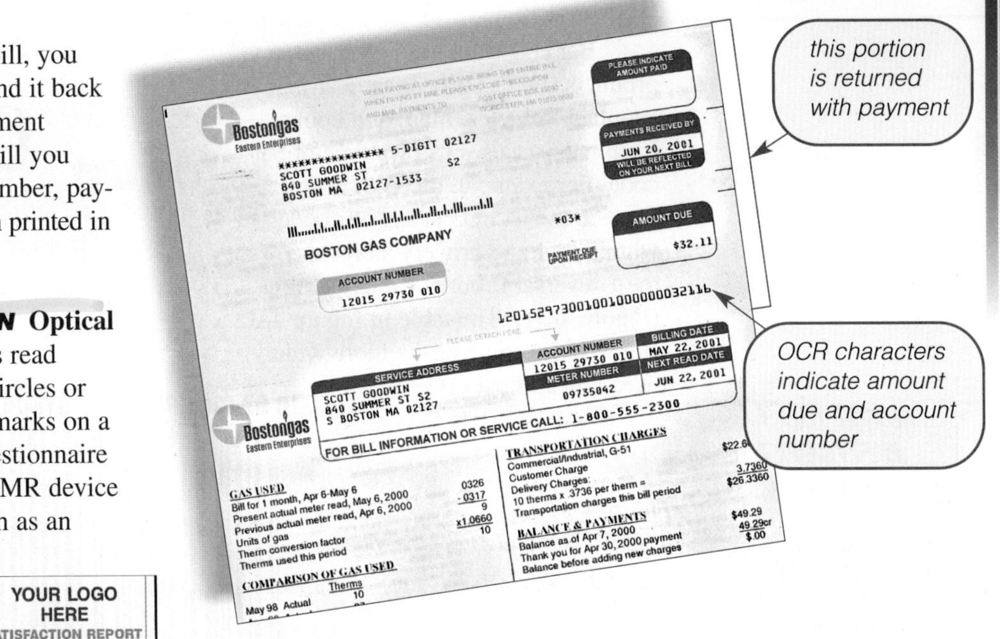
Bostongas
Eastern Enterprises
PLEASE INDICATE AMOUNT PAID
PAYMENTS RECEIVED BY
JUN 20, 2001
WILL BE REFLECTED ON YOUR NEXT BILL
**************** 5-DIGIT 02127
SCOTT GOODWIN S2
840 SUMMER ST
BOSTON MA 02127-1533
03
AMOUNT DUE
$32.11
BOSTON GAS COMPANY
PAYMENT DUE UPON RECEIPT
ACCOUNT NUMBER
12015 29730 010
1201529730010010000000032116
SERVICE ADDRESS
SCOTT GOODWIN
840 SUMMER ST S2
S BOSTON MA 02127
ACCOUNT NUMBER
12015 29730 010
METER NUMBER
09735042
BILLING DATE
MAY 22, 2001
NEXT READ DATE
JUN 22, 2001
Bostongas
Eastern Enterprises
FOR BILL INFORMATION OR SERVICE CALL: 1-800-555-2300

GAS USED	
Bill for 1 month, Apr 6-May 6	
Present actual meter read, May 6, 2000	0326
Previous actual meter read, Apr 6, 2000	- 0317
Units of gas	9
Therm conversion factor	x1.0660
Therms used this period	10

TRANSPORTATION CHARGES	
Commercial/Industrial, G-51	
Customer Charge	$22.6
Delivery Charges: 10 therms x .3736 per therm =	3.7360
Transportation charges this bill period	$26.3360

BALANCE & PAYMENTS	
Balance as of Apr 7, 2000	$49.29
Thank you for Apr 30, 2000 payment	49.29cr
Balance before adding new charges	$.00

COMPARISON OF GAS USED
Therms
May 98 Actual 10

Figure 4-28 OCR is used frequently with turn-around documents. With this gas bill, you tear off the top portion and return it with your payment.

OPTICAL MARK RECOGNITION Optical **mark recognition (OMR)** devices read hand-drawn marks such as small circles or rectangles. A person places these marks on a form, such as a test, survey, or questionnaire answer sheet (Figure 4-29). The OMR device first reads a master document, such as an answer key sheet for a test, to record correct answers based on patterns of light; the remaining documents then are passed through the OMR device and their patterns of light are matched against the master document.

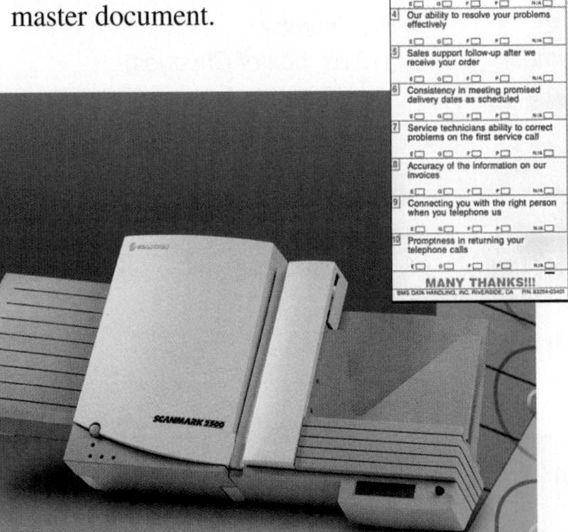
YOUR LOGO HERE
SATISFACTION REPORT
Please rate our performance in each section by marking one box with a #2 pencil
1. Our overall performance in dealing with you as a customer
2. The quality of our products
3. Getting product support information to you in a timely manner
4. Our ability to resolve your problems effectively
5. Sales support follow-up after we receive your order
6. Consistency in meeting promised delivery dates as scheduled
7. Service technicians ability to correct problems on the first service call
8. Accuracy of the information on our invoices
9. Connecting you with the right person when you telephone us
10. Promptness in returning your telephone calls
MANY THANKS!!!

Figure 4-29 OMR devices commonly are used to scan test, survey, or questionnaire answer sheets.

BAR CODE SCANNER A **bar code scanner** uses laser beams to read bar codes (Figure 4-30). A **bar code** is an identification code that consists of a set of vertical lines and spaces of different widths. The bar code, which represents some data that identifies the item, is printed on a product's package or on a label that is affixed to a product so it can be read by a bar code scanner. The bar code scanner uses light patterns from the bar code lines to identify the item.

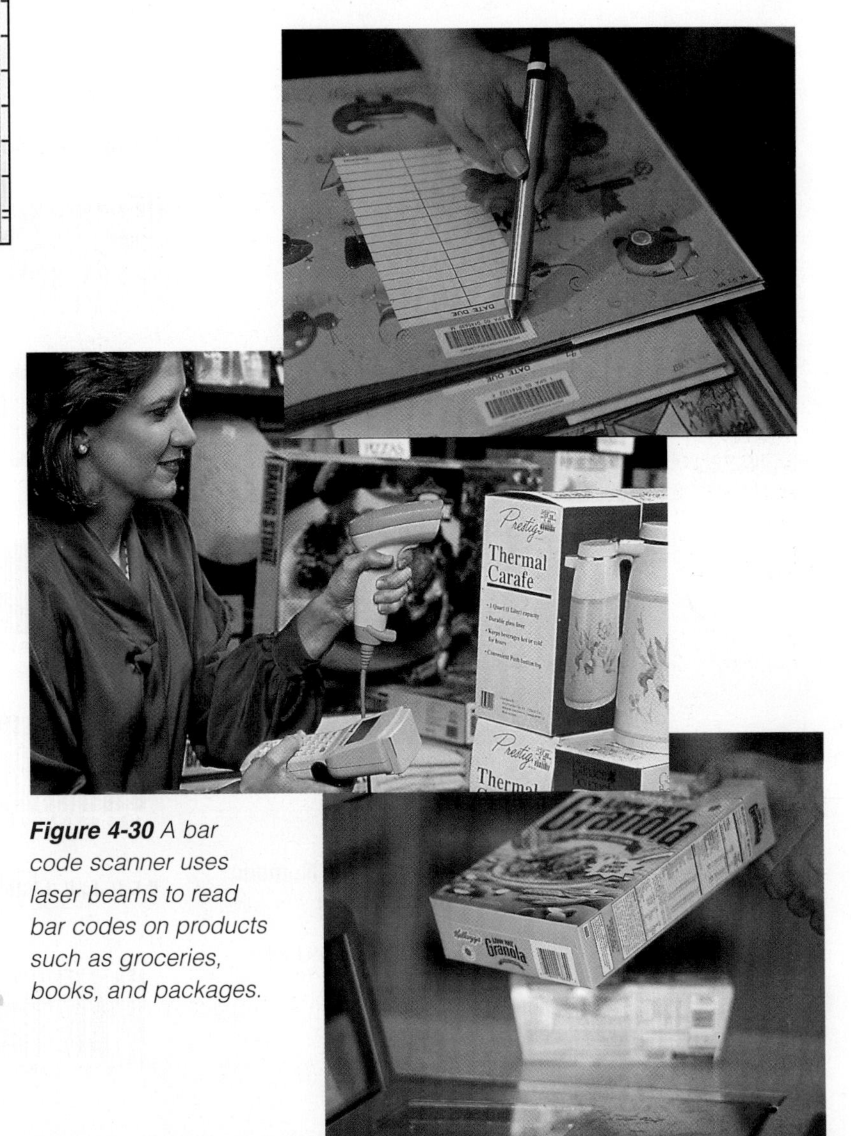

Figure 4-30 A bar code scanner uses laser beams to read bar codes on products such as groceries, books, and packages.

Bar codes are used on a variety of products such as groceries, pharmacy supplies, vehicles, mail, and books. Each industry uses its own type of bar code. For example, the U.S. Postal Service uses a POSTNET bar code, while retail and grocery stores use the Universal Product Code, or UPC (Figure 4-31). The table in Figure 4-32 summarizes some of the more widely used types of bar codes.

WEB INFO

For more information on bar codes, visit the Discovering Computers 2001 Chapter 4 WEB INFO page (**www.scsite.com/dc2001/ch4/webinfo.htm**) and click Bar Codes.

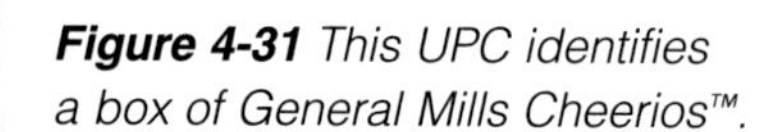

Figure 4-31 *This UPC identifies a box of General Mills Cheerios™.*

Figure 4-32 *Some of the more widely used types of bar codes.*

TYPES OF BAR CODES

Bar Code Name	Sample Bar Code	Primary Market
Codabar	 Codabar A 1 2 3 4 5 6 7 8 9 0 1 2 A	Libraries, blood banks, and air parcel carriers.
Code 39	C O D E 3 9	Nonretail applications such as manufacturing, military, and health applications requiring numbers and letters in the bar code.
EAN – European Article Numbering	 EAN-13 1 234567 890128	Similar to UPC, except used in Europe. A variation of EAN is used for ISBNs on books.
Interleaved 2 of 5	 Interleaved 2 of 5 1 2 3 4 5 6 7 8 9 0 1 2	Nonretail applications requiring only numbers in the bar code.
POSTNET – Postal Numeric Encoding Technique		U.S. Postal Service to represent a postal code or delivery point code.
UPC – Universal Product Code	 UPC-A with Supplemental 0 12345 67890 5	Supermarkets, convenience, and specialty stores used to identify manufacturers and products.

Magnetic Ink Character Recognition Reader

A **magnetic-ink character recognition (MICR)** reader is used to read text printed with magnetized ink (Figure 4-33). MICR is used almost exclusively by the banking industry for check processing. Each check in your checkbook has precoded MICR characters on the lower-left edge; these characters represent the bank number, your account number, and the check number.

When a check is presented for payment, the bank uses an MICR inscriber to print the amount of the check in MICR characters in the lower-right corner (Figure 4-34). The check then is sorted or routed to the customer's bank, along with thousands of others. Each check is inserted into an MICR reader, which sends the check information – including the amount of the check – to a computer for processing. When you balance your checkbook, you should verify that the amount printed in the lower-right corner is the same as the amount written on the check; otherwise, your statement will not balance.

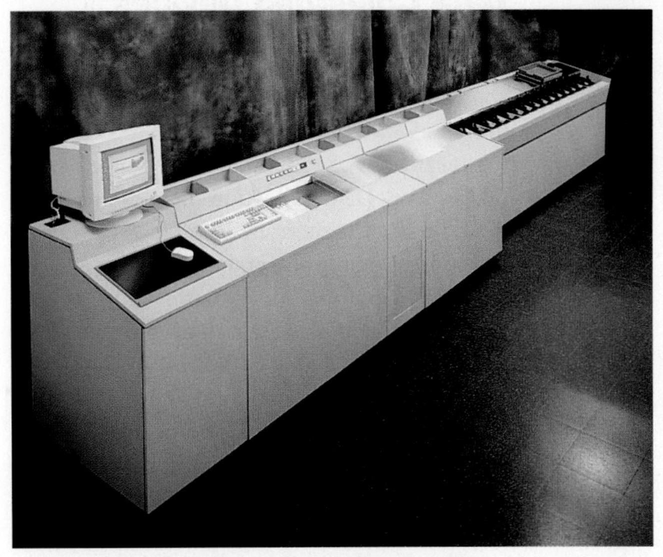

Figure 4-33 *An MICR reader is used almost exclusively by the banking industry for check processing.*

The banking industry has established an international standard not only for bank numbers, but also for the font of the MICR characters. This standardization makes it possible for you to write checks in another country.

Figure 4-34 *The MICR characters preprinted on the check represent the bank number, your account number, and the check number. The amount of the check in the lower-right corner is added after you write the check.*

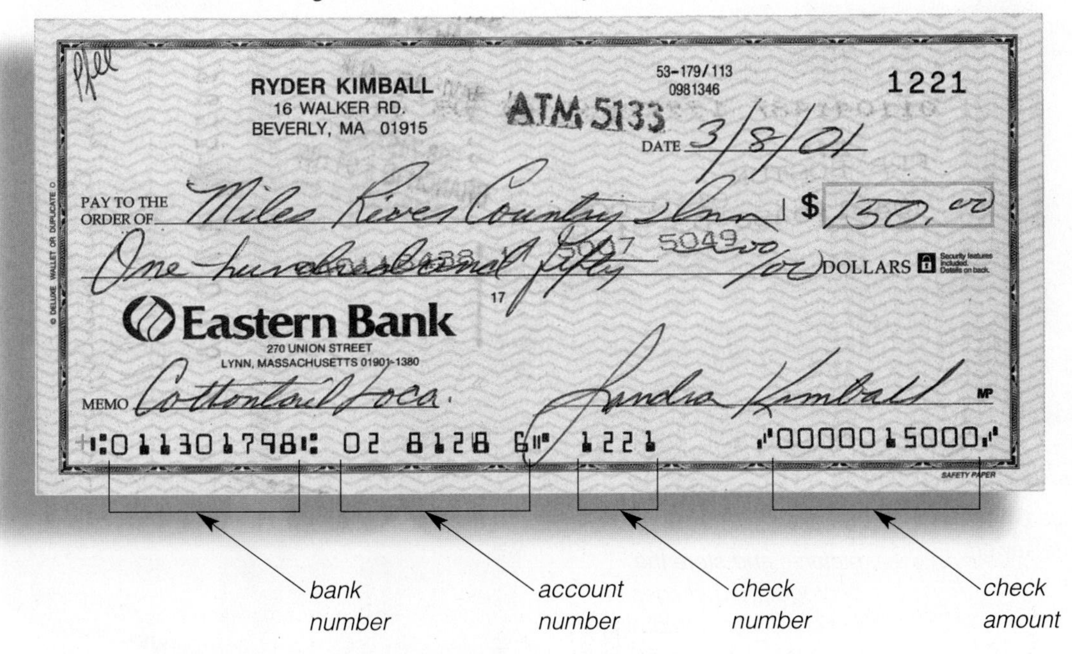

Data Collection Devices

Instead of reading or scanning data from a source document, **data collection devices** are designed and used to obtain data directly at the location where the transaction or event takes place. Data collection devices are used in factories, warehouses, or other locations where heat, humidity, and cleanliness are difficult to control. An example of this type of environment is a researcher who must be outside in the elements when collecting the data (points, lines, and area features) for a geographic information system (GIS). The data collection devices used to gather data for a GIS thus are rugged and durable, allowing researchers to create maps, analyze and interpret data for the maps, and capture images from the air or the ground (Figure 4-35).

DIGITAL CAMERAS

A **digital camera** allows you to take pictures and store the photographed images digitally instead of on traditional film (Figure 4-36). With some digital cameras, you **download**, or transfer a copy of, the stored pictures to your computer by connecting a cable between the digital camera and your computer and using special software included with the camera. With other digital cameras, the pictures are stored directly on storage media such as a floppy disk, PC Card, or flash card. You then copy the pictures to your computer by inserting the floppy disk into a disk drive or the card into a card slot or reader. Once the pictures are on your computer, they can be edited with photo-editing software, printed, faxed, sent via electronic mail,

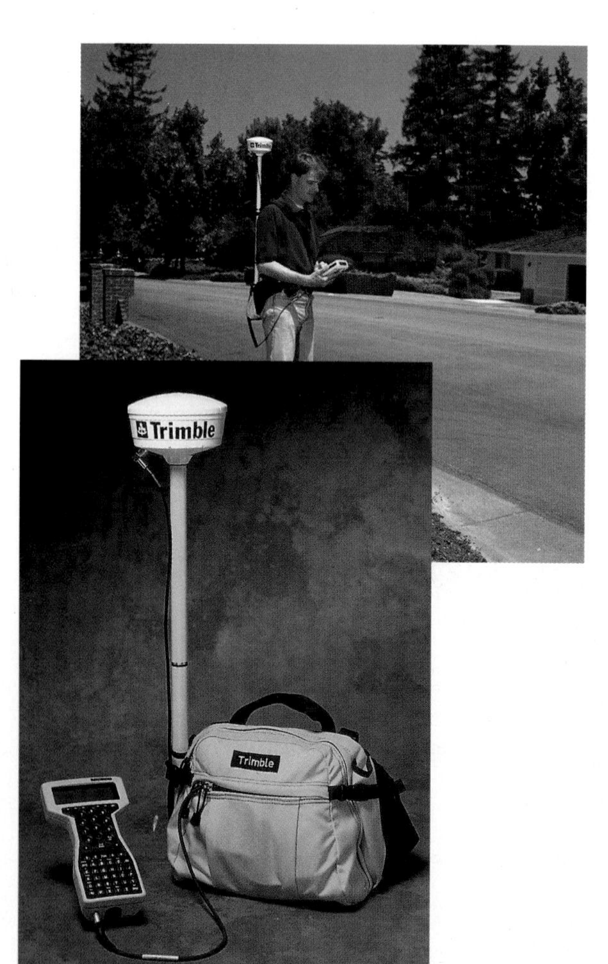

Figure 4-35 *Civil engineers use GIS data collection device to locate sewers, roads, and public buildings.*

Figure 4-36a *(digital camera)*

Figure 4-36b *(digital camera with flash card)*

Figure 4-36 *A digital camera is used to take pictures and store the photographed images on the computer. Some digital cameras can save images directly onto a floppy disk or PC Card.*

Figure 4-36c *(digital camera with floppy disk)*

included in another document, or posted to a Web site for everyone to see (Figure 4-37). Instead of copying the images to your computer, many of today's digital cameras allow you to review and edit the images directly in the camera, as well as connect the camera to a television or printer.

The three basic types of digital cameras are studio cameras, field cameras, and point-and-shoot cameras. The most expensive and highest quality of the three, a **studio camera** is a stationary camera used for professional studio work. Often used by photojournalists, a **field camera** is a portable camera that has many lenses and other attachments; like the studio camera, a field camera can be quite expensive. A **point-and-shoot camera** is more affordable and lightweight and provides acceptable quality photographic images for the home or small business user. These cameras often include features such as flash, zoom, automatic focus, and special effects. You can use a point-and-shoot camera to add pictures to personalized greeting cards, a computerized photo album, a family newsletter, certificates, awards, or your own Web site. The point-and-shoot camera also is ideal for mobile users such as real estate agents, insurance agents, and general contractors.

As with a scanner, the quality of a digital camera is measured by the number of bits it stores in a dot and the number of dots per inch, or resolution. The higher each number, the better quality, but the more expensive the camera. Most of today's point-and-shoot digital cameras are at least 24-bit with a resolution ranging from 640 x 480 to 1280 x 960. Other features of digital cameras are discussed in Chapter 13.

WEB INFO

For more information on digital cameras, visit the Discovering Computers 2001 Chapter 4 WEB INFO page (**www.scsite.com/dc2001/ch4/webinfo.htm**) and click Digital Cameras.

Figure 4-37
HOW A DIGITAL CAMERA WORKS

Step 1: *Point to the image to photograph. Light passes into the lens of the camera.*

Step 2: *The image is focused on a chip called a charge-coupled device (CCD).*

Step 3: *The CCD generates an analog signal that represents the image.*

Step 4: *The analog signal is converted to a digital signal by an analog-to-digital converter (ADC).*

Step 5: *A digital signal processor (DSP) adjusts the quality of the image and stores the digital image on a PC Card, floppy disk, or other memory in the camera.*

Step 6: *Images are transferred to a computer by plugging one end of the cable into a camera and the other end of the cable into a computer; or a PC Card or floppy disk is inserted into the computer and the images are copied to the hard disk.*

Step 7: *Using software supplied with the camera, the images are viewed on the screen.*

AUDIO AND VIDEO INPUT

Although characters (text and numbers) are still the primary form of input into a computer, the use of other types of input such as images, audio, and video is increasing. In the previous sections, you learned about a variety of ways to enter image data. The next sections discuss methods used to enter audio and video data into a computer. Other techniques for using images, audio, and video are discussed in Chapter 13.

Audio Input

Audio input is the process of entering (recording) music, speech, or sound effects. To record high quality sound, your personal computer must have a sound card. (Most new computers today come equipped with a sound card.) Sound is entered via a device such as a microphone, tape player, or audio CD player, each of which plugs into a port on the sound card. External MIDI devices such as an electric piano keyboard also can connect to the sound card for audio input (Figure 4-38).

Figure 4-38 An electronic keyboard is an external MIDI device that can be used to play music, which can be stored in the computer.

Recall that, in addition to being a port, MIDI (musical instrument digital interface) is the electronic music industry's standard that defines how sounds are represented electronically by digital musical devices. Software programs that conform to the MIDI standard allow you to compose and edit music and other sounds. For example, you can change the speed, add notes, or rearrange the score to produce an entirely new sound.

With a microphone plugged into the microphone port on the sound card, you can record sound using the Windows Sound Recorder (Figure 4-39). Windows stores audio files as **waveforms**, which are called **WAV** files and have a .wav extension. Once you save the sound in a file, you can play it using the Sound Recorder, or edit it using music-editing software that conforms to the MIDI standard. You also can attach the audio file to an e-mail message or include it in a document such as a word processing report or presentation graphics slide show.

WAV files often are large – requiring more than 1 MB of storage space for a single minute of audio. For this reason, WAV files often are compressed so they take up less storage space.

Figure 4-39 The waveform shown represents a portion of the word, Hello, as spoken into a microphone.

Speech Recognition

Another use for a microphone is speech recognition. **Speech recognition**, also called **voice recognition**, is the computer's capability of distinguishing spoken words. Speech recognition programs do not understand speech; they only recognize a vocabulary of certain words. The vocabulary of speech recognition programs can range from two words (such as Yes and No) to more than sixty thousand words (Figure 4-40).

Speech recognition programs are either speaker dependent or speaker independent. With **speaker-dependent software**, the computer makes a profile of your voice, which means you have to train the computer to recognize your voice. To train the computer, you must speak each of the words in the vocabulary into the computer repeatedly. After *hearing* the spoken word repeatedly, the program develops and stores a digital pattern for the word. When you later speak a word, the program compares the spoken word to those stored. **Speaker-independent software** has a built-in set of word patterns, so you do not have to train a computer to recognize your voice.

Some speech recognition software requires **discrete speech**, which means you have to speak slowly and separate each word with a short pause. Higher-quality speech recognition software allows you to speak in a flowing conversational tone, called **continuous speech**. Several continuous-speech systems are available for personal computers, and advances in speech recognition continue to be made.

Speech recognition systems often are used in specialized applications in which a user's hands are occupied or disabled, or by users such as reporters and attorneys. Instead of typing or using a pointing device, the user speaks into a microphone to dictate words, issue commands, or perform other tasks. The table in Figure 4-41 summarizes three types of speech recognition applications and examples of their uses.

- ☑ **Continuous speech—increased productivity.** Speak to your computer without pausing between words. At speeds faster than that of an average typist, you can create documents more quickly than ever before. ViaVoice automatically spells each word correctly, so there's no need to spend time looking for spelling mistakes.
- ☑ **Dictate directly into Microsoft Word, Microsoft Excel, and Microsoft Outlook.** Save precious time by speaking directly into these popular programs. No need to cut and paste... dictate right into the software!
- ☑ **Speech-enabled Web features.** Talk directly into most Web browsers and chat rooms.
- ☑ **No initial training.** No initial enrollment is necessary (use ViaVoice out of the box). Just spend a few minutes with the Dictation Trainer to learn how to dictate, and you're off and running.
- ☑ **Adapts to the way you work.** As you use the system, it learns your vocabulary and usage patterns. The more you use the system, the more accurate it becomes. You can have any number of users and guests on a single system.
- ☑ **Shortcuts.** Use a keyword to automatically trigger commonly used phrases, paragraphs, or pages. By simply dictating "T-letter" the entire text of a thank you letter can appear on your screen.
- ☑ **Accuracy.** Industry-leading technology. IBM's multi-patented breakthrough technology means you can spend less time correcting and more time creating.
- ☑ **Context recognition.** ViaVoice uses context to distinguish between similar sounding words and phrases (for example: there, their and they're; or whirled peas and world peace).
- ☑ **Text to speech.** Listen to your own documents or imported documents. In a rush? Let ViaVoice read aloud your e-mail as you prepare for that important meeting. Or, just give your tired eyes a rest.
- ☑ **Large vocabulary.** The base vocabulary of 260,000 words can be expanded to 2 million words to accommodate your own personal preferences. ViaVoice also provides a 200,000 word back-up dictionary.
- ☑ **Expand your vocabulary.** Analyze text documents for new words with the option of adding them to your vocabulary. This is a quick and easy way to customize your vocabulary to the words used most often.
- ☑ **Ease of correction.** Many other programs insist that you correct errors as they occur. IBM ViaVoice allows you to complete your thoughts and make corrections at a convenient time or defer the corrections to a colleague. Simply click on a word to play back that word. This feature makes corrections a breeze and ensures your original intent and accuracy.
- ☑ **Includes noise-canceling headset microphone.** ViaVoice includes a top-of-the-line headset microphone.

Figure 4-40 Speech recognition programs recognize a vocabulary of words. ViaVoice, for example, has a 260,000 word base vocabulary that can be expanded to 2 million words.

SPEECH RECOGNITION APPLICATIONS

Speech Recognition Application	Explanation	Example Uses
Command	Controls equipment	• Issue instructions to personal computer • Dial a cellular telephone • Route calls in an automated telephone system
Dictation/Data Entry	Types spoken words	• Office employees dictate letters • Doctors update hospital patient records • Reporters write stories • Lawyers develop briefs • Bankers transfer funds among bank accounts
Information Access	Enables access to products and services	• Access credit card account information • Access product information

Figure 4-41 Summary of the three types of speech recognition applications and examples of their uses.

Video Input

Video input or **video capture** is the process of entering a full-motion recording into a computer and storing the video on a hard disk or some other medium. To capture video, you plug a video camera, VCR, or other video device into a **video capture card**, which is an expansion card that converts the analog video signal into a digital signal that a computer can understand. (Most new computers are not equipped with a video capture card.) Once the video device is connected to the video capture card, you can begin recording. After you save the video on a hard disk, you can play it or edit it using video-editing software.

Just as with audio files, video files can require tremendous amounts of storage space. A three-minute segment, or clip, of high-quality video, for example, can take an entire gigabyte of storage (equal to approximately 50 million pages of text). To decrease the size of the files, video often is compressed. A popular video compression standard is defined by the **Moving Picture Experts Group (MPEG)**. DVD-ROMs use the MPEG standard to compress video data.

If you do not want to save an entire video clip on your computer, you can use a **video digitizer** to capture an individual frame from a video and then save the still picture in a file. To do this, you plug the recording device such as a video camera, VCR, or television into the video digitizer, which then usually connects to a parallel port on the system unit (Figure 4-42). As you watch the video using special software, you can stop it and capture any single frame. The resulting files are similar to those generated with a digital camera.

Figure 4-42
HOW A VIDEO DIGITIZER WORKS

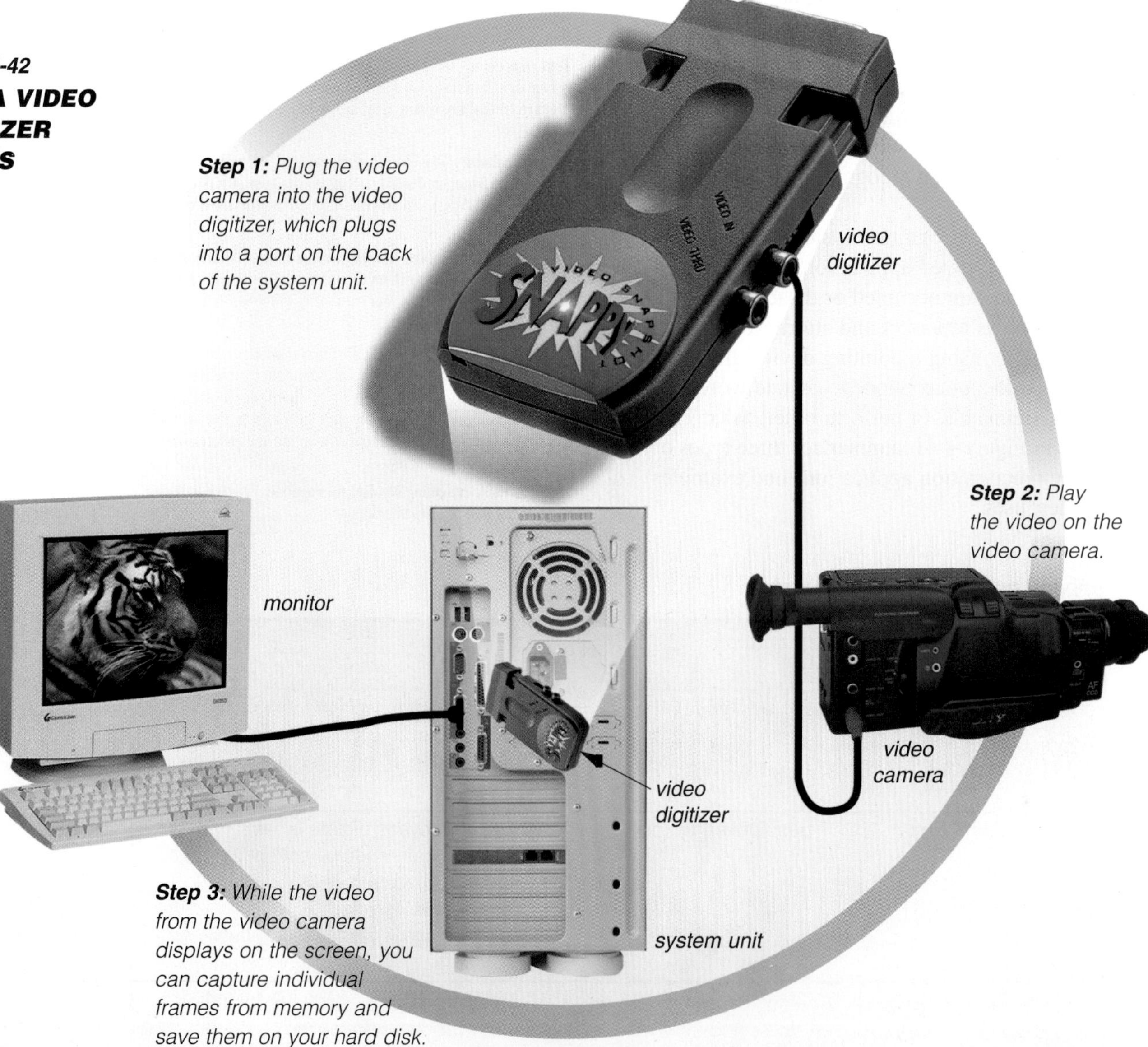
Step 1: *Plug the video camera into the video digitizer, which plugs into a port on the back of the system unit.*
VIDEO IN
VIDEO THRU
video digitizer
Step 2: *Play the video on the video camera.*
monitor
video camera
video digitizer
system unit
Step 3: *While the video from the video camera displays on the screen, you can capture individual frames from memory and save them on your hard disk.*

Videoconferencing

A **videoconference** is a meeting between two or more geographically separated individuals who use a network or the Internet to transmit audio and video data. To participate in a videoconference, you must have a microphone, speakers, and a video camera mounted on your computer (Figure 4-43). As you speak, members of the meeting hear your voice on their speakers. Any image in front of the video camera, such as a person's face, displays in a window on each participant's screen.

Another window on the screen that displays notes and drawings simultaneously on all the participants' screens, called a **whiteboard**, provides multiple users with an area on which they can write or draw. As the costs of videoconferencing hardware and software continue to decrease, more people and companies are taking advantage of this cost-effective way to conduct business meetings, corporate training, and educational classes.

Many home computers now also include a video camera as standard equipment. These cameras, often called **PC cameras** or **Webcams,** allow the home user to see people at the same time they communicate on the Internet. These cameras are discussed in more depth in Chapter 13.

WEB INFO

For more information on videoconferencing, visit the Discovering Computers 2001 Chapter 4 WEB INFO page (**www.scsite.com/dc2001/ch4/webinfo.htm**) and click Videoconference.

***Figure 4-43** As you speak, members of the videoconference hear your voice on their speakers. With the video camera facing you, an image of your face displays in a window on each participant's screen.*

INPUT DEVICES FOR PHYSICALLY CHALLENGED USERS

The growing presence of computers in everyone's lives has generated an awareness of the need to address computing requirements for those with physical limitations. Today, the **Americans with Disabilities Act (ADA)** requires that any company with 15 or more employees make reasonable attempts to accommodate the needs of physically challenged workers. Whether at work or at home, you may find it necessary to obtain input devices that address physical limitations. Besides speech recognition, which is ideal for blind or visually impaired users, several other input devices are available.

Users with limited hand mobility that wish to use a keyboard have several options. A **keyguard**, which is placed over the keyboard, allows you to rest your hand on the keyboard without accidentally pressing any keys; a keyguard also guides your finger or pointing device so you press only one key at a time (Figure 4-44).

Figure 4-44 *A keyguard allows you to rest your hand on the keyboard without accidentally pressing any keys and guides your finger or pointing device onto a key so you press only a single key at a time.*

Keyboards with larger keys also are available. Still another option is the screen-displayed keyboard, in which a graphic of a standard keyboard displays on the user's screen. Using a pointing device, the individual presses the keys on the screen-displayed keyboard.

Various pointing devices are available for users with motor disabilities. Small trackballs that can be controlled with a thumb or one finger can be attached to a table, mounted to a wheelchair, or held in a user's hand. People with limited hand movement can use a **head-mounted pointer** to control the pointer or insertion point (Figure 4-45). To simulate the functions of a mouse button, you use a single-switch scanning display. The switch might be a pad you press with your hand, a foot pedal, a receptor that detects facial motions, or a pneumatic instrument controlled by puffs of air.

Figure 4-45 *A head-mounted pointer can be used to control the pointer on a screen-displayed keyboard, which is a graphical image of the standard keyboard.*

PUTTING IT ALL TOGETHER

When you purchase a computer, you should have an understanding of the input devices included with the computer, as well as those you may need that are not included. Many factors influence the type of input devices you may use: the type of input desired, the hardware and software in use, and the desired cost. The type of input devices you require depends on your intended use. Figure 4-46 outlines several suggested input devices for specific computer users.

SUGGESTED INPUT DEVICES BY USER

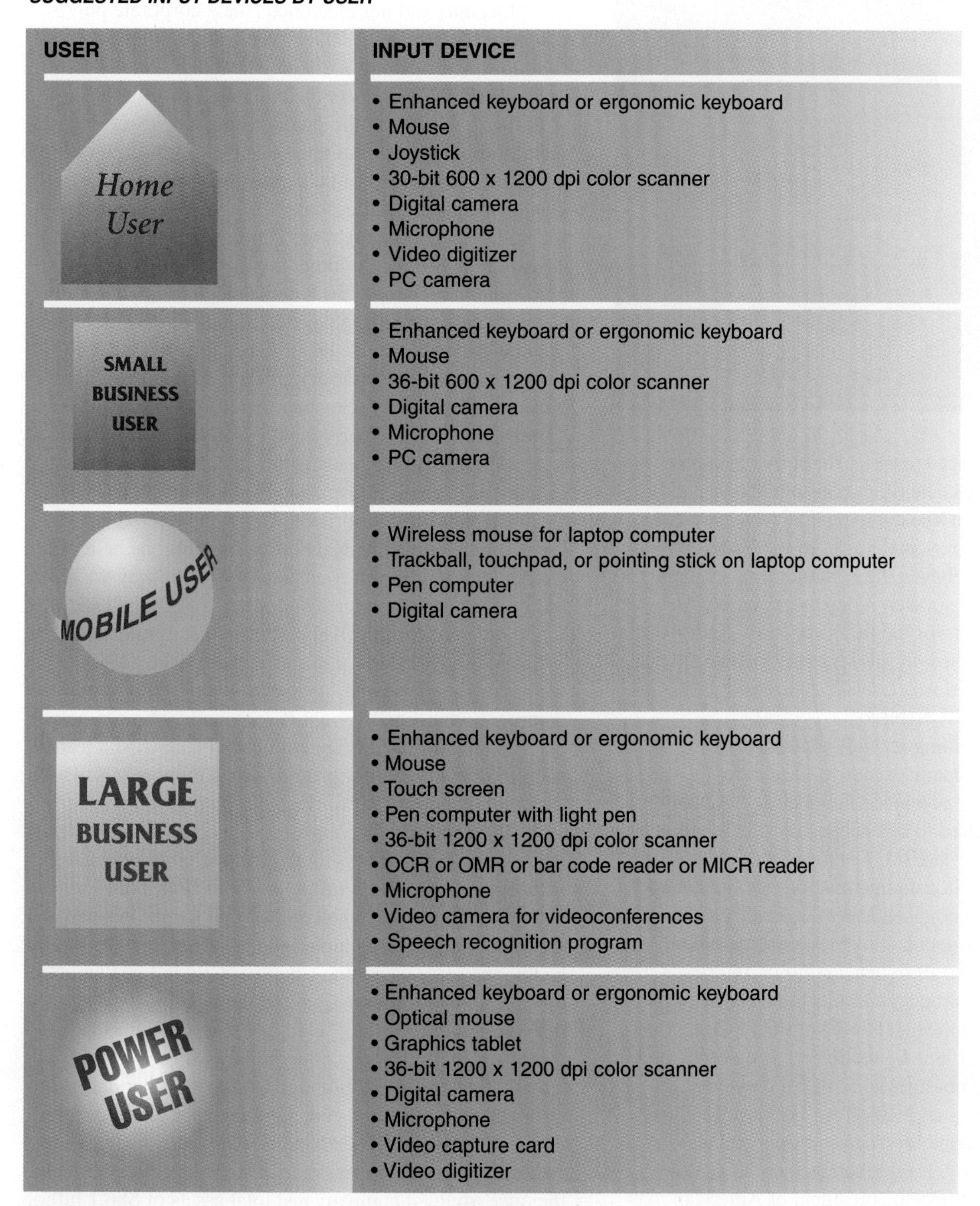

USER	INPUT DEVICE
Home User	• Enhanced keyboard or ergonomic keyboard • Mouse • Joystick • 30-bit 600 x 1200 dpi color scanner • Digital camera • Microphone • Video digitizer • PC camera
SMALL BUSINESS USER	• Enhanced keyboard or ergonomic keyboard • Mouse • 36-bit 600 x 1200 dpi color scanner • Digital camera • Microphone • PC camera
MOBILE USER	• Wireless mouse for laptop computer • Trackball, touchpad, or pointing stick on laptop computer • Pen computer • Digital camera
LARGE BUSINESS USER	• Enhanced keyboard or ergonomic keyboard • Mouse • Touch screen • Pen computer with light pen • 36-bit 1200 x 1200 dpi color scanner • OCR or OMR or bar code reader or MICR reader • Microphone • Video camera for videoconferences • Speech recognition program
POWER USER	• Enhanced keyboard or ergonomic keyboard • Optical mouse • Graphics tablet • 36-bit 1200 x 1200 dpi color scanner • Digital camera • Microphone • Video capture card • Video digitizer

Figure 4-46 *This table recommends suggested input devices.*

TECHNOLOGY TRAILBLAZER

LOU GERSTNER

In 1993, IBM faced a crisis. Beset by changing customer relationships, a shifting consumer focus, and increasing competition, the company had lost more than $8 billion. IBM's future – perhaps even its existence – was in doubt; some believed the computer industry giant should be broken into smaller, independent enterprises. In search of a solution, for the first time in its eighty-two year history IBM reached outside its ranks for a corporate leader. The man tapped was Louis V. Gerstner, chief executive officer at RJR Nabisco.

Why would an ailing technological colossus turn to a tobacco/food company executive for a cure? Surprisingly, Gerstner combined the qualities, education, and experience required to be chairman of the board and chief executive officer at IBM. Gerstner learned accountability and discipline in a competitive private high school, where grades were announced and rules strictly enforced. (Years later, Gerstner detailed his views on education in *Reinventing Education: Entrepeneurship in America's Public Schools.*) He earned a bachelor's degree in engineering as a scholarship student at Dartmouth College. Colleagues remember him as intelligent, purposeful, and demanding. "He did not tolerate fools," a classmate recalls. After receiving an MBA from Harvard Business School, Gerstner became the youngest principal at McKinsey & Company, Inc., a management consulting firm. He moved on to become president of American Express Company, where he showed the ability to invigorate an established business. At RJR Nabisco, his next stop, Gerstner demonstrated the willingness to make tough decisions in a variety of crises.

Gerstner did not announce a new corporate philosophy when he arrived at IBM. "The last thing IBM needs right now," he said, "is a vision." Instead, Gerstner took practical steps to stabilize the company: cutting the workforce, rebuilding the product line, and instituting cost reductions. He brought a no-nonsense approach to an organization where internal dissent was commonplace. Gerstner was an exacting boss: "I'm intense, competitive, focused, blunt, and tough, yes." Yet, he permitted the more casual dress common at other companies, rejecting the blue-suited conformity once demanded at *Big Blue*. Gerstner cast aside IBM's geographic divisions, reorganizing along industry lines. He railed against the fanatic perfectionism that characterized IBM's labs. "You don't launch products here," Gerstner complained to technicians. "They escape." Gerstner speeded product development and, in 1996, IBM PCs using new technology were first to market.

Perhaps most surprising, Gerstner rejected calls to break up the company. Network computing, he believed, would "change the way we do business, the way we teach our children, communicate and interact as individuals." In a networked world, IBM's ability to combine hardware, software, and service was an important asset. "This technology stuff is hard," Gerstner said, "and if I'm a company, what I really want is somebody who can help me implement it." Gerstner set out to make IBM a leading source for the technology, hardware, software, installation, and maintenance of networked computing systems.

IBM still had to reestablish the customer confidence that once was taken for granted. An early adage claimed, *Nobody was ever fired for buying from IBM*, but in a changing marketplace the company no longer inspired that assurance. Technology alone was not enough. "We should be impressed by technology, but we shouldn't be distracted by it, or fooled into thinking that technology, unto itself, is the solution to anything," Gerstner maintained. Restoring customer faith required cooperation with each individual client. "We start with an understanding of what our customers need," Gerstner said, "and then we work together to fashion a solution." Gerstner seems on the way to fashioning his own solution at IBM. At the close of 1998, the once troubled company had total assets of $86.1 billion.

COMPANY ON THE CUTTING EDGE

COMPANY ON THE CUTTING EDGE

IBM

What do meat and cheese slicers, commercial scales, industrial time recorders, tabulators, and punched cards have in common? These were some of the first products offered by IBM. It was more than thirty years before the company created its first computer. Despite the delayed start, today IBM, nicknamed Big Blue, is a leader in the computer industry with a reputation for innovative products.

In 1911, the Computing Scale Company of America, Herman Hollerith's Tabulating Machine Company, and the International Time Recording Company merged and incorporated to form the Computing – Tabulating – Recording Company (C-T-R). The company sold a variety of business-related gadgets, but Hollerith's tabulating machine presaged its future. The machine, which used punched cards to catalogue data, helped complete the 1890 census in record time and was an ancestor of the modern computer. Nine years after C-T-R was created, its name was changed to the International Business Machines Corporation (IBM).

IBM's first electronic computer, the Mark I, was completed in 1944. Designed to perform long computations automatically, the huge Mark I took about twelve seconds to divide – far slower than today's pocket calculators. In 1952, IBM produced its first vacuum tube computer. This IBM 701 executed 17,000 instructions per second and soon was employed for business applications. Five years later, IBM introduced the first computer disk storage system and FORTRAN, a scientific programming language. In 1959, IBM presented the IBM 7090, a mainframe that could perform 229,000 calculations per second.

In the 1960s and 1970s, IBM changed how computers were marketed and used. The System/360 was the first family of computers. Interchangeable software and peripherals allowed users to upgrade processors without having to replace the entire system. In 1969, IBM unbundled components (hardware, software, and services), and began to sell each separately – thus giving birth to the software and service industries. Three devices, now taken for granted, were invented at IBM in the 1970s: the supermarket price scanner, the automatic teller machine (ATM), and a new type of laser printer.

In 1981, the IBM Personal Computer (PC) revolutionized the computer industry. With 16 KB of RAM, a floppy disk drive, an optional color monitor, and prices starting at $1,565, more than 136,000 IBM PCs were sold in the first year and a half. For the first time, IBM had turned to outside companies for components – an Intel processor and a Microsoft operating system. As a result, a number of IBM PC-compatible clones with the same components were built by other companies. In 1992, IBM's new ThinkPad notebook computer combined two inventive input devices – a butterfly-like keyboard that expanded when the computer case was opened and a unique pointing device, called a TrackPoint, positioned between keys. The ThinkPad collected more than 300 awards for its creative design.

As purchasing decisions moved down the corporate ladder, customer relationships changed and many buyers turned to less-expensive IBM-compatible clones. With declining sales, some industry insiders advocated breaking up the company. New CEO Louis Gerstner, however, recognized that the rise of the Internet and network computing dovetailed with an IBM strength – the ability to provide integrated solutions. Building on this advantage, revenues climbed. Acquisitions such as Lotus Development Corporation in the mid 1990s increased IBM's market value by $50 billion.

In 1997, IBM had 369,465 employees with revenues of $78.5 billion and net earnings of more than $6 billion. The company also experienced a public relations coup when Deep Blue, an IBM supercomputer, defeated World Chess Champion Gary Kasparov in a six-game match, becoming the first computer to beat a reigning champion. The unprecedented success of Deep Blue is emblematic of Big Blue's determination to remain at the forefront of the computer revolution.

Student Exercises
WEB INFO
IN BRIEF
KEY TERMS
AT THE MOVIES
CHECKPOINT
AT ISSUE
MY WEB PAGE
HANDS ON
NET STUFF
Special Features
TIMELINE 2001
GUIDE TO WWW SITES
MAKING A CHIP
E-COMMERCE 2001
BUYER'S GUIDE 2001
CAREERS 2001
TRENDS 2001
LEARNING GAMES
CHAT
INTERACTIVE LABS
NEWS
HOME

CHAPTER 1 2 3 4 5 6 7 8 9 10 11 12 13 14 INDEX

IN BRIEF

www.scsite.com/dc2001/ch4/brief.htm

WEB INSTRUCTIONS: *To display this page from the Web, launch your browser and enter the URL, www.scsite.com/dc2001/ch4/brief.htm. Click the links for current and additional information. To listen to an audio version of this IN BRIEF, click the Audio button to the right of the title, IN BRIEF, at the top of the page. To play the audio, RealPlayer must be installed on your computer (download by clicking here).*

1. What Are the Four Types of Input?

Input is any data or instructions entered into the memory of a computer. Four types of input are data, programs, commands, and user responses. **Data** is a collection of unorganized facts that can include words, numbers, pictures, sounds, and video. A **program** is a series of instructions that tell a computer how to process data into information. **Commands** are instructions given to a computer program. **User responses** are instructions a user issues to the computer by responding to questions posed by a computer program. Any component used to enter data, programs, commands, and user responses into a computer is an **input device**.

2. What Are the Characteristics of a Keyboard?

The **keyboard** is a primary input device on a computer. All keyboards have a typing area used to type letters of the alphabet, numbers, punctuation marks, and other basic characters. A keyboard also may include a **numeric keypad** designed to make it easier to enter numbers, **function keys** programmed to issue commands and accomplish certain tasks, **arrow keys** used to move the **insertion point**, and **toggle keys** that can be switched between two different states.

3. Describe the Various Types of Keyboards

A standard computer keyboard sometimes is called a QWERTY keyboard because of the layout of its typing area. The **Dvorak keyboard** places the most frequently typed letters in the middle of the typing area. **Enhanced keyboards** have function keys, CTRL keys, ALT keys, and a set of arrow and additional keys. **Wireless keyboards** transmit data via infrared light waves. Laptop and handheld computer keyboards sometimes contain smaller and fewer keys.

4. What Are Various Types of Pointing Devices?

Pointing devices control the movement of a pointer on the screen. A **mouse** is a pointing device that is moved across a flat surface, controls the movement of the pointer on the screen, and is used to make selections from the screen. A **trackball** is a stationary pointing device with a ball mechanism on its top. A **touchpad** is a flat, rectangular pointing device that is sensitive to pressure and motion. A **pointing-stick** is a pressure-sensitive pointing device shaped like a pencil eraser. Other pointing devices include a **joystick** (a vertical lever mounted on a base), a **touch screen** (a monitor with a touch-sensitive panel on the screen), a **light-pen** (a handheld device that contains a light source or can detect light), and a **graphics tablet** (an electronic plastic board used to input graphical data).

5. How Does a Mouse Work?

The bottom of a mouse is flat and contains a multidirectional mechanism, either a small ball or an optical sensor, which detects movement of the mouse. As the mouse is moved across a flat surface, electronic circuits in the mouse translate the movement into signals that are sent to the computer, and the pointer on the screen also moves.

www.scsite.com/dc2001/ch4/brief.htm

6. What Are Different Mouse Types?

A **mechanical mouse** has a rubber or metal ball on its underside. An optical mouse uses devices that emit light to detect the mouse's movement. A **cordless mouse**, or **wireless mouse**, relies on battery power and uses infrared or radio waves to communicate with a receiver.

7. How Do Scanners and Other Reading Devices Work?

A **scanner** is a light-sensing input device that reads printed text and graphics and then translates the results into a form the computer can use. When a **source document** is scanned, a bright light moves across the document and an image is reflected into a series of mirrors. The light image is converted into an analog electrical current, which is converted into a digital signal that is sent to software in the computer. The scanned results are stored in rows and columns of dots, called a **bitmap**. An **optical reader** uses a light source to read characters, marks, and codes and converts them into digital data that can be processed by a computer. Three types of optical readers are **optical character recognition (OCR)**, **optical mark recognition (OMR)**, and **bar code scanners.**

8. Why Is a Digital Camera Used?

A digital camera is used to take pictures and store the photographed images digitally. The photographed image can be **downloaded**, or transferred, to a computer by a connecting cable; or it can be stored on a floppy disk, PC Card, or flash card and copied to a computer through a disk drive, card slot, or reader. Once on a computer, pictures can be edited with photo-editing software, printed, faxed, sent via electronic mail, included in another document, or posted to a Web site.

9. What Are Various Techniques Used for Audio and Video Input?

Audio input is the process of entering (recording) music, speech, or sound effects. Sound is entered via a microphone, tape player, audio CD player, or external MIDI (musical instrument digital interface) device such as an electric piano keyboard. Each plugs into a port on a computer's sound card. Audio files can be played, edited, attached to an e-mail message, or included in a document. **Video input** is the process of entering a full-motion recording into a computer and storing the video on a hard disk or some other medium. To capture video, a video camera is plugged into a video capture card, which is an expansion card that converts the analog video signal into a digital signal. A **video digitizer** can be used to capture an individual frame from a video and save the still picture in a file.

10. What Are Some Alternative Input Devices for Physically Challenged Users?

Speech recognition, or the computer's capability of distinguishing spoken words, is ideal for blind or visually impaired computer users. A **keyguard**, which is placed over the keyboard, allows people with limited hand mobility to rest their hands on the keyboard and guides a finger or pointing device so that only one key is pressed. Keyboards with larger keys and screen-displayed keyboards on which keys are pressed using a pointing device also can help. Pointing devices such as small trackballs that can be controlled with a thumb or one finger and **head-mounted pointers** also are available for users with motor disabilities.

KEY TERMS

www.scsite.com/dc2001/ch4/terms.htm

WEB INSTRUCTIONS: *To display this page from the Web, launch your browser and enter the URL, www.scsite.com/dc2001/ch4/terms.htm. Scroll through the list of terms. Click a term to display its definition and a picture. Click KEY TERMS on the left to redisplay the KEY TERMS page. Click the TO WEB button for current and additional information about the term from the Web. To see animations, Shockwave and Flash Player must be installed on your computer (download by clicking here).*

Americans with Disabilities Act (ADA) (4.26)
arrow keys (4.5)
audio input (4.22)
bar code (4.17)
bar code scanner (4.17)
bitmap (4.14)
click (4.8)
command (4.2)
continuous speech (4.23)
cordless mouse (4.10)
data (4.2)
data collection devices (4.20)
digital camera (4.20)
digitizer (4.13)
digitizing tablet (4.13)
discrete speech (4.23)
dots per inch (dpi) (4.15)
download (4.20)
Dvorak keyboard (4.6)
enhanced keyboards (4.6)
ergonomics (4.6)
field camera (4.21)
function keys (4.4)
graphical user interface (4.3)
graphics tablet (4.13)
handwriting recognition software (4.13)
head-mounted pointer (4.26)
image processing (4.16)
image processing system (4.16)
imaging (4.16)
input (4.2)
input device (4.3)
insertion point (4.5)
joystick (4.12)
keyboard (4.3)
keyguard (4.26)
keyword (4.2)

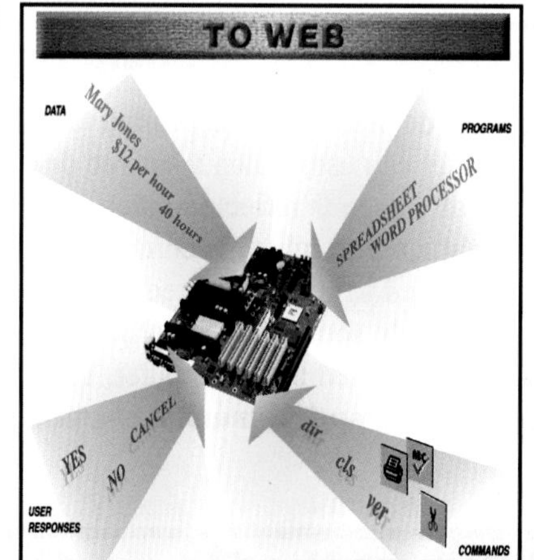

INPUT: Any data or instructions entered into the memory of a computer. Once in memory, the CPU can access it and process it into output. Four types of input are data, programs, commands, and user responses. (4.2)

light pen (4.12)
magnetic-ink character recognition (MICR) (4.19)
mechanical mouse (4.9)
menu-driven (4.3)
mouse (4.7)
mouse pad (4.7)
mouse pointer (4.7)
Moving Picture Experts Group (MPEG) (4.24)
numeric keypad (4.4)
OCR devices (4.16)
optical character recognition (OCR) (4.16)
optical character recognition (OCR) software (4.16)
optical mark recognition (OMR) (4.17)
optical mouse (4.9)
optical reader (4.16)
optical scanner (4.14)
PC cameras (4.25)
pen (4.13)
point-and-shoot camera (4.21)
pointer (4.7)
pointing device (4.7)
pointing stick (4.11)
program (4.2)
puck (4.13)
QWERTY keyboard (4.6)
resolution (4.15)
scanner (4.14)
source document (4.13)
speaker-dependent software (4.23)
speaker-independent software (4.23)
speech recognition (4.22)
studio camera (4.21)
stylus (4.13)
toggle keys (4.5)
touch screen (4.12)
touchpad (4.11)
trackball (4.10)
trackpad (4.11)
turnaround documents (4.16)
user response (4.3)
video capture (4.24)
video capture card (4.24)
video digitizer (4.24)
video input (4.24)
videoconference (4.24)
voice recognition (4.22)
WAV (4.22)
waveforms (4.22)
Webcams (4.25)
whiteboard (4.25)
wireless keyboards (4.6)
wireless mouse (4.10)

Student Exercises
WEB INFO
IN BRIEF
KEY TERMS
AT THE MOVIES
CHECKPOINT
AT ISSUE
MY WEB PAGE
HANDS ON
NET STUFF
Special Features
TIMELINE 2001
GUIDE TO WWW SITES
MAKING A CHIP
E-COMMERCE 2001
BUYER'S GUIDE 2001
CAREERS 2001
TRENDS 2001
LEARNING GAMES
CHAT
INTERACTIVE LABS
NEWS
HOME

AT THE MOVIES www.scsite.com/dc2001/ch4/movies.htm

WELCOME to VIDEO CLIPS from CNN

WEB INSTRUCTIONS: *To display this page from the Web, launch your browser and enter the URL, www.scsite.com/dc2001/ch4/movies.htm. Click a picture to view a video. After watching the video, close the video window and then complete the exercise by answering the questions about the video. To view the videos, RealPlayer must be installed on your computer (download by clicking here).*

1 Walk Through Computer

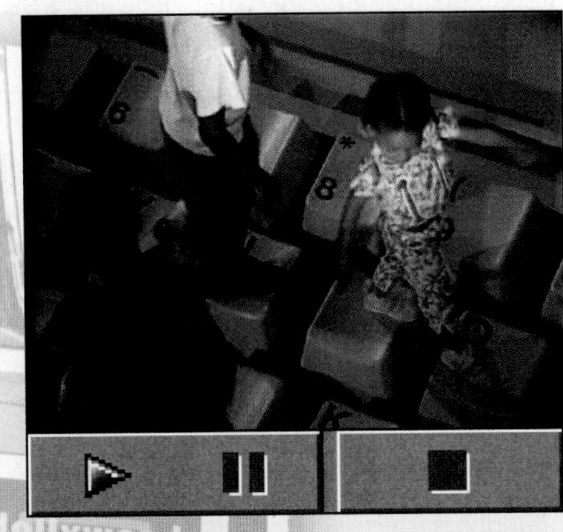

What is the idea behind the oversized computer exhibit at the Computer Museum in Boston? How does a museum visitor interact with the computer? What input devices are included in this larger-than-life computer? Do you think that being inside a computer could help you understand how computers work? To learn more about the museum's latest exhibits, visit the Web site.

2 Joystick Review

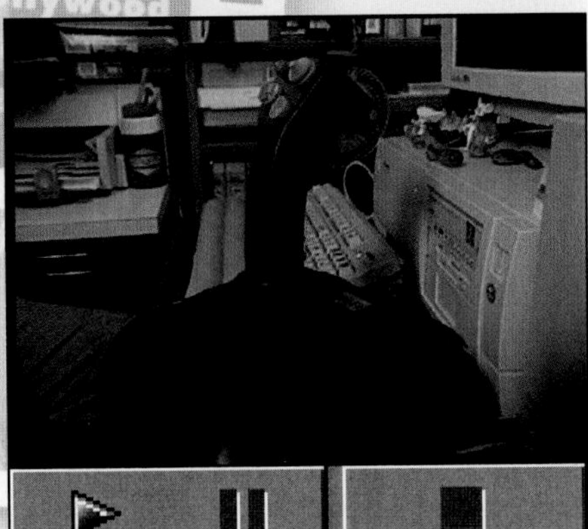

Advances in joystick technology have created exciting new possibilities. Where would you expect to use a joystick? What two types of joysticks are described in the video? Describe the features of each. How do they differ from other input devices, such as a mouse? How are they the same? In relation to the joystick, what is meant by feedback technology? Which model of joystick would you purchase?

3 Faces in the Crowd

What are some applications of face-recognition technology? What types of input devices are used in a face-recognition system? In what segment of society are the privacy issues raised by use of these face-recognition systems? If you were recognized as someone else and detained, would you be upset? What can be done to make sure this technology is not abused? Who do you think is accountable for the responsible use of face-recognition technology – government or the private sector?

CHECKPOINT www.scsite.com/dc2001/ch4/check.htm

WEB INSTRUCTIONS: *To display this page from the Web, launch your browser and enter the URL, www.scsite.com/dc2001/ch4/check.htm. Click the links for current and additional information. To experience the animation and interactivity, Shockwave and Flash Player must be installed on your computer (download by clicking here).*

Label the Figure

Instructions: *Identify these areas or keys on a typical desktop computer keyboard.*

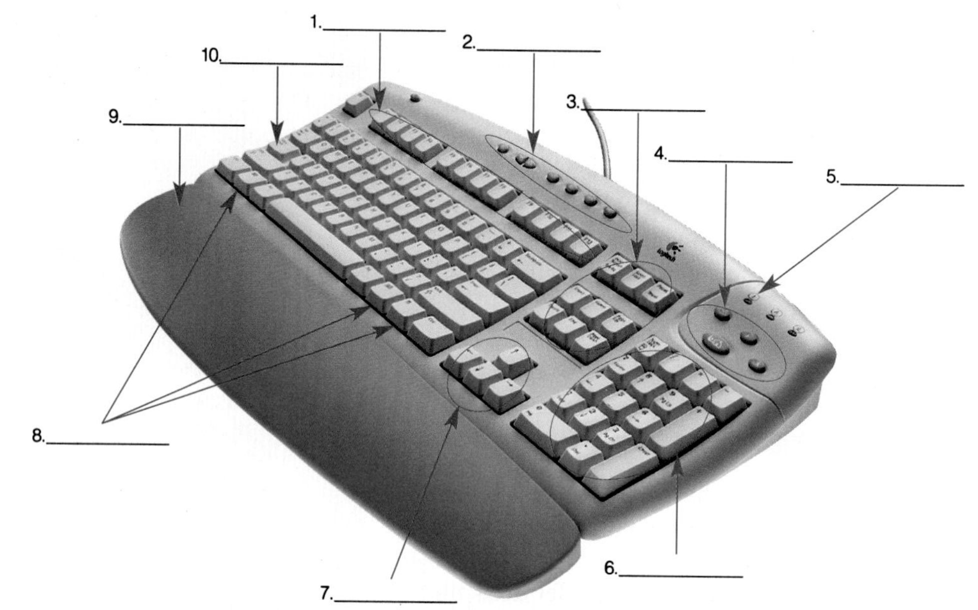

Matching

Instructions: *Match each key name from the column on the left with the best description from the column on the right.*

_________ 1. ENTER

_________ 2. BACKSPACE

_________ 3. ESC

_________ 4. CTRL

_________ 5. NUM LOCK

a. Toggle key that causes numeric keypad keys to function like a calculator.
b. Erases the character to the right of the insertion point.
c. Used at the end of a command to direct the computer to process the command.
d. When pressed in combination with another key(s), usually issues a command.
e. Toggle key that shifts alphabetic letters to uppercase.
f. Often used to quit a program or operation.
g. Erases a character to the left of the insertion point.

Short Answer

Instructions: *Write a brief answer to each of the following questions.*

1. Why is resolution important when using a scanner? __________ How is resolution typically measured and stated? __________
2. How is optical character recognition different from optical mark recognition? __________
3. What is a bar code? __________ How are bar codes read? __________ On what products are they used? __________
4. How is speaker-dependent software different from speaker-independent software? __________ How is discrete speech recognition different from continuous speech recognition? __________
5. What is videoconferencing? __________ How does a whiteboard enhance videoconferencing? __________

AT ISSUE www.scsite.com/dc2001/ch4/issue.htm

Student Exercises
WEB INFO
IN BRIEF
KEY TERMS
AT THE MOVIES
CHECKPOINT
AT ISSUE
MY WEB PAGE
HANDS ON
NET STUFF

Special Features
TIMELINE 2001
GUIDE TO WWW SITES
MAKING A CHIP
E-COMMERCE 2001
BUYER'S GUIDE 2001
CAREERS 2001
TRENDS 2001
LEARNING GAMES
CHAT
INTERACTIVE LABS
NEWS
HOME

WEB INSTRUCTIONS: *To display this page from the Web, launch your browser and enter the URL, www.scsite.com/dc2001/ch4/issue.htm. Click the links for current and additional information.*

ONE

More than 10,000 World Wide Web sites show images from Webcams. These sites feature real-time (often 24-hour) views from a variety of venues, including beaches, buildings, classrooms, dorm rooms, baby bassinets, fish tanks, taxicab dashboards, even inside a refrigerator. As entertainment, the appeal of Webcam Web sites is questionable. With minute-by-minute displays of the most mundane moments, many Webcam Web sites seem, well, boring. Yet, the sites are increasingly popular, with some experiencing almost 4.5 million hits a day. What motivates cyber-voyeurs? Why would someone want to see what are, in essence, a stranger's home movies? In what, if any, type of Webcam Web site would you be interested? Why?

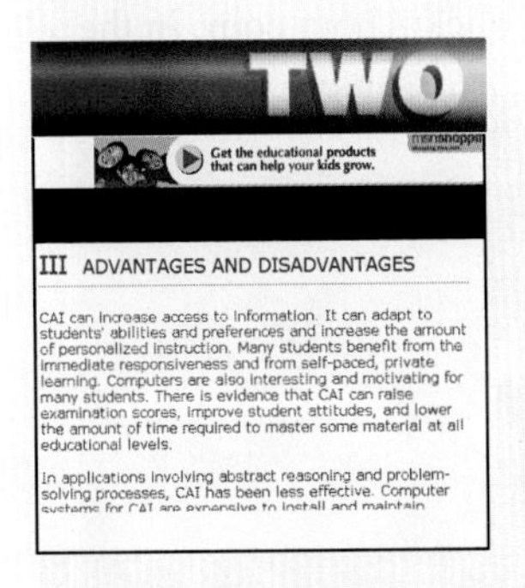

In addition to data, programs, commands, and user responses, researchers are experimenting with a fifth type of input – human emotions. A development called affective computing uses input devices such as video cameras and skin sensors with software similar to speech recognition programs to allow a computer to read a user's emotions. For example, a furrowed brow and sweaty palms might indicate frustration. Emotional input could be invaluable in conjunction with computer-aided instruction, letting a computer know whether to speed up or slow down tutorials. Some people, however, see affective computing as an invasion of privacy. How do you feel? In what areas might affective computing be useful? Why? Would you be comfortable with a computer knowing how you feel? Why or why not?

THREE

Although input varies, in one way input devices do not; most are encased in a bland, beige, plastic shell. A California company is changing that by offering keyboards and mouse units in oak, cherry, or maple. As a fitting adjunct, another company is selling mouse pads that replicate traditional Oriental or Persian carpets. Upgrading appearance is not cheap. A wooden keyboard costs more than $600, a wooden mouse more than $300, and the rodent rug almost $20. Yet, some people insist these devices provide a more attractive and, in the long run, more productive work setting. How much money would you spend to upgrade the appearance of your computer equipment? Why? Would this kind of upgrade increase productivity? Why or why not?

FOUR

Experts agree that speech recognition capability represents the future of software. Experts do not agree, however, that the future is now. The best speech recognition programs are ninety to ninety-five percent accurate. Yet, advocates admit that this assessment is based on expected speech and vocabulary. When confronted with unusual dialogue, accuracy drops; one speech recognition program translated the line, "I'll make him an offer he can't refuse" (from *The Godfather*) into "I held make them that off or he can refuse." Even a ninety percent accuracy rate means one out of ten words will be wrong. How accurate must speech recognition software be before it can be used effectively? When would speech recognition be an advantage? Might it ever be a disadvantage? Why?

FIVE

Getting ready to study for an exam can take longer than studying. By the time notes written on scraps of paper, food wrappers, book covers, and so on have been collected and arranged, little study time may be left. Fortunately, a new device called CrossPad can assist less-than-organized scholars. The portable digital notepad allows students to jot notes with a digital pen (storing up to one hundred pages), download the notes to a PC, and then organize the information by date or keywords. How would a portable digital notepad change your study habits? If every student had a portable digital notepad, how might it change the way teachers teach? In what occupations could people benefit from a portable digital notepad? Why?

MY WEB PAGE www.scsite.com/dc2001/ch4/myweb.htm

WEB INSTRUCTIONS: *The icons to the left of the numbered activities on this page indicate the learning system availability. If you are using WebCT, follow the instructions in the exercises below. If you are using the textbook Web site or CyberClass, launch your browser, enter the URL www.scsite.com/dc2001/ch4/myweb.htm, and then click the corresponding icon to the left of the exercise.*

1. Practice Test: Click My Web Page and then click Chapter 4. Click Practice Test. Answer each question and then click the Save Answer button. When completed, click the Finish button. Click the OK button to submit the quiz for grading. Click the View Results button, and then make a note of any missed answers.

2. Web Guide: Click My Web Page and then click Chapter 4. Click Web Guide to display the Guide to WWW Sites page. Click Reference and then click About.com. In the Find It Now text box, type `digital camera`. Scroll through the results and then click a link of your choice. Use your word processing program to prepare a brief report on your findings and submit your assignment to your instructor.

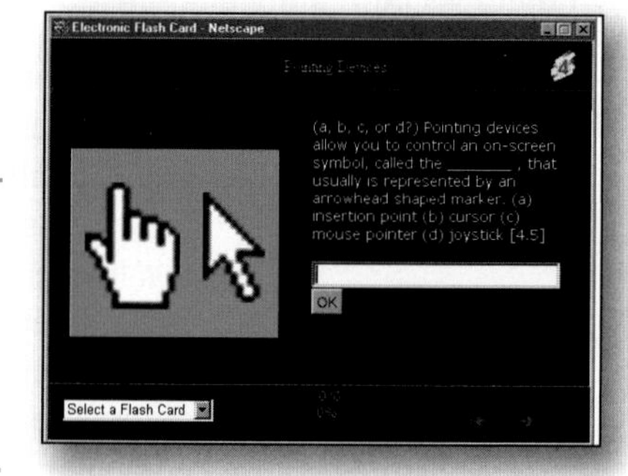

3. Scavenger Hunt: Click My Web Page and then click Chapter 4. Click Scavenger Hunt. Print a copy of the Scavenger Hunt page; use this page to write down your answers as you search the Web. Submit your completed page to your instructor.

4. Who Wants to Be a Computer Genius?: Click My Web Page and then click Chapter 4. Click Computer Genius to find out if you are a computer genius. For directions, click the How to Play button. When you are ready to play, click the Game button. Submit your score to your instructor.

5. Wheel of Terms: Click My Web Page and then click Chapter 4. Click Wheel of Terms to reinforce important terms you learned in this chapter by playing the Shelly Cashman Series version of this popular game. For directions, click the How to Play button. When you are ready to play, click the Game button. Submit your score to your instructor.

6. Career Corner: Click My Web Page and then click Chapter 4. Click Career Corner to display the Career Magazine page. Review this page. Click the links that you find interesting. Write a brief report on three topics you found to be the most interesting. Submit the report to your instructor.

7. Search Sleuth: Click My Web Page and then click Chapter 4. Click Search Sleuth to learn search techniques that will help make you a research expert. Submit the completed assignment to your instructor.

8. Crossword Puzzle Challenge: Click My Web Page and then click Chapter 4. Click Crossword Puzzle Challenge. Complete the puzzle to reinforce skills you learned in this chapter. For directions, click the How to Play button. When you are ready to play, click the Game button. Submit the completed puzzle to your instructor.

9. Flash Cards: This exercise uses CyberClass. Follow the instructions at the top of this page and then click the CyberClass icon to the left; or ask your instructor for logon instructions. Click Flash Cards on the Main Menu. Click the plus sign. Answer all the questions in any two subjects of your choice. Click the Close button in the upper-right corner to close all windows. Notice the many other exercises at this site. Complete all other exercises as assigned by your instructor.

HANDS ON

www.scsite.com/dc2001/ch4/hands.htm

WEB INSTRUCTIONS: *To display this page from the Web, launch your browser and enter the URL, www.scsite.com/dc2001/ch4/hands.htm. Click the links for current and additional information.*

SHELLY CASHMAN SERIES®
DISCOVERING COMPUTERS 2001

Student Exercises
WEB INFO
IN BRIEF
KEY TERMS
AT THE MOVIES
CHECKPOINT
AT ISSUE
MY WEB PAGE
HANDS ON
NET STUFF

Special Features
TIMELINE 2001
GUIDE TO WWW SITES
MAKING A CHIP
E-COMMERCE 2001
BUYER'S GUIDE 2001
CAREERS 2001
TRENDS 2001
LEARNING GAMES
CHAT
INTERACTIVE LABS
NEWS
HOME

One

About Your Computer

This exercise uses Window 95 or Windows 98 procedures. Your computer probably has more than one input device. To learn a little bit about the input devices on your computer, right-click the My Computer icon on the desktop. Click Properties on the shortcut menu. When the System Properties dialog box displays, click the Device Manager tab. Click View devices by type. Under Computer, a list of hardware device categories displays. What input devices appear in the list? Click the plus sign next to each category. What specific input devices in each category are connected to your computer? Click the Cancel button in the System Properties dialog box.

Two

Customizing the Keyboard

The Windows operating system provides several ways to customize the keyboard for people with physical limitations. Some of these options are StickyKeys, FilterKeys, and ToggleKeys. To find out more about each option, click the Start button, point to Settings on the Start menu, and then click Control Panel on the Settings submenu. Double-click Accessibility Options in the Control Panel window. Click the Keyboard tab in the Accessibility Properties dialog box. Click the question mark in the title bar, click StickyKeys, read the information in the pop-up window, and then click the pop-up window to close it. Repeat this process for FilterKeys and ToggleKeys. What is the purpose of each option? How might each option benefit someone with a physical disability?

Three

Using the Mouse and Keyboard to Interact with an Online Program

See your instructor for the location of the Loan Payment Calculator program. Click the Start button on the taskbar, and then click Run on the Start menu to display the Run dialog box. In the Open text box, type the path and file name of the program. For example, type `a:loancalc.exe` and then press the ENTER key to display the Loan Payment Calculator window. Type `12500` in the LOAN AMOUNT text box. Click the YEARS right scroll arrow or drag the scroll box until YEARS equals 15. Click the APR right scroll arrow or drag the scroll box until APR equals 8.5. Click the Calculate button. Write down the monthly payment and sum of payments. Click the Clear button. What are the monthly payment and sum of payments for each of these loan amounts, years, and APRs: (1) 28000, 5, 7.25; (2) 98750, 30, 9; (3) 6000, 3, 8.75; (4) 62500, 15, 9.25. Close the Loan Payment Calculator.

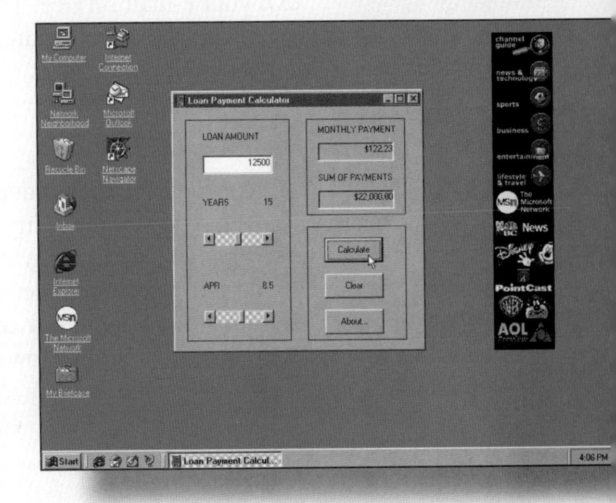

Four

MouseKeys

A graphical user interface allows you to perform many tasks with just the point and click of a mouse. Yet, what if you do not have, or cannot use, a mouse? The Windows operating system covers this possibility with an option called MouseKeys. When the MouseKeys option is turned on, you can use numeric keypad keys to move the mouse pointer, click, right-click, double-click, and drag. To find out how, click the Start button and click Help on the Start menu. Click the Index tab. Type `MouseKeys` in the text box and click the Display button. To answer each of the following questions, click an appropriate topic in the Topics Found dialog box, click the Display button, and read the Help information. To display a different topic, click the topic and then click the Display button.

- How do you turn on MouseKeys?
- How do you use MouseKeys to move the mouse pointer?
- How do you perform each of these operations using MouseKeys: click, right-click, double-click, drag?

Click the Close button to close the Windows Help dialog box.

Student Exercises
WEB INFO
IN BRIEF
KEY TERMS
AT THE MOVIES
CHECKPOINT
AT ISSUE
MY WEB PAGE
HANDS ON
NET STUFF
Special Features
TIMELINE 2001
GUIDE TO WWW SITES
MAKING A CHIP
E-COMMERCE 2001
BUYER'S GUIDE 2001
CAREERS 2001
TRENDS 2001
LEARNING GAMES
CHAT
INTERACTIVE LABS
NEWS
HOME

CHAPTER 1 2 3 4 5 6 7 8 9 10 11 12 13 14 INDEX

NET STUFF www.scsite.com/dc2001/ch4/net.htm

WEB INSTRUCTIONS: *To display this page from the Web, launch your browser and enter the URL, www.scsite.com/dc2001/ch4/net.htm. To use the Scanning Documents lab from the Web, Shockwave and Flash Player must be installed on your computer (download by clicking here).*

SCANNING DOCUMENTS LAB

1. Shelly Cashman Series Scanning Documents Lab

Follow the appropriate instructions in NET STUFF 1 on page 1.48 to start and use the Scanning Documents lab. If you are running from the Web, enter the URL, www.scsite.com/sclabs/menu.htm; or display the NET STUFF page (see instructions at the top of this page) and then click the SCANNING DOCUMENTS LAB button.

DIGITAL CAMERAS

2. Digital Cameras

Digital cameras, which record photographs in the form of digital data, have fans and critics. Fans claim that digital cameras make it easy to store, organize, edit, and transmit photographs. Critics argue that digital cameras are difficult to use, produce poor-quality images, and cost too much money. Click the DIGITAL CAMERAS button and complete this exercise to learn more about digital cameras, see a live digital camera in action, and perhaps form your own opinion about their value.

SENDING E-MAIL

3. Sending E-Mail

In Chapter 2, you set up an e-mail account. Now, you can use that account to send messages across the world, across the country, or across the room. To send an e-mail message, click the SENDING E-MAIL button to display your e-mail service. Log in to your e-mail service. When you are finished, send the message. Next, read any messages you have received. When you are finished, exit your e-mail service. Then, follow the procedure to compose a message. The subject of the message should be input devices. Enter the e-mail address of one of your classmates. In the message itself, type something your classmates should know about input devices.

IN THE NEWS

4. In the News

Input devices can enhance the productivity of users and increase the number of potential users. The U.S. Army recently discovered this by replacing the many buttons used to operate a tank's onboard computer with a joystick and just three buttons. To the Army's delight, tank-driver performance improved, and even individuals who scored poorly on Army intelligence tests handled the tanks effectively. Click the IN THE NEWS button and read a news article about a new or improved input device, an input device being used in a new way, or an input device being made more available. What is the device? Who is promoting it? How will it be used? Will the input device change the number, or effectiveness, of potential users? If so, why?

WEB CHAT

5. Web Chat

Although satisfactory for Western languages, the keyboard is inadequate for many Asian languages. While English uses 26 letters, even the simpler version of written Chinese used in mainland China has almost 7,000 characters. This presents a major roadblock to the introduction of computers in the world's largest country. Three solutions have been offered for the problem of Chinese input. The first, called pinyin, uses English letters to express the sound of Chinese words. Pinyin is difficult to learn and slow to use, however. The second remaps keyboard keys with strokes used to draw Chinese characters. As keys are pressed, software suggests a list of characters that incorporate those strokes. The program is easy to learn, but also is fairly slow. The third is speech recognition software. As users speak words into a microphone, the computer produces the Chinese characters. The software must be reprogrammed for regional dialects, requires a powerful computer, and is less than 100 percent accurate. What do you think is the best method to handle input in China? Why? Can you suggest a better approach? Click the WEB CHAT button to enter a Web Chat discussion related to this topic.

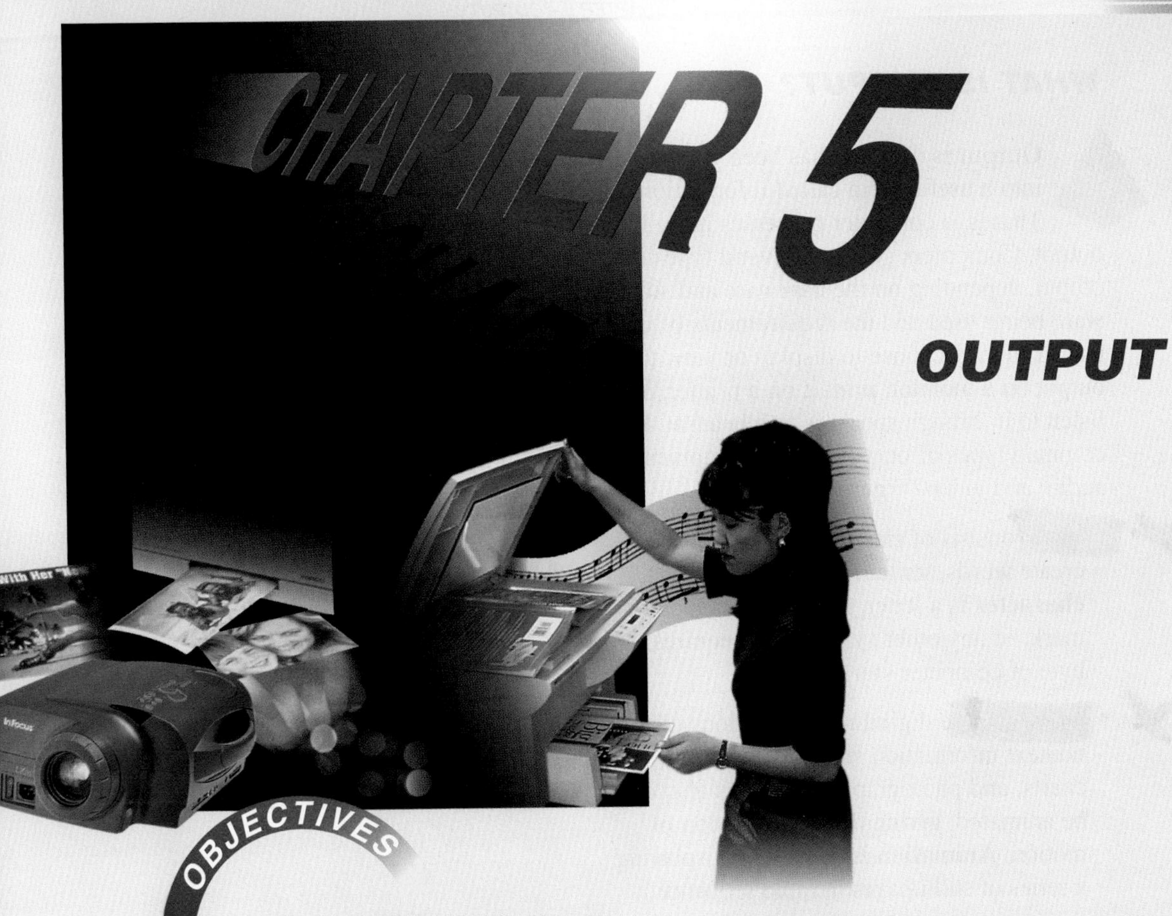

CHAPTER 5

OUTPUT

OBJECTIVES

After completing this chapter, you will be able to:

- Define the four types of output
- Identify the different types of display devices
- Describe factors that affect the quality of a monitor
- Understand the purpose of a video card
- Identify monitor ergonomic issues
- Explain the differences among various types of printers
- List various types of audio output devices
- Identify the purpose of data projectors, fax machines, and multifunction devices
- Explain how a terminal is both an input and output device
- Identify output options for physically challenged users

Data is a collection of unorganized items that can include words, numbers, images, and sounds. Computers process and organize data into information, which has meaning and is useful. Output devices such as printers, monitors, and speakers are used to convey that information to a user. This chapter describes the various methods of output and several commonly used output devices.

WHAT IS OUTPUT?

Output is data that has been processed into a useful form called information. That is, a computer processes input into output. Computers generate several types of output, depending on the hardware and software being used and the requirements of the user. You may choose to display or view this output on a monitor, print it on a printer, or listen to it through speakers or a headset. Four common types of output are text, graphics, audio, and video (Figure 5-1).

- **Text** consists of characters that are used to create words, sentences, and paragraphs. A **character** is a letter, number, punctuation mark, or any other symbol that requires one byte of computer storage space.
- **Graphics** are digital representations of nontext information such as drawings, charts, and photographs. Graphics also can be animated, giving them the illusion of motion. **Animation** is created by displaying a series of still images in rapid sequence.

  Many of today's software programs support graphics. For example, you can include a photograph in a word processing document or create a chart of data in a spreadsheet program.

  Some software packages are designed specifically to create and edit graphics. Paint programs, for instance, allow you to create graphics that can be used in brochures, newsletters, and Web pages. **Image editing software** allows you to alter graphics by including enhancements such as blended colors, animation, and other special effects.
- **Audio** is music, speech, or any other sound. Recall that sound waves, such as the human voice or music, are analog. To store such sounds, a computer converts the sounds from a continuous analog signal into a digital format. Most output devices require that the computer convert the digital format back into analog signals.

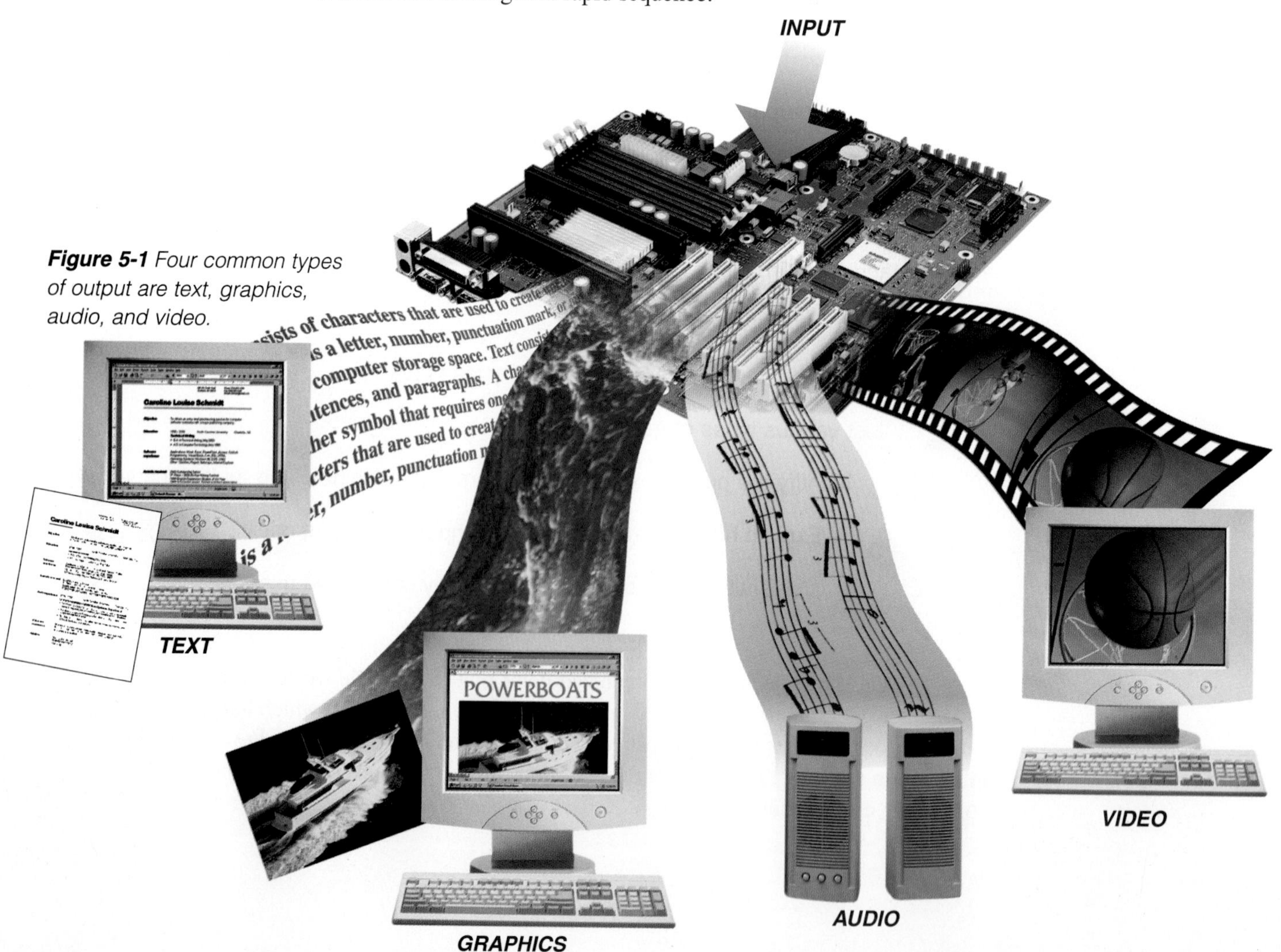

Figure 5-1 *Four common types of output are text, graphics, audio, and video.*

- **Video** consists of images that are played back at speeds that provide the appearance of full motion. Video often is captured with a video input device such as a video camera or VCR. Most video signals are analog; however, some video devices record the video images digitally.

  A video capture card converts an analog video signal into a digital signal that a computer can understand. The digital signal then is stored on the computer's hard disk. Some output devices accept the digital signal, while others require that the computer convert the digital signals back into analog signals.

Figure 5-2 *A color monitor displays text, graphics, and video information in color.*

WHAT ARE OUTPUT DEVICES?

An **output device** is any computer component capable of conveying information to a user. Commonly used output devices include display devices, printers, speakers, headsets, data projectors, facsimile machines, and multifunction devices. Each of these output devices is discussed in the following pages.

DISPLAY DEVICES

A **display device** is an output device that visually conveys text, graphics, and video information. Information shown on a display device often is called **soft copy**, because the information exists electronically and is displayed for a temporary period of time.

Display devices include CRT monitors, flat-panel displays, and high-definition televisions.

CRT Monitors

A **CRT monitor**, or **monitor**, is a display device that consists of a screen housed in a plastic or metal case. A **color monitor** displays text, graphics, and video information in color (Figure 5-2). Color monitors are used widely with all types of computers because most of today's software is designed to display information in color.

Monitors that display only one color are considered monochrome. A **monochrome monitor** displays text, graphics, and video information in one color (usually white, amber, or green) on a black background. Because monochrome monitors are less expensive than color monitors, some organizations use them for applications that do not require color or detailed graphics, such as order entry.

To enhance the quality of their graphics display, some monochrome monitors use **gray scaling**, which involves using many shades of gray from white to black to form the images.

Like a television set, the core of a CRT monitor is a large glass tube called a **cathode ray tube (CRT)** (Figure 5-3). The **screen**, which is the front of the tube, is coated with

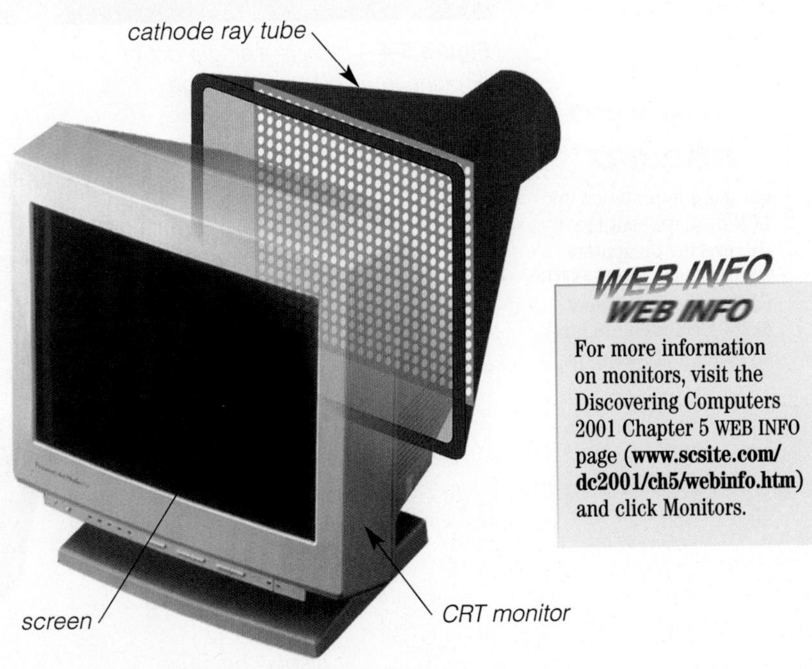

Figure 5-3 *The core of most desktop monitors is a cathode ray tube.*

WEB INFO

For more information on monitors, visit the Discovering Computers 2001 Chapter 5 WEB INFO page (**www.scsite.com/dc2001/ch5/webinfo.htm**) and click Monitors.

tiny dots of phosphor material that glow when electrically charged. Inside the CRT, an electron beam moves back and forth across the back of the screen, causing the dots to glow, which produces an image on the screen.

Each dot, called a **pixel** (short for *pic*ture *el*ement), is a single point in an electronic image (Figure 5-4). Monitors consist of hundreds, thousands, or millions of pixels arranged in rows and columns that can be used to create images. The pixels are so close together that they appear connected.

CRT monitors are used with a variety of computers. The CRT monitors used with desktop computers are available in a number of sizes, with the more common being 15, 17, 19, 20, and 21 inches. The size of a monitor is measured diagonally, from corner to corner.

Most monitors are referred to by their **viewable size**, which is the diagonal measurement of the cathode ray tube inside the monitor and is larger than the actual viewing area provided by the monitor. A monitor listed as a 19-inch monitor, for example, may have a viewable size of only 17.9 inches. Manufacturers are required to list a monitor's viewable size in any advertisement.

Determining what size monitor to use depends on your intended use. A large monitor allows you to view more information on the screen at once, but usually is more expensive. If you work on the Web or use multiple applications at one time, however, you may want to invest in at least a 19-inch monitor. If you use your computer for intense graphing applications such as desktop publishing and engineering, you may want an even larger monitor.

Figure 5-4 A pixel is a single dot of color, or point, in an electronic image.

Flat-Panel Displays

A **flat-panel display** is a lightweight, thin screen that consumes less power than a CRT monitor. Two common types of flat-panel displays are LCD and gas plasma.

LCD DISPLAYS LCD displays commonly are used in laptop computers, handheld computers, digital watches, and calculators because they are thinner and more lightweight than CRT monitors (Figure 5-5). While most LCD displays are color, even those on newer models of handheld computers, some devices use monochrome LCD displays to save battery power (Figure 5-6).

WEB INFO

For more information on LCD displays, visit the Discovering Computers 2001 Chapter 5 WEB INFO page (**www.scsite.com/dc2001/ch5/webinfo.htm**) and click LCD Displays.

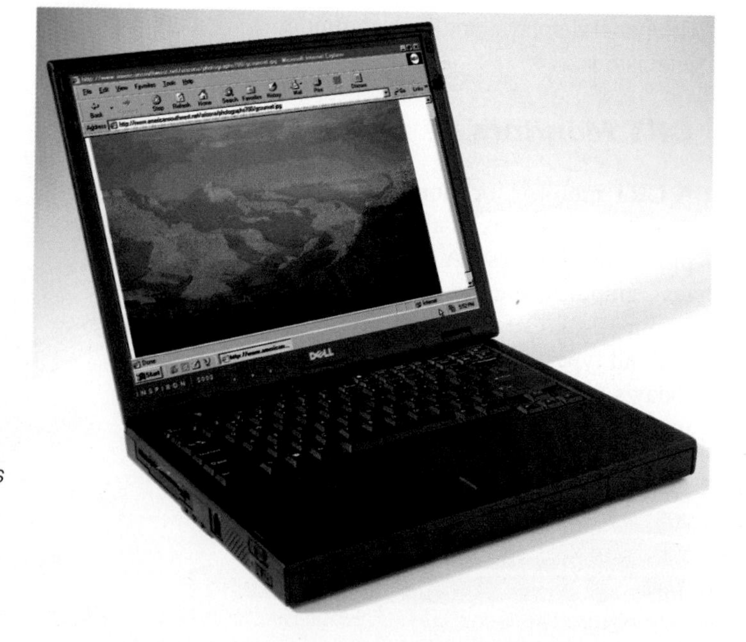

Figure 5-5 Most laptop computers use an LCD display because it is lightweight and thin.

Unlike a CRT monitor, an **LCD display** does not use a cathode ray tube (CRT) to create images on the screen; it instead uses a liquid crystal display (LCD). A **liquid crystal display (LCD)** has special molecules (called liquid crystals) deposited between two sheets of material. When an electric current passes through them, the molecules twist, causing some light waves to be blocked and allowing others to pass through, which then creates the desired images on the screen.

Like CRT monitors, the LCD displays used with laptop computers are available in a variety of sizes, with the more common being 14.1, 15.0, and 15.4 inches. LCD displays produce color using either passive matrix or active matrix technology. An **active-matrix display** uses a separate transistor for each color pixel and thus can display high-quality color that is viewable from all angles. Active-matrix displays sometimes are called **TFT displays**, named after the thin-film transistor (TFT) technology they use. Because they use many transistors, active-matrix displays require a lot of power.

A **passive-matrix display** uses fewer transistors and requires less power, but its color display often is not as bright as an active-matrix display. Images on a passive-matrix display can be viewed best when you work directly in front of the display. Passive-matrix displays are less expensive than active-matrix displays.

While LCD displays are used most often with laptops, stand-alone LCD monitors also can be used with desktop computers (Figure 5-7). Like CRT monitors, LCD monitors are available in a variety of sizes, with the more common being 15, 18, and 20 inches. (The term monitor typically is used for desktop display devices, while the term display is used for display devices on portable computers.) LCD monitors require less power and take up less desk space than traditional CRT monitors, making them ideal for users with space limitations. Some LCD monitors even can be mounted on the wall for increased space savings. Stand-alone LCD monitors, however, are more expensive than CRT monitors.

Figure 5-7 *Desktop applications that have space and weight limitations sometimes use an LCD monitor.*

Figure 5-6 *Some devices use color displays, while others use monochrome displays to save battery power.*

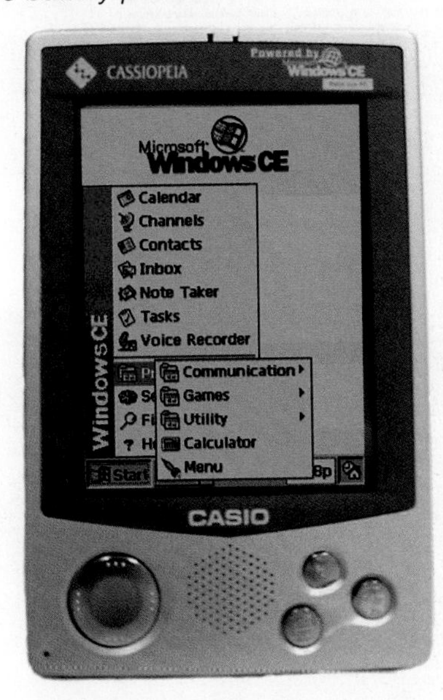

Figure 5-6a *(handheld computer)*

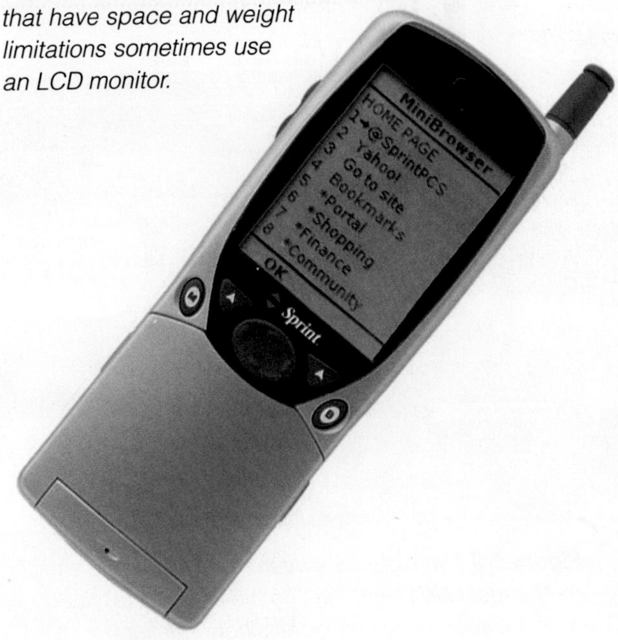

Figure 5-6b *(smart phone)*

GAS PLASMA MONITORS For even larger displays, some large business or power users prefer gas plasma monitors, which can measure more than 42 inches and hang directly on a wall (Figure 5-8). **Gas plasma monitors** use gas plasma technology, which substitutes a layer of gas for the liquid crystal material in an LCD monitor. When voltage is applied, the gas glows and produces the pixels that form an image. Gas plasma monitors offer larger screen sizes and higher display quality than LCD monitors but are much more expensive.

Figure 5-8 Large gas plasma displays can measure more than 42 inches and hang directly on a wall.

Monitor Quality

The quality of a monitor's display depends largely on its resolution, dot pitch, and refresh rate. The **resolution**, or sharpness and clarity, of a monitor is related directly to the number of pixels it can display. Resolution is expressed as two separate numbers: the number of columns of pixels and the number of rows of pixels a monitor can display.

A screen with a 640 × 480 (pronounced 640 by 480) resolution, for example, can display 640 columns and 480 rows of pixels (or a total of 307,200 pixels). Most modern monitors can display 800 × 600 and 1024 × 768 pixels; some high-end monitors can display 1280 × 1024, 1600 × 1200, or even 1920 × 1440 pixels.

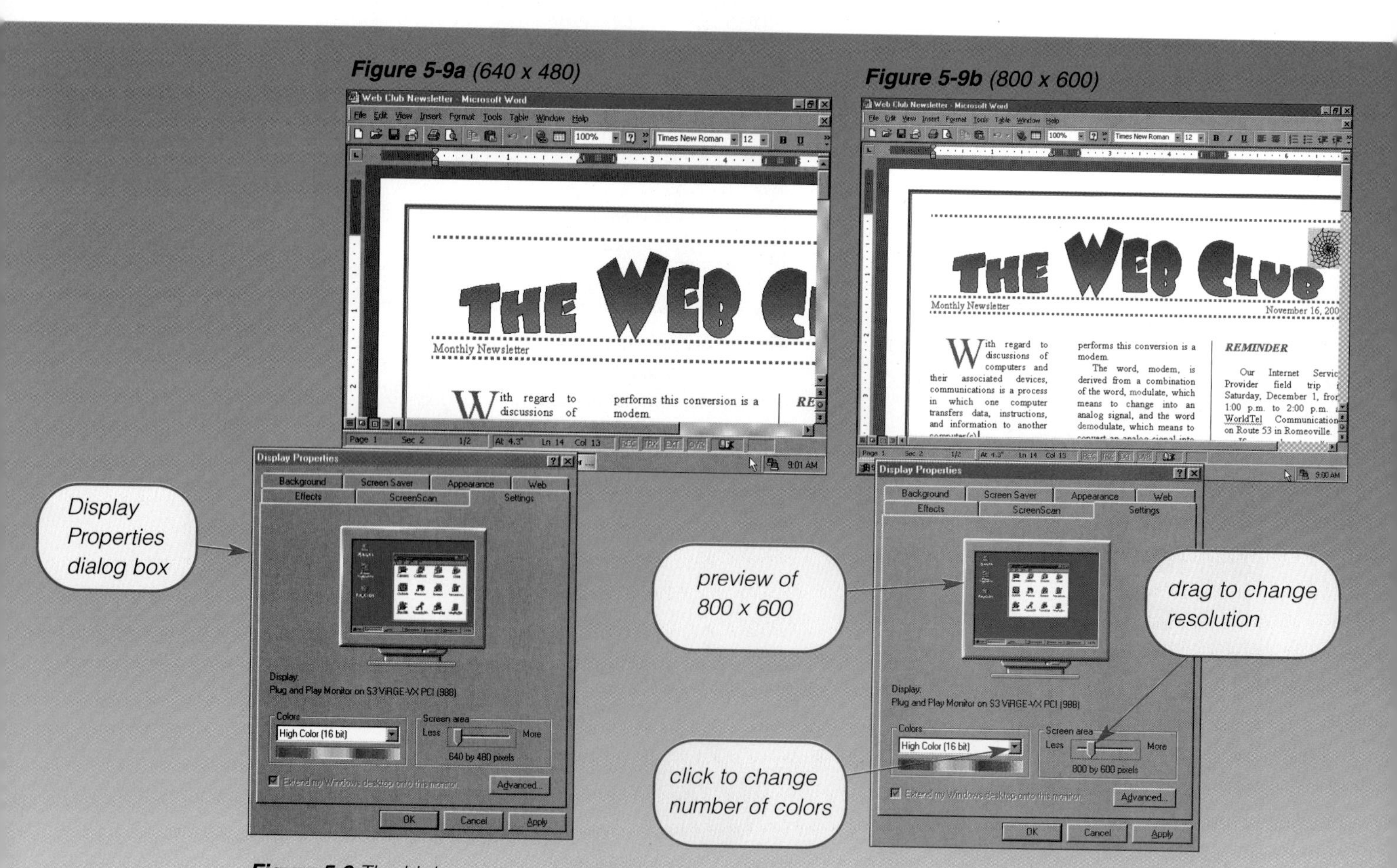

Figure 5-9 The higher a screen's resolution, the smaller the images display on the screen. As the resolution increases from 640 x 480 (a) to 800 x 600 (b) to 1024 x 768 (c), the image on the screen becomes increasingly smaller.

A monitor with a higher resolution displays a greater number of pixels, which provides a smoother image. A higher resolution, however, also causes images to display smaller on the screen. For this reason, you would not use a high resolution on a small monitor, such as a 14-inch monitor, because the small characters would be difficult to read. In Windows, you can change the resolution and other monitor characteristics using the Display Properties dialog box (Figure 5-9).

The ideal monitor resolution to use is a matter of preference. A higher resolution is desirable for graphics applications; a lower resolution usually is satisfactory for applications such as word processing.

Another factor that determines monitor quality is **dot pitch**, which is a measure of image clarity. The dot pitch is the vertical distance between each pixel on a monitor. The smaller the distance between the pixels, the sharper the displayed image. For example, as shown in Figure 5-10, text created with a smaller dot pitch is easier to read. To minimize eye fatigue, you should use a monitor with a dot pitch of .28 millimeters or smaller.

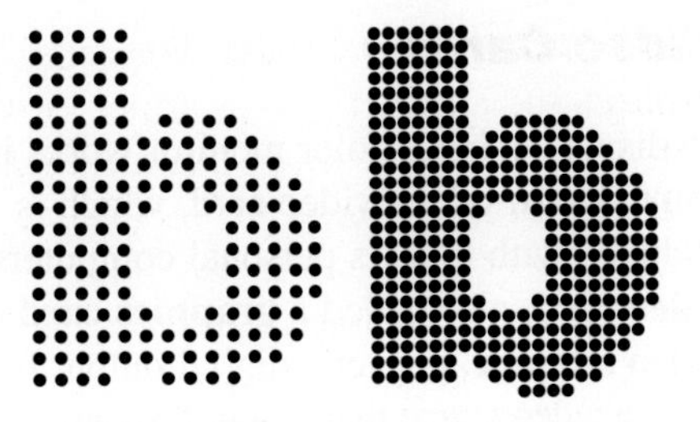

Figure 5-10 *In this figure, the letter b on the right is easier to read because it has a smaller dot pitch.*

Figure 5-9c *(1024 x 768)*

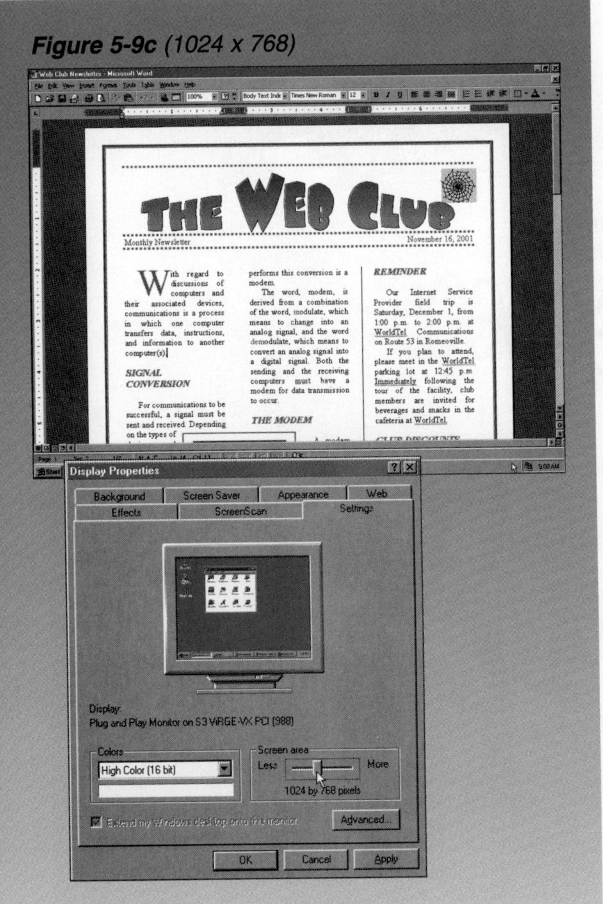

Recall that images on a CRT monitor are drawn on the screen as an electron beam moves back and forth across the back of the screen and causes pixels on the screen to glow. These pixels, however, glow for only a fraction of a second before beginning to fade. The monitor thus must redraw the picture many times per second so the image does not fade.

The speed that the monitor redraws images on the screen is called the **refresh rate**. Ideally, a monitor's refresh rate should be fast enough to maintain a constant, flicker-free image. A slower refresh rate causes the image to fade and then flicker as it is redrawn, which can cause headaches for users. Refresh rate is measured according to **hertz**, which is the number of times per second the screen is redrawn. Although most people can tolerate a refresh rate of 60 hertz, a high-quality monitor will provide a refresh rate of at least 75 hertz.

Some older monitors refresh images using a technique called interlacing. With **interlacing**, the electron beam draws only half the horizontal lines with each pass (for example, all odd-numbered lines on one pass and all even-numbered lines on the next pass). Because it happens so quickly, your brain perceives the two images as a single image. Interlacing originally was developed for monitors with slow refresh rates. Most of today's monitors are *noninterlaced*, and they provide a much better, flicker-free image than interlaced monitors.

Video Cards

To display color, a color monitor works in combination with a video card, which is included with today's personal computers. A **video card**, also called a **graphics card** or **video adapter**, converts digital output into an analog video signal that is sent through a cable to the monitor (Figure 5-11). The monitor separates the video signal into red, green, and blue signals. Electron guns then fire the three color signals to the front of the monitor. These three dots – one red, one green, and one blue – are combined to make up each single pixel.

WEB INFO

For more information on video cards, visit the Discovering Computers 2001 Chapter 5 WEB INFO page (**www.scsite.com/dc2001/ch5/webinfo.htm**) and click Video Cards.

The number of colors that a video card can display is determined by the number of bits it uses to store information about each pixel. For example, an 8-bit video card (also called 8-bit color) uses 8 bits to store information about each pixel and thus can display 256 different colors (computed as 2^8 or $2 \times 2 \times 2 \times 2 \times 2 \times 2 \times 2 \times 2$); a 24-bit video card uses 24 bits to store information about each pixel and can display 16.7 million colors (Figure 5-12). In Windows, you can change the number of displayed colors in the Display Properties dialog box (see Figure 5-9 on page 5.6).

Figure 5-11

HOW VIDEO TRAVELS FROM THE CPU TO YOUR MONITOR

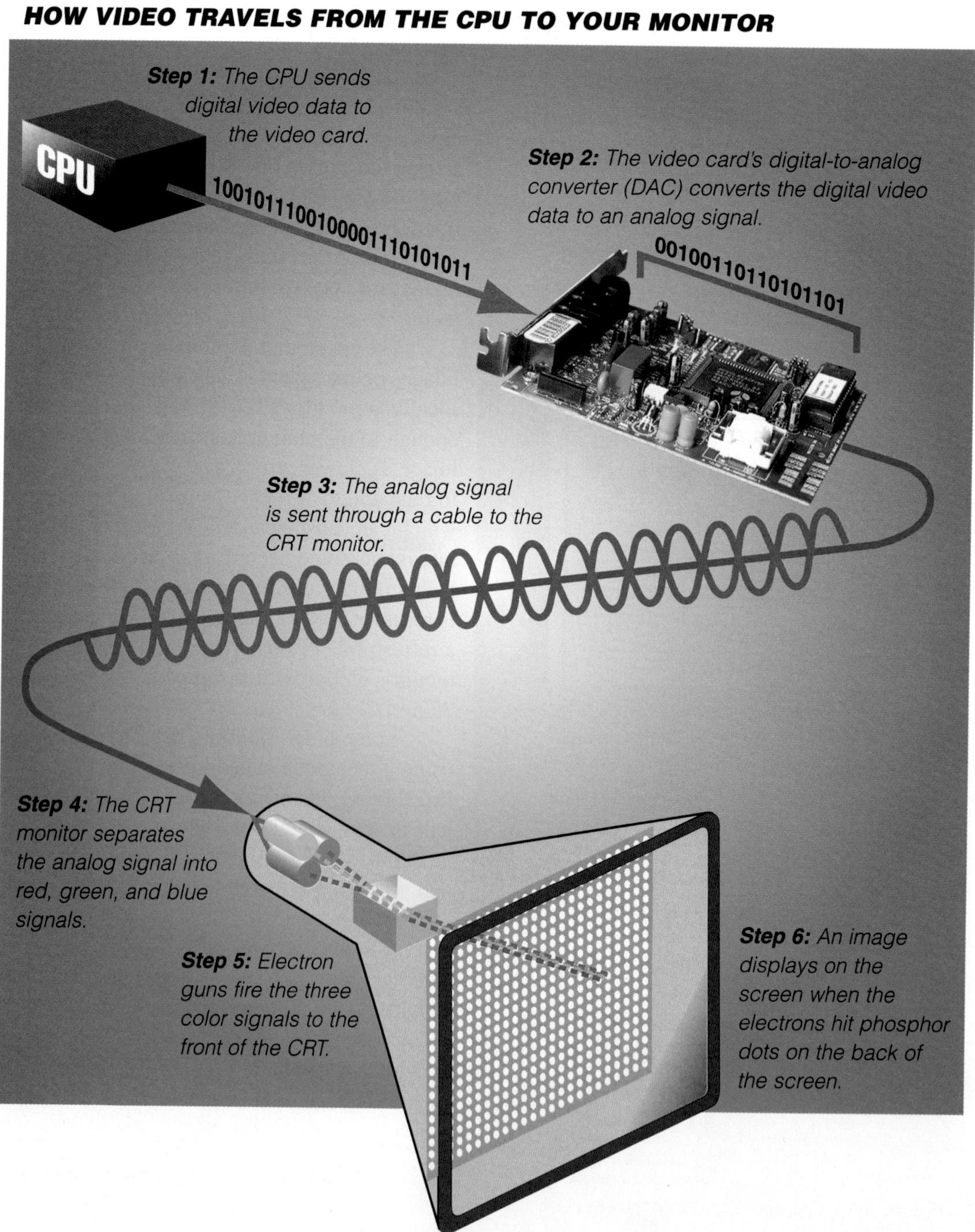

Step 1: The CPU sends digital video data to the video card.
CPU
1001011100100001110101011
Step 2: The video card's digital-to-analog converter (DAC) converts the digital video data to an analog signal.
0010011011010101101
Step 3: The analog signal is sent through a cable to the CRT monitor.
Step 4: The CRT monitor separates the analog signal into red, green, and blue signals.
Step 5: Electron guns fire the three color signals to the front of the CRT.
Step 6: An image displays on the screen when the electrons hit phosphor dots on the back of the screen.

Over the years, several video standards have been developed to define the resolution, number of colors, and other properties for various types of monitors. Today, just about every monitor supports the **super video graphics array (SVGA)** standard, which also supports resolutions and colors in the VGA standard. The table in Figure 5-13 outlines the suggested resolution and number of displayed colors in the MDA, VGA, XGA, and SVGA standards.

For a monitor to display images using the resolution and number of colors defined by a video standard, the monitor must support the video standard and the video card must be capable of communicating appropriate signals to the monitor. That is, the video card and the monitor must support the video standard to generate the desired resolution and number of colors.

Most video cards are equipped with memory, which is used to store information about each pixel. While some use dynamic RAM (DRAM), higher-quality video cards use **video RAM** or **VRAM** (pronounced *VEE*-ram) to improve the quality of graphics. As with other types of memory, VRAM is measured in megabytes.

VIDEO CARD DISPLAY DESIGNATIONS

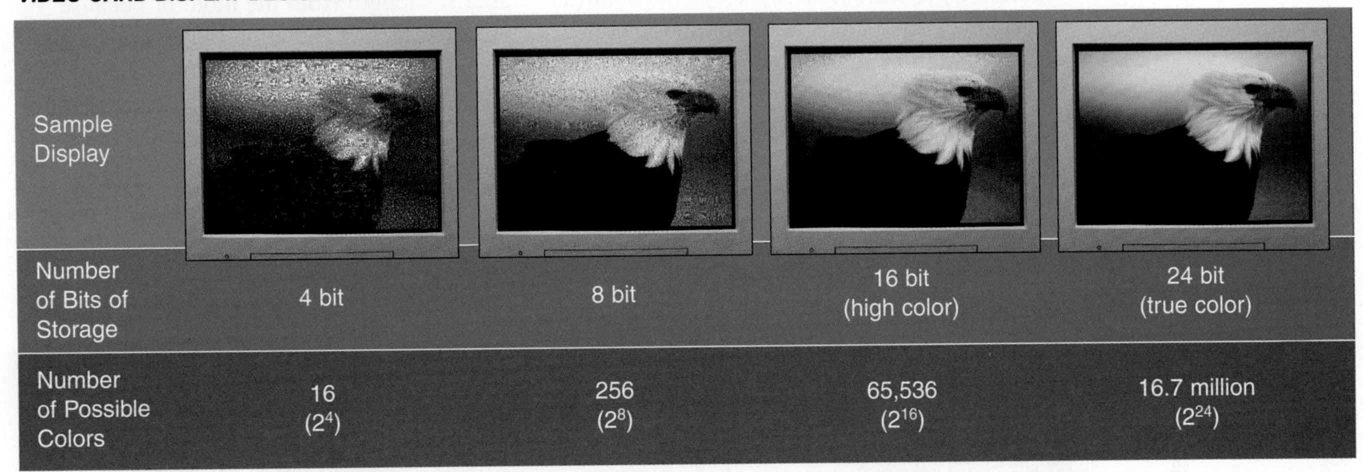

Sample Display
Number of Bits of Storage
4 bit
8 bit
16 bit (high color)
24 bit (true color)
Number of Possible Colors
16 (2^4)
256 (2^8)
65,536 (2^16)
16.7 million (2^24)

Figure 5-12 *The number of colors that can be displayed on a monitor is determined by the numbers of bits it uses to store information about each pixel. The greater the number of bits, the more colors. Most monitors today use at least 8-bit color.*

VIDEO STANDARDS

Standard	Suggested Resolution	Possible Simultaneous Colors
Monochrome Display Adapter (MDA)	720 x 350	1
Video Graphics Array (VGA)	640 x 480	16
	320 x 200	256
Extended Graphics Array (XGA)	1024 x 768	256
	640 x 480	65,536
Super Video Graphics Array (SVGA)	800 x 600	16 million
	1024 x 768	16 million
	1280 x 1024	16 million
	1600 x 1200	16 million

Figure 5-13 *The various video standards.*

Your video card must have enough memory to generate the resolution and number of colors you want to display. The table in Figure 5-14 outlines the amount of VRAM suggested for various screen resolutions and color displays. For example, if you wanted an 800 × 600 resolution with 16-bit color (65,536 colors), then your video card should have at least 1 MB of VRAM.

Monitor Ergonomics

Recall that the goal of ergonomics is to incorporate comfort, efficiency, and safety into the design of items in the workplace. Many monitors have features that help address ergonomic issues, such as the controls that allow you to adjust the brightness, contrast, positioning, height, and width of images. These controls usually are positioned on the front of the monitor for easy access. Newer monitors have digital controls that allow you to fine-tune the display in small increments (Figure 5-15).

Another advantage of digital controls is you quickly can return to the default settings. Many monitors have a tilt-and-swivel base, so you can adjust the angle of the screen to minimize neck strain and reduce glare from overhead lighting.

Because they use electricity, monitors produce a small amount of **electromagnetic radiation (EMR)**, which is a magnetic field that travels at the speed of light. Although no solid evidence exists to prove that EMR poses a health risk, an established set of standards, known as MPR II, define acceptable levels of EMR for a monitor. All high-quality monitors should comply with MPR II standards. To protect yourself even further, you should sit at arm's length from the monitor because electromagnetic radiation only travels a short distance. Also, electromagnetic radiation is greatest on the sides and back of the monitor.

VRAM REQUIREMENTS

Number of Colors	640 x 480	800 x 600	1024 x 768	1280 x 1024	1600 x 1200
16 (4 bit)	256 KB	256 KB	512 KB	1 MB	1 MB
256 (8 bit)	512 KB	512 KB	1 MB	2 MB	2 MB
65,536 (16 bit)	1 MB	1 MB	2 MB	4 MB	4 MB
16.7 million (24 or 32* bit)	1 MB	2 MB	4 MB	4 MB	6 MB

Figure 5-14 *The amount of video RAM required for various screen resolutions.*
**Some video cards use 32 bits to store information about each pixel for true color.*

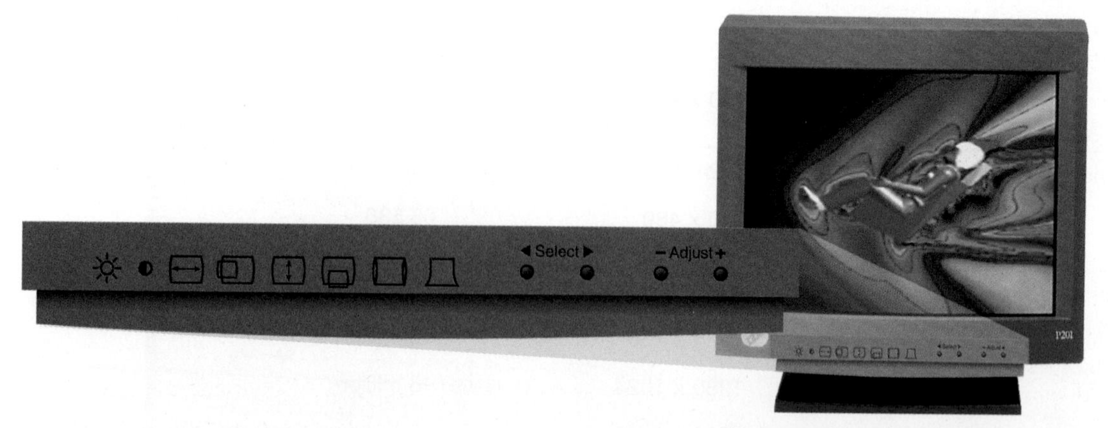

Figure 5-15 *Digital controls on many monitors allow a user to fine-tune the display in small increments.*

To help reduce the amount of electricity used by monitors and other computer components, the U.S. Department of Energy (DOE) and the U.S. Environmental Protection Agency (EPA) developed the **ENERGY STAR program**. This program encourages manufacturers to create energy-efficient devices that require little power when they are not in use. Monitors and devices that meet ENERGY STAR guidelines display an ENERGY STAR® label (Figure 5-16).

Figure 5-16 *Products with an ENERGY STAR® label are energy efficient as defined by the Environmental Protection Agency (EPA).*

High-Definition Television

High-definition television (HDTV) is a type of television set that works with digital broadcasting signals and supports a wider screen and higher resolution display than a standard television set. When you use a standard television set as a monitor for your computer, the output must be converted to an analog signal that can be displayed by the television set.

With HDTV, the broadcast signals are digitized when they are sent. Digital television signals provide two major advantages over analog signals. First, digital signals produce a higher-quality picture. Second, many programs can be broadcast on a single digital channel, whereas only one program can be broadcast on an analog channel. Currently, only a few U.S. television stations broadcast digital signals. By 2006, all stations are to have converted from analog to digital. As the cost of HDTV becomes more reasonable, home users will begin to use it as their computer's display device.

PRINTERS

A **printer** is an output device that produces text and graphics on a physical medium such as paper or transparency film. Printed information is called **hard copy** because the information exists physically and is a more permanent form of output than that presented on a display device (soft copy).

Hard copy, also called a **printout**, can be printed in portrait or landscape orientation. A page with **portrait orientation** is taller than it is wide, with information printed across the shorter width of the paper; a page with **landscape orientation** is wider than it is tall, with information printed across the widest part of the paper (Figure 5-17). Letters, reports, and books typically are printed in portrait orientation; spreadsheets, slide shows, and graphics often are printed in landscape orientation.

WEB INFO

For more information on HDTV, visit the Discovering Computers 2001 Chapter 5 WEB INFO page (**www.scsite.com/dc2001/ch5/webinfo.htm**) and click High-Definition Television.

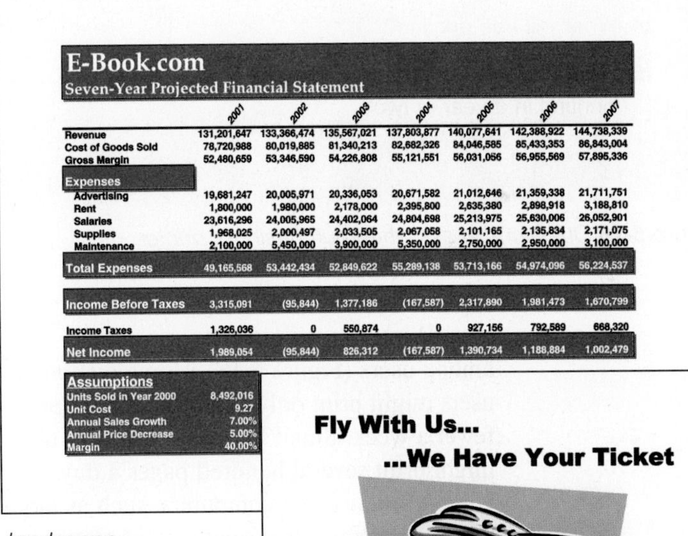

E-Book.com
Seven-Year Projected Financial Statement

	2001	2002	2003	2004	2005	2006	2007
Revenue	131,201,647	133,366,474	135,567,021	137,803,877	140,077,641	142,388,922	144,738,339
Cost of Goods Sold	78,720,988	80,019,885	81,340,213	82,682,326	84,046,585	85,433,353	86,843,004
Gross Margin	52,480,659	53,346,590	54,226,808	55,121,551	56,031,056	56,955,569	57,895,336
Expenses							
Advertising	19,681,247	20,005,971	20,336,053	20,671,582	21,012,646	21,359,338	21,711,751
Rent	1,800,000	1,980,000	2,178,000	2,395,800	2,635,380	2,898,918	3,188,810
Salaries	23,616,296	24,005,965	24,402,064	24,804,698	25,213,975	25,630,006	26,052,901
Supplies	1,968,025	2,000,497	2,033,505	2,067,058	2,101,165	2,135,834	2,171,075
Maintenance	2,100,000	5,450,000	3,900,000	5,350,000	2,750,000	2,950,000	3,100,000
Total Expenses	49,165,568	53,442,434	52,849,622	55,289,138	53,713,166	54,974,096	56,224,537
Income Before Taxes	3,315,091	(95,844)	1,377,186	(167,587)	2,317,890	1,981,473	1,670,799
Income Taxes	1,326,036	0	550,874	0	927,156	792,589	668,320
Net Income	1,989,054	(95,844)	826,312	(167,587)	1,390,734	1,188,884	1,002,479

Assumptions	
Units Sold in Year 2000	8,492,016
Unit Cost	9.27
Annual Sales Growth	7.00%
Annual Price Decrease	5.00%
Margin	40.00%

Figure 5-17 *Portrait orientation is taller than it is wide; landscape orientation is wider than it is tall.*

PRINTING REQUIREMENTS QUESTION SHEET

1. How fast must my printer print?
2. Do I need a color printer?
3. What is the cost per page for printing?
4. Do I need multiple copies?
5. Do I need to print graphics?
6. Do I need to print photographic-quality images?
7. What types of paper does the printer use?
8. What sizes of paper does the printer accept?
9. How much paper can the printer tray hold?
10. Will the printer work with my computer and software?
11. How much do supplies such as ink and paper cost?
12. Can the printer print on envelopes and transparencies?
13. What is my budget?
14. What will I be printing?
15. How much do I print now, and what will I be printing in a year or two?

Figure 5-18 *Questions to ask when purchasing a printer.*

Printing requirements vary greatly among users (Figure 5-18). Home computer users might print only a hundred pages or fewer a week. Small business computer users might print several hundred pages a day. Users of mainframe computers, such as large utility companies that send printed statements to hundreds of thousands of customers each month, require printers that are capable of printing thousands of pages per hour. These different needs have resulted in the development of printers with varying speeds, capabilities, and printing methods.

Generally, printers can be grouped into two categories: impact and nonimpact. Printers in each of these categories are discussed in the following sections.

Impact Printers

An **impact printer** forms characters and graphics on a piece of paper by striking a mechanism against an ink ribbon that physically contacts the paper. Because of the striking activity, impact printers generally are noisy.

Many impact printers do not provide letter quality print. **Letter quality (LQ)** output is a quality of print acceptable for business letters. Many impact printers produce **near letter quality (NLQ)** print, which is slightly less clear than letter quality. NLQ impact printers are used for jobs that require only near letter quality, such as printing mailing labels, envelopes, or invoices.

Impact printers also are ideal for printing multipart forms because they easily can print through many layers of paper. Finally, impact printers are used in many factories and at retail counters because they can withstand dusty environments, vibrations, and extreme temperatures.

Two commonly used types of impact printers are dot-matrix printers and line printers. Each of these printers is discussed in the following sections.

DOT-MATRIX PRINTERS A **dot-matrix printer** is an impact printer that produces printed images when tiny wire pins on a print head mechanism strike an inked ribbon (Figure 5-19). When the ribbon presses against the paper, it creates dots that form characters and graphics.

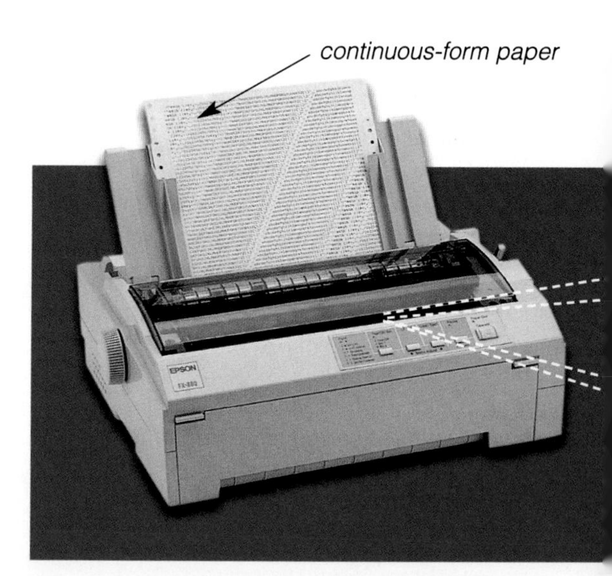

Figure 5-19 *A dot-matrix printer produces printed images when tiny pins strike an inked ribbon.*

Most dot-matrix printers use **continuous-form paper**, in which each sheet of paper is connected together. The pages generally have holes punched along two opposite sides so the paper can be fed through the printer. Perforations along the inside of the punched holes and at each fold allow the sheets to be separated into standard-sized sheets of paper, such as 8½ x 11 inches. One advantage of continuous-form paper is that it does not have to be changed very often because thousands of pages are connected together. Many dot-matrix printers also can be adjusted to print pages in either portrait or landscape orientation.

The print head mechanism on a dot-matrix printer can contain nine to twenty-four pins, depending on the manufacturer and the printer model. A higher number of pins means more dots are printed, which results in higher print quality.

The speed of a dot-matrix printer is measured by the number of **characters per second (cps)** it can print. The speed of dot-matrix printers ranges from 50 to 700 characters per second (cps), depending on the desired print quality.

LINE PRINTERS A **line printer** is a high-speed impact printer that prints an entire line at a time (Figure 5-20). The speed of a line printer is measured by the number of lines per minute (lpm) it can print. Capable of printing up to 3,000 lines per minute (lpm), these printers often are used with mainframes, minicomputers, or with a network in applications such as manufacturing, distribution, or shipping. Line printers typically use 11 x 17-inch continuous-form paper, such as the greenbar paper used to print reports from mainframe computers.

Two popular types of line printers used for high-volume output are band and shuttle-matrix. A **band printer** prints fully-formed characters when hammers strike a horizontal, rotating band that contains shapes of numbers, letters of the alphabet, and other characters. A **shuttle-matrix printer** works more like a dot-matrix printer; the difference is the shuttle-matrix printer moves a series of print hammers back and forth horizontally at incredibly high speeds. Unlike a band printer, a shuttle-matrix printer can print characters in various fonts and font sizes.

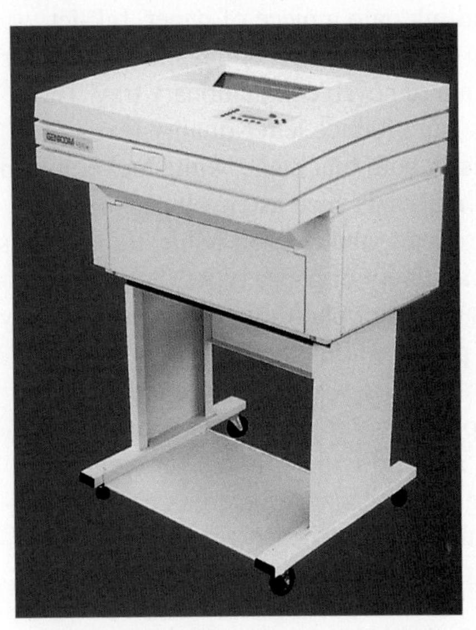

Figure 5-20 *A line printer is a high-speed printer often connected to a mainframe, minicomputer, or network.*

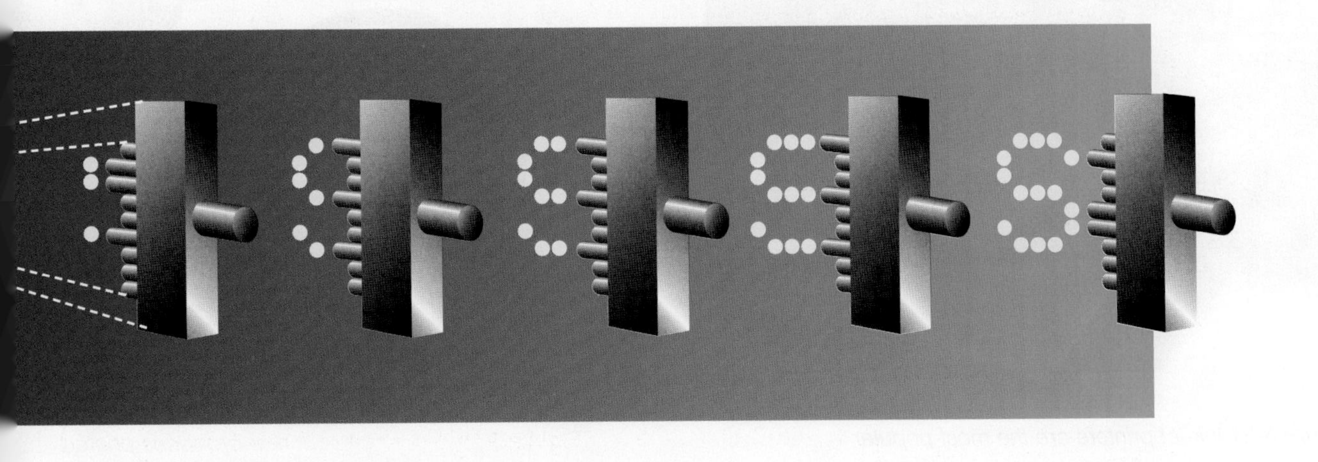

Nonimpact Printers

A **nonimpact printer** forms characters and graphics on a piece of paper without actually striking the paper. Some spray ink, while others use heat and pressure to create images. Because these printers do not strike the paper, they are much quieter than the previously discussed impact printers.

Three commonly used types of nonimpact printers are ink-jet printers, laser printers, and thermal printers. Each of these printers is discussed in the following sections.

WEB INFO

For more information on ink-jet printers, visit the Discovering Computers 2001 Chapter 5 WEB INFO page (**www.scsite.com/dc2001/ch5/webinfo.htm**) and click Ink-Jet Printers.

INK-JET PRINTERS An **ink-jet printer** is a type of nonimpact printer that forms characters and graphics by spraying tiny drops of liquid ink onto a piece of paper. Ink-jet printers usually use individual sheets of paper stored in a removable or stationary tray. These printers can produce letter-quality text and graphics in both black-and-white and color on various materials such as envelopes, labels, transparencies, and iron-on t-shirt transfers, as well as a variety of paper types. Some ink-jet printers can print photo-quality images on any type of paper, while other ink-jet printers require a heavier weight premium paper for better-looking color documents. Many ink-jet printers are sold with software for creating greeting cards, banners, business cards, letterheads, and transparencies (Figure 5-21).

Because of their reasonable cost and letter-quality print, ink-jet printers have become the most popular type of color printer for use in the home. You can purchase an ink-jet printer of reasonable quality for a few hundred dollars.

One factor that determines the quality of an ink-jet printer is its resolution, or sharpness and clarity. Printer resolution is measured by the number of **dots per inch** (**dpi**) a printer can output. As shown in Figure 5-22, the higher the dpi, the better the print quality. With an ink-jet printer, a dot is a drop of ink. A higher dpi means the drops of ink are smaller, which provides a higher quality image. Most ink-jet printers have a dpi that ranges from 300 to 1,440 dpi. Typically, printers with a higher dpi are more expensive.

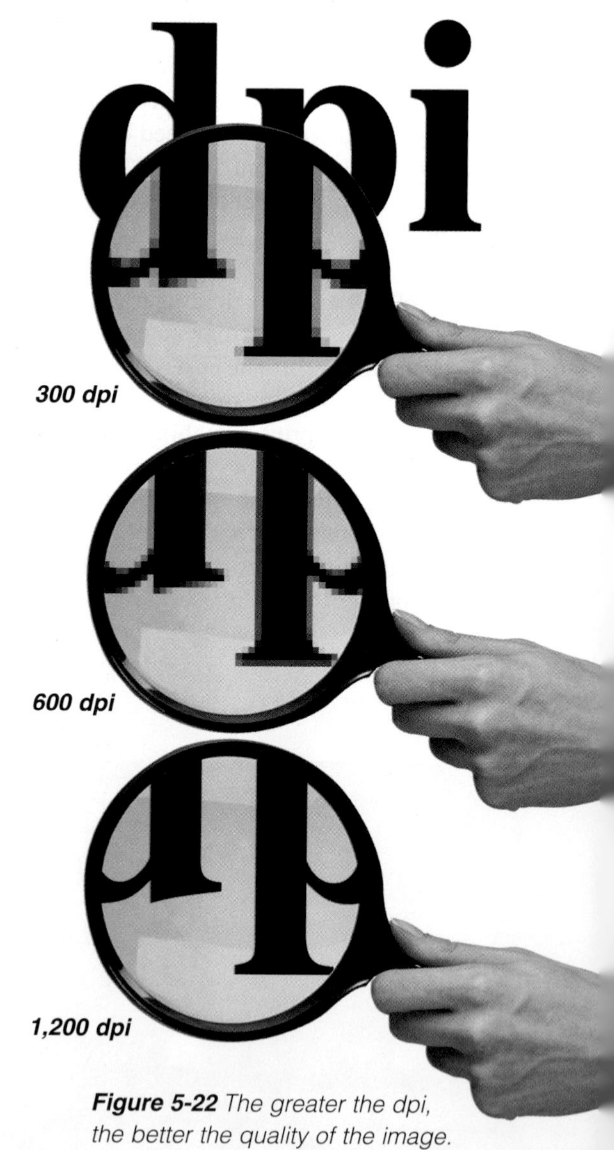

Figure 5-22 The greater the dpi, the better the quality of the image.

Figure 5-21 Ink-jet printers are the most popular type of color printer for use in the home.

The speed of an ink-jet printer is measured by the number of **pages per minute (ppm)** it can print. Most ink-jet printers print from one to twelve pages per minute (ppm). Graphics and colors print at the slower rate.

The print head mechanism of an ink-jet printer contains ink-filled print cartridges, each with fifty to several hundred small ink holes, or nozzles. The steps in Figure 5-23 illustrate how a drop of ink appears on a page. Each nozzle in the print cartridge is similar to an individual pin on a dot-matrix printer. Just as any combination of dot-matrix pins can be activated, ink can be propelled by heat or pressure through any combination of the nozzles to form a character or image on the paper.

When the print cartridge runs out of ink, you simply replace the cartridge. Most ink-jet printers have at least two print cartridges: one containing black ink and the other(s) containing colors.

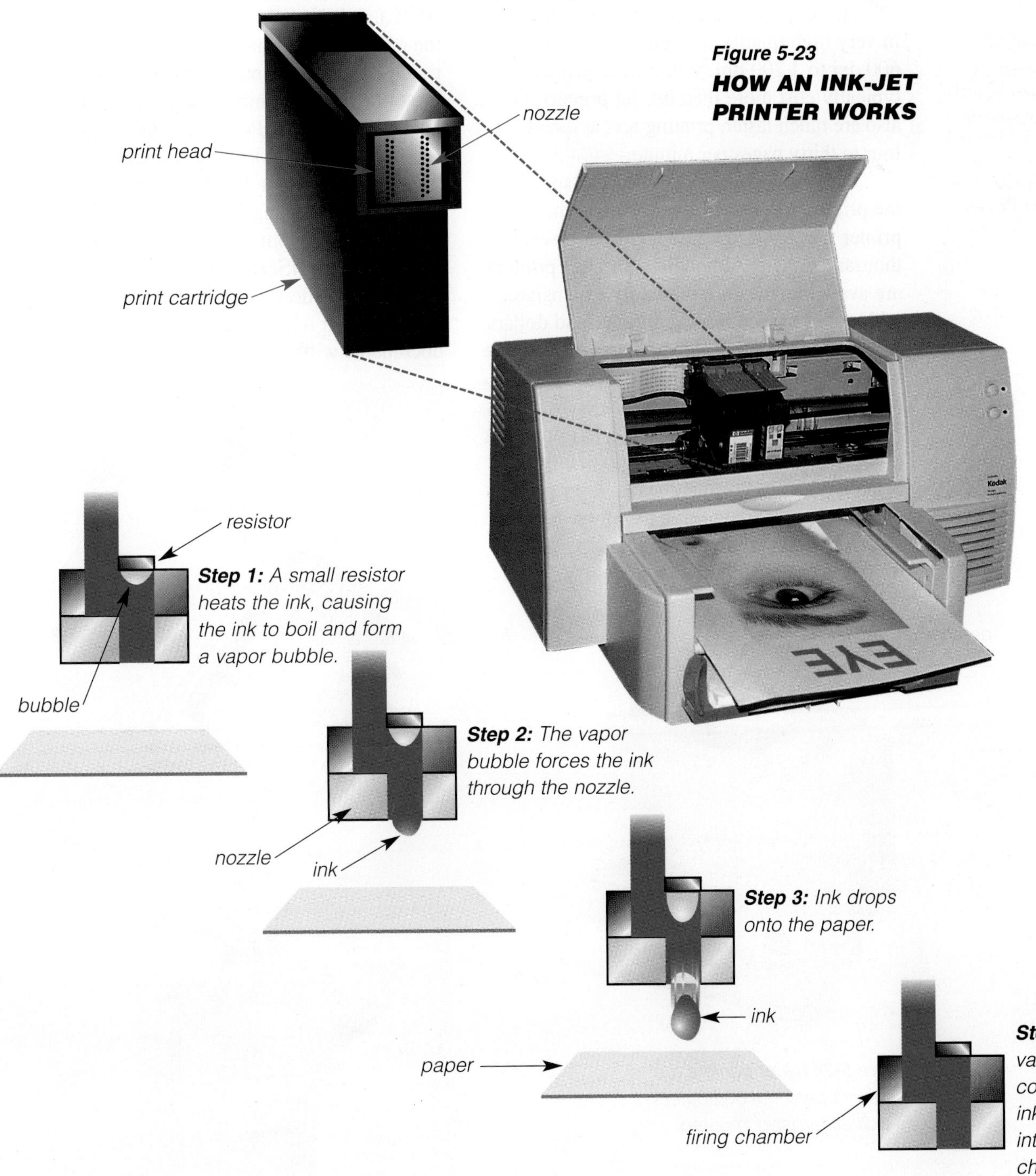
nozzle
print head
print cartridge
EYE
resistor
Step 1: A small resistor heats the ink, causing the ink to boil and form a vapor bubble.
bubble
Step 2: The vapor bubble forces the ink through the nozzle.
nozzle
ink
Step 3: Ink drops onto the paper.
ink
paper
Step 4: As the vapor bubble collapses, fresh ink is drawn into the firing chamber.
firing chamber
ink dot

Figure 5-23
HOW AN INK-JET PRINTER WORKS

LASER PRINTERS A **laser printer** is a high-speed, high-quality nonimpact printer (Figure 5-24). Laser printers for personal computers usually use individual sheets of paper stored in a removable tray that slides into the printer case. Some laser printers have trays that can accommodate different sizes of paper, while others require separate trays for letter- and legal-sized paper. Most laser printers have a manual feed slot where you can insert individual sheets and envelopes. You also can print transparencies on a laser printer.

Laser printers can print text and graphics in very high quality resolutions, ranging from 600 dpi to 1,200 dpi. While laser printers typically cost more than ink-jet printers, they also are much faster, printing text at speeds of four to thirty pages per minute.

Depending on the quality and speed of the printer, the cost of a black-and-white laser printer ranges from a few hundred to several thousand dollars. Although color laser printers are available, they are relatively expensive, with prices exceeding several thousand dollars. The higher the resolution and speed, the more expensive the printer. High-end fast laser printers are used with mainframe computers.

When printing a document, laser printers process and store the entire page before they actually print it. For this reason, laser printers sometimes are called **page printers**. Storing a page before printing requires laser printers to have a certain amount of memory.

Depending on the amount of graphics you intend to print, a laser printer can have up to 200 MB of memory. To print a full-page 600-dpi picture, for instance, you might need 8 MB of memory on the printer. If your printer does not have enough memory to print the picture, it either will print as much of the picture as its memory will allow or it will display an error message and not print any of the picture.

Laser printers use software that enables them to interpret a **page description language (PDL)**. A PDL tells the printer how to layout the contents of a printed page. When you purchase a laser printer, it comes with at least one of two common page description languages: PCL or PostScript. Developed by Hewlett-Packard, a leading printer manufacturer, **PCL (Printer Control Language)** is a standard printer language designed to support the fonts and layout used in standard office documents. **PostScript** commonly is used in fields such as desktop publishing and graphic art because it is designed for complex documents with intense graphics and colors.

WEB INFO

For more information on laser printers, visit the Discovering Computers 2001 Chapter 5 WEB INFO page (**www.scsite.com/dc2001/ch5/webinfo.htm**) and click Laser Printers.

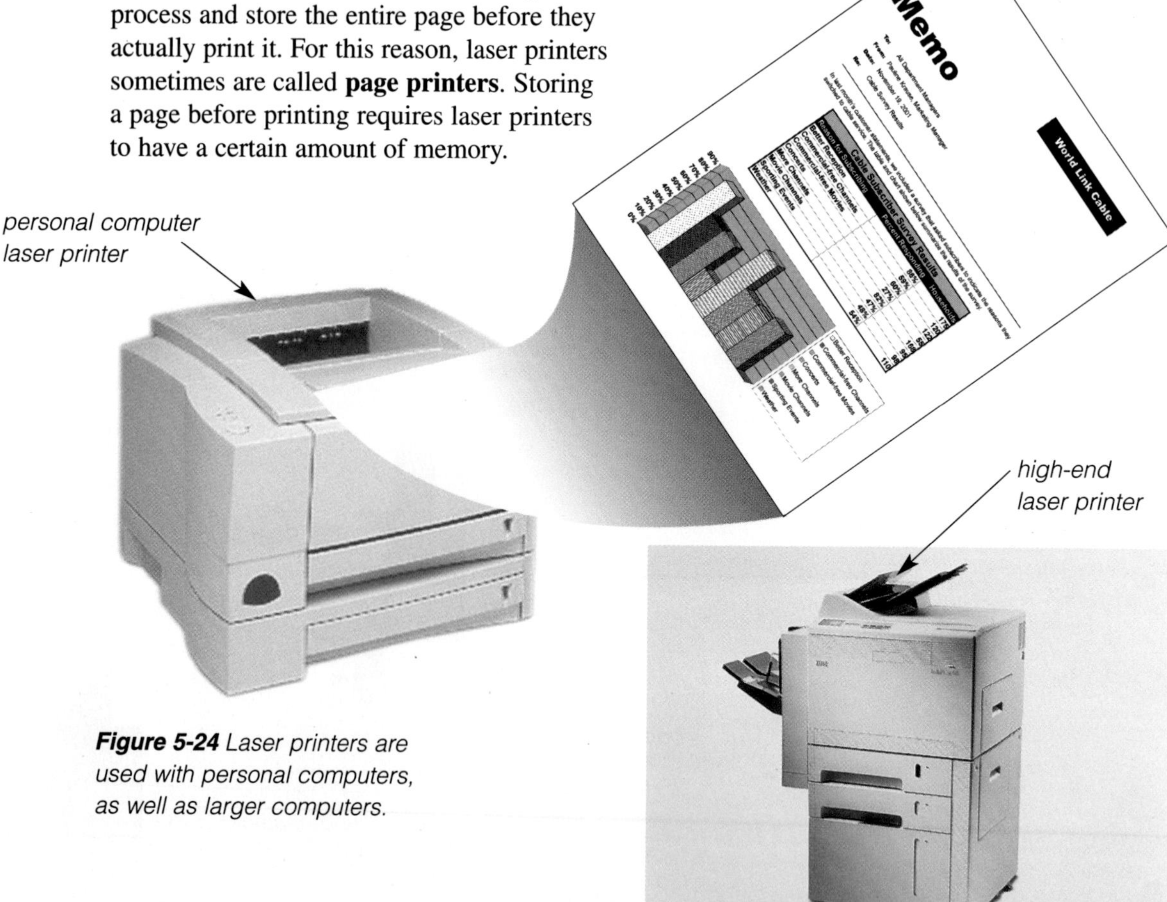

Figure 5-24 *Laser printers are used with personal computers, as well as larger computers.*

Operating in a manner similar to a copy machine, a laser printer creates images using a laser beam and powdered ink, called **toner**, which is packaged in a cartridge. The laser beam produces an image on a special drum inside the printer. The light of the laser alters the electrical charge on the drum wherever it hits. When this occurs, the toner sticks to the drum and then is transferred to the paper through a combination of pressure and heat (Figure 5-25). When the toner runs out, you can replace the toner cartridge (Figure 5-26).

THERMAL PRINTERS A **thermal printer** generates images by pushing electrically heated pins against heat-sensitive paper. Standard thermal printers are inexpensive, but the print quality is low and the images tend to fade over time. Thermal printers are, however, ideal for use in small devices such as adding machines.

Figure 5-25

HOW A LASER PRINTER WORKS

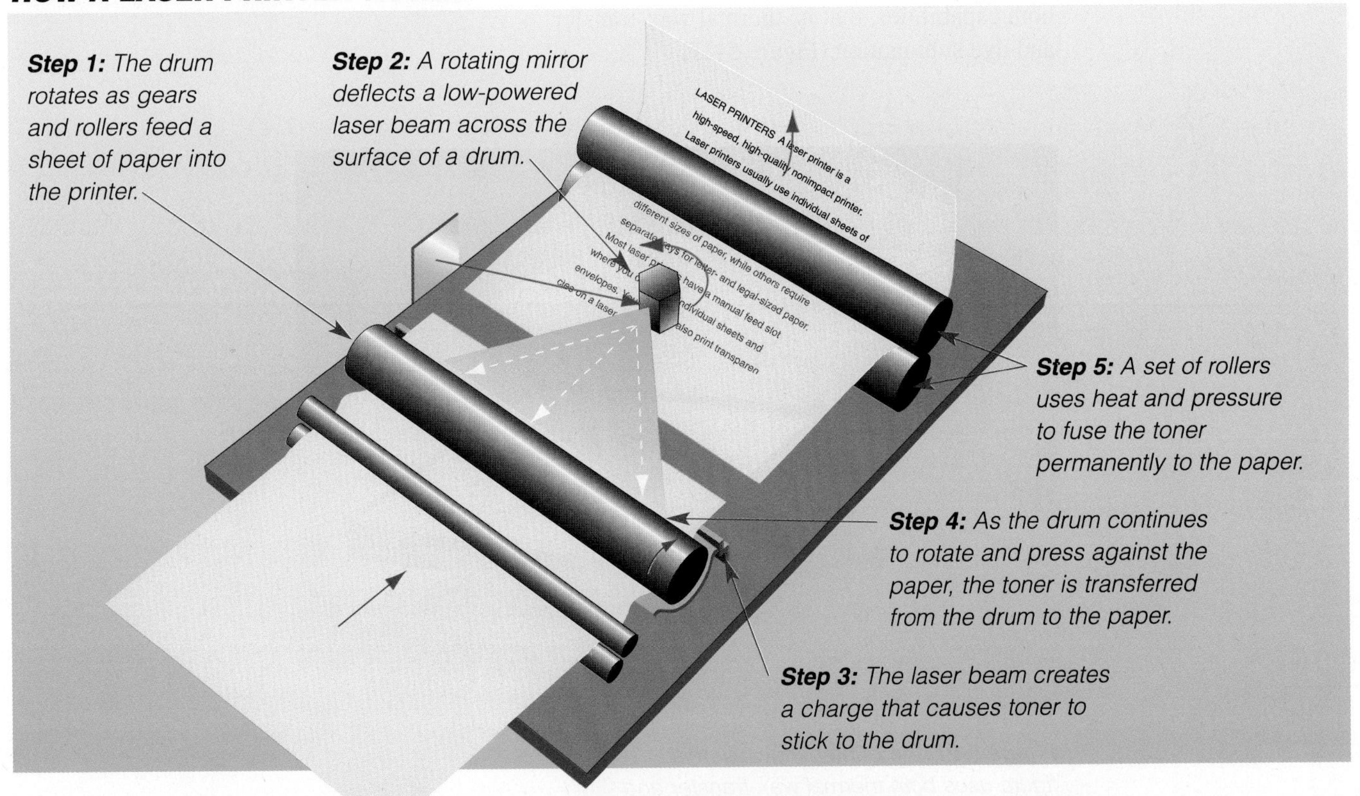

Step 1: The drum rotates as gears and rollers feed a sheet of paper into the printer.
Step 2: A rotating mirror deflects a low-powered laser beam across the surface of a drum.
Step 3: The laser beam creates a charge that causes toner to stick to the drum.
Step 4: As the drum continues to rotate and press against the paper, the toner is transferred from the drum to the paper.
Step 5: A set of rollers uses heat and pressure to fuse the toner permanently to the paper.

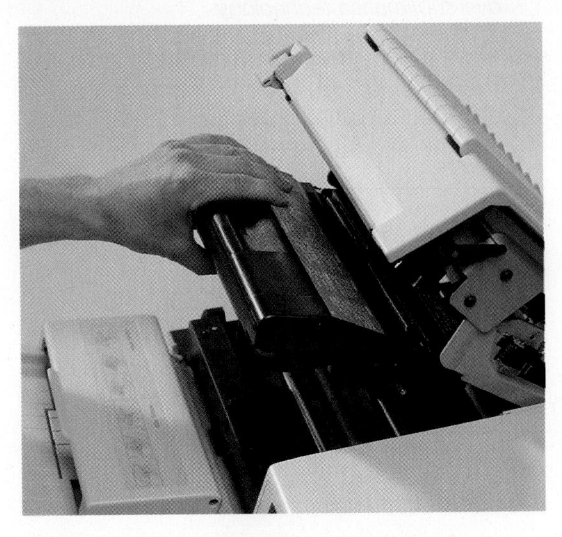

Figure 5-26 *Replacing the toner cartridge in a laser printer.*

Two special types of thermal printers have a much higher print quality. A **thermal wax-transfer printer**, also called a **thermal transfer printer**, generates rich, nonsmearing images by using heat to melt colored wax onto heat-sensitive paper. Thermal wax-transfer printers are more expensive than ink-jet printers, but less expensive than many color laser printers. A **dye-sublimation printer**, also called a **thermal dye transfer printer**, uses heat to transfer colored dye to specially coated paper. While among the more expensive types of printers, dye-sublimation printers can create images that are of photographic quality. Some manufacturers offer a printer for a few thousand dollars that have both capabilities, that is, thermal wax-transfer and dye sublimation (Figure 5-27).

WEB INFO

WEB INFO

For more information on portable printers, visit the Discovering Computers 2001 Chapter 5 WEB INFO page **(www.scsite.com/dc2001/ch5/webinfo.htm)** and click Portable Printers.

Portable Printers

A **portable printer** is a small, lightweight printer that allows a mobile user to print from a laptop or handheld computer while traveling (Figure 5-28). Barely wider than the paper on which they print, portable printers easily can fit in a briefcase alongside a laptop computer.

Some portable printers use ink-jet technology, while others are thermal or thermal wax-transfer. Portable ink-jet printers provide better output quality than portable thermal printers, but usually are larger. Many of these printers connect to a parallel port; others have a built-in infrared port through which they communicate with the computer.

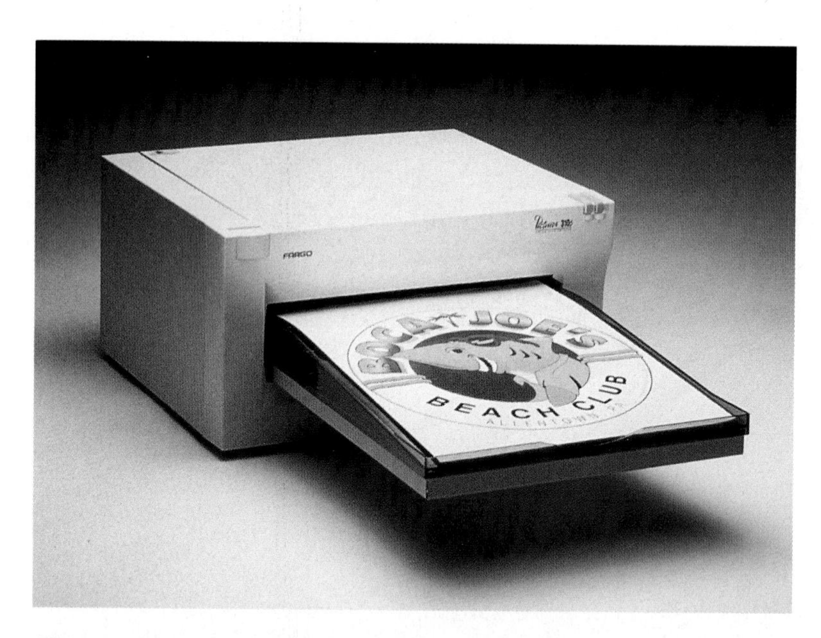

Figure 5-27 *The printer shown in this figure uses both thermal wax-transfer and dye-sublimation technology.*

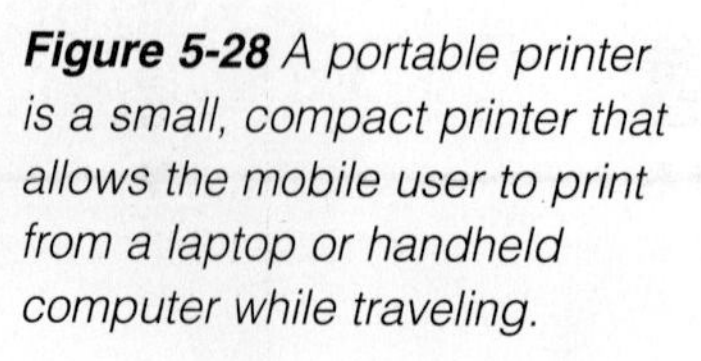

Figure 5-28 *A portable printer is a small, compact printer that allows the mobile user to print from a laptop or handheld computer while traveling.*

Plotters and Large-Format Printers

Plotters and large-format printers are sophisticated printers used to produce high-quality drawings such as blueprints, maps, circuit diagrams, and signs (Figure 5-29). Because blueprints, maps, and other such drawings can be quite large, these printers typically can handle paper with widths up to 60 inches. Some plotters and large-format printers use individual sheets of paper, while others take large rolls. These printers are used in specialized fields such as engineering, drafting, and graphic art and usually are very costly.

Two basic types of plotters are pen plotters and electrostatic plotters. A **pen plotter** uses one or more colored pens, light beams, or a scribing device to draw on paper or transparencies. Pen plotters differ from other printers in that they produce continuous lines, whereas most printers generate lines by printing a closely spaced series of dots.

An **electrostatic plotter** uses a row of charged wires (called styli) to draw an electrostatic pattern on specially coated paper and then fuses toner to the pattern. The printed image is composed of a series of very small dots, which provide high-quality output.

Operating like an ink-jet printer, but on a much larger scale, a **large-format printer** creates photo-realistic quality color prints. Used by graphic artists, these high performance printers are used for signs, posters, and other displays.

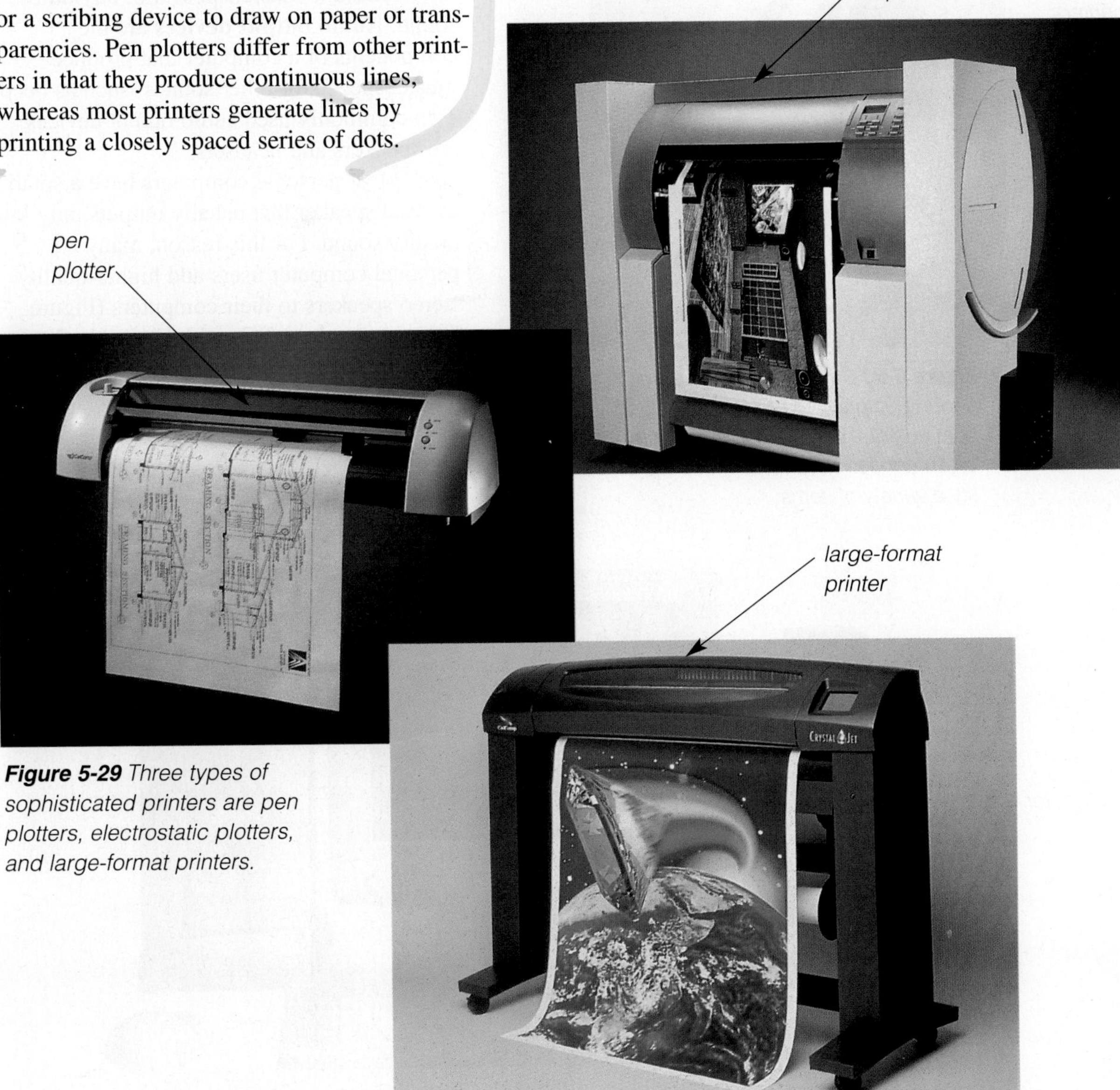

Figure 5-29 Three types of sophisticated printers are pen plotters, electrostatic plotters, and large-format printers.

Special-Purpose Printers

In addition to the printers just discussed, other printers have been developed for special purposes (Figure 5-30). A **photo printer** is a color printer designed to produce photo lab quality pictures directly from a digital camera. A **label printer** is a small printer that prints on an adhesive-type material that can be placed on a variety of items such as envelopes, disks, audiocassettes, photographs, and toys. Many label printers are used to print bar codes.

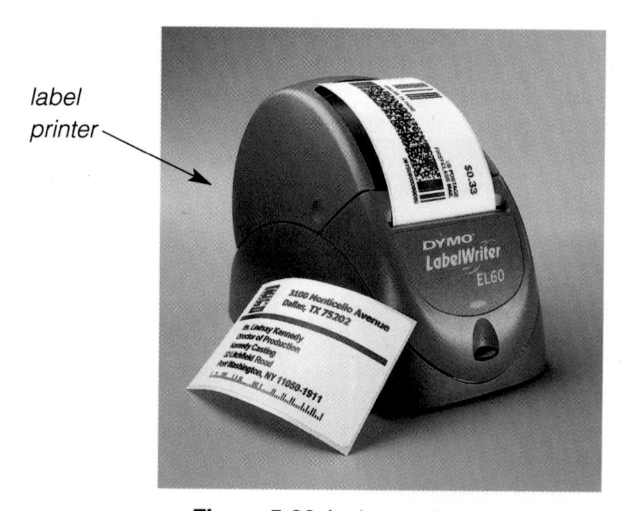

Figure 5-30 *A photo printer is a color printer designed to print photos directly from a digital camera. A label printer prints on an adhesive-type material used to identify a variety of items.*

AUDIO OUTPUT

Audio is music, speech, or any other sound. **Audio output devices** are the components of a computer that produce music, speech, or other sounds, such as beeps. Two commonly used audio output devices are speakers and headsets.

Most personal computers have a small internal speaker that usually outputs only low-quality sound. For this reason, many personal computer users add higher-quality stereo speakers to their computers (Figure 5-31) or purchase PCs with larger speakers built into the sides of the monitor (see Figure 5-2 on page 5.3).

Figure 5-31 *Many personal computer users add high-quality stereo speakers and a woofer to their computers.*

To boost the low bass sounds, you can add a woofer (also called a subwoofer). The stereo speakers and woofer are connected to ports on the sound card. Most speakers have tone and volume controls so you can adjust these settings.

When using speakers, anyone within listening distance can hear the output. If you are in a computer laboratory or some other crowded environment, speakers might not be practical; instead, you can plug a headset into a port on the sound card (Figure 5-32). With the **headset**, only you can hear the sound from the computer.

Figure 5-32 *In a crowded environment where speakers are not practical, you can use a headset for audio output.*

OTHER OUTPUT DEVICES

Although monitors, printers, and speakers are the more widely used output devices, many other output devices are available for particular uses and applications. These include data projectors, facsimiles, and multifunction devices. Each of these devices is discussed in the following sections.

Data Projectors

A **data projector** takes the image that displays on a computer screen and projects it onto a screen so that an audience of people can see the image clearly (Figure 5-33). Data projectors can be large devices attached to a ceiling or wall in an auditorium, or they can be small portable devices. Two types of smaller, lower-cost units are LCD projectors and DLP projectors.

An **LCD projector**, which uses liquid crystal display technology, attaches directly to a computer and uses its own light source to display the information shown on the computer screen. Because LCD projectors tend to produce lower-quality images, some users prefer to use a DLP projector for sharper, brighter images.

A **digital light processing (DLP) projector** uses tiny mirrors to reflect light, producing crisp, bright, colorful images that remain in focus and can be seen clearly even in a well-lit room.

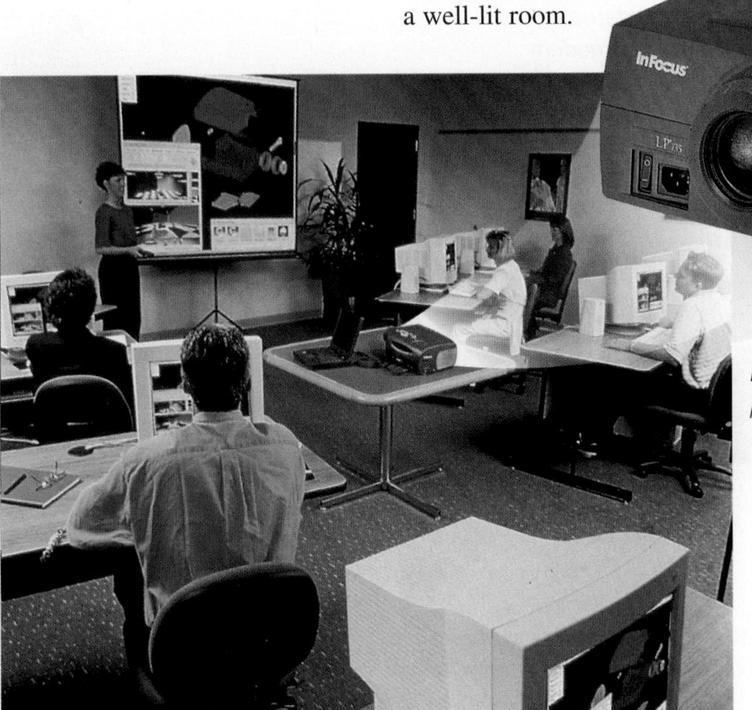

Figure 5-33 *An LCD data projector.*

Facsimile (Fax) Machine

WEB INFO

For more information on fax modems, visit the Discovering Computers 2001 Chapter 5 WEB INFO page (**www.scsite.com/dc2001/ch5/webinfo.htm**) and click Fax Modems.

A **facsimile** (**fax**) **machine** is a device that transmits and receives documents over telephone lines. The documents can contain text, drawings, or photographs, or can be handwritten. When sent or received via a fax machine, these documents are known as faxes. A stand-alone fax machine scans the original document, converts the image into digitized data, and transmits the digitized image (Figure 5-34). A fax machine at the receiving end reads the incoming data, converts the digitized data into an image, and prints or stores a copy of the original image.

Figure 5-34 *A stand-alone fax machine.*

Fax capability also can be added to your computer using a fax modem. A **fax modem** is a communications device that allows you to send (and sometimes receive) electronic documents as faxes (Figure 5-35). A fax modem transmits computer-prepared documents, such as a word processing letter, or documents that have been digitized with a scanner or digital camera. A fax modem is like a regular modem except that it is designed to transmit documents to a fax machine or to another fax modem.

When a computer (instead of a fax machine) receives a fax, you can view the document on the screen or print it using special fax software. The quality of the viewed or printed fax is less than that of a word processing document because the fax actually is a large image. If you have optical character recognition (OCR) software, you also can edit the document.

A fax modem can be an external peripheral that plugs into a port on the back of the system unit or an internal card that is inserted into an expansion slot on the motherboard. In addition, most fax modems function as regular modems.

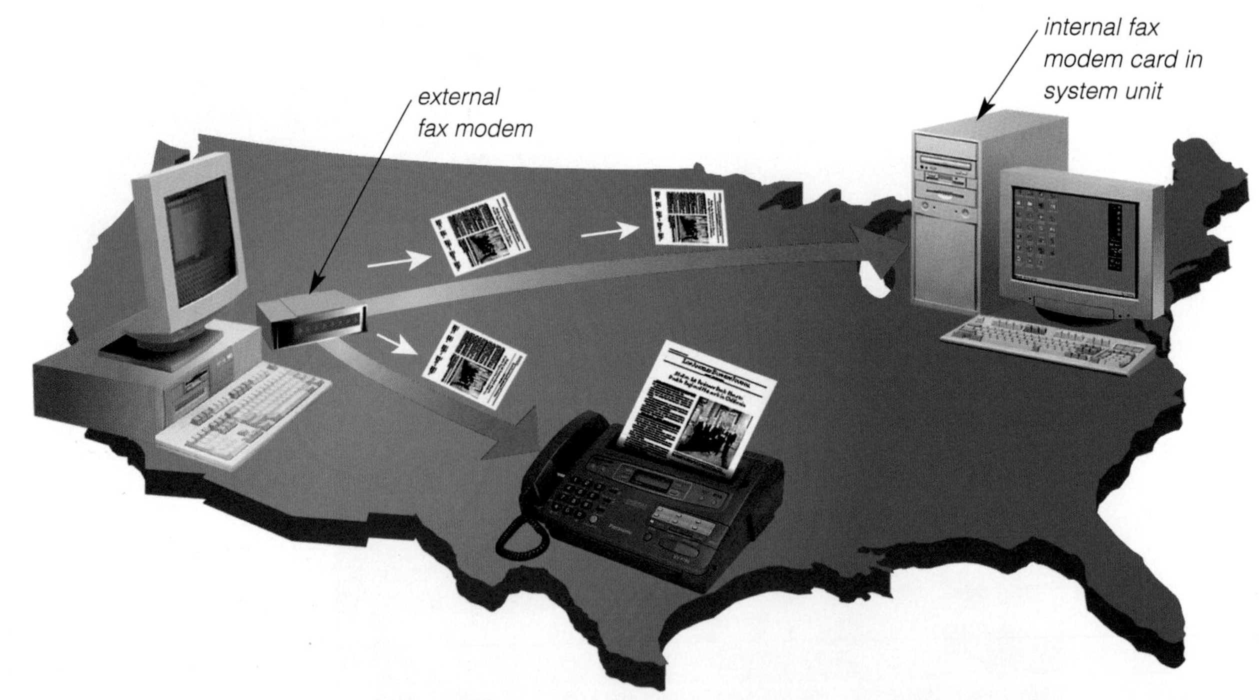

Figure 5-35 *A fax modem allows you to send (and sometimes receive) electronic documents as faxes to a fax machine or another computer.*

Multifunction Devices

A **multifunction device** (**MFD**) is a single piece of equipment that looks like a copy machine but provides the functionality of a printer, scanner, copy machine, and perhaps a fax machine (Figure 5-36). Sometimes called a multifunction peripheral, the features of multifunction devices vary widely. For example, some use color ink-jet printer technology, while others include a black-and-white laser printer.

Small offices and home offices use multifunction devices because they take up less space than having a separate printer, scanner, copy machine, and fax machine. Another advantage of an MFD is that it is significantly less expensive than if you purchased each device separately. The primary disadvantage of an MFD is that if the machine breaks down you lose all four functions. Given the advantages, however, increasingly more users are bringing multifunction devices into their offices and homes.

Figure 5-36 *This OfficeJet by Hewlett-Packard is a color printer, scanner, and copy machine all in one device.*

TERMINALS

A **terminal** is a device that performs both input and output because it consists of a monitor (output), a keyboard (input), and a video card. Terminals fall into three basic categories: dumb terminals, intelligent terminals, and special-purpose terminals.

A **dumb terminal** has no processing power, and thus, cannot function as an independent device (Figure 5-37). A dumb terminal is used to enter and transmit data to, or receive and display information from, a computer to which it is connected. Dumb terminals are connected to a **host computer** that performs the processing and then sends the output back to the dumb terminal. The host computer usually is a minicomputer, mainframe computer, or supercomputer.

In addition to a monitor and keyboard, an **intelligent terminal** also has memory and a processor that has the capability of performing some functions independent of the host computer. Intelligent terminals sometimes are called **programmable terminals** because they can be programmed by the software developer to perform basic tasks. In recent years, personal computers have replaced many intelligent terminals.

Figure 5-37 *Dumb terminals have no processing power and usually are connected to larger computer systems.*

WEB INFO

For more information on MFDs, visit the Discovering Computers 2001 Chapter 5 WEB INFO page (**www.scsite.com/dc2001/ch5/webinfo.htm**) and click Multifunction Devices.

WEB INFO

For more information on POS terminals, visit the Discovering Computers 2001 Chapter 5 WEB INFO page (**www.scsite.com/dc2001/ch5/webinfo.htm**) and click Point-of-Sale Terminals.

WEB INFO

For more information on ATMs, visit the Discovering Computers 2001 Chapter 5 WEB INFO page (**www.scsite.com/dc2001/ch5/webinfo.htm**) and click Automatic Teller Machines.

Other special-purpose terminals perform specific tasks and contain features uniquely designed for use in a particular industry (Figure 5-38). Two of these special-purpose terminals are point-of-sale terminals and automatic teller machines.

A **point-of-sale (POS) terminal** is used to record purchases at the point where the consumer purchases the product or service. The POS terminal used in a grocery store, for example, is a combination of an electronic cash register and bar code reader. When the bar code on the food product is scanned, the price of the item displays on the monitor, the name of the item and its price print on a receipt, and the item being sold is recorded so the inventory can be updated.

As indicated by this example, POS terminals serve as input to other computers to maintain sales records, update inventory, verify credit, and perform other activities associated with the sales transactions that are critical to running the business.

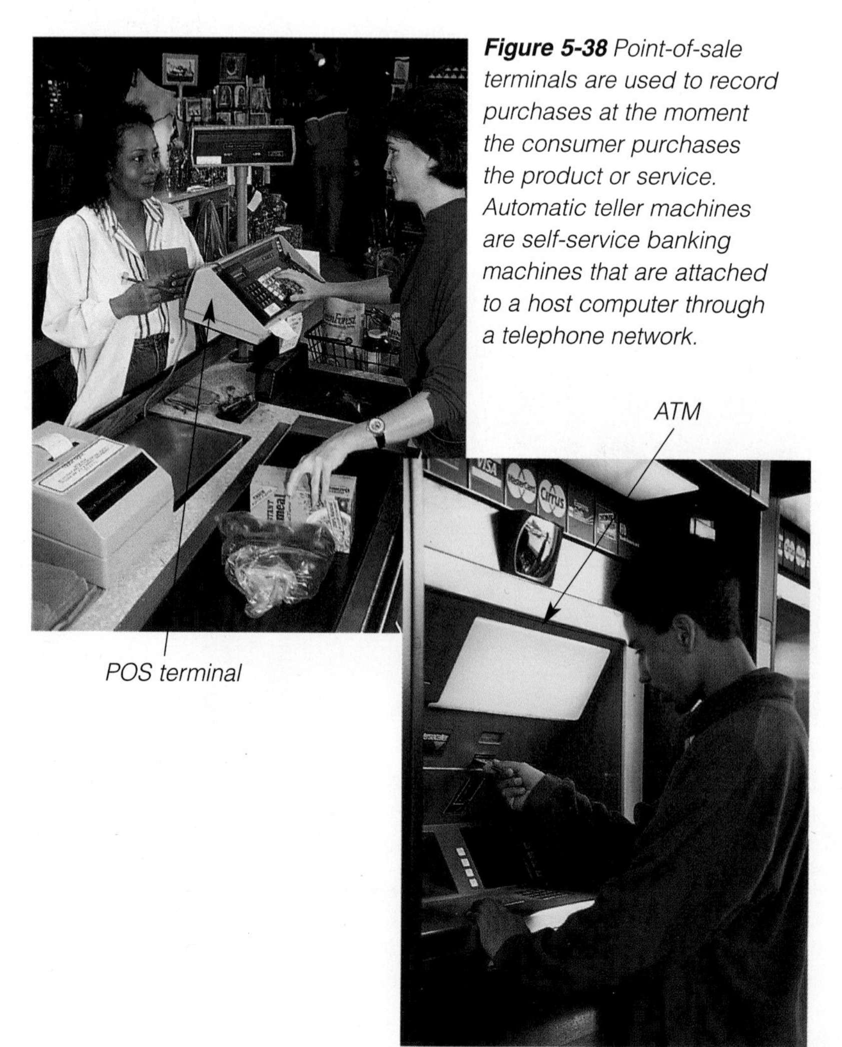

Figure 5-38 *Point-of-sale terminals are used to record purchases at the moment the consumer purchases the product or service. Automatic teller machines are self-service banking machines that are attached to a host computer through a telephone network.*

An **automatic teller machine (ATM)** is a self-service banking machine attached to a host computer through a telephone network. You insert a plastic bankcard with a magnetic strip into the ATM and enter your password, called a personal identification number (PIN), to access your bank account.

Some ATMs have touch screens, while others have special keyboards for input. Using an ATM, you can withdraw cash, deposit money, transfer funds, or inquire about an account balance.

OUTPUT DEVICES FOR PHYSICALLY CHALLENGED USERS

As discussed in Chapter 4, the growing presence of computers in everyone's lives has generated an awareness of the need to address computing requirements for those with physical limitations. For users with mobility, hearing, or vision disabilities, many different types of output devices are available. Hearing-impaired users, for example, can instruct programs to display words instead of sounds. With Windows, such users also can set options in the **Accessibility Properties dialog box** to instruct Windows to display visual signals in situations where normally it would make a sound (Figure 5-39).

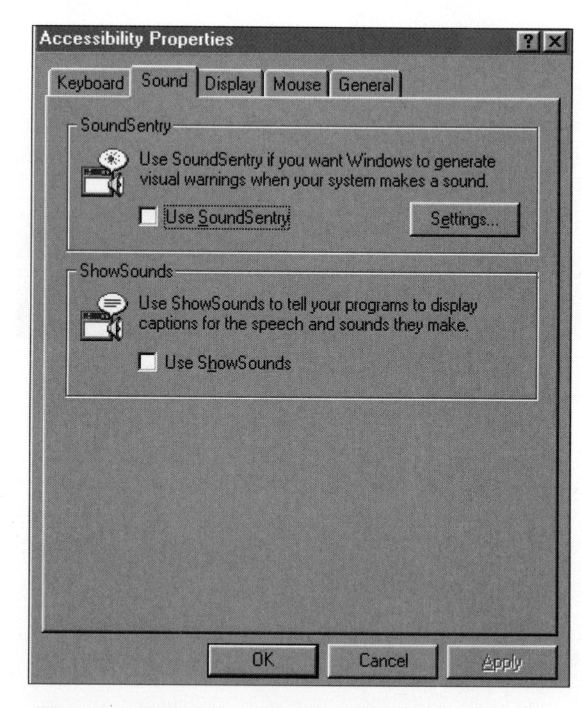

Figure 5-39 *Setting options in the Accessibility Properties dialog box makes Windows easier to use for physically challenged individuals.*

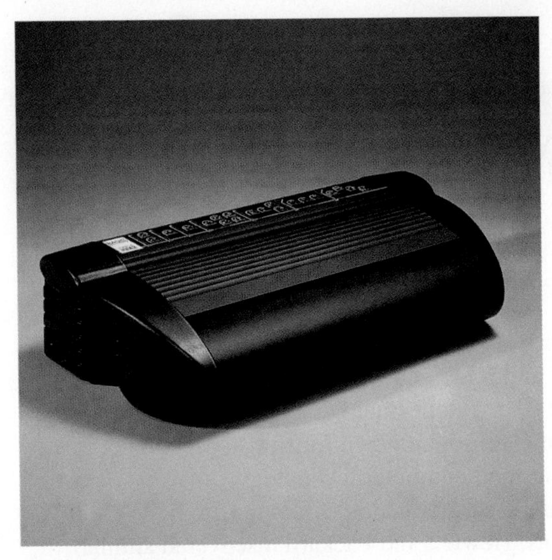

Figure 5-40 A Braille printer.

Visually impaired users can change Windows settings such as increasing the size or changing the color of the text to make the words easier to read. Instead of using a monitor, blind users can utilize speech output, where the computer *reads* the information that displays on the screen. Another alternative is a **Braille printer**, which outputs information in Braille onto paper (Figure 5-40).

PUTTING IT ALL TOGETHER

Many factors influence the type of output devices you should use: the type of output desired, the hardware and software in use, and the desired cost. Figure 5-41 outlines several suggested monitors, printers, and other output devices for various types of computer users.

SUGGESTED OUTPUT DEVICES BY USER

User	Monitor	Printer	Other
Home User	• 17- or 19-inch color CRT monitor	• Ink-jet color printer • Photo printer • Label printer	• Speakers • Headset
SMALL BUSINESS USER	• 19- or 21-inch color CRT monitor • Color LCD display for a handheld computer (PDA)	• Multifunction device; *or* • Ink-jet color printer; *or* • Laser printer, black and white • Label printer	• Fax machine • Speakers
MOBILE USER	• 15.4-inch color LCD display with a laptop computer • 19-inch color CRT monitor for a laptop docking station • Color LCD display for a handheld computer (PDA)	• Portable printer • Ink-jet color printer; *or* • Laser printer, black and white, for in-office use • Photo printer	• Fax modem • Headset • LCD or DLP data projector
LARGE BUSINESS USER	• 19- or 21-inch color CRT monitor • 15.4-inch color LCD display for a laptop computer • Color LCD display for a handheld computer (PDA)	• Laser printer, black and white • Line printer (for large reports from a mainframe) • Label printer	• Fax machine *or* fax modem • Speakers • LCD or DLP data projector • Dumb terminal
POWER USER	• 21-inch color monitor (for multimedia, desktop publishing, and engineering)	• Laser printer, black and white • Plotter • Snapshot printer • Dye-sublimation printer • Large format printer	• Fax machine *or* fax modem • Speakers • Headset

Figure 5-41 This table recommends suggested output devices for various types of users.

TECHNOLOGY TRAILBLAZERS

WILLIAM HEWLETT AND DAVID PACKARD

William Hewlett, David Packard

William Hewlett's children have formed a group called SPPW – the Society to Prove Papa Wrong. The group's objective is no small task. Hewlett's knowledge ranges from medicine, botany, and education to mountain climbing, fishing, and ranching. The subject that must be avoided most, however, is computer technology. Some sixty years ago William Hewlett, along with David Packard, co-founded Hewlett-Packard Company, the world's second largest manufacturer of computers and computer peripherals.

Hewlett and Packard met at Stanford University, where they both were students of Fred Terman, an inspirational teacher and author of a book on radio engineering, one of Packard's passions. Terman would become a mentor to both Hewlett and Packard. In the midst of the depression, Terman's electronics lab at Stanford was in the attic of a building badly in need of renovation. With no repair money, students scattered tar-paper-lined wooden baskets to catch water from a leaky roof. Yet, enthusiasm never waned; one winter, Hewlett stocked the baskets with goldfish to add a homey touch. Terman made up for a limited faculty by encouraging seminars in which students taught each other. To remedy Stanford's financial problems, Terman convinced the school to lease university lands to high-tech companies. He conducted tours of the labs and factories, pointing out that many were led by people with little conventional schooling. Just think, Terman mused, how successful someone with a formal engineering education and a little business training could be. Terman's insights and advice colored decisions made by Hewlett and Packard and other entrepreneurs. Today, Terman often is called "The Intellectual Father of Silicon Valley," that region of California known for firms specializing in electronics and computer technology.

On a post-graduation camping trip in the Colorado mountains, Hewlett and Packard discovered similar attitudes and cemented a lasting friendship. Hewlett went on to MIT, while Packard accepted a job with General Electric. Eventually, both returned to Stanford to complete their graduate work. With Terman's encouragement, in 1938 they started their own electronics company, Hewlett-Packard Company, in Packard's garage. Terman checked on his ex-students' progress whenever he drove past. "If the car was in the driveway," he recalled, "business was good." The car often was in the driveway, for that first year started an astonishing run. "We made $1,536 in profits," Packard noted. "We would show a profit every year thereafter." To avoid competitors, the company created original products. "Never take a fortified position unless you have to," Hewlett said. "The world is a big place." Hewlett-Packard products were accepted enthusiastically by engineers and scientists. Sales representatives were hired, and World War II fired an explosion of government orders. From electronic components, the company expanded into medical equipment, chemical analysis, test and measurement instruments, and computer equipment, which today forms the bulk of the company's sales. Hewlett-Packard earned a reputation for innovation. Packard proudly stated, "HP continually strives to develop products that represent true advancements."

Innovative products were accompanied by an inventive managerial style. Perhaps reminiscent of Terman's student-led seminars, a recurring theme is the idea of sharing. Hewlett-Packard employees share setting and achieving goals, share professional development, share ownership (through stock purchases), and share in the company's highs and lows. In his book, *The HP Way*, Packard described the decentralized decision making, team orientation, and hands-on leadership that characterize working at Hewlett-Packard. A key principle is "management by walking around (MBWA)," which maintains that managers learn more in the plant than they do from behind a desk. *The HP Way* emphasizes seeking out, and hearing, what others have to say. Packard died in 1996, but listening remains a priority at Hewlett-Packard. As Hewlett points out, "I learned how to listen…What you learn is you have a great memory."

COMPANY ON THE CUTTING EDGE

COMPANY ON THE CUTTING EDGE

HEWLETT-PACKARD

In a one-car garage in 1938, two 26-year-old engineers started a business with a collection of simple tools (bench, vise, drill press, screwdriver, file, soldering iron, hacksaw, and some electrical supplies) and $538 in working capital. It was a humble beginning, but the business begun by William Hewlett and David Packard would become Hewlett-Packard (HP) Company, one of the largest companies in the computer industry, with 121,000 employees worldwide and revenues of more than $42.8 billion. Today, the Palo Alto garage is a California State Historical Landmark, a Silicon Valley milestone.

The Hewlett-Packard partnership was formed officially in 1939. A coin toss determined the company's name. The first product was an audio oscillator, which is an instrument that employed a light bulb to help test sound equipment. Walt Disney ordered eight oscillators to use in developing sound effects for the movie, *Fantasia.* In 1942, the company built its first Hewlett-Packard owned building — a 10,000-square-foot office/laboratory/factory. Business was good, but the founders designed the new facility so it easily could be converted into a grocery store if the electronics business failed.

Instead of failing, Hewlett-Packard grew. By 1962, the company had earned a place on *Fortune* magazine's list of the top 500 U.S. companies. During its first two decades, Hewlett-Packard introduced a variety of products, including microwave signal generators, high-speed frequency counters, and other instruments. Through a number of acquisitions, the company expanded into other fields. Today, Hewlett-Packard markets and services electronic components, test and measurement supplies, medical electronic equipment, chemical analysis instruments, and computer products.

Hewlett-Packard entered the computer field in 1966 with a computer designed as a controller for some of the company's test and measurement instruments. A flurry of computer-related products followed, including personal computers, electronic mail systems, printers, palmtop PCs, and handheld organizers. The company also achieved a number of *firsts*, such as the first desktop scientific calculators (presaging workstations) and the first scientific handheld calculators (making slide rules obsolete). By 1997, computer-related sales made up 82 percent of the company's revenues.

Hewlett-Packard's most successful product is the LaserJet printer. Interestingly, the company developed *two* new printers in the early 1980s: the HP Thinkjet ink-jet printer and the HP LaserJet laser printer. Some companies would have held back one of the competing printers for later release. Hewlett-Packard decided, however, that it would rather compete with itself than with another printer manufacturer. When the printers were introduced, Hewlett-Packard sales skyrocketed. By 1993, the company had sold 20 million printers, and it is estimated that 60 percent of printers sold bear the Hewlett-Packard logo.

Despite its well-earned reputation for innovation among industry insiders, Hewlett-Packard's image among home consumers is less inspiring. Recent company surveys found that if Hewlett-Packard was a person, it would be seen as a "trustworthy but boring computer geek wearing a turtleneck and sandals." Hewlett-Packard is working to change that picture, with a new marketing campaign, a new corporate slogan ("Expanding Possibilities"), and new services (such as automated printer-supply vending machines). The company hopes this fresh approach, coupled with the unique ability to combine measurement, computer, and communication technologies, will give Hewlett-Packard the recognition it deserves both inside and outside the computer industry.

IN BRIEF  www.scsite.com/dc2001/ch5/brief.htm

WEB INSTRUCTIONS: *To display this page from the Web, launch your browser and enter the URL, www.scsite.com/dc2001/ch5/brief.htm. Click the links for current and additional information. To listen to an audio version of this IN BRIEF, click the Audio button to the right of the title, IN BRIEF, at the top of the page. To play the audio, RealPlayer must be installed on your computer (download by clicking here).*

1. What Are the Four Types of Output?

Output is data that has been processed into a useful form, called information. Four types of output are text, graphics, audio, and video. **Text** consists of **characters** that are used to create words, sentences, and paragraphs. **Graphics** are digital representations of nontext information such as drawings, charts, and photographs. **Audio** is music, speech, or any other sound. **Video** consists of images that provide the appearance of full motion.

2. What Are Different Types of Output Devices?

An **output device** is any computer component capable of conveying information to a user. A **display device** is an output device that visually conveys text, graphics, and video information. A printer is an output device that produces text and graphics on a physical medium such as paper or transparency film. An **audio output device** produces music, speech, or other sounds. Other output devices include **data projectors**, **facsimile (fax) machines**, and **multifunction devices**.

3. What Factors Affect the Quality of a Monitor?

A **monitor** is a display device that consists of a screen housed in a plastic or metal case. The **screen** is coated with tiny dots of phosphor material, called pixels, that glow when electrically charged. The quality of the resulting image depends on a monitor's resolution, dot pitch, and refresh rate. **Resolution**, or sharpness, is related to the number of pixels a monitor can display. Higher resolution means a greater number of pixels display, providing a smoother image. **Dot pitch**, a measure of image clarity, is the vertical distance between each pixel. The smaller the dot pitch, the clearer the displayed image. **Refresh rate** is the speed with which a monitor redraws images on the screen. Refresh rate should be fast enough to maintain a constant, flicker-free image.

4. Why Is a Video Card Used?

A **video card** converts digital output into an analog video signal that is sent through a cable to the monitor. The monitor separates the video signal into red, green, and blue signals. Electron guns fire the signals to the front of the monitor, and these three colored dots are combined to make up each single pixel. Several standards have been developed to define resolution, the number of colors, and other monitor properties. Today, most monitors and video cards support the **super video graphics array** (SVGA) standard.

5. What Are Monitor Ergonomic Issues?

Features that address monitor ergonomic issues include controls to adjust the brightness, contrast, positioning, height, and width of images. Many monitors have a tilt and swivel base so the angle of the screen can be altered to minimize neck strain and glare. Monitors produce a small amount of **electromagnetic radiation (EMR)**, which is a magnetic field that travels at the speed of light. High-quality monitors should comply with MPR II, which is a standard that defines acceptable levels of EMR for a monitor.

 www.scsite.com/dc2001/ch5/brief.htm

6. How Are Various Types of Printers Different?

Printers can be grouped in two categories: impact and nonimpact. **Impact printers** form characters and graphics by striking a mechanism against an ink ribbon that physically contacts the paper. A **dot-matrix printer** is an impact printer that prints images when tiny wire pins on a print head mechanism strike an inked ribbon. A **line printer** is an impact printer that prints an entire line at one time. **Nonimpact printers** form characters and graphics without actually striking the paper. An **ink-jet printer** is a nonimpact printer that sprays drops of ink onto a piece of paper. A **laser printer** is a nonimpact printer that operates in a manner similar to a copy machine. A **thermal printer** is a nonimpact printer that generates images by pushing electrically heated pins against heat-sensitive paper.

7. What Are Various Types of Audio Output Devices?

Two commonly used audio output devices are speakers and headsets. Most personal computers have an internal speaker that outputs low-quality sound. Many users add high-quality stereo speakers or purchase PCs with larger speakers built into the sides of the monitor. A woofer can be added to boost low bass sounds. A **headset** plugged into a port on the sound card allows only the user to hear sound from the computer.

8. Why Are Data Projectors, Fax Machines, and Multifunction Devices Used?

A **data projector** takes the image on a computer screen and projects it onto a large screen so an audience of people can see the image. A **facsimile (fax) machine** transmits and receives documents over telephone lines. A **fax modem** is a communications device that sends (and sometimes receives) electronic documents as faxes. A **multifunction device (MFD)** is a single piece of equipment that looks like a copy machine but provides the functionality of a printer, scanner, copy machine, and sometimes a fax machine.

9. How Is a Terminal Both an Input and Output Device?

A **terminal** is a device that consists of a keyboard (input), a monitor (output), and a video card. A terminal is used to input and transmit data to, or receive and output information from, a **host computer**. Three basic categories of terminals are **dumb terminals**, **intelligent terminals**, and special-purpose terminals.

10. What Are Output Options for Physically Challenged Users?

Hearing-impaired users can instruct programs to display words instead of sounds. With Windows, the **Accessibility Properties dialog box** can be used to instruct Windows to display visual signals in situations where normally it would make a sound. Visually impaired users can change the size or color of text to make words easier to read. Blind users can utilize speech output, where the computer reads information that displays on the screen. A **Braille printer** outputs information in Braille onto paper.

KEY TERMS

www.scsite.com/dc2001/ch5/terms.htm

WEB INSTRUCTIONS: *To display this page from the Web, launch your browser and enter the URL, www.scsite.com/dc2001/ch5/terms.htm. Scroll through the list of terms. Click a term to display its definition and a picture. Click KEY TERMS on the left to redisplay the KEY TERMS page. Click the TO WEB button for current and additional information about the term from the Web. To see animations, Shockwave and Flash Player must be installed on your computer (download by clicking here).*

Accessibility Properties dialog box (5.24)
active-matrix display (5.5)
animation (5.2)
audio (5.2, 5.20)
audio output devices (5.20)
automatic teller machine (ATM) (5.24)
band printer (5.13)
Braille printer (5.25)
cathode ray tube (CRT) (5.3)
character (5.2)
characters per second (cps) (5.13)
color monitor (5.3)
continuous-form paper (5.13)
CRT monitor (5.3)
data projector (5.21)
digital light processing (DLP) projector (5.21)
display device (5.3)
dot pitch (5.7)
dot-matrix printer (5.12)
dots per inch (dpi) (5.14)
dumb terminal (5.23)
dye-sublimation printer (5.18)
electromagnetic radiation (EMR) (5.10)
electrostatic plotter (5.19)
ENERGY STAR program (5.11)
facsimile (fax) machine (5.22)
fax modem (5.22)
flat-panel display (5.4)
gas plasma monitors (5.6)
graphics (5.2)
graphics card (5.8)
gray scaling (5.3)
hard copy (5.11)
headset (5.21)
hertz (5.7)
high-definition television (HDTV) (5.11)
host computer (5.23)

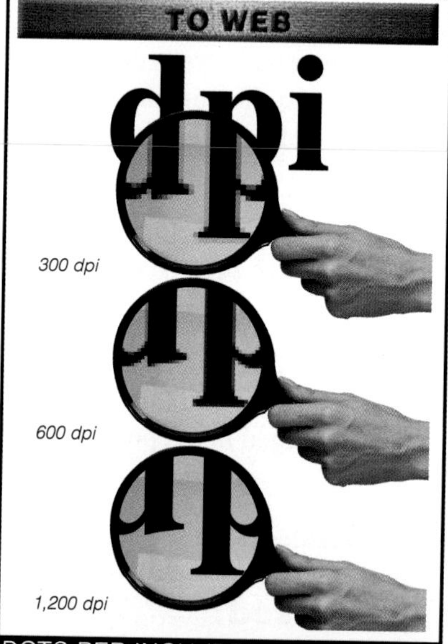

DOTS PER INCH (DPI): Measure of printer resolution; the higher the dpi, the better the quality of the print image. (5.14)

image editing software (5.2)
impact printer (5.12)
ink-jet printer (5.14)
intelligent terminal (5.23)
interlacing (5.7)
label printer (5.20)
landscape orientation (5.11)
large-format printer (5.19)
laser printer (5.16)
LCD display (5.5)
LCD projector (5.21)
letter quality (LQ) (5.12)
line printer (5.13)
liquid crystal display (LCD) (5.5)
monitor (5.3)
monochrome monitor (5.3)
multifunction device (MFD) (5.23)
near letter quality (NLQ) (5.12)
nonimpact printer (5.14)
output (5.2)
output device (5.3)
page description language (PDL) (5.16)
page printers (5.16)
pages per minute (ppm) (5.15)
passive-matrix display (5.5)
PCL (Printer Control Language) (5.16)
pen plotter (5.19)
photo printer (5.20)
pixel (5.4)
point-of-sale (POS) terminal (5.24)
portable printer (5.18)
portrait orientation (5.11)
PostScript (5.16)
printer (5.11)
printout (5.11)
programmable terminals (5.23)
refresh rate (5.7)
resolution (5.6)
screen (5.3)
shuttle-matrix printer (5.13)
soft copy (5.3)
super video graphics array (SVGA) (5.9)
terminal (5.23)
text (5.2)
TFT displays (5.5)
thermal dye transfer printer (5.18)
thermal printer (5.17)
thermal transfer printer (5.18)
thermal wax-transfer printer (5.18)
toner (5.17)
video (5.3)
video adapter (5.8)
video card (5.8)
video RAM (5.9)
viewable size (5.4)
VRAM (5.9)

AT THE MOVIES

www.scsite.com/dc2001/ch5/movies.htm

WELCOME to VIDEO CLIPS from CNN

WEB INSTRUCTIONS: *To display this page from the Web, launch your browser and enter the URL, www.scsite.com/dc2001/ch5/movies.htm. Click a picture to view a video. After watching the video, close the video window and then complete the exercise by answering the questions about the video. To view the videos, RealPlayer must be installed on your computer (download by clicking here).*

1 IP Faxville Press

The use of the facsimile (fax) machine has changed many aspects of political life. Do you think the change is for the better? Why? Does the use of a fax give one candidate an advantage over another candidate? Is this good? What other activities have been altered by the use of fax machines? Describe what you see as problems with this instant communications device. Do you see any way of solving the problems or must we put up with whatever we are sent on a fax machine? Would you support "fax regulation?" Why?

The United States Postal Service is the technology for testing e-stamps. What do you think of e-stamps? How are e-stamps placed on envelopes? Do you need a special printer to print the e-stamps? Can a person handwrite the address on an e-stamped envelope? What are the advantages and disadvantages of e-stamps? Will you use e-stamps? Why?

3 Biz Apple 20th

New versatile features of Apple laptop computers parallel trends in PC laptops. What are some of the new features offered on laptop models? What new types of input and output devices are used with laptops? Do you think laptops will look the same ten years from now? If not, how might laptops be different?

Student Exercises
WEB INFO
IN BRIEF
KEY TERMS
AT THE MOVIES
CHECKPOINT
AT ISSUE
MY WEB PAGE
HANDS ON
NET STUFF
Special Features
TIMELINE 2001
GUIDE TO WWW SITES
MAKING A CHIP
E-COMMERCE 2001
BUYER'S GUIDE 2001
CAREERS 2001
TRENDS 2001
LEARNING GAMES
CHAT
INTERACTIVE LABS
NEWS
HOME

CHECKPOINT

www.scsite.com/dc2001/ch5/check.htm

WEB INSTRUCTIONS: *To display this page from the Web, launch your browser and enter the URL, www.scsite.com/dc2001/ch5/check.htm. Click the links for current and additional information. To experience the animation and interactivity, Shockwave and Flash Player must be installed on your computer (download by clicking here).*

Label the Figure

Instructions: *Complete these steps to show how video travels from the CPU to the monitor.*

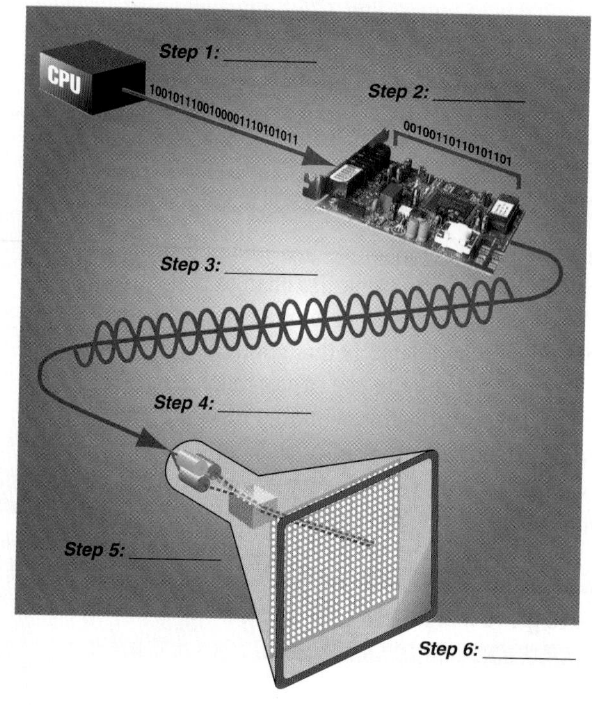

Matching

Instructions: *Match each term from the column on the left with the best description from the column on the right.*

_______ 1. LCD (liquid crystal display)
_______ 2. EMR (electromagnetic radiation)
_______ 3. NLQ (near letter quality)
_______ 4. PDL (page description language)
_______ 5. ATM (automatic teller machine)

a. Has special molecules deposited between two sheets of material.
b. Used by higher-quality video cards to improve the quality of graphics.
c. Print that is slightly less clear than letter quality but ideal for mailing labels and envelopes.
d. A magnetic field that travels at the speed of light.
e. A standard printer language designed to support the fonts and layouts used in standard documents.
f. Tells the printer how to layout the contents of a printed page.
g. A self-service banking machine attached to a host computer through a telephone network.

Short Answer

Instructions: *Write a brief answer to each of the following questions.*

1. How are color monitors, monochrome monitors, and monitors that use gray scaling different? __________ When are color monitors and monochrome monitors most widely used? __________
2. How is an active matrix display different from a passive matrix display? __________ What is a gas plasma monitor? __________
3. What is high-definition television (HDTV)? __________ What advantages does HDTV provide over analog signals? __________
4. How is hard copy different from soft copy? __________ How is portrait orientation different from landscape orientation? __________
5. How is a dumb terminal different from an intelligent terminal? __________ For what purpose is a point-of-sale terminal used? __________

AT ISSUE

www.scsite.com/dc2001/ch5/issue.htm

WEB INSTRUCTIONS: *To display this page from the Web, launch your browser and enter the URL, www.scsite.com/dc2001/ch5/issue.htm. Click the links for current and additional information.*

ONE

The era of "Internet wearables" soon will be upon us. These devices consist of small monitors or speakers coupled with wearable items (such as broaches, eyeglasses, watches, or hearing aids) that offer wireless access to the Internet. By issuing commands through a tiny microphone attached to the device, wearers can send and receive e-mail, obtain business information and stock quotes, monitor health conditions with medical databases, surf the Web, or perform any other Internet-related task. But, are people ready for Internet wearables? Any-time Internet access could prove an irresistible distraction for drivers in cars, students in classes, workers at job sites, or even people in conversations. Are Internet wearables a good, or bad, idea? Why? What, if anything, could be done to limit the potential problems?

Short of cash? Some college students found a solution to the problem: they made some. With a scanner, personal computer, and color printer, the students produced bogus bills and passed more than $1,000 in counterfeit currency before they were caught. As printer quality improves, police have arrested counterfeiters ranging from high school students to senior citizens, and the problem is growing. Fake money usually can be spotted with a close inspection, but most people never look that closely. Once a counterfeit bill is accepted, even as change, it is yours. Should printer manufacturers take some responsibility for this problem? Why or why not? What, if anything, could printer makers do? Can you offer any other possible solutions?

THREE

Computer screens appear in unlikely places, but perhaps no spot is more improbable than a cemetery. Nevertheless, a company called Leif Technologies is changing that by offering computerized grave markers. The headstones have a 5-inch by 4-inch screen and a computer that stores up to 85 pages of information about the deceased person. Visitors can access records, stories, poems, and even photographs. Buyers think these modern memorials will prove invaluable to future generations, but traditionalists feel the technological tombstones are unseemly in such a solemn setting. Are computerized grave markers appropriate? Why or why not? At a cost of $5,000, would you consider an interactive plaque? Why?

FOUR

Many interior decorators feel CRT monitors are the most unattractive and inefficient component of the modern office. Perched on a desktop, these boxy devices add little to office allure and subtract much from available workspace. Liquid crystal display (LCD) screens, however, soon may remedy this problem. Modern LCDs provide resolution, screen sizes, and clarity equivalent to or better than the best CRTs and occupy much less space. Designer LCDs also are available in a range of attractive colors and designs. Yet, high-quality LCDs can cost more than $2,000, while a top-quality CRT monitor can be purchased for only $500. Is the extra expense worth it? When would you consider buying a designer LCD instead of a standard CRT monitor? Why?

At one time, people predicted paperless offices in which all information was processed electronically, and hard copy was nonexistent. Instead, paper use has risen more than forty percent over the past thirty years, and the largest increase is in the use of office paper. Experts explain this phenomenon by pointing out that paper is less expensive and paper documents are easier to create than in the past. Every office has computer printers, copying machines, and facsimile machines that reproduce documents at an ever-increasing rate. In addition, paper documents help sift the glut of electronic information. Can, and should, people and organizations try to reduce paper use? Why or why not? What can be done to make, if not a paperless office, at least a less-paper office?

MY WEB PAGE

www.scsite.com/dc2001/ch5/myweb.htm

WEB INSTRUCTIONS: *The icons to the left of the numbered activities on this page indicate the learning system availability. If you are using WebCT, follow the instructions in the exercises below. If you are using the textbook Web site or CyberClass, launch your browser, enter the URL www.scsite.com/dc2001/ch5/myweb.htm, and then click the corresponding icon to the left of the exercise.*

1. Practice Test: Click My Web Page and then click Chapter 5. Click Practice Test. Answer each question and then click the Save Answer button. When completed, click the Finish button. Click the OK button to submit the quiz for grading. Click the View Results button, and then make a note of any missed answers.

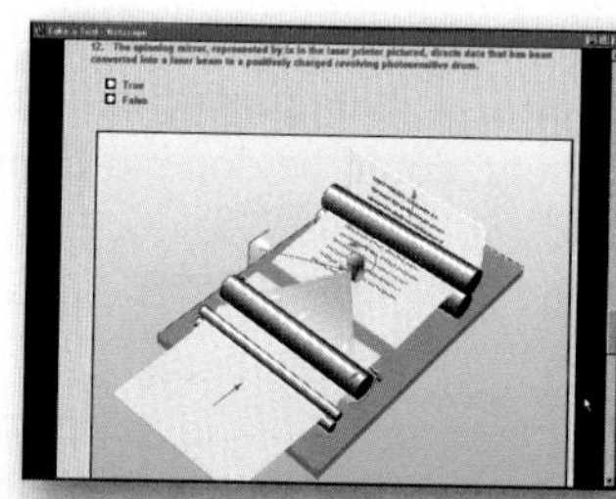

2. Web Guide: Click My Web Page and then click Chapter 5. Click Web Guide to display the Guide to WWW Sites page. Click Computers and Computing. Click PC Guide and search for Monitors. Click Monitors and then review the information. Use your word processing program to prepare a brief report on your findings and submit your assignment to your instructor.

3. Scavenger Hunt: Click My Web Page and then click Chapter 5. Click Scavenger Hunt. Print a copy of the Scavenger Hunt page; use this page to write down your answers as you search the Web. Submit your completed page to your instructor.

4. Who Wants to Be a Computer Genius?: Click My Web Page and then click Chapter 5. Click Computer Genius to find out if you are a computer genius. For directions, click the How to Play button. When you are ready to play, click the Game button. Submit your score to your instructor.

5. Wheel of Terms: Click My Web Page and then click Chapter 5. Click Wheel of Terms to reinforce important terms you learned in this chapter by playing the Shelly Cashman Series version of this popular game. For directions, click the How to Play button. When you are ready to play, click the Game button. Submit your score to your instructor.

6. Career Corner: Click My Web Page and then click Chapter 5. Click Career Corner to display the MSN Career page. Review this page. Click a link about Interviewing or Resumes. Write a brief report on what you discovered. Submit the report to your instructor.

7. Search Sleuth: Click My Web Page and then click Chapter 5. Click Search Sleuth to learn search techniques that will help make you a research expert. Submit the completed assignment to your instructor.

8. Crossword Puzzle Challenge: Click My Web Page and then click Chapter 5. Click Crossword Puzzle Challenge. Complete the puzzle to reinforce skills you learned in this chapter. For directions, click the How to Play button. When you are ready to play, click the Game button. Submit the completed puzzle to your instructor.

9. Flash Cards: This exercise uses CyberClass. Follow the instructions at the top of this page and then click the CyberClass icon to the left; or ask your instructor for logon instructions. Click Flash Cards on the Main Menu. Click the plus sign. Answer all the questions in any two subjects of your choice. Click the Close button in the upper-right corner to close all windows. Notice the many other exercises at this site. Complete all other exercises as assigned by your instructor.

HANDS ON

www.scsite.com/dc2001/ch5/hands.htm

WEB INSTRUCTIONS: *To display this page from the Web, launch your browser and enter the URL, www.scsite.com/dc2001/ch5/hands.htm. Click the links for current and additional information.*

One

About Your Computer

This exercise uses Windows 95 or Windows 98 procedures. Your computer probably has more than one output device. To learn a little bit about the output devices on your computer, right-click the My Computer icon on the desktop. Click Properties on the shortcut menu. When the System Properties dialog box displays, click the Device Manager tab. Click View devices by type. Below Computer, a list of hardware device categories displays. What output devices display in the list? Click the plus sign next to each category. What specific output devices in each category are connected to your computer? Close the System Properties dialog box.

Two

Accessibility Options

This exercise uses Windows 95 or Windows 98 procedures. The Windows operating system offers several output options for people with hearing or visual impairments. Three of these options are SoundSentry, ShowSounds, and High Contrast. To find out more about each option, click the Start button, point to Settings on the Start menu, and then click Control Panel on the Settings submenu. Double-click the Accessibility Options icon in the Control Panel window. Click the Sound tab in the Accessibility Properties dialog box. Click the question mark button on the title bar, click Use SoundSentry, read the information in the pop-up window, and then click the pop-up window to close it. Repeat this process for Use ShowSounds. Click the Display tab. Click the question mark button on the title bar and then click Use High Contrast. Read the information in the pop-up window, and then click the pop-up window to close it. What is the purpose of each option? Click the Cancel button. Click the Close button to close the Control Panel window.

Three

Self-Portrait

Windows includes a drawing program called Paint. The quality of graphics produced with this program depends on a variety of factors, including the quality of your printer, your understanding of the software, and (to some extent) your artistic talent. In this exercise, you will use Paint to create a self-portrait. To access Paint, click the Start button, point to Programs on the Start menu, point to Accessories on the Programs submenu, and then click Paint on the Accessories submenu. When the Paint window displays, you will see the Paint toolbar on the left side of the window. Point to a toolbar button to see a tool's name; click a button to use that tool. Use the tools and colors available in Paint to draw a picture of yourself. If you make a mistake, you can click Undo on the Edit menu to undo your most recent action, you can erase part of your picture using the Eraser/Color Eraser tool, or you can clear the entire picture by clicking Clear Image on the Image menu. When your self-portrait is finished, print it by clicking Print on the File menu. Close Paint.

Four

Microsoft Magnifier

This exercise uses Windows 98 or Windows 2000 procedures. Microsoft Magnifier is a Windows utility for the visually impaired. To find out about the Microsoft Magnifier capabilities, click the Start button and then click Help on the Start menu. Click the Index tab. Type `magnifier` in the text box and then click the Display button. Click the Using Microsoft Magnifier topic or click overview and then click the Display button in the Windows Help window. Read the information and answer these questions:

- How does Microsoft Magnifier make the screen more readable for the visually impaired?
- What viewing options does Microsoft Magnifier have?
- What tracking options does Microsoft Magnifier have?

Click the Close button to close the Windows Help window.

Student Exercises
WEB INFO
IN BRIEF
KEY TERMS
AT THE MOVIES
CHECKPOINT
AT ISSUE
MY WEB PAGE
HANDS ON
NET STUFF
Special Features
TIMELINE 2001
GUIDE TO WWW SITES
MAKING A CHIP
E-COMMERCE 2001
BUYER'S GUIDE 2001
CAREERS 2001
TRENDS 2001
LEARNING GAMES
CHAT
INTERACTIVE LABS
NEWS
HOME

NET STUFF

www.scsite.com/dc2001/ch5/net.htm

WEB INSTRUCTIONS: *To display this page from the Web, launch your browser and enter the URL, www.scsite.com/dc2001/ch5/net.htm. To use the Setting Up to Print lab or the Configuring Your Display lab from the Web, Shockwave and Flash Player must be installed on your computer (download by clicking here).*

SETTING UP TO PRINT LAB

1. Shelly Cashman Series Setting Up to Print Lab

Follow the appropriate instructions in NET STUFF 1 on page 1.48 to start and use the Setting Up to Print lab. If you are running from the Web, enter the URL, www.scsite.com/sclabs/menu.htm; or display the NET STUFF page (see instructions at the top of this page) and then click the SETTING UP TO PRINT LAB button.

CONFIGURING YOUR DISPLAY LAB

2. Shelly Cashman Series Configuring Your Display Lab

Follow the appropriate instructions in NET STUFF 1 on page 1.48 to start and use the Configuring Your Display lab. If you are running from the Web, enter the URL, www.scsite.com/sclabs/menu.htm; or display the NET STUFF page (see instructions at the top of this page) and then click the CONFIGURING YOUR DISPLAY LAB button.

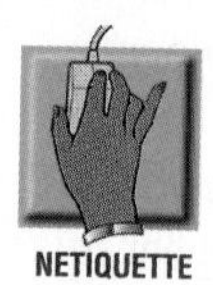

NETIQUETTE

3. Newsgroup Etiquette

In Chapter 3, you learned about newsgroups. Before getting involved in a newsgroup, however, you should learn a little about newsgroup etiquette. Perfectly acceptable communication in e-mail messages to a friend may be highly inappropriate in postings to newsgroup readers. Click the NETIQUETTE button for an informative, amusing, common-sense guide to newsgroup etiquette. This site's Emily Post-like approach presents etiquette questions with unlikely answers. As you explore the guide, you will see that the actual answer to each question usually is the opposite of what the given answer suggests. Make a list of five etiquette questions presented. Using the satirical answer given and your own common sense, write a genuine reply to each question.

IN THE NEWS

4. In the News

Monitors continue to grow clearer and thinner. A newly introduced 50-inch gas plasma display presents near-photographic images and is less than four inches thick. At a cost of $25,000, the monitors probably will be seen first at stadiums, in airports, and as touch screens in stores. Yet, as prices fall, consumers surely will purchase the monitors for HDTV and crystal-clear Internet access. Click the IN THE NEWS button and then read a news article about a new or improved output device. What is the device? Who manufactures it? How is the output device better than, or different from, earlier devices? Who do you think is most likely to use the device? Why?

WEB CHAT

5. Web Chat

A photomosaic is a large image, often a portrait, made up of many small photographs. The result is a remarkable representation that is enchanting from a distance and stunning from up close. Photomosaics include an Abraham Lincoln composed of Civil War photographs, a Vincent Van Gogh comprised of nature scenes, and a Bill Gates consisting of various world currencies. Robert Silvers created the software to form photomosaics while a graduate student at MIT. The software digitizes a large image and then selects the smaller photographs from special collections or a huge database of stock pictures. The software then sorts the photographs based on color, tone, and discernible shapes. Finally, the small photographs are digitally arranged to create the large image. A photomosaic costs more than $75,000, but is it art? Silvers is not sure. He claims the entire process could be done by hand, although over a much longer period of time. Critics argue that photomosaics are little more than computer wizardry, but admirers insist photomosaics may express emotion, an essential characteristic of art, even better than conventional artwork. Would you consider a photomosaic a work of art? Why or why not? Click the WEB CHAT button to enter a Web Chat discussion concerning photomosaics.

STORAGE

OBJECTIVES

After completing this chapter, you will be able to:

- Differentiate between storage and memory
- Identify various types of storage media and storage devices
- Explain how data is stored on a floppy disk
- Understand how to care for a floppy disk
- Describe how a hard disk organizes data
- List the advantages of using disks
- Explain how data is stored on compact discs
- Understand how to care for a compact disc
- Differentiate between CD-ROMs and DVD-ROMs
- Identify uses of tapes, PC Cards, smart cards, microfilm, and microfiche

Storage refers to the media on which data, instructions, and information are kept, as well as the devices that record and retrieve these items. This chapter explains various storage media and storage devices. Following completion of this chapter, you will have an understanding of all four operations in the information processing cycle: input, processing, output, and storage.

MEMORY VERSUS STORAGE

It is important to understand the difference between memory, which was discussed in Chapter 3, and storage, the focus of this chapter. To clarify the differences, the next section reviews memory and then discusses basic storage concepts.

Memory

While performing a processing operation, the CPU needs a place to temporarily hold instructions to be executed and data to be used with those instructions. Memory, which is composed of one or more chips on the motherboard, holds data and instructions while they are being processed by the CPU.

The two basic types of memory are volatile and nonvolatile. The contents of **volatile memory**, such as RAM, are lost (erased) when the power to the computer is turned off. The contents of **nonvolatile memory**, however, are not lost when power is removed from the computer. For example, once instructions have been recorded onto a nonvolatile ROM chip, they usually cannot be erased or changed, and the contents of the chip are not erased when power is turned off.

Storage

Storage, also called **secondary storage**, **auxiliary storage**, or **mass storage**, holds items such as data, instructions, and information for future use.

Think of storage as a filing cabinet used to hold file folders, and memory as the top of your desk. When you need to work with a file, you remove it from the filing cabinet (storage) and place it on your desk (memory). When you are finished with the file, you return it to the filing cabinet (storage).

Storage is nonvolatile, which means that items in storage are retained even when power is removed from the computer (Figure 6-1). A **storage medium** (media is the plural) is the physical material on which items are kept. One commonly used storage medium is a **disk**, which is a round, flat piece of plastic or metal with a magnetic coating on which items can be written. A **storage device** is the mechanism used to record and retrieve items to and from a storage medium.

AN ILLUSTRATION OF VOLATILITY

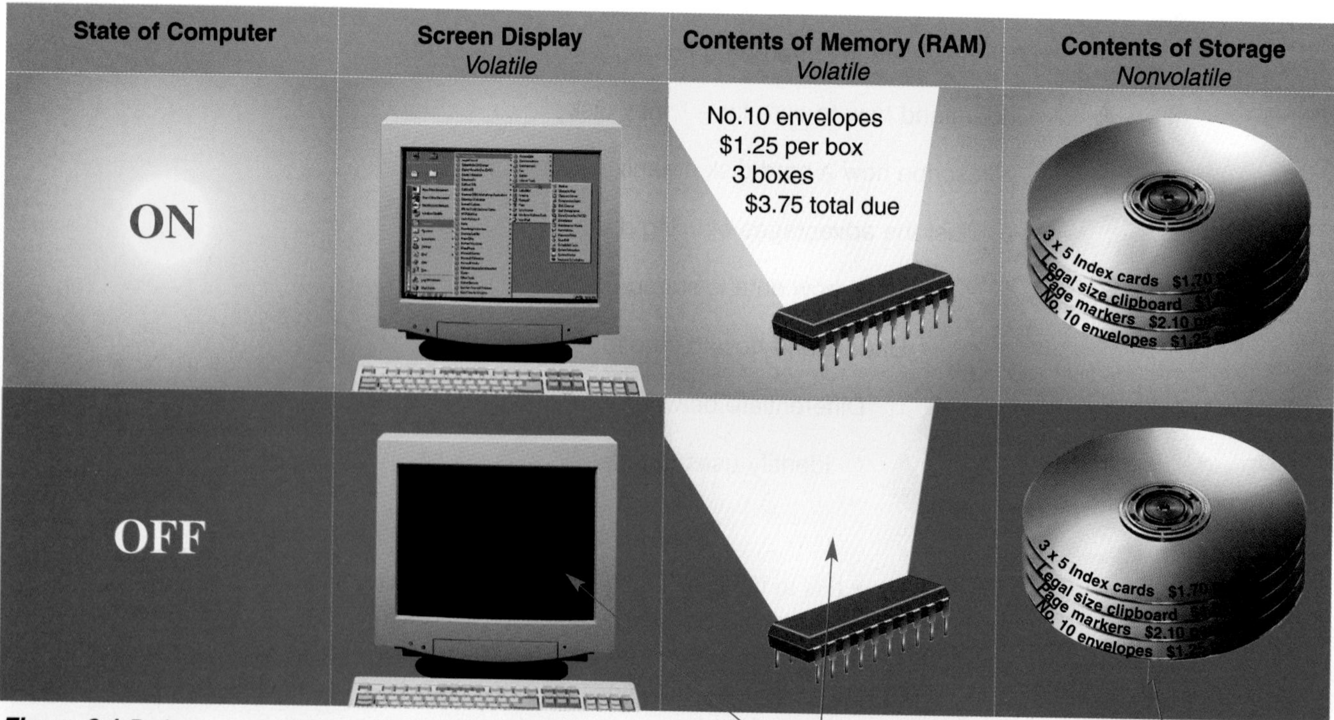

Figure 6-1 *Both a screen display and RAM are volatile; that is, their contents are erased when power is removed from the computer. Storage, in contrast, is nonvolatile – its contents are retained when power is off.*

Storage devices can function as sources of input and output. For example, each time a storage device transfers data, instructions, and information from a storage medium into memory — a process called **reading** — it functions as an input source. When a storage device transfers these items from memory to a storage medium — a process called **writing** — it functions as an output source.

The speed of a storage device is defined by its **access time**, which is the minimum time it takes the device to locate a single item on a disk. Compared to memory, storage devices are slow. The access time of memory devices is measured in nanoseconds (billionths of a second), while the access time of storage devices is measured in milliseconds (thousandths of a second).

The size, or **capacity**, of a storage device, is measured by the number of bytes (characters) it can hold. Figure 6-2 lists the terms used to define the capacity of storage devices. For example, a typical floppy disk can store 1.44 MB of data (approximately 1,440,000 bytes) and a typical hard disk can store 8 GB of data (approximately 8,000,000,000 bytes).

Storage requirements among users vary greatly. Users of smaller computers, such as small business users, might need to store a relatively small amount of data. For example, a field sales representative might have a list of names, addresses, and telephone numbers of 50 customers, which he or she uses on a daily basis. Such a list might require no more than several thousand bytes of storage. Users of larger computers, such as banks, libraries, or insurance companies, process data for millions of customers and thus might need to store trillions of bytes worth of historical or financial records in their archives.

To meet the needs of a wide range of users, numerous types of storage media and storage devices exist. Figure 6-3 shows how different types of storage media and memory compare in terms of relative cost and speed. The storage media included in the pyramid are discussed in this chapter.

STORAGE TERMS

Storage Term	Abbreviation	Number of Bytes
Kilobyte	KB	1 thousand
Megabyte	MB	1 million
Gigabyte	GB	1 billion
Terabyte	TB	1 trillion

Figure 6-2 *The capacity of a storage device is measured by the number of bytes it can hold.*

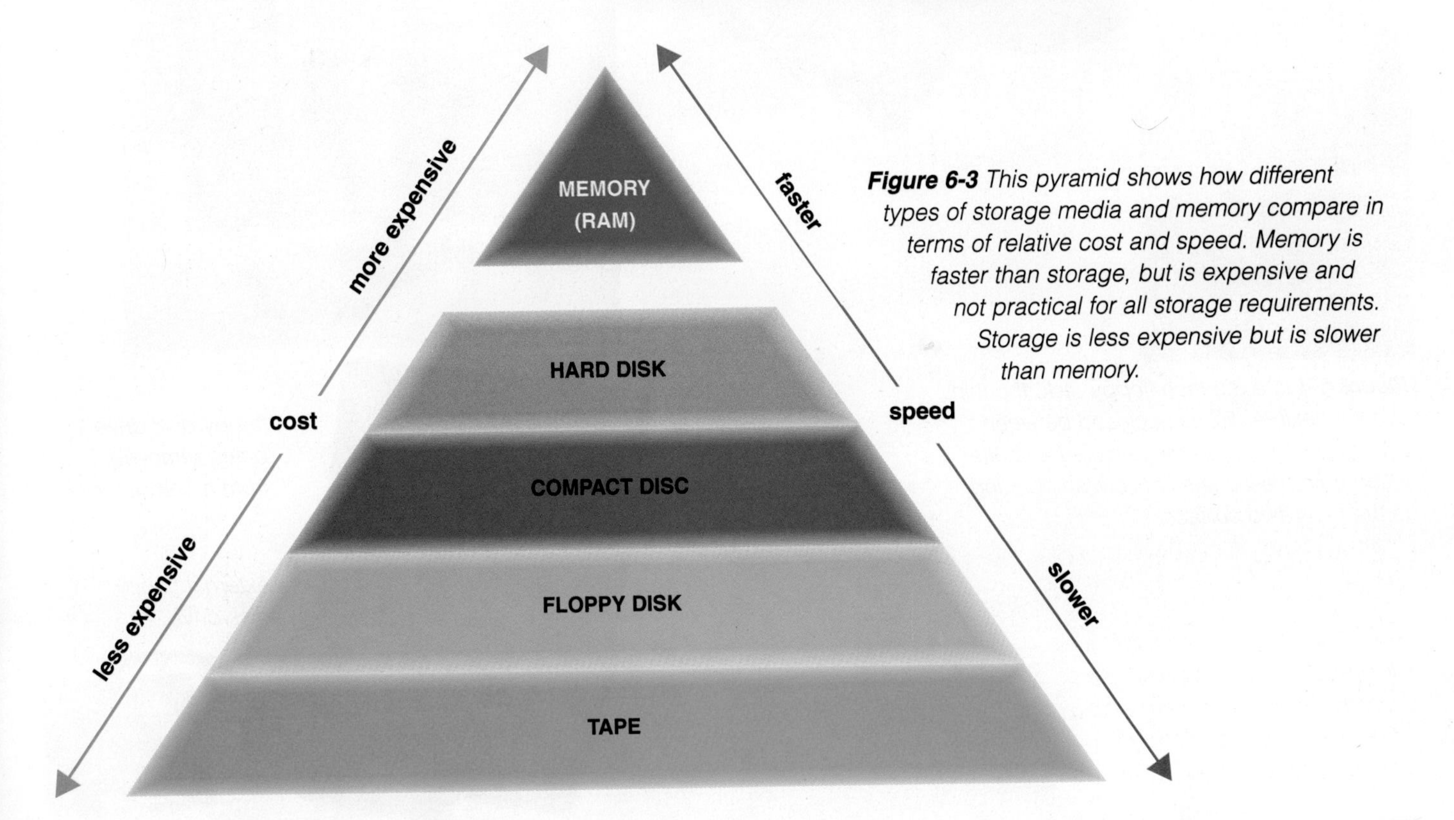

Figure 6-3 *This pyramid shows how different types of storage media and memory compare in terms of relative cost and speed. Memory is faster than storage, but is expensive and not practical for all storage requirements. Storage is less expensive but is slower than memory.*

WEB INFO

For more information on floppy disks, visit the Discovering Computers 2001 Chapter 6 WEB INFO page (**www.scsite.com/dc2001/ch6/webinfo.htm**) and click Floppy Disks.

FLOPPY DISKS

A **floppy disk**, or **diskette**, is a portable, inexpensive storage medium that consists of a thin, circular, flexible plastic disk with a magnetic coating enclosed in a square-shaped plastic shell (Figure 6-4). In the early 1970s, IBM introduced the floppy disk as a new type of storage. Because these early 8-inch wide disks had flexible plastic covers, many users referred to them as floppies. The next generation of floppies looked much the same, but were only 5.25-inches wide.

Today, the most widely used floppy disk is 3.5-inches wide. The flexible cover of the earlier floppy disks has been replaced with a rigid plastic outer cover. Thus, although today's 3.5-inch disks are not at all floppy, the term floppy disk still is used.

As noted, a floppy disk is a portable storage medium. When discussing a storage medium, the term portable means you can remove the medium from one computer and carry it to another computer. For example, you can insert a floppy disk into and remove it from a floppy disk drive on many types of computers (Figure 6-5). A floppy disk drive is a device that can read from and write to a floppy disk.

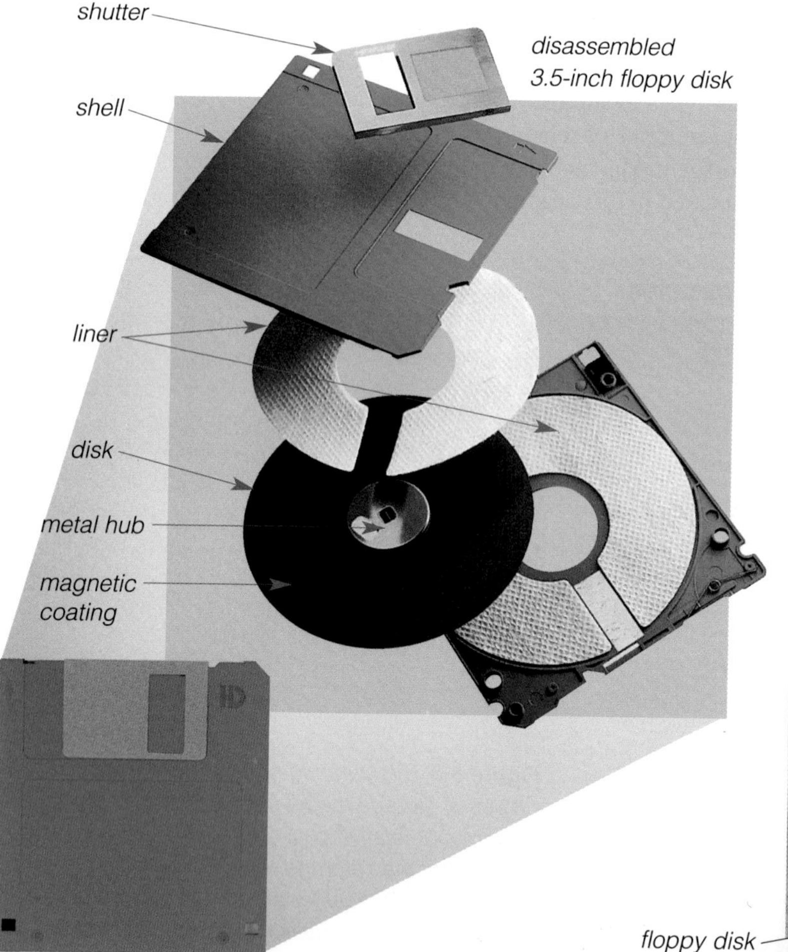

Figure 6-4 *In a 3.5-inch floppy disk, the thin circular flexible disk is enclosed between two liners. A piece of metal called a shutter in the rigid plastic shell covers an opening to the recording surface.*

Figure 6-5 *Various types of floppy disk drives.*

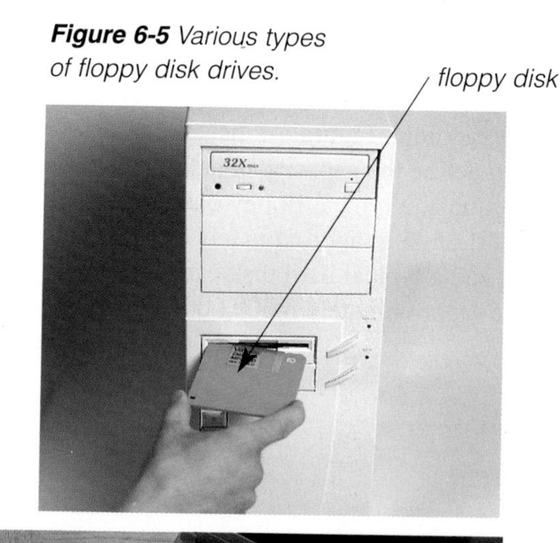

Characteristics of Magnetic Media

A floppy disk is a type of magnetic media, which means it uses magnetic patterns to store items such as data, instructions, and information on the disk's surface. Most magnetic disks are **read/write** storage media; that is, you can access (read) data from and place (write) data on a magnetic disk any number of times, just as you can with an audiocassette tape.

A new, blank floppy disk has nothing stored on it. Before you can write on a new floppy disk, it must be formatted.

Formatting is the process of preparing a disk (floppy disk or hard disk) for reading and writing by organizing the disk into storage locations called tracks and sectors (Figure 6-6). A **track** is a narrow recording band that forms a full circle on the surface of the disk. The disk's storage locations then are divided into pie-shaped sections, which break the tracks into small arcs called sectors. A **sector** is capable of holding 512 bytes of data. A typical floppy disk stores data on both sides and has 80 tracks on each side of the recording surface with 18 sectors per track.

Sometimes, a sector is damaged or has a flaw and cannot store data. A sector that cannot be used due to a physical flaw on the disk is called a bad sector. When you format a disk, the operating system marks these bad sectors as unusable. If a sector that contains data is damaged, you may be able to use special software to recover the data.

For reading and writing purposes, sectors are grouped into clusters. A **cluster** consists of two to eight sectors (the number varies depending on the operating system). A cluster is the smallest unit of space used to store data. Even if a file consists of only a few bytes, an entire cluster is used for storage. Although each cluster holds data from only one file, one file can be stored in many clusters.

A disk's storage capacity is determined by the density of the disk. A higher density means that the disk has a larger storage capacity. Disk **density** is computed by

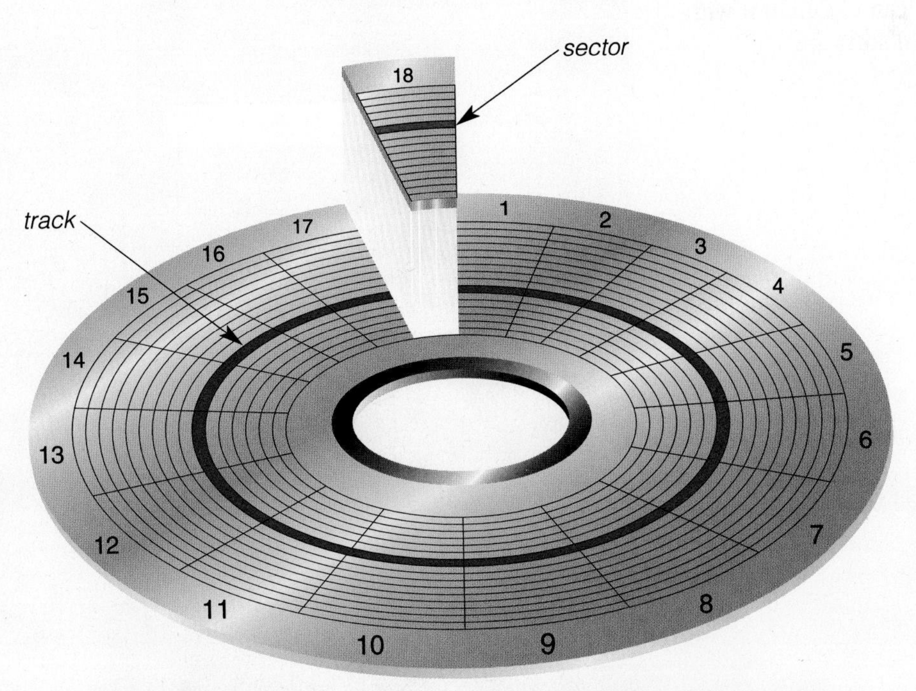

Figure 6-6 A track is a narrow recording band that forms a full circle on the surface of a disk. The disk's storage locations then are divided into pie-shaped sections, which break the tracks into small arcs called sectors. A sector can store 512 bytes of data.

multiplying together the number of sides on the disk, the number of tracks on the disk, the number of sectors per track, and the number of bytes in a sector. For example, for a typical 3.5-inch floppy disk, disk density is computed as follows: 2 (sides) × 80 (tracks) × 18 (sectors per track) × 512 (bytes per sector) = 1,474,560 bytes, or approximately 1.44 MB. The table in Figure 6-7 shows the number of sides, tracks, sectors per track, and bytes per sector for a 3.5-inch high-density floppy disk, which is the most common size of floppy disk used today.

If you are using the Windows operating system, the formatting process also defines the **file allocation table (FAT)**, which is a table of information used to locate files on a disk. The FAT is like a library card catalog for your disk that contains a listing of all files, file types, and locations. If you format a disk that already contains data, instructions, or information, the formatting process erases the file location information and redefines the file allocation table for these items. The actual files on the disk, however, are not erased. For this reason, if you accidentally format a disk, you often can *unformat* it with special software.

Characteristics of a Floppy Disk

To protect them from accidentally being erased, floppy disks have a write-protect notch. A **write-protect notch** is a small opening in the corner of the floppy disk with a tab that you slide to cover or expose the notch (Figure 6-8). The write-protect notch works much like the recording tab on a VHS tape: if the recording tab is removed, a VCR cannot record onto the VHS tape.

On a floppy disk, if the write-protect notch is exposed, or open, the drive cannot write on the floppy disk. If the write-protect notch is covered, or closed, the drive can write on the floppy disk. The write-protect notch only affects the floppy disk drive's capability of writing on the disk; a floppy disk drive can read from a floppy disk whether the write-protect notch is open or closed. Some floppy disks have a second opening on the opposite side of the disk that does not have the small tab; this opening identifies the disk as a high-density floppy disk.

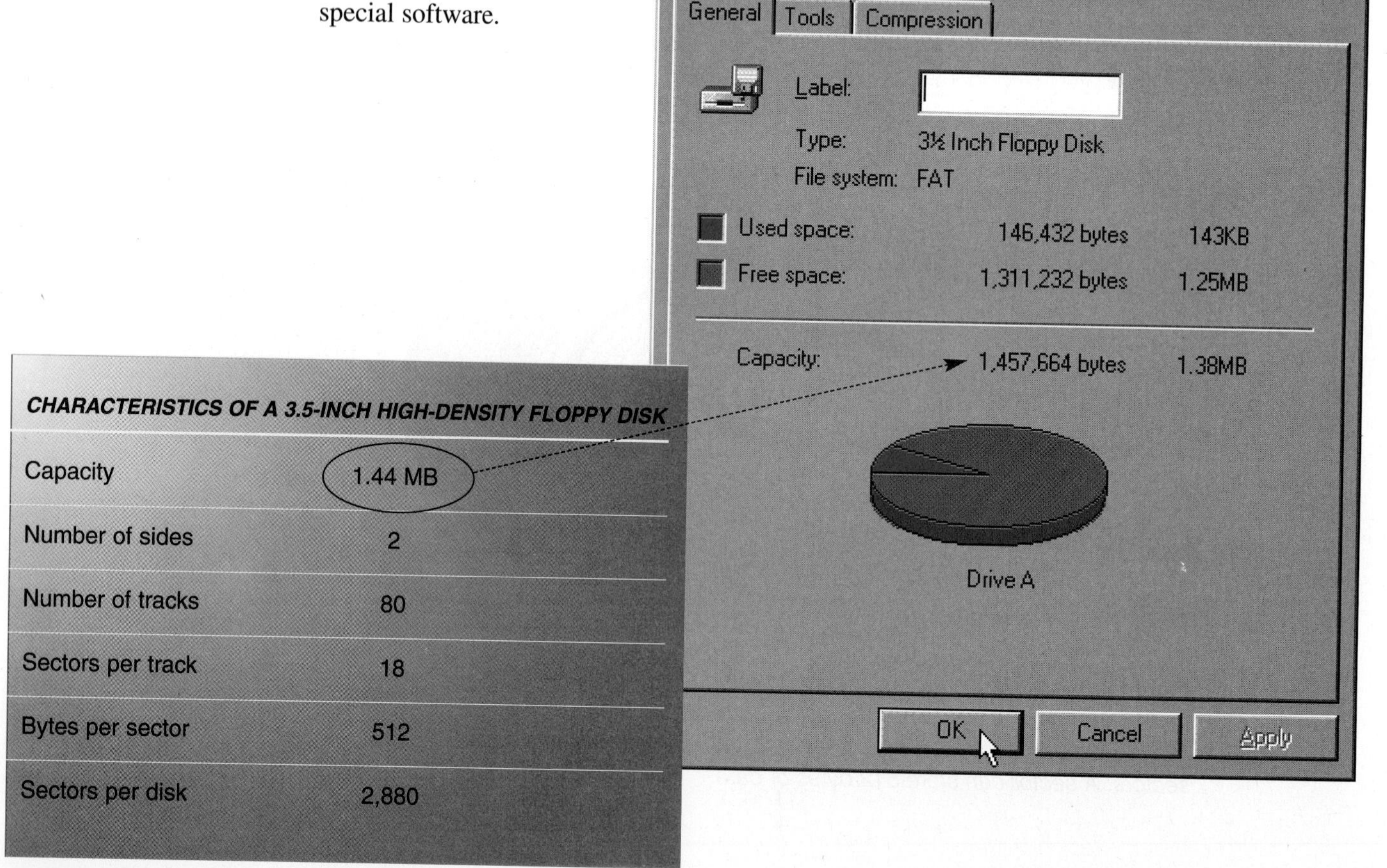

CHARACTERISTICS OF A 3.5-INCH HIGH-DENSITY FLOPPY DISK	
Capacity	1.44 MB
Number of sides	2
Number of tracks	80
Sectors per track	18
Bytes per sector	512
Sectors per disk	2,880

Figure 6-7 *Most of today's personal computers use high-density disks.*

Figure 6-8 To protect data from being erased accidentally, floppy disks have a write-protect notch. By sliding a small tab, you can either cover or expose the notch.

Floppy Disk Drives

As noted, a **floppy disk drive (FDD)** is a device that can read from and write on a floppy disk. Desktop personal computers usually have a floppy disk drive installed inside the system unit. Many laptop computers have removable floppy disk drives that can be replaced with other types of drives or devices, or they use an external floppy disk drive that plugs into the laptop (see Figure 6-5 on page 6.4).

If a computer has one floppy disk drive, the drive usually is designated *drive A*; if the computer has two floppy disk drives, the second one usually is designated *drive B*.

To read from or write on a floppy disk, a floppy disk drive must support that floppy disk's density. That is, to use a high-density floppy disk, you must have a high-density floppy disk drive. Floppy disk drives are **downward compatible**, which means they recognize and can use earlier media. Floppy disk drives are not **upward compatible**, however, which means they cannot recognize

Most floppy disks are preformatted by the disk's manufacturer. If you must format a floppy disk yourself, you do so by issuing a formatting command to the operating system (Figure 6-9). Because PC-compatible computers using the Windows operating system format floppy disks differently than Macintosh computers, a Macintosh computer cannot use a PC formatted floppy disk without special equipment or software. A disk drive such as the Apple Macintosh SuperDrive, however, can read from and write on both Macintosh and PC formatted floppy disks.

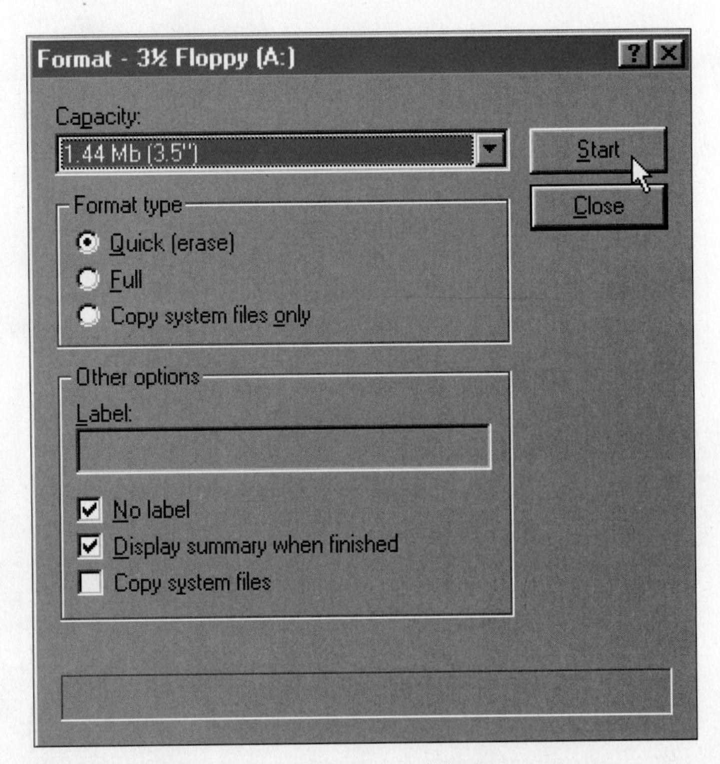

Figure 6-9 The Format dialog box in Windows contains options to format a floppy disk.

newer media. For example, a lower-density floppy disk drive cannot read from or write on a high-density floppy disk.

On any 3.5-inch floppy disk, a piece of metal called the **shutter** covers an opening in the rigid plastic shell. When you insert a floppy disk into a floppy disk drive, the drive slides the shutter to the side to expose a portion of both sides of the floppy disk's recording surface.

The **read/write head** is the mechanism that actually reads items from or writes items on the floppy disk. Figure 6-10 illustrates the steps for reading from and writing on a floppy disk.

The average access time for current floppy disk drives to locate an item on the disk is 84 ms, or approximately 1/12 of a second.

On the front of most floppy disk drives is a light emitting diode (LED) that lights up when the drive is accessing the floppy disk. You should not remove a floppy disk when the floppy disk drive is accessing the disk.

Sometimes, a floppy disk drive will malfunction when it is attempting to access a floppy disk and will display an error message on the computer's monitor screen. If the same error occurs with multiple floppy disks, the read/write heads in the floppy disk drive may have a buildup of dust or dirt. In this case, you can try cleaning the read/write heads using a floppy disk cleaning kit.

Figure 6-10
HOW A FLOPPY DISK DRIVE WORKS

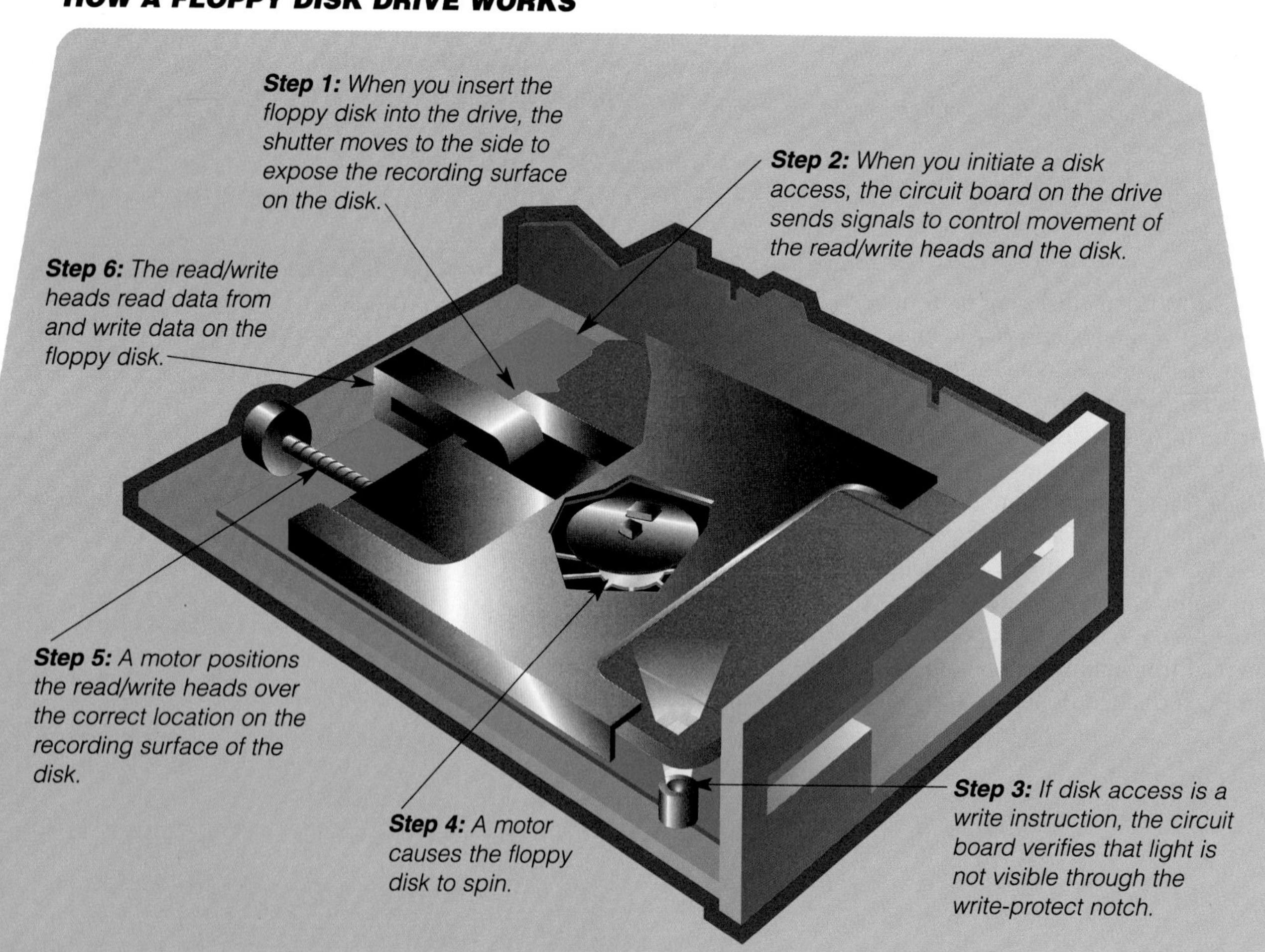

Step 1: When you insert the floppy disk into the drive, the shutter moves to the side to expose the recording surface on the disk.
Step 2: When you initiate a disk access, the circuit board on the drive sends signals to control movement of the read/write heads and the disk.
Step 6: The read/write heads read data from and write data on the floppy disk.
Step 5: A motor positions the read/write heads over the correct location on the recording surface of the disk.
Step 4: A motor causes the floppy disk to spin.
Step 3: If disk access is a write instruction, the circuit board verifies that light is not visible through the write-protect notch.

Care of Floppy Disks

With reasonable care, floppy disks can last at least seven years – providing an inexpensive and reliable form of storage. When handling a floppy disk, you should avoid exposing it to heat, cold, magnetic fields, and contaminants such as dust, smoke, or salt air. Exposure to any of these elements could damage or destroy the data, instructions, and information stored on the floppy disk. Figure 6-11 outlines some guidelines for the proper care of floppy disks.

High-Capacity Floppy Disks

Several manufacturers have high-capacity floppy disk drives that use disks with capacities of 100 MB and greater. With these high-capacity disks, you can store large files containing graphics, audio, or video; transport a large number of files from one computer to another; or make a backup of all of your important files. A **backup** is a duplicate of a file, program, or disk that can be used if the original is lost, damaged or destroyed.

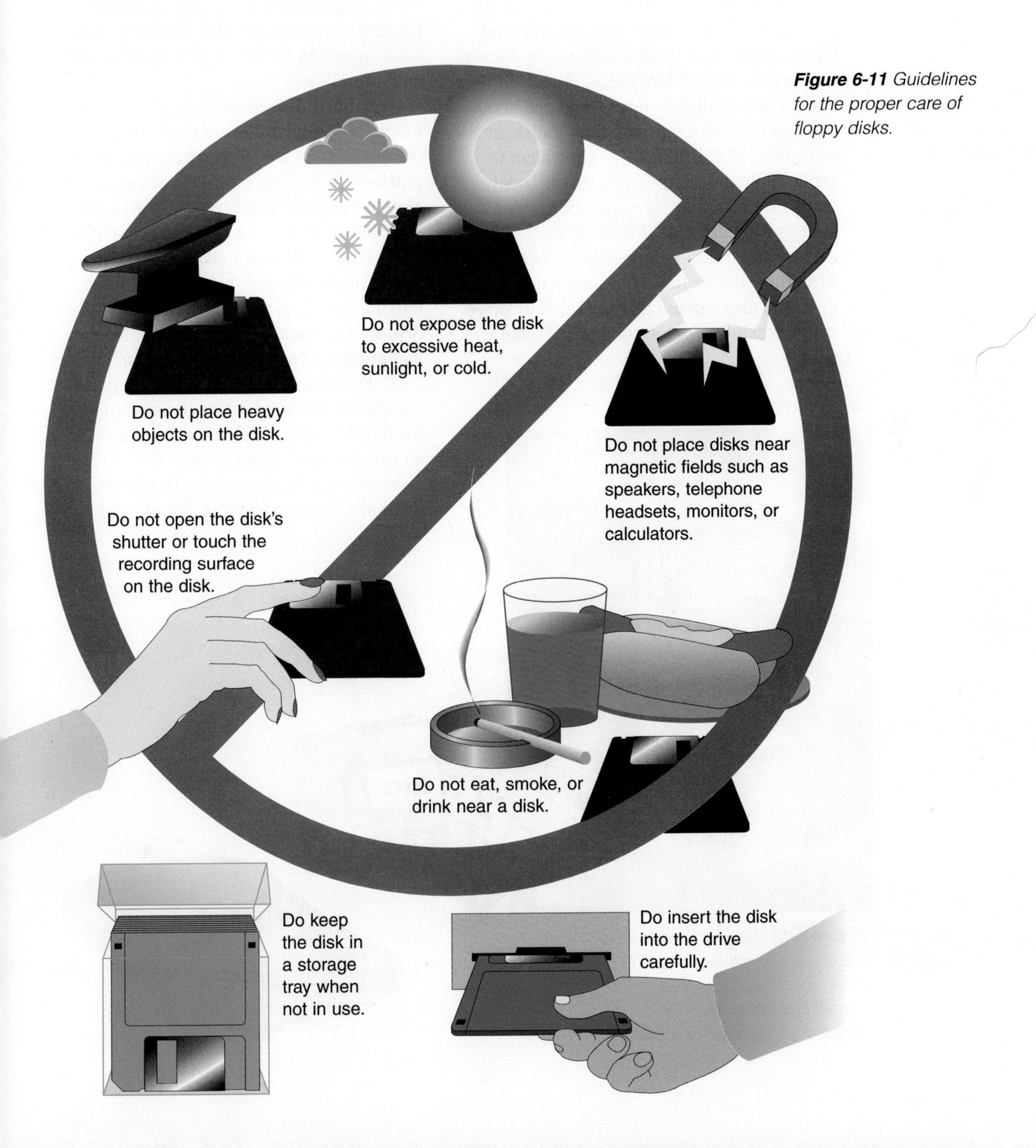

***Figure 6-11** Guidelines for the proper care of floppy disks.*

A **SuperDisk™ drive** is a high-capacity disk drive developed by Imation that reads from and writes on a 120 MB SuperDisk™ floppy disk. Sony Electronics Inc. has developed **HiFD™ (High-Capacity FD)**, a high-capacity floppy disk drive that reads from and writes on a 200 MB HiFD™ floppy disk. Both the SuperDisk™ drive and the HiFD™ drive are downward compatible; that is, they can read from and write on standard 3.5-inch floppy disks as well as their own high-capacity disks.

Another type of high-capacity disk drive is the Zip® drive. A **Zip® drive** is a special high-capacity disk drive developed by Iomega Corporation. Zip® drives use a 3.5-inch **Zip® disk**, which is slightly larger than and about twice as thick as a 3.5-inch floppy disk, and can store 250 MB of data — equivalent to about 70 high-density floppy disks. Today, many new computers are equipped with a built-in Zip® drive, so that using Zip® disks to store and transport large files is as easy as storing smaller files on a standard floppy disk. You also can use an external Zip® drive with a desktop or laptop computer (Figure 6-12).

WEB INFO

For more information on a Zip® drive, visit the Discovering Computers 2001 Chapter 6 WEB INFO page (**www.scsite.com/dc2001/ch6/webinfo.htm**) and click Zip® Drives.

HARD DISKS

When personal computers were introduced, software programs and their related files required small amounts of storage and fit easily on floppy disks. As software became more complex and included graphical user interfaces and multimedia, file sizes and storage requirements increased. Today, hard disks — which provide far larger storage capacities and much faster access times than floppy disks — are one of the primary media for storing software programs and files. Current personal computer hard disks can store from 4 to 50 GB of data, instructions, and information.

A **hard disk** usually consists of several inflexible, circular disks, called platters, on which items are stored electronically. A **platter** in a hard disk is made of aluminum, glass, or ceramic and is coated with a material that allows items to be magnetically recorded on its surface. On hard disks, the platters, the read/write heads, and the mechanism for moving the heads across the surface of the disk are enclosed in an airtight, sealed case that protects the platters from contamination.

The hard disk in most desktop personal computers is housed inside the system unit. Such hard disks, which are not portable, are considered **fixed disks** (Figure 6-13). Hard disks also can be removable. Removable hard disks are discussed later in this chapter.

Figure 6-12 *Many new computers are equipped with a built-in Zip® drive. External Zip® drives sometimes are attached to desktop and laptop computers.*

Characteristics of a Hard Disk

Like a floppy disk, a hard disk is a type of magnetic media that stores items using magnetic patterns. Hard disks also are read/write storage media; that is, you can both read from and write on a hard disk any number of times.

Hard disks undergo two formatting steps, and possibly a third process, called partitioning. The first format, called a **low-level format**, organizes both sides of each platter into tracks and sectors to define where items will be stored on the disk. Because a hard disk often has some bad sectors, the hard disk manufacturer usually performs the low-level format.

After low-level formatting is complete, the hard disk can be divided into separate areas called **partitions** by issuing a special operating system command. Each partition functions as if it were a separate hard disk drive. Partitioning often is performed to make hard disks more efficient (faster) or to allow you to install multiple operating systems on the same hard disk.

If a hard disk has only one partition, the hard disk usually is called, or designated, drive C. If the hard disk is divided into two partitions, the first partition is designated drive C and the second partition is designated drive D, and so on. Unless specifically requested by the consumer, most manufacturers define a single partition (drive C) on the hard disk.

After low-level formatting and partitioning, a **high-level format** command is issued through the operating system to define, among other items, the file allocation table (FAT) for each partition. Recall that the FAT is a table of information used to locate files on a disk. As with the low-level format, most hard disk manufacturers perform the high-level format for the consumer.

You can partition a hard disk yourself using special operating system commands. You then must issue a high-level format command for each partition.

WEB INFO

For more information on hard disks, visit the Discovering Computers 2001 Chapter 6 WEB INFO page (**www.scsite.com/dc2001/ch6/webinfo.htm**) and click Hard Disks.

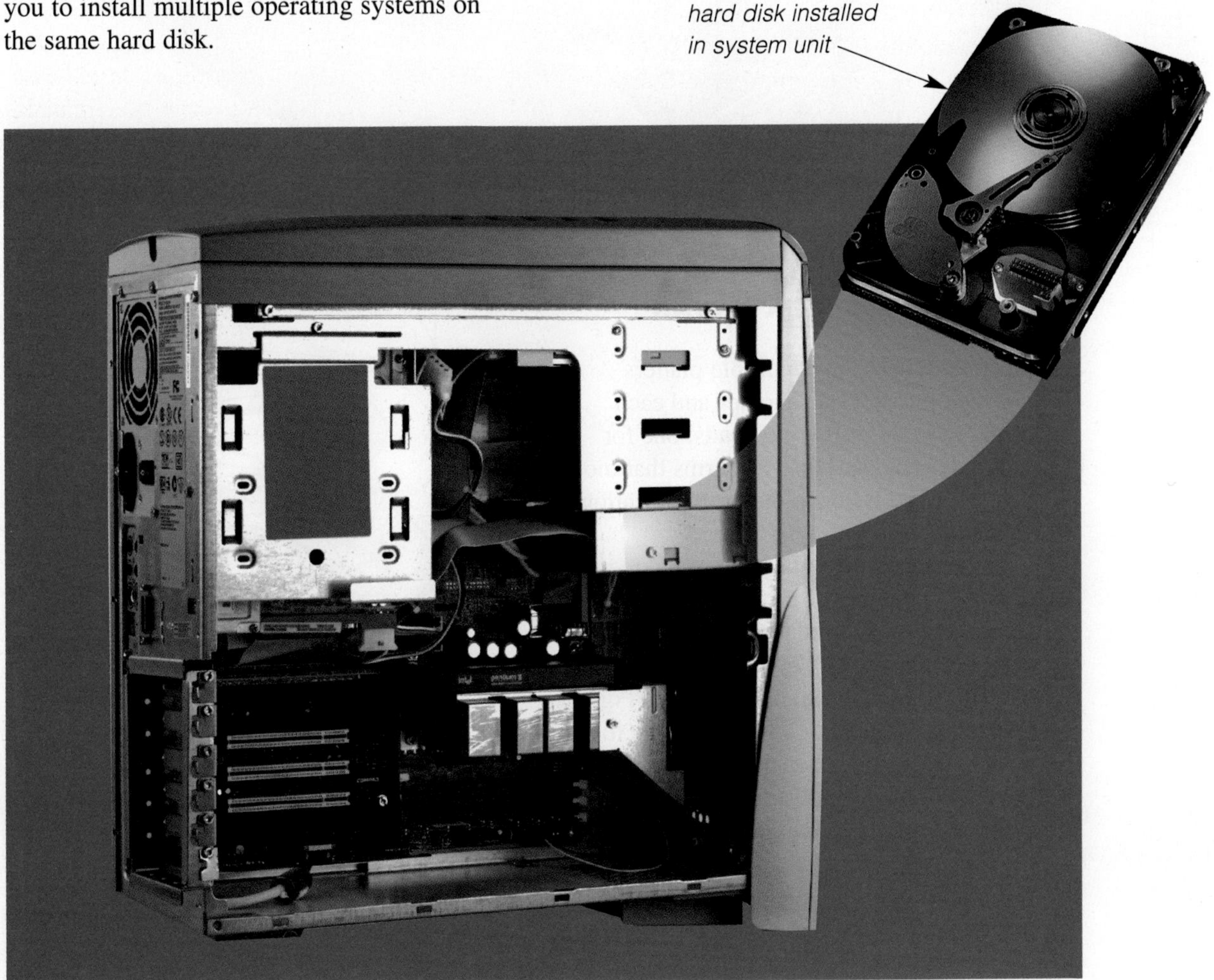

Figure 6-13 *The hard disk in a desktop personal computer normally is housed permanently inside the system unit; that is, it is not portable.*

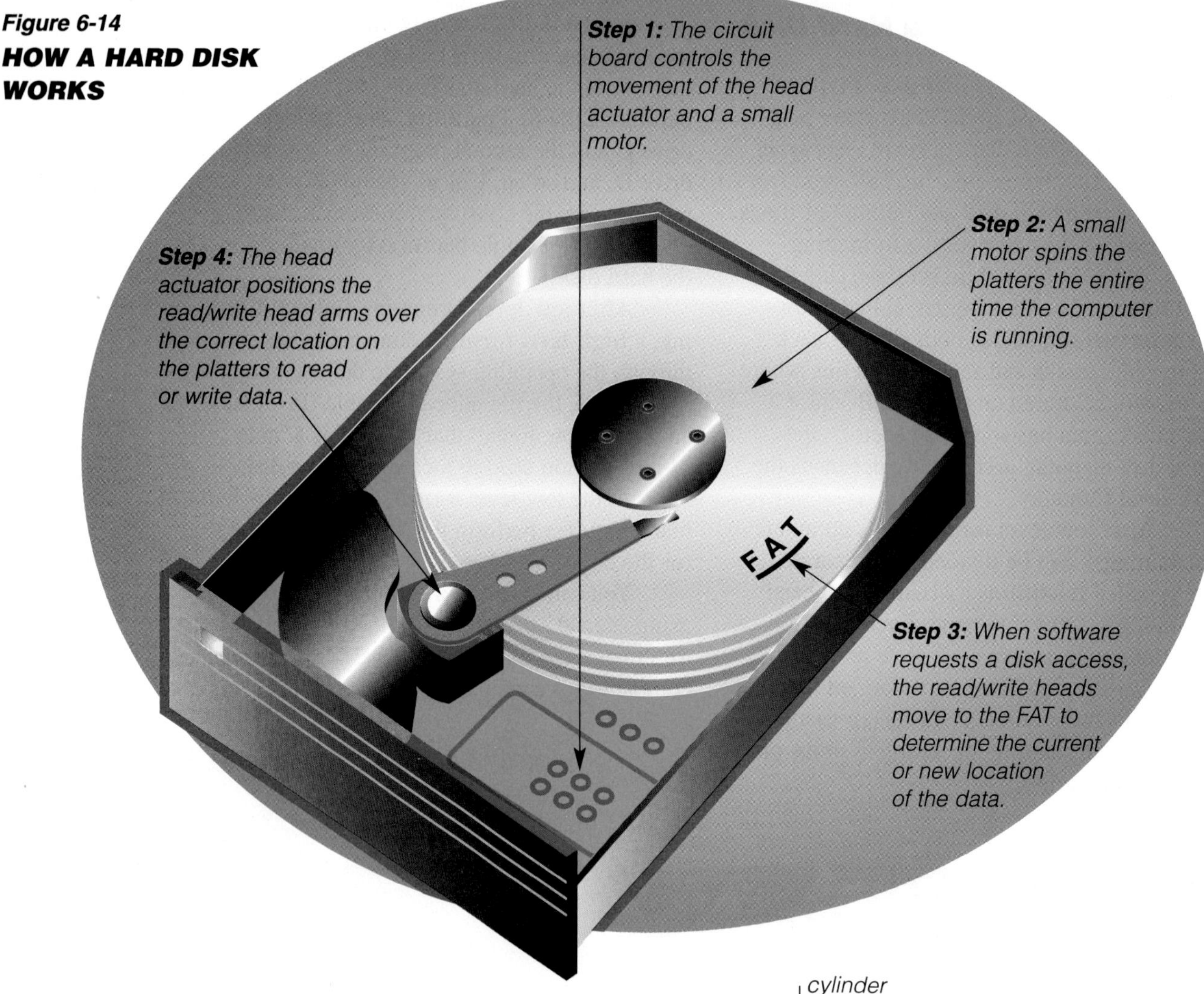

Step 1: The circuit board controls the movement of the head actuator and a small motor.
Step 2: A small motor spins the platters the entire time the computer is running.
Step 3: When software requests a disk access, the read/write heads move to the FAT to determine the current or new location of the data.
Step 4: The head actuator positions the read/write head arms over the correct location on the platters to read or write data.
FAT

Figure 6-14
HOW A HARD DISK WORKS

How a Hard Disk Works

Most hard disks have multiple platters stacked on top of one another and each platter has two read/write heads, one for each side. The hard disk has arms that move the read/write heads to the proper location on the platter (Figure 6-14).

Because of the stacked arrangement of the platters, the location of the read/write heads often is referred to by its cylinder instead of its track. A **cylinder** is the location of a single track through all platters (Figure 6-15). For example, if a hard disk has four platters (eight sides), each with 1,000 tracks, then it will have 1,000 cylinders with each cylinder consisting of eight tracks (two for each platter).

While your computer is running, the platters in the hard disk rotate at a high rate of speed, usually 5,400 to 7,200 revolutions per minute. The platters continue spinning until power is removed from the computer.

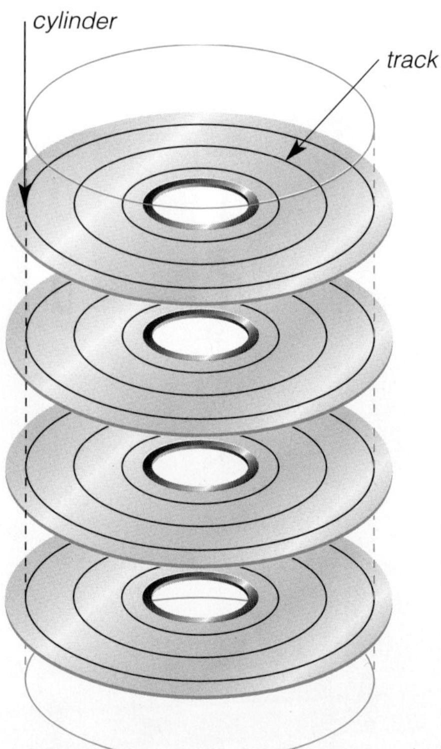

Figure 6-15 *A cylinder is the location of a single track through all platters on a hard disk.*

The spinning motion creates a cushion of air between the platter and its read/write head so the read/write head floats above the platter instead of making direct contact with the platter surface. The distance between the read/write head and the platter is approximately two millionths of an inch.

As shown in Figure 6-16, this close clearance leaves no room for any type of contamination. If contamination is introduced, the hard disk can have a head crash. A **head crash** occurs when a read/write head touches the surface of a platter, usually resulting in a loss of data or sometimes loss of the entire drive. Today's hard disks are built to withstand shocks and are sealed tightly to keep out contaminants, which means head crashes are less likely to occur.

Access time for today's hard disks ranges from five to eleven milliseconds. Access time for a hard disk is significantly faster than for a floppy disk for two reasons: (1) a hard disk spins much faster than a floppy disk and (2) a hard disk spins constantly, while a floppy disk starts spinning only when it receives a read or write command.

Some computers are able to improve the hard disk access time by using disk caching. **Disk cache** is a portion of memory that the CPU uses to store frequently accessed items (Figure 6-17). Disk cache works similarly to memory cache. When a program needs data, instructions, or information, the CPU checks the disk cache. If the item is located in disk cache, the CPU uses that item and completes the process. If the CPU does not find the requested item in the disk cache, then the CPU must wait for the hard disk drive to locate and transfer the item from the disk to the CPU.

Some disk caching systems also attempt to predict what data, instructions, or information might be needed next and place them into cache before they are requested. Because disk caching significantly improves disk access times, almost all new disk drives work with some amount of disk cache.

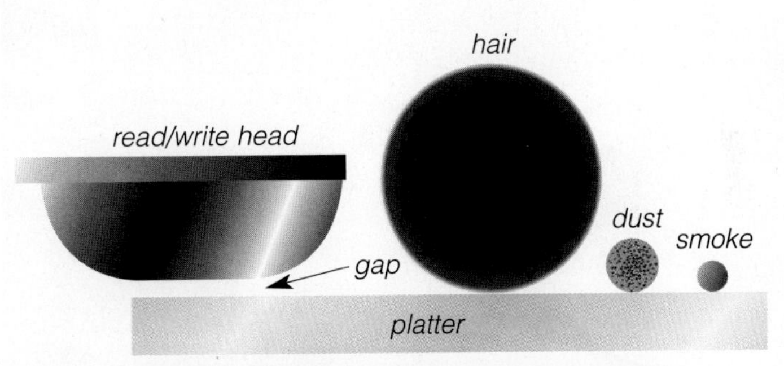

Figure 6-16 *Because the gap between a disk read/write head and the platter is so small, contaminants such as a smoke particle, dust particle, or human hair could render the drive unusable.*

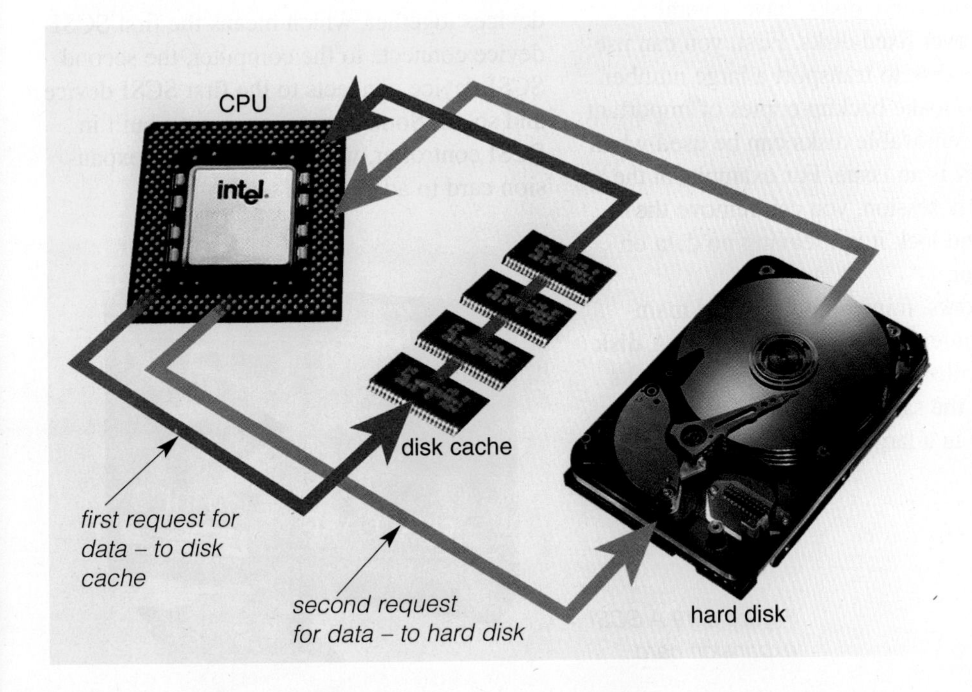

Figure 6-17 *When a program needs an item, the CPU checks the disk cache. If the item is located, the CPU uses it. If the CPU does not find the item in the disk cache, then the CPU must wait for the disk drive to locate and transfer the data from the disk.*

Removable Hard Disks

As noted, some hard disks are removable, which means they can be inserted and removed from a hard disk drive, much like a floppy disk. A **removable hard disk** or **disk cartridge** is a disk drive in which hard disks are enclosed in plastic or metal cases so they can be removed from the drive (Figure 6-18). A popular, reasonably priced, removable hard disk is the **Jaz® disk** by Iomega, which can store up to two gigabytes (GB) of data, instructions, and information.

Figure 6-18 *The Jaz® disk by Iomega is a removable hard disk with a capacity of up to 2 GB.*

Portable media, such as floppy disks and other removable disks, have several advantages over fixed disks. First, you can use a removable disk to transport a large number of files or to make backup copies of important files. Also, removable disks can be used when data security is an issue. For example, at the end of a work session, you can remove the hard disk and lock it up, leaving no data on the computer.

Networks, minicomputers, and mainframe computers often use disk packs. A **disk pack** is a collection of removable hard disks mounted in the same cabinet as the computer or enclosed in a large stand-alone cabinet.

Hard Disk Controllers

The flow of data, instructions, and information to and from a disk is managed by a special-purpose chip and its associated electronic circuits called the **disk controller**. Because a disk controller controls the transfer of items from the disk to the rest of the computer, it often is referred to as a type of interface.

A controller for a hard disk is called a **hard disk controller** (**HDC**). On a personal computer, the hard disk controller either is built into the disk drive or is a separate expansion card that plugs into an expansion slot. Two types of hard disk controllers for personal computers are IDE and SCSI.

IDE One of the most widely used controllers, or interfaces, for hard disks is the Integrated Drive Electronics (IDE) controller. **Integrated Drive Electronics** (**IDE**) controllers support up to four hard disks and can transfer data, instructions, and information to the disk at rates of up to 66 MB per second. IDE controllers also are referred to as **ATA**, short for the **AT Attachment**, that integrates the controller into the disk drive. Many versions of ATA exist, including ATA, ATA-4, Ultra ATA, Ultra DMA, and ATA/66.

SCSI **Small computer system interface**, or **SCSI**, (pronounced scuzzy) controllers can support multiple disk drives, as well as other peripherals such as scanners high-capacity disk drives, CD-ROM/DVD-ROM drives, tape drives, and printers. When using SCSI devices, you can daisy chain as many as 30 devices together, which means the first SCSI device connects to the computer, the second SCSI device connects to the first SCSI device, and so on. Some computers have a built in SCSI controller, while others use an expansion card to add a SCSI controller

Figure 6-19 *A SCSI expansion card.*

(Figure 6-19). SCSI controllers are faster than EIDE controllers, providing up to 160 MB per second transfer rates. SCSI controllers, however, are more expensive than EIDE controllers. As with ATA, many versions of SCSI exist, including SCSI-3, Wide SCSI, Fast SCSI, and Ultra2 SCSI.

RAID

For networks and other applications that depend on reliable data access, it is crucial that the data is available when a user attempts to access it. For these applications, some manufacturers developed a type of hard disk system that connects several smaller disks into a single unit that acts like a single large hard disk. A group of two or more integrated hard disks is called a **RAID** (**redundant array of independent disks**). Although quite expensive, RAID is more reliable than traditional disks and thus often is used with network and Internet servers (Figure 6-20).

Reliability is improved with RAID through the duplication of data, instructions, and information. This duplication is implemented in different ways, depending on the RAID storage design, or level, used. (These levels are not hierarchical; that is, higher levels are not necessarily better than lower levels.) The simplest RAID storage design is **level 1**, called **mirroring**, which has one backup disk for each disk (Figure 6-21a). A level 1 configuration enhances system reliability because, if a drive should fail, a duplicate of the requested item is available elsewhere within the array of disks.

Levels beyond level 1 use a technique called **striping**, which splits data, instructions, and information across multiple disks in the array (Figure 6-21b). Striping improves disk access times, but does not offer data duplication. For this reason, some RAID levels combine both mirroring and striping.

Figure 6-20 A group of two or more integrated hard disks, called a RAID (redundant array of independent disks), often is used with network servers.

WEB INFO

For more information on RAID, visit the Discovering Computers 2001 Chapter 6 WEB INFO page (**www.scsite.com/dc2001/ch6/webinfo.htm**) and click RAID.

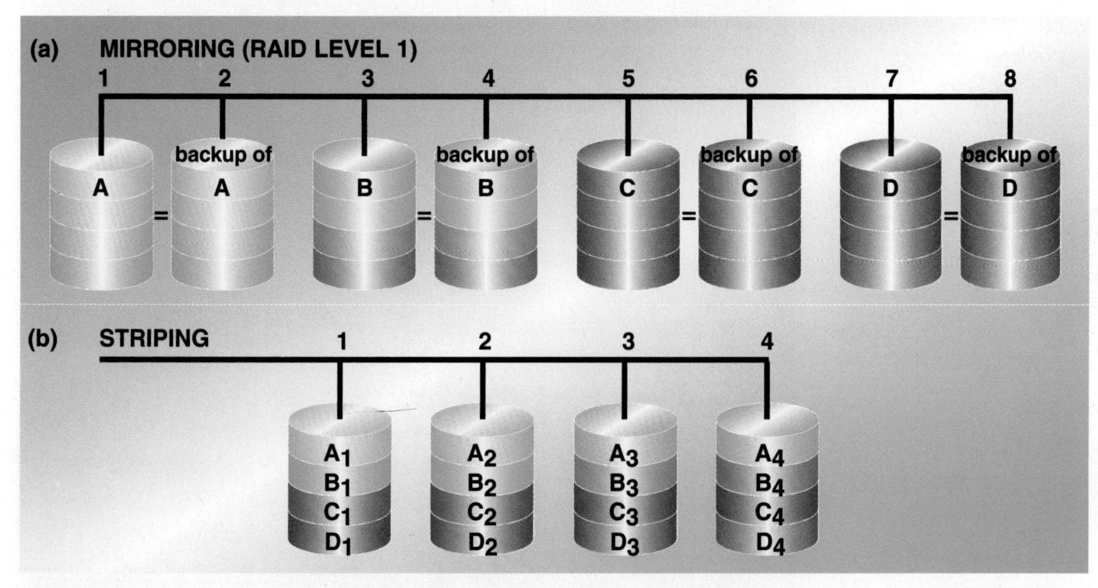

Figure 6-21 In RAID level 1, called mirroring, a backup disk exists for each disk. Higher RAID levels use striping; that is, portions of each disk are placed on multiple disks.

Maintaining Data Stored on a Hard Disk

Most manufacturers guarantee their hard disks to last somewhere between three and five years, although many last much longer with proper care. To prevent the loss of items stored on a hard disk, you should perform preventative maintenance such as defragmenting or scanning the disk for errors. As outlined in the table in Figure 6-22, operating systems such as Windows provide many maintenance and monitoring utilities. These and other utilities are discussed in more depth in Chapter 8.

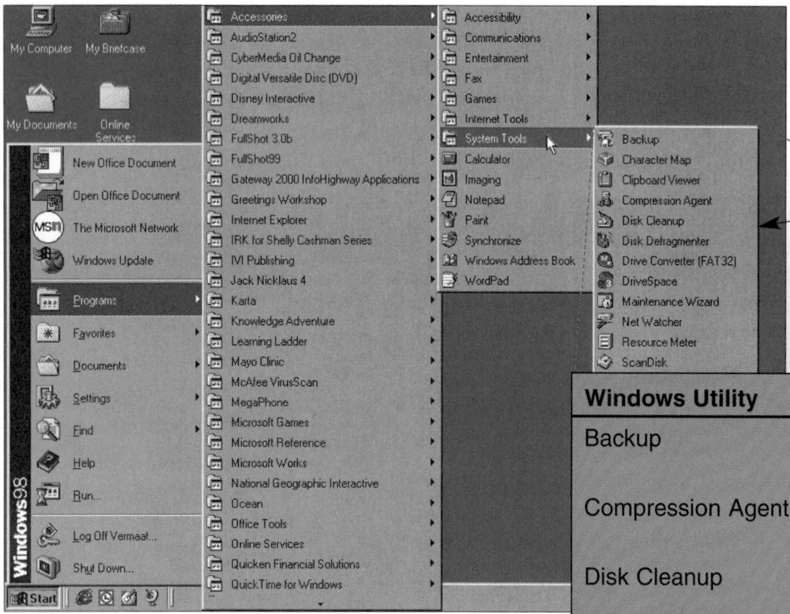

Windows Utility	Function
Backup	Creates a copy of files on a hard disk in case the original is damaged or destroyed.
Compression Agent	Recompresses files according to settings in DriveSpace (see below).
Disk Cleanup	Frees up space on a hard disk by listing files that can be deleted safely.
Disk Defragmenter	Reorganizes files and unused space on a hard disk so programs run faster.
Drive Converter (FAT 32)	Improves the FAT method of storing data, which frees up hard disk space and makes programs load faster.
DriveSpace	Compresses a hard disk or floppy disk to create free space on the disk.
Maintenance Wizard	Runs utilities that optimize your computer's performance.
Net Watcher	Monitors users and disk/file usage when computers are networked.
Resource Meter	Monitors system, user, and graphics resources being used by programs.
ScanDisk	Detects errors on a disk and then repairs the damaged areas.
Scheduled Tasks	Automatically runs a utility at a specified time.
System Monitor	Monitors disk access, the processor, memory, and network usage.

***Figure 6-22** Windows provides many maintenance and monitoring utilities on the System Tools submenu. The table in this figure briefly describes some of these utilities.*

COMPACT DISCS

In the past, when you purchased off-the-shelf software, you received one or more floppy disks that contained the files needed to install or run the software program. As software programs became more and more complex, the number of floppy disks required to store the programs increased, sometimes exceeding thirty disks. These more complex programs required a larger storage medium, which is why many of today's software programs are distributed on compact discs.

A **compact disc** (**CD**) is a flat, round, portable, metal storage medium that usually is 4.75 inches in diameter and less than one-twentieth of an inch thick (Figure 6-23). Compact disks store items such as data, instructions, and information by using

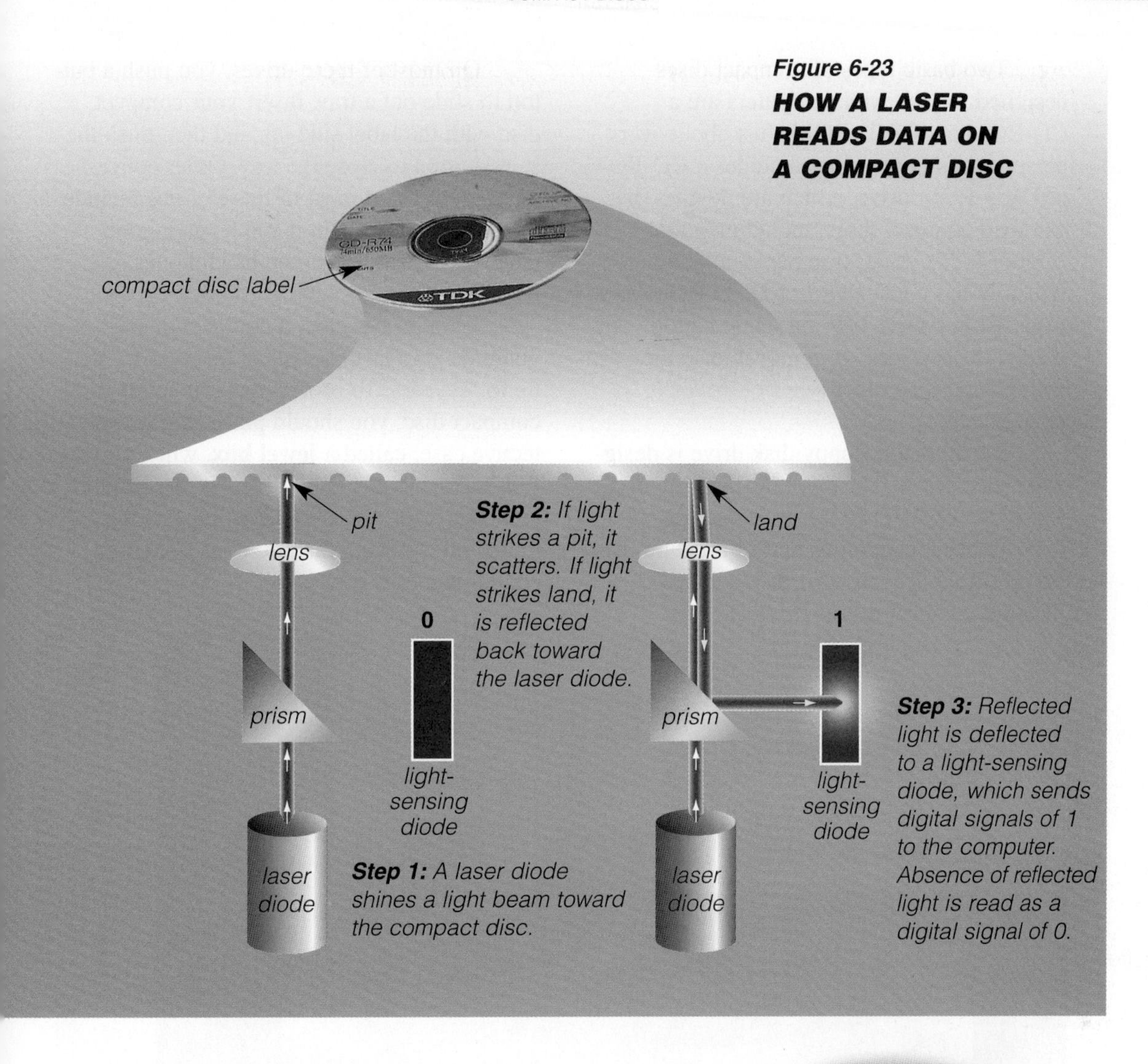
Figure 6-23
HOW A LASER READS DATA ON A COMPACT DISC
compact disc label
pit
land
lens
lens
0
1
Step 2: If light strikes a pit, it scatters. If light strikes land, it is reflected back toward the laser diode.
prism
prism
light-sensing diode
light-sensing diode
Step 3: Reflected light is deflected to a light-sensing diode, which sends digital signals of 1 to the computer. Absence of reflected light is read as a digital signal of 0.
laser diode
laser diode
Step 1: A laser diode shines a light beam toward the compact disc.

microscopic pits (indentations) and land (flat areas) that are in the middle layer of the disc. (Most manufacturers place a silk-screened label on the top layer of the disc so you can identify it.) A high-powered laser light creates the pits. A lower-powered laser light reads items from the compact disc by reflecting light through the bottom of the disc, which usually is either solid gold or silver in color. The reflected light is converted into a series of bits that the computer can process. Land causes light to reflect, which is read as binary digit 1. Pits absorb the light; this absence of light is read as binary digit 0.

A compact disc stores items in a single track that spirals from the center of the disc to the edge of the disc. As with a hard disk, this single track is divided into evenly sized sectors in which items are stored (Figure 6-24).

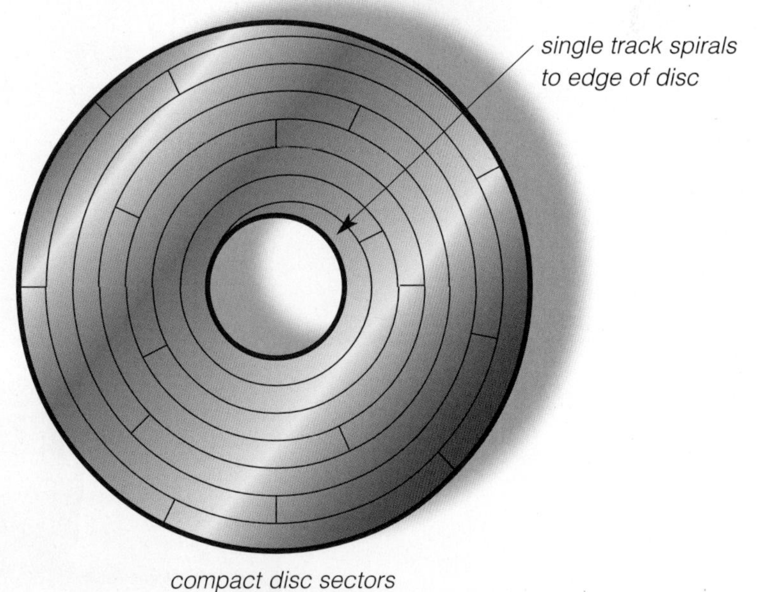
single track spirals to edge of disc
compact disc sectors

Figure 6-24 *The data on a compact disc is stored in a single track that spirals from the center of the disc to the edge of the disc; this track is divided into evenly sized sectors.*

Two basic types of compact discs designed for use with computers are a CD-ROM and DVD-ROM. Just about every personal computer today includes a CD-ROM or DVD-ROM drive, which are devices that can read compact discs, including audio CDs (Figure 6-25). A desktop personal computer typically has a CD-ROM or DVD-ROM drive installed in a drive bay; on many laptop computers, these drives are removable so they can be replaced with other types of drives or devices.

Recall that a floppy disk drive is designated as drive A. The drive designation of a CD-ROM or DVD-ROM drive usually follows alphabetically after that of the hard disk. For example, if your hard disk is drive C, then the compact disc is drive D.

On most of these drives, you push a button to slide out a tray, insert your compact disc with the label side up, and then push the same button to close the tray. Other convenient features on most of these drives include a volume control button and a headphone jack so you can use stereo headphones to listen to audio.

With proper care, a compact disc is guaranteed to last five years, but could last up to 50 years. To protect data on any type of compact disc, you should place it in its protective case, called a **jewel box**, when you are finished using it (Figure 6-26). When handling compact discs, you should avoid stacking them and exposing them to heat, cold, and contaminants.

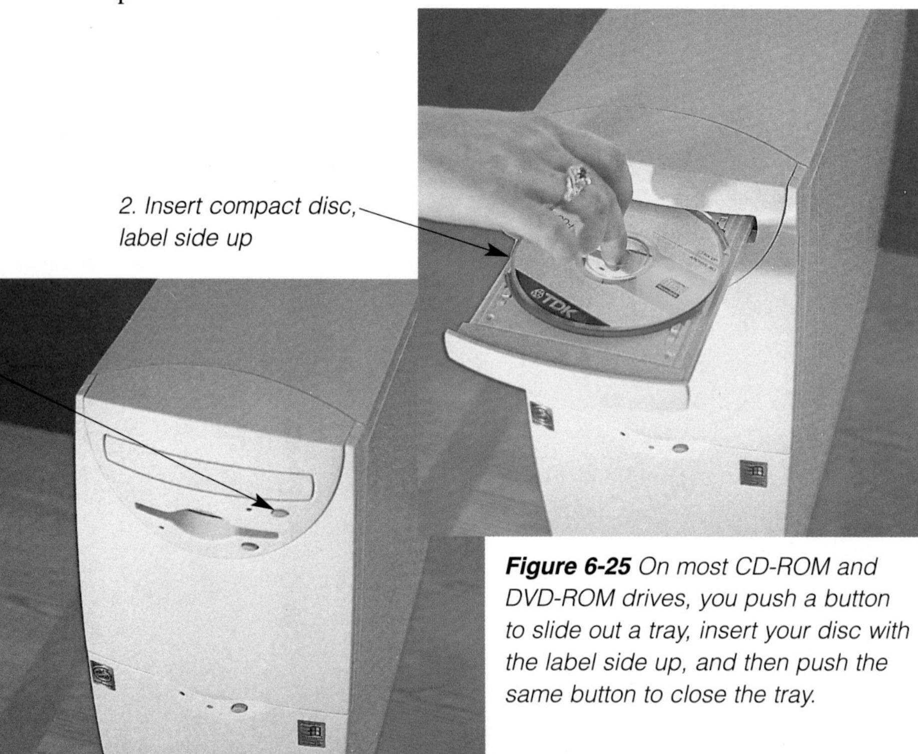

Figure 6-25 *On most CD-ROM and DVD-ROM drives, you push a button to slide out a tray, insert your disc with the label side up, and then push the same button to close the tray.*

Figure 6-26 *To protect data on a compact disc, you should place it in a jewel box when you are finished using it.*

Figure 6-27 outlines some guidelines for the proper care of compact discs. You can clean the bottom surface of a compact disc with a soft cloth and warm water or a specialized CD cleaning kit, but never clean the label side because this may destroy the data. You also can repair scratches on the bottom surface with a specialized CD repair kit.

Compact discs are available in a variety of formats. The following sections discuss two basic types used with personal computers: CD-ROMs and DVD-ROMs.

CD-ROMs

A **CD-ROM** (pronounced SEE-DEE-rom, which is an abbreviation for **compact disc read-only memory**) is a silver-colored compact disc that uses the same laser technology as audio CDs for recording music. Unlike an audio CD, a CD-ROM can contain text, graphics, and video, as well as sound. The contents of standard CD-ROMs are written, or **recorded**, by the manufacturer and only can be read and used. That is, they cannot be erased or modified — hence, the name read-only.

WEB INFO

For more information on CD-ROMs, visit the Discovering Computers 2001 Chapter 6 WEB INFO page (**www.scsite.com/dc2001/ch6/webinfo.htm**) and click CD-ROMs.

Figure 6-27 *Guidelines for the proper care of compact discs.*

For a computer to read items on a CD-ROM, you must place it into a **CD-ROM drive** or **CD-ROM player**. Because audio CDs and CD-ROMs use the same laser technology, you also can use your CD-ROM drive to listen to an audio CD while working on your computer.

A CD-ROM can hold up to 700 MB of data, instructions, and information, or about 450 times that which can be stored on a high-density 3.5-inch floppy disk. Because CD-ROMs have such high storage capacities, they are used to store and distribute today's complex software (Figure 6-28). Some programs even require that the disc be in the drive each time you use the program.

Figure 6-28a (encyclopedia)

Figure 6-28b (word processing)

Figure 6-28c (flight simulator)

Figure 6-28 *CD-ROMs are used to store and distribute multimedia and other complex software.*

CD-ROM DRIVE SPEED The speed of a CD-ROM drive is extremely important when viewing animation or video such as those found in multimedia encyclopedias and games. A slower CD-ROM drive will result in choppy images or sound. A CD-ROM drive's speed is measured by its **data transfer rate**, which is the time it takes the drive to transmit data, instructions, and information from the CD-ROM to another device.

The original CD-ROM drive was a single-speed drive with a data transfer rate of 150 KB per second. All subsequent CD-ROM drives have been measured relative to this first CD-ROM drive and use an X to denote the original transfer rate of 150 KB per second. For example, a 40X CD-ROM drive has a data transfer rate of 6,000 (16 × 150) KB per second or 6 MB per second. Current CD-ROM drives have speeds ranging from 40X to 75X. The higher the number, the faster the CD-ROM drive, which results in smoother playback of images and sounds. Faster CD-ROM drives, however, are more expensive than slower drives.

PHOTOCD, CD-R, AND CD-RW Three variations of the standard CD-ROM are the PhotoCD, the recordable CD, and the rewriteable CD. These CD-ROMs typically are **multisession**, which means additional data, instructions, and information can be written to the disc at a later time. (Most standard CD-ROMs are called **single-session** because all items must be written to the disc at the time it is manufactured.)

- Based on a file format developed by Eastman Kodak, a **PhotoCD** is a compact disc that only contains digital photographic images saved in the PhotoCD format. You can purchase PhotoCDs that already contain pictures or you can have your own pictures or negatives recorded on a PhotoCD so that you have digital versions of your photographs. The images on a PhotoCD can be printed, faxed, sent via electronic mail, included in another document, or posted to a Web site for everyone to see. Many film developers offer this service, often called a **Picture CD** for the consumer, when you drop off film to be developed; that is, in addition to printed photographs and negatives, you also receive a CD containing your pictures (Figure 6-29).

WEB INFO

For more information on PhotoCDs, visit the Discovering Computers 2001 Chapter 6 WEB INFO page (**www.scsite.com/dc2001/ch6/webinfo.htm**) and click PhotoCDs.

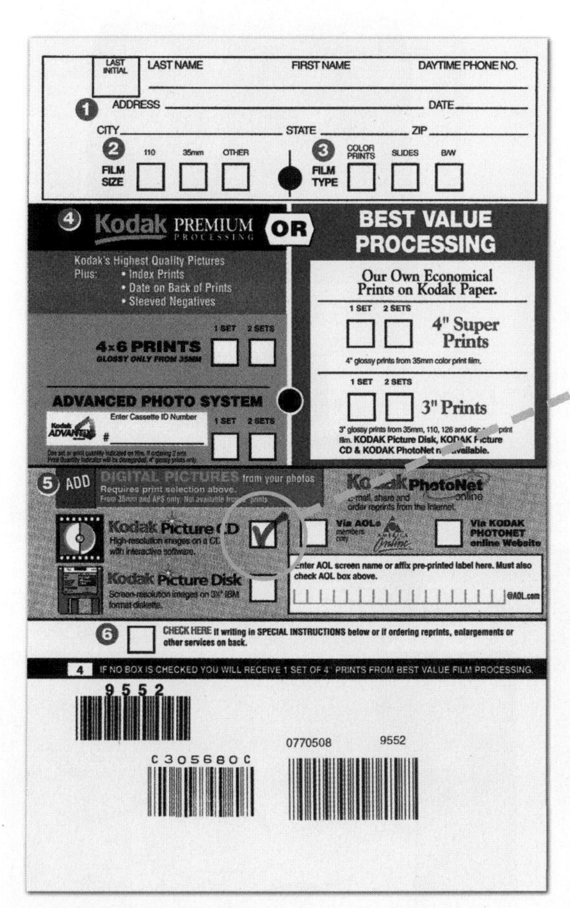
LAST INITIAL LAST NAME FIRST NAME DAYTIME PHONE NO.
1 ADDRESS DATE
CITY STATE ZIP
2 110 35mm OTHER FILM SIZE
3 COLOR PRINTS SLIDES B/W FILM TYPE
4 Kodak PREMIUM PROCESSING OR BEST VALUE PROCESSING
Kodak's Highest Quality Pictures Plus: • Index Prints • Date on Back of Prints • Sleeved Negatives
Our Own Economical Prints on Kodak Paper.
4x6 PRINTS 1 SET 2 SETS
4" Super Prints 1 SET 2 SETS
ADVANCED PHOTO SYSTEM
3" Prints 1 SET 2 SETS
5 ADD DIGITAL PICTURES from your photos
Kodak PhotoNet online
Kodak Picture CD
Via AOL
Via KODAK PHOTONET online Website
Kodak Picture Disk
Enter AOL screen name or affix pre-printed label here. Must also check AOL box above.
6 CHECK HERE if writing in SPECIAL INSTRUCTIONS below or if ordering reprints, enlargements or other services on back.
4 IF NO BOX IS CHECKED YOU WILL RECEIVE 1 SET OF 4" PRINTS FROM BEST VALUE FILM PROCESSING.
9552
C305680C
0770508 9552

Figure 6-29 *Many film developers offer a PhotoCD service when you drop off film to be developed.*

WEB INFO

For more information on DVD-ROMs, visit the Discovering Computers 2001 Chapter 6 WEB INFO page (**www.scsite.com/dc2001/ch6/webinfo.htm**) and click DVD-ROMs.

- **CD-R (compact disc-recordable)** is a technology that allows you to write on a compact disc using your own computer. Whereas the disc's manufacturer records the data, instructions, and information on a standard CD-ROM, you record your own items such as text, graphics, and audio, onto a CD-R (compact disc-recordable). You can write on the disc in stages – writing on part of it one time and writing on another part at a later time. You can, however, write on each part only once and you cannot erase it. Once you have recorded the CD-R, you can read from it as many times as you wish. In order to write on a CD-R, you must have CD-R software and a CD-R drive. A **CD-R drive** can read and write both audio CDs and standard CD-ROMs with read speeds of up to 24X and write speeds of up to 8X. Manufacturers often list the write speed first, for example, as 8/24. While CD-R drives are somewhat more expensive than standard CD-ROM drives, their price continues to drop and many computers today are equipped with CD-R drives.

- A **CD-RW (compact disc-rewritable)** is an erasable disc that you can write on multiple times. CD-RW overcomes one of the disadvantages of CD-R disks — that you can write on them only once. With CD-RW, the disc acts like a floppy or hard disk, allowing you to write and rewrite data, instructions, and information onto it multiple times. To write on a CD-RW disc, you must have CD-RW software and a **CD-RW drive**. The read speed of these drives is up to 32X, write speed up to 8X, and rewrite speed up to 4X. One problem with CD-RW is that these discs cannot be read by all CD-ROM drives.

DVD-ROMs

Although CD-ROMs have huge storage capacities, even a CD-ROM is not large enough for many of today's complex programs. Some software, for example, is sold on five or more CD-ROMs. To meet these tremendous storage requirements, some software moved from CD-ROMs to the larger DVD-ROM format — a technology that can be used to store video items, such as motion pictures (Figure 6-30). A **DVD-ROM (digital video disc-ROM)** is an extremely high capacity compact disc capable of storing from 4.7 GB to 17 GB — more than enough to hold a telephone book containing every resident in the United States. Not only is the storage capacity of a DVD-ROM greater than a CD-ROM, a DVD-ROM's quality also far surpasses that of a CD-ROM. In order to read a DVD-ROM, you must have a **DVD-ROM drive** or DVD player. These drives have speeds up to 40X and can read most types of CD-ROMs.

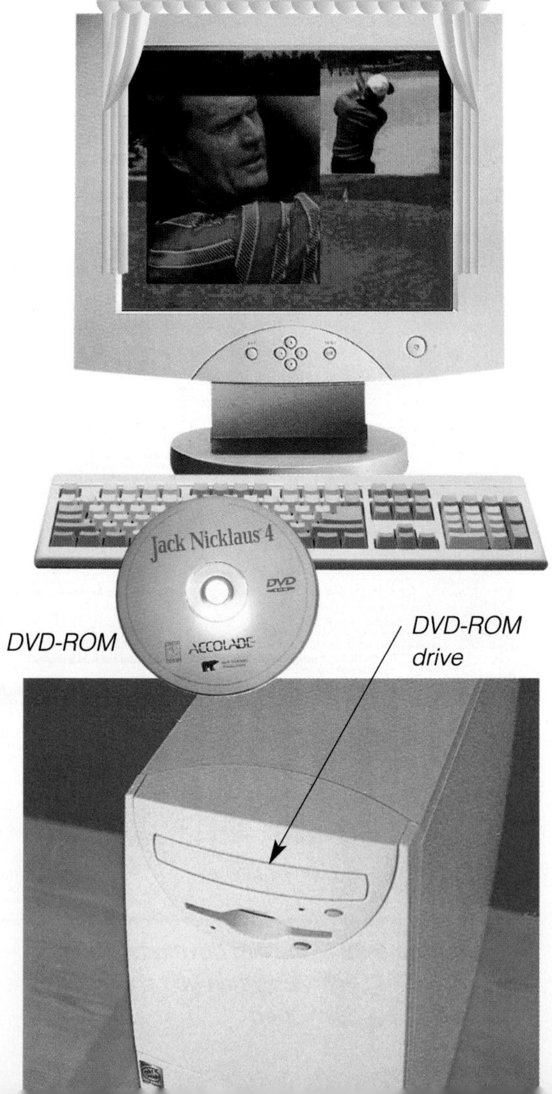

Figure 6-30 *A DVD-ROM is an extremely high capacity compact disc capable of storing 4.7 GB to 17 GB.*

At a glance, a DVD-ROM looks just like a CD-ROM. Although the size and shape are similar, a DVD-ROM stores data, instructions, and information in a slightly different manner and thus achieves a higher storage capacity. A DVD-ROM uses one of three storage techniques. The first technique involves making the disc more dense by packing the pits closer together. A second technique involves using two layers of pits. For this technique to work, the lower layer of pits is semitransparent so the laser can read through it to the upper layer. This technique doubles the capacity of the disc. Finally, some DVD-ROMs are double-sided, which means you must remove the DVD-ROM and turn it over to read the other side. The storage capacities of various types of DVD-ROMs are shown in the table in Figure 6-31.

DVD VARIATIONS DVDs are available in a variety of formats, one of which stores video such as digital movies and/or audio such as music. To view a movie that has been stored on a DVD, you insert the DVD into a DVD player that is connected to your television set, somewhat like the VCR you use to play VHS tapes. Movies on DVD have near-studio-quality video, which far surpasses VHS tapes. When music is stored on a DVD, it includes surround sound and has a much better quality than audio CDs.

As with compact discs, you can obtain recordable and rewritable versions of DVD. With a **DVD-R (DVD-recordable)**, you can write once on it and read (play) it many times. DVD-R drives have a read speed of up to 32X and a write speed of up to 8X. With a rewritable version of a DVD, called **DVD-RAM**, you can erase and record on it multiple times. DVD-RAM drives typically can read DVD-ROM, DVD-R, and all CD-ROM media. As the cost of DVD technologies becomes more reasonable, many industry professionals expect that DVD eventually will replace compact discs.

TAPES

One of the first storage media used with mainframe computers was **magnetic tape**, a magnetically coated ribbon of plastic capable of storing large amounts of data and information at a low cost. Tape storage requires **sequential access**, which refers to reading or writing data consecutively. Like a music tape, you must forward or rewind the tape to a specific point to access a specific piece of data.

DVD-ROM STORAGE CAPACITIES

Number of sides	1	1	2	2
Number of layers	1	2	1	2
Storage capacity	4.7 GB	8.5 GB	9.4 GB	17 GB

Figure 6-31 *Storage capacities of DVD-ROMs.*

For example, to access item W, you must pass sequentially through points A through V. Floppy disks, hard disks, and compact discs all use **direct access**, or **random access**, which means you can locate a particular data item or file immediately, without having to move consecutively through items stored in front of the desired data item or file. Because sequential access is much slower than direct access, tapes are no longer used as a primary method of storage. Instead, tapes are used most often for long-term storage and backup.

Similar to a tape recorder, a **tape drive** is used to read from and write data and information onto a tape. Although older computers used reel-to-reel tape drives, today's tape drives use tape cartridges. A **tape cartridge** is a small, rectangular, plastic housing for tape (Figure 6-32). Tape cartridges containing one-quarter-inch wide tape are slightly larger than audiocassette tapes and frequently are used for personal computer backup.

WEB INFO

For more information on tapes, visit the Discovering Computers 2001 Chapter 6 WEB INFO page (**www.scsite.com/dc2001/ch6/webinfo.htm**) and click Tapes.

Figure 6-32 *Tape cartridges are an inexpensive, reliable form of storage often used for personal computer backup.*

Some personal computers have permanently mounted tape drives, while others have external units (Figure 6-33). On larger computers, tape cartridges are mounted in a separate cabinet (Figure 6-34).

Three common types of tape drives are QIC, DAT, and DLT, which is the fastest and most expensive of the three. The table in Figure 6-35 summarizes each of these tapes.

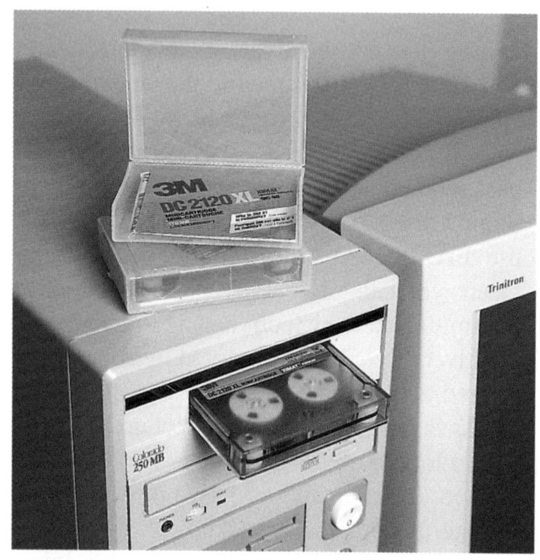

Figure 6-33 *Some personal computers have permanently mounted tape drives, while others have external units.*

POPULAR TYPES OF TAPES

Name	Abbreviation	Storage Capacity
Quarter-inch cartridge	QIC	40 MB to 5 GB
Digital audio tape	DAT	2 to 40 GB
Digital linear tape	DLT	20 to 80 GB

Figure 6-35 *Common types of tapes.*

PC CARDS

As discussed in Chapter 3, a **PC Card** is a thin, credit card-sized device that fits into a PC Card expansion slot on a laptop or other personal computer. Different types and sizes of PC Cards are used to add storage, additional memory, communications, and sound capabilities to a computer. PC Cards most often are used with laptops and other portable computers (Figure 6-36).

PC Cards are available in three types, which are designated Type I, Type II, and Type III. The thicker **Type III cards** are used to house hard disks and currently have storage capacities of more than 520 MB (Figure 6-37). The advantage of a PC Card hard disk is

Figure 6-34 *On larger computers, tape cartridges are mounted in a separate cabinet.*

Figure 6-36 *PC Cards normally are used with laptops and other portable computers.*

portability; that is, you easily can transport large amounts of data, instructions, and information from one machine to another. Type I and Type II cards, used to add memory or communications capabilities to a computer, are discussed in other chapters.

Some digital cameras also use a matchbook-size card, sometimes called a **picture card** or **compact flash card**, to store pictures, which are then transferred to a computer by inserting the card into a card reader or slot. These compact flash cards have storage capacities ranging from 2 MB to 256 MB.

Figure 6-37 *This Type III PC Card is a hard disk with 1 GB of storage space.*

OTHER TYPES OF STORAGE

Although the majority of data, instructions, and information are stored on floppy disk, hard disk, compact disc, tape, and PC Cards, other more specialized means for storing these items also are used. These include smart cards and microfilm and microfiche. Each of these media is discussed in the following sections.

Smart Cards

A **smart card**, which is similar in size to a credit card or ATM card, stores data on a thin microprocessor embedded in the card (Figure 6-38). Two types of smart cards exist: intelligent and memory. An **intelligent smart card** contains a CPU and has input, process, output, and storage capabilities. In contrast, a **memory card** has only storage capabilities. When the smart card is inserted into a specialized card reader, the information on the smart card is read and, if necessary, updated.

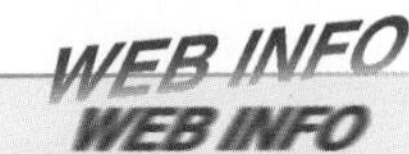

For more information on smart cards, visit the Discovering Computers 2001 Chapter 6 WEB INFO page (**www.scsite.com/dc2001/ch6/webinfo.htm**) and click Smart Cards.

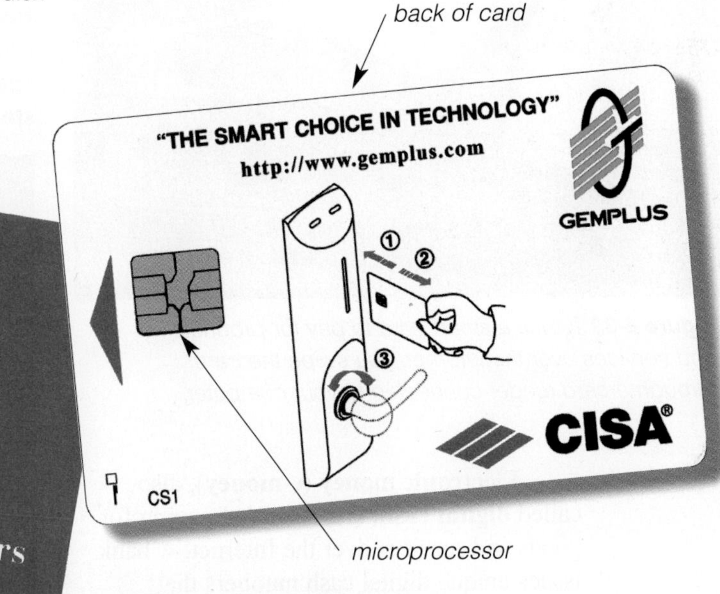

Figure 6-38 *Many hotels issue smart cards instead of keys to hotel guests. With the smart card, guests gain access to their rooms as well as hotel services such as cafeterias, swimming pools, lockers, and parking lots.*

WEB INFO

For more information on e-money, visit the Discovering Computers 2001 Chapter 6 WEB INFO page (**www.scsite.com/ dc2001/ch6/webinfo.htm**) and click E-money.

One popular use of smart cards is to store a prepaid dollar amount, as in a prepaid telephone calling card. You receive the card with a specific dollar amount stored in the microprocessor. Each time you use the card, the available amount of money is reduced. Using these cards provides convenience to the caller, eliminates the telephone company's need to collect coins from telephones, and reduces vandalism of pay telephones. Other uses of smart cards include storing patient records, vaccination data, and other health-care information; tracking information such as customer purchases or employee attendance; and storing a prepaid amount, such as electronic money.

Figure 6-39 *To use a smart card to pay for products and services over the Internet, you swipe the card through a card reader connected to your computer.*

Electronic money (**e-money**), also called **digital cash**, is a means of paying for goods and services over the Internet. A bank issues unique digital cash numbers that represent an amount of money. When you purchase digital cash, the amount of money is withdrawn from your bank account. One implementation of e-money places the digital cash on a smart card. To use the card, you swipe it through a card reader on your computer or one that is attached to your computer (Figure 6-39).

Microfilm and Microfiche

Microfilm and microfiche are used to store microscopic images of documents on roll or sheet film (Figure 6-40). **Microfilm** uses a 100- to 215-foot roll of film. **Microfiche** uses a small sheet of film, usually about four inches by six inches. The images are recorded onto the film using a device called a **computer output microfilm** (**COM**) **recorder**. The stored images are so small they can be read only with a microfilm or microfiche reader.

Applications of microfilm and microfiche are widespread. Libraries use these media to store back issues of newspapers, magazines, and genealogy records. Large organizations use microfilm and microfiche to archive inactive files. Banks, for example, use it to store transactions and cancelled checks, and the U.S. Army uses it to store personnel records. Using microfilm and microfiche provides a number of advantages: it greatly reduces the amount of paper firms must handle; it is inexpensive; and it has the longest life of any storage medium (Figure 6-41).

Enterprise Storage Systems and Data Warehouses

Because data is such an important asset to an organization, many storage sellers now offer enterprise storage systems. An **enterprise storage system** is a strategy that focuses on

Figure 6-40 *Microfilm and microfiche are used to stor microscopic images of documents on roll or sheet film*

the availability, protection, organization, and backup of storage in a company. The goal of an enterprise storage system is to consolidate storage so that operations run as efficiently as possible. Many software and hardware techniques are used to implement an enterprise storage system, from RAID to tape libraries to storage area networks. A **storage area network (SAN)** is a high-speed network that connects storage devices. Some companies manage an enterprise storage system in house, while others elect to offload all (or at least the backup) storage management to an outside organization or online service.

Another growing strategy, called a **data warehouse**, centralizes the computing environment, in which large megaservers store data, information, and programs, and less powerful client devices connect to the megaservers to access these items.

PUTTING IT ALL TOGETHER

Many factors influence the type of storage devices you should use: the amount of data, instructions, and information to be stored; the hardware and software in use; and the desired cost. The table in Figure 6-42 outlines several suggested storage devices for various types of computer users.

MEDIA LIFE EXPECTANCIES

Media Type	Guaranteed Life Expectancy	Potential Life Expectancy
Tape	2 to 5 years	20 years
Compact disc	5 years	50-100 years
Microfilm	100 years	200 years

Figure 6-41 Microfilm is the medium with the longest life.

SUGGESTED STORAGE DEVICES BY USER

USER	STORAGE DEVICE
Home User	• 3.5-inch high-density floppy disk drive • 250 MB Zip® drive • 20 GB hard disk • 40X CD-ROM drive; DVD-ROM drive and/or CD-RW drive
SMALL BUSINESS USER	• 3.5-inch high-density floppy disk drive • 27 GB hard disk • 48X CD-ROM drive; DVD-ROM drive and/or CD-RW drive • 2 GB Jaz® drive
MOBILE USER	• 3.5-inch high-density floppy disk drive • 10 GB hard disk • 1 GB PC Card hard disk • 24X CD-ROM drive; DVD-ROM drive and/or CD-RW drive
LARGE BUSINESS USER	• 3.5-inch high-density floppy disk drive • Tape drive • 34 GB hard disk • 48X CD-ROM drive; DVD-ROM drive and/or CD-RW drive • Smart card reader • Microfilm or microfiche • RAID
POWER USER	• 3.5-inch high-density floppy disk drive • 50 GB hard disk • 48X CD-ROM drive; DVD-ROM drive and/or CD-RW drive • 2 GB Jaz® drive

Figure 6-42 This table recommends suggested storage devices.

TECHNOLOGY TRAILBLAZER

SCOTT McNEALY

Scott McNealy — chairman of the board, president, and CEO of Sun Microsystems — is a sought-after speaker. His talks are laced with humor, insight, and controversy, causing listeners to rethink their opinions about the computer industry. In short, McNealy's orations reflect the same motto that he applies at Sun — to work hard, but have fun.

The son of an automobile company executive, McNealy learned important lessons growing up in Detroit. His first job at a car dealership taught him the value of organization and customer satisfaction. He also discovered that, in addition to superior technology, successful companies need vision and the knowledge of how to market their products. These lessons would be guiding principles for McNealy, who claims he would trade all he was taught in college for everything he learned working at the car dealership in the Motor City. A wealthy man today, McNealy still drives "good ol' Deeeeetroit iron."

Despite what he says, it is obvious that McNealy reaped the benefits of his schooling — earning a bachelor's degree at Harvard and an MBA at Stanford, with a concentration in manufacturing. After graduation, he worked in middle management at several companies. In 1982, former schoolmate Vinod Khosla asked McNealy to join him in forming a company to build engineering workstations in the network computer model. Together with Andreas Bechtolsheim and Bill Joy, Khosla and McNealy founded Sun Microsystems, Inc., a company that would work closely with the Stanford University Network (hence, the *SUN* in Sun Microsystems).

Company growth was rapid. The first Sun system, SUN 1, was built with inexpensive, readily available components and the UNIX operating system. In 1983, Sun and Computervision began to work together to develop and manufacture new workstation products.

Although McNealy had little technical expertise, he could envision and sell ideas and products. When appointed company president in 1984, McNealy championed his latest vision for Sun: "the network is the computer." According to McNealy, "[the personal computer] is a personal activity generator," McNealy says. "It was a brilliant stroke to call it a personal productivity tool." As an example, McNealy cites his own belief that productivity increased when presentation graphics software was banned at Sun.

Workers concentrated on what they were producing instead of how to present it. He further states that, even if personal computers *can* improve individual productivity, networked workstations allow groups of people to work together, thus increasing overall productivity.

It is an idea that is working. By 1988, Sun became the fastest-growing computer company in history, reaching $1 billion in revenues just six years after its founding. By 1992, Sun was the world's second most profitable company. Today, Sun employs more than 17,000 people and has revenues of more than $8.5 billion.

One of Sun's key innovations is Java. Java combines two of Sun's founding principles: network computing and open systems that make software and information portable. Java is the first universal programming language designed to let developers write user-friendly applications that will run on any computer. "The problem with personal computers today," insists McNealy, "is that incompatible systems make it difficult to share information." According to McNealy, "Java is the way for everybody to share the technology they invent, to share the information they publish, and to provide an interactivity that has never been there before."

McNealy knows that not everyone accepts his, or Sun's, vision. "I want Sun to be controversial. If everyone believes in your strategy you have zero chance of profit." In the future, McNealy thinks network computers will replace the world of personal computers we know today. "I am a firm believer that my little boy is going to come to me some day and say, 'Daddy, you *really* had a computer in your home? Why?'"

COMPANY ON THE CUTTING EDGE

COMPANY ON THE CUTTING EDGE

SILICON GRAPHICS

The oldest and most celebrated event in international sailing is the series of races known as the America's Cup. Since 1851, when the sloop *America* brought home the first United States victory, U.S. ships have dominated the America's Cup. Now, a team from New Zealand has joined the short list of non-American winners using a new tool — networked computer workstations from Silicon Graphics, Inc.

Designing a winning racing yacht has been an exacting, expensive, and time-consuming process of trial and error. Team New Zealand, however, saved time and money by using Silicon Graphics's workstations and servers to test yacht designs under a range of conditions, do structural analysis simulations, and provide design corroboration before prototypes were built. "Our yachts are going faster than ever before because of this program," says the syndicate head for Team New Zealand. "New design ideas can be tested literally overnight and implemented as early as the next series of races."

Silicon Graphics, Inc. is a leading manufacturer of high-performance visual computing systems ranging from desktop workstations to servers and supercomputers. Dr. James Clark, an associate professor at Stanford University, and six graduate students founded the company in 1982. The firm was based on Clark's idea for a chip that would speed a computer's capability of displaying three-dimensional graphics. In 1983, Silicon Graphics shipped its first graphics terminals. The company released its first workstations one-year later and introduced its first RISC workstations in 1987.

In the 1990s, mergers resulted in two new Silicon Graphics subsidiaries, thus expanding the company's horizons. Alias|Wavefront became a leading innovator in graphics technology, creating software for markets including film, video games, interactive media, and industrial design. Cray Research, Inc. placed its focus on supercomputing technologies used for weather forecasts, molecular modeling, new car design, and hazardous waste site cleanup.

Silicon Graphics's first venture into the general consumer marketplace took place in 1993 — the result of an agreement with Nintendo to create the Nintendo 64™ video game system. The *Reality Immersion Technology* combined Silicon Graphics's digital media and computing technologies with MIPS microprocessor technology to power a highly realistic gaming experience. In 1998, Silicon Graphics expanded its presence in the consumer marketplace through an alliance with Microsoft Corp. aimed at increasing graphics capabilities for a variety of customers.

Over the years, Silicon Graphics has pioneered four core technologies: graphics innovations for workstations, servers, and supercomputers; symmetric multiprocessing that optimizes the performance of applications; RISC CPUs based on the MIPS architecture; and digital media that integrates 3-D graphics, animation, and text with video, audio, and videoconferencing capabilities.

Today, Silicon Graphics is developing Internet and Web technologies. The company offers systems for Web server setup and Web page authoring packages designed for use with workstations. Silicon Graphics hopes to challenge Sun Microsystems's Solaris™ software (another proprietary version of UNIX) Internet servers for a larger portion of the Web server market.

With 10,300 employees worldwide and revenues of $2.7 billion for fiscal year 1999, Silicon Graphics is a major player in the computer industry. The company name may be less well known to the general public, but in recent years, its creations have become uncomfortably familiar to millions of movie-goers with its production of the life-like, computer-generated monsters in movies including *The Lost World* and *Godzilla*.

IN BRIEF

www.scsite.com/dc2001/ch6/brief.htm

WEB INSTRUCTIONS: *To display this page from the Web, launch your browser and enter the URL, www.scsite.com/dc2001/ch6/brief.htm. Click the links for current and additional information. To listen to an audio version of this IN BRIEF, click the Audio button to the right of the title, IN BRIEF, at the top of the page. To play the audio, RealPlayer must be installed on your computer (download by clicking here).*

1. How Is Storage Different from Memory?

Memory, which is composed of one or more chips on the motherboard, holds data and instructions while they are being processed by the CPU. Storage holds items such as data, instructions, and information for future use.

2. What Are Storage Media and Storage Devices?

A **storage medium** (media is the plural) is the physical material on which items are kept. A **storage device** is the mechanism used to record and retrieve items to and from a storage medium. When a storage device transfers items from a storage medium into memory – a process called **reading** – it functions as an input device. When a storage device transfers items from memory to a storage medium – a process called **writing** – it functions as an output device.

3. How Is Data Stored on a Floppy Disk?

A **floppy disk** is a portable, inexpensive storage medium that consists of a thin, circular, flexible disk with a plastic coating enclosed in a square-shaped plastic shell. **Formatting** prepares a disk for reading and writing by organizing the disk into storage locations called **tracks** and **sectors**. Most floppy disks are preformatted by the disk's manufacturer. When data is stored, a formatted floppy disk is inserted in a **floppy disk drive**, and the drive slides the shutter to the side to expose a portion of both sides of the floppy disk's recording surface. A circuit board on the drive sends signals to control the movement of the **read/write head**, the mechanism that reads items from or writes items on the floppy disk. The circuit board on the drive verifies that light is not visible through the disk's **write-protect notch**. A motor causes the floppy disk to spin and positions the read/write head over the correct position on the recording surface. The read/write head then writes data on the floppy disk.

4. How Do You Care for a Floppy Disk?

Floppy disks should not be exposed to excessive heat, sunlight, cold, magnetic fields, or contaminants such as dust, smoke, or salt air. Heavy objects should not be placed on a disk. The disk's shutter should not be opened and the recording surface should not be touched. Floppy disks should be inserted carefully into the disk drive and kept in a storage tray when not in use.

5. How Does a Hard Disk Organize Data?

A **hard disk** usually consists of several inflexible, circular disks called platters on which items are stored electronically. Hard disks undergo two formatting steps. A **low-level format** organizes both sides of each platter into tracks and sectors to define where items will be stored on the disk. After low-level formatting is complete, a hard disk can be divided into separate areas called **partitions**; each partition functions as if it were a separate hard disk drive. Partitioning makes the hard disk more efficient (faster). A **high-level format** defines, among other items, the **file allocation table (FAT)** for each partition, which is a table of information used to locate files on the disk.

6. What Are the Advantages of Using Disks?

Disks are a nonvolatile, relatively inexpensive means for storing data. Floppy disks can be used to transport a large number of files or to make backup copies of important files. Removable disks can be used when security is an issue.

7. How Is Data Stored on Compact Discs?

A **compact disc (CD)** is a flat, round, portable metal storage medium that usually is 4.75 inches in diameter and less than one-twentieth of an inch thick. Compact discs store items in microscopic pits (indentations) and land (flat areas) that are under the printed label on the disc. A high-powered laser light creates the pits in a single track, divided into evenly spaced sectors, that spirals from the center of the disc to the edge of the disc. A low-powered laser reads items from the compact disc by reflecting light through the bottom of the disc surface. The reflected light is converted into a series of bits that the computer can process.

8. How Do You Care for a Compact Disc?

Compact discs should not be stacked or exposed to heat, cold, and contaminants. The label side should not be written on or the underside touched. A compact disc should be held by its edges and placed in its protective case, called a **jewel box**, when it is not being used. The bottom surface of the compact disc can be cleaned with a soft cloth and warm water or a specialized CD cleaning kit, but the label side should not be cleaned because this is where data is stored.

9. How Are CD-ROMs and DVD-ROMs Different?

A **CD-ROM** is a silver-colored compact disc that uses the same laser technology as audio CDs for recording music. A CD-ROM can hold up to 700 MB of data, instructions, and information. A DVD-ROM is an extremely high-capacity compact disc capable of storing from 4.7 GB (4,700 MB) to 17 GB (17,000 MB). Both the storage capacity and quality of a DVD-ROM surpass that of a CD-ROM. A DVD-ROM stores data in a different manner than a CD-ROM, making the disc more dense by packing pits closer together, using two layers of pits, or using both sides of the disc.

10. How Are Tapes, PC Cards, Smart Cards, Microfilm, and Microfiche Used?

Tape is a magnetically-coated ribbon of plastic capable of storing large amounts of data and information at a low cost. Tape storage requires **sequential access** which refers to reading or writing data consecutively and is used for long-term storage and backup. A **PC Card** is a thin, credit card-sized device that fits into a PC Card expansion slot on a personal computer. PC Cards are used to add storage, memory, communications, and sound capabilities to laptops and other portable computers. A smart card stores data on a thin microprocessor embedded in the card similar in size to an ATM card. Smart cards are used to store prepaid dollar amounts such as electronic money, patient records in the health-care industry, and tracking information such as customer purchases. **Microfilm** and **microfiche** store microscopic images of documents on roll (microfilm) or sheet (microfiche) film. Libraries and large organizations use microfilm and microfiche to archive relatively inactive documents and files.

KEY TERMS

www.scsite.com/dc2001/ch6/terms.htm

WEB INSTRUCTIONS: *To display this page from the Web, launch your browser and enter the URL, www.scsite.com/dc2001/ch6/terms.htm. Scroll through the list of terms. Click a term to display its definition and a picture. Click KEY TERMS on the left to redisplay the KEY TERMS page. Click the TO WEB button for current and additional information about the term from the Web. To see animations, Shockwave and Flash Player must be installed on your computer (download by clicking here).*

access time (6.3)
AT Attachment (ATA) (6.14)
auxiliary storage (6.2)
backup (6.9)
capacity (6.3)
CD-R drive (6.22)
CD-ROM drive (6.20)
CD-ROM player (6.20)
CD-RW (6.22)
cluster (6.5)
compact disc read-only memory (CD-ROM) (6.19)
compact disc-recordable (CD-R) (6.22)
compact disc-rewritable (CD-RW) (6.22)
compact disc (CD) (6.16)
compact flash card (6.25)
computer output microfilm (COM) recorder (6.26)
cylinder (6.12)
data transfer rate (6.21)
data warehouse (6.27)
density (6.5)
digital cash (6.26)
digital video disc-ROM (DVD-ROM) (6.22)
direct access (6.23)
disk (6.2)
disk cache (6.13)
disk cartridge (6.14)
disk controller (6.14)
disk pack (6.14)
diskette (6.4)
downward compatible (6.7)
DVD-R (DVD-recordable) (6.23)
DVD-RAM (6.23)
DVD-ROM drive (6.22)
electronic money (e-money) (6.26)
enterprise storage system (6.26)
file allocation table (FAT) (6.6)

TO WEB

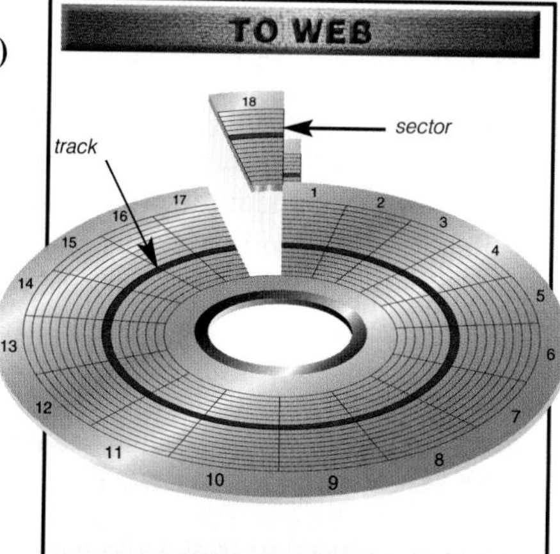

TRACK: Narrow recording band that forms a full circle on the surface of a disk. The disk's storage locations are then divided into pie-shaped sections, which break the tracks into small arcs called sectors. A sector can store 512 bytes of data. (6.5)

fixed disks (6.10)
floppy disk (6.4)
floppy disk drive (FDD) (6.7)
formatting (6.5)
hard disk (6.10)
hard disk controller (HDC) (6.14)
head crash (6.13)
HiFD™ (High-Capacity FD) (6.10)
high-level format (6.11)
Integrated Drive Electronics (IDE) (6.14)
intelligent smart card (6.25)
Jaz® disk (6.14)
jewel box (6.18)
level 1 (6.15)
low-level format (6.11)
magnetic tape (6.23)
mass storage (6.2)
memory card (6.25)
microfiche (6.26)
microfilm (6.26)
mirroring (6.15)
multisession (6.21)
nonvolatile memory (6.2)
optical disc (6.16)
partitions (6.11)
PC Card (6.24)
PhotoCD (6.21)
picture card (6.24)
Picture CD (6.21)
platter (6.10)
random access (6.23)
reading (6.3)
read/write (6.5)
read/write head (6.8)
recorded (6.19)
RAID (redundant array of independent disks) (6.15)
removable hard disk (6.14)
secondary storage (6.2)
sector (6.5)
sequential access (6.23)
shutter (6.8)
single-session (6.21)
small computer system interface (SCSI) (6.14)
smart card (6.25)
storage (6.2)
storage area network (SAN) (6.27)
storage device (6.2)
storage medium (6.2)
striping (6.15)
SuperDisk™ drive (6.10)
tape cartridge (6.23)
tape drive (6.23)
track (6.5)
Type III cards (6.25)
upward compatible (6.7)
volatile memory (6.2)
write-protect notch (6.6)
writing (6.3)
Zip® disk (6.10)
Zip® drive (6.10)

AT THE MOVIES

www.scsite.com/dc2001/ch6/movies.htm

WELCOME to VIDEO CLIPS from CNN

WEB INSTRUCTIONS: *To display this page from the Web, launch your browser and enter the URL, www.scsite.com/dc2001/ch6/movies.htm. Click a picture to view a video. After watching the video, close the video window and then complete the exercise by answering the questions about the video. To view the videos, RealPlayer must be installed on your computer (download by clicking here).*

1 Biz IBM Storage

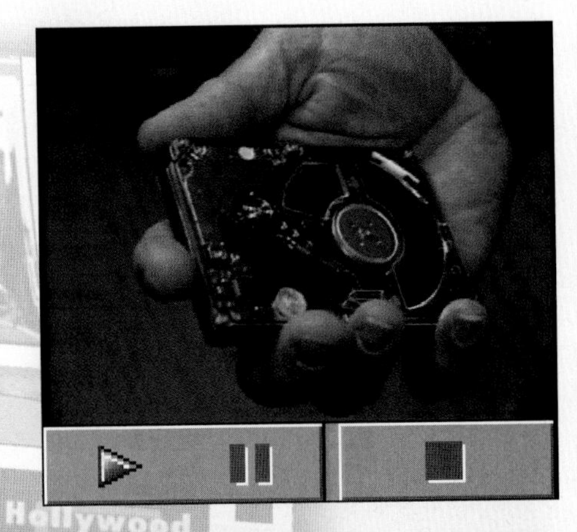

Based on what you learned in this video and in this chapter, will increased storage change the way you use a computer? Will it change the types of applications that can run on personal computers? What size storage do you think will be available in ten years? Why? How much has the cost of 1 MB of storage fallen since 1991? In the future, storage capacities will continue to increase. How much data will higher-density storage being developed by IBM allow you to store on 1 square inch of disk space? Why is greater storage capacity important? Does IBM intend to share this new technology with competitors?

2 SBT Online/NetFlix

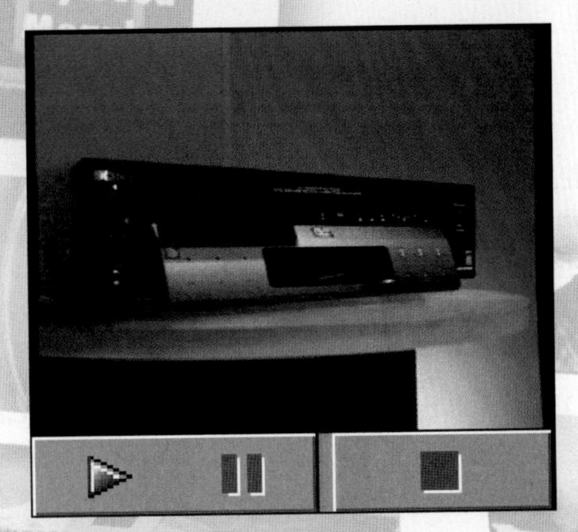

Popularity is growing for sites like the NetFlix movie Web site where you can search for your favorite movie and then either rent or buy it in DVD-ROM format. What would it cost to rent a DVD movie online and how would you receive and return the movie? What are the advantages and disadvantages of an online DVD movie rental? What movie companies are strong players in the DVD movie market? Are you excited about renting DVD movies online? Why?

3 Internet Commerce

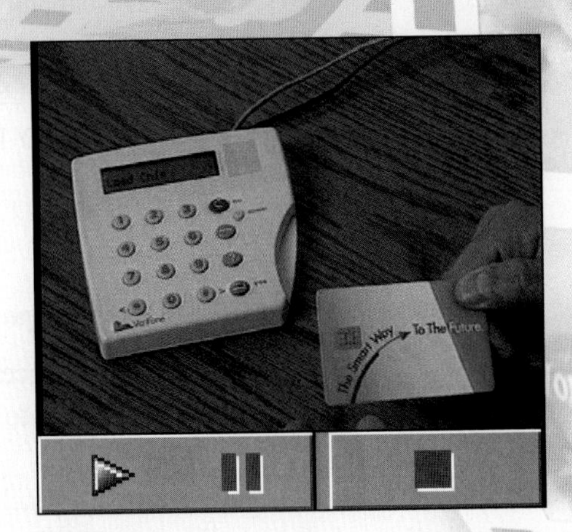

One practice soon to become commonplace is the use of bank-assigned smart cards. What is a smart card? How does it work? What kind of password-protection is used? What are the advantages of having a "personal ATM"? Are there disadvantages? Would you use a personal ATM rather than cash? Why? What are the most likely uses of your smart card. The least likely uses? Why?

CHECKPOINT

www.scsite.com/dc2001/ch6/check.htm

WEB INSTRUCTIONS: *To display this page from the Web, launch your browser and enter the URL, www.scsite.com/dc2001/ch6/check.htm. Click the links for current and additional information. To experience the animation and interactivity, Shockwave and Flash Player must be installed on your computer (download by clicking here).*

Label the Figure

Instructions: *Identify each part of a 3.5-inch floppy disk.*

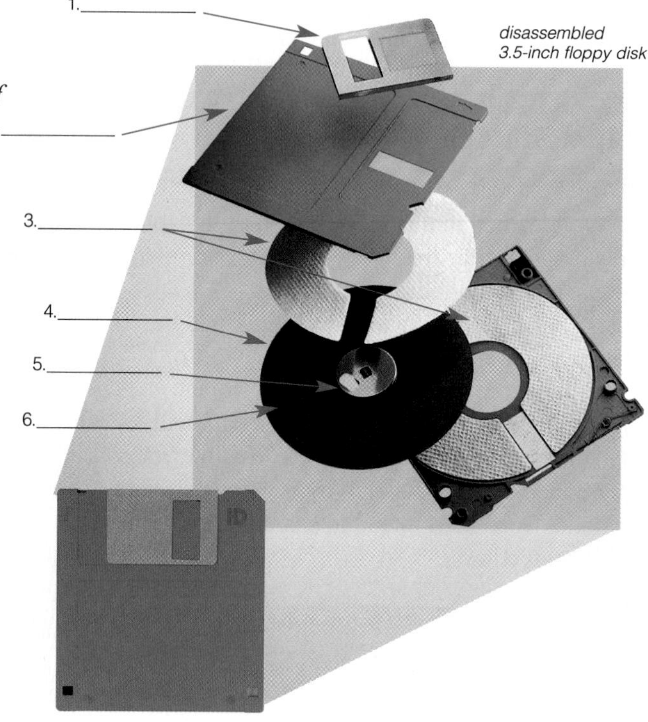

disassembled 3.5-inch floppy disk

Matching

Instructions: *Match each storage medium or storage device from the column on the left with the best description from the column on the right.*

_______ 1. SuperDisk™ drive

_______ 2. HiFD™ (High FD)

_______ 3. Zip® drive

_______ 4. Zip® disk

_______ 5. Jaz® disk

a. Floppy disk drive technology that replaces existing disk drives; uses 200 MB high-capacity floppy disks.

b. Silver-colored optical disk that uses the same laser technology as audio CDs.

c. Slightly larger than and about twice as thick as a 3.5-inch floppy disk and can store 100 MB of data.

d. High-capacity disk drive developed by Imation that uses 120 MB floppy disks.

e. Special high-capacity disk drive developed by Iomega Corporation that is built into many new computers.

f. Used to read a DVD-ROM, and is downward compatible with standard CD-ROMs.

g. Popular, reasonably priced, removable hard disk by Iomega, which can store up to 2 GB of data.

Short Answer

Instructions: *Write a brief answer to each of the following questions.*

1. What is access time? __________ Why is hard disk access time faster than floppy disk access time? __________
2. What is disk density? __________ What does it mean to say that floppy disk drives are downward compatible but not upward compatible? __________
3. What is a head crash? __________ How does a disk cache improve hard disk access time? __________
4. What is a disk controller? __________ How are EIDE controllers and SCSI controllers different? __________
5. How are multisession CD-ROMs different from single-session CD-ROMs? __________ What are three variations of the standard CD-ROM? __________

CHAPTER 1 2 3 4 5 6 7 8 9 10 11 12 13 14 INDEX

Student Exercises
WEB INFO
IN BRIEF
KEY TERMS
AT THE MOVIES
CHECKPOINT
AT ISSUE
MY WEB PAGE
HANDS ON
NET STUFF
Special Features
TIMELINE 2001
GUIDE TO WWW SITES
MAKING A CHIP
E-COMMERCE 2001
BUYER'S GUIDE 2001
CAREERS 2001
TRENDS 2001
LEARNING GAMES
CHAT
INTERACTIVE LABS
NEWS
HOME

AT ISSUE www.scsite.com/dc2001/ch6/issue.htm

WEB INSTRUCTIONS: *To display this page from the Web, launch your browser and enter the URL, www.scsite.com/dc2001/ch6/issue.htm. Click the links for current and additional information.*

ONE

Instead of supplying potential customers with printed brochures, an increasing number of companies now offer marketing CD-ROMs for customers with personal computers. Although the CDs are more expensive than conventional advertisements, the companies feel the storage medium's flexibility is worth the cost. Some automobile manufacturers, for example, offer a CD-ROM that introduces new vehicles to prospective buyers. The CD shows photographs, statistics, option packages, and pricing information. Using a mouse, automobile shoppers can "walk around and kick the tires," viewing the car from different angles, obtaining close-ups of special features, and even going on a figurative test drive. What products are best suited to promotion on CD-ROM? Why? What products are least suited? Why? Will marketing CD-ROMs even replace printed advertising materials? Why or why not?

TWO

Up to 75 percent of today's data is "born digital" and has never existed on paper. Although written documents can be read hundreds of years after they were created, rapid changes in computer technology can make digital records almost inaccessible in just one decade. Without the necessary hardware or software, information stored on once-popular 5¼-inch floppy disks can be impossible to access. Pennsylvania State University recently admitted that 2,986 student and school files could not be accessed due to lost or outdated software. Is the potential unavailability of digital data a problem? Why or why not? What can be done to keep today's digital information available in the future?

THREE

Apple Computer's iMac shocked traditionalists. The new computer's translucent case and distinctive shape alone made it unlike conventional desktop computers, but perhaps its most controversial feature was the absence of a 3½-inch floppy disk drive. Steve Jobs, acting CEO of Apple Computer, claimed Apple was "leading the way" by abandoning an outmoded and superfluous technology. Is Jobs correct? Will larger hard drives, increased Internet access, and new developments make 3½-inch floppy disks as antiquated as the 8-inch floppy disks used in the early 1980s? Why? Would you buy a computer without a 3½-floppy disk drive? Why or why not?

FOUR

Nothing lasts forever. ***This aphorism is true even in regards to computer storage.*** NASA discovered almost 20 percent of the data collected during the Viking mission was lost on decaying magnetic tape. Veterans' files, census statistics, and toxic-waste records also have been lost on deteriorating storage media. One computer scientist admits, "Digital information lasts forever or five years – whichever comes first." A major problem with digital data is that, unlike the visible deterioration in a faded document, the extent of decay on a storage medium such as a CD-ROM may be invisible until it is too late. If you were the leader of an information-intensive organization, what medium would you choose to store your records? Why? What steps would you take to ensure the records were intact ten years from now? Twenty years from now?

FIVE

Today's sports cards are different from those you once traded, and not only are the players new. Topps has introduced a series of plastic baseball cards, each of which has four seconds of game footage from Kodak. Eventually, Kodak hopes to add sound chips to the $10 cards. Donruss already offers CD-ROM trading cards. When the rectangular (yes, rectangular) $20 cards are inserted into a CD-ROM drive, they display a host of statistics, present personal information, and even show video highlights analyzing individual playing styles. Will the new sports cards eventually replace traditional trading cards? Why or why not? In light of what you have learned regarding computer storage, how else might sports cards be different in the future?

Student Exercises
WEB INFO
IN BRIEF
KEY TERMS
AT THE MOVIES
CHECKPOINT
AT ISSUE
MY WEB PAGE
HANDS ON
NET STUFF
Special Features
TIMELINE 2001
GUIDE TO WWW SITES
MAKING A CHIP
E-COMMERCE 2001
BUYER'S GUIDE 2001
CAREERS 2001
TRENDS 2001
LEARNING GAMES
CHAT
INTERACTIVE LABS
NEWS
HOME

CHAPTER 1 2 3 4 5 6 7 8 9 10 11 12 13 14 INDEX

MY WEB PAGE

www.scsite.com/dc2001/ch6/myweb.htm

WEB INSTRUCTIONS: *The icons to the left of the numbered activities on this page indicate the learning system availability. If you are using WebCT, follow the instructions in the exercises below. If you are using the textbook Web site or CyberClass, launch your browser, enter the URL www.scsite.com/dc2001/ch6/myweb.htm, and then click the corresponding icon to the left of the exercise.*

1. Practice Test: Click My Web Page and then click Chapter 6. Click Practice Test. Answer each question and then click the Save Answer button. When completed, click the Finish button. Click the OK button to submit the quiz for grading. Click the View Results button, and then make a note of any missed answers.

2. Web Guide: Click My Web Page and then click Chapter 6. Click Web Guide to display the Guide to WWW Sites page. Click Reference and then click Webopedia. Search for Optical Disk. Click one of the Optical Disk links. Use your word processing program to prepare a brief report on your findings and submit your assignment to your instructor.

3. Scavenger Hunt: Click My Web Page and then click Chapter 6. Click Scavenger Hunt. Print a copy of the Scavenger Hunt page; use this page to write down your answers as you search the Web. Submit your completed page to your instructor.

4. Who Wants to Be a Computer Genius? Click My Web Page and then click Chapter 6. Click Computer Genius to find out if you are a computer genius. For directions, click the How to Play button. When you are ready to play, click the Game button. Submit your score to your instructor.

5. Wheel of Terms: Click My Web Page and then click Chapter 6. Click Wheel of Terms to reinforce important terms you learned in this chapter by playing the Shelly Cashman Series version of this popular game. For directions, click the How to Play button. When you are ready to play, click the Game button. Submit your score to your instructor.

6. Career Corner: Click My Web Page and then click Chapter 6. Click Career Corner to display the Career Campus page. Click a link of an area of interest and review the information. Write a brief report on what you discovered. Submit the report to your instructor.

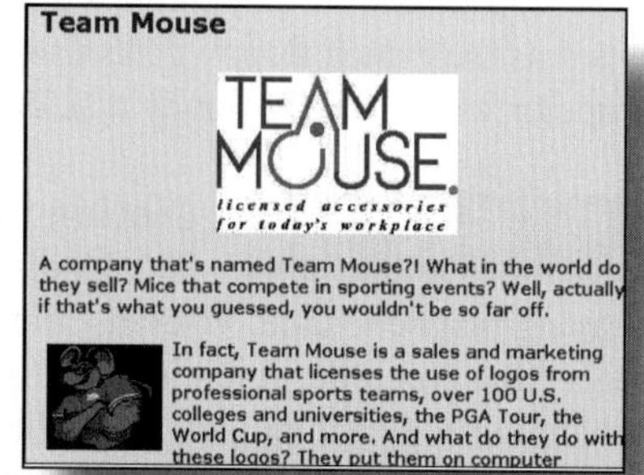

Team Mouse
TEAM MOUSE
licensed accessories for today's workplace
A company that's named Team Mouse?! What in the world do they sell? Mice that compete in sporting events? Well, actually if that's what you guessed, you wouldn't be so far off.
In fact, Team Mouse is a sales and marketing company that licenses the use of logos from professional sports teams, over 100 U.S. colleges and universities, the PGA Tour, the World Cup, and more. And what do they do with these logos? They put them on computer

7. Search Sleuth: Click My Web Page and then click Chapter 6. Click Search Sleuth to learn search techniques that will help make you a research expert. Submit the completed assignment to your instructor.

8. Crossword Puzzle Challenge: Click My Web Page and then click Chapter 6. Click Crossword Puzzle Challenge. Complete the puzzle to reinforce skills you learned in this chapter. For directions, click the How to Play button. When you are ready to play, click the Game button. Submit the completed puzzle to your instructor.

9. Flash Cards: This exercise uses CyberClass. Follow the instructions at the top of this page and then click the CyberClass icon to the left; or ask your instructor for logon instructions. Click Flash Cards on the Main Menu. Click the plus sign. Answer all the questions in any two subjects of your choice. Click the Close button in the upper-right corner to close all windows. Notice the many other exercises at this site. Complete all other exercises as assigned by your instructor.

HANDS ON

www.scsite.com/dc2001/ch6/hands.htm

WEB INSTRUCTIONS: *To display this page from the Web, launch your browser and enter the URL, www.scsite.com/dc2001/ch6/hands.htm. Click the links for current and additional information.*

One

Examining My Computer

How many disk drives does your computer have? What letter is used for each? To find out more about the disk drives on your computer, right-click the My Computer icon in the upper-left corner on the desktop. Click Open on the shortcut menu. What is the drive letter for the floppy disk drive on your computer? What letter(s) are used for the hard disk drives on your computer? If you have a CD-ROM drive, what letter is used for it? Double-click the Hard disk (C:) drive icon in the My Computer window. The Hard disk (C:) window shows the file folders (yellow folder icons) stored on your hard disk. How many folders are on the hard disk? Click the Close button to close the Hard disk (C:) window.

Two

Working with Files

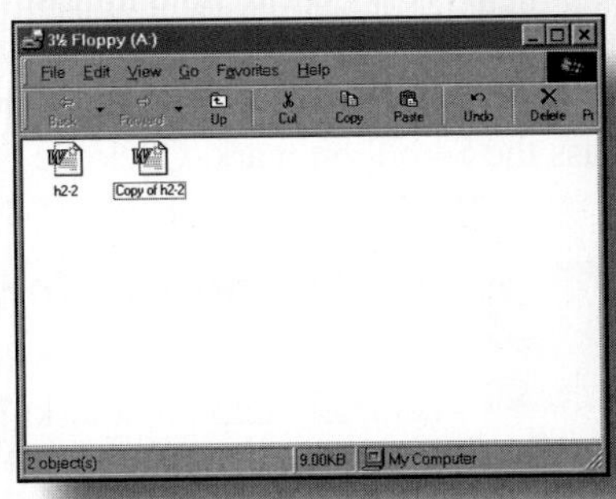

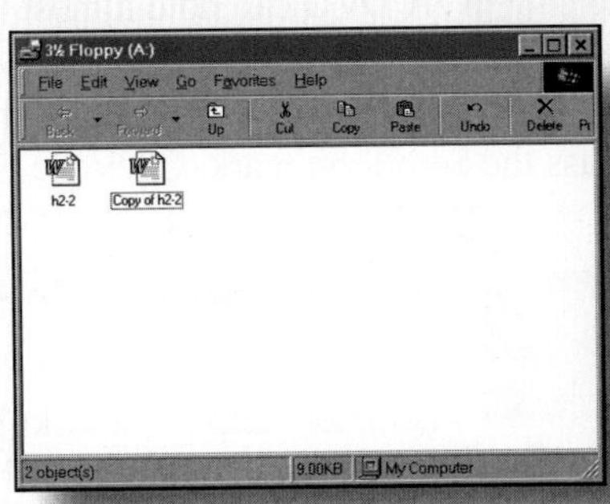

Insert the Discover Data Disk into drive A. If you do not have the Discover Data Disk, see the inside back cover of this book or your instructor. Double-click the My Computer icon on the desktop. When the My Computer window displays, right-click the 3½ Floppy (A:) icon. Click Open on the shortcut menu. Click View on the menu bar and then click Large Icons. Right-click the h2-2 icon. If h2-2 is not on the floppy disk, ask your instructor for a copy. Click Copy on the shortcut menu. Click Edit on the menu bar and then click Paste. How has the 3½ Floppy (A:) window changed? Right-click the new icon (Copy of h2-2) and then click Rename on the shortcut menu. Type `h6-2` and then press the ENTER key. Right-click the h6-2 icon and then click Print on the shortcut menu. Close the 3½ Floppy (A:) window.

Three

Learning About Your Hard Disk

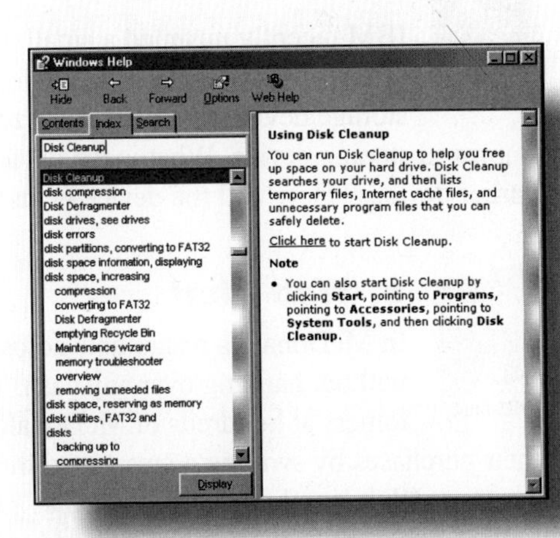

What are the characteristics of your hard disk? To find out, right-click the My Computer icon in the upper-left corner on the desktop. Click Open on the shortcut menu. Right-click the Hard disk (C:) icon in the My Computer window. Click Properties on the shortcut menu. If necessary, click the General tab and then answer the following questions:

- What Label is on the disk?
- What Type of disk is it?
- How much of the hard disk is Used space?
- How much of the hard disk is Free space?
- What is the total Capacity of the hard disk?

Close the Hard disk (C:) dialog box and the My Computer window.

Four

Disk Cleanup

This exercise uses Windows 98 or Windows 2000 procedures. Just as people maintain they never can have too much money, computer users insist that you never can have too much hard disk space. Fortunately, Windows includes a utility program called Disk Cleanup that can increase available hard disk space. To find out more about Disk Cleanup, click the Start button and then click Help on the Start menu. Click the Index tab in the Windows Help window and then type `Disk Cleanup` in the Index sheet text box. Click the Display button. Click Using Disk Cleanup or click overview in the Topics Found dialog box and then click the Display button. How does Disk Cleanup or click overview help to free up space on the hard disk? How do you start Disk Cleanup using the Start button? Click the Close button to close the Windows Help window.

NET STUFF

www.scsite.com/dc2001/ch6/net.htm

WEB INSTRUCTIONS: *To display this page from the Web, launch your browser and enter the URL, www.scsite.com/dc2001/ch6/net.htm. To use the Maintaining Your Hard Drive lab from the Web, Shockwave and Flash Player must be installed on your computer (download by clicking here).*

MAINTAINING YOUR HARD DRIVE LAB

1. Shelly Cashman Maintaining Your Hard Drive Lab

Follow the appropriate instructions in NET STUFF 1 on page 1.48 to start and use the Maintaining Your Hard Drive lab. If you are running from the Web, enter the URL, www.scsite.com/sclabs/menu.htm; or display the NET STUFF page (see instructions at the top of this page) and then click the MAINTAINING YOUR HARD DRIVE LAB button.

DVD

2. Digital Video Disk (DVD)

A DVD can hold almost twenty-five times more data than a CD. This translates into richer sound and images than ever seen or heard before. The quality of DVD storage is beginning to have a major impact on the market. Some expect the sales of DVD optical drives will soon pass the $4 billion mark. Click the DVD button and complete this exercise to learn more about DVDs.

PIM

3. Personal Information Management

Tired of forgetting birthdays, missing meetings, overlooking appointments, or neglecting to complete important tasks? If so, then personal information management software may be perfect for you. Click the PIM button to find out about a free, Internet-based calendar. How could this calendar help you organize your life? How might the calendar help you have more fun? After reading the information you may sign up to create your own Internet-based calendar.

IN THE NEWS

4. In the News

IBM recently unveiled a small disk drive, about the size of a quarter, that is capable of storing 340 megabytes of information, as much as 230 floppy disks. The drive will be used in devices such as digital cameras. What other storage devices are on the horizon? Click the IN THE NEWS button and read a news article about a new or improved storage device. What is the device? Who manufactures it? How is the storage device better than, or different from, earlier devices? How will the device be used? Why?

WEB CHAT

5. Web Chat

In McDonald's restaurants across Germany, you can get a Big Mac without handing over any cash. The sandwiches are not free, but customers at hundreds of McDonalds Deutschland, Inc., now can make their purchases by swiping a smart card through terminals at the restaurants' counters. With simple touch screens that guide them through the process, customers also can use the terminals to add value to their smart cards by downloading money electronically from their bank accounts. Is McDonald's experiment a forerunner for the future? Smart cards offer greater security and provide more control (card-owners can restrict the type of purchases that can be made) than conventional currency. On the other hand, because transactions are recorded, smart cards eliminate anonymity in purchasing and may lead to invasion of privacy. What is the greatest advantage of smart cards? What is the greatest disadvantage? Do you think smart cards will someday replace money? Why or why not? Click the WEB CHAT button to enter a Web Chat discussion related to this topic.

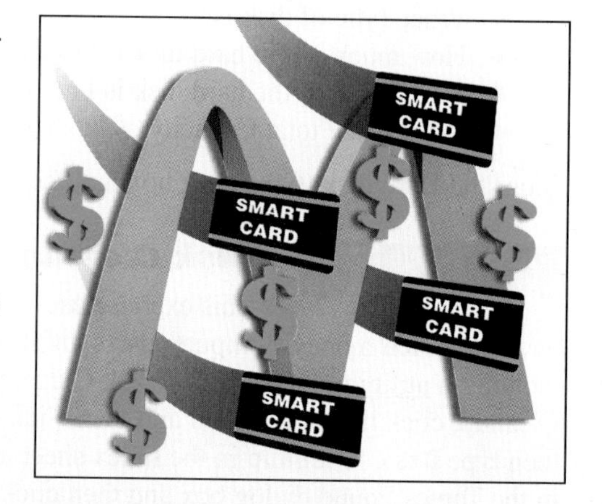

THE INTERNET

OBJECTIVES

After completing this chapter, you will be able to:

- Describe how the Internet works
- Recognize how graphics, animation, audio, video, and virtual reality are used on the World Wide Web
- Identify the tools required for Web publishing
- Describe the uses of electronic commerce (e-commerce)
- Explain how e-mail, FTP, Telnet, newsgroups, mailing lists, and chat rooms work
- Identify the rules of netiquette
- Understand security precautions for the Internet
- Describe how to access the Internet wirelessly

Today, one of the major reasons business, home, and other users purchase computers is for Internet access. Many companies and organizations assume the public is familiar with the Internet. Web addresses appear on television, in radio broadcasts, in printed newspapers magazines, and other forms of advertising. Software companies use their Web sites as a place for you to download upgrades or enhancements to software products. The government publishes thousands of informational Web pages to provide individuals with material such as legislative updates, tax forms, and e-mail addresses for Congress members. To be successful today, you must have an understanding of the Internet. Without it, you are missing a tremendous resource for goods, services, and information. This chapter discusses the history and structure of the Internet and the various services available on the Internet.

THE INTERNET

You have learned that a **network** is a collection of computers and devices connected together via communications devices and media such as cables, telephone lines, modems, and satellites.

The world's largest network is the **Internet**, which is a worldwide collection of networks that links together millions of businesses, government offices, educational institutions, and individuals (Figure 7-1). Each of these networks provides resources that add to the abundance of goods, services, and information accessible via the Internet.

The many networks that comprise the Internet, also called the **Net**, are local, regional, national, and international. Although each network that constitutes the Internet is owned by a public or private organization, no single organization owns or controls the Internet. Each organization on the Internet is responsible only for maintaining its own network.

Today more than 225 million users around the world connect to the Internet for a variety of reasons. Some of the uses of the Internet are as follows:

- To access a wealth of information, news, research, and educational material
- To conduct business or complete banking and investing transactions
- To access sources of entertainment and leisure such as online games, magazines, and vacation planning guides

Figure 7-1 *The world's largest network is the Internet, which is a worldwide collection of networks that links together millions of businesses, the government, educational institutions, and individuals.*

- To shop for goods and services
- To meet and converse with people around the world in discussion groups or chat rooms
- To access other computers and exchange files
- To send messages to or receive messages from other connected users
- To provide information, photographs, audio, or video
- To take a course or access other educational material

Figure 7-2 shows Web pages that illustrate some of these uses.

To support these and other activities, the Internet provides a variety of services including the World Wide Web, electronic mail (e-mail), FTP, Telnet, newsgroups, mailing lists, chat rooms, instant messaging, and portals. These services, along with a discussion of the history of the Internet and how the Internet works, are explained in the following pages.

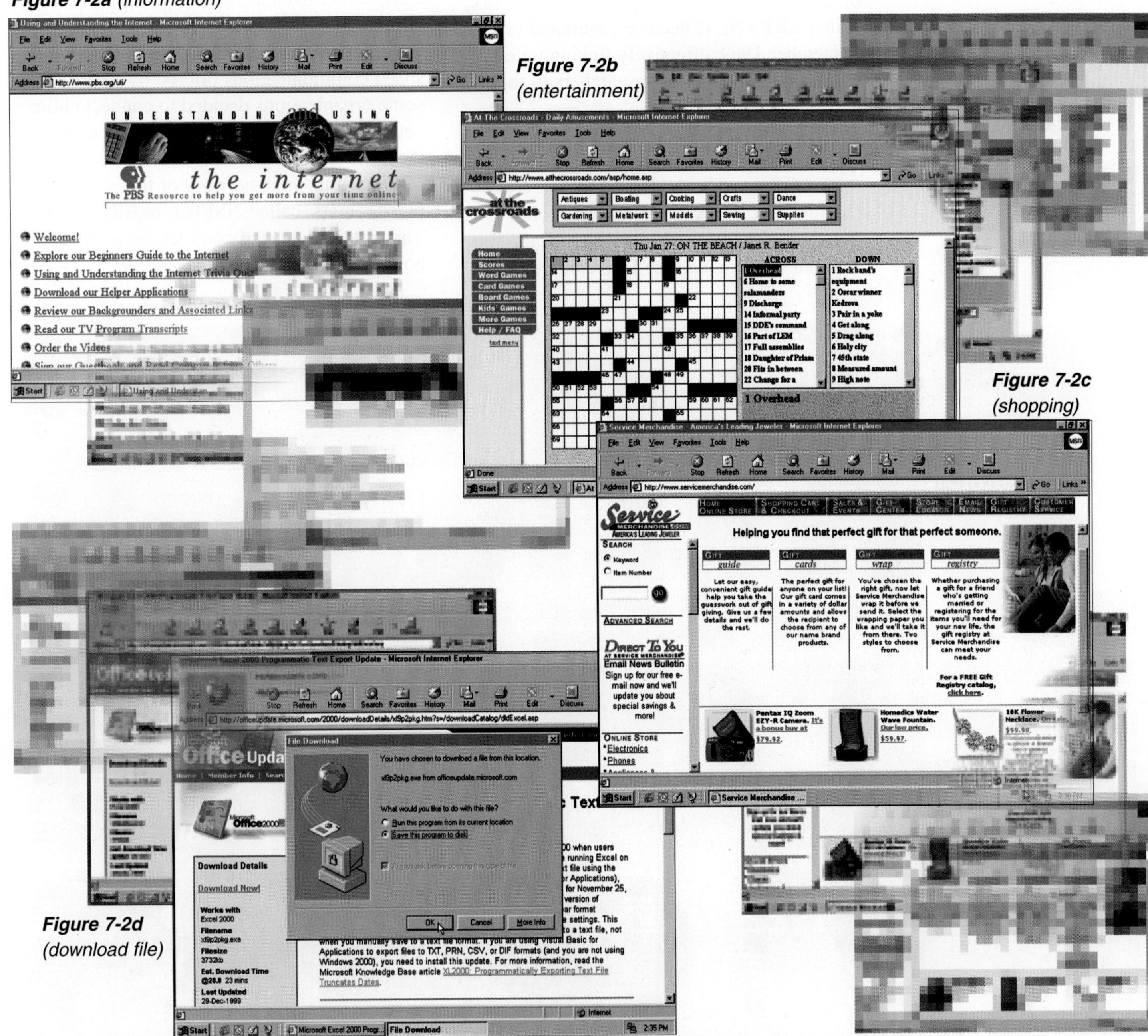

Figure 7-2a (information)

Figure 7-2b (entertainment)

Figure 7-2c (shopping)

Figure 7-2d (download file)

Figure 7-2 *Today, more than 225 million users around the world connect to the Internet for a variety of reasons.*

HISTORY OF THE INTERNET

Although the history of the Internet is relatively short, its growth has been explosive (Figure 7-3). The Internet has it roots in a networking project started by the Pentagon's **Advanced Research Projects Agency (ARPA)**, which is an agency of the U.S. Department of Defense. ARPA's goal was to build a network that (1) would allow scientists at different locations to share information and collaborate on military and scientific projects and (2) could function even if part of the network were disabled or destroyed by a disaster, such as a nuclear war. That network, called **ARPANET**, became functional in September 1969, effectively linking together scientific and academic researchers in the United States.

The original ARPANET was a wide area network consisting of four main computers, one each located at the University of California at Los Angeles, the Stanford Research Institute, the University of California at Santa Barbara, and the University of Utah. Each of these four computers served as the network's host nodes. In a network, a **host node**, or **host**, is any computer directly connected to the network. A host often stores and transfers data and messages on high-speed communications lines and provides network connections for other computers. Hosts and communications lines are discussed in more depth in Chapter 9.

As researchers and others realized the great benefit of using ARPANET's electronic mail to share information and notes, ARPANET underwent phenomenal growth. By 1984, ARPANET had more than 1,000 individual computers linked as hosts. (Today, more than 70 million hosts are connected to the Internet.)

To take further advantage of the high-speed communications offered by ARPANET, some organizations decided to connect entire networks to ARPANET. In 1986, for example, the National Science Foundation (NSF) connected its huge network of five supercomputer centers, called **NSFnet**, to ARPANET. This configuration of complex networks and hosts became known as the Internet.

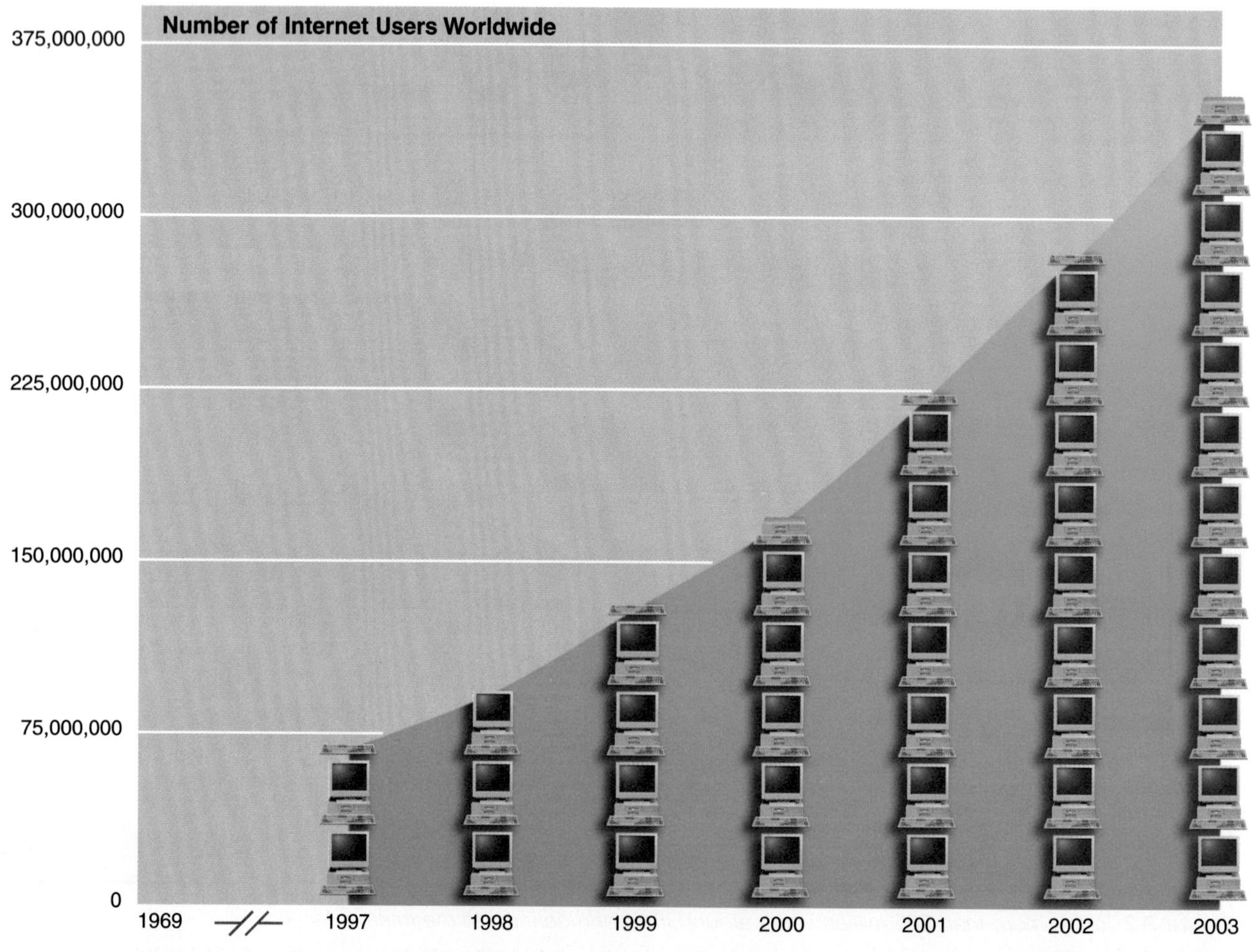

***Figure 7-3** The growth of the Internet has been explosive.*

Because of its advanced technology, NSFnet served as the major backbone network on the Internet until 1995. A **backbone** is a high-speed network that connects regional and local networks to the Internet; other computers then connect to these regional and local networks to access the Internet. A backbone thus handles the bulk of the communications activity, or **traffic**, on the Internet.

In 1995, NSFnet terminated its backbone network on the Internet to return its status to a research network. Since then, a variety of corporations, commercial firms, and other companies run backbone networks that provide access to the Internet.

These backbone networks, along with telephone companies, cable and satellite companies, and the government all contribute toward the internal structure of the Internet. Many donate resources, such as servers, communication lines, and technical specialists — making the Internet truly collaborative.

Even as it grows, the Internet remains a public, cooperative, and independent network. Although no single person, company, institution, or government agency controls or owns the Internet, several organizations contribute toward its success by advising, defining standards, and addressing other issues. The table in Figure 7-4 outlines the functions of some of these organizations.

Two Internet-related research and development projects that have emerged in recent years are the Internet2 and Next Generation Internet. **Internet2** (**I2**) is an extremely high-speed network whose purpose is to develop and test advanced Internet technologies for research, teaching, and learning. The result of collaboration among more than 170

WEB INFO

For more information on the Internet backbone, visit the Discovering Computers 2001 Chapter 7 WEB INFO page (**www.scsite.com/dc2001/ch7/webinfo.htm**) and click Internet Backbone.

INTERNET ADVISORY GROUPS AND ORGANIZATIONS

Organization	Abbreviation	Composition	Function
World Wide Web Consortium	W3C	Commercial and educational institutions	Oversees research and sets standards for many areas of the Web
Internet Society	ISOC	Individuals, corporations, nonprofit organizations, foundations, and government agencies	Concerned with use, maintenance, and development of Internet; oversees other boards and task forces; publishes Internet Society News; coordinates annual Internet conference called INET
Internet Architecture Board	IAB	Advisory group to the ISOC	Defines the architecture of the Internet – including backbone and all attached networks; resolves standards' disputes
Internet Engineering Steering Group	IESG	Body of the ISOC	Responsible for Internet standards process; approves final Internet standards specifications
Internet Engineering Task Force	IETF	Body of the IESG	Studies technical problems and recommends solutions to the IAB and IESG
Internet Assigned Numbers Authority	IANA	Body of the IAB	Assigns and controls numeric designations on the Internet, such as IP addresses and protocols
Internet Network Information Center	InterNIC	National Science Foundation, AT&T, General Atomics, and Network Solutions, Inc.	Registers domain names and IP addresses; distributes information about the Internet
Internet Engineering and Planning Group	IEPG	Internet service providers	Coordinates technical efforts on the Internet; promotes usage of the Internet
Internet Research Task Force	IRTF	Volunteers	Makes recommendations about the Internet to the IAB

Figure 7-4 *The functions of the various organizations that define standards and make recommendations for Internet issues.*

universities in the United States, along with several industry and government partners, the goal of I2 is eventually to deploy its advanced technologies to the Internet. **Next Generation Internet** (**NGI**) is a program similar to that of I2, except that it is funded and led by the United States federal government.

HOW THE INTERNET WORKS

For more information on ISPs, visit the Discovering Computers 2001 Chapter 7 WEB INFO page (**www.scsite.com/dc2001/ch7/webinfo.htm**) and click ISP.

Data sent over the Internet travels via networks and communications lines owned and operated by many companies. Various ways to connect to these networks are presented in the following sections.

Internet Service Providers and Online Services

An **Internet service provider (ISP)** is an organization that has a permanent Internet connection and provides temporary connections to individuals and companies for free or for a fee. The most common ISP fee arrangement is a fixed amount, usually about $10 to $20 per month for an individual account. For this amount, many ISPs provide unlimited Internet access, while others specify a set number of access hours per month, such as 100 hours. If you spend more time on the Internet than the allotted access hours, you are charged an additional amount, based on an hourly rate.

Two types of ISPs exist: local and national (Figure 7-5). A **local ISP** usually provides one or more local telephone numbers to provide access to the Internet. A **national ISP** is a larger business that provides local telephone numbers in most major cities and towns nationwide; some also provide a toll-free telephone number. Because of their size, national ISPs offer more services and generally have a larger technical support staff than local ISPs. The most important consideration when selecting an ISP is to be sure that it provides a local telephone number for Internet access, called a **point of presence**

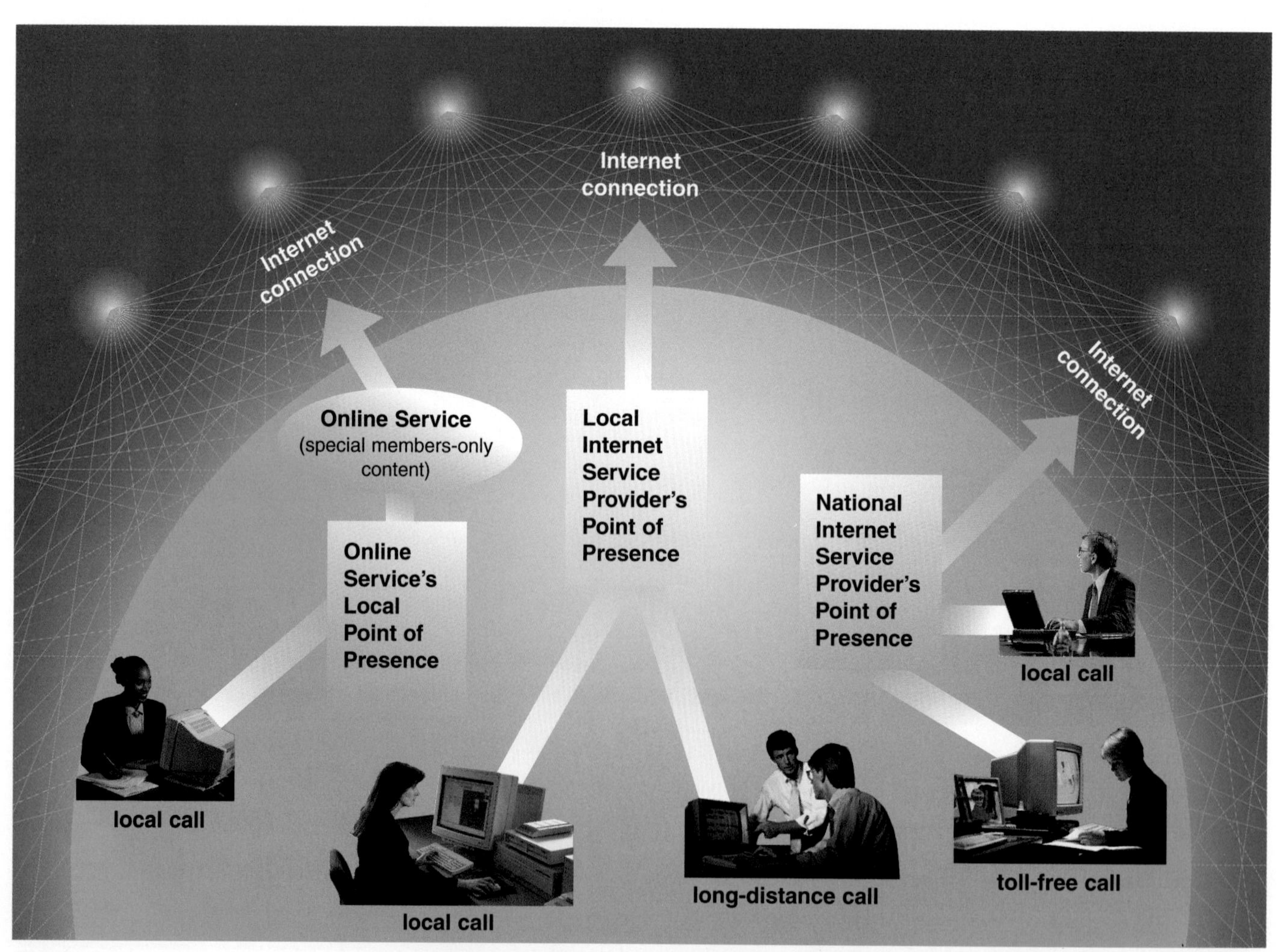

Figure 7-5 *Common ways to access the Internet are through an online service or a local or national Internet service provider.*

(**POP**); otherwise, you pay long-distance telephone bills for the time you are connected to the Internet.

Like an ISP, an **online service** provides Internet access, but such online services also have members-only features that offer a variety of special content and services such as news; weather; legal information; financial data; hardware and software guides; games; and travel guides. For this reason, the fees for using an online service usually are slightly higher than fees for an ISP. Online services such as America Online and The Microsoft Network (MSN) usually have thousands of POPs all over the world and large customer and technical support staffs.

Connecting to the Internet

Many users connect to the Internet through a business or school network. In this instance, their computers usually are part of a local area network (LAN) that is connected to an ISP through a high-speed connection line leased from the local telephone company. Instead of connecting via a modem, a personal computer connects to the LAN using a network interface card (NIC), which is an expansion card that allows the computer to be networked.

When connecting from home or while traveling, individuals often use dial-up access to connect to the Internet. With **dial-up access**, you might use your computer and a modem to dial into an ISP or online service over a regular telephone line. The computer at the receiving end, whether at an ISP or online service, also may use a modem. Dial-up access provides an easy way for mobile and home users to connect to the Internet to check e-mail, read the news, and access research material. Because dial-up access uses regular telephone lines, however, the speed of the connection is limited.

Newer high-speed technologies such as Integrated Services Digital Network (ISDN), digital subscriber lines (DSL), and cable services provide an alternative to dial-up access over regular telephone lines. These technologies, along with modems, network interface cards, and other types of communications equipment and media are discussed in more detail in Chapter 9.

Some users connect to the Internet wirelessly; that is, without using any physical cables or lines. Wireless Internet access is discussed at the end of this chapter.

How Data Travels the Internet

Computers connected to the Internet work together to transfer data and information around the world using servers and clients. Recall that a **server** is a computer that manages the resources on a network and provides a centralized storage area for software programs and data. A **client** is a computer that can access the contents of the storage area on the server, including programs, data, and other resources. On the Internet, for example, your computer is a client that can access files and services on a variety of servers, which are called host computers.

When a client computer sends data over the Internet, the data is divided into small pieces, called **packets**. The data in a packet might be part of an e-mail message, a file, a document, or a request for a file. Each packet contains the data, as well as the recipient (destination), origin (sender), and sequence information used to reassemble the data at the destination. These packets are sent along the fastest available path to the recipient's computer via devices called **routers**. If the most direct path to the destination is overloaded or not operating, the routers send the packets along an alternate path.

If necessary, each packet can be sent over a different path to the destination. If the packets arrive out of sequence, the destination computer uses the sequence information contained in each packet to reassemble the original message, file, document, or request. This technique of breaking a message into individual packets, sending the packets along the best route available, and then reassembling the data is called **packet switching**.

For a technique such as packet switching to work, all of the devices on the network must follow certain standards, or protocols. A **communications protocol** specifies the rules that define how devices connect to each other and transmit data over a network. The protocol used to define packet switching on the Internet is a communications protocol known as **TCP/IP** (**transmission control protocol/Internet protocol**).

WEB INFO

For more information on online services, visit the Discovering Computers 2001 Chapter 7 WEB INFO page (**www.scsite.com/dc2001/ch7/webinfo.htm**) and click Online Services.

The inner structure of the Internet works much like a transportation system. Just as highways connect major cities and carry the bulk of the automotive traffic across the country, several main communications lines carry the heaviest amount of traffic on the Internet. These communications lines are referred to collectively as the Internet **backbone**.

In the United States, the communications lines that make up the Internet backbone exchange data at several different locations. These locations, which function like a highway interchange, are one of two basic types: Network Access Points or Metropolitan Area Exchanges (Figure 7-6). Network Access Points (NAPs) and Metropolitan Area Exchanges (MAEs) are located in major cities and use high-speed equipment to transfer data packets from one network to another.

National ISPs, sometimes called **backbone providers**, use dedicated lines to connect directly to the Internet backbone at one or more NAPs or MAEs. Smaller regional networks and local ISPs, by contrast, lease lines from local telephone companies to connect to national ISPs. These smaller, slower-speed regional and local networks extend out from the backbone into regions and local communities. Figure 7-7 illustrates how these components of the Internet work together to transfer data over the Internet to and from your computer.

Internet Addresses

The Internet relies on an addressing system much like the postal service to send data to a computer at a specific destination. Each computer location on the Internet has a numeric address called an **IP** (**Internet protocol**) **address**. The IP address consists of four groups of numbers, each separated by a period. The number in each group is between 0 and 255. For example, the numbers, 198.112.168.223, are an IP address. In general, the first portion of each IP address identifies the network and the last portion identifies the specific computer.

Because these all-numeric IP addresses are difficult to remember and use, the Internet supports the use of a text name that represents one or more IP addresses. The text version of an IP address is called a **domain name**. Figure 7-8 shows an IP address and its associated domain name. Like an IP address,

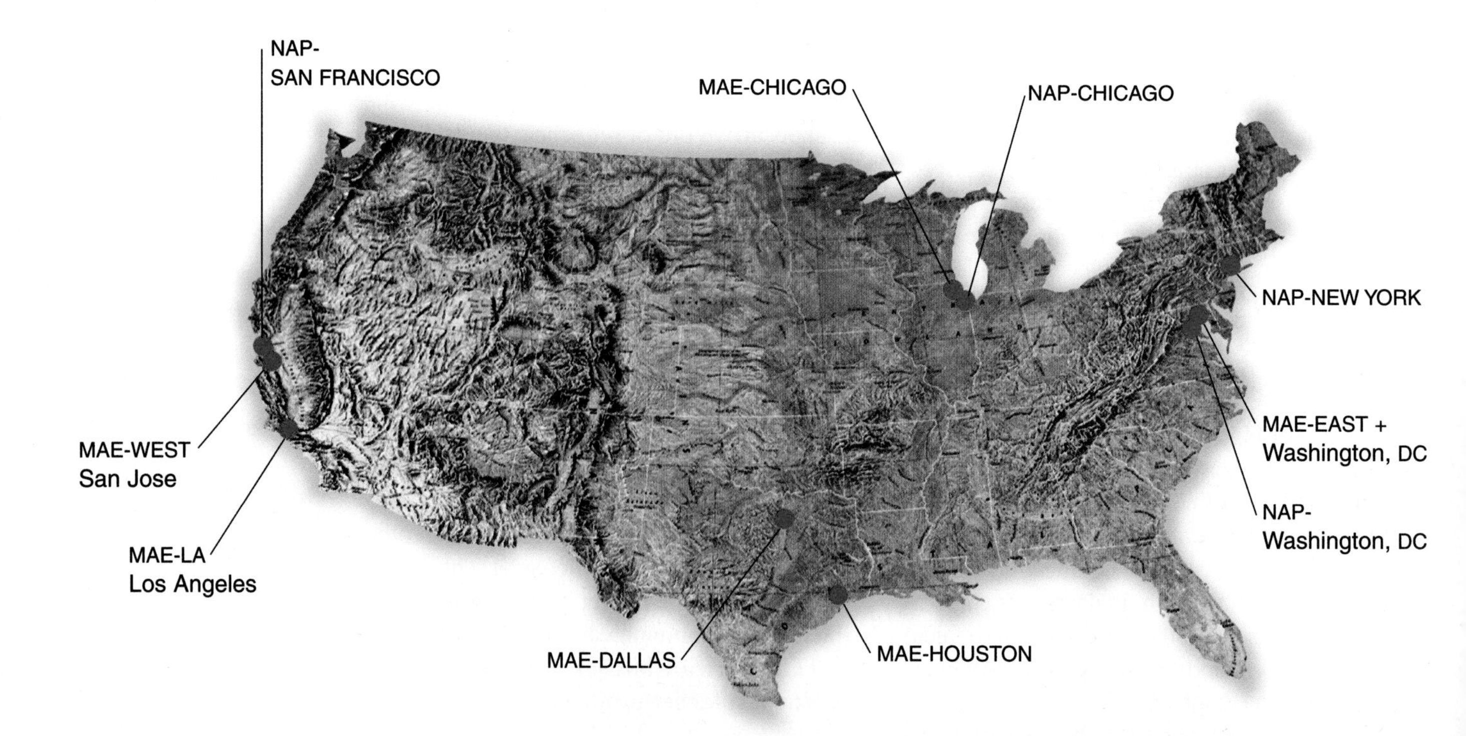

Figure 7-6 *The map shows where the MAEs and NAPs currently are located in the U.S.*

Figure 7-7
HOW DATA MIGHT TRAVEL THE INTERNET

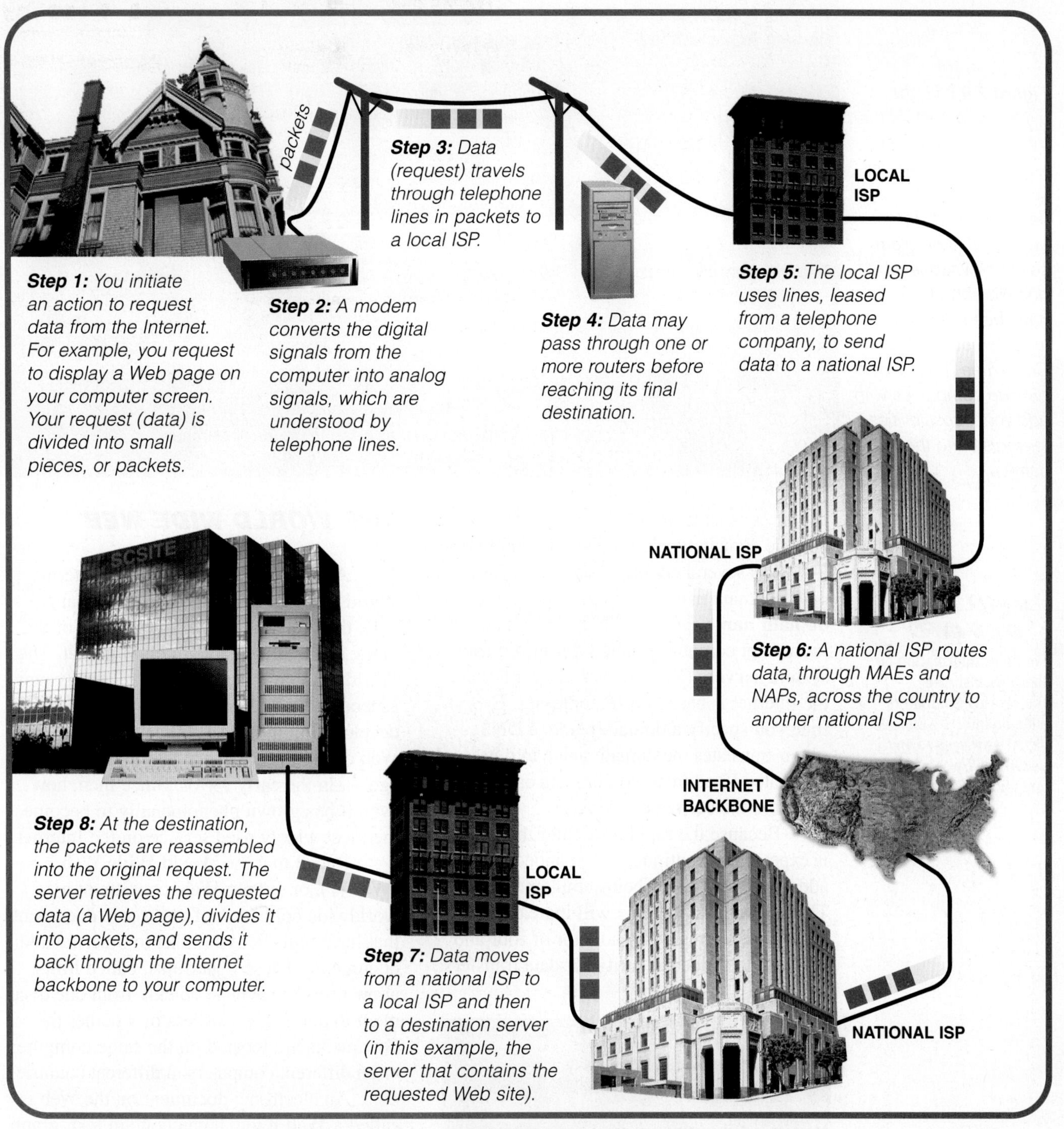

the components of a domain name are separated by periods.

packets
Step 1: You initiate an action to request data from the Internet. For example, you request to display a Web page on your computer screen. Your request (data) is divided into small pieces, or packets.
Step 2: A modem converts the digital signals from the computer into analog signals, which are understood by telephone lines.
Step 3: Data (request) travels through telephone lines in packets to a local ISP.
Step 4: Data may pass through one or more routers before reaching its final destination.
LOCAL ISP
Step 5: The local ISP uses lines, leased from a telephone company, to send data to a national ISP.
NATIONAL ISP
Step 6: A national ISP routes data, through MAEs and NAPs, across the country to another national ISP.
INTERNET BACKBONE
NATIONAL ISP
LOCAL ISP
Step 7: Data moves from a national ISP to a local ISP and then to a destination server (in this example, the server that contains the requested Web site).
SCSITE
Step 8: At the destination, the packets are reassembled into the original request. The server retrieves the requested data (a Web page), divides it into packets, and sends it back through the Internet backbone to your computer.

Every domain name contains a **top-level domain** (**TLD**) abbreviation that identifies the type of organization that operates the site. In Figure 7-8, for example, the abbreviation, com, represents a top-level domain. For international Web sites, the domain name also includes a country code, which usually is placed at the end of a domain name — for

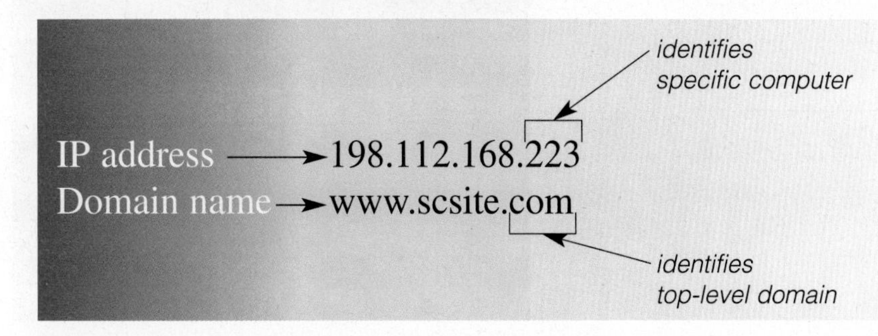

Figure 7-8 *The IP address and domain name for the Shelly Cashman Series® Instructional Web site.*

TOP-LEVEL DOMAIN ABBREVIATIONS

Original Top-Level Domain Abbreviations	Type of Organization
com	Commercial organizations, businesses, and companies
edu	Educational institutions
gov	Government institution
mil	Military organizations
net	Network provider
org	Non-profit organizations

Newer Top-Level Domain Abbreviations	Type of Organization
arts	Arts and cultural-oriented entities
firm	Other businesses or firms
info	Information services
nom	Individuals or families
rec	Recreation/entertainment sources
store	Merchants, businesses offering goods to purchase
web	Parties emphasizing Web activities

Figure 7-9 *With the explosive growth of the Internet over the last few years, the competition for domain names has increased. To address this problem, seven new top-level domain abbreviations have been created by the Internet Ad Hoc Committee (IAHC) to provide companies with additional possibilities for registering their names.*

countries outside the United States. Figure 7-9 lists the domain type abbreviations, and Figure 7-10 lists several country code abbreviations.

Domain names are registered in the **domain name system (DNS)** and are stored in Internet computers called **domain name system servers (DNS servers)**. Recall that the Internet is based on IP addresses. Every time you specify a domain name, a DNS server translates the domain name into its associated IP address, so that data can be routed to the correct computer.

Because the rapid growth of the Internet is expected to continue, an expanded IP addressing scheme is being implemented. The new address scheme will increase the number of addresses by a factor of four and will provide added security for data transfers.

WEB INFO

For more information on the DNS, visit the Discovering Computers 2001 Chapter 7 WEB INFO page (**www.scsite.com/dc2001/ch7/webinfo.htm**) and click DNS.

Country Code Abbreviations	Country
au	Australia
ax	Antarctica
ca	Canada
de	Germany
dk	Denmark
fr	France
jp	Japan
nl	Netherlands
se	Sweden
th	Thailand
uk	United Kingdom
us	United States

Figure 7-10 *A partial listing of country code abbreviations. The us code usually is omitted.*

THE WORLD WIDE WEB

Although many people use the terms World Wide Web and Internet interchangeably, the World Wide Web is just one of the many services available on the Internet. The World Wide Web actually is a relatively new aspect of the Internet. While the Internet was developed in the late 1960s, the World Wide Web came into existence less than a decade ago — in the early 1990s. Since then, however, it has grown phenomenally to become the most widely used service on the Internet.

Recall that the **World Wide Web (WWW)**, or simply **Web**, consists of a worldwide collection of electronic documents that have built-in hyperlinks to other related documents. These hyperlinks, called **links**, allow users to navigate quickly from one document to another, regardless of whether the documents are located on the same computer or on different computers in different countries.

An electronic document on the Web is called a **Web page**; it can contain text, graphics, sound, and video, as well as links to other Web pages. A collection of related Web pages that you can access electronically is called a **Web site**. Most Web sites have a starting point, called a **home page**, which is similar to a book cover or table of contents for the site and provides information about the site's purpose and content.

Each Web page on a Web site has a unique address, called a **Uniform Resource Locator** (**URL**). As shown in Figure 7-11, a URL consists of a protocol, a domain name, and sometimes the path to a specific Web page or location in a Web page. Most Web page URLs begin with **http**://, which stands for **hypertext transfer protocol**, the communications protocol used to transfer pages on the Web. You access and view Web pages using a software program called a **Web browser**, or **browser**. The more widely used Web browsers today are Microsoft Internet Explorer and Netscape Navigator.

The Web pages that comprise a Web site are stored on a server, called a Web server. A **Web server** is a computer that delivers (serves) requested Web pages. For example, when you enter the URL, http://www.sportsline.com/tennis/index.html in your browser, your browser sends a request to the server that stores the Web site of www.sportsline.com. The server then fetches the page named index.html in the tennis path and sends it to your browser.

Multiple Web sites can be stored on the same Web server. For example, many Internet service providers grant their subscribers storage space on a Web server for their personal or company Web sites.

A **Webmaster** is the individual responsible for developing Web pages and maintaining a Web site. Webmasters and other Web page developers create and format Web pages using hypertext markup language (HTML), which is a set of special codes that define the placement and format of text, graphics, video, and sound on a Web page. Because HTML can be difficult to learn and use, many user-friendly tools exist for Web publishing, which is the development and maintenance of Web pages. HTML and other Web publishing techniques are discussed later in this chapter.

WEB INFO

For more information on URLs, visit the Discovering Computers 2001 Chapter 7 WEB INFO page (**www.scsite.com/dc2001/ch7/webinfo.htm**) and click Uniform Resource Locators.

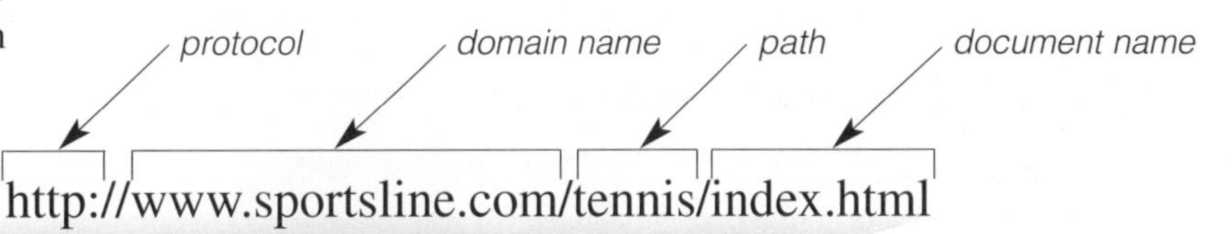

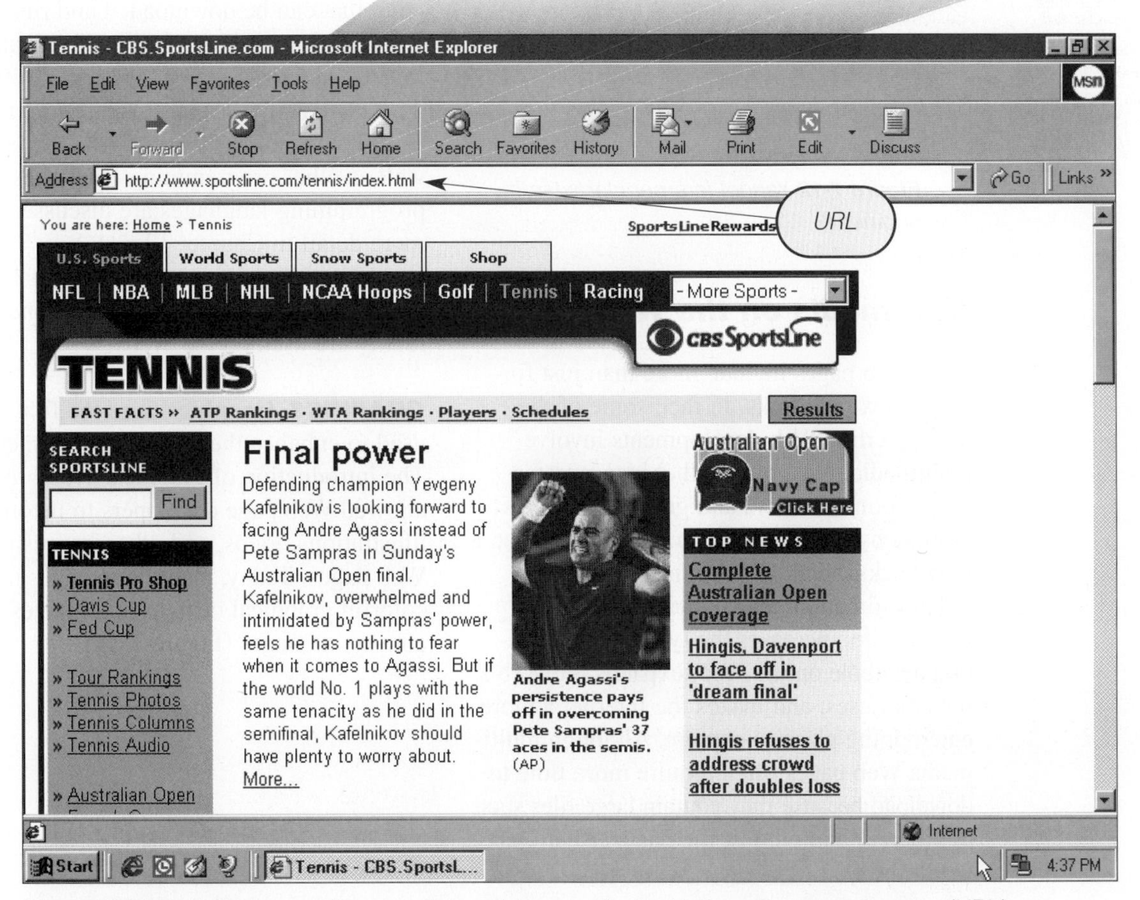

Figure 7-11 *Each Web page has a unique address, called a Uniform Resource Locator (URL).*

WEB INFO

For more information on plug-ins, visit the Discovering Computers 2001 Chapter 7 WEB INFO page (**www.scsite.com/dc2001/ch7/webinfo.htm**) and click Plug-ins.

Search Engines

No single organization controls additions, deletions, and changes to Web sites, which means no central menu or catalog of Web site content and addresses exists. Several companies, however, maintain organized directories of Web sites to help you find information on specific topics. The companies provide a software program, called a **search engine**, which helps you locate Web sites, Web pages, and Internet files (Figure 7-12). For additional information on search engines, see Chapter 2.

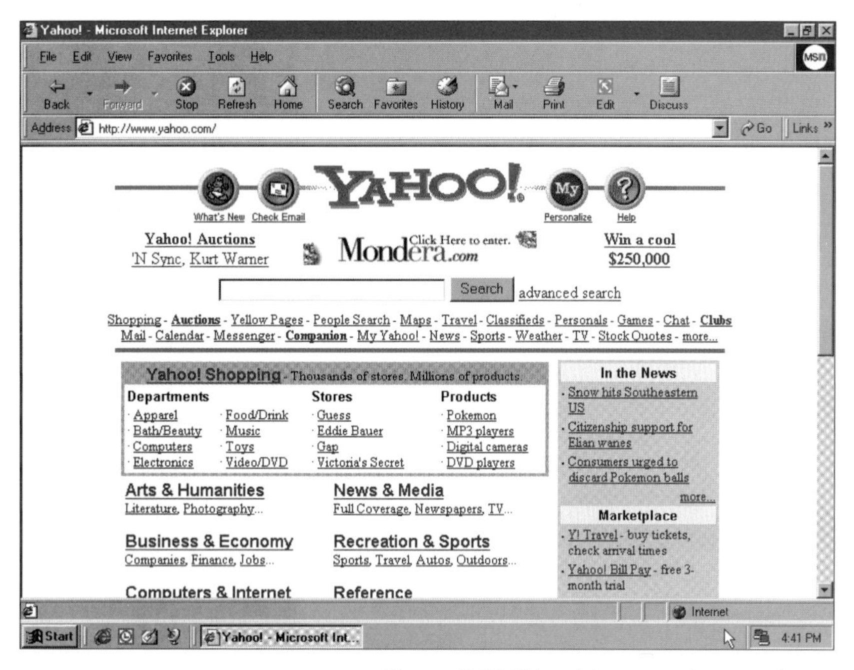

Figure 7-12 *Yahoo! is a popular search engine.*

Multimedia on the Web

Most Web pages include more than just formatted text and links. In fact, some of the more exciting Web developments involve multimedia. A Web page that incorporates color, sound, motion, and graphics with text has much more appeal than one with text on a gray background. Combining text, audio, video, animation, and sound brings a Web page to life; increases the types of information available on the Web; expands the Web's potential uses; and makes the Internet a more entertaining place to explore. Although multimedia Web pages often require more time to download because they contain large files such as video or audio clips, the pages usually are worth the wait.

Most browsers have the capability of displaying basic multimedia elements on a Web page. Sometimes, however, your browser might need an additional program, called a plug-in or helper application, which extends the capability of the browser. A **plug-in** runs multimedia elements within the browser window, while a **helper application** runs multimedia elements in a window separate from the browser. Plug-ins and helper applications can be downloaded, or copied, at no charge from many sites on the Web (Figure 7-13). In fact, Web pages that use multimedia elements often include links to Web sites that contain the required plug-in or helper. Some browsers include commonly used plug-ins, such as Shockwave, which is required for viewing many multimedia Web pages.

Some of the multimedia on the Web is developed in **Java**, which is a programming language specifically designed by Sun Microsystems for use on the Internet. Developers use Java to create small programs called **applets** that can be downloaded and run in a browser window. Applets can be used by just about any type of computer. Similar to an applet, an **ActiveX control** is a small program that can be downloaded and run in a browser, thus adding multimedia capabilities to Web pages. Although Microsoft initially developed ActiveX controls, programmers can develop ActiveX controls using Java as well as other programming languages. These programming languages are discussed in more detail in Chapter 12.

The following sections discuss how graphics, audio, animation, video, and virtual reality are used on the Web.

GRAPHICS **Graphics** were the first media used to enhance the text-based Internet. The introduction of graphical Web browsers allowed Web page developers to incorporate illustrations, logos, and other images into Web pages. Today, many Web pages use colorful graphical designs and images to convey messages (Figure 7-14).

POPULAR PLUG-IN/HELPER APPLICATIONS

Plug-In/Helper Application		Description	Web Site URL
Get Acrobat Reader	Acrobat Reader	View, navigate, and print Portable Document Format (PDF) files — documents formatted to look just as they look in print	www.adobe.com
Cosmo Player 2.1 The Universal VRML 2.0 Client	Cosmo Player	View 3-D and other virtual reality applications written in Virtual Reality Modeling Language (VRML)	cosmosoftware.com
macromedia FLASH PLAYER	Flash Player	View dazzling graphics and animation, hear outstanding sound and music; display Web pages across entire screen	www.macromedia.com
download liquid player	Liquid Player	Listen to and purchase CD-quality music tracks and audio CDs over the Internet; access MP3 files	www.liquidaudio.com
	QuickTime	View animation, music, audio, video, and virtual reality panoramas and objects directly in a Web page	www.apple.com
FREE real jukebox	RealJukebox	Play MP3 files; create music CDs	www.real.com
real player 7 FREE	RealPlayer	Live and on-demand near-CD-quality audio and newscast-quality video; stream audio and video content for faster viewing	www.real.com
macromedia SHOCKWAVE	Shockwave	Experience dynamic interactive multimedia, graphics, and streaming audio	www.macromedia.com
stamps.com	Stamps.com	Print postage from your computer onto envelopes or labels	www.stamps.com

Figure 7-13 *Most of these popular plug-ins and helper applications can be downloaded free from the Web.*

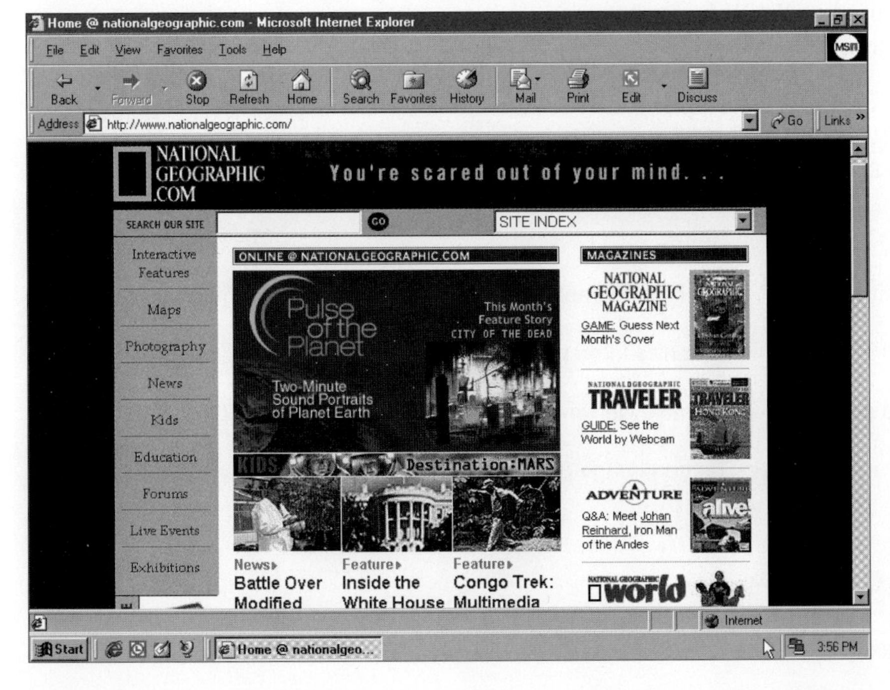

Figure 7-14 *Many Web pages use colorful graphic designs and images to convey their messages.*

Graphics files on the Web must be saved in a certain format. The two more common file formats for graphical images on the Web are JPEG and GIF. A **JPEG** (pronounced JAY-peg) file, which stands for **Joint Photographic Experts Group**, is a graphical image saved using compression techniques to reduce the file size for faster downloading from the Web. When you create a JPEG image, you can specify the image quality to reach a balance between image quality and file size. The JPEG format often is used for scanned photographs, artwork, and other images that include smooth color variations.

A graphical image saved as a **GIF** (pronounced jiff or giff) file, which stands for **Graphics Interchange Format**, also is saved using compression techniques to reduce its file size for downloading. The GIF format works best for images with only a few distinct colors, such as line drawings, single-color borders, and simple cartoons. The technique used to compress GIF files (called LZW compression), however, is patented, which means companies that make products using the GIF format must obtain a license. (Most Web users or businesses that include GIFs in their Web pages are not required to obtain a license.)

A patent-free replacement for the GIF, the PNG format, has been developed and approved by the World Wide Web Consortium as an Internet graphics standard. The **PNG** (pronounced ping) format, which stands for **portable network graphics**, also is a compressed file format that supports multiple colors and resolutions. These and other graphics formats used on the Web are shown in the table in Figure 7-15.

The Web contains thousands of image files on countless subjects, many of which can be downloaded free and used for non-commercial purposes. Because graphics files can be time consuming to download, some Web sites use thumbnails on their pages. A **thumbnail** is a small version of a larger graphical image that you usually can click to display the full-sized image (Figure 7-16).

ANIMATION **Animation** is the appearance of motion that is created by displaying a series of still images in rapid sequence. Animated graphics can make Web pages more visually interesting or draw attention to important information or links. For example, text that is animated to scroll across the screen, called a **marquee** (pronounced mar-KEE), can serve as a ticker to display stock updates, news, sports scores, or weather (Figure 7-17). Animation often is used in Web-based games; some animations even contain links to a different page.

One popular type of animation, called an **animated GIF**, is created using computer animation and graphics software to combine several images into a single GIF file. You also can create applets or ActiveX controls that include animation, or you simply can download many already-developed animations from the Web.

GRAPHICS FORMATS USED ON THE INTERNET

Acronym	Name	File Extension
JPEG (pronounced JAY-peg)	Joint Photographic Experts Group	.jpg
GIF (pronounced jiff)	Graphics Interchange Format	.gif
PNG (pronounced ping)	Portable Network Graphics	.png
TIFF	Tagged Image File Format	.tif
PCX	PC Paintbrush	.pcx
BMP	Bitmap	.bmp

Figure 7-15 *Graphics formats used on the Internet.*

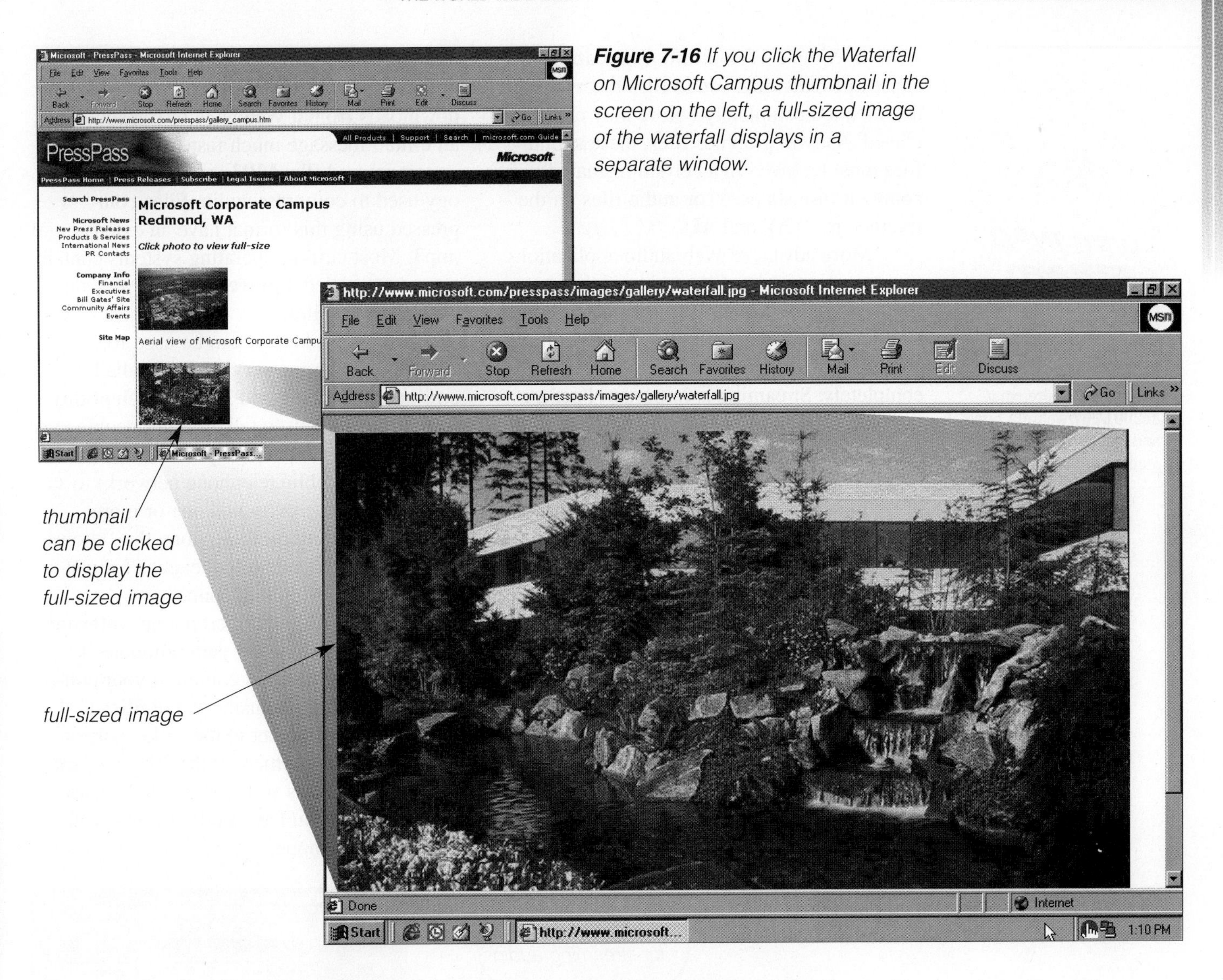

Figure 7-16 If you click the Waterfall on Microsoft Campus thumbnail in the screen on the left, a full-sized image of the waterfall displays in a separate window.

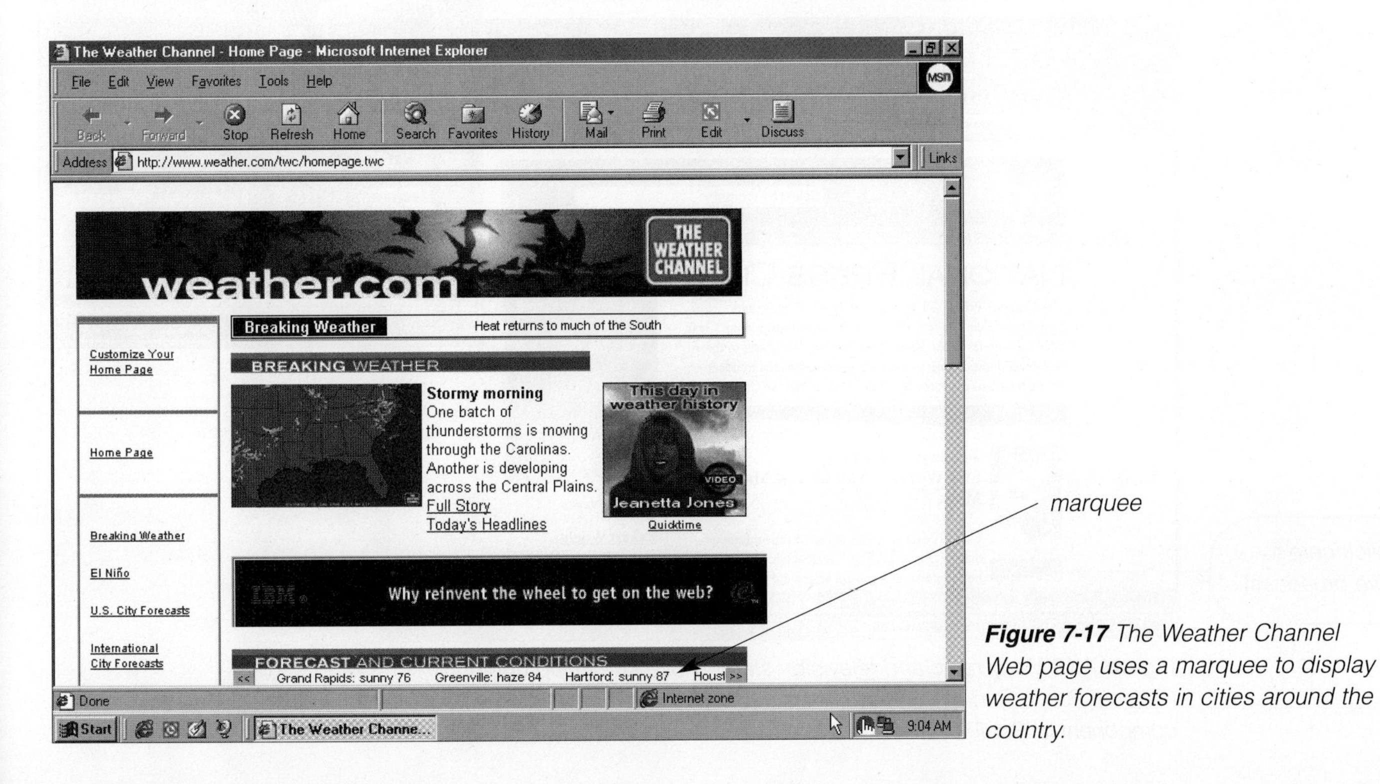

Figure 7-17 The Weather Channel Web page uses a marquee to display weather forecasts in cities around the country.

AUDIO Simple **Web audio applications** consist of individual sound files that must be downloaded completely before they can be played. As with graphics files, these sound files must be saved in a certain format. Two common formats used for audio files on the Internet are **WAV** and **AU**.

More advanced Web audio applications use streaming audio. **Streaming** is the process of transferring data in a continuous and even flow, which allows users to access and use a file before it has been transmitted completely. Streaming is important because most users do not have fast enough Internet connections to download a large multimedia file quickly. **Streaming audio**, enables you to listen to the sound (the data) as it downloads to your computer. Many radio and television stations use streaming audio to broadcast music, interviews, talk shows, sporting events, music videos, news, live concerts, and other segments (Figure 7-18). One accepted standard for transmitting audio data on the Internet is **RealAudio**, which is supported by most current Web browsers.

If, however, you download the audio completely before listening to it, the audio more than likely has been compressed using a file compression utility. Because the size of the compressed file is much smaller than its uncompressed version, a compressed file downloads from a Web page or transmits in an e-mail message much faster than an uncompressed file. **MP3** is a popular technology used to compress audio. Files compressed using this format have an extension of .mp3. Most current operating systems contain a program called a **player** that can play the audio in MP3 files.

Web-based audio can be used for **Internet telephone service**, also called **audioconferencing** and **Internet telephony**, which enables you to talk to other people over the Web. Internet telephony uses the Internet (instead of a public telephone network) to connect a calling party and one or more called parties. Internet telephony thus allows you to talk to friends or colleagues for just the cost of your Internet connection. As you speak into a computer microphone, **Internet telephone software** and your computer's sound card digitize and compress your conversation and then transmit the digitized audio over the Internet to the called parties. Software and equipment at the receiving end reverse the process so the receiving parties can hear what you have said, just as if you were on a telephone.

WEB INFO

For more information on streaming media, visit the Discovering Computers 2001 Chapter 7 WEB INFO page (**www.scsite.com/dc2001/ch7/webinfo.htm**) and click Streaming Media.

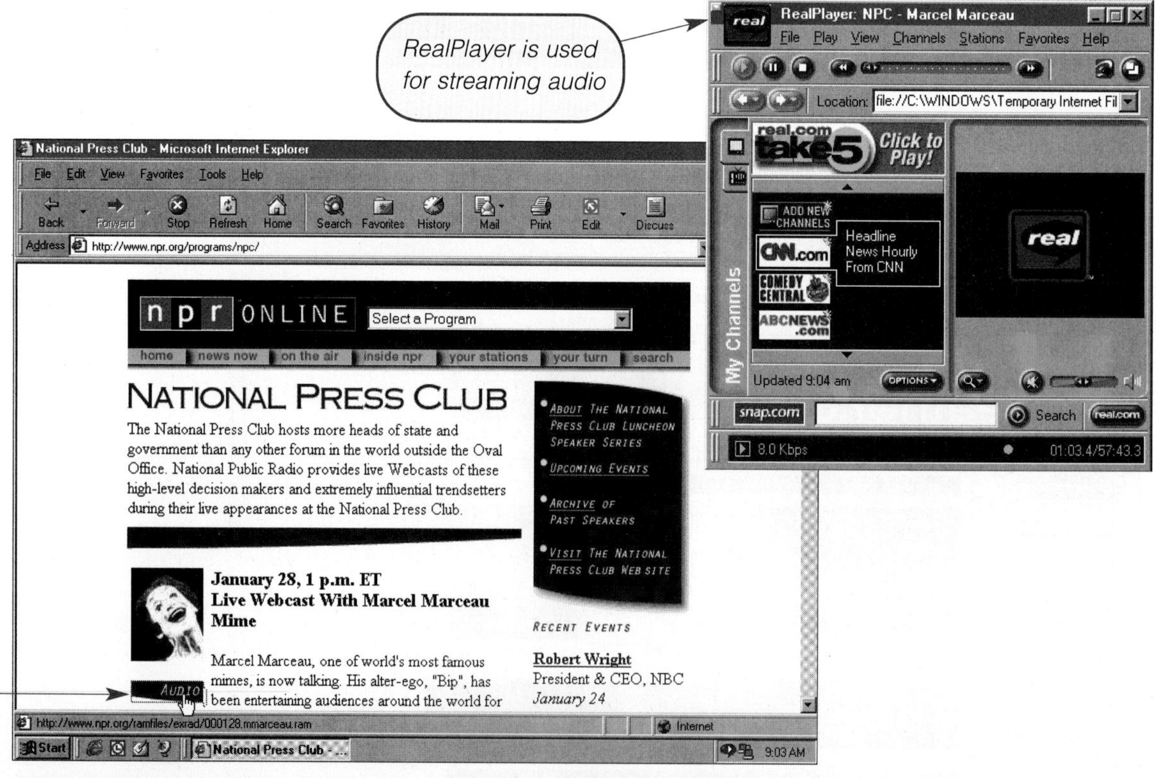

Figure 7-18 *Many radio and television stations use streaming audio. National Public Radio (NPR) transmits audio data in RealAudio, which is a component of RealPlayer.*

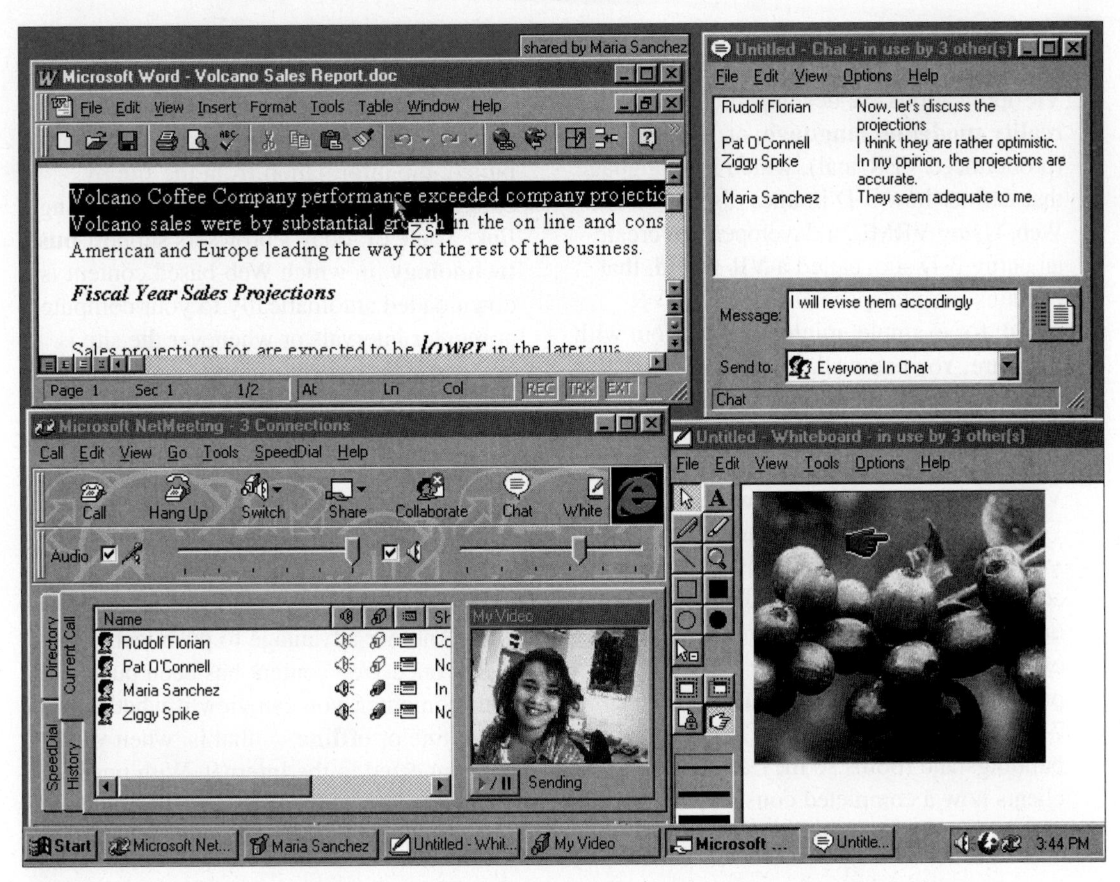

Figure 7-19 *NetMeeting, which is included with Internet Explorer, offers Internet telephony.*

Some of today's Web browsers include software such as CoolTalk and Microsoft NetMeeting, which supports Internet telephony (Figure 7-19). In addition to Internet telephony, these products typically offer additional services such as a whiteboard to display drawings, diagrams, and other graphics; chat tools to type text messages; and even videoconferencing so you can see images of the meeting participants.

VIDEO Like audio, simple **Web video applications** consist of individual video files, such as movie or television clips, that must be downloaded completely before they can be played on your computer. Because video files often are large and can take a long time to download, these video clips usually are quite short and compressed. A popular video compression standard is defined by the **Moving Picture Experts Group** (**MPEG**). Files in the MPEG format typically have an .mpg extension.

As with streaming audio, **streaming video** allows you to view longer or live video images as they are downloaded to your computer. A widely used standard for transmitting video data on the Internet is **RealVideo**. Like RealAudio, RealVideo, is a component of RealPlayer and is supported by most current Web browsers.

Streaming video also allows you to conduct Internet videoconferences, which work much like Internet telephony. As you are filmed by a video camera, videoconferencing software and your computer's video capture card digitize and compress the images and sounds.

This compressed data is divided into packets and sent over the Internet. Equipment and software at the receiving end assemble the packets, decompress the data, and present the image and sound as video. As with traditional videoconferencing, live Internet videoconferences can be choppy and blurry depending on the speed of the slowest communications link. As mentioned earlier, many products that support Internet telephony also have videoconferencing capabilities, thus allowing for face-to-face conversations over the Internet (see Figure 7-19).

VIRTUAL REALITY **Virtual reality (VR)** is the simulation of a real or imagined environment that appears as a three-dimensional (3-D) space. On the Web, VR involves the display of 3-D images that you can explore and manipulate interactively. Most Web-based VR applications are developed using **virtual reality modeling language**, or **VRML** (pronounced VER-mal), which is a language that defines how 3-D images display on the Web. Using VRML, a developer can create an entire 3-D site, called a **VR world**, that contains infinite space and depth. A VR world, for example, might show a room with furniture. You can walk through such a VR room by moving your pointing device forward, backward, or to the side. To view a VR world, you need a VRML browser or a VRML plug-in to a Web browser.

VR often is used for games, but it has many practical applications as well. Science educators can create VR models of molecules, organisms, and other structures for students to examine. Companies can use VR to showcase products or create advertisements (Figure 7-20). Architects can create VR models of buildings and rooms so they can show their clients how a completed construction project will look before it is built.

WEBCASTING

Until recently, when you wanted information from a Web site, you requested, or pulled, the information from the site by entering a URL in your browser or clicking a link. Some of today's browsers support **push technology**, in which Web-based content is downloaded automatically to your computer at regular intervals or whenever the site is updated. You can choose to have the entire site or just a portion of it, such as the latest news, pushed to your computer (Figure 7-21). You can choose to view the information immediately or access it later. Push technology thus saves you time by delivering information at regular intervals, without your having to request it.

Another advantage to push technology is that once Web content has been pushed to your computer, you can view it whether you are online or **offline** — that is, when you are not connected to the Internet. With push technology, the contents of one or more Web sites are downloaded to your hard disk while you

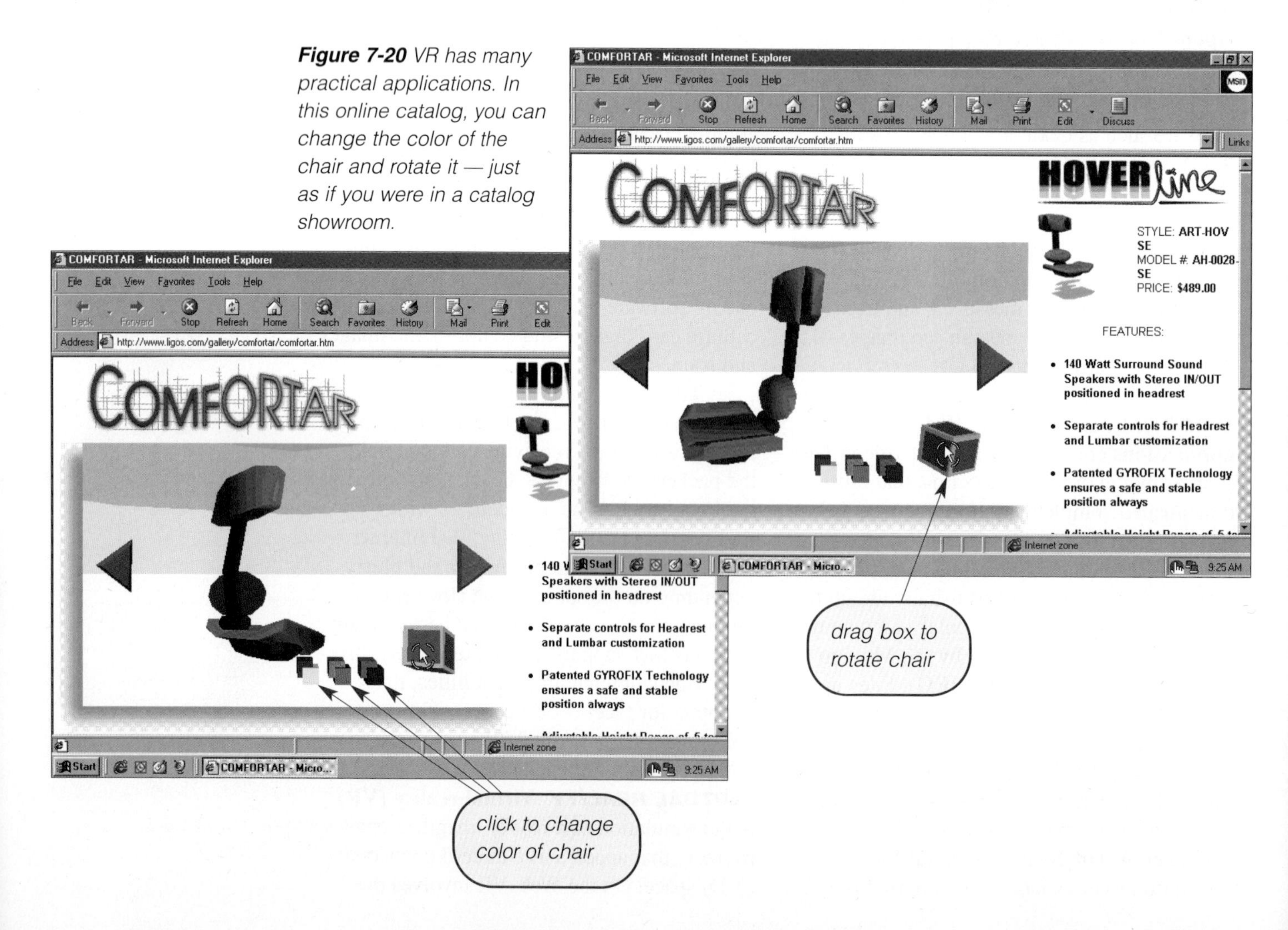

Figure 7-20 *VR has many practical applications. In this online catalog, you can change the color of the chair and rotate it — just as if you were in a catalog showroom.*

you are online, thus making them available for browsing while you are offline. Offline browsing is ideal for mobile users because they do not always have access to the Internet.

The concept of using push and pull technologies to broadcast audio, video, and text information is called **webcasting**. With webcasting, you receive customized Web content that is updated regularly and automatically. The information you receive is based on the preferences you choose, such as selecting certain channels and setting times for updates.

ELECTRONIC COMMERCE

When you conduct business activities online, you are participating in **electronic commerce**, also known as **e-commerce** (Figure 7-22). These commercial activities include shopping, investing, and any other venture that represents a business transaction or uses electronic data interchange.

One of the earliest forms of e-commerce was **electronic data interchange** (**EDI**), which is the transmission of business documents or data over communication lines.

Today, three types of e-commerce exist: business to consumer, consumer to consumer, and business to business. **Business to consumer** (**B-to-C** or **B2C**) **e-commerce** consists of the sale of goods to the general public. Just about any product or service you purchase from a traditional storefront also can be purchased on the Internet. For example, instead of visiting a computer retailer to purchase a computer, you can order one designed to your specifications directly from the manufacturer's Web site. With B2C e-commerce, transactions can occur instantaneously and globally, thereby saving money for both the business and the consumer.

A customer (consumer) visits an online business through its an electronic storefront. An **electronic storefront** contains descriptions, graphics, and a shopping cart. The **shopping cart** allows the customer to collect purchases. When ready to complete the sale, the customer often enters personal and credit card data through a secure Web connection. Instead of purchasing from a business, consumers can purchase from each other. When one consumer sells directly to another, such as in an online auction, this type of

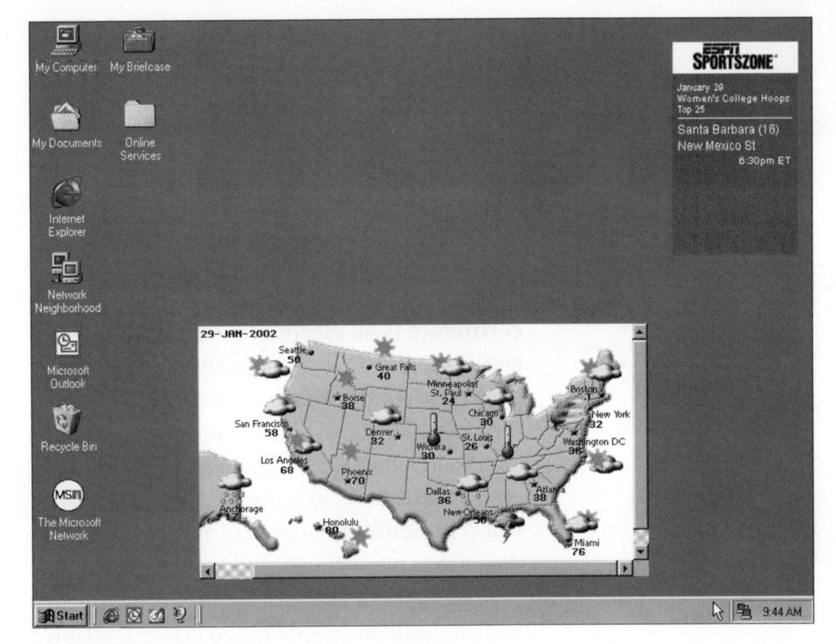

Figure 7-21 On this screen, the weather map and sports headlines are pushed onto the desktop.

e-commerce sometimes is called **consumer to consumer** (**C-to-C** or **C2C**).

Most e-commerce, though, actually takes place between businesses, which is called **business to business** (**B-to-B** or **B2B**) **e-commerce**. Businesses typically provide goods and services to other businesses, such as online advertising, recruiting, credit, sales, market research, technical support, and training.

WEB INFO

For more information on e-commerce, visit the Discovering Computers 2001 Chapter 7 WEB INFO page (**www.scsite.com/dc2001/ch7/webinfo.htm**) and click E-commerce.

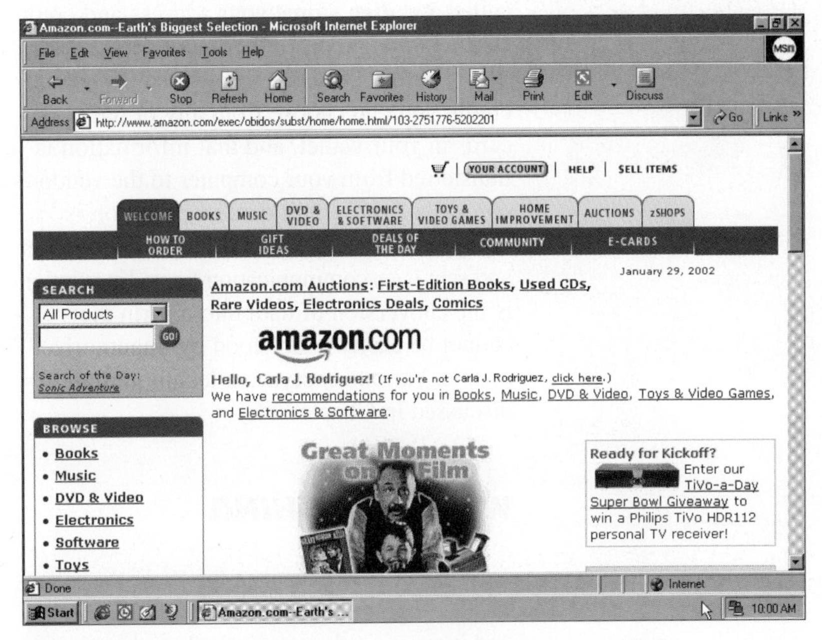

Figure 7-22 Online shopping is a popular e-commerce activity. Web sites, such as this one, allow you to purchase a range of products over the Internet.

If you are uncomfortable sending your credit card information over the Internet, some businesses accept **electronic money (e-money)**, also called **digital cash** or **e-cash**, as payment for goods and services. With one method, you purchase digital certificates from a **certificate authority** (**CA**), which is an authorized company or person that issues and verifies digital certificates. A **digital certificate** is an electronic credential that ensures a user is legitimate and the transfer of confidential materials is secure. Many banks, for example, allow you to purchase digital certificates that represent cash. Each digital cash certificate is assigned a unique number that represents an amount of money, just like the unique numbers on real currency.

Once you have the digital cash certificate, you can purchase goods or services over the Web. To do this, you transfer the digital cash certificate to the vendor, who then deposits the certificate number in a bank or transfers it to another vendor. Just like paper money, digital cash is reusable and anonymous; that is, the vendor has no information about the buyer. Because of this, many users prefer to pay with e-money instead of a credit card. Currently, no standard exists for e-money, and many companies offer various electronic payment schemes for use on the Internet.

For more information on Web publishing, visit the Discovering Computers 2001 Chapter 7 WEB INFO page (**www.scsite.com/dc2001/ch7/webinfo.htm**) and click Web Publishing.

A variation on e-money is **electronic credit** or an **electronic wallet**, which is a credit-card payment scheme for use on the Web. With electronic credit, a small program, called a **wallet**, stores your address and credit card information on your computer's hard disk. When you purchase something using electronic credit, you can choose from any of the credit cards in your wallet, and that information is transferred from your computer to the vendor's computer. Your credit card information is encrypted when it is stored on your computer and sent over communication lines. Encryption is the conversion of data into a form that cannot be easily understood by unauthorized people. Various encryption techniques are discussed in Chapter 14.

WEB PUBLISHING

Before the advent of the Web, the means to share opinions and ideas with others easily and inexpensively was limited to classroom, work, or social environments. Generating an advertisement or publication that could reach a massive audience required a lot of expense. Today, businesses and individuals can convey information to millions of people by using Web pages. Individual Web pages are sometimes called **personal Web pages**.

As mentioned earlier in this chapter, Web pages are created and formatted using **hypertext markup language (HTML)**, which is a set of special codes used to format a file for use as a Web page. These codes, called **tags**, specify how the text and other elements display in a browser and where the links lead. Figure 7-23 shows the HTML document used to create the Web page shown in Figure 7-24. Your Web browser translates the document with HTML tags into a functional Web page.

The development and maintenance of Web pages, called **Web publishing**, is fairly easy as long as you have the proper tools.

- To incorporate pictures in your Web pages, you could use a digital camera to take digital photographs or a scanner to convert your existing photographs and other graphics into digital format. It also would be beneficial to have a collection of clip art and/or other images, which you can download from the Web or purchase on CD-ROM or DVD-ROM.
- With a sound card, you can add sounds to your Web pages. A microphone allows you to include your voice in a Web page.
- To incorporate videos, you could use a PC camera or a video capture card and a video camera. Or, you can purchase a video digitizer to capture still photographs from videos.
- If you are comfortable using HTML tags, you can create an HTML document using any text editor or standard word processing software. You must save the HTML document as an ASCII file with an **.htm** or **.html** extension, instead of as a formatted word processing document.

To develop a Web page, however, you do not have to be a computer programmer — or even learn more than basic HTML. Instead, you can generate HTML tags with Web page authoring software — or just your word processing software. Many current word processing packages include Web page authoring features that help you to

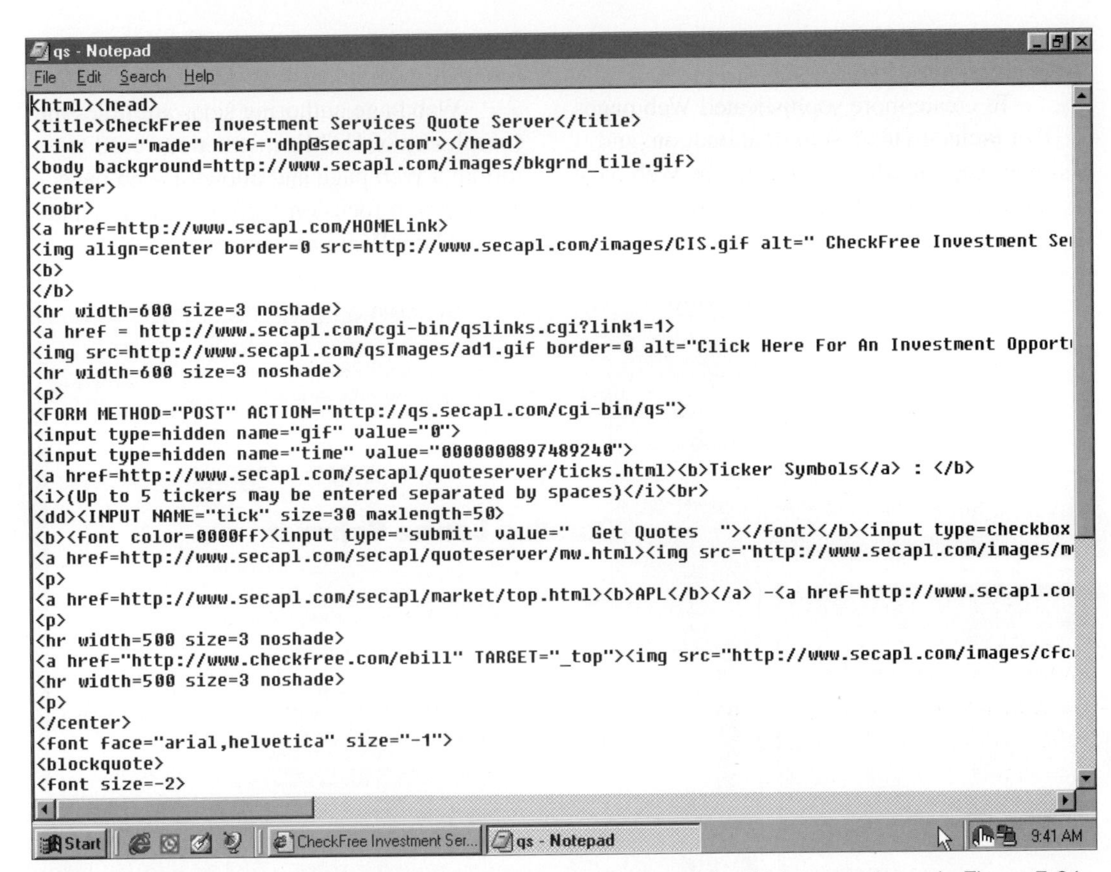

```
<html><head>
<title>CheckFree Investment Services Quote Server</title>
<link rev="made" href="dhp@secapl.com"></head>
<body background=http://www.secapl.com/images/bkgrnd_tile.gif>
<center>
<nobr>
<a href=http://www.secapl.com/HOMELink>
<img align=center border=0 src=http://www.secapl.com/images/CIS.gif alt=" CheckFree Investment Se
<b>
</b>
<hr width=600 size=3 noshade>
<a href = http://www.secapl.com/cgi-bin/qslinks.cgi?link1=1>
<img src=http://www.secapl.com/qsImages/ad1.gif border=0 alt="Click Here For An Investment Opport
<hr width=600 size=3 noshade>
<p>
<FORM METHOD="POST" ACTION="http://qs.secapl.com/cgi-bin/qs">
<input type=hidden name="gif" value="0">
<input type=hidden name="time" value="0000000897489240">
<a href=http://www.secapl.com/secapl/quoteserver/ticks.html><b>Ticker Symbols</a> : </b>
<i>(Up to 5 tickers may be entered separated by spaces)</i><br>
<dd><INPUT NAME="tick" size=30 maxlength=50>
<b><font color=0000ff><input type="submit" value="  Get Quotes  "></font></b><input type=checkbox
<a href=http://www.secapl.com/secapl/quoteserver/mw.html><img src="http://www.secapl.com/images/m
<p>
<a href=http://www.secapl.com/secapl/market/top.html><b>APL</b></a> -<a href=http://www.secapl.co
<p>
<hr width=500 size=3 noshade>
<a href="http://www.checkfree.com/ebill" TARGET="_top"><img src="http://www.secapl.com/images/cfc
<hr width=500 size=3 noshade>
<p>
</center>
<font face="arial,helvetica" size="-1">
<blockquote>
<font size=-2>
```

Figure 7-23 *This HTML document represents the top portion of the Web page shown in Figure 7-24. Web browser software interprets the HTML tags and displays the text, graphics, and links accordingly.*

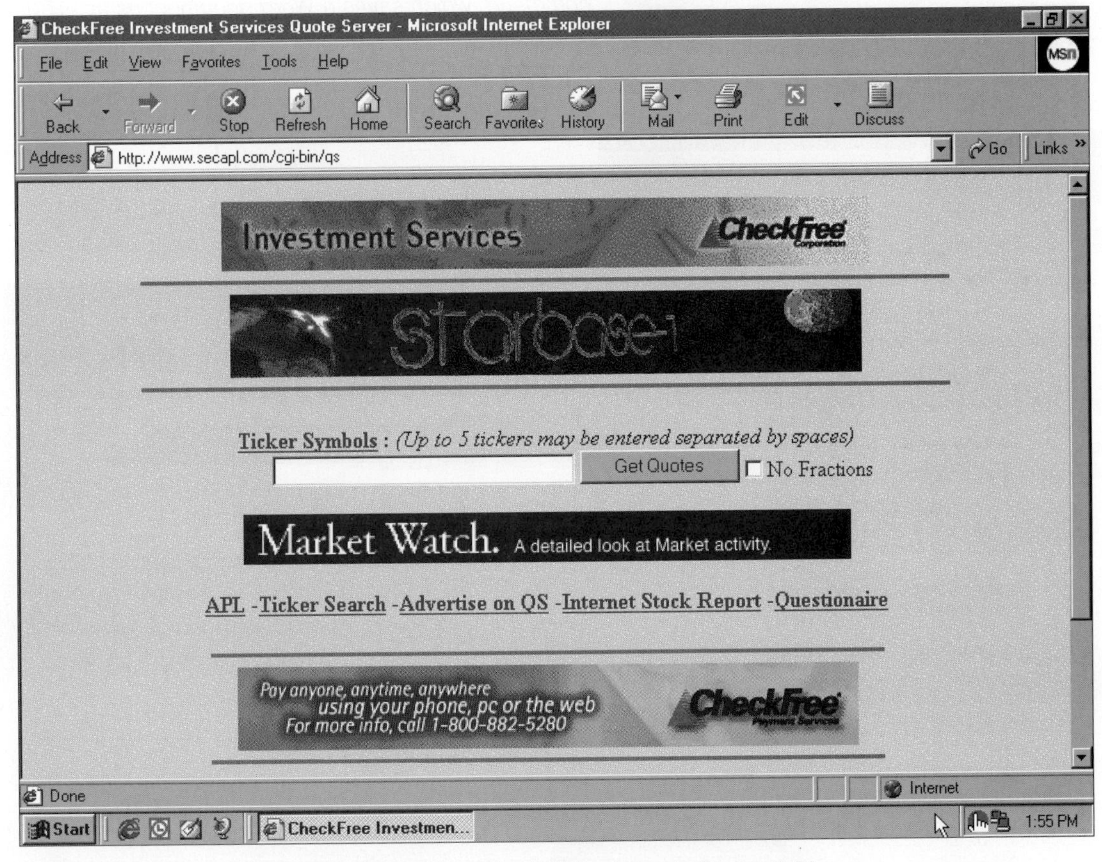

Figure 7-24 *The Web page generated from the HTML shown in Figure 7-23.*

create basic Web pages that contain text and graphics (Figure 7-25).

To create more sophisticated Web pages that include video, sound, animation, and other special effects, you can use Web page authoring software such as Adobe PageMill or Microsoft FrontPage (Figure 7-26). Both new and experienced users can create fascinating Web sites with this software.

Web page authoring software hides the complexity of HTML instructions used to format a Web page and allows a Web page developer to focus on Web page design and style. Many Web page authoring products

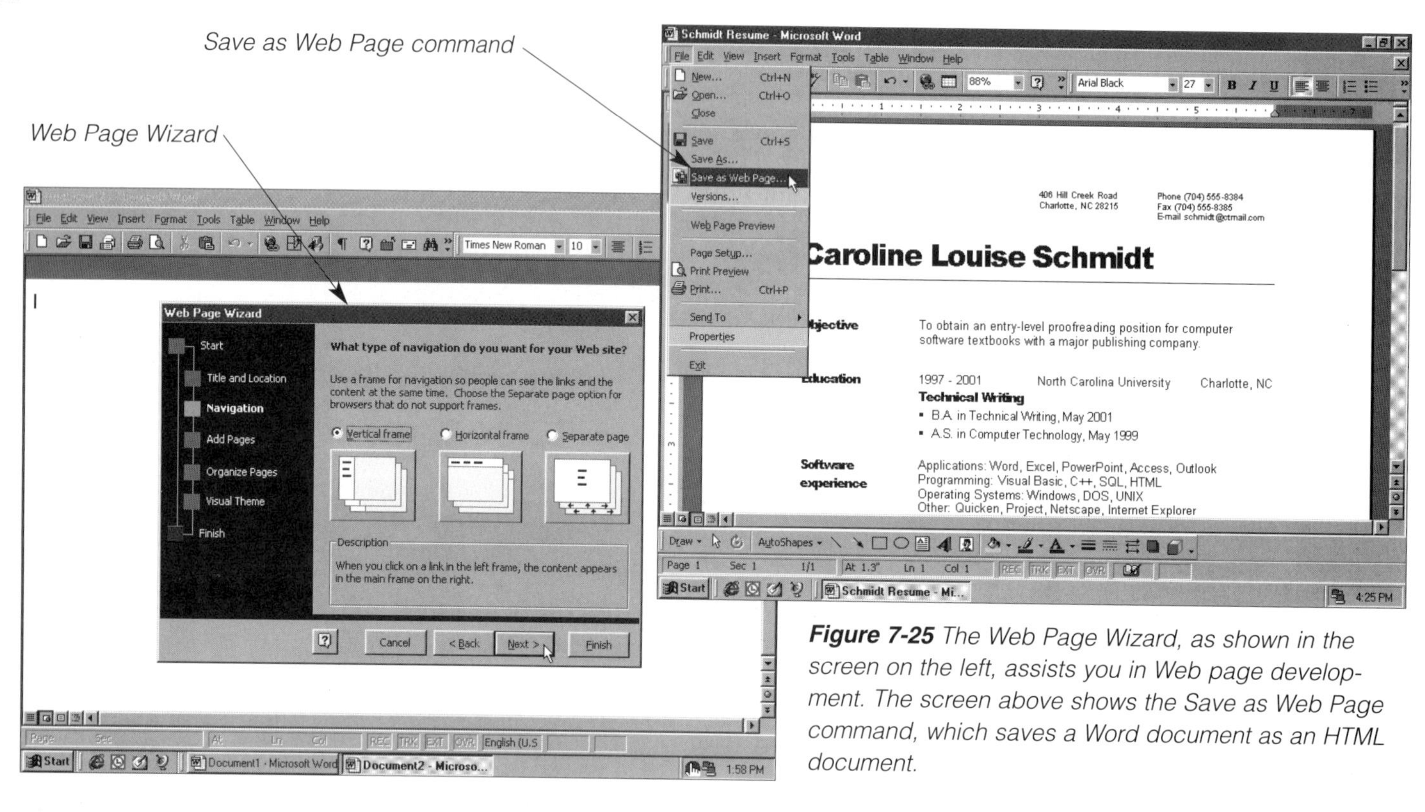

Figure 7-25 The Web Page Wizard, as shown in the screen on the left, assists you in Web page development. The screen above shows the Save as Web Page command, which saves a Word document as an HTML document.

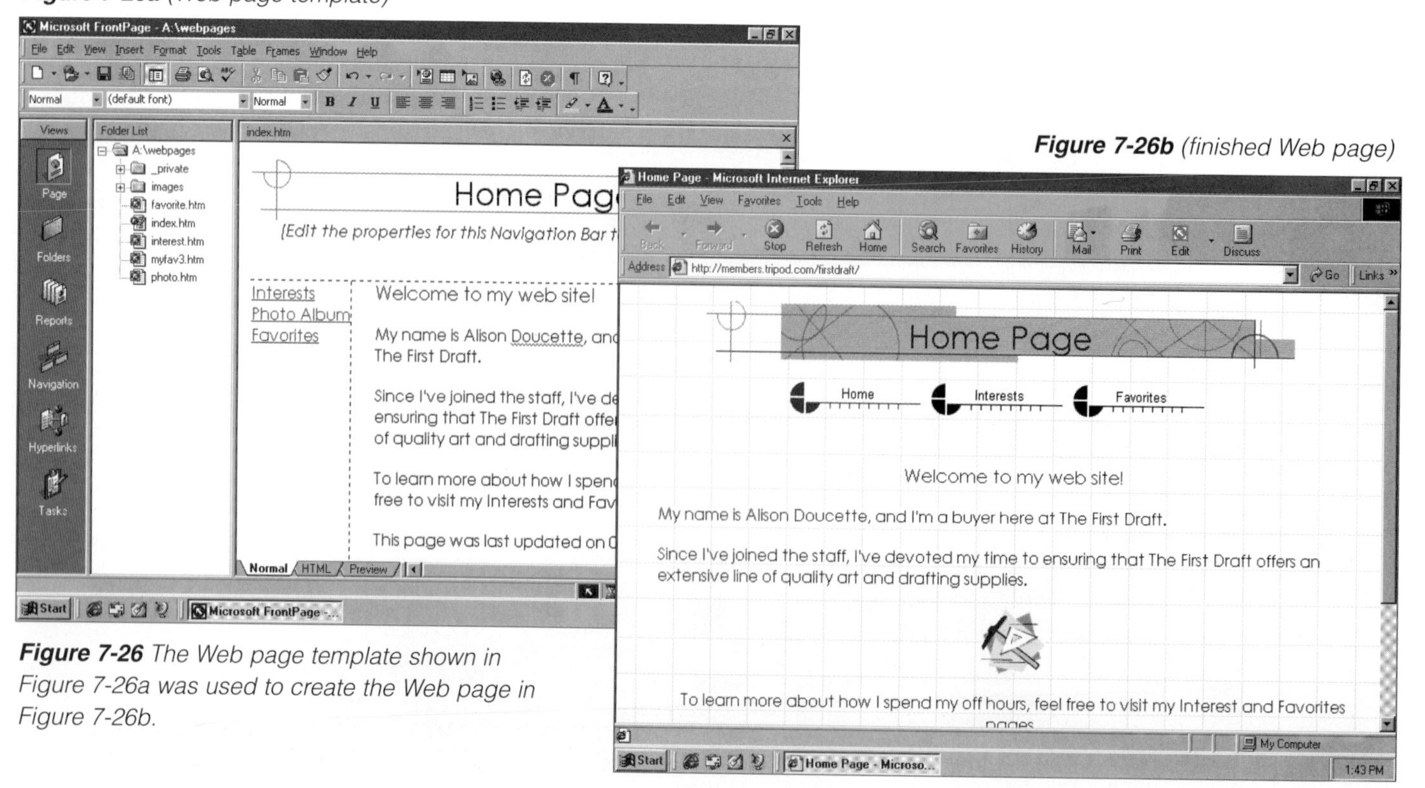

Figure 7-26a (Web page template)
Figure 7-26b (finished Web page)

Figure 7-26 The Web page template shown in Figure 7-26a was used to create the Web page in Figure 7-26b.

provide templates and wizards, along with collections of design elements such as bullets, backgrounds, patterns, fonts, and graphics, to assist you with Web page development.

After your Web pages are created, you store them on a Web server. Many ISPs and online services provide their customers with a Web address and 1 to 10 MB of storage on a Web server without an additional charge. If your ISP does not include this service, companies called **Web hosting services** provide storage for your Web pages for a reasonable monthly fee. The fee charged by a Web hosting service varies based on factors such as the amount of storage your Web pages require, whether your pages use streaming or other multimedia, and whether the pages are personal or for business use.

Once you have created a Web page and located a Web server to store it, you need to **upload** the Web page, or copy it, from your computer to the Web server. A common procedure used to upload files is FTP, which is discussed later in this chapter.

If you created the Web page using an Office 2000 program, such as Word, Excel, or FrontPage, you can use the Save As command on the File menu to save it to a Web server using FTP or Web folders. A **Web folder** is an Office 2000 shortcut to a Web server. You must contact the network administrator or technical support staff at your ISP to determine if the Web server supports Web folders, FTP, or both. You also must obtain necessary permissions to access the Web server.

To help others locate your Web site, you should register it with various search engines. Doing so ensures that your site will appear in the results returned for searches on keywords related to your site. Many search engines allow you to register your URL and keywords without cost.

Registering your site with the various search engines, however, can be an extremely time-consuming task. Instead, you can use a **submission service**, which is a Web-based business that usually offers free registration of your site with several search engines or a registration package in which you pay to register with hundreds of search engines (Figure 7-27).

In addition to supplying a title for your site, the URL, and a site description, the submission service probably will require you to identify several features of your site, such as whether the site is commercial or personal; a category and subcategory; and search keywords. For example, if your Web site business sells greeting cards, you could register under the Products and Services subcategory in the Business and Economy category, and specify keywords such as greeting cards, birthday cards, and anniversary cards.

Most Web page authoring software packages provide basic Web site management tools, allowing you to add and modify Web pages within the Web site. For more advanced features such as managing users, passwords, chat rooms, and e-mail, you need to purchase specialized **Web site management software**.

OTHER INTERNET SERVICES

Although the World Wide Web is the most talked about service on the Internet, many other Internet services are used widely. These include e-mail, FTP, Telnet, newsgroups, mailing lists, chat rooms, instant messaging, and portals. Each of these services is discussed in the following sections.

E-mail

E-mail (**electronic mail**) is the transmission of messages and files via a computer network. E-mail was one of the original services on the Internet, enabling scientists and researchers working on government-sponsored projects to communicate with colleagues at other locations. Today, e-mail quickly is becoming a primary communication method for both personal and business use.

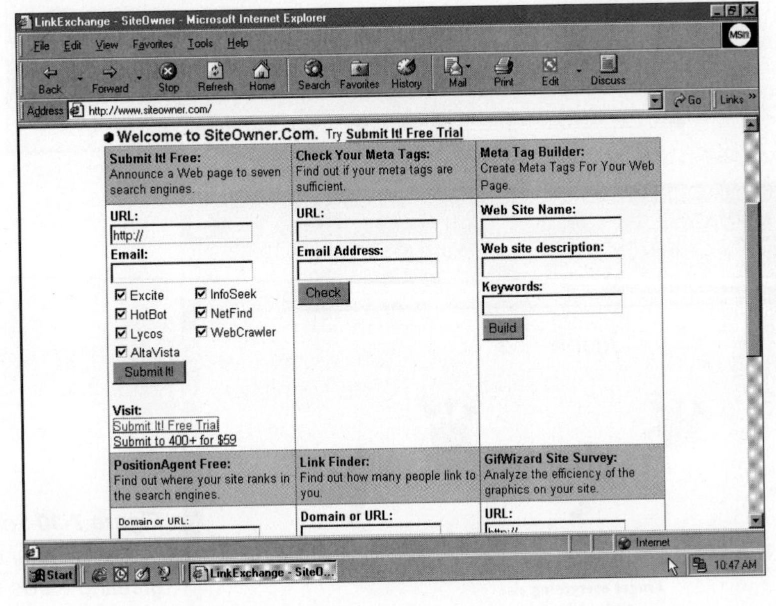

Figure 7-27 *Submit It! is a popular submission service that offers free registration to several search engines or a complete registration to hundreds of search engines for a fee.*

Using an **e-mail program**, you can create, send, receive, forward, store, print, and delete messages (Figure 7-28). To receive messages, you need an **e-mail address**, which is a combination of a user name and a domain name that identifies a user (Figure 7-29). When you receive an e-mail message, the message is placed in your mailbox. A **mailbox** is a storage location usually residing on the computer that connects you to the Internet, such as the server operated by your Internet service provider (ISP). The server that contains the mailboxes often is called a **mail server**. Most ISPs and online services provide an Internet e-mail program and a mailbox on a mail server as a standard part of their Internet access services.

Figure 7-28 *Outlook is an e-mail program that allows you to create, send, receive, forward, store, print, and delete messages.*

Some Web sites provide e-mail services free of charge. To use these Web-based e-mail programs, you connect to the Web site and set up an e-mail account, which typically includes an e-mail address and a password. Instead of sending e-mail messages, several Web sites provide services that allow you to send other items such as online invitations and greetings (Figure 7-30). These Web sites have a server that stores your messages, invitations, and greetings.

When you send a message, the message is transmitted according to a communications protocol called **SMTP** (**simple mail transfer protocol**). The mail server uses SMTP to determine how to route the message through the Internet and then sends the message. When the message arrives at the recipient's mail server, the message is transferred to a POP or POP3 server. **POP** (**Post Office Protocol**) is a communications protocol used to retrieve e-mail from a mail server. The POP server holds the message until the recipient retrieves it with his or her e-mail software (Figure 7-31). The newest version of POP is **POP3**, or **Post Office Protocol 3**.

Most e-mail programs allow you to send messages that contain graphics, audio and video clips, and computer files, as attachments. These attachments must be converted, or encoded, to binary format so they can be sent over the Internet and then decoded when the recipient retrieves them. Most e-mail

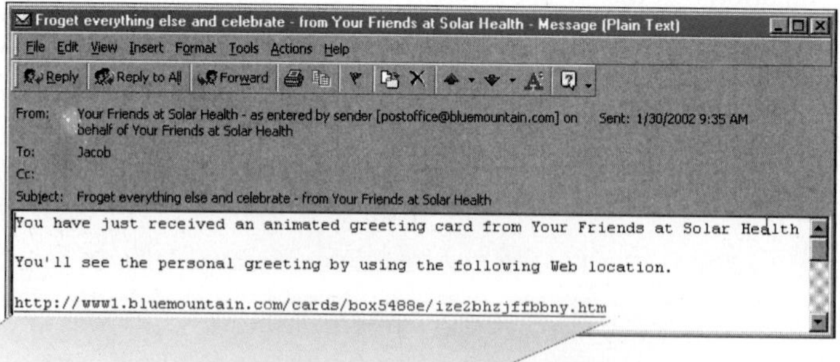

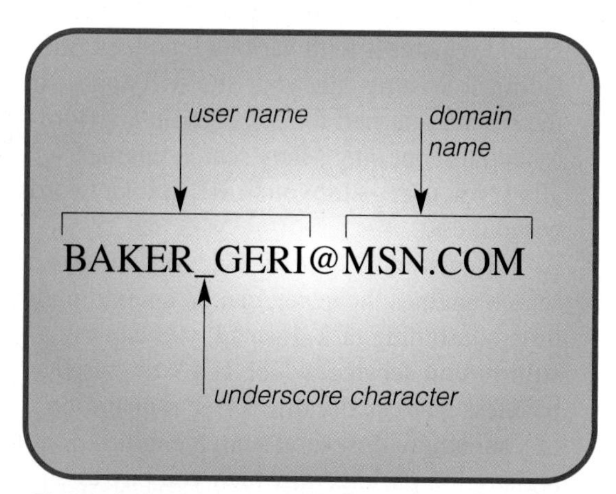

Figure 7-29 *An e-mail address is a combination of a user name and a domain name. The underscore character (_) often is used to separate sections of the user name.*

Figure 7-30 *Some Web sites, such as Blue Mountain Arts, allow you to send electronic greeting cards to any e-mail address.*

Figure 7-31
HOW AN E-MAIL MESSAGE TRAVELS FROM THE SENDER TO THE RECEIVER

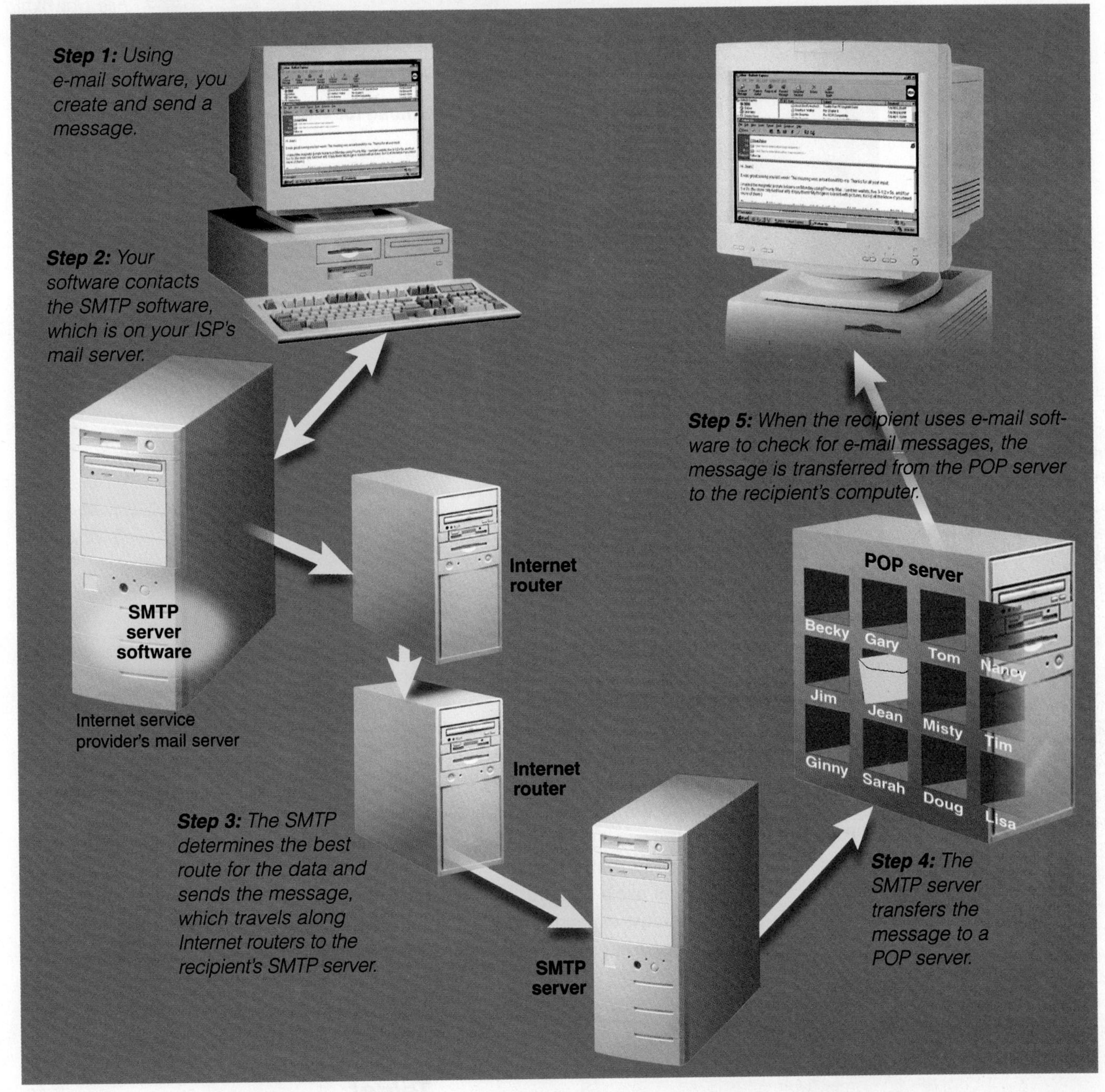

software includes **encoding schemes** that make it possible for attachments to arrive at the recipient's computer in the same format in which they left the sender's computer.

FTP

FTP (file transfer protocol) is an Internet standard that allows you to exchange files with other computers on the Internet. For example, if you click a link on a Web page that begins to download a file to your hard disk, you probably are using FTP (Figure 7-32 on the next page).

An **FTP server** is a computer that allows users to upload and download files using FTP. An FTP server contains one or more FTP sites. An **FTP site** is a collection of files including text, graphics, audio, video, and program files that reside on an FTP server. Some FTP sites limit file transfers to individuals who have authorized accounts (user names and passwords) on the FTP server. Many FTP sites allow **anonymous FTP**, whereby anyone can transfer some, if not all, available files. Microsoft, for example, has an FTP site that uses anonymous FTP to allow customers to download software updates, manuals, and

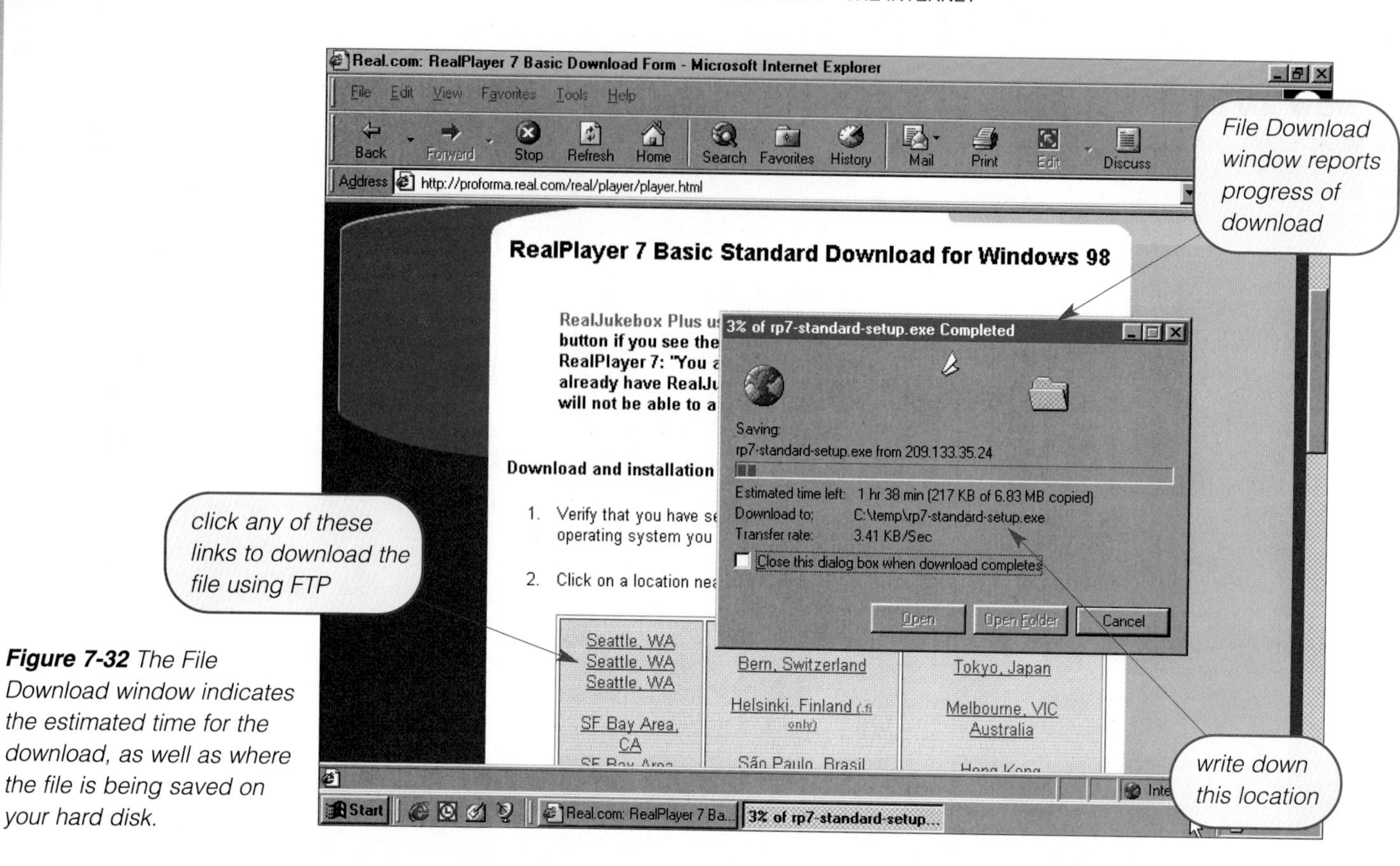

Figure 7-32 *The File Download window indicates the estimated time for the download, as well as where the file is being saved on your hard disk.*

other files. Many program files on anonymous FTP sites are freeware or public domain software, while others are shareware.

To view or use a file on an FTP site, you first must download it to your computer. In most cases, you click the file name to begin the download procedure. Large files on FTP sites often are compressed to reduce storage space and download transfer time.

Before you use a compressed file, you must expand it with a decompression program, such as WinZip. Such programs usually are available for download from the FTP site. Compression and decompression programs are discussed in more detail in Chapter 8.

In some cases, you may want to upload a file to an FTP site. For example, if you create personal Web pages, you will want to post them on a Web server. To do this, many Web servers require you to upload the files using FTP. To upload files from your computer to an FTP site, you use an **FTP program** (Figure 7-33). Many Internet service providers include an FTP program as part of their Internet access service; you also can download FTP programs from the Web.

Figure 7-33 *You use an FTP program such as WS_FTP to upload a file to an FTP site.*

contents of your computer

contents of FTP site

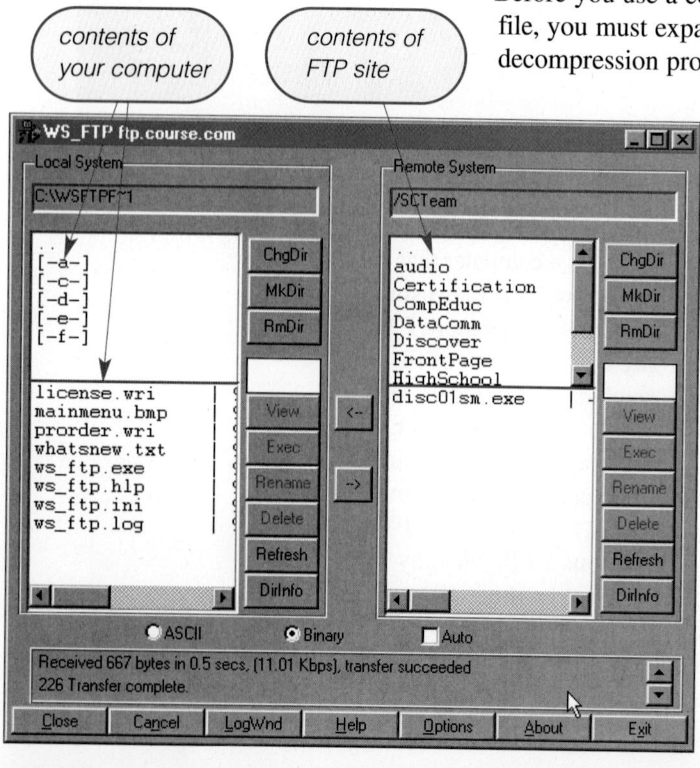

Telnet

Telnet is a program or command that enables you to connect to a remote computer on the Internet. To make a Telnet connection to the remote computer, you enter a user name and password. Once connected, your computer acts like a terminal directly linked to the remote computer. Telnet access to many remote computers is free, while others are accessed for a fee. For convenience, some remote computers provide the Telnet program.

Some uses of Telnet include connecting to a remote computer to access databases, directories, and library catalogs. Online databases may contain research results, and directories containing lists of information such as people and organizations.

The widespread use of Internet service providers and Web browsers, however, has reduced the need to log in directly to remote computers using Telnet. Many libraries, for example, have converted their library databases to allow Web-based access.

Newsgroups and Message Boards

A **newsgroup** is an online area in which users conduct written discussions about a particular subject. To participate in a discussion, a user sends a message to the newsgroup, and other users in the newsgroup read and reply to the message. The entire collection of Internet newsgroups is called **Usenet**, which contains thousands of newsgroups on a multitude of topics. Some major topic areas include news, recreation, business, science, and computers.

A computer that stores and distributes newsgroup messages is called a **news server**. Many universities, corporations, ISPs, online services, and other large organizations have a news server. Some newsgroups require you to enter your user name and password to participate in the discussion. These types of newsgroups are used when the messages on the newsgroup are to be viewed only by authorized members, such as students taking a college course.

To participate in a newsgroup, you use a program called a **newsreader**, which is included with most browsers. The newsreader enables you to access a newsgroup to read a previously entered message, called an **article**. You also can add an article of your own, a process called **posting**. A newsreader also keeps track of which articles you have and have not read.

Newsgroup members frequently post articles as a reply to another article — either to answer a question or to comment on material in the original article. These replies may cause the author of the original article, or others, to post additional articles related to the original article. The original article and all subsequent related replies are called a **thread** or **threaded discussion** (Figure 7-34). A thread can be short-lived or continue for some time, depending on the nature of the topic and the interest of the participants.

Using a newsreader, you can search for newsgroups discussing a particular subject, such as a type of musical instrument, brand of sports equipment, or employment opportunities. To help you determine what topics are discussed in a particular newsgroup, newsgroups are identified using a hierarchical naming system, with a major category divided into one or more subcategories. Each subcategory is separated by a period. Figure 7-35 on the next page lists major categories for newsgroups. If you like the discussion in a particular newsgroup, you can **subscribe** to it, which means its location is saved in your newsreader so you can access it easily in the future.

In some newsgroups, when you post an article, it is sent to a moderator instead of immediately displaying on the newsgroup. The **moderator** reviews the contents of the article and then posts it, if appropriate. Called a **moderated newsgroup**, the moderator decides if the article is relevant to the discussion. The moderator may choose to edit or discard inappropriate articles. For this reason, the content of a moderated newsgroup is considered more valuable.

WEB INFO

For more information on newsgroups, visit the Discovering Computers 2001 Chapter 7 WEB INFO page (**www.scsite.com/dc2001/ch7/webinfo.htm**) and click Newsgroups.

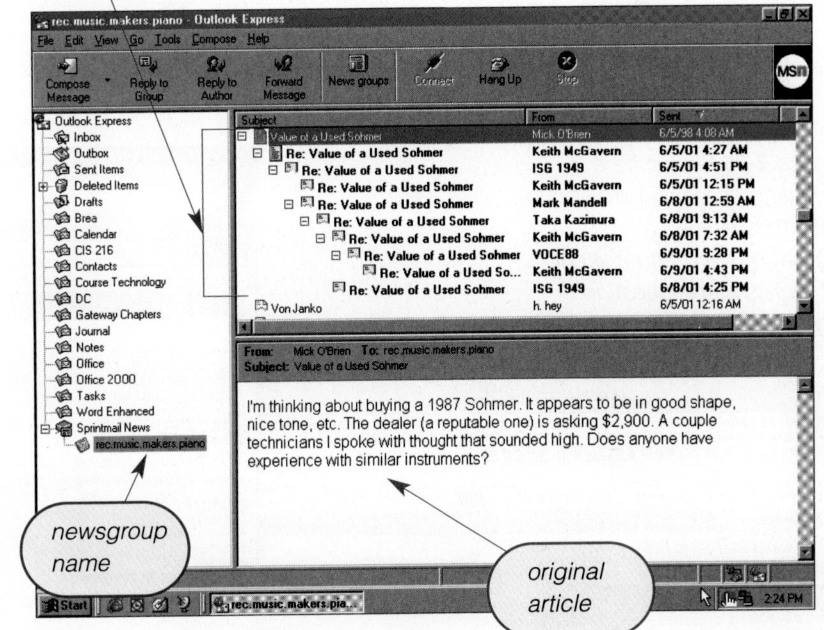

Figure 7-34 *A newsgroup is an online area in which users conduct written discussions about a particular subject. Shown is a thread about the value of a used piano model.*

NEWSGROUP CATEGORIES

Category	Description	Examples
alt	Alternative	alt.family-names.hendricks alt.music.paul-simon
biz	Business	biz.jobs biz.marketplace.computers
comp	Computer	comp.graphics.animation comp.sys.handhelds
news	Newsgroups	news.announce.newsgroups news.groups.reviews
rec	Recreation	rec.autos.antique rec.travel.misc
soc	Social	soc.college.gradinfo soc.culture.spain
talk	Talk	talk.environment talk.politics.misc

Figure 7-35 *Major categories of newsgroups.*

A popular Web-based type of discussion group that does not require a newsreader is a message board (Figure 7-36). Many Web sites provide a **message board**, also called a **discussion board**, service free of charge; that is, you can post messages, reply to messages, search for messages by keyword, and browse messages by category or topic. Message boards typically are easier to use than newsgroups.

Mailing Lists

A **mailing list** is group of e-mail names and addresses given a single name. When a message is sent to a mailing list, every person on the list receives a copy of the message in his or her mailbox. To add your e-mail name and address to a mailing list, you **subscribe** to it; to remove your name, you **unsubscribe** from the mailing list. Some mailing lists are called

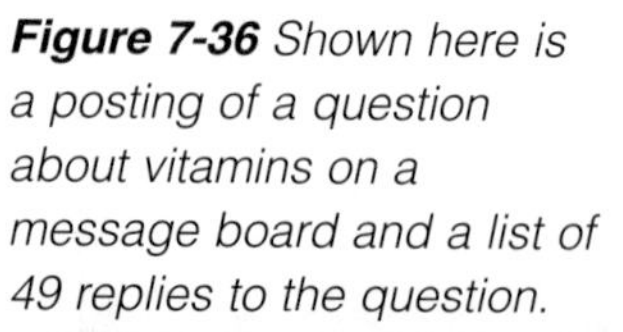

Figure 7-36 *Shown here is a posting of a question about vitamins on a message board and a list of 49 replies to the question.*

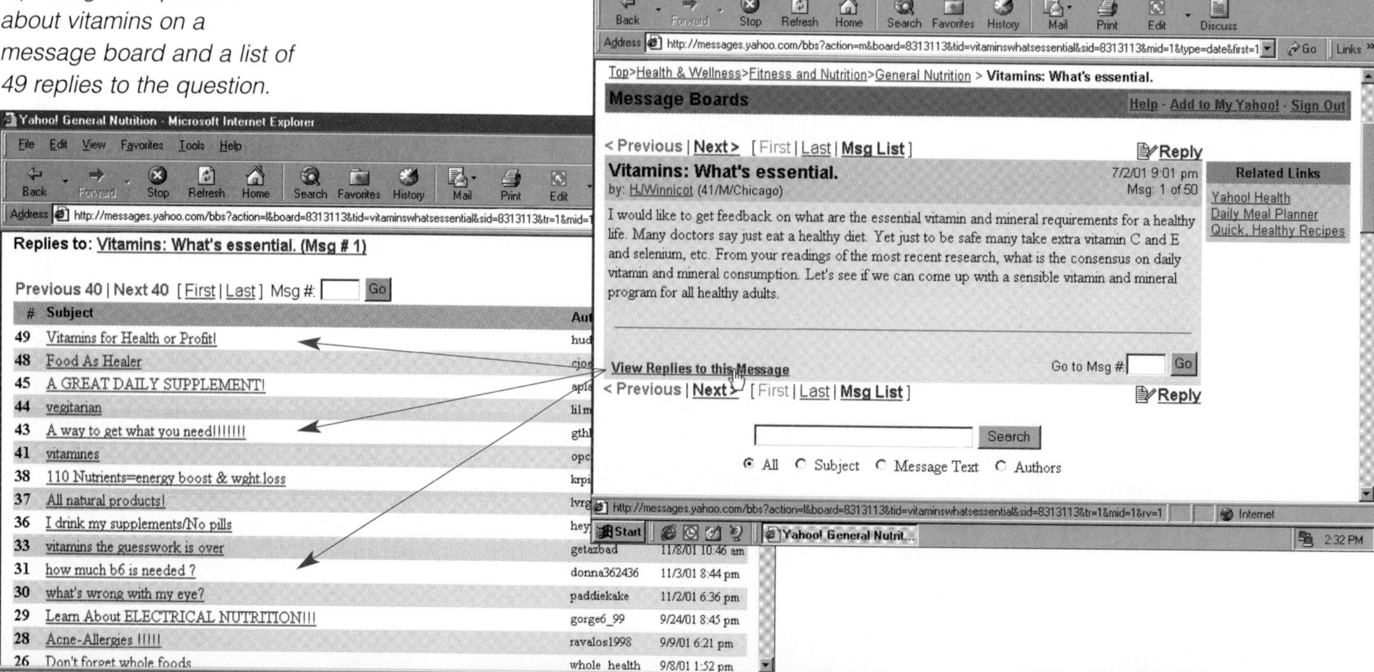

LISTSERVs, named after a popular mailing list software product.

Thousands of mailing lists exist on a variety of topics in areas of entertainment, business, computers, society, culture, health, recreation, and education. To locate a mailing list dealing with a particular topic, you can search for the keywords, mailing lists or LISTSERVs, using your Web browser.

Chat Rooms

A **chat** is a real-time typed conversation that takes place on a computer. **Real-time** means that you and the people with whom you are conversing must be online at the same time. When you enter a line of text on your computer screen, your words display on one or more participant's screens.

A **chat room** refers to the communications medium, or channel, that permits users to chat with each other. Anyone on the channel can participate in the conversation, which usually is specific to a particular topic. Each discussion is assigned a different channel. Some chat rooms support **voice chats** and **video chats**, where you hear or see others and they can hear or see you as you chat.

To start a chat session, you connect to a chat server through a **chat client**, which is a program on your computer. Today's browsers usually include a chat client. If yours does not, you can download a chat client from the Web. Some chat clients are text-based, such as IRC (Internet relay chat) and ichat, while others, such as Microsoft Chat, support graphical and text-based chats.

Once you have installed a chat client, you then can create or join a conversation on the chat server to which you are attached. The channel name should indicate the topic of discussion. The person who creates a channel acts as the channel operator and has responsibility for monitoring the conversation and disconnecting anyone whom becomes disruptive. Operator status can be shared or transferred to someone else.

Several Web sites exist for the purpose of conducting chats. Most chat sites allow participants to assume the role or appearance of a fictitious character.

Instant Messaging

Instant messaging (IM) is a service that notifies you when one or more people are online and then allows you to exchange messages or files with them or join a private chat room with them. Many services also can alert you to personal information such as calendar appointments, stock quotes, weather, or sports scores. IM can be used on desktop or laptop computers, as well as wireless devices such as handheld computers and smart phones.

Portals

A **portal** is a Web site designed to offer a variety of Internet services from a single, convenient location (Figure 7-37). Most portals offer the following free services: search engine; local, national, and worldwide news, sports, and weather; free personal Web pages; reference such as yellow pages, stock quotes, and maps; shopping malls and auctions; e-mail; instant messaging, message boards, calendars, and chat rooms. Some portals also provide Internet access and Web communities. A **Web community** is a Web site geared toward a specific group of people with similar interests or relationships. These communities usually offer a message board, chat room, e-mail, and photo albums to facilitate communications among members.

For more information on portals, visit the Discovering Computers 2001 Chapter 7 WEB INFO page (**www.scsite.com/dc2001/ch7/webinfo.htm**) and click Portals.

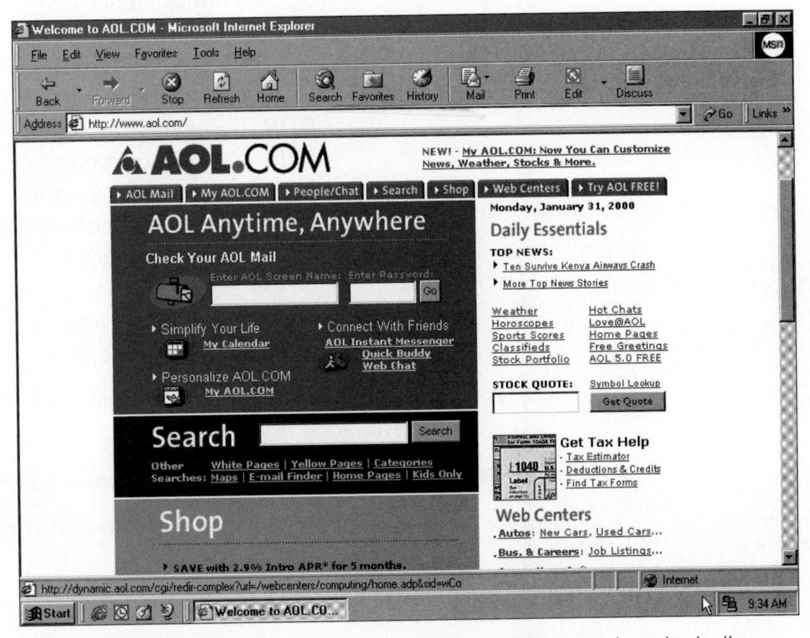

Figure 7-37 *A portal typically offers a variety of free services including a search engine, new, sports, weather, stock quotes, yellow pages, maps, and other reference information, shopping and auctions, calendars, e-mail, message boards, instant messaging, and chat rooms.*

Popular portals include AltaVista, AOL.COM, Disney's GO Network, Excite, InfoSeek, Lycos, MSN.com, Netscape Netcenter, Snap, and Yahoo!. The goal of these portals is to be designated as your browser's starting Web page.

NETIQUETTE

Netiquette, which is short for *Internet etiquette*, is the code of acceptable behaviors users should follow while on the Internet; that is, the conduct expected of individuals while online. Netiquette includes rules for all aspects of the Internet, including the World Wide Web, e-mail, FTP, Telnet, newsgroups, instant messaging, message boards, and chat rooms. Figure 7-38 outlines the rules of netiquette.

Netiquette

Golden Rule: *Treat others as you would like them to treat you.*

1. In e-mail, newsgroups, and chat rooms:
   - Keep messages brief, using proper grammar and spelling.
   - Be careful when using sarcasm and humor, as it might be misinterpreted.
   - Be polite. Avoid offensive language.
   - Avoid sending or posting **flames**, which are abusive or insulting messages. Do not participate in **flame wars**, which are exchanges of flames.
   - Avoid sending spam, which is the Internet's version of junk mail. **Spam** is an unsolicited e-mail message or newsgroup posting sent to many recipients or newsgroups at once.
   - Do not use all capital letters, which is the equivalent of SHOUTING!
   - Use **emoticons** to express emotion. Popular emoticons include

:)	Smile
:(	Frown
:l	Indifference
:\	Undecided
:o	Surprised

   - Use abbreviations and acronyms for phrases such as

BTW	by the way
FYI	for your information
FWIW	for what it's worth
IMHO	in my humble opinion
TTFN	ta ta for now
TYVM	thank you very much

   - Clearly identify a **spoiler**, which is a message that reveals a solution to a game or ending to a movie or program.
2. Read the **FAQ** (frequently asked questions), if one exists. Many newsgroups and Web sites have an FAQ.
3. Use your user name for personal purposes only.
4. Do not assume material is accurate or up to date. Be forgiving of other's mistakes.
5. Never read someone's private e-mail.

Figure 7-38 *The rules of netiquette.*

USING THE INTERNET: COOKIES AND SECURITY

While it is a vast and exciting resource, the Internet also is a public place, and as with all other public places, you should use common sense while there. The following sections explain guidelines for the use of cookies and security precautions for your consideration while using the Internet.

Cookies

Webcasting, e-commerce, and other Web applications often rely on cookies to track information about viewers, customers, and subscribers. A **cookie** is a small file that a Web server stores on your computer. Cookie files typically contain data about you, such as your user name or viewing preferences. Some Web sites send a cookie to your browser, which stores it on your computer's hard disk. The next time you visit the Web site, your browser retrieves the cookie from your hard disk and sends the data in the cookie to the Web site. Web sites use cookies for a variety of purposes.

- Web sites that allow for personalization often use cookies to track user preferences (Figure 7-39). On such sites, you may be asked to fill in a form requesting personal information, such as your name and site preferences. A news site, for example, might allow you to customize your viewing preferences to display business and sports news only. Your preferences are stored in a cookie on your hard disk.
- Online shopping sites generally use cookies to keep track of items in your shopping cart. This way, you can start an order during one Web session and finish it on another day in another session.
- Some Web sites use cookies to track how regularly you visit a site and the Web pages you visit while at the site.
- Web sites may use cookies to target advertisements. Your interests and browsing habits are stored in the cookie.

Although many believe that cookies allow other Web sites to read information on your computer, a Web site can read data only from its own cookie file; that is, it cannot access or view any other data on your hard

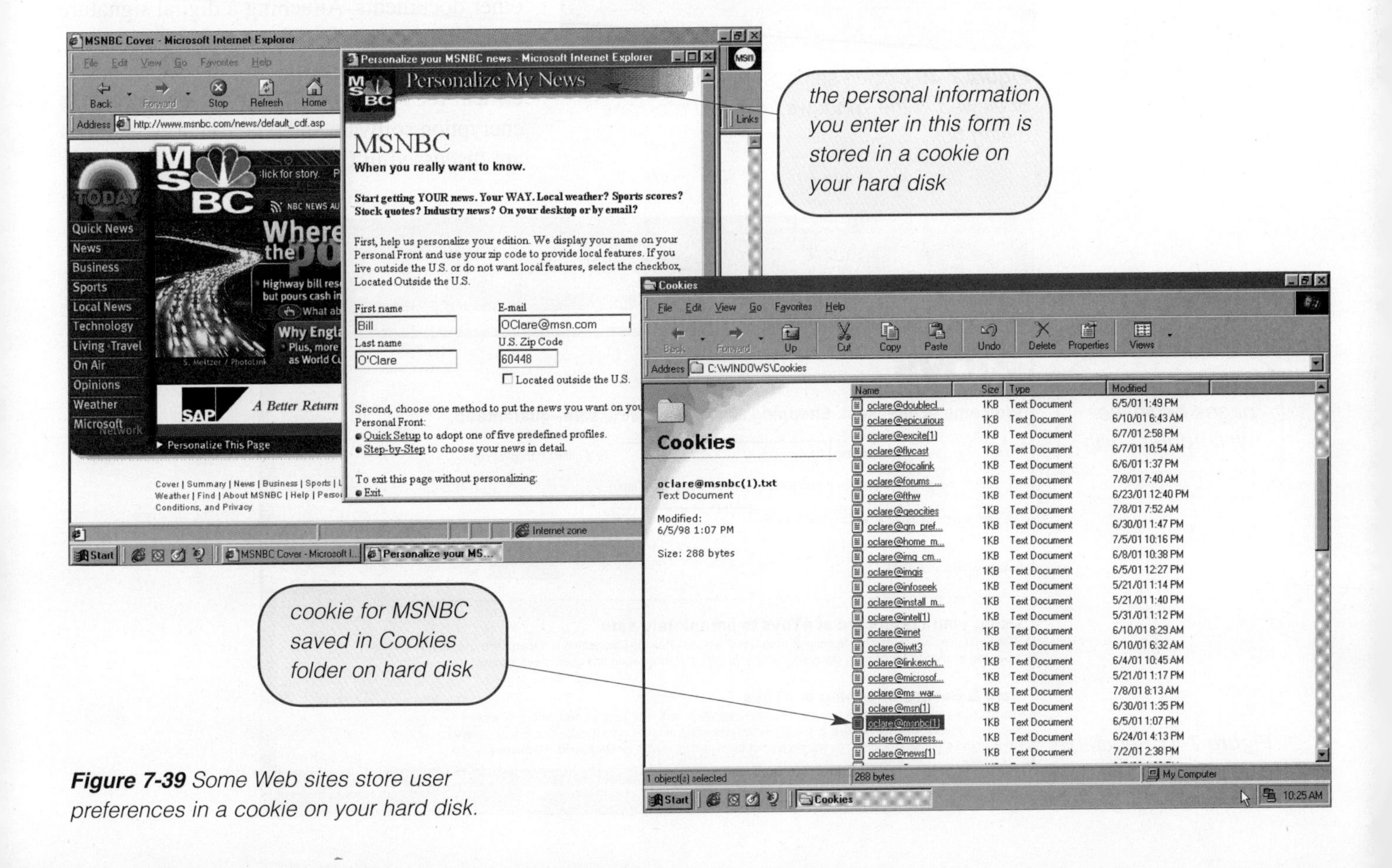

Figure 7-39 *Some Web sites store user preferences in a cookie on your hard disk.*

disk — including another cookie file. Some Web sites do, however, sell or trade infor-mation stored in your cookie to advertisers — a practice many believe to be unethical. If you do not want your personal information being distributed, you should limit the amount of information you provide to a Web site. You can set your browser to accept cookies automatically, prompt you if you wish to accept a cookie, or disable cookie use altogether (Figure 7-40).

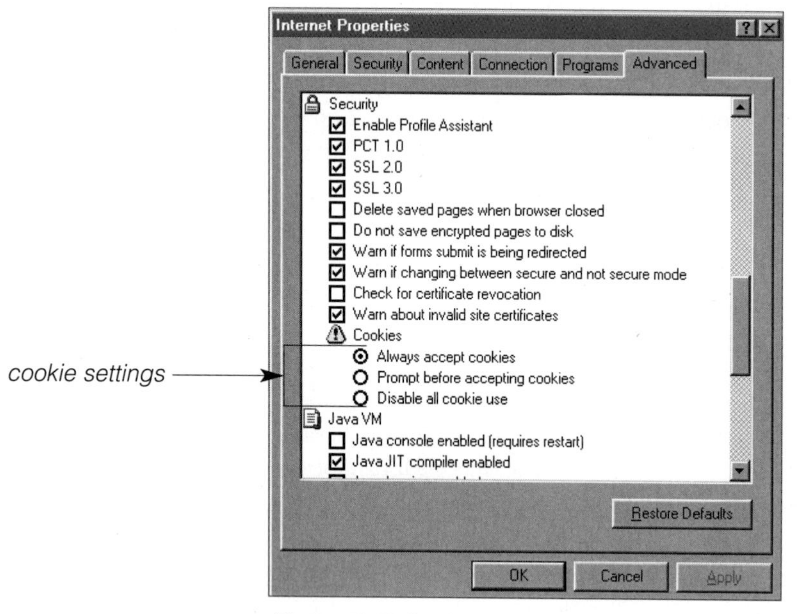

Figure 7-40 *Cookie settings are changed or viewed in the Windows Internet Properties dialog box.*

Internet Security

Even with netiquette guidelines, the Internet opens up the possibility for improper behaviors and content. For example, amidst the wealth of information and services on the Internet, some content may be inappropriate for certain people. Some Web sites, newsgroups, or chat rooms, for instance, contain content or discus-sions that are unsuitable for children. To assist parents with these types of issues, many browsers include software that can screen out unacceptable content. You also can purchase Internet **filtering software**, which allows parents, teachers, and others to block access to certain materials on the Internet.

Confidentiality is another important consideration on the Internet. For example, when shopping online, you should be sure that confidential or personal information such as your credit card number is encrypted during transmission. Reputable companies have secure servers that automatically encrypt this type of information while it is being transmitted.

One way to identify a secure Web page is to see if its URL begins with https://, instead of http:// (Figure 7-41). Many browsers today also include encryption software that allows you to encrypt e-mail messages or other documents. Attaching a digital signature to a document or message verifies your identity to the recipient, which is especially critical for e-commerce. Various types of encryption software and other security methods are discussed in Chapter 14.

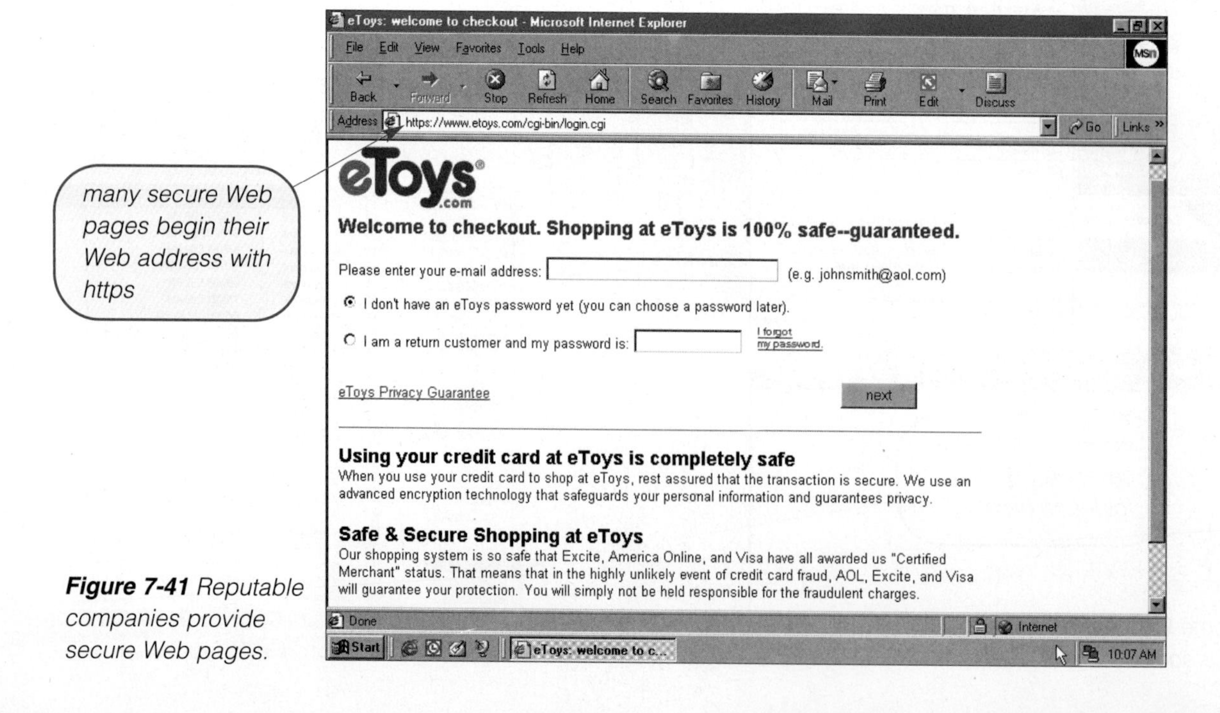

Figure 7-41 *Reputable companies provide secure Web pages.*

NETWORK COMPUTERS

For some applications, including many of the Internet services, a personal computer has more capability than the application requires. Jobs that primarily involve entering transactions or looking up information in a database — even viewing pages on the Web — do not require floppy disks, CD-ROMs, or large hard disks. These extra components contribute to both the cost and complexity of a personal computer; the more complex a computer is, the more expensive it is to maintain. As the costs of owning, operating, and maintaining a personal computer increase, many users are turning to network computers.

A **network computer** (**NC**), sometimes called a **thin-client computer**, is a less expensive version of a personal computer designed specifically to connect to a network, especially the Internet. Most network computers cannot operate as a stand-alone computer; that is, they must be connected to a network to be functional. Network computers typically rely on the network for storage and, therefore, usually do not have a hard disk or CD-ROM drive (Figure 7-42).

A specific type of network computer used in business applications is the **network personal computer**, or **NetPC**, which was designed cooperatively by Microsoft and Intel. A NetPC primarily relies on the server for software and storage but does have a hard disk for storing some data and programs. A NetPC can run Java and other programs such as Microsoft Windows applications.

WEB APPLIANCES

Many consumers today use Web appliances at home or on the road. A **Web appliance**, sometimes called an **Internet appliance**, is a device designed specifically to connect to the Internet. Among the variety of Web appliances on the market, four of the more popular are the set-top boxes, smart phones, smart pagers, and Web-enabled PDAs.

A **set-top box**, such as WebTV™, sits on top of your television set and allows you to access the Internet and navigate Web pages using a device that looks like a remote control (Figure 7-43). Set-top boxes have become

WEB INFO

For more information on set-top boxes, visit the Discovering Computers 2001 Chapter 7 WEB INFO page (**www.scsite.com/dc2001/ch7/webinfo.htm**) and click Set-top Boxes.

Figure 7-42 *This thin-client computer is used in schools, businesses, medical facilities, and other applications designed specifically to connect to a network, especially the Internet.*

Figure 7-43 *With a device such as WebTV™, you can access the Internet and navigate Web pages using a device that looks like a remote control.*

popular in hotel rooms, allowing users to check out, order food, rent movies, and access the Internet without leaving the room.

A **smart phone** is a cellular phone that, in addition to other features, allows you to send and receive messages on the Internet and browse Web sites specifically configured for display on a phone (Figure 7-44). A little larger than a standard pager, a **smart pager** is a two-way pager whose features include sending and receiving e-mail and receiving news alerts from the Web. Several handheld computers and Personal Digital Assistants (PDAs) on the market today are Web-enabled, which means you also can access the Web through these portable computers.

Some of these devices, such as the smart phone, smart pager, and Web-enabled PDAs communicate with the Web wirelessly. Wireless Web communications are discussed in the next section.

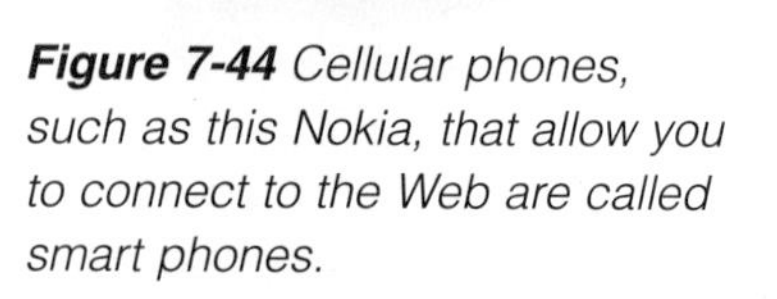

Figure 7-44 *Cellular phones, such as this Nokia, that allow you to connect to the Web are called smart phones.*

WIRELESS WEB COMMUNICATIONS

Today, many users connect to the Web wirelessly. That is, instead of using wires and cables for Internet connections, they typically use cellular or satellite services. This section discusses how devices connect wirelessly to the Web and the features of some of these Web-enabled devices, which includes portable computers; handheld devices such as PDAs, smart pagers, and smart phones; and hands-free (voice activated) Internet devices found in automobiles. Chapter 9 discusses the specifics of cellular and satellite communications.

To connect to the Internet using wireless technology, you use an Internet service provider (ISP) that supplies wireless connections to the Internet. When you purchase a wireless device or modem, it often includes information about a wireless provider that can communicate with the device or modem (Figure 7-45). Once you subscribe to this service, the wireless provider then communicates with an antenna on your Web-enabled device.

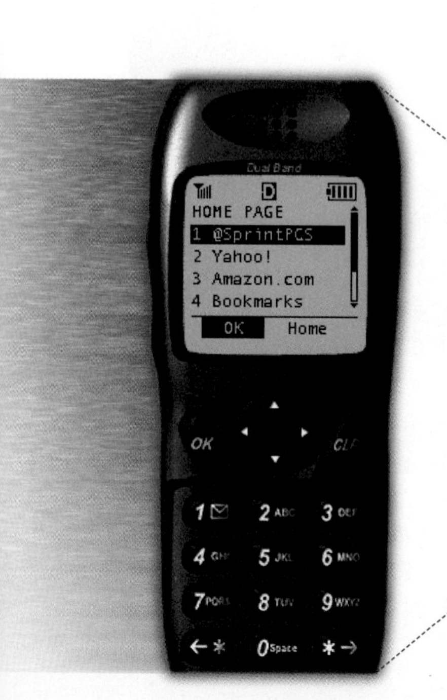

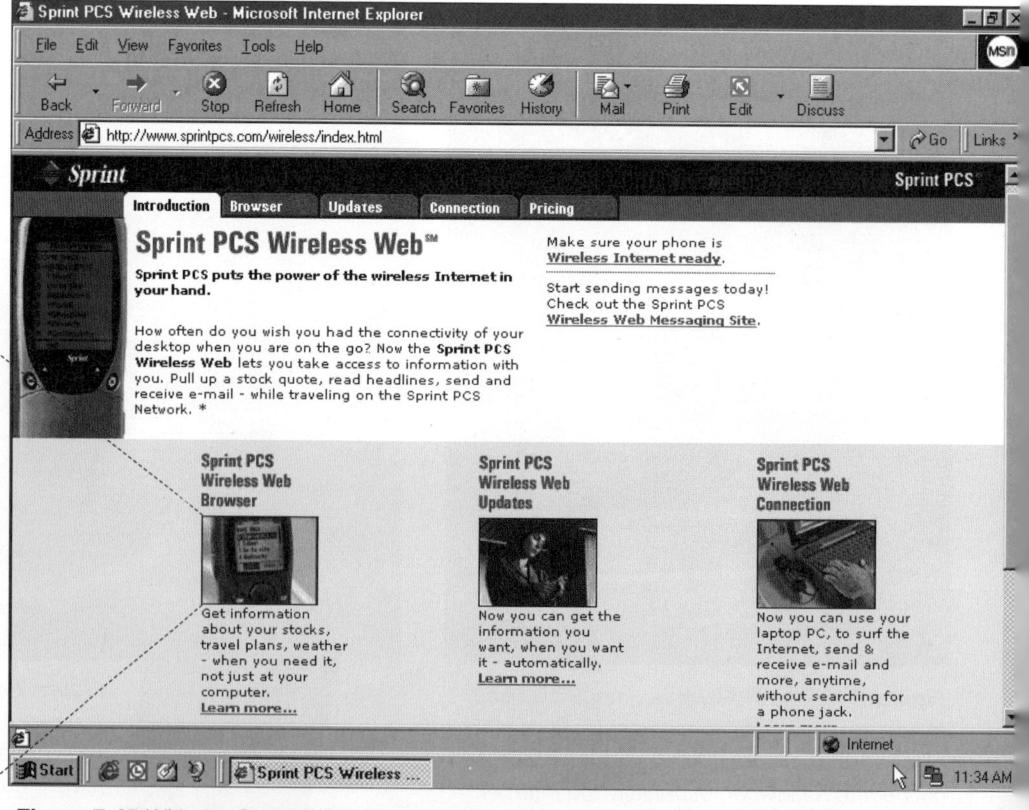

Figure 7-45 *With the Sprint PCX Wireless Web, you can access the Web directly on the smart phone or you can connect the phone to your laptop computer – providing wireless access from your laptop computer.*

A development specifically designed for Web-enabled devices is the **wireless portal**, which is a single Web site that attempts to provide all information a wireless user might require. These wireless portals offer services such as search engines, news, stock quotes, weather, maps, e-mail, calendar, instant messaging, and shopping.

Because of the slow download times of many Web sites, you do not browse the Web with these devices; instead, you use Web clippings. A **Web clipping** is information you request from a Web site that is targeted specifically for display on a Web-enabled device. Figure 7-46, for example, illustrates a Web clipping on a Palm Computing device that requests travel information. Many Web sites have developed Web clippings for use on Web-enabled devices.

The future of wireless Web communications continues to grow rapidly. For example, you soon will be able to send digital cash wirelessly using devices such as smart phones and Web-enabled PDAs. This service will give mobile users the freedom to transfer cash from areas not equipped with standard telephone service such as an auction, a restaurant, or a coffee shop.

Figure 7-46

HOW TO USE A TRAVEL WEB CLIPPING APPLICATION ON A PALM COMPUTING DEVICE

Step 1: *Raise the device's antenna and then tap the MapQuest icon.*

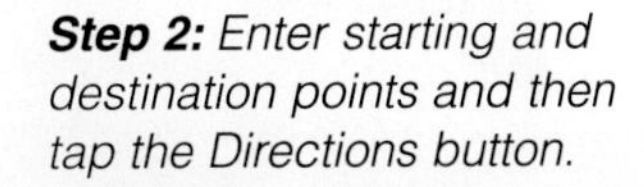

Step 2: *Enter starting and destination points and then tap the Directions button.*

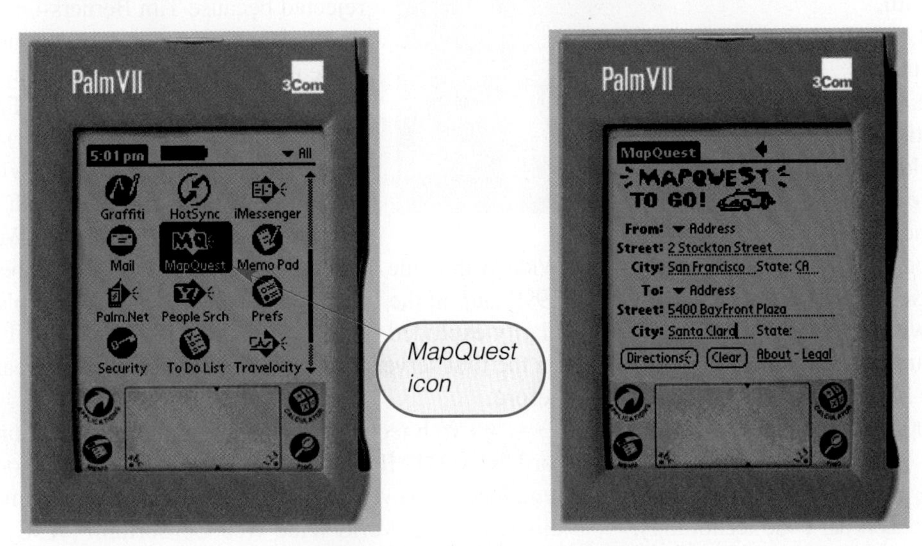

Step 3: *Scroll through directions. Read total distance and estimated travel time below directions.*

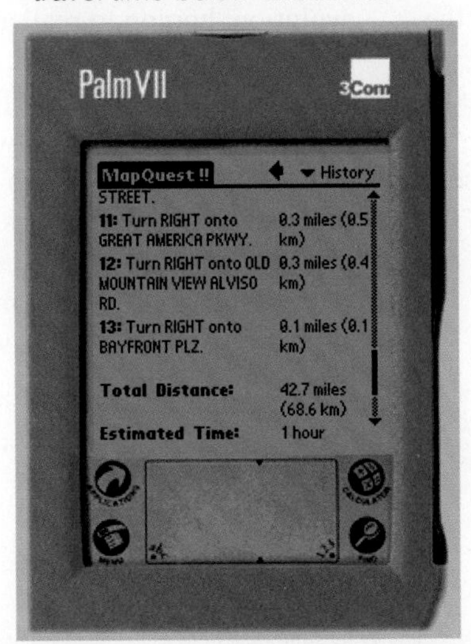

TECHNOLOGY TRAILBLAZER

TIM BERNERS-LEE

Tim Berners-Lee is the creator of the World Wide Web, the innovation that revolutionized the way people use computers and obtain information. Millions access the Web every day, and many wonder why he did not commercialize his brainchild. Berners-Lee responds, "It's a strange question. By asking the question, people are suggesting that they respect people as a function of their net worth. That's worrying. It's not an assumption I was brought up with."

Tim Berners-Lee learned his values growing up in London, the child of mathematicians who met while working with the Ferranti Mark I, one of the first computers sold commercially. Berners-Lee had an early interest in both mathematics and electronics, playing with five-hole paper tape and making toy computers out of cardboard boxes. He attended Oxford University's Queen's College where, as a compromise between mathematics and electronics, he studied physics. While at Oxford, he also made his first working computer with a soldering iron, an M6800 processor, and an old television.

After graduating in 1976, Berners-Lee worked with transaction systems, message relays, bar code technology, typesetting software, multitasking operating systems, real-time control firmware, generic macro languages, and graphics and communications software. Looking back, his most significant work may have been a software program, called Enquire, that he wrote for his own use at CERN, the European Laboratory for Particle Physics based in Geneva, Switzerland. The program was inspired by the English edition of a book titled, *Enquire Within Upon Everything*. Enquire, which was intended as a resource you could use to learn about anything, foreshadowed the World Wide Web.

Computers always had stored information and worked with it mechanically, in tables and hierarchies. "One of the things computers have not done," Berners-Lee wrote, "is to be able to store random associations between disparate things, although this is something the brain has always done relatively well." Enquire not only stored text but also stored *random associations* in the form of hypertext links that pointed to related documents. Enquire's associations were limited, but in 1989, Berners-Lee suggested a universal system to let people everywhere share knowledge in a collection of hypertext documents. Associations would be sweeping and links could point to any document — primitive or polished, personal or public, everyday or exotic.

Early names for the project were rejected because Tim Berners-Lee was uncomfortable with their acronyms — Mine of Information (MOI) seemed self-centered ("moi" is "my" in French), and The Information Mine (TIM) was even worse. Finally, the World Wide Web (WWW) was chosen as both descriptive and memorable. The World Wide Web made its debut within CERN in December of 1990 and on the Internet at large in 1991. The Web's popularity exploded — each year from 1991 to 1994, the load on the first server was ten times greater than the year before.

As Web technology spread, initial specifications for URLs, HTML, and HTTP were refined. Questions remained, however, and in 1994 Berners-Lee became the director of the World Wide Web Consortium (W3C) based at the Massachusetts Institute of Technology (MIT). The Consortium consisted of, in 1999, more than 300 organizations — including Microsoft, IBM, Hewlett-Packard, and Lotus — that consider issues in the Web's evolution, hoping to realize its full potential and ensure its stability.

"We have to be careful," Berners-Lee warns, "because the sort of Web we end up with and the society we end up building on top of it will be determined by the decisions we make." Users shape the Web's future. The places people browse, the sites they setup, and the links they establish affect how the Web develops. "People have to be aware of this. We have the answers in our own hands."

COMPANY ON THE CUTTING EDGE

YAHOO!

One of the more recognizable brands on the World Wide Web has one of the more unusual names: Yahoo!. Yahoo!, a leading global Internet media, is visited by millions of people worldwide every day. Supposedly, the name is an acronym for Yet Another Hierarchical Officious Oracle, but the co-founders claim they simply consider themselves yahoos (or tough guys). For millions of grateful users, however, the name is the cry of joy they utter when Yahoo! (www.yahoo.com) guides them to an elusive Web site. Whatever the origin, today Yahoo! is one of the more popular destinations.

Jerry Yang and David Filo, Ph.D. candidates in electrical engineering at Stanford University, started Yahoo! in 1994. Envisioned as a way of organizing their personal interests on the Internet, they began with collections of lists. As the lists became cumbersome, Yang and Filo transformed them into a tree-style database. Thousands of people soon were accessing the database, and software was developed to find, catalog, and edit material. Yang maintained the database on his student workstation, while Filo kept the search engine on his computer. The machines were dubbed *Akebono* and *Konishiki* in honor of two celebrated Hawaiian sumo wrestlers. Grappling with the Web went from a part-time pursuit to a full-time fixation, and Filo and Yang took a leave of absence from their doctoral work. In 1995, Marc Andreessen asked Yang and Filo to bring their operation to the large computers at Netscape Communications Corporation.

Yahoo!'s navigational guide and directory is a tailor-made database that runs on the UNIX platform and provides links to other Web sites. Links come from users who submit sites by clicking, Add URL, on the Yahoo! menu bar and from automated robots that seek out new sites. The database's renown stems from its simple indexing. Every site is hand placed in an appropriate category by human beings. A value in parentheses following a category name indicates the number of entries in that category, and an @ symbol at the end of a category means the heading appears in several different places in the Yahoo! hierarchy.

What differentiates the Yahoo! navigational guide from other online guides is that the directory is built by people. Every site is visited and evaluated by a Yahoo! staff member, or surfer. Site creators suggest the category under which a site belongs, but the Yahoo! surfers ultimately decide where each site is placed. To do this, staff members ride the surfboard of a typical user. While a book on building a chair, for example, may seem appropriate for the furniture category, most people surfing that category are interested in buying, not making, chairs. Therefore, Yahoo! surfers would place a site advertising the book in the how-to category. These decisions often are made in spontaneous staff meetings, which can take several days.

Yahoo! is the largest guide to the Web in terms of traffic and user reach. Services offered include Yahoo! Mail (http://mail.yahoo.com), Yahoo! Chat (http://chat.yahoo.com), and My Yahoo! (http://my.yahoo.com); and features such as What's New, What's Cool, and What's Popular. Yahoo! was incorporated in 1995 and became a publicly owned company in 1996. Today, Yahoo! employs more than 800 people. It costs nothing to use Yahoo!. Revenue comes from advertising, including banner advertisements, sponsorships, promotion, keywords, and from electronic commerce, hosting, and licensing. People using this service may be asked for such information as their name, e-mail address, or other personal information that Yahoo! uses to target advertising at specific audiences based on their demographics, geographic location, interests, or other factors. Consolidated statistics also may be used to describe Yahoo! audiences to potential advertisers.

In 1999, Yahoo! recorded net revenues of more than $588 million. With more than 120 million unique users and traffic increased to over 160 million page views per day, Yahoo! corporate shareholders have good reason to shout — well, you know.

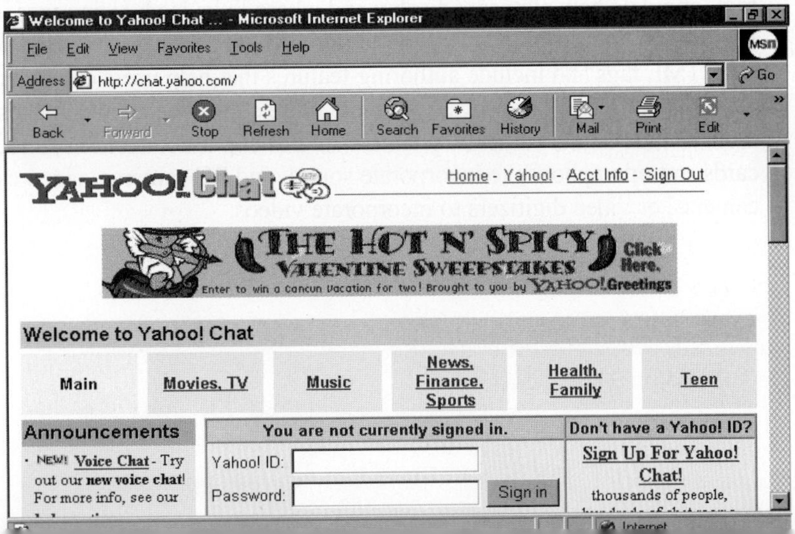

IN BRIEF

www.scsite.com/dc2001/ch7/brief.htm

WEB INSTRUCTIONS: *To display this page from the Web, launch your browser and enter the URL, www.scsite.com/dc2001/ch7/brief.htm. Click the links for current and additional information. To listen to an audio version of this IN BRIEF, click the Audio button to the right of the title, IN BRIEF, at the top of the page. To play the audio, RealPlayer must be installed on your computer (download by clicking here).*

1. How Does the Internet Work?

The **Internet** is a worldwide collection of networks that links millions of businesses, government offices, educational institutions, and individuals. An **Internet service provider (ISP)** provides temporary Internet connections to individuals and companies. **Online services** also provide Internet access, in addition to a variety of special services. At a business or school, many users connect to the Internet through a local area network that is connected to an ISP. At home, individuals often use their computers and a modem to dial into an ISP or online service over a regular telephone line. Data is transferred over the Internet using **servers**, which are computers that manage network resources and provide centralized storage areas, and **clients**, which are computers that can access the contents of the storage areas. The data is divided into small pieces, called **packets**, that move over communications lines to the destination, where they are reassembled. Each computer destination has a numeric address called an **IP (Internet protocol) address**, the text version of which is called a **domain name**.

2. How Are Graphics, Animation, Audio, Video, and Virtual Reality Used on the World Wide Web?

The **World Wide Web (WWW)** consists of a collection of electronic documents, called **Web pages**, that have built-in **links** to related Web pages. A Web page can contain graphics, animation, audio, video, and virtual reality. **Graphics** were the first media used to enhance the text-based Internet. **Animation** is the appearance of motion that is created by displaying a series of still images in rapid sequence. Simple **Web audio applications** and **Web video applications** consist of individual sound and video files that must be downloaded completely before they can be played on your computer. **Virtual reality (VR)** is the simulation of a real or imagined environment that appears as a three-dimensional (3-D) space.

3. What Tools Are Required for Web Publishing?

Web publishing is the development and maintenance of Web pages. Web pages are created and formatted using a set of codes called **hypertext markup language (HTML)**. Special codes, called **tags**, stipulate how elements display and where links lead. Using tags, developers can create an HTML document with any text editor or word processor. Many word processing packages generate HTML tags and include authoring features that help users create basic Web pages. Web page authoring software can be used to create more sophisticated Web pages. Other Web publishing tools include digital cameras, scanners, and/or clip art collections to incorporate pictures; sound cards and microphones to incorporate sound; and PC cameras, video capture cards and video cameras, or video digitizers to incorporate videos.

IN BRIEF

 www.scsite.com/dc2001/ch7/brief.htm

4. How Is Electronic Commerce Used?

Electronic commerce (e-commerce) is the performance of business activities online. Today, there are three types of e-commerce. **Business to consumer (B-to-C** or **B2C) e-commerce** consists of the sale of goods to the general public. **Consumer to consumer (C-to-C** or **C2C) e-commerce** occurs when one consumer sells directly to another, such as in an online auction. **Business to business (B-to-B** or **B2B) e-commerce**, which is the most prevalent type of e-commerce, takes place between businesses, with businesses typically providing services to other businesses.

5. What Internet Services Are Available?

A variety of services are widely used on the Internet. **E-mail (electronic mail)**, which is the transmission of messages and files via a computer network, is becoming a primary method of communication. **FTP (file transfer protocol)** is an Internet standard that allows you to exchange files with other computers. **Telnet** is a program or command that enables you to connect to a remote computer. **Newsgroups** are online areas in which users conduct written discussions about a particular subject. **Message boards** are Web-based discussion groups where you can post messages, reply to messages, search for messages, and browse through messages. **Mailing lists** are groups of e-mail names and addresses given a single name. **Chat** is real-time typed conversation that takes place on a computer through a **chat room**, or communications medium. **Instant messaging (IM)** is a service that notifies you when one or more people are online and then allows you to exchange messages or join a private chat room.

6. What Are the Rules of Netiquette?

Netiquette is the code of acceptable behaviors when using the Internet. Rules for e-mail, newsgroups, and chat rooms include: keep messages short and polite; avoid sarcasm, **flames** (abusive messages), and **spam** (unsolicited junk mail); and read the **FAQ** (frequently asked questions). Do not assume all material is accurate or up-to-date, and never read private e-mail.

7. Why Are Security Precautions Necessary on the Internet?

Because the Internet is a public place, you should use common sense while there. Security precautions include understanding cookies, handling improper behavior or content, and ensuring confidentiality. **Cookies** are small files containing data about you that a Web server stores on your computer. You can instruct your browser to accept or disable cookie use. You can use Internet **filtering software** to block access to certain material on the Internet, such as sites with content unsuitable for children. You can enhance confidentiality by dealing with secure Web pages, using a browser that encrypts (codes) messages, and attaching digital signals to messages.

8. How Can You Access the Internet Wirelessly?

People who connect to the Web wirelessly typically use satellites. The satellites send and receive signals to and from earth stations, which include reflective dishes, a variety of portable computers, handheld devices (such as PDAs, smart pagers, and smart phones), and Internet devices found in automobiles. Using an antenna, a wireless device communicates with an ISP that supplies wireless connections to the Internet. The **wireless portal**, which is a single Web site that provides all information a wireless user might require, is a development specifically designed for Web-enabled devices.

KEY TERMS

www.scsite.com/dc2001/ch7/terms.htm

WEB INSTRUCTIONS: *To display this page from the Web, launch your browser and enter the URL, www.scsite.com/dc2001/ch7/terms.htm. Scroll through the list of terms. Click a term to display its definition and a picture. Click KEY TERMS on the left to redisplay the KEY TERMS page. Click the TO WEB button for current and additional information about the term from the Web. To see animations, Shockwave and Flash Player must be installed on your computer (download by clicking here).*

ActiveX control (7.12)
Advanced Research Projects Agency (ARPA) (7.4)
animation (7.14)
animated GIF (7.14)
anonymous FTP (7.25)
applets (7.12)
ARPANET (7.4)
article (7.27)
AU (7.16)
audioconferencing (7.16)
backbone (7.5, 7.8)
backbone providers (7.8)
browser (7.11)
business to business (B-to-B or B2B) e-commerce (7.19)
business to consumer (B-to-C or B2C) e-commerce (7.19)
certificate authority (CA) (7.20)
chat (7.29)
chat room (7.29)
chat client (7.29)
client (7.7)
communications protocol (7.7)
consumer to consumer (C-to-C or C2C) e-commerce (7.19)
cookie (7.31)
dial-up access (7.7)
digital cash (7.20)
digital certificate (7.20)
discussion board (7.28)
domain name (7.8)
domain name system (DNS) (7.10)
domain name system servers (DNS servers) (7.10)
e-cash (7.20)
electronic commerce (e-commerce) (7.19)
electronic credit (7.20)
electronic data interchange (EDI) (7.19)
electronic money (e-money) (7.20)
electronic storefront (7.19)
electronic wallet (7.20)
e-mail address (7.24)
e-mail (electronic mail) (7.23)
emoticons (7.30)
encoding schemes (7.25)
FAQ (7.30)
filtering software (7.32)
flames (7.30)
flame wars (7.30)
FTP (file transfer protocol) (7.25)
FTP server (7.25)
FTP site (7.25)
graphics (7.12)
Graphics Interchange Format (GIF) (7.14)
helper application (7.12)
home page (7.10)
host (7.4)
host node (7.4)
.htm (7.20)
.html (7.20)
hypertext markup language (HTML) (7.20)
hypertext transfer protocol (http:) (7.11)
instant messaging (IM) (7.29)
Internet (7.2)
Internet2 (I2) (7.5)
Internet appliance (7.33)
Internet protocol (IP) address (7.8)

TO WEB

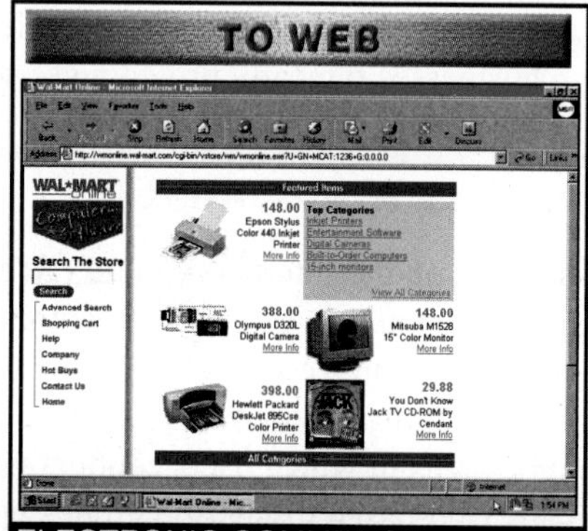

ELECTRONIC COMMERCE (E-COMMERCE): Conducting business activities online, including shopping, investing, and any other venture that uses either electronic money (e-money) or electronic data interchange. Using online shopping allows users to purchase a range of products over the Internet. (7.19)

Internet service provider (ISP) (7.6)
Internet telephone service (7.16)
Internet telephone software (7.16)
Internet telephony (7.16)
Java (7.12)
Joint Photographic Experts Group (JPEG) (7.14)
links (7.10)
LISTSERVs (7.29)
local ISP (7.6)
mail server (7.24)
mailbox (7.24)
mailing list (7.28)
marquee (7.14)
message board (7.28)
moderated newsgroup (7.27)
moderator (7.27)
Moving Picture Experts Group (MPEG) (7.17)
MP3 (7.16)
national ISP (7.6)
Net (7.2)
netiquette (7.30)
NetPC (7.33)
network (7.2)
network computer (NC) (7.33)
network personal computer (NetPC) (7.33)
news server (7.27)
newsgroup (7.27)
newsreader (7.27)
Next Generation Internet (NGI) (7.6)
NSFnet (7.4)
offline (7.18)
online service (7.7)
packets (7.7)
packet switching (7.7)
personal Web pages (7.20)
player (7.16)
plug-in (7.12)
point of presence (POP) (7.6)
portable network graphics (PNG) (7.14)
portal (7.29)
Post Office Protocol (POP) (7.24)
Post Office Protocol 3 (POP3) (7.24)
posting (7.27)
push technology (7.18)
RealAudio (7.16)
real-time (7.29)
RealVideo (7.17)
routers (7.7)
search engine (7.12)
server (7.7)
set-top box (7.33)
shopping cart (7.19)
simple mail transfer protocol (SMTP) (7.24)
smart pager (7.34)
smart phone (7.34)
spam (7.30)
spoiler (7.30)
streaming (7.16)
streaming audio (7.16)
streaming video (7.17)
submission service (7.23)
subscribe (7.28)
tags (7.20)
Telnet (7.26)
thin-client computer (7.33)
thread (7.27)
threaded discussion (7.27)
thumbnail (7.14)
top-level domain (TLD) (7.9)
traffic (7.5)
transmission control protocol/Internet protocol (TCP/IP) (7.7)
Uniform Resource Locator (URL) (7.11)
unsubscribe (7.28)
upload (7.23)
Usenet (7.27)
video chats (7.29)
virtual reality (VR) (7.17)
virtual reality modeling language (VRML) (7.18)
voice chats (7.29)
VR world (7.18)
wallet (7.20)
WAV (7.16)
Web (7.10)
Web appliance (7.33)
Web audio applications (7.16)
Web browser (7.11)
Web clipping (7.35)
Web community (7.29)
Web folder (7.23)
Web hosting services (7.23)
Web page (7.10)
Web publishing (7.20)
Web server (7.11)
Web site (7.10)
Web site management software (7.23)
Web video applications (7.17)
webcasting (7.19)
Webmaster (7.11)
wireless portal (7.35)
World Wide Web (WWW) (7.10)
Yahoo! (7.37)

AT THE MOVIES

www.scsite.com/dc2001/ch7/movies.htm

WELCOME to VIDEO CLIPS from CNN

WEB INSTRUCTIONS: *To display this page from the Web, launch your browser and enter the URL, www.scsite.com/dc2001/ch7/movies.htm. Click a picture to view a video. After watching the video, close the video window and then complete the exercise by answering the questions about the video. To view the videos, RealPlayer must be installed on your computer (download by clicking here).*

1 New Internet

The U.S. government and businesses are partnering to make I-2, "The Next Generation Internet," happen. What is I-2? How fast will it be? What is meant by "World Wide Wait"? Who can use I-2 now? Who will be able to use it in the future? What are I-2's advantages? What is the expected growth in Internet traffic over the next few years? Is the World Wide Web too slow for you now? If I-2 is significantly faster, will your use of the Web change? How?

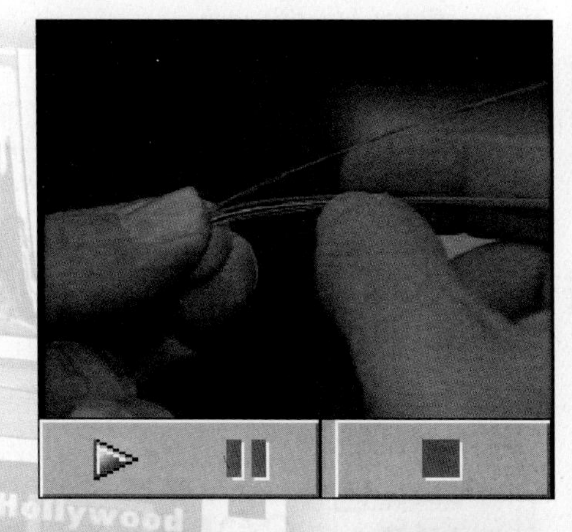

2 Yahoo! Millionaires

At the time two ingenious college students started Yahoo!, the success it now enjoys was inconceivable. Its founders became so wealthy that they now are million dollar benefactors of Stanford University. Who are the founders of Yahoo? How old are they? What services does Yahoo! offer? Have you ever used Yahoo!? What is the corporate culture like at Yahoo! Would you want to work at a firm like Yahoo!?

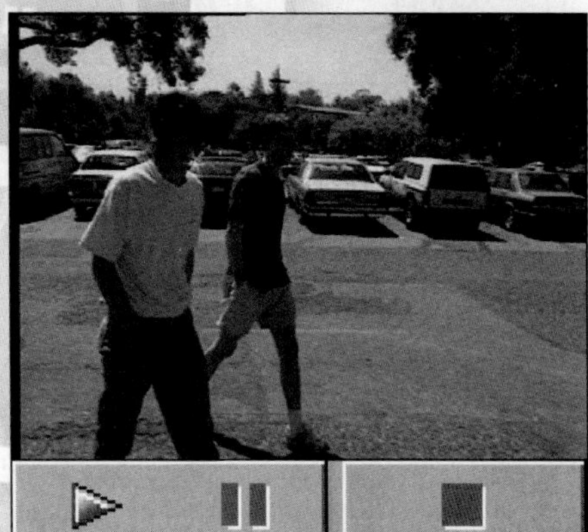

3 Chat Goes to Work

Chat is becoming a valuable business tool. Based on the video, what kinds of companies are using chat on their Web sites to support customers? Why do companies want to use chat in this manner? What are the advantages and disadvantages? Do you think using chat is more or less effective than using telephone and/or e-mail support? Would you prefer chat? If chat was the only support you received for a product, would you be more inclined, or less inclined to buy the product? Why? What other uses might chat have in a business environment?

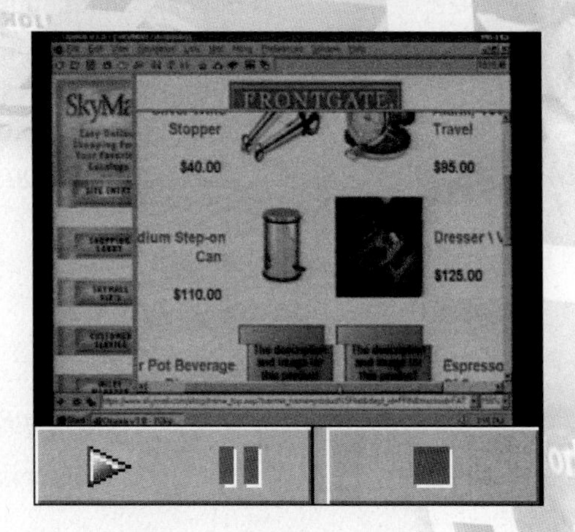

Student Exercises
WEB INFO
IN BRIEF
KEY TERMS
AT THE MOVIES
CHECKPOINT
AT ISSUE
MY WEB PAGE
HANDS ON
NET STUFF
Special Features
TIMELINE 2001
GUIDE TO WWW SITES
MAKING A CHIP
E-COMMERCE 2001
BUYER'S GUIDE 2001
CAREERS 2001
TRENDS 2001
LEARNING GAMES
CHAT
INTERACTIVE LABS
NEWS
HOME

CHECKPOINT

www.scsite.com/dc2001/ch7/check.htm

WEB INSTRUCTIONS: *To display this page from the Web, launch your browser and enter the URL, www.scsite.com/dc2001/ch7/check.htm. Click the links for current and additional information. To experience the animation and interactivity, Shockwave and Flash Player must be installed on your computer (download by clicking here).*

Label the Figure

Instructions: *Identify each part of the URL and e-mail address.*

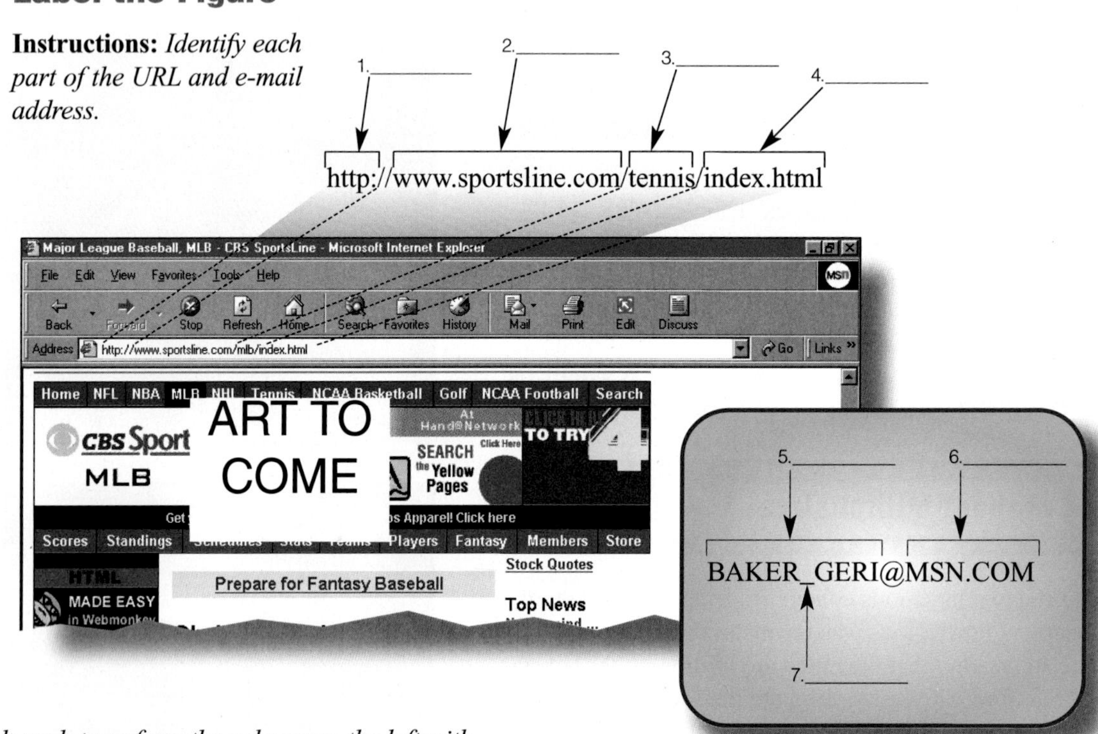

Matching

Instructions: *Match each term from the column on the left with the best description from the column on the right.*

_______ 1. TCP/IP

_______ 2. HTML

_______ 3. JPEG

_______ 4. VRML

_______ 5. SMTP

a. Set of special codes used to format a file for use as a Web page.
b. Popular standard used in compression of video and audio files.
c. Language that defines how 3-D images display on the Web.
d. Protocol used to define packet-switching on the Internet.
e. File format that uses compression techniques to reduce the size of graphical files.
f. Communications protocol used when e-mail messages are transmitted.
g. Communications protocol used to transfer pages on the Web.

Short Answer

Instructions: *Write a brief answer to each of the following questions.*

1. What is packet switching? __________ How do routers and communications protocols enable packet switching? __________
2. How are a Web page, Web site, and home page different? __________ What is a URL? __________
3. What is a search engine? __________ How is a plug-in different from a helper application? __________
4. What does it mean to subscribe to a newsgroup? __________ What is a thread? __________
5. Why do Web sites use cookies? __________ What cookie-related practice often is considered unethical? __________

AT ISSUE

www.scsite.com/dc2001/ch1/issue.htm

WEB INSTRUCTIONS: *To display this page from the Web, launch your browser and enter the URL, www.scsite.com/dc2001/ch1/issue.htm. Click the links for current and additional information.*

ONE

In 1992, the Supreme Court ruled that buyers had to pay state and local sales taxes only when merchants had a physical presence, such as a store or office building, in the buyer's state. Because Internet merchants often do not have a physical presence in a buyer's state, Internet purchases usually are sales tax free. Some believe this tax free status has a negative impact on local businesses (who must charge sales tax), state and local tax coffers (one estimate claims $60 billion in revenues is lost annually), and lower-income families (who are less likely to buy online). Yet, others feel any tax on e-commerce would be unmanageable (forcing vendors to adjust to varying sales tax rates) and unjustified. Should there be a sales tax on Internet purchases? Why or why not? How can the problems of taxing, or not taxing, Internet purchases be addressed?

TWO

A recent study of Pittsburgh families discovered a disturbing trend – the more time subjects spent online, the more depressed and lonely they tended to be. The report concluded that one hour a week of Internet use led to an average increase of 1% on the depression scale and 0.04% on the loneliness scale. Researchers think time spent on the Internet may be deducted from contact with real people. Participants in a study reported a decline in interaction with family members and a reduction in numbers of friends. No matter how heartfelt, e-mail, chat rooms, and newsgroups have an ephemeral quality compared to real human relationships. How does Internet communication affect mental health? Why? In terms of mental health, what do you think is the best way to use the Internet?

THREE

A father sat down at a computer with his young child, typed what he thought was the URL for a site of national interest, and was surprised to encounter pornographic material. "I should have paid closer attention to the (URL) suffix," he admitted. Stealth URLs – addresses similar to those of other Web pages – attract visitors and potential subscribers. Some Web pages adopt the URLs of popular Web sites, with minor changes in spelling or domain name. Critics claim this misleads consumers and weakens the value of the original name. Defendants counter that restrictions on URLs would violate rights to free speech. Do you think URLs should be regulated? Why or why not? How else can people deal with the problem of stealth URLs?

FOUR

Choosing a college once involved poring over catalogs, mailing letters, completing applications, and anxiously awaiting letters of acceptance. Today, a college selection Web site allows students to answer a few questions, and the site presents links to appropriate schools. Click a link to tour the college's Web page and complete an online application. Though the process is great for students, some admissions offices have reservations. Colleges have little control over what schools are promoted, and cannot prevent chat-room slander by grumpy students. Online applications requires less commitment than paper forms, and admissions personnel see an increasing number of incomplete or fake online applications. What, if anything, do you think can be done to make Internet college selection a benefit for both candidates and colleges?

FIVE

Almost 80 percent of America's public schools have Internet access. Although a boon to students and teachers, Internet use could prove a headache for school administrators. Legally, a school's responsibility for Internet use is unclear. No one is sure of a school's liability if students access objectionable material, hack into other computers, or violate copyrights. So far, most schools have relied on filtering software and permission slips. Yet, filtering software can screen too much (students might be unable to research sextuplets), and permission slips have questionable legal value. Besides, neither teaches responsible Internet use. What can, and should, schools do to control Internet use? Do you think schools should be required to take any steps? Why or why not? Where should responsibility for in-school Internet use ultimately lie? Why?

MY WEB PAGE

www.scsite.com/dc2001/ch7/myweb.htm

WEB INSTRUCTIONS: *The icons to the left of the numbered activities on this page indicate the learning system availability. If you are using WebCT, follow the instructions in the exercises below. If you are using the textbook Web site or CyberClass, launch your browser, enter the URL www.scsite.com/dc2001/ch7/myweb.htm, and then click the corresponding icon to the left of the exercise.*

1. Practice Test: Click My Web Page and then click Chapter 7. Click Practice Test. Answer each question and then click the Save Answer button. When completed, click the Finish button. Click the OK button to submit the quiz for grading. Click the View Results button, and then make a note of any missed answers.

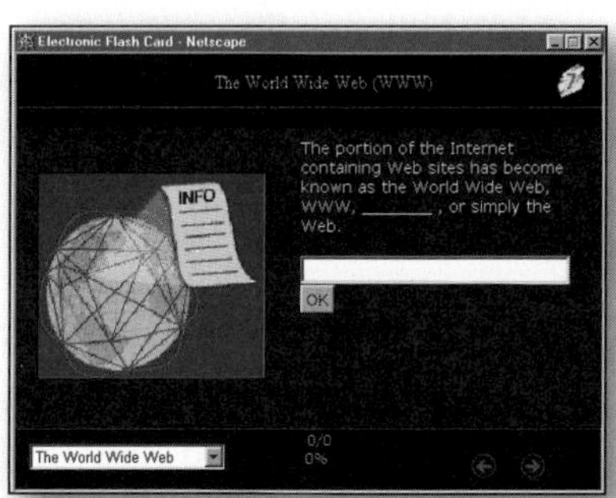

2. Web Guide: Click My Web Page and then click Chapter 7. Click Web Guide to display the Guide to WWW Sites page. Click Reference and then click AskJeeves. Ask Jeeves about the history of the Internet. Click an answer of your choice. Use your word processing program to prepare a brief report on what you discovered and submit your assignment to your instructor.

3. Scavenger Hunt: Click My Web Page and then click Chapter 7. Click Scavenger Hunt. Print a copy of the Scavenger Hunt page; use this page to write down your answers as you search the Web. Submit your completed page to your instructor.

4. Who Wants to Be a Computer Genius?: Click My Web Page and then click Chapter 7. Click Computer Genius to find out if you are a computer genius. For directions, click the How to Play button. When you are ready to play, click the Game button. Submit your score to your instructor.

5. Wheel of Terms: Click My Web Page and then click Chapter 7. Click Wheel of Terms to reinforce important terms you learned in this chapter by playing the Shelly Cashman Series version of this popular game. For directions, click the How to Play button. When you are ready to play, click the Game button. Submit your score to your instructor.

6. Career Corner: Click My Web Page and then click Chapter 7. Click Career Corner to display the USA TODAY page. Click Career Center. Write a brief report on what you discovered. Submit the report to your instructor.

7. Search Sleuth: Click My Web Page and then click Chapter 7. Click Search Sleuth to learn search techniques that will help make you a research expert. Submit the completed assignment to your instructor.

8. Crossword Puzzle Challenge: Click My Web Page and then click Chapter 7. Click Crossword Puzzle Challenge. Complete the puzzle to reinforce skills you learned in this chapter. For directions, click the How to Play button. When you are ready to play, click the Game button. Submit the completed puzzle to your instructor.

9. Flash Cards: This exercise uses CyberClass. Follow the instructions at the top of this page and then click the CyberClass icon to the left; or ask your instructor for logon instructions. Click Flash Cards on the Main Menu. Click the plus sign. Answer all the questions in any two subjects of your choice. Click the Close button in the upper-right corner to close all windows. Notice the many other exercises at this site. Complete all other exercises as assigned by your instructor.

Student Exercises
WEB INFO
IN BRIEF
KEY TERMS
AT THE MOVIES
CHECKPOINT
AT ISSUE
MY WEB PAGE
HANDS ON
NET STUFF
Special Features
TIMELINE 2001
GUIDE TO WWW SITES
MAKING A CHIP
E-COMMERCE 2001
BUYER'S GUIDE 2001
CAREERS 2001
TRENDS 2001
LEARNING GAMES
CHAT
INTERACTIVE LABS
NEWS
HOME

HANDS ON

www.scsite.com/dc2001/ch7/hands.htm

WEB INSTRUCTIONS: *To display this page from the Web, launch your browser and enter the URL, www.scsite.com/dc2001/ch7/hands.htm. Click the links for current and additional information.*

One

Online Services

This exercise uses Windows 98 procedures. What online services are available on your computer? Right-click the Online Services icon on the desktop and then click Open on the shortcut menu. What online services have shortcut icons in the Online Services window? Right-click each shortcut and then click Properties on each shortcut menu. Click the General tab. When was each shortcut created? When was it modified? Close the dialog box and then click the Close button to close the Online Services window.

Two

Understanding Internet Properties

Right-click an icon for a Web browser that displays on your desktop. Click Properties on the shortcut menu. When the Internet Properties dialog box displays, click the General tab. Click the question mark button on the title bar and then click one of the buttons. Read the information in the pop-up window and then click the pop-up window to close it. Repeat the process for other areas of the dialog box. Click the Cancel button.

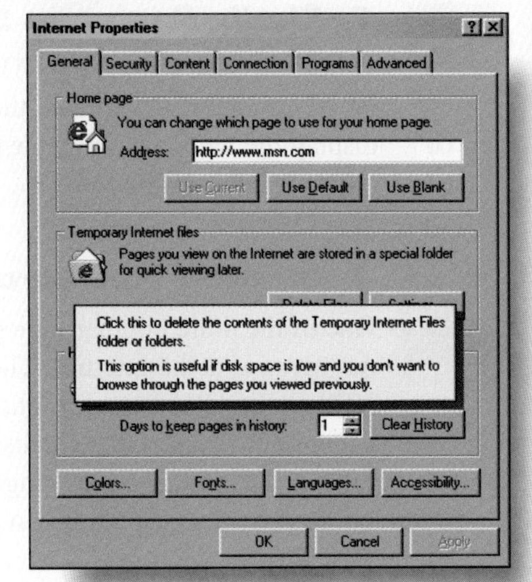

Three

Determining Dial-Up Networking Connections

This exercise uses Windows 98 procedures. Click the Start button on the taskbar. Point to Programs on the Start menu, point to Accessories on the Program submenu, point to Communications on the Accessories submenu, and then click Dial-Up Networking on the Accessories submenu. When the Dial-Up Networking window displays, right-click a connection displayed in the window and then click Connect on the shortcut menu. Write down the User name and the Phone number. Close the connect to dialog box and the Dial-Up Networking window.

Four

Using Help to Understand the Internet

This exercise uses Windows 98 procedures. Click the Start button on the taskbar and then click Help on the Start menu. Click the Contents tab. Click the Exploring the Internet book and then click the Explore the Internet topic. Click the Click here link to find out more about Internet Explorer. Answer the following questions:

- How can you update your favorite Web sites and view them at your leisure?
- How can you move around the Web faster and easier with the Explorer bar?
- How can you browse the Web safely?
- How can you view Web pages in other languages?

Close the Internet Explorer Help window and the Windows Help window.

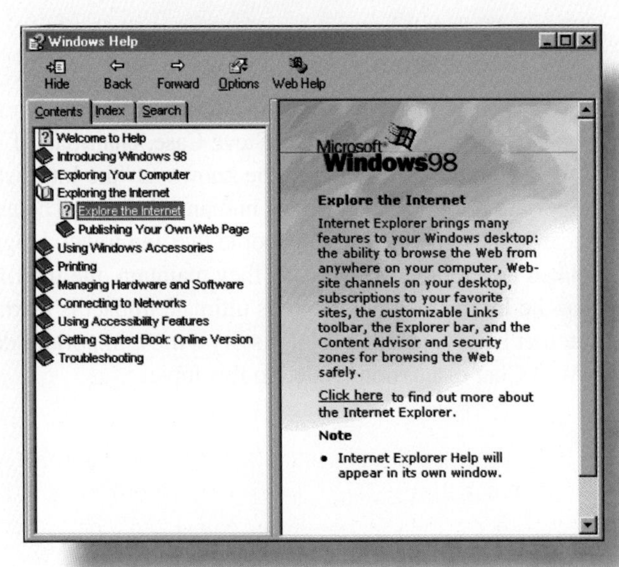

NET STUFF

www.scsite.com/dc2001/ch7/net.htm

WEB INSTRUCTIONS: *To display this page from the Web, launch your browser and enter the URL, www.scsite.com/dc2001/ch7/net.htm. To use the Connecting to the Internet lab or the World Wide Web lab from the Web, Shockwave and Flash Player must be installed on your computer (download by clicking here).*

CONNECTING TO THE INTERNET LAB

1. Shelly Cashman Series Connecting to the Internet Lab

Follow the instructions in NET STUFF 1 on page 1.48 to start and use the Connecting to the Internet lab. If you are running from the Web, enter the URL www.scsite.com/sclabs/menu.htm; or display the NET STUFF page (see instructions at the top of this page) and then click the CONNECTING TO THE INTERNET LAB button.

WORLD WIDE WEB LAB

2. Shelly Cashman Series World Wide Web Lab

Follow the instructions in NET STUFF 1 on page 1.46 to start and use the World Wide Web lab. If you are running from the Web, enter the URL, www.scsite.com/sclabs/menu.htm; or display the NET STUFF page (see instructions at the top of this page) and then click the WORLD WIDE WEB LAB button.

NEWSGROUP

3. Internet Newsgroups

One of the more popular topics for Internet newsgroups is the Internet. Click the NEWSGROUP button for a list of newsgroups. Find one or more newsgroups that discuss something about the Internet. Read the newsgroup postings and briefly summarize the topic under discussion. If you like, post a reply to a message.

IN THE NEWS

4. In the News

In her book, *Caught in the Net*, Kimberly S. Young argues that the Internet can be addictive. Young's methodology and conclusions have been questioned by several critics, but Young remains resolute. She points out that at one time no one admitted the existence of alcoholism. Click the IN THE NEWS button and read a news article about the impact of Internet use on human behavior. What affect did the Internet have? Why? In your opinion, is the Internet's influence positive or negative? Why?

WEB CHAT

5. Web Chat

The importance of the Internet is a hotly debated topic. Some claim that the Internet is one of history's most significant developments. Steve Case, chairman of America Online, believes the Internet's magnitude is so striking that the 21st century will be known as "the Internet century." Yet, not everyone agrees. Many insist an innovation is truly important only if it changes the human condition. The automobile, medical advances, and electricity were major innovations because they changed where people live and travel, lengthened life expectancies, and expanded the usable day (not to mention related developments). To date, they maintain, the Internet has not had a similar impact. Internet enthusiasts counter, however, that the Internet is young and its ultimate impact is unknown. How has the Internet changed the human condition? Will the Internet be considered one of history's most important developments? Why or why not? Click the WEB CHAT button to enter a Web Chat discussion related to this topic.

2001 E-Commerce

A Revolution in Merchandising

From a business perspective, the Internet means opportunity. The Internet provides companies with avenues to save money, to reduce the costs of procurement and communication between organizations, and to increase human resource productivity. It opens a world market to small and large businesses, giving each an equal footing in a new era of electronic trade. The Internet has stimulated the growth of many support industries that establish, maintain, and expand Internet connectivity and assist businesses in using the technology.

Forrester Research predicts that the amount of business conducted over the Internet will increase from $130 billion in 1999 to $1.5 trillion by 2003. Although e-trade on the Internet is still in its infancy, these figures indicate a strong confidence in the Internet as a business transaction tool. The Internet is an environment where imaginative vision and reality can merge to create unique and unexpected opportunities.

WHAT IS E-COMMERCE?

Electronic commerce, also known as **e-commerce**, is business activity that takes place over an electronic network. Broadly speaking, e-commerce includes transactions that utilize any computing or communication technology. E-commerce thus includes commercial activities involving e-mail, an online information service, a bulletin board system (BBS), and Electronic Data Interchange (EDI) systems. The most well-known and powerful medium for e-commerce is the Internet. For this reason e-commerce sometimes is called **Internet commerce** or **iCommerce.**

E-commerce has changed the way businesses do business. Organizations that do not have a Web site have been labeled **brick-and-mortar** businesses, while those with an online presence are sometimes termed **click-and-order** businesses. Businesses that have both a physical and an online presence are called **click-and-mortar.** E-commerce virtually eliminates the barriers of time and distance that slow traditional business dealings. Now, with e-commerce, transactions can occur instantaneously and globally, saving money for participants on both ends. Business can be taken out of the ordinary workplace. Transactions can occur in any location with access to a computer and an Internet connection (Figure 1).

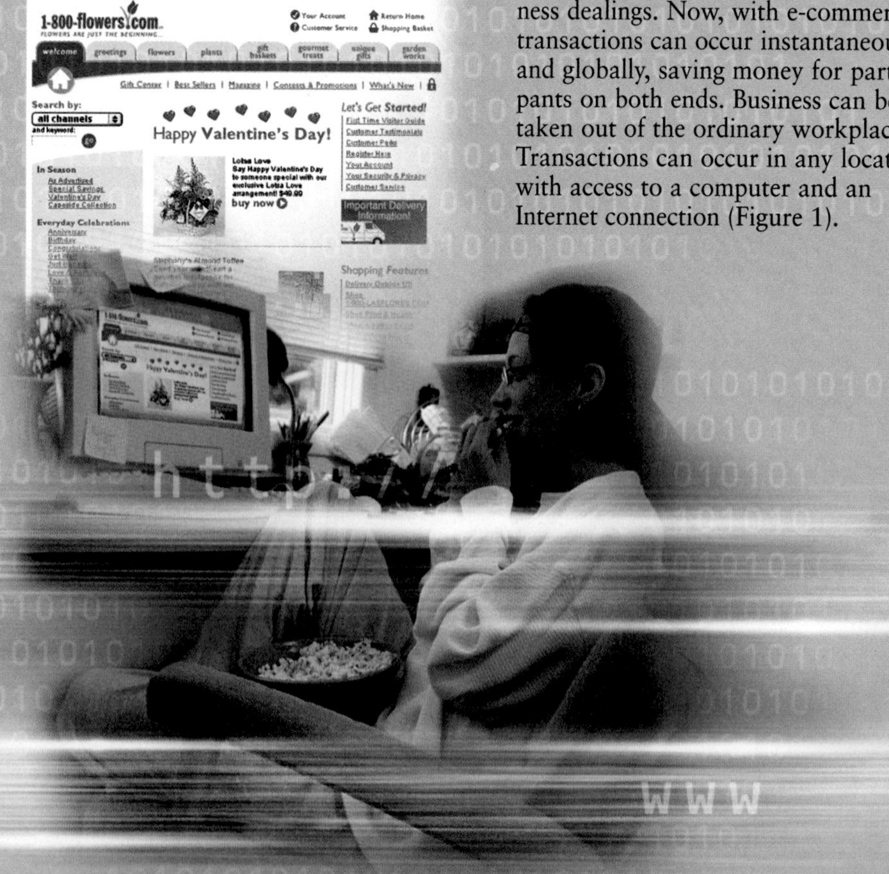

Figure 1 E-commerce allows business transactions such as shopping to occur outside of traditional brick-and-mortar settings.

The Growth of E-Commerce

Research experts such as Forrester Research, ActivMedia, and Jupiter Communications report that total worldwide e-commerce for 1999 exceeded $100 billion. They estimate that in 2003 this number will escalate to between $1.3 and $3.2 trillion. While no one actually knows where e-commerce will be in the future, these estimates indicate that it will continue to grow at an incredible rate (Figure 2).

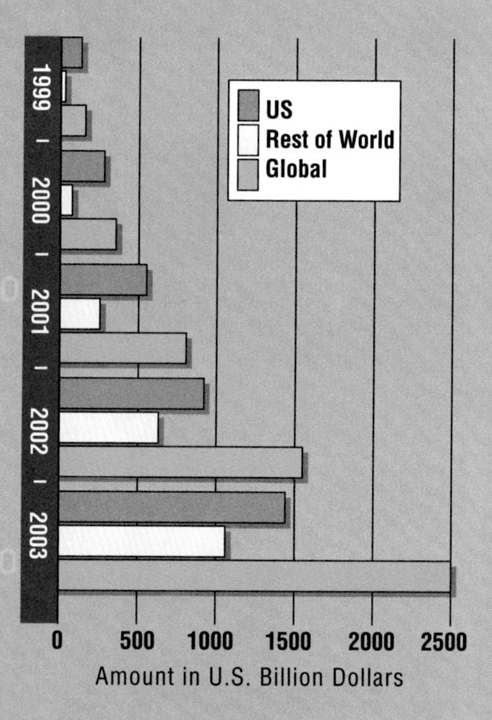

Figure 2 This estimate indicates projected e-commerce growth between 1999 and 2003.

(Source: Forrester Research)

E-commerce has developed along with improvements in communications and computing channels. One of the earliest forms of electronic commerce was Electronic Data Interchange (EDI). EDI originally was created to eliminate paperwork and increase response time in business interactions. It is a set of standards to control the transfer of business information among computers both within and between companies. Now these standards are being applied to communications on the Internet to link industry partners. Another precursor to the present form of e-commerce was the

Automatic Teller Machine (ATM). This device allows bank customers to transfer funds.

Radical change came to e-commerce when the Internet was opened for commercial use in 1991. At this time, most consumers knew little about the Internet, much less could they conceive of any profitable use for it. Within the year, however, consumers were using the Web to look up product information. Now e-commerce almost is synonymous with the World Wide Web. The growth of one enhances the other. Advances in technology required for encryption, security, and personal interaction all have been swiftly incorporated into commercial Web sites. Figure 3 lists some advantages of doing business on the Internet that have fueled the growth of e-commerce.

Global market with no geographic boundaries and access to 200 million people
Customers have access to multiple suppliers and prices
Internet is available 24 hours a day, 7 days a week
Interactive environment minimizes delays
Feedback from customers is immediate
Changing information is available quickly, with short turnaround time
Feedback through supply chain channel is instantaneous
Provides customer support and answers—FAQs
Ability to gather customer information, analyze it, and react
Cost-effective, one-to-one marketing
New and traditional approaches to generating revenue
Manufacturers sell and buy directly, avoiding the cost of the middleman
Distribution costs for information are reduced or eliminated
Reduces cost of paperwork, with savings in human resource hours

Figure 3 Advantages of doing business on the Internet.

E-Commerce Revenue Streams

A **revenue stream** is the method used by a business to generate income. Many levels of involvement exist in e-commerce. Some brick-and-mortar companies use their Web sites as electronic billboards to advertise and provide information about their product (Figure 4a). Sites such as this have acquired the name **brochureware** because they are not interactive and could be displayed equally well on paper.

Other companies incorporate Web features, although the site does not generate income directly (Figure 4b). These organizations may use their Web pages to facilitate communication or to provide **informational** resources, such as product support. This provides savings for the company by reducing the need for physical and human resources.

The possibilities for e-commerce expand when a company uses its site to **produce revenue** actively (Figure 4c). Some use traditional methods such as direct sales to generate profits. The nature of the Internet, however, has led to the development of revenue streams unique to the Web.

As you visit sites on the Web, you can observe how different companies earn money. Just a few methods you might see are advertising, transaction fees, commissions, data analysis, subscriptions, and development fees. As the Web develops, new opportunities for generating income are developing with it.

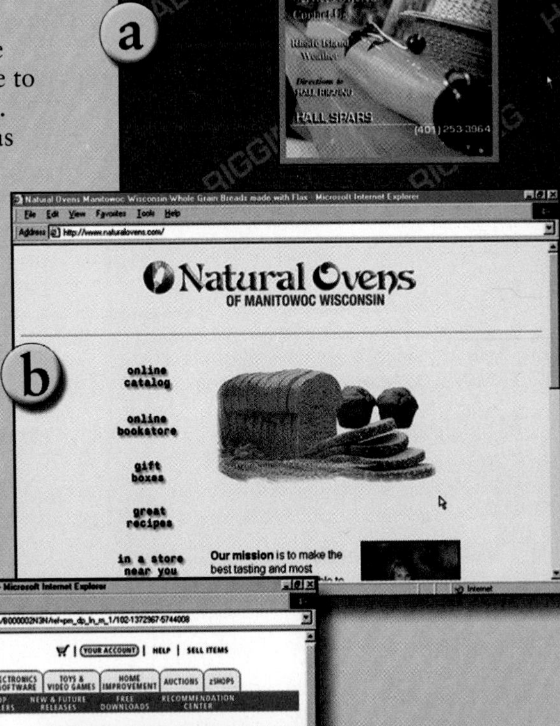

Figure 4 Involvement in e-commerce varies from brochureware (a), to informational (b), to revenue producing (c).

E-COMMERCE INTERACTIONS

Issue:

Disintermediation, Reintermediation, and Channel Conflict

As B-to-B commerce increases on the Internet, traditional **distribution channels** — the paths of goods from manufacturer to consumer — are being redefined. The Internet allows producers to communicate directly with both suppliers and consumers, eliminating the need for intermediaries at stages throughout the process. This displacement is called **disintermediation.**

As e-commerce develops, however, new channels are emerging to facilitate business interactions. The process is creating new intermediaries, and thus, is called **reintermediation.** Producers risk damaging relationships with traditional business partners because of this displacement. Companies doing business on the Internet are challenged to find new ways to resolve this channel conflict. To learn more about disintemediation, reintermediation, and channel conflict, visit www.scsite.com/dc2001/ch7/e-commerce.htm and then click Disintermediation or Channel Conflict.

The Internet e-commerce community contains as much diversity as business in the physical world. Most people are aware of the exchanges between businesses and consumers in retail trade. While e-retail is highly visible, it is only the tip of the e-commerce iceberg. The other major type of e-commerce interaction occurs between businesses.

Business-to-Consumer E-Commerce

Most business-to-consumer (B-to-C or B2C) e-commerce consists of the sale of goods to the general public. Goods can be physical objects such as books, flowers, or computers. They also can be intangible products delivered online. Consumers can subscribe to online magazines, download software, and participate in online classes. In addition, B-to-C companies provide services. Financial and investment companies, travel agents, and Internet service providers (ISPs) are a few examples of service companies. In the United States, the business-to-consumer market is expected to grow from $20 billion in 1999 to $110 billion by 2003. Figure 5 shows a breakdown of the types of products and services available in the business-to-consumer market.

Business-to-consumer companies maximize benefits for the sellers. In addition, consumers derive benefits from the e-commerce ability to access a variety of products and services without the constraints of time or distance. Customers easily can comparison shop to find the best buy. To attract and retain customers, sellers increasingly are providing services such as access to product reviews, chat rooms and forums, and product-related information. Many businesses are using Web-tracking technology to determine customer needs and personalize their offerings to a customer's profile.

The term consumer-to-consumer e-commerce (C-to-C or C2C) sometimes is used to refer to a small segment of B-to-C e-commerce. Consumer-to-consumer business consists of sites such as online auctions where individuals can connect with other individuals.

Business-to-Business E-Commerce

Only one fourth of all e-commerce revenue comes from business-to-consumer interactions. Most e-commerce actually takes place between businesses (Figure 6). The tremendous potential of business-to-business e-commerce, also known as B-to-B or B2B, is attracting global attention. The B-to-B market is expanding at a much faster rate than the B-to-C market. In the United States, B-to-B e-commerce is expected to grow from $110 billion in 1999 to more than $1.3 trillion in 2003. Businesses are rushing to participate in the anticipated business-to-business boom.

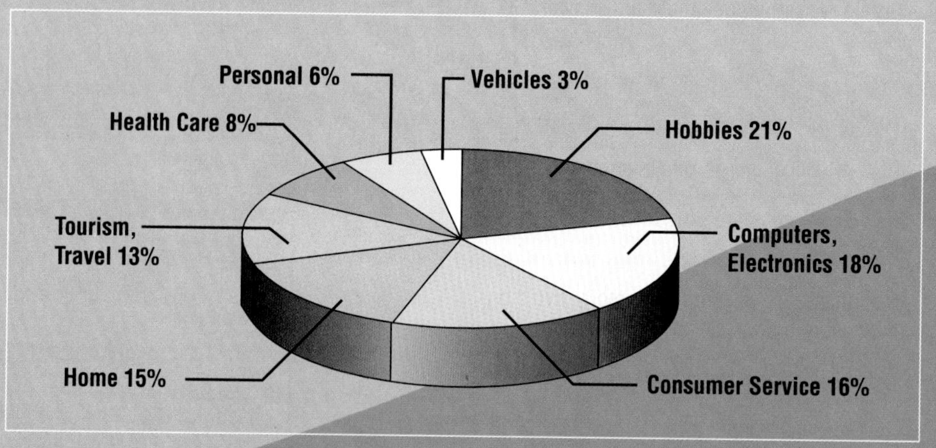

Figure 5 Business-to-consumer products and services.

Source: *Chicago Tribune*

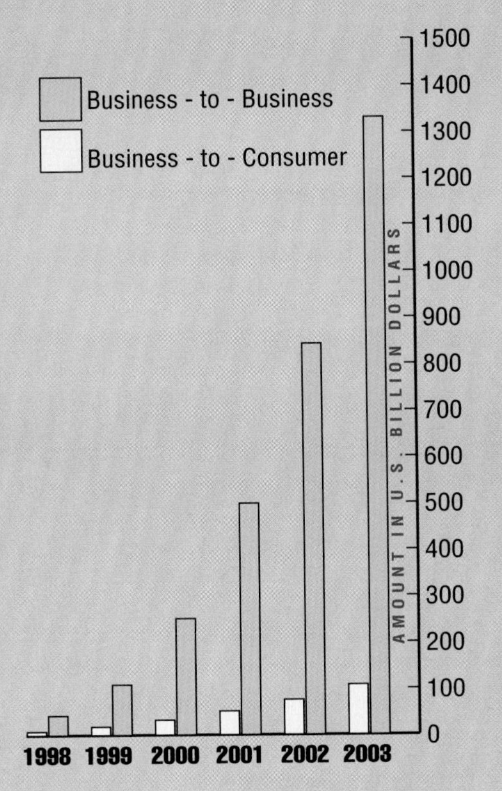

Figure 6 The projection for the growth of both business-to-consumer and business-to-business e-commerce in the United States is an indication of expectations for e-commerce worldwide.

Source: Forrester Research

Figure 7 illustrates numerous B-to-B opportunities. Many companies are utilizing the unique advantages of the Internet to communicate with business partners. The Internet is used to relay information among participants in the supply chain. The **supply chain** is the interrelated network of suppliers that work together to create a product. Interdependent B-to-B communities are being created where partners can share information and react immediately to changing demands. To aid communication, some organizations are revamping their EDI systems for Internet use. In addition, other systems are being created to facilitate business interactions.

The earliest participants in Internet B-to-B commerce provided Internet access and support. These companies supply the infrastructure that maintains and expands the Internet. This area includes hardware developers, software developers, IT support businesses, and service providers that market to businesses.

B-to-B includes a number of organizations that provide services to businesses. These range from entirely online services such as those offered by financial institutions to physical services such as warehousing and shipping. Other organizations specialize in providing information resources to businesses. They collect data, organize it, and market it to businesses to aid in the decision-making process.

Many companies are becoming B-to-B **infomediaries**, or **information intermediaries**. The infomediary establishes an electronic hub to match and interface among buyers and sellers in a particular industry segment. These hubs produce benefits for both buyers and sellers. They reduce search costs, transaction costs, and allow business to find better buyer and seller matches because of the increased pool of participants.

Business-to-business e-commerce demands that companies provide accurate and reliable services. Today's economy depends on the complex systems handling B-to-B exchanges. Establishing a B-to-B solution requires a significant investment of both time and money. New standards for communication are being established to address the needs of B-to-B interactions.

Intrabusiness E-Commerce

In addition to commerce transactions among businesses, e-commerce can include exchanges within an organization. **Intrabusiness e-commerce** refers to the use of Web-based technology to handle electronic transactions that take place within a business. Intrabusiness e-commerce does not generate revenue, but it allows savings as an organization operates more efficiently.

Primary Focus

Computers 27%
Retail Distribution 21%
Professional Services 17%
Manufacturing 17%
Marketing 6%
Production 5%
Office Equipment and Supplies 4%
Training and Education 3%

Figure 7 Primary product or service offered by business-to-business Web sites.

Source: *Chicago Tribune*

MARKET SECTORS AND THE INTERNET

The Internet is complementing and supplanting old businesses and creating new ones. The market sectors affected most visibly by the Internet are retail, finance, supply chain, business services, media, and entertainment. The Internet itself has developed a supporting infrastructure that has become a new revenue generating market. Figure 8 lists a few well-known participants in each sector.

Retailing and Personal Services

In retail, sales of goods and services are made directly to a consumer. Retail on the Internet often is called **e-retail** or **e-tail.** Research analysts show that Internet sales constitute .08 percent of the total sales revenue generated in the United States in 1999. Although this is a comparatively small amount, no one underestimates the meaning of this percentage point in real dollar earnings. E-retail companies constantly are challenging the old ways of doing business as they bring new products and services to market.

Finance

Financial institutions — traditional brick-and-mortar institutions such as banks, brokerage firms, and insurance companies — are conducting business on the Internet. In 1999, **e-finance** accounted for about $103 billion, or five percent of the total revenue attributed to the United States finance segment of business.

The banking industry can expect to see more demands for online services by customers. Direct online trading of stocks continues to grow as customers take control of their own stock portfolios and at the same time avoid the higher commissions charged by brokers.

Business Services

Many businesses are service providers to other businesses. Public relations, online advertising, direct mail, recruiting, credit, sales, market research, technical support, training, software consulting, and travel represent a few of the areas of support. These types of services made up $2.8 trillion of the United States economy. In 1999, the business service market on the Internet accounted for less than one percent of that figure, or $22 billion.

Supply Chain

Many businesses are developing services that help develop products through the stages from raw materials to customer delivery. Some companies provide services to help a manufacturer locate and do business with suppliers. This facet of the supply chain market is a $126 billion a year business. Of that amount, $3.9 billion involved Internet-based solutions. Some software development firms are creating software that provides integrated online transaction services for use by large manufacturing firms. Other development companies are creating Web host sites that place buyers and sellers together for a particular manufacturing process. Many suppliers have created catalogs of their products and opened Web stores.

Entertainment, Media, and Advertising

The technology behind the Web has enabled entertainment, media presentations, and advertising to take many forms. Entertainment companies such as Disney and Time Warner have a lot to gain by penetrating the Internet and setting up shop. Music, videos, live entertainment, sports events, and multiplayer gaming are a growing part of the Web's future. **Media** involves the use of the Internet relay information. Newspapers, television, and radio generally are considered media. The Internet technology currently supports live radio broadcasting due to companies such as RealAudio combined with streaming technology. Newsprint on the Web is not replacing the newspaper, but enhancing it and reaching different populations.

Advertising firms are discovering both overt and covert methods to gain information about the people who visit the Web. They use the knowledge to help people and business connect. All businesses realize that name recognition is important. The Web is so big, that companies that ignore it only are allowing competition to establish a presence first.

Internet Access and Infrastructure

Companies such as America Online, Earthlink, Juno Online, and WorldGate provide bandwidth for user access to the Internet. More than ten million people use the Internet through America Online. Its merger with Time Warner has rasied questions about services that will be accessible to members of America Online that will not be available to other ISP users. The building of the Internet and its maintenance and support rests in the hands of hardware manufacturers and specialized software development companies. Companies such as Cisco Systems manufacture the critical equipment that allows so many people to connect to the Internet at the same time. Other companies such as VeriSign are important for protecting the transactions that take place on the Internet.

http://
www.3com.com
dell.com
amazon.com
travelocity.com
www.1800flowers.com
disney.com
autoweb.com
www.aol.com
www.etrade.com
drugstore.com
www.arriba.com
schwab.com
www.ameritrade.com

Category/Company	URL	Comments
Retail		
1-800-flowers	www.1800flowers.com	Gift and delivery service
Amazon.com	www.amazon.com	Book and entertainment media sales
Autoweb.com	www.autoweb.com	Automobile sales
Dell	www.dell.com	Computer sales
drugstore.com	www.drugstore.com	Subscription drugs, over-the-counter medication, cosmetics, and health product sales
eBay	www.ebay.com	World's largest online auction site
EMusic.com	www.emusic.com	Digital music Sales
Finance		
Ameritrade, Inc.	www.ameritrade.com	Securities broker
Charles Schwab	www.schwab.com	Securities broker
Citicorp	www.citi.com	Banking and credit card services
E-LOAN	www.eloan.com	Consumer credit
E*TRADE	www.etrade.com	Securities broker
New York Life Insurance	www.newyorklife.com	Insurance sales
TD Waterhouse	www.waterhouse.com	Brokerage, banking, and mutual funds
Business Services		
AllBusiness.com	www.allbusiness.com	Administrative and professional functions support
Beyond.com	www.beyond.com	E-commerce Web site setup and management
BizBuyer	www.bizbuyer.com	Hub for buying and purchasing business products and services
Industrial Developments International (IDI)	www.idi.com	Real estate development and listing
Onvia.com	www.onvia.com	Hub for buying business services
Travelocity.com	www.travelocity	Online travel service
Supply Chain		
Ariba, Inc	www.ariba.com	Integrated B2B solutions for manufacturers and suppliers
EqualFooting.com	www.equalfooting.com	Wholesale supplier for small businesses
Grainger.com	www.grainger.com	Maintenance, repair, and operating supply parts distributor
NECX	webexchange.necx.com	Global electronic exchange for semiconductors
Entertainment, Media, and Advertising		
Disney.com	www.disney.com	Entertainment portal for Disney products
DoubleClick	www.doubleclick.com	Advertising company
National Public Radio (NPR Online)	www.npr.com	News and entertainment
RealNetworks	realguide.real.com	Streaming technology live broadcast portal
USA TODAY	www.usatoday.com	Online newspaper
VerticalNet.com	www.verticalnet.com	Hub for vertical trade communities and software supplier
GameSpy.com	www.gamespy.com	Online multiplayer action game hub
Infrastructure, Support, and Internet Access		
America Online	www.aol.com	Online service
MCIWorldcom	www.mci.com	Online service
Cisco Systems	www.cisco.com	Manufacturer of routers, switches, and hubs
Lucent	www.lucent.com	Fiber optics supplier
3Com	www.3com.com	Network management and control
VeriSign, Inc.	www.verisign.com	Internet trust and security
For an updated list, visit www.scsite.com/dc2001/ch7/e-commerce.htm.		

Figure 8 E-commerce companies facing the challenge of an e-market economy.

RETAILING ON THE WEB

The smallest to largest e-retail businesses operate in a similar way. Figure 9 illustrates how an e-commerce transaction might occur. A customer (1) visits an online business at the Web-equivalent of a showroom: the **electronic storefront**, or **online catalog**. This is the Web site where an e-retail business displays its products (2). The **storefront** contains descriptions, graphics, and even product reviews.

After browsing through the merchandise, the customer makes a selection. This activates a second area of the store known as the shopping cart (3). The **shopping cart** allows the customer to collect purchases. Items in the cart can be added, deleted, or even saved for a future visit. When ready to complete the sale, the customer goes to another software component, the **checkout** (4). Through a secure Web connection, the customer interacts with the **transaction server**. This is a safe system where the customer enters personal, e-cash, and/or credit card data (5). This data automatically is verified at a banking Web site (6). If the financial data is accepted (7), the customer side of the transaction is completed. The customer returns to the merchant's Web server (8) and sees a message that the purchase is on its way (9).

Now, the order proceeds to the merchant to be filled. In a fully automated system, a **back office system** processes the order (10). The order then is sent to a **fulfillment** center (11) where it is packaged and shipped (12). Inventory systems are updated and, if needed, and order is placed with the factory (12a). The fulfillment center sends a report to the back office system where records are updated (12b). The bank is notified of the shipment and payment is sent via electronic channels (12c). Shipping information is posted on the Internet, so the customer can track the order (13). The customer typically receives the order in a few days (14).

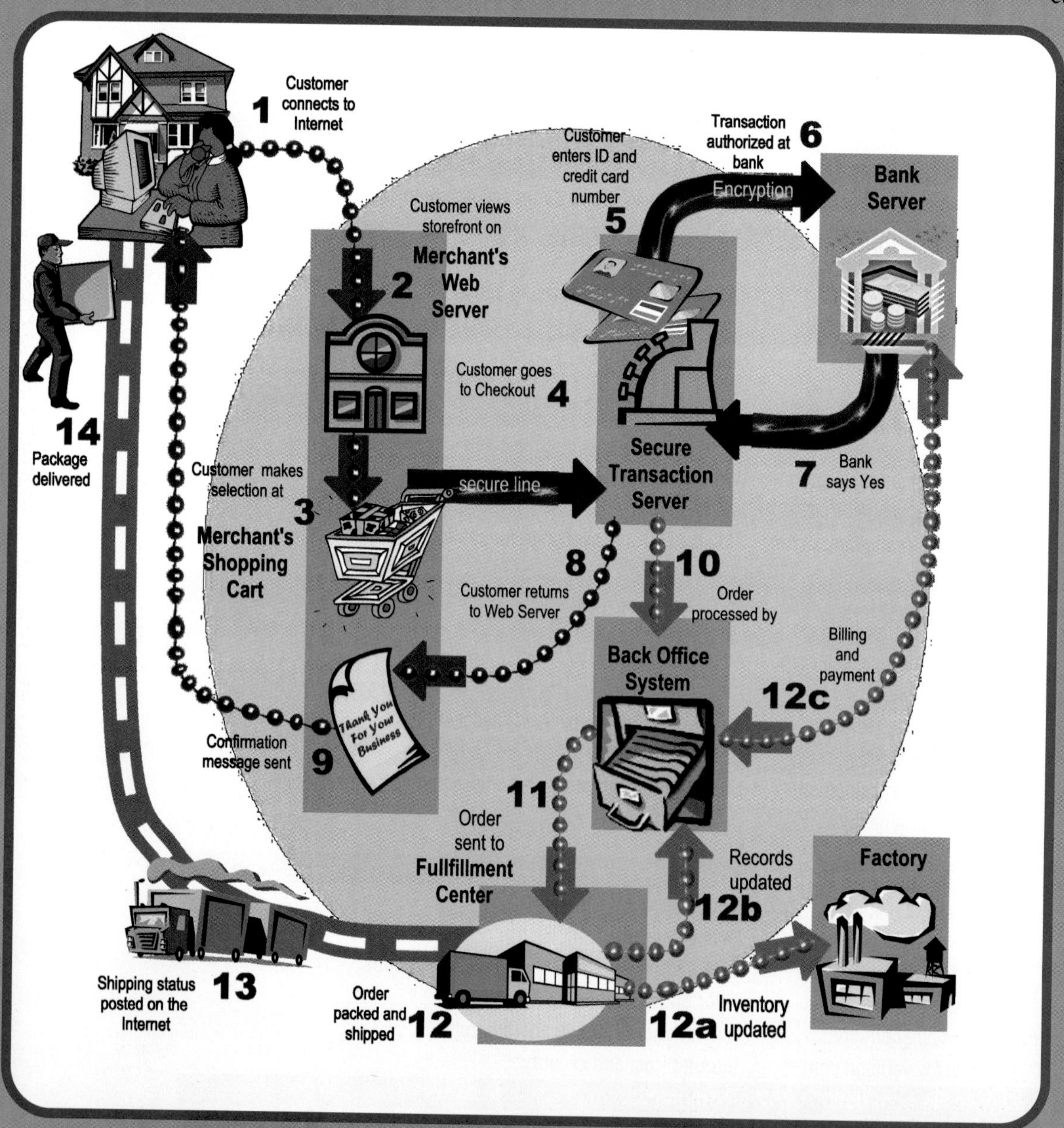

Figure 9 A model of an e-retail transaction.

CREATING AN ONLINE STORE

A variety of approaches exist to establish an online presence. Some merchants choose to start an e-retail store from the ground up. Others establish an electronic storefront as an extension of an existing brick-and-mortar business. Some expand an informational Web site into a full-featured e-commerce site. No matter what their scope, all e-commerce businesses must address some common concerns. To provide for e-commerce, a company must plan how to:

- Handle hardware and software needs
- Accept payment and provide security
- Manage product delivery
- Design a site that attracts customers and keeps them returning
- Promote the site

Building a Storefront

One of the more important decisions facing e-retailers is the choice of the software and hardware to build a storefront. The e-retail storefront must inform customers about the business, its products, and its services. It must provide for purchases, and supply feedback to the retailer. The software/hardware solution is influenced by the time and money available and the required complexity of the site. Merchants must decide whether to purchase and maintain their own system or to lease all or a portion of the system from an outside company. Costs can range from a few hundred dollars to hundreds of thousands of dollars.

Many businesses prefer to use an outside company, called a **Web host**, to provide hardware and software needs. Figure 10 shows two Web hosting sites. A Web host allows individuals or companies to use the host's server to store Web pages.

Many Internet service providers (ISPs), telecommunications providers, content portals, and online malls provide

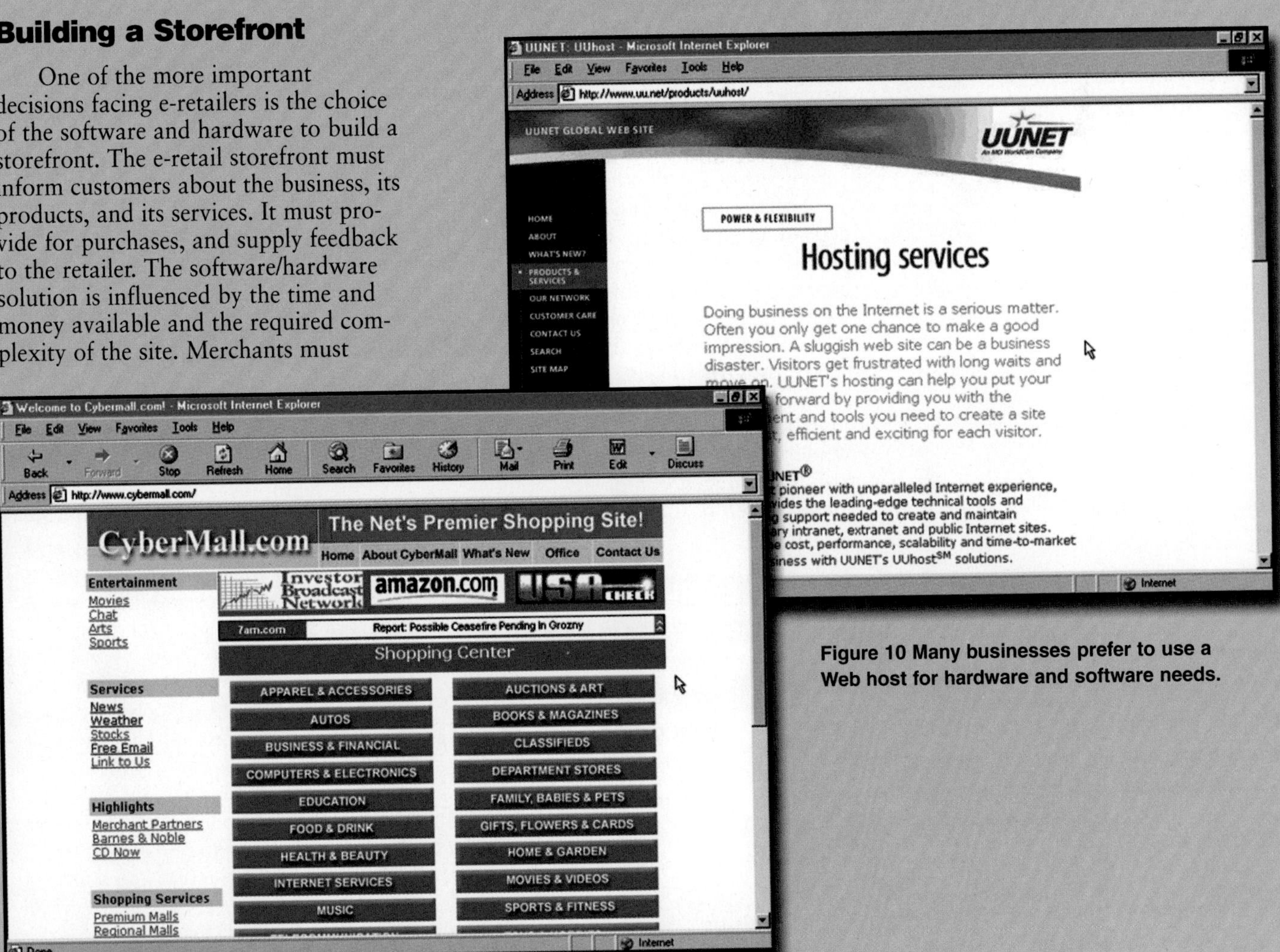

Figure 10 Many businesses prefer to use a Web host for hardware and software needs.

Web Host	Primary Service	Examples
Internet Service Provider	Provide Internet access and Web hosting	GTE Internetworking www.bbn.com AT&T CERFnet www.cerf.net UUNET www.uu.net
Telecommunications Provider	Communications businesses that extend their enterprises to include Web hosting and Internet connectivity	AT&T Easy World Wide Web www.att.com MCI Web Commerce www.wcom.com Sprint www.sprint.com
Content Portal	Search engine providing multiple Internet services	Yahoo! Shopping shopping.Yahoo.com
Online Mall (also called Virtual Mall or Internet Mall)	Mall shopping: buyers enter the mall through the front door, the mall's URL; retailers store their individual Web pages under this URL	ShopNow.com www.shopnow.com Choice Mall www.choicemall.com

Figure 11 Numerous Web hosts are available.

commerce software packages. The table in Figure 11 describes some of the Web hosting services. Web site development services also may be available. Hosting services allow entry into the e-retail business with a minimum investment. The hosting service charges a monthly fee, or may take a percentage of the sales income.

Software capabilities vary widely, depending on the host. Some solutions provide little more than a product billboard, while others allow merchants to set up elaborate, customized sites. With some hosts, the merchant operates independently. Other hosts, particularly Internet malls, may set restrictions and controls over the store's operation. While they are very popular and offer many advantages, companies should research Web hosting services before spending the time and resources to establish a site.

A business that wants total control over its e-retail site can purchase and maintain the software and hardware. This option requires a larger financial outlay. Businesses usually hire an expert in Web design and development to implement this type of system. Many e-commerce software packages are available with a range of features and pricing. The table in Figure 12 lists various software packages with their pricing and URLs. E-commerce software allows the merchant to set up a storefront with a product database, combined with a shopping cart. In addition, the software should provide a secure environment to process order transactions. Statistical tracking features may be included. A sophisticated system is capable of integrating with a business's other systems to provide for billing, inventory management, and back office activities such as orders to suppliers. Many systems allow the use of third-party plug-ins. A **plug-in** is add-on software used to provide additional custom features to a site.

Hybrid solutions also are available. Some companies use their own in-house server to store parts of their Web site and use an outside host for other components. Often companies elect to outsource the transaction services that require a secure server. A **secure server** is one that prevents access to the

Software	Price	URL
Calico eSales Suite	$250	www.calicotechnology.com
freemerchant.com	Free	www.freemerchant.com
IBM Net.Commerce	$5,000-$20,000	www.ibm.com/e-business
iCat Electronic Commerce Suite	$3,500-$10,000	www.icat.com
INTERSHOP Communications	$5,000	www.intershop.com
Maestro Commerce Suite	$1,000-15,000	www.maestrocommerce.com
Microsoft Site Server	$4,500-$5,500	www.microsoft.com/siteserver

Figure 12 Web commerce software packages.

system by unauthorized users. The storefront and shopping cart are connected to the transaction services with hyperlinks. This means that a customer browses for products at a merchant's in-house site. When ready to order, the customer clicks an order button and moves seamlessly to the transaction area on the hosting company's server. After placing the order on the secure system, the customer returns to the original business site.

Managing Payments

Before a business can open on the Internet, it needs to be able to accept customer payments. Although some companies accept digital cash and electronic wallets, credit cards are the most popular method of payment on the Web. They are used for 98 percent of all Internet purchases. To accept credit cards safely from consumers, a business must complete three steps: (1) obtain a merchant account; (2) provide a secure ordering form; and (3) utilize payment-processing software.

Merchants typically apply to a bank to obtain an e-commerce merchant account. A **merchant account** is used to accept and hold money until an online transaction is complete. The merchant pays a monthly fee to the bank to maintain this account plus commissions on each transaction.

Businesses use an order form to collect orders and credit card information from the customer. The form is stored on a secure server, provided either by the hosting service or the merchant. A protocol such as **Secure Sockets Layer (SSL)** or **Secure Electronics Transaction (SET)** is used to **encrypt**, or code, the transaction. In addition, the merchant should obtain a digital certificate from an organization such as VeriSign. The digital certificate is a method to verify the source of financial transaction messages.

Finally, the merchant must arrange to use a payment-processing service. These services provide software to manage the transaction between the merchant and the bank. The merchant pays a monthly fee for this service. Several well-known, reputable payment-processing services are Authorize.net, Cybercash, IC Verify, and PC Authorize.

Often, hosting services provide some or all of the payment and security needs as part of the overall e-commerce software package. In addition, new systems are being offered to provide secure and reliable management of the entire ordering and payment process. Merchants usually pay the service a percentage of transactions. A few of these companies are Verza, the Bag Boy, PaySystems, iBill, and CcNow.

The entire payment process is set up to protect against fraud. Most experts agree that all of these safeguards make credit card use safer over the Internet than in many face-to-face transactions. Consumers should verify that a merchant provides secure transactions before using a credit card on the Internet. Secure sites have URLs that begin with https:// rather than http://. Figure 13 lists several organizations that guard against Internet fraud and offer practical advice for both merchants and consumers.

National Fraud Consumers League National Fraud Information Center and Internet Fraud Watch
www.fraud.org/welmes.htm

The American Bar Association Safe Shopping Page
www.safeshopping.org

United States Federal Trade Commission (FTC)
Consumer Protection for E-Commerce & Internet
www.ftc.gov/bcp/menu-internet.htm

Figure 13 Monitoring organizations offer advice about Internet fraud.

Fulfillment

Before a storefront can be opened, the merchant must have a workable, deliverable product. Merchandise does not transfer equally well to the Web. For example, clothing sales often work better when customers can browse physically and try on items in a physical store. A major challenge for online businesses is determining how a product will be delivered. Increased Internet sales mean increased business for traditional delivery services such as Federal Express and UPS (Figure 14). In addition, new service niches are being created for delivery of goods and services that the traditional delivery services cannot provide.

A business must consider how it will manage and store inventory, package and ship products, and maintain records of all transactions. Existing retailers already have a system in place to handle this part of the business. Sophisticated e-commerce software packages can integrate and help automate existing business functions. As with other aspects of the e-commerce world, however, outsourcing solutions are available. **Fulfillment companies** can provide warehousing and inventory management, product assembly, order processing, packing, shipping, return processing, and online reporting.

Figure 14 Internet retail sales make up 20 percent of all packages delivered by services such as Federal Express and UPS.

Attracting and Retaining Customers

A successful Web site attracts customers and keeps them returning to the site. The best storefronts plan for convenience, efficiency, and ease of navigation. Studies indicate that Web customers will click to another site if they must wait more than eight seconds for a page to download. Consumers want to navigate easily through a site. Instructions should be clear and easy to follow. Too many special effects can slow downloading and clutter a site. In addition, the fewer clicks it takes for a customer to find a product and place an order, the more sales the store will make.

Successful businesses incorporate features to take advantage of the capabilities of the Internet. These features enhance a customer's experience and move the store beyond simply being a catalog. The features should coincide with the store's function. For example, a store selling music can provide audio previews. Discussion groups, newsletters, and informational articles related to the store's product could draw visitors to the site and generate goodwill.

Issue:

Security

Just as in real-world business, both merchants and consumers must guard against fraud. Merchants are legally responsible for ensuring security at their sites. They should investigate the security measures they use thoroughly, whether installed in-house or out sourced. To learn more about merchant security, visit www.scsite.com/dc2001/ch7/e-commerce.htm and then click Security.

Consumers can protect themselves best by becoming educated about Internet fraud and avoiding unsafe practices. To learn more about consumer protection, visit www.scsite.com/dc2001/ch7/e-commerce.htm and then click Consumer Protection.

Issue:

Taxes

Should sales taxes be charged for Internet sales? If so, how can these be regulated when sales are made to a global marketplace? Currently United States law requires that Web businesses pay sales taxes in states where the business has a physical presence. When a business does not have a physical presence, and it out sources many of its functions, it is difficult to determine where the company is required to pay sales tax.

Brick-and-mortar businesses are placed at a disadvantage when they must compete with online companies that do not charge tax. Sales tax revenue traditionally is used for schools, transportation, and community services. As more of the economy moves to online sales, the potential becomes greater of reducing the resources needed to serve local communities. To learn more about issues concerning taxation, visit www.scsite.com/dc2001/ch7/e-commerce.htm and then click Tax.

Issue:

Trust

Because Internet interactions are faceless, a business has to generate trust to turn window shoppers into customers. Displaying an actual address and telephone number indicates that a business is established and willing to be contacted. Merchants can join an Internet protection group such as Netcheck, TRUSTe, or the Better Business Bureau Online (BBBOnLine) to show that they are reputable and safe businesses. Stores often display a security or privacy statement to encourage customer confidence. To learn more about trust, visit www.scsite.com/dc2001/ch7/e-commerce.htm and then click Trust.

Site Management

By monitoring Web site use, businesses can collect data and use it to improve their sites. Most e-commerce software packages include features to monitor site use and collect statistics. The programs count the numbers of **hits**, or downloads, for each page. They can track a customer's progress from a previous location, and follow the path taken through the site. Merchants use **personalization** or **one-to-one marketing** to adapt a site to individual customers.

The programs often store **cookies**, or small files about site use, on the customer's computer. Cookies are used to store information such as customer preferences and the contents of shopping carts. When a customer returns to the site, the program accesses the cookies and uses the data to personalize the interaction. **Intelligent agents**, sometimes called **bots**, are software programs that analyze a customer's preferences and tailor the storefront to match.

Businesses need to consider carefully how to provide service after a sale. Surveys indicate that a large percentage of customers are dissatisfied with customer service at online businesses. E-retailers may improve communication by using automatic e-mail to confirm orders, frequently asked questions (FAQs), and surveys for customer feedback. Customer queries should be answered quickly and accurately. Returns policies should allow customers to make returns and exchanges conveniently. Follow up after a sale can generate return business and recommendations.

Issue:

Privacy

An enormous potential exists for abuse as companies increasingly collect data and profile customers. Who owns and has control over personal and collective data, the individuals or the data collector, are unresolved questions that courts may have do decide. Many merchants are adding a privacy policy to their sites to notify customers how collected information will be used. Watchdog organizations such as TRUSTe, the Online Privacy Alliance, and the Federal Trade Commission (FTC) monitor privacy issues. To learn more about privacy, visit www.scsite.com/dc2001/ch7/e-commerce.htm and then click Privacy.

Naming a Storefront

The first decision in promoting a site actually should be made before the storefront is set up: the company must have a name. Choosing a name for a site and an associated domain name can be a crucial decision. A company may use the brick-and-mortar business name, create a Web variation of the name, or coin an entirely new title. **The domain name**, or **Web address**, is the registered name used to identify and locate a Web site. It is like a street address on the Internet. A customer types the domain name to visit a Web page, so domain names can have a great influence on the number of hits a Web page receives. Ideally, a domain name corresponds to the business name or the function of the business.

Issue:

Cybersquatting

Cybersquatting is the practice of domain name speculation. Cybersquatters register domain names they think will be popular, and then attempt to resell the rights to each name to the highest bidder. Some organizations have paid millions of dollars for a single domain name. For example, the name business.com sold for $7.5 million and autos.com for $2.2 million. Alternately, the cybersquatter may use a name to divert Web surfers from legitimate sites with similar names.

Cybersquatters have registered names of famous people, company names, and products. They also register modifications of these. The World Intellectual Property Organization (WIPO) and The Internet Corporation for Assigned Names and Numbers (ICANN) are establishing guidelines to help regulate rights to domain names. To learn more about cybersquatting, visit www.scsite.com/dc2001/ch7/e-commerce.htm and then click Cybersquatter.

Promoting a Site

Most people locate Web sites using search engines and Web portals. Businesses can register domain names at each search engine individually. Alternately, the business can hire a promotion service such as Promote It!, Register It!, or Submit It! to register the site. Online magazines and informational sites related to the product often maintain lists where businesses can register. In addition, a company can trade links with similar sites or join a Web ring that links groups of related sites together. Again, services exist to associate similar Web sites.

A more expensive option is to use online advertisements at another site. These often are called **banner ads**. The ads display a brief message and are hyperlinked to the advertiser's Web site. They can be personalized to match a customer's interests. The advertiser usually pays based on the number of click-throughs. A **click-through** occurs when a visitor clicks an ad to move to the advertiser's Web page.

Some businesses consider using unsolicited advertising on newsgroups and e-mail. This electronic form of junk mail is called **spamming**. In the Internet world, spamming generates antagonism rather than sales. A better method is to promote goodwill by providing information or services for groups and individuals. They then may reciprocate by promoting the site.

Issue:

Attention

With global access to millions of Internet sites, people now are bombarded with information. A major challenge to a business's success is the capability of attracting consumer attention. Businesses now compete for this resource. Thomas Mandel and Gerard Van der Leun wrote in their 1996 book *Rules of the Internet* that, "Attention is the hard currency of cyberspace." Because it is becoming a scarce commodity, attention can be considered the basis of the future economy. To learn more about Attention, visit www.scsite.com/dc2001/ch7/e-commerce.htm and then click Attention.

PUTTING IT ALL TOGETHER: MASKPARADE AT YAHOO! SHOPPING

Putting it altogether introduces you to Yahoo!, Yahoo! Shopping, and Yahoo! Store. These three Web sites provide a solution for Web customers who want to find services and products and merchants who want to establish a presence on the Web. The steps in this section illustrate the process of creating an inexpensive storefront and some of the special features included in such a storefront.

Many businesses, both small and large, perceive selling a service or product on the Web a difficult process. Yahoo!, one of the more popular Web sites on the Internet, however, offers a convenient, easy-to-use, and inexpensive storefront solution (Figure 15). Yahoo! has evolved from a stepping stone to other destinations into a **content portal** where dozens of free services such as chat rooms, e-mail, news, stock market reports, and mall shopping are available.

Yahoo! visitors consist of 120 million unique users, and more than 450 million page views occur there each day. Yahoo!'s home page provides access to Yahoo! Shopping (Figure 16). Here, Web surfers can visit an online shopping mall with more than 8,000 storefronts where millions of products are sold.

To display a storefront in the shopping mall, all a merchant needs to know is the name of his or her store and the items for sale. Such technical features as a shopping cart, secure lines, and encryption for credit card transactions are handled by Yahoo! Anyone can try out the services at Yahoo! Store for a trial period of ten days. During the ten-day trial period, a storefront can be built and items entered to sell on the Web. To keep the store for more than ten days, Yahoo! must be notified and the site must be registered as a B-to-C storefront. Presence at Yahoo! Shopping costs $100 per month if your store lists up to 50 items for sale and $300 per month if your sales list falls between 51 to 1,000 items. After registering the site with Yahoo, the business appears in the shopping mall.

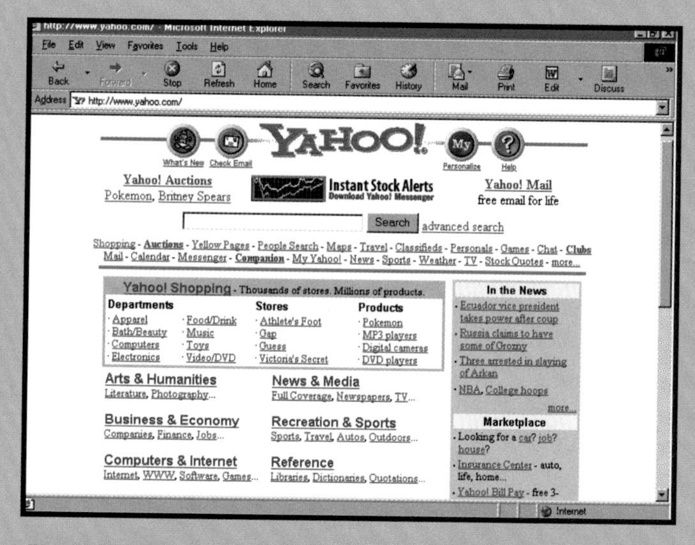

Figure 15 Yahoo! home page at www.yahoo.com.

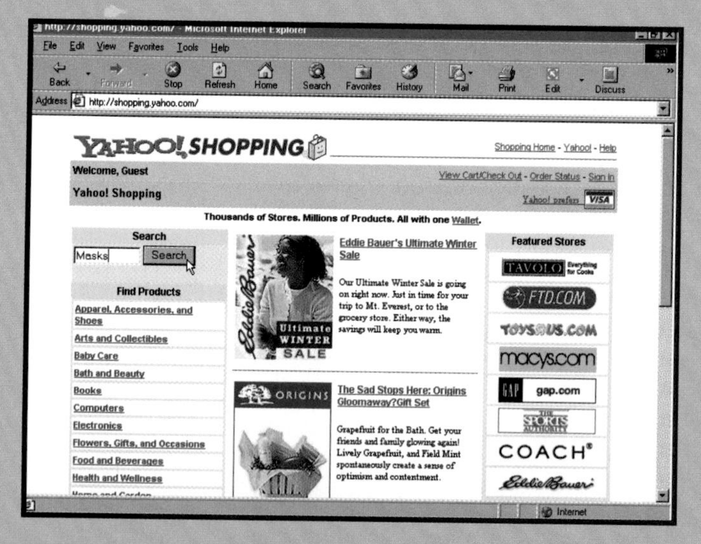

Figure 16 Yahoo! Shopping page at shopping.yahoo.com.

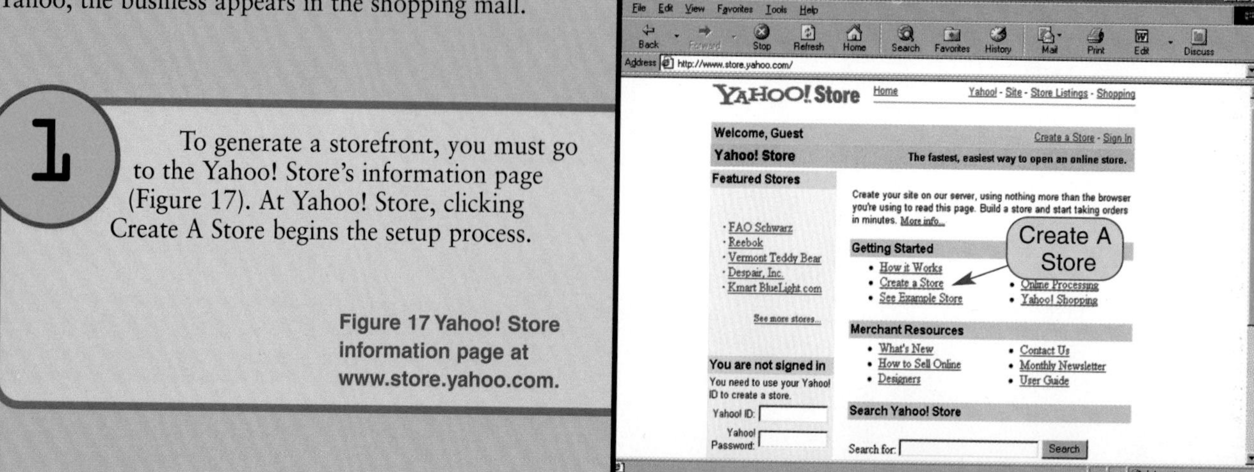

1

To generate a storefront, you must go to the Yahoo! Store's information page (Figure 17). At Yahoo! Store, clicking Create A Store begins the setup process.

Figure 17 Yahoo! Store information page at www.store.yahoo.com.

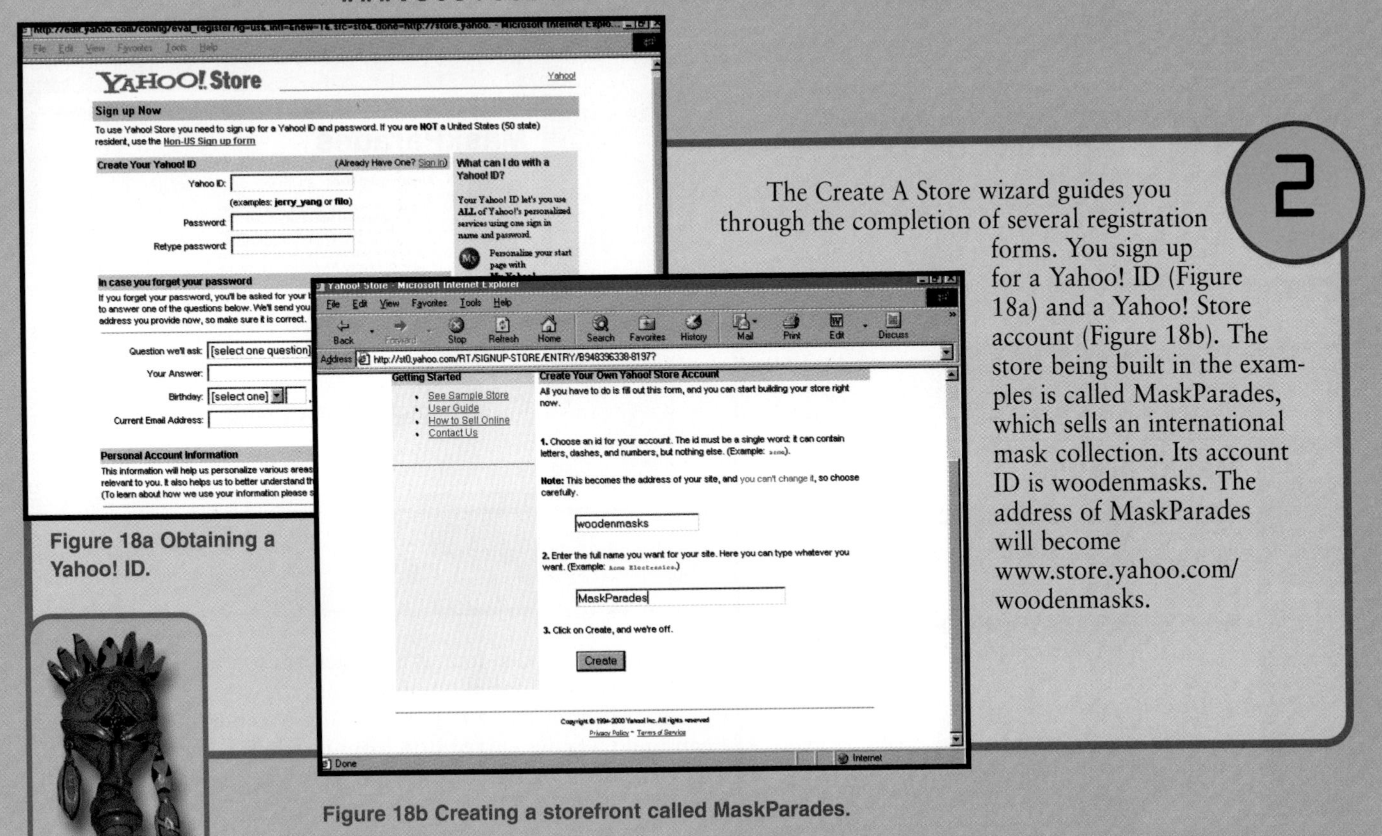

Figure 18a Obtaining a Yahoo! ID.

Figure 18b Creating a storefront called MaskParades.

2

The Create A Store wizard guides you through the completion of several registration forms. You sign up for a Yahoo! ID (Figure 18a) and a Yahoo! Store account (Figure 18b). The store being built in the examples is called MaskParades, which sells an international mask collection. Its account ID is woodenmasks. The address of MaskParades will become www.store.yahoo.com/woodenmasks.

3

After completing the forms, you are directed to take a tour of the working area where the storefront is developed. This tour actually is a tutorial that shows you how to construct your storefront. Later you can use the program's editing features to modify the store's content and appearance. The first stop in your guided tour is the new home page, or storefront (Figure 19). Notice that the program automatically generates the left panel and its buttons. An Edit bar always is visible during storefront construction.

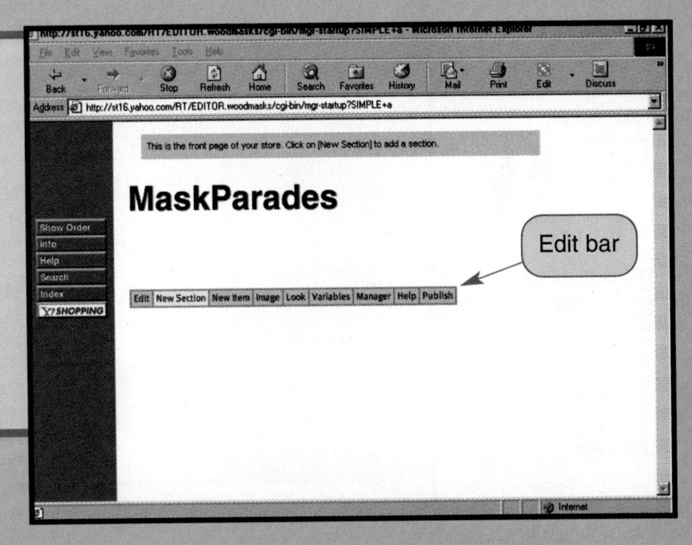

Figure 19 MaskParades storefront.

Figure 20 MaskParades store.

4

As you work your way through the tour, you are shown how to build sections, or rooms, in the store, how to add buttons on the storefront panel, and how to enter sales items. Each item is identified with a name, price, graphic, code number, and description. In the MaskParades store, two items, an Indonesian Garuda Mask and an African Spirit Mask, have been entered in the Buy International Masks section (Figure 20).

5 Each item has its own display case. Clicking an item enlarges its view and displays any pertinent information associated with it. In Figure 21, the thumbnail of the African Spirit Mask has been selected and the item's descriptive text, product code, and enlarged picture display.

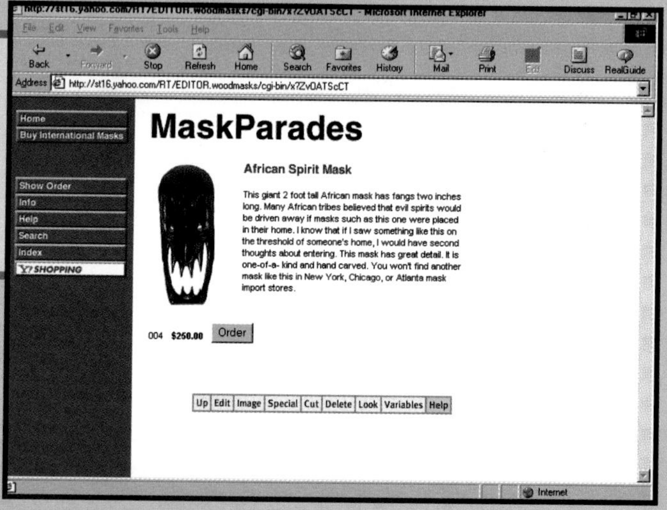

Figure 21 Display case.

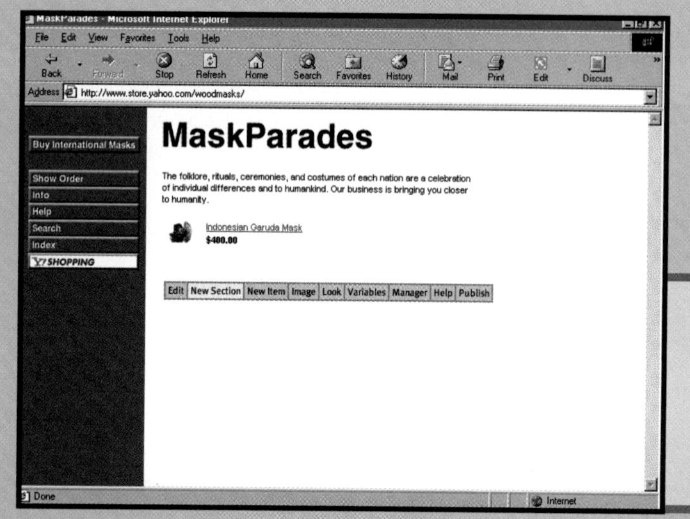

Figure 22 MaskParades storefront after completion of tutorial.

6 Yahoo!'s guided tour is not intended to be a comprehensive tutorial on how to use the program. It demonstrates that a few simple commands can build a rudimentary storefront in a few minutes. After completing the tutorial, the MaskParades storefront displays (Figure 22).

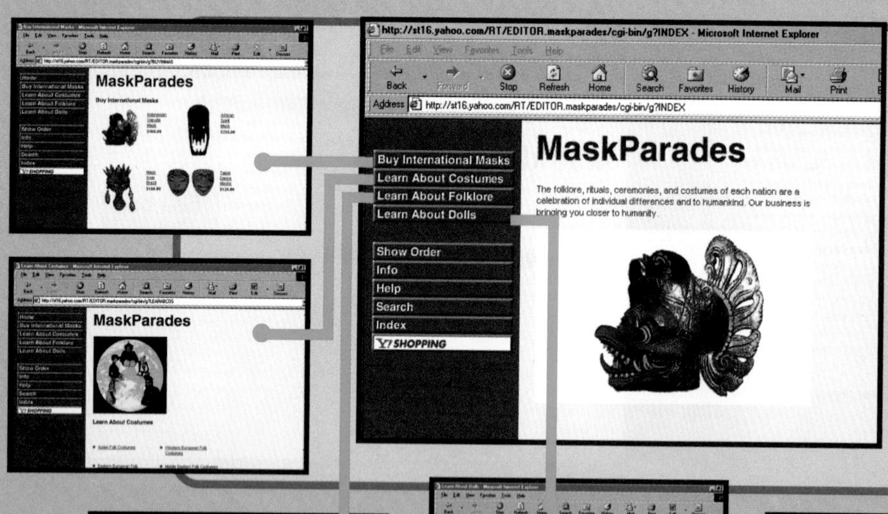

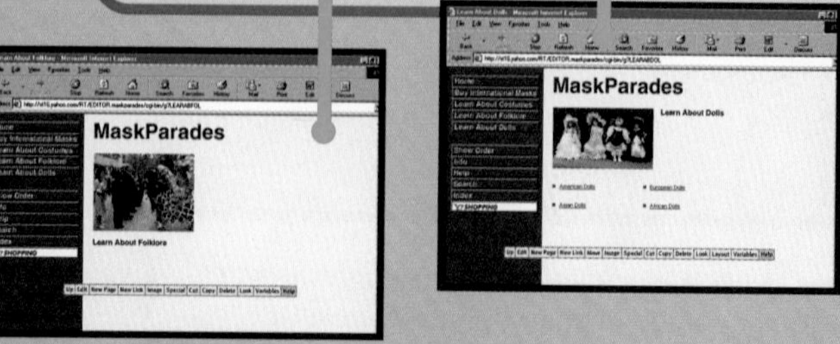

Figure 23 Expanded MaskParades site.

7 Editing the site and making it more attractive requires time and planning. As with any retail store, a Web storefront ultimately must attract customers and provide reasons for a customer to return to the site. To this end, the MaskParade site has been expanded beyond the simple storefront completed in the tutorial (Figure 23). In addition to its primary product, the sale of unique masks, MaskParades offers visitors an international museum of folk costumes, traditional dolls, and folklore.

8 After the store is completed, it must be tested for accuracy before a public opening (Figure 24). The e-retailer enters the store as a customer, orders some products, and verifies that the shopping cart and customer order form work properly.

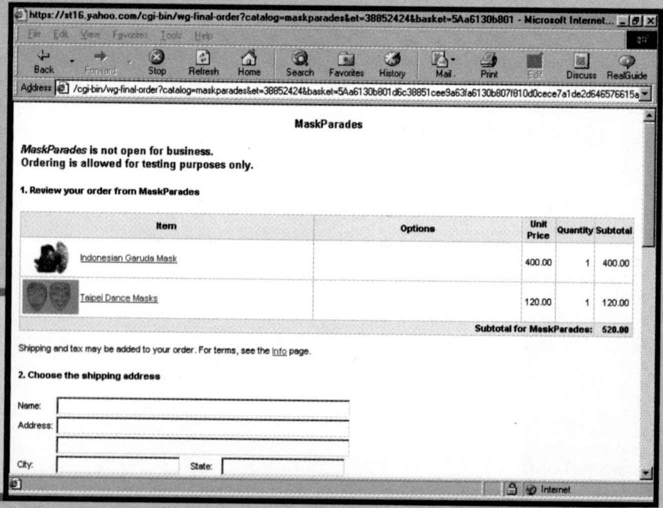

Figure 24 MaskParades test page to ensure accuracy.

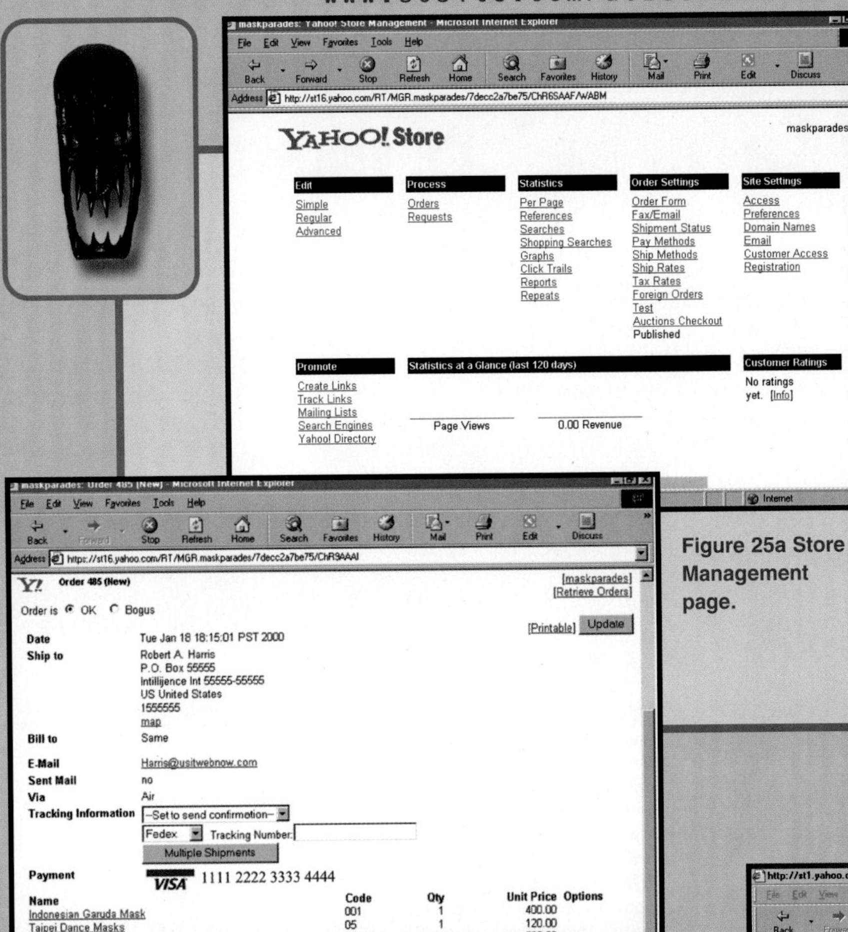

Figure 25a Store Management page.

Figure 25b Customer Order Status Form.

9

The e-retailer can verify that the order was received by returning to the Store Management screen at the editing site (Figure 25a). The Store Management screen accesses all of the back office operations necessary to maintain the site and communicate with customers. The Orders link in the Process category accesses the customer orders and related information. One type of automated feedback you can receive is the Customer Order Status Form (Figure 25b). It displays all of the pertinent information about a purchase including the shipping date and tracking number.

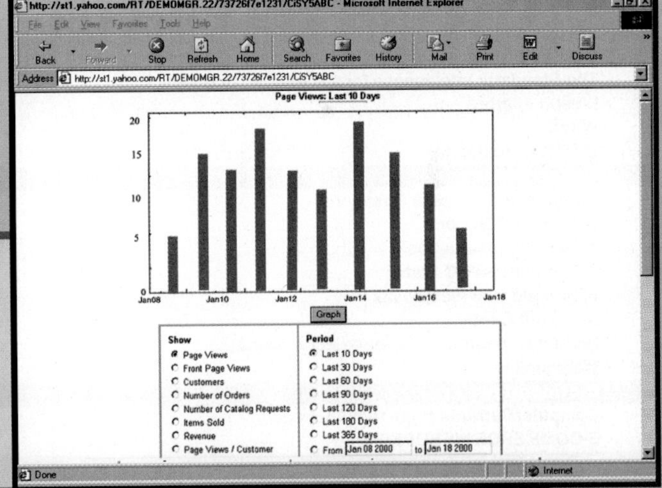

10

Sales and customer tracking statistics are generated immediately and are accessible through the Statistics option on Yahoo!'s Store Management screen. You receive instantaneous reports showing the number of customers visiting the site, the pages visited, repeat customers, graphs, and more (Figure 26).

Figure 26 Sales and customer tracking.

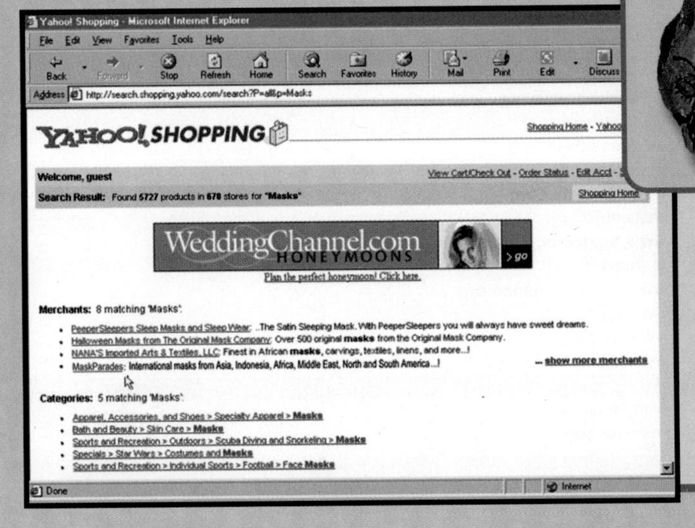

11

Your store must be accessible to the e-consumers visiting Yahoo! Shopping. A customer at Yahoo! Shopping can search for a product using the product category, the product name, or the company name. When a search is made for a product you sell, your store's name or product will be listed among all of the other stores that sell similar products (Figure 27).

Figure 27 Merchants and categories.

LEARNING MORE ABOUT E-COMMERCE

As the e-commerce boom continues, businesses rush to go online. E-commerce constantly is making headlines. Computer industry and business-oriented publications regularly publish e-commerce articles. General news magazines and newspapers have added e-commerce reports to their financial sections. Stock indexes such as Dow Jones have added Internet Indexes to measure the financial standings of online companies. Every year, increasingly more books, available at local libraries and bookstores, are being published about e-commerce.

Many **e-zines**, which are publications that exist only online, sponsor discussion groups and forums on e-commerce issues. Hardware and software manufacturers often provide news and general information at their sites. Online guides to e-commerce provide resources, tutorials, and links.

E-commerce has created opportunities for related careers. Many universities now offer courses in e-commerce as part of the business curriculum. Some schools even offer a master's degree in e-commerce or an MBA with an e-commerce concentration.

Government agencies, professional organizations, and nonprofit groups maintain many Web sites. These provide free information to support the growth of e-commerce. Some organizations act as industry watchdogs. Often these groups sponsor forums and workshops. Figure 28 lists many online sources for news and information about e-commerce. Using the popular Web portals such as Yahoo!, AltaVista, and others, you will find links to thousands of other e-commerce sites.

Name	URL
Online Magazines	
Business 2.0	www.business2.com
CNET News.com	news.cnet.com
ComputerCurrents	computercurrents.com/magazine/national
Electronic Commerce Today, or EC Today	www.ectoday.com
eMarketer	www.emarketer.com
InformationWeek Online	www.informationweek.com
InternetWeek Online	www.internetwk.com
internet.com: E-Commerce Guide	ecommerce.internet.com
New York Times: E-Commerce Times	www.ecommercetimes.com
The Standard: Intelligence for the Internet Economy	www.thestandard.com
Wilson Internet	www.webcommercetoday.com
Wired	www.wired.com
ZD Net E-Business	www.zdnet.com/enterprise/e-business
Online Guides	
About.com Electronic Commerce	ccommerce.miningco.com/business/ecommerce
CNet BUILDER.com	www.builder.com
eCommerce Guidebook	www.online-commerce.com
E-Commerce@NC State	ecommerce.ncsu.edu
Electronic Commerce FAQ	cism.bus.utexas.edu/resources/ecfaq.html
Net Profit Center	www.net-profit-center.net
Nightcat's Beginners Guide to Ecommerce	www.nightcats.com/sales/free.html
Webmonkey	hotwired.lycos.com/webmonkey/e-business/tutorials/tutorials3.htm
Dictionaries	
ComputerCurrents High-Tech Dictionary	computercurrents.com/resources/dictionary
E-COMMERCE WEBOPEDIA	e-comm.webopedia.com
Education And Careers	
About.com Career In Electronic Commerce	ecommerce.about.com/business/ecommerce/library/weekly/aa121399c.htm
About.com Guide to Business Majors	businessmajors.about.com
IS WorldNet: Electronic Commerce Course Syllabi Page	dossantos.cbpa.louisville.edu/ISNET/Ecomm
LifeLongLearning.com	www.lifelonglearning.com
Government and Nonprofit	
Association for Electronic Commerce Professionals (AECPII)	www.aecpii.com
BBBOnline	www.bbbonline.org
Data Interchange Standards Association (DISA)	www.disa.org
EC Institute: Professional E-Commerce Accreditation	www.ecschool.org
Electronic Commerce and the European Union	www.ispo.cec.be/ecommerce/Welcome.htm
Electronic Commerce Resource Center (ECRC)	www.ecrc.ctc.com
Electronic Privacy Information Center (EPIC)	www.epic.org
FTC Consumer Protection	www.ftc.gov/bcp/menu-internet.htm
InterNIC	rs.internic.net
Online Privacy Alliance	www.privacyalliance.org
The Internet Corporation for Assigned Names and Numbers (ICANN)	www.icann.org
TRUSTe	www.truste.org
United States Government Electronic Commerce Policy	www.ecommerce.gov
Online Bookstores	
Amazon.com	amazon.com
Barnes and Noble Booksellers	www.bn.com

For an updated list, visit www.scsite.com/dc2001/ch7/e-commerce.htm.

Figure 28 Online sources for information about e-commerce.

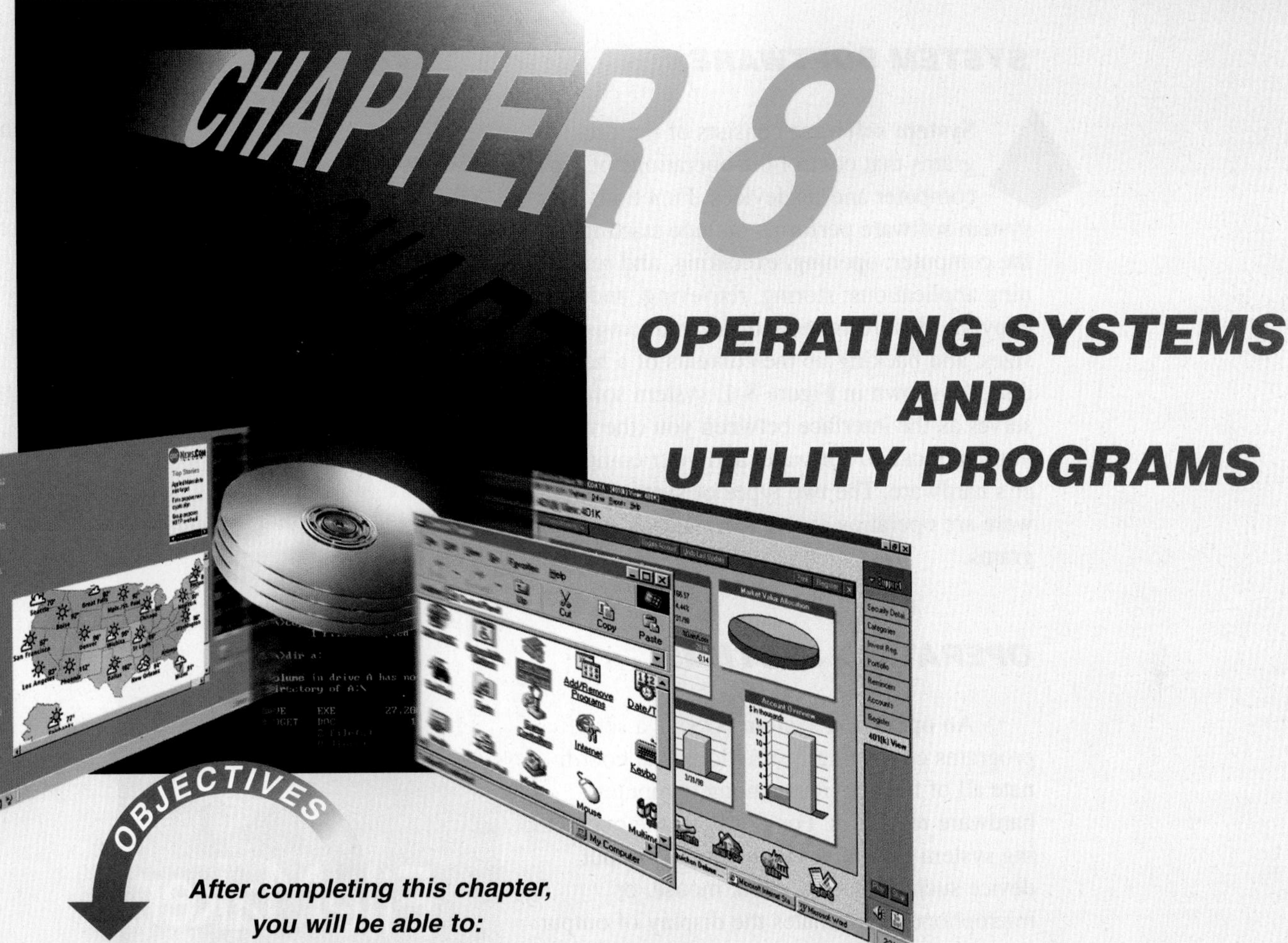

OPERATING SYSTEMS AND UTILITY PROGRAMS

OBJECTIVES

After completing this chapter, you will be able to:

- Identify the various types of system software
- Differentiate between an operating system and utility program
- Describe the features of operating systems
- Describe the functions of an operating system
- Identify and briefly describe popular operating systems used today
- Explain the startup process for a personal computer
- Discuss the purpose of the following utilities: viewer, file compression, diagnostic, disk scanner, defragmenter, uninstaller, backup, antivirus, and screen saver

Like most computer users, you probably are familiar with application software, such as word processing, spreadsheet, e-mail, and a Web browser. To run any application software, also called an application, your computer also must be running another type of software -- an operating system. In addition to an operating system, modern computers also contain several utility programs. Together, operating systems and utility programs comprise a category of software called system software. This chapter discusses the operating system and its functions, as well as several utility programs used with today's personal computers.

SYSTEM SOFTWARE

System software consists of the programs that control the operations of the computer and its devices. Functions that system software performs include starting up the computer; opening, executing, and running applications; storing, retrieving, and copying files; formatting disks; reducing file sizes; and backing up the contents of a hard disk. As shown in Figure 8-1, system software serves as the interface between you (the user), your application software, and your computer's hardware. The two types of system software are operating systems and utility programs.

OPERATING SYSTEMS

An **operating system (OS)** is a set of programs containing instructions that coordinate all of the activities among computer hardware resources. For example, the operating system recognizes input from an input device such as the keyboard, mouse, or microphone; coordinates the display of output on the monitor; instructs a printer how and when to print information; and manages data, instructions, and information stored on disk. A computer cannot function without an operating system.

An operating system also contains instructions that allow you to run application software. Application software is written to run with particular operating systems. Thus, the operating system installed on your computer directly determines which application software you can and cannot run.

The operating system used on a computer sometimes is called the **software platform**. When you purchase application software, the package states the software platform (operating system) on which it runs.

icons indicate software platforms

FileMaker Pro 5

Manage your data from desktop to Web.

Figure 8-2 *Some application software can run on multiple software platforms (operating systems) on which it runs.*

Figure 8-1 *System software serves as the interface between the user, application programs, and the computer's hardware.*

Applications that run identically on more than one operating system are referred to as **cross-platform**. FileMaker Pro, for example, runs on several operating systems, including several versions of Microsoft Windows and the Macintosh operating system (Figure 8-2).

An operating system usually is stored on the computer's hard disk. The core of an operating system, called the **kernel**, is responsible for managing memory, files, and devices; maintaining the computer's clock, which contains the current date and time; starting applications; and assigning the computer's resources, such as hardware devices, software programs, data, and information. Each time you turn on the computer, the kernel and other frequently used instructions in the operating system are copied from the hard disk (storage) to the computer's memory.

Any program or command that remains in memory while the computer is running is called **memory-resident**. This includes the operating system kernel and programs such as calendars and calculators that you access frequently or need to access quickly.

User Interfaces

A **user interface** is the part of the software with which you interact; it controls how data and instructions are entered and information is presented on the screen. Two types of user interfaces are command-line and graphical (Figure 8-3). Many operating systems use a combination of these types of user interfaces to define how you interact with your computer.

With a **command-line interface**, you type keywords or press special keys on the keyboard to enter data and instructions. Recall that a keyword is a special word, phrase, or code that a program understands as an instruction. Some keyboards also include keys that send a command to a program when you press them. When working with a command-line interface, the set of commands you use to interact with the computer is called the **command language**.

Figure 8-3a *(command-line)*

```
C:\>dir budget.*

 Volume in drive C has no label
 Volume Serial Number is 1B7B-07D0
 Directory of C:\

BUDGET   DOC              11  07-16-01 11:54a
         1 file(s)              11 bytes
         0 dir(s)      107,708,416 bytes free

C:\>copy budget.doc a:
        1 file(s) copied

C:\>dir a:

 Volume in drive A has no label
 Directory of A:\

SAVE     EXE          27,206  08-12-01  2:50p
BUDGET   DOC              11  07-16-01 11:54a
         2 file(s)          27,217 bytes
         0 dir(s)          701,440 bytes free
```

Figure 8-3b *(graphical)*

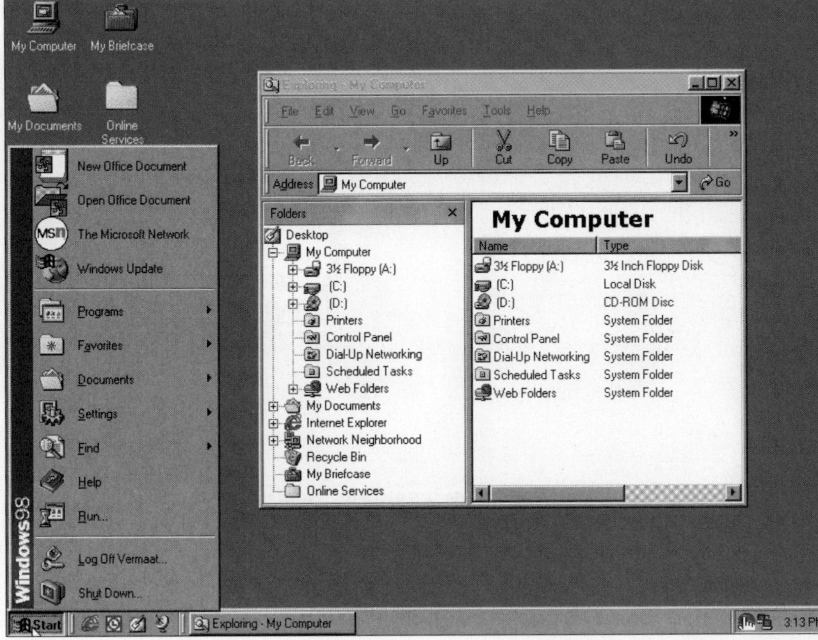

Figure 8-3 *Examples of command-line and graphical user interfaces.*

WEB INFO

For more information on graphical user interfaces, visit the Discovering Computers 2001 Chapter 8 WEB INFO page (**www.scsite.com/dc2001/ch8/webinfo.htm**) and click Graphical User Interface.

As noted in Chapter 1, a **graphical user interface** allows you to use menus and visual images such as icons, buttons, and other graphical objects to issue commands.

Instead of requiring you to remember keywords or special keys, a **menu** displays a set of available commands or options from which you choose one or more. You can use a keyboard, mouse, or any other pointing device to select items on the menu. Because they do not require you to know a command language, menus typically are easier to learn and use than commands.

An **icon** is a small image that represents an item such as a program, an instruction, or a file.

Although any input device can be used with a graphical user interface, the mouse and other pointing devices are used more commonly. Of all the interfaces, a graphical user interface is the easiest to learn and work with, characteristics described as being **user-friendly**.

Today, many graphical user interfaces incorporate Web browser-like features, which increase their ease of use. In these browser-like graphical user interfaces, icons function like Web links, toolbar buttons look like those used in Web browsers, and Web pages can be delivered automatically to your computer (Figure 8-4).

Features of Operating Systems

Depending on its intended use, an operating system will support just one user running one program or thousands of users running multiple programs. These various capabilities of operating systems are described as single user, multiuser, multiprocessing, multitasking. A single operating system may support one or all of these capabilities.

A **single user**, also called **single tasking** operating system allows only one user to run one program at a time. Suppose, for example, you are typing a memorandum in a word processing program and decide to browse the Web for more information. If you are working with a single user operating system, you must quit the word processing program before you can run the Web browser. You then must close the Web browser before you restart the word processing program to finish the memorandum. Early systems were single user; however, most operating systems today are multitasking and multiuser.

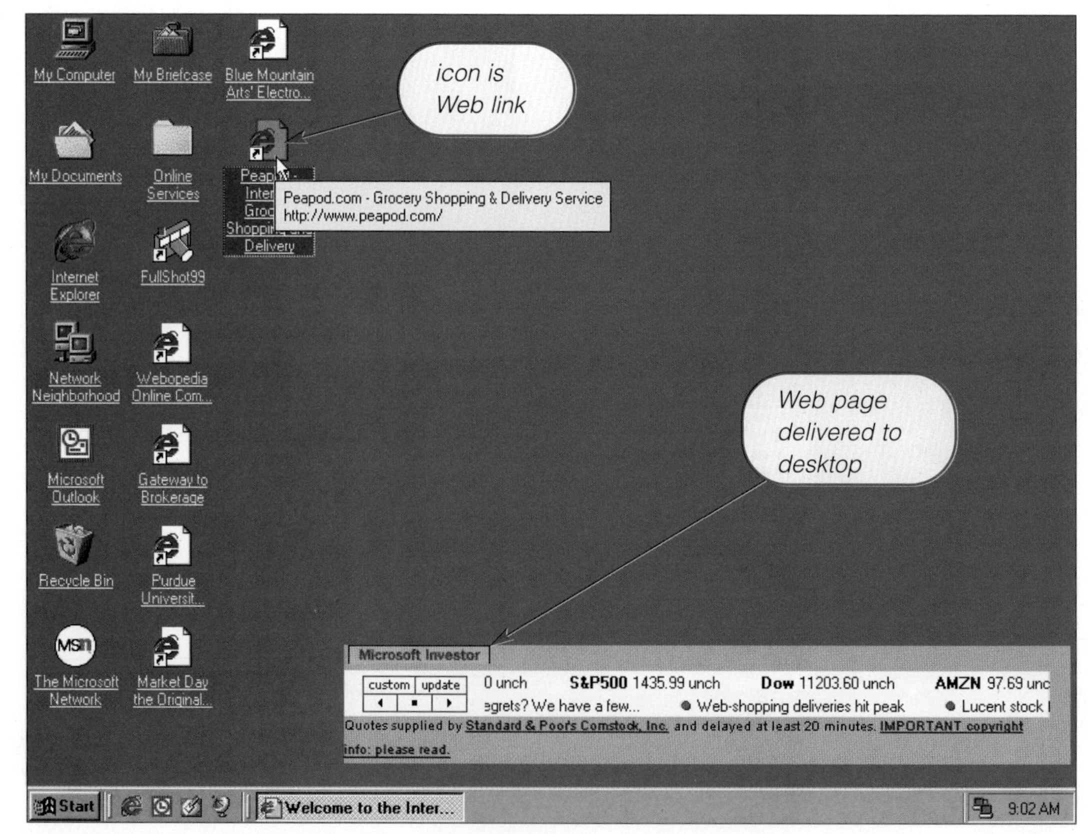

Figure 8-4 *This graphical user interface incorporates icons that function like Web links and Web pages that are pushed onto the desktop.*

A **multitasking** operating system allows a single user to work on two or more applications that reside in memory at the same time. Using the example just cited, if you are working with a multitasking operating system, you do not have to quit the word processing program to run a Web browser; that is, both programs can run concurrently. When you are running multiple applications, the one that you currently are working with is in the **foreground**, and the others that are running but not being used are in the **background** (Figure 8-5).

A **multiuser** operating system enables two or more users to run a program simultaneously. Networks, minicomputers, mainframes, and supercomputers allow hundreds to thousands of users to be connected at the same time, and thus are multiuser.

A **multiprocessing** operating system can support two or more CPUs running programs at the same time. Multiprocessing works much like parallel processing, which was discussed in Chapter 3, and involves the coordinated processing of programs by more than one CPU. As with parallel processing, multiprocessing increases a computer's processing speed.

A computer with separate CPUs also can serve as a **fault-tolerant computer**; that is, one that continues to operate even if one of its components fails. Fault-tolerant computers are built with duplicate components such as CPUs, memory, and disk drives. If any one of these components fails, the computer switches to the duplicate component and continues to operate. Fault-tolerant computers are used for airline reservation systems, communications networks, bank teller machines, and other systems that are of critical importance and must be operational at all times.

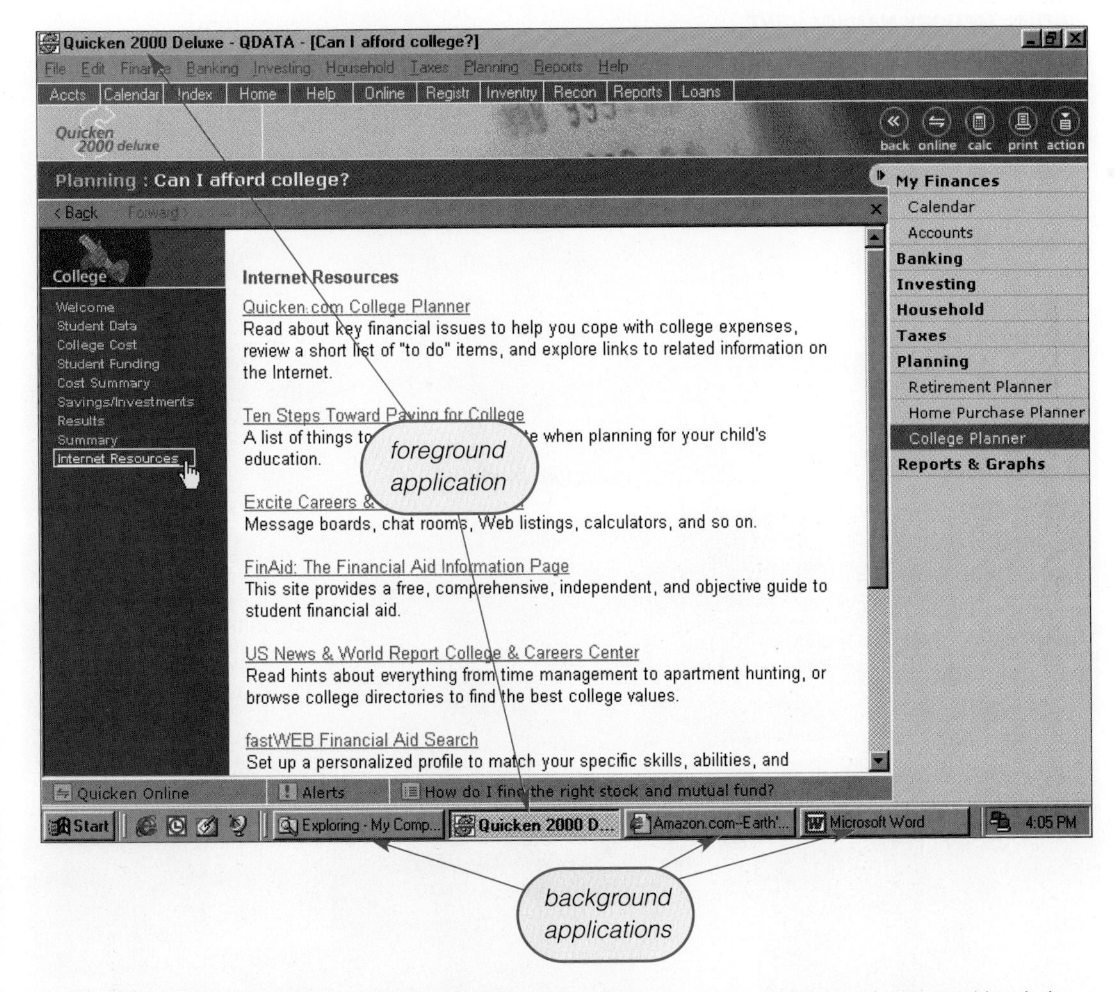

Figure 8-5 *When running multiple applications, the one with which you currently are working is in the foreground; others that are running but not being used are in the background. In this screen, the buttons in the taskbar button area indicate that Quicken Deluxe is in the foreground and Windows Explorer, Microsoft Internet Explorer, and Microsoft Word are in the background.*

Functions of an Operating System

An operating system performs a number of basic functions that enable you and the application software to interact with the computer: managing memory, spooling print jobs, configuring devices, monitoring system performance, administering security, and managing storage media and files. The following paragraphs describe these activities.

MEMORY MANAGEMENT The purpose of **memory management** is to optimize use of random access memory (RAM). Recall that RAM, often simply called memory, is one or more chips on the motherboard that temporarily hold items such as data and instructions while they are being processed by the CPU. The operating system has the responsibility to allocate, or assign, these items to an area of memory while they are being processed; to monitor carefully the contents of these items in memory; and to clear these items from memory when they are no longer required by the CPU. For example, the operating system manages areas of memory or storage called buffers. A **buffer** is an area of memory or storage in which data and information is placed while waiting to be transferred to or from an input or output device. The contents of the buffers are managed by the operating system.

Some operating systems use virtual memory to optimize RAM. With **virtual memory (VM)**, the operating system allocates a portion of a storage medium, usually the hard disk, to function as additional RAM (Figure 8-6). The area of the hard disk used for virtual memory is called a **swap file**

VIRTUAL MEMORY MANAGEMENT

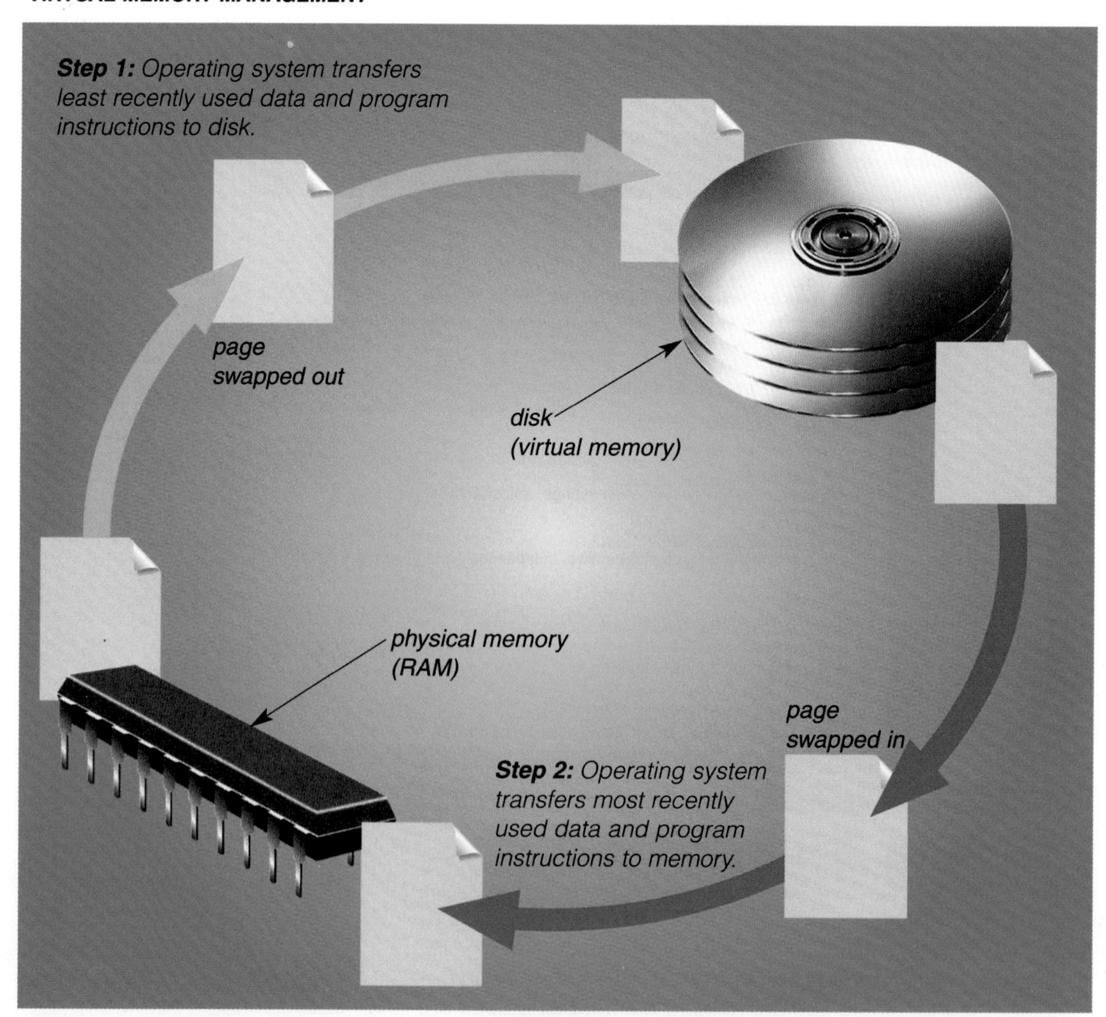
Step 1: Operating system transfers least recently used data and program instructions to disk.
page swapped out
disk (virtual memory)
physical memory (RAM)
page swapped in
Step 2: Operating system transfers most recently used data and program instructions to memory.

Figure 8-6 *With virtual memory (VM), the operating system allocates a portion of a storage medium, usually the hard disk, to function as additional RAM.*

because it is used to swap (exchange) data and program instructions between memory and storage. The amount of data and program instructions exchanged at a given time is called a **page**. Thus, the technique of swapping items between memory and storage often is called **paging**.

When an operating system spends much of its time paging, instead of executing application software, it is said to be **thrashing**. If application software, such as a Web browser, has stopped responding and the LED for your hard disk is blinking repeatedly, the operating system probably is thrashing. To stop the thrashing, you should quit the application that stopped responding. If thrashing occurs frequently, you may need to install more RAM in your computer.

SPOOLING PRINT JOBS When you instruct an application to print a document, such as a memorandum, newsletter, photograph, or e-mail message, the document you are printing is called a **print job**. Because the CPU sends print jobs to the printer at a rate much faster than the printer can print, operating systems typically use a technique called spooling to increase printer efficiency.

With **spooling**, the print jobs are placed in a buffer instead of being sent immediately to the printer. Recall that a buffer is an area of memory or storage that holds data and information waiting to be transferred from one device to another. In the case of print spooling, the buffer holds the information waiting to print while the printer prints from the buffer at its own printing rate. As soon as the print job is placed in the buffer, the CPU is available to process the next instruction, usually in less than a few seconds. Thus, once the print job is in the buffer, you can use your computer for other tasks.

Spooling allows you to send a second job to the printer without waiting for the first job to finish printing. Multiple print jobs are **queued**, or lined up, in the buffer. The program that manages and intercepts print jobs and places them in the queue is called the **print spooler** (Figure 8-7).

CONFIGURING DEVICES To communicate with each device in the computer, the operating system relies on device drivers. A **device driver** is a small program that accepts commands from another program and then converts these commands into commands that the device understands. Each device on a computer, such as the mouse, keyboard,

WEB INFO

For more information on device drivers, visit the Discovering Computers 2001 Chapter 8 WEB INFO page (**www.scsite.com/dc2001/ch8/webinfo.htm**) and click Device Drivers.

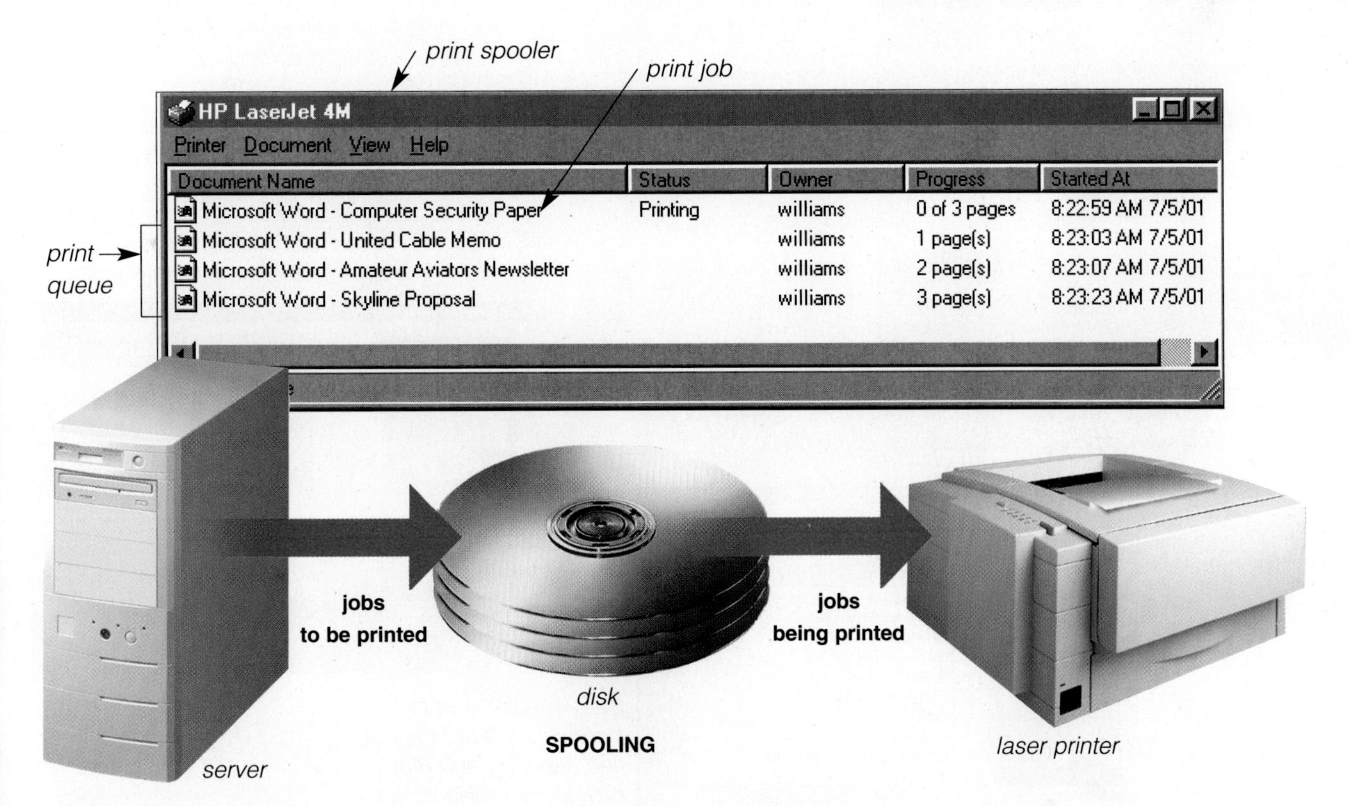

Figure 8-7 *Spooling increases both CPU and printer efficiency by writing print jobs to a disk before they are printed. In this figure, three jobs are in the queue and one is printing.*

monitor, and printer, has its own specialized set of commands and thus requires its own device driver, also called a **driver**. These devices will not function unless the correct device driver is installed on the computer. In Windows environments, most device drivers have a .drv extension.

If you add a new device to your computer, such as a printer or scanner, its driver must be installed before the device will be operational. Windows 98 provides a wizard to guide you through the installation steps (Figure 8-8). For many devices, your computer's operating system already may include the necessary device drivers. If it does not, you can install the drivers from the disk included with the device upon purchase. If, for some reason, you need a driver for your device and do not have the original disk, you can obtain the driver by contacting the vendor that sold you the device or contacting the manufacturer directly. Many manufacturers also post device drivers on their Web sites for anyone to download.

Figure 8-8

HOW TO INSTALL NEW HARDWARE IN WINDOWS 98

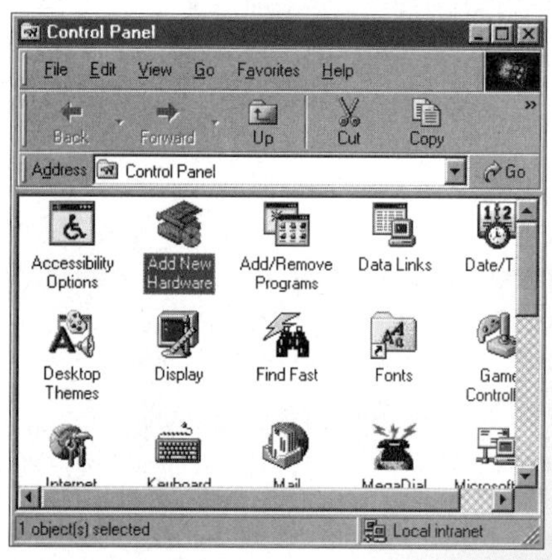

Step 1: *Open the Control Panel window.*

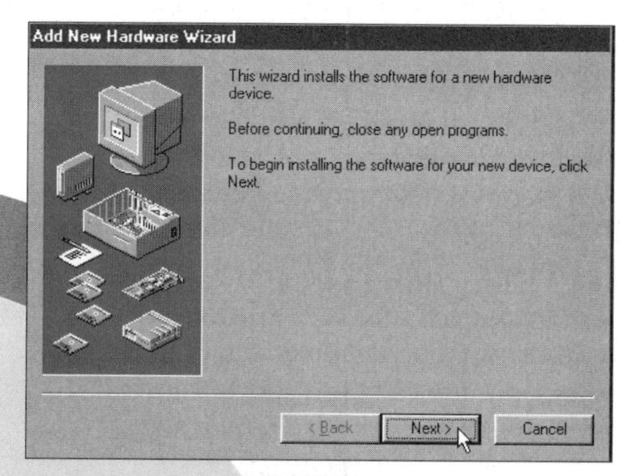

Step 2: *Start the Add New Hardware Wizard by clicking the Add New Hardware icon. Follow the on-screen instructions.*

Step 3: *The Add New Hardware Wizard searches for Plug and Play devices on your system. If it finds any such devices, it installs them.*

Step 4: *If the Add New Hardware Wizard cannot find any Plug and Play devices, you can select the type of device you want to install.*

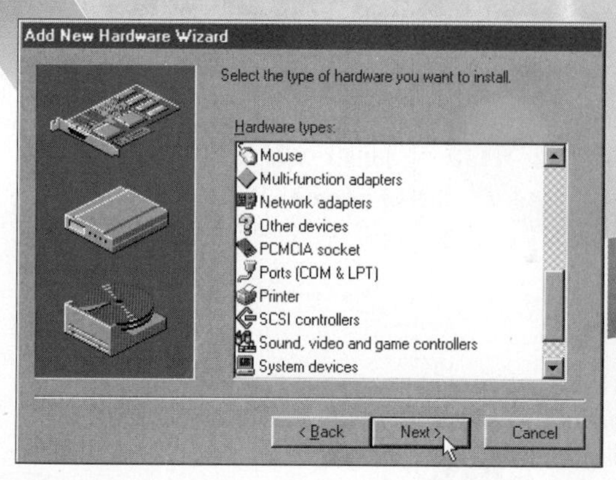

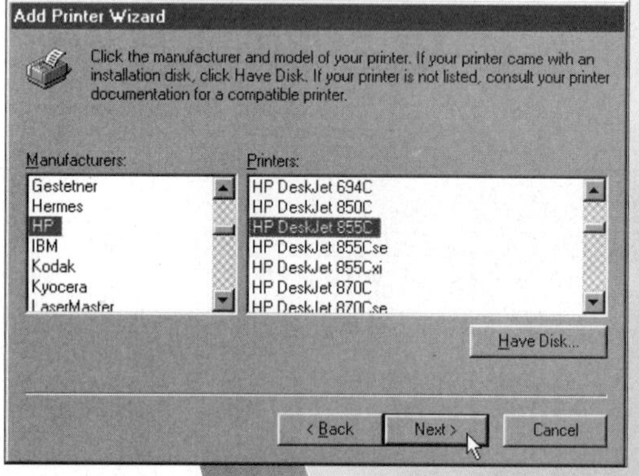

Step 5: *You may be requested to insert the floppy disk, CD-ROM, or DVD-ROM that contains necessary driver files to complete the installation of the device.*

In the past, installing a new device often required setting switches and other elements on the motherboard. Installation of current devices is easier because most devices and operating systems support Plug and Play. As previously noted, **Plug and Play** is the computer's capability of recognizing any new device and assisting in the installation of the device by loading the necessary drivers automatically and checking for conflicts with other devices. Having Plug and Play support means a user can plug in a device, turn on the computer, and then use, or play, the device without having to configure the system manually.

When installing some components, occasionally you have to know which interrupt request the device should use for communications. An **interrupt request (IRQ)** is a communications line between a device and the CPU. Most computers have 15 IRQs (Figure 8-9). With Plug and Play, the operating system determines the best IRQ to use for these communications. If your operating system uses an IRQ that already is assigned to another device, an IRQ conflict will occur and the computer will not work properly. If an IRQ conflict occurs, you will have to obtain the correct IRQ for the device, which usually is specified in the installation directions that accompany the device.

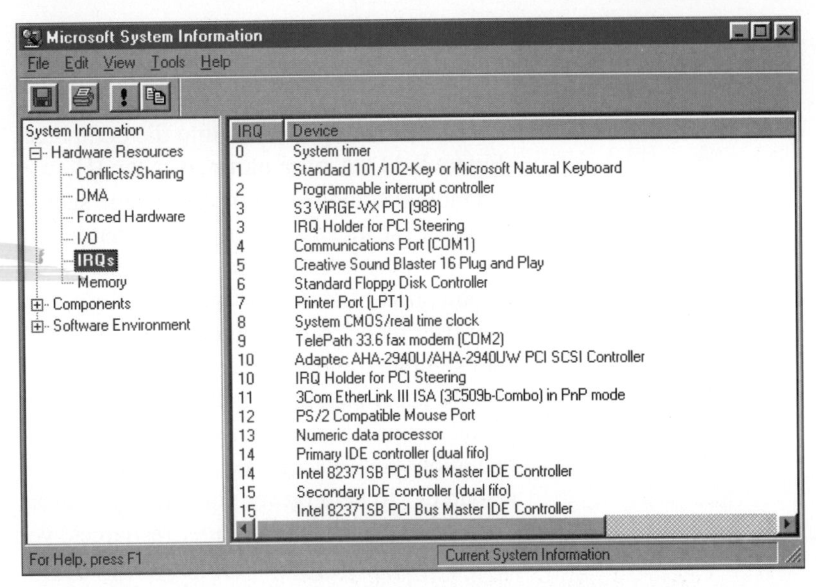

Figure 8-9 To display IRQs in Windows 98, click the Start button, point to Programs, point to Accessories, point to System Tools, and click System Information. When the Microsoft System Information window opens, click the Hardware Resources icon and then click IRQs.

WEB INFO

For more information on Plug and Play, visit the Discovering Computers 2001 Chapter 8 WEB INFO page (**www.scsite.com/dc2001/ch8/webinfo.htm**) and click Plug and Play.

MONITORING SYSTEM PERFORMANCE

Operating systems typically contain a **performance monitor**, which is a program that assesses and reports information about various system resources and devices (Figure 8-10). For example, you can monitor the CPU, disks, memory, and network usage, as well as the number of times a file is read or written. The information in these reports can help you identify problems with resources so you can attempt to resolve the problem. For example, if your computer is running extremely slow and you determine that the computer's memory is utilized to its maximum, then you might consider installing additional memory.

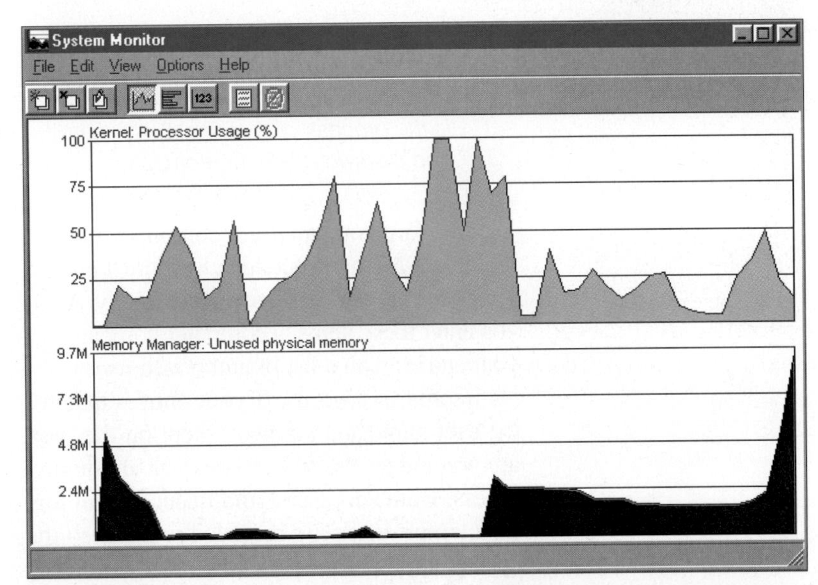

Figure 8-10 A performance monitor is a program that assesses and reports information about various system resources and devices. The System Monitor above is tracking the processor usage and the amount of unused physical memory.

ADMINISTERING SECURITY Most multiuser operating systems allow each user to **log on**, which is the process of entering a user name and a password into the computer (Figure 8-11). A **user name**, or **user ID**, is a unique combination of characters, such as letters of the alphabet or numbers, that identifies one specific user. Many users select a combination of their first and last names as their user name. A user named Rick Williams, for example, might choose, rwilliams, as his user name. A **password** is a combination of characters associated with your user name that allow you to access certain computer resources. To help prevent unauthorized users from accessing those computer resources, you should keep your password confidential. As you enter your password, most computers hide the actual password characters by displaying some other characters, such as asterisks (*). Guidelines for selecting good passwords are discussed in Chapter 14.

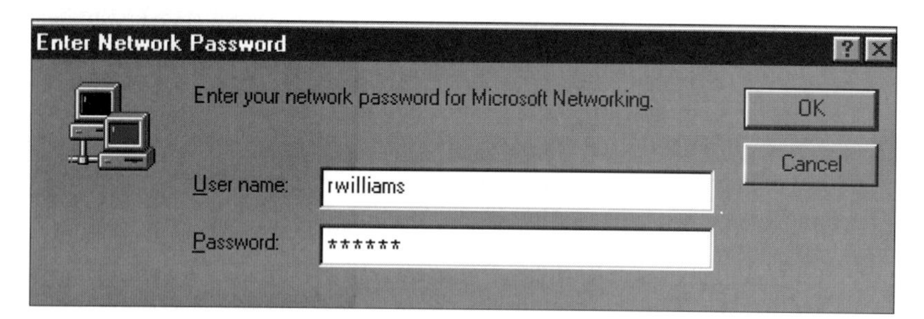

Figure 8-11 *Most multiuser operating systems allow each user to log on, which is the process of entering a user name and password into the computer.*

Before you can use a computer that requires a log on procedure, you must enter your user name and password correctly. As you enter these items, a computer compares your entries with a list of authorized user names and passwords. If your entries match the user name and password kept on file, you are granted access; otherwise, you are denied access. Both successful and unsuccessful log on attempts often are recorded in a file so the system administrator can review who is using or attempting to use the computer. System administrators also use these files to monitor computer usage.

Some operating systems also allow you to assign passwords to files so that only authorized users can open them.

MANAGING STORAGE MEDIA AND FILES Operating systems also contain a type of program called a **file manager**, which performs functions related to storage and file management (Figure 8-12). Some of the storage and file management functions performed by a file manager are formatting and copying disks; displaying a list of files on a storage medium; checking the amount of used or free space on a storage medium; and copying, renaming, deleting, moving, and sorting files.

POPULAR OPERATING SYSTEMS

Many of the first operating systems were **device dependent**; that is, they were developed by manufacturers specifically for the computers in their product line. Software that is privately owned and limited to a specific vendor or computer model is called **proprietary software**. When manufacturers introduced a new computer or model, they often produced an improved and different proprietary operating system. Problems arose, however, when a user wanted to switch computer models or manufacturers. Because applications were designed to work with a specific operating system, the user's application software often would not work on the new computer.

Although some operating systems still are device dependent, the trend today is toward **device-independent** operating systems that will run on many manufacturers' computers. The advantage of device independent operating systems is that, even if you change computer models or vendors, you can retain existing application software and data files, which generally represent a sizable investment in time and money.

New versions of an operating system usually are downward compatible. A **downward-compatible** operating system is one that recognizes and works with application software that was written for an earlier version of the operating system. The application software, by contrast, is **upward compatible**; meaning it was written for an earlier version of the operating system, but runs under the new version.

The following sections discuss some of the more popular operating systems.

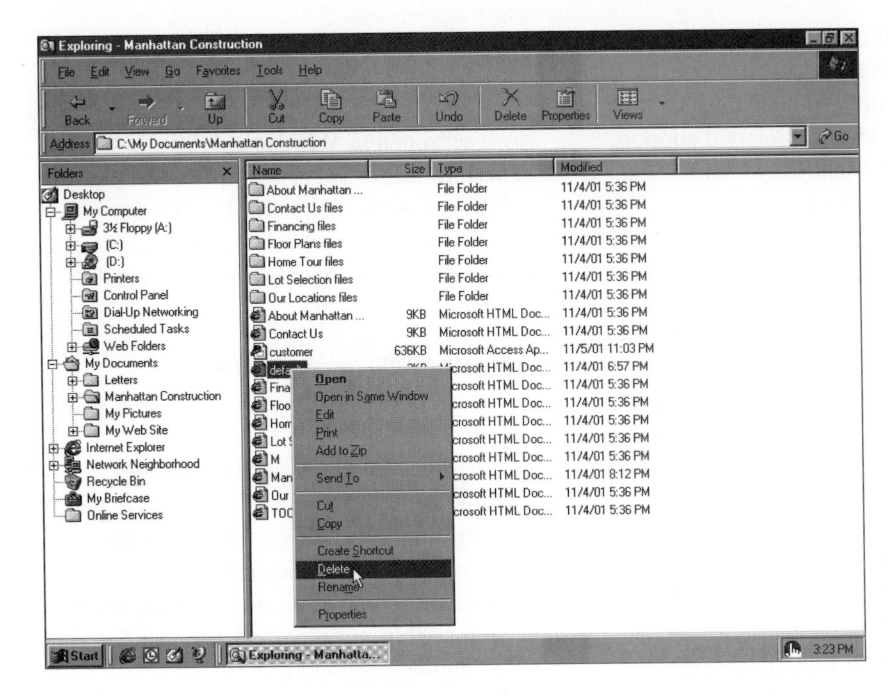

Figure 8-12 Windows Explorer is a file manager. A file manager performs functions related to storage and file management.

DOS

The term **DOS (Disk Operating System)** refers to several single user operating systems that were developed in the early 1980s for PCs. The two more widely used versions of DOS were PC-DOS and MS-DOS. Microsoft Corporation developed both PC-DOS and MS-DOS; the two operating systems were essentially the same. Microsoft developed PC-DOS (Personal Computer DOS) for IBM, and IBM installed and sold PC-DOS on its computers. At the same time, Microsoft marketed and sold MS-DOS (Microsoft DOS) to makers of IBM-compatible personal computers.

When first developed by Microsoft, DOS used a command-line interface. Later versions of DOS included both command-line and menu-driven user interfaces, as well as improved memory and disk management.

At its peak, DOS was a widely used operating system, with an estimated 70 million computers running it. Today, DOS no longer is widely used because it does not offer a graphical user interface and it cannot take full advantage of modern 32-bit microprocessors. Although it still has some users, many PC users prefer to use the graphical user interface of Windows platforms.

Windows 3.x

To meet the need for an operating system that had a graphical user interface, Microsoft developed **Windows**. **Windows 3.x** refers to three early versions of Microsoft Windows: Windows 3.0, Windows 3.1, and Windows 3.11. These Windows 3.x versions were not operating systems; instead they were operating environments. An **operating environment** is a graphical user interface that works in combination with an operating system to simplify its use. Windows 3.x was designed to work as an operating environment for DOS.

Windows 95

With **Windows 95**, also referred to as **Win95**, Microsoft developed a true multitasking operating system – not an operating environment like early versions of Windows. Windows 95 thus did not require DOS to run, although it included some DOS and Windows 3.x features to allow for downward compatibility.

One advantage of Windows 95 was its improved graphical user interface, which made working with files and programs easier than the earlier versions. In addition, most programs ran faster under Windows 95

WEB INFO

For more information on Windows, visit the Discovering Computers 2001 Chapter 8 WEB INFO page (**www.scsite.com/dc2001/ch8/webinfo.htm**) and click Windows.

because it was written to take advantage of 32-bit processors (versus 16-bit processors) and supported a more efficient form of multitasking. Windows 95 also included support for networking, Plug and Play technology, longer file names, and e-mail.

Windows NT

Microsoft **Windows NT**, also referred to as **NT**, was an operating system designed for client-server networks. The server used Windows NT Server, and the clients connected to the server used Windows NT Workstation, which had a Windows 95 interface.

Windows 98

Microsoft developed an upgrade to the Windows 95 operating system, called Windows 98. The **Windows 98** operating system, also called **Win98**, was easier to use than Windows 95 and was more integrated with the Internet (Figure 8-13). For example, Windows 98 included **Microsoft Internet Explorer**, a popular Web browser. The Windows 98 file manager, called **Windows Explorer**, also had a Web browser look and feel. With Windows 98, you could have an Active Desktop™ interface, which allowed you to set up Windows so icons on the desktop and file names in Windows Explorer worked like Web links.

Windows 98 also provided faster system startup and shutdown, better file management, and support for new multimedia technologies such as DVD and WebTV. Windows 98 supported the Universal Serial Bus (USB) so you easily could add and remove devices on your computer. The table in Figure 8-14 lists features of Windows 98. Like Windows 95, Windows 98 could run 16- and 32-bit software, which means it could run software designed for DOS and earlier versions of Windows.

Windows 2000

Microsoft Windows 2000 is an upgrade to the Windows 98 and Windows NT operating systems. Like Windows 98 and Windows NT, Windows 2000 is a complete multitasking operating system (not an operating environment) that has a graphical user interface. Two basic versions of Windows 2000 exist: the **Windows 2000 Server** family, which includes Windows 2000 Server, Windows 2000 Advanced Server, and Windows 2000 Datacenter Server, for various levels of network servers and **Windows 2000 Professional** for stand-alone business desktop or laptop computers, as well as for computers connected to the network, called the clients. Windows 2000 includes these additional features:

Windows 2000 Server Family

- Various networking versions to meet your needs:
  - Windows 2000 Server for all sizes of business
  - Windows 2000 Advanced Server for e-commerce applications
  - Windows 2000 Datacenter Server for demanding, time-critical applications such as data warehousing
- Capability of hosting and managing Web sites
- Windows Distributed interNet Applications (DNA) Architecture provides a tool for easy application development across platforms
- Deliver and manage multimedia across intranets and the Internet
- Enable users to store documents in Web folders
- Manage information about network users and resources with Active Directory™
- Supports clients using Windows 2000 Professional, Windows NT, Windows 98, Windows 95, Windows 3.x, Macintosh, and Unix

Windows 2000 Professional

- Reliable operating system for desktop and laptop business computers
- Windows File Protection safeguards operating system files from being overwritten during installation of applications
- Certified device drivers safeguard drivers against tampering
- Windows Installer Service guides you through installation or upgrade of applications
- Faster performance than Windows 98
- Adapts Start menu to display applications you use most frequently
- Preview multimedia files in Windows Explorer before opening them
- Enhanced technology to increase efficiency of mobile users

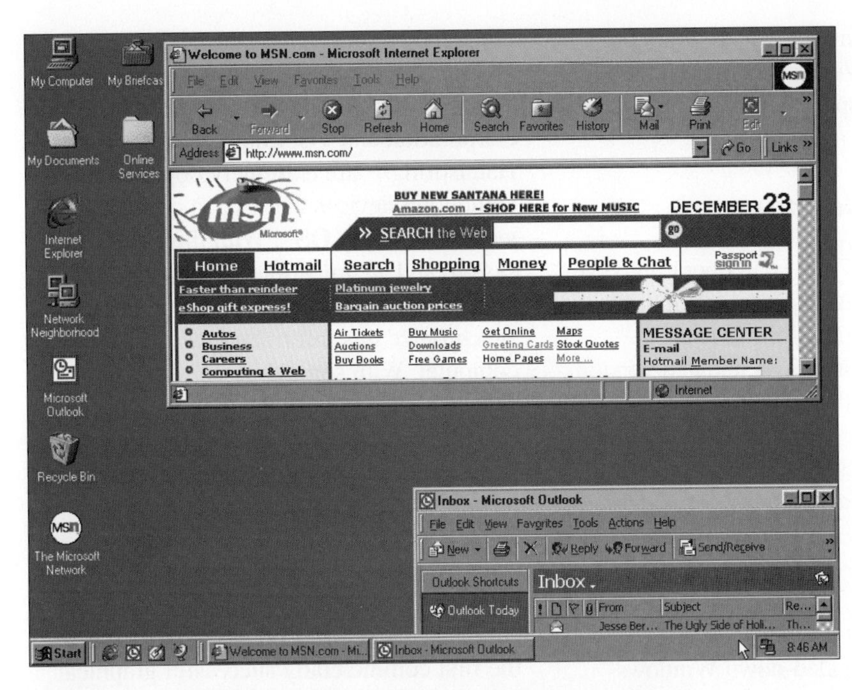

Figure 8-13 *Microsoft Windows 98 is easy to use, fast, and integrated with the Internet. This desktop view is called the Classic style; the desktop view shown in Figure 8-4 on page 8.4 is called the Web style.*

FEATURES COMMON TO MOST WINDOWS OPERATING SYSTEMS

Feature	Description
1. Active Desktop™	Allows you to set up Windows so icons on the desktop and file names in Windows Explorer work like Web links (single-click) and create real-time windows that display television-style news or an animated ticker that provides stock updates, news, or other information.
2. Taskbar/toolbars	Several new toolbars can be added to the taskbar by right-clicking the taskbar. These toolbars make it easier to use your computer.
3. Web browser look and feel in Web-style interface	Several Web browser tools have been added to the Web-style interface. For example, in Windows Explorer, your hard drive is viewed as an extension of the World Wide Web. Back and Forward buttons allow you easily to revisit folders you have selected previously. A Favorites menu allows you to view quickly your favorite folders. You also can view folder contents in Web-page format.
4. Increased speed	Faster startup and shutdown of Windows. Also, loads 32-bit applications faster.
5. Tune-Up Wizard	Makes your programs run faster, checks for hard disk problems, and frees up hard disk space.
6. Multiple display support	Makes it possible for you to use several monitors at the same time to increase the size of your desktop, run different programs on separate monitors, and run programs or play games with multiple views.
7. Universal Serial Bus Support	Add devices to your computer easily without having to restart.
8. Accessibility Settings Wizard	Accessibility options, such as StickyKeys, ShowSounds, and MouseKeys, are Wizards designed to help users with specific disabilities make full use of the computer.
9. Update Wizard	Reviews device drivers and system software on your computer, compares findings with a database on the Web, and then recommends and installs updates specific to your computer.
10. Registry Checker	A system maintenance program that finds and fixes registry problems.
11. FAT32	An improved version of the File Allocation Table (file system) that allows hard drives larger than two gigabytes to be formatted as a single drive.
12. New hardware support	Supports a variety of new hardware devices, such as DVD, digital audio speakers, and recording devices. Improved Plug and Play capabilities make installing new hardware easier than early versions of Windows.

Figure 8-14 *The more recent Windows operating systems have many common features.*

WEB INFO

For more information on Mac OS, visit the Discovering Computers 2001 Chapter 8 WEB INFO page (**www.scsite.com/dc2001/ch8/webinfo.htm**) and click Mac OS.

Because it is more complex than previous versions of Windows, Windows 2000 requires more disk space, memory, and faster processors.

Windows Millennium

Whereas Windows 2000 is an operating system designed for business users, Windows Millennium is designed for home users. **Windows Millennium** is an updated version of Windows 98 for the consumer that uses a computer to surf the Internet or for entertainment (playing games, working with digital photographs, listening to music, or watching videos).

Windows CE

Windows CE is a scaled-down Windows operating system designed for use on wireless communications devices and smaller computers such as handheld computers, in-vehicle devices, and network computers. Because it is designed for use on smaller computing devices, Windows CE requires little memory. On most of these devices, the Windows CE interface incorporates many elements of the Windows graphical user interface including color and sound. It also has multitasking, e-mail, and Internet capabilities.

Many applications, such as Microsoft Word and Microsoft Excel, have scaled-down versions that run under Windows CE.

Recently, Microsoft introduced the **Auto PC**, which is a device mounted onto a vehicle's dashboard that is powered by Windows CE (Figure 8-15). Using an automobile equipped with Auto PC, the driver can obtain information such as driving directions, traffic conditions, and weather; access e-mail; listen to the radio or a CD; and share information with a handheld computer. Because the Auto PC is directed through voice commands, it is ideal for the mobile user.

WEB INFO

For more information on OS/2, visit the Discovering Computers 2001 Chapter 8 WEB INFO page (**www.scsite.com/dc2001/ch8/webinfo.htm**) and click OS/2.

Figure 8-15 *Auto PC is powered by Windows CE.*

Palm OS

The Palm Computing devices from 3Com Corporation and Visor devices from Handspring™ are extremely popular mobile computing devices that use an operating system called **Palm OS®**. With this operating system and a compatible device, you can manage schedules and contacts and easily synchronize this information with a desktop computer. With some devices, you also have wireless access to the Internet and your e-mail. In addition to containing PIM software, these devices contain handwriting recognition software called Graffiti®.

Mac OS

Apple's **Macintosh operating system** was the first commercially successful graphical user interface. It was released with Macintosh computers in 1984; since then, it has set the standard for operating system ease of use and has been the model for most of the new graphical user interfaces developed for non-Macintosh systems.

In recent years Apple changed the name of the operating system to **Mac OS**. The Mac OS is available only on computers manufactured by Apple. Figure 8-16 shows a screen of the latest version of Mac OS. This version includes the two more popular Web browsers: Netscape Navigator and Microsoft Internet Explorer. It also has the capability of opening, editing, and saving files created using the Windows and DOS platforms. Other features of the latest version of Mac OS include multitasking, built-in networking support, electronic mail, online shopping, enhanced speech recognition, and enhanced multimedia capabilities.

OS/2

OS/2 (pronounced OH-ESS too) is IBM's multitasking graphical user interface operating system designed to work with 32-bit microprocessors (Figure 8-17). In addition to its capability of running programs written specifically for OS/2, the operating system also can run programs written for DOS and most Windows 3.x programs. The latest version of OS/2 is called **OS/2 Warp**.

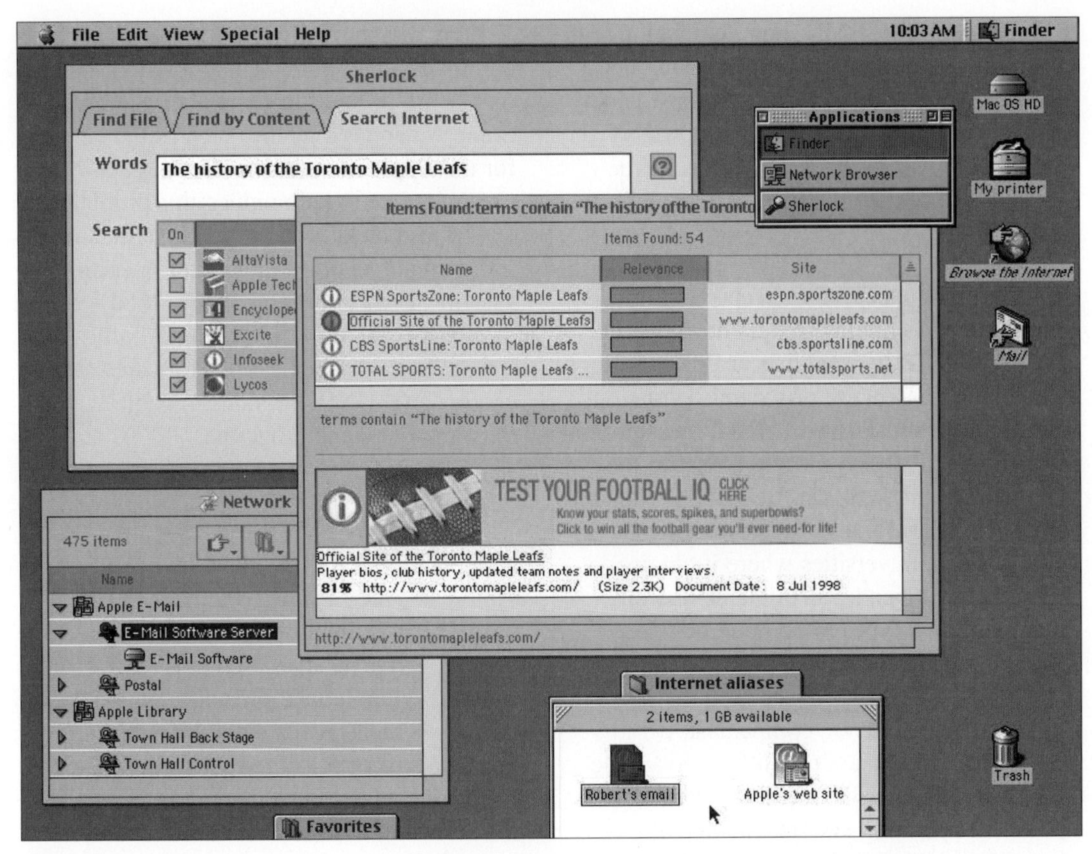

Figure 8-16 *Mac OS is the operating system used with Apple Macintosh computers.*

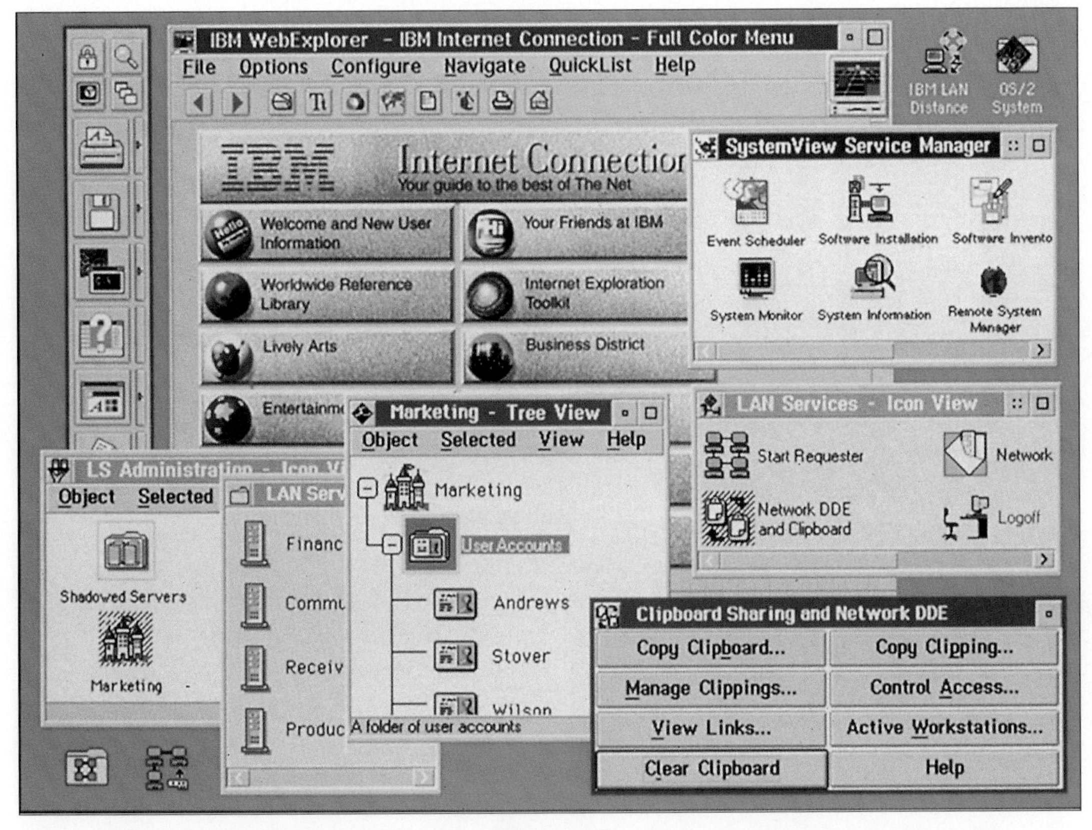

Figure 8-17 *OS/2 is IBM's multitasking graphical user interface operating system designed to work with 32-bit microprocessors.*

WEB INFO

For more information on UNIX, visit the Discovering Computers 2001 Chapter 8 WEB INFO page (**www.scsite.com/dc2001/ch8/webinfo.htm**) and click UNIX.

For more information on Linux, visit the Discovering Computers 2001 Chapter 8 WEB INFO page (**www.scsite.com/dc2001/ch8/webinfo.htm**) and click Linux.

Because of IBM's long association with business computing and OS/2's strong networking support, OS/2 has been most widely used by businesses. As with Windows NT, a version of OS/2 exists for use on a server.

UNIX

UNIX (pronounced YOU-nix) is a multiuser, multitasking operating system developed in the early 1970s by scientists at Bell Laboratories. Because of federal regulations, Bell Labs (a subsidiary of AT&T) was prohibited from actively promoting UNIX in the commercial marketplace. Bell Labs instead licensed UNIX for a low fee to numerous colleges and universities where it obtained a wide following and was implemented on many different types of computers. After deregulation of the telephone companies in the 1980s, UNIX was licensed to many hardware and software companies.

Today, a version of UNIX is available for most computers of all sizes. UNIX is a powerful operating system, capable of handling a high volume of transactions in a multiuser environment and working with multiple CPUs using multiprocessing. UNIX thus is used most often on workstations and servers.

A weakness of UNIX is that it has a command-line interface, and many of its commands are difficult to remember and use. Some versions of UNIX, such as the version for the Apple Macintosh, offer a graphical user interface to help reduce this problem. UNIX also lacks some of the system administration features offered by other operating systems. Finally, several widely used versions of UNIX exist, each of which is slightly different. To move application software from one of these UNIX versions to another, you must rewrite some programs.

Linux

A popular, free, UNIX-like GUI operating system is called **Linux** (pronounced LINN-uks). Linux is not proprietary software like the operating systems discussed thus far. Instead, Linux is **open-source software**, which means its code is made available to the public. Promoters of open-source software state two main advantages: customers can personalize the software to meet their needs, and users that modify the software share their improvements with others.

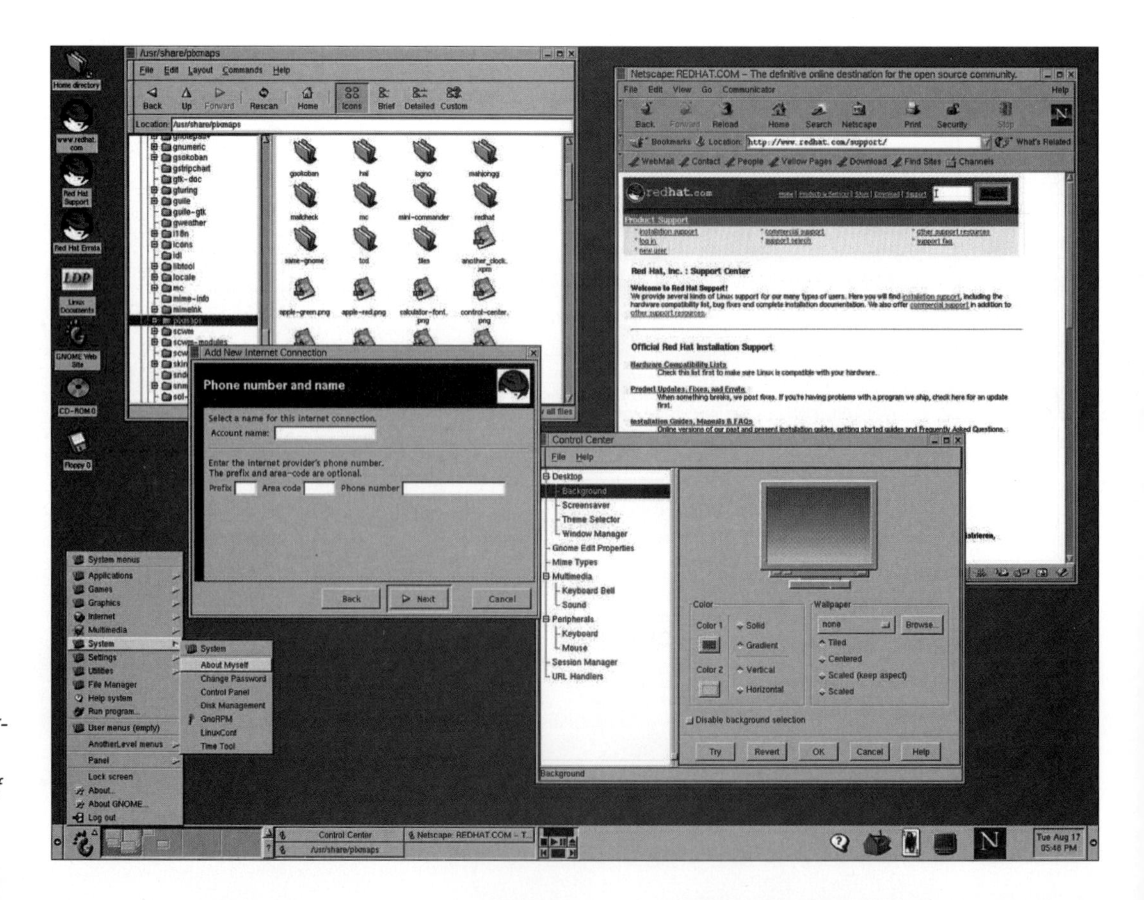

Figure 8-18 Red Hat Software provides a version of Linux designed to run on many types of computers, including PCs.

Many software applications run on Linux. Red Hat Software (Figure 8-18) and Corel Corporation, for example, sell products and services specifically developed for Linux.

NetWare

Novell's **NetWare** is a widely used network operating system designed for client-server networks. NetWare has a server portion that resides on the network server and a client portion that resides on each client computer connected to the network. The server portion of NetWare allows you to share hardware devices attached to the server (such as a printer), as well as any files or application software stored on the server. The client portion of NetWare communicates with the server. Client computers also have a local operating system, such as Windows 98.

STARTING A COMPUTER

The process of starting or resetting a computer involves loading an operating system into memory – a process called **booting** the computer. When you turn on a computer after it has been powered off completely, you are performing a **cold boot**. A **warm boot**, by contrast, is the process of restarting, or resetting, a computer that already is on. When using Windows, you typically can perform a warm boot, also called a **warm start**, by pressing a combination of keyboard keys (CTRL+ALT+DEL), selecting options from a menu, or pressing a Reset button on your computer.

When you boot a computer, information displays on the screen (Figure 8-19). The actual information displayed varies depending on the make of the computer and the equipment installed. The boot process, however, is similar for large and small computers.

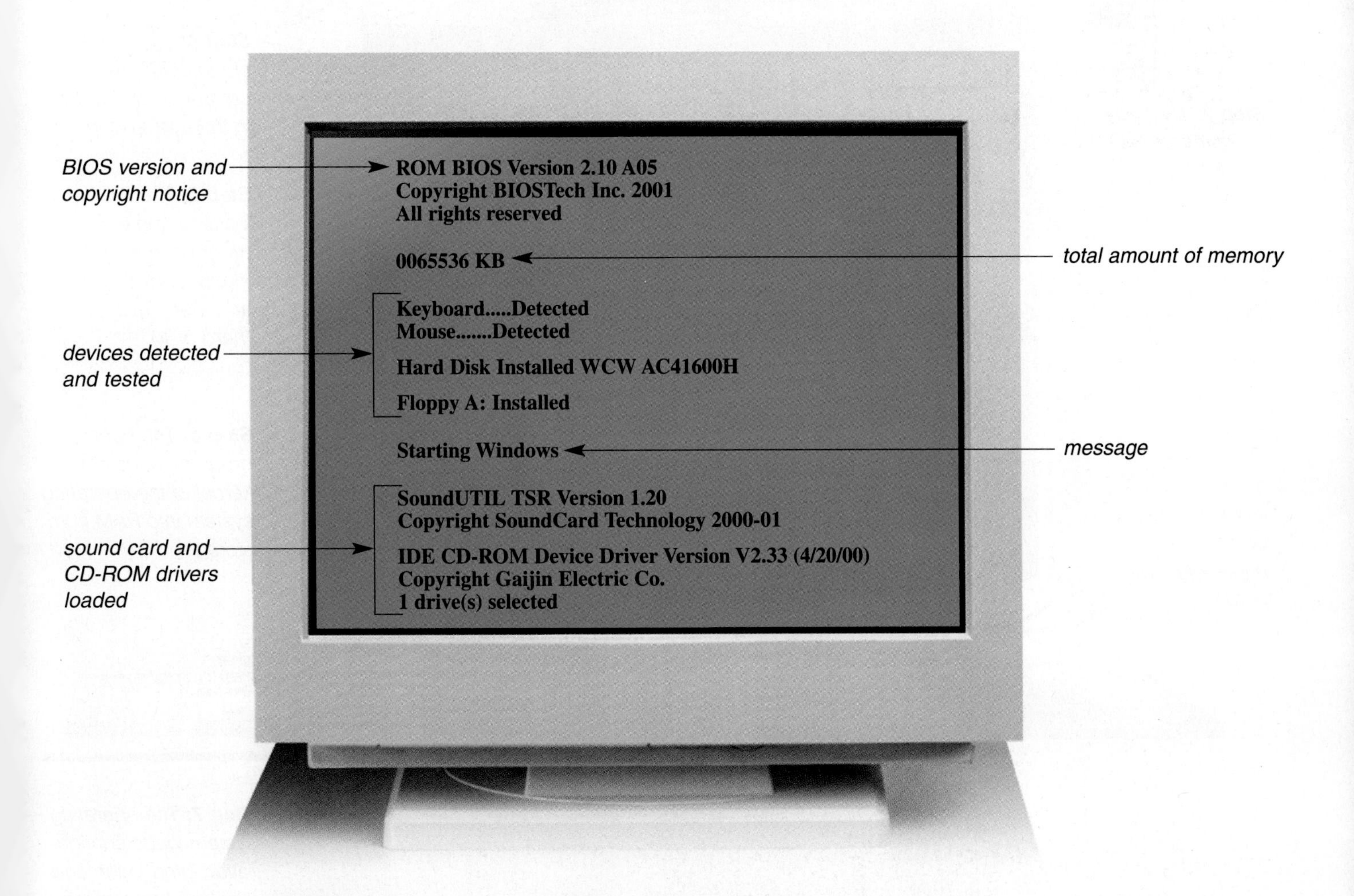

Figure 8-19 *When you boot a computer, information displays on the screen. The actual information displayed varies depending on the make of the computer and the equipment installed.*

The following steps explain what occurs during a cold boot for a personal computer using the Windows operating system (Figure 8-20).

1. When you turn on your computer, the power supply sends an electrical signal to the motherboard and the other devices located in the system unit.
2. The surge of electricity causes the CPU chip to reset itself and look for the ROM chip(s) that contains the BIOS. The **BIOS** (pronounced BYE-oss), which stands for **basic input/output system**, is firmware that contains the computer's startup instructions. Recall that firmware consists of ROM chips that contain permanently written instructions and data.

Figure 8-20
HOW SEVERAL COMPONENTS ARE ACCESSED DURING THE BOOT PROCESS

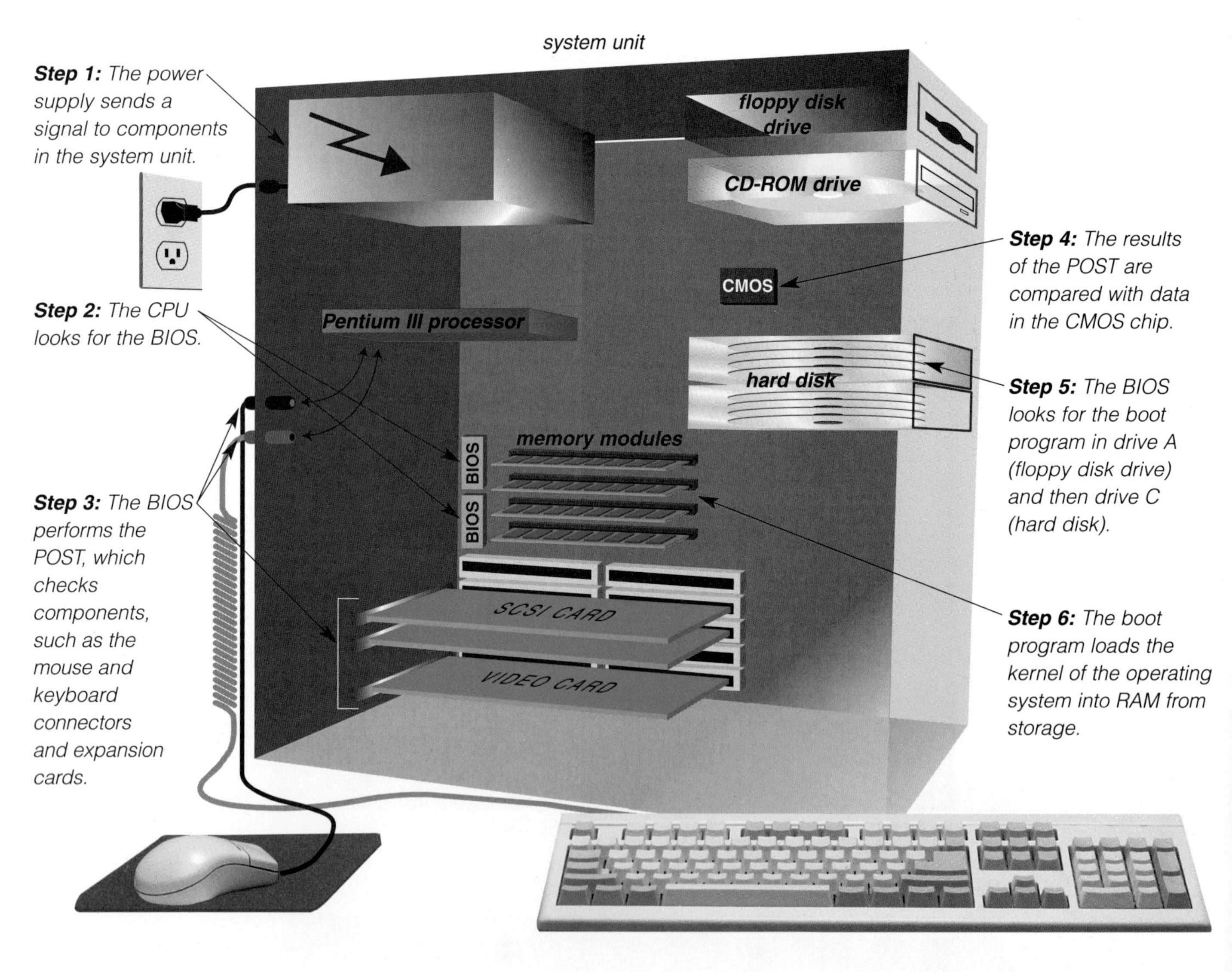
system unit
Step 1: The power supply sends a signal to components in the system unit.
floppy disk drive
CD-ROM drive
CMOS
Pentium III processor
Step 2: The CPU looks for the BIOS.
hard disk
memory modules
BIOS
BIOS
Step 3: The BIOS performs the POST, which checks components, such as the mouse and keyboard connectors and expansion cards.
SCSI CARD
VIDEO CARD
Step 4: The results of the POST are compared with data in the CMOS chip.
Step 5: The BIOS looks for the boot program in drive A (floppy disk drive) and then drive C (hard disk).
Step 6: The boot program loads the kernel of the operating system into RAM from storage.

Step 7: The operating system loads configuration information and displays the desktop on the screen.

3. The BIOS begins by executing a series of tests to make sure the computer hardware is connected properly and operating correctly. The tests, collectively called the **power-on self test** (**POST**), check the various system components such as the buses, system clock, expansion cards, RAM chips, keyboard, floppy disk drive(s), and hard disk. As the POST is performed, LEDs flicker on devices such as the disk drives and keyboard, several beeps sound, and messages display on the monitor's screen.

4. The results of the POST are compared with data in a CMOS chip on the motherboard. Recall that the CMOS chip stores configuration information about the computer, such as the amount of memory; type of disk drives, keyboard, and monitor; the current date and time; and other startup information needed when the computer is turned on. The CMOS chip is updated whenever new components are installed. If any problems are found, the computer may beep, display error messages, or cease operating — depending on the severity of the problem.

5. If the POST is completed successfully, the BIOS looks for the boot program that loads the operating system. Usually, it first looks in drive A (the designation for a floppy disk drive). If an operating system disk is not inserted into drive A, the BIOS looks in drive C, which is the designation usually given to the first hard disk. If neither drive A nor drive C contain the boot program, some computers look to the CD-ROM drive.

6. Once located, the boot program is loaded into memory and executed. The boot program then loads the kernel of the operating system into RAM. The operating system takes control of the computer.

7. The operating system loads system configuration information. In Windows, system configuration information is contained in several files called the **registry**. Windows constantly accesses the registry during the computer's operation for information such as installed hardware and software devices and individual user preferences for mouse speed, passwords, and other user-specific information.

   For each hardware device identified in the registry, such as the sound card, a CD-ROM or DVD-ROM drive, or a scanner, the operating system loads a device driver. Recall that a device driver is a program that tells the operating system how to communicate with a device.

   The remainder of the operating system is loaded into RAM and the desktop and icons display on the screen. The operating system executes programs in the StartUp folder, which contains a list of programs that open automatically when you boot your computer.

Boot Disk

Under normal circumstances, the drive from which your computer boots, called the **boot drive**, is drive C (the hard disk). If you cannot boot from the hard disk, for example if it is damaged or destroyed, you can boot from a boot disk. A **boot disk** is a floppy disk that contains certain operating system commands that will start the computer. For this reason, it is crucial you have a boot disk available — ready for use at any time.

When you install an operating system, one of the installation steps involves making a boot disk. You may not, however, have a boot disk because the operating system was pre-installed by the computer's manufacturer when you purchased the computer. If you do not have a boot disk, you should create one and keep it in a safe place. The steps in Figure 8-21 show how to create a boot disk in Windows 98.

Figure 8-21
HOW TO CREATE A BOOT DISK IN WINDOWS 98

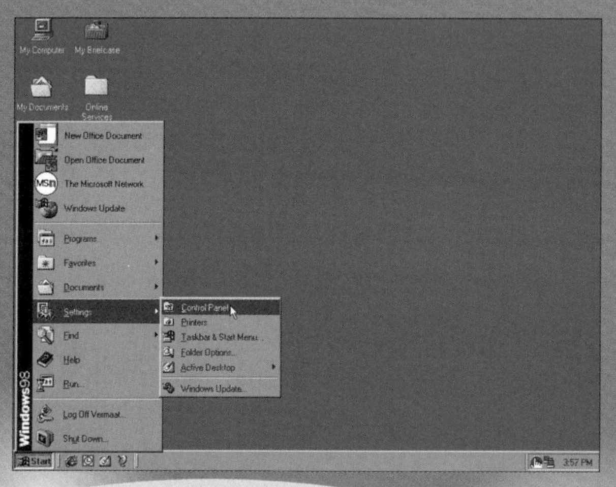

Step 1: *Click the Start button on the taskbar, point to Settings on the Start menu, point to Control Panel on the Settings submenu.*

Step 2: *Click Control Panel on the Settings submenu to open the Control Panel window.*

Step 3: *Double-click the Add/Remove Programs icon in the Control Panel window to display the Add/Remove Programs Properties dialog box.*

Step 4: *Click the Startup Disk tab and then click the Create Disk button to create the boot disk. Follow the on-screen instructions.*

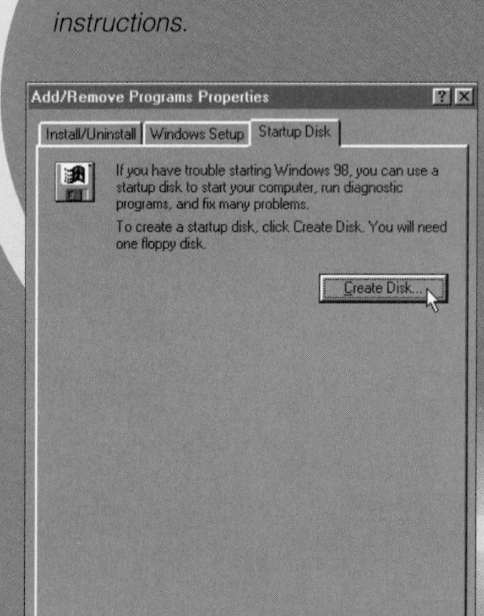

UTILITY PROGRAMS

A **utility program**, also called a **utility**, is a type of system software that performs a specific task, usually related to managing a computer, its devices, or its programs. Most operating systems include several utility programs. As shown in Figure 8-22, for example, Windows 98 provides access to many utility programs through the System Tools submenu. You also can buy stand-alone utilities that offer improvements over those supplied with the operating system.

Popular utility programs perform these functions: viewing files, compressing files, diagnosing problems, scanning disks, defragmenting disks, uninstalling software, backing up files and disks, checking for viruses, and displaying screen savers. Each of these utilities is briefly discussed in the following paragraphs.

- A **file viewer** is a utility that displays the contents of a file. An operating system's file manager often includes a file viewer. In Windows 98, for example, Windows Explorer has two viewers: one called **Quick View** to display the contents of text files and another called **Imaging Preview** for graphics files (Figure 8-23). The title bar of the file viewer window displays the name of the file being viewed.

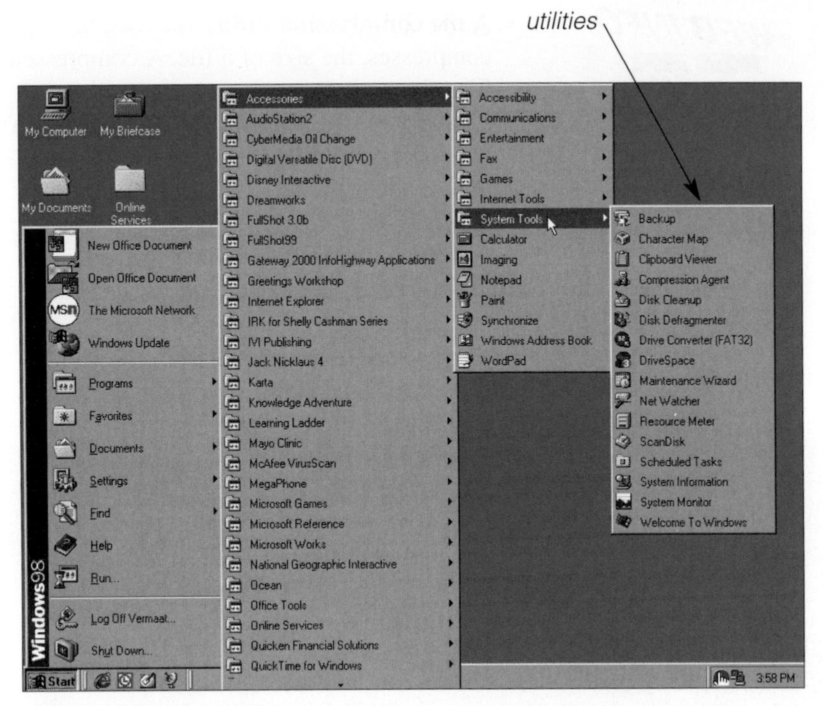

Figure 8-22 Many Windows 98 utility programs are accessed on the System Tools submenu.

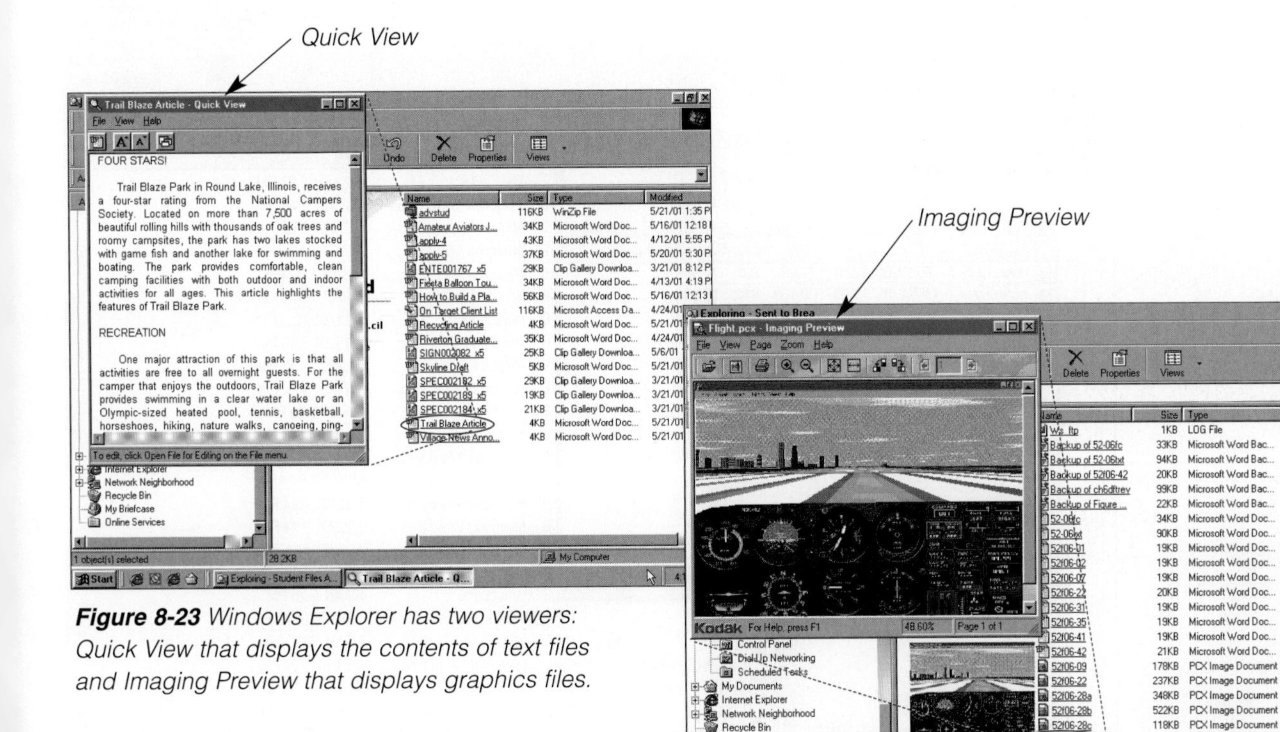

Figure 8-23 Windows Explorer has two viewers: Quick View that displays the contents of text files and Imaging Preview that displays graphics files.

WEB INFO

For more information on file compression utilities, visit the Discovering Computers 2001 Chapter 8 WEB INFO page (**www.scsite.com/dc2001/ch8/webinfo.htm**) and click File Compression Utilities.

- A **file compression utility** reduces, or compresses, the size of a file. A compressed file takes up less storage space on a hard disk or floppy disk, which frees up room on the disk and improves system performance. Files available for download from the Internet often are compressed to reduce download time. Compressing files attached to e-mail messages also reduces the time needed for file transmission.

  Because compressed files usually have a .zip extension, compressed files sometimes are called **zipped files**. When you receive a compressed file, you must **uncompress**, or **unzip**, the file — or restore it to its original form. Two popular stand-alone file compression utilities are PKZIP™ and WinZip®. WinZip® is shown in Figure 8-24.

- A **diagnostic utility** compiles technical information about your computer's hardware and certain system software programs and then prepares a report outlining any identified problems. For example, Windows 98 includes the diagnostic utility, Dr. Watson, which diagnoses problems as well as suggests courses of action (Figure 8-25).

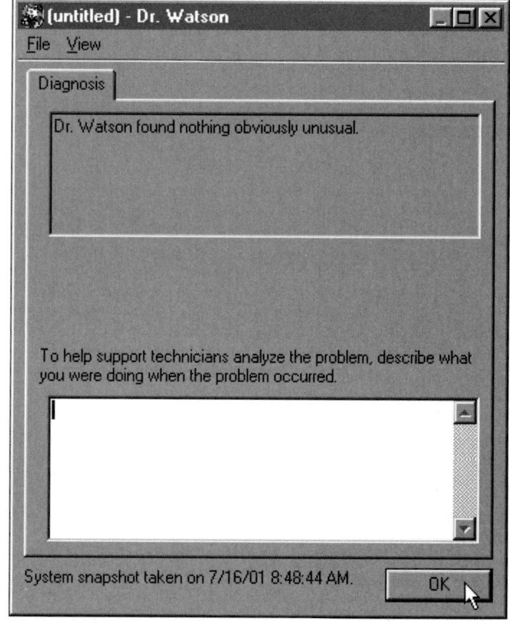

Figure 8-25 *Dr. Watson is a diagnostic utility included with Windows 98.*

Figure 8-24 *WinZip® is a popular stand-alone file compression utility. This zipped file (ch1notes) contains 5 files for a total of 79 KB. Without being zipped, these files consume 478 KB. Thus, zipping the files reduced the amount of storage by 399 KB.*

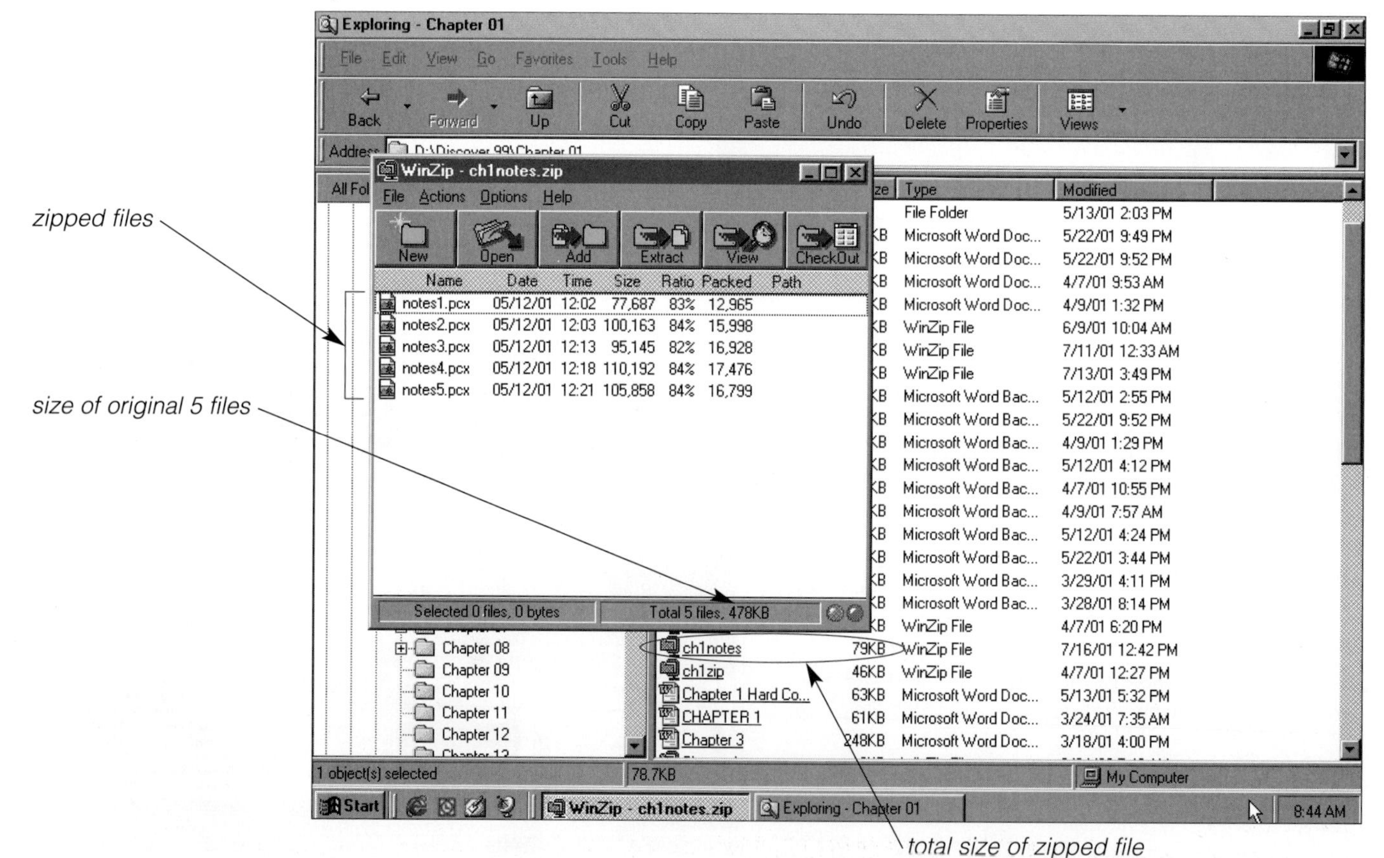

- A **disk scanner** is a utility that (1) detects and corrects both physical and logical problems on a hard disk or floppy disk and (2) searches for and removes unwanted files. A physical problem is one with the media, such as a scratch on the surface of the disk. A logical problem is one with the data, such as a corrupted file allocation table (FAT). Windows 98 includes two disk scanner utilities: **ScanDisk**, which detects and corrects problems, and **Disk Cleanup**,which searches for and removes unnecessary files such as temporary files (Figure 8-26).
- A **disk defragmenter** is a utility that reorganizes the files and unused space on a computer's hard disk so data can be accessed more quickly and programs can run faster. When a computer stores data on a disk, it places the data in the first available sector on the disk. Although the computer attempts to place data in sectors that are contiguous (next to each other), this is not always possible. When the contents of a file are scattered across two or more noncontiguous sectors, the file is **fragmented**. Fragmentation slows down disk access and thus the performance of the entire computer. The process of **defragmentation** — that is, reorganizing the disk so the files are stored in contiguous sectors — solves this problem (Figure 8-27). Windows 98 includes a disk defragmenter, called **Disk Defragmenter**.

Figure 8-26a (ScanDisk utility)

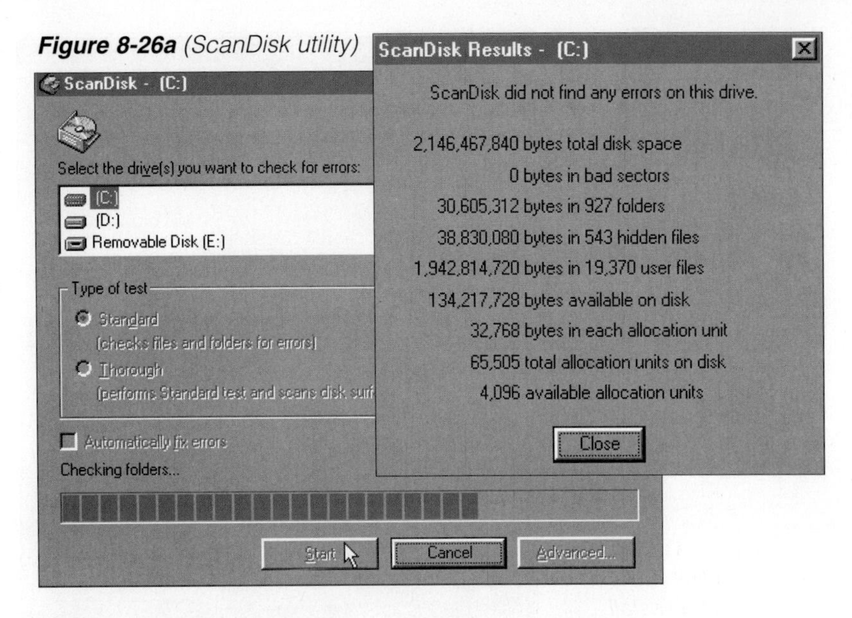

Figure 8-26b (Disk Cleanup utility)

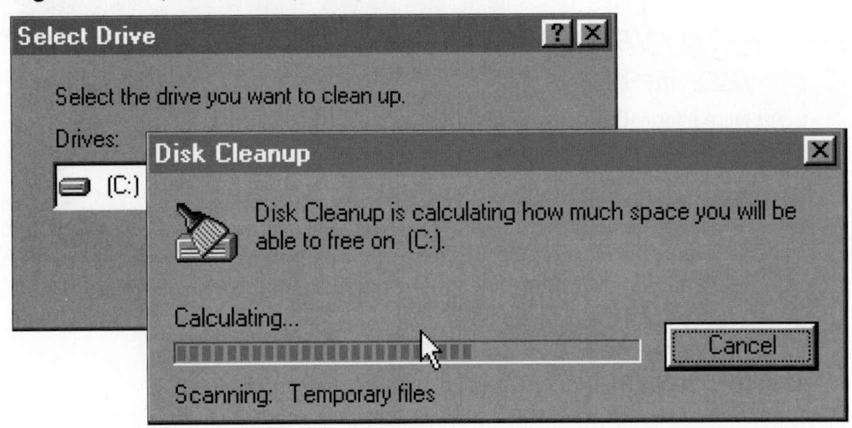

Figure 8-26 *ScanDisk detects and corrects problems on a disk, and Disk Cleanup searches for and removes unnecessary files.*

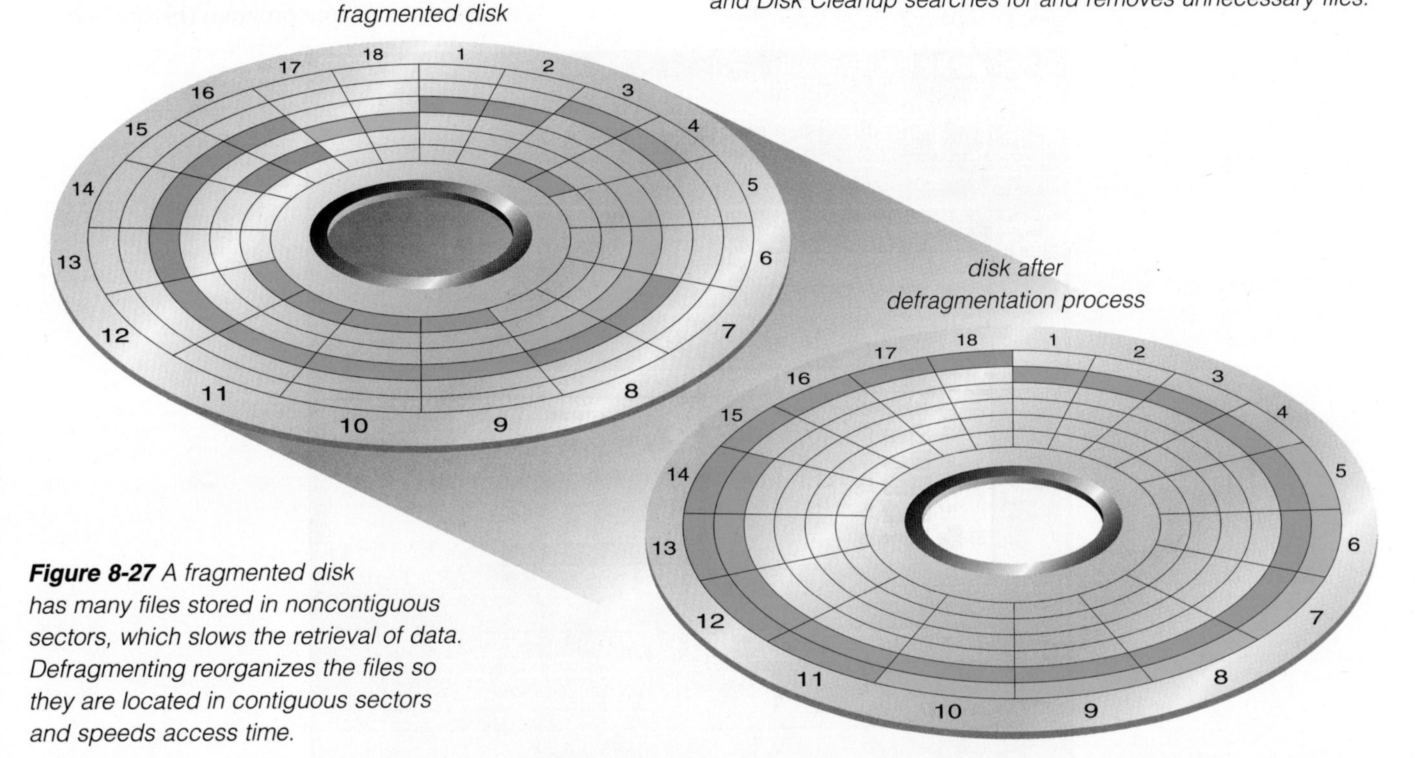

Figure 8-27 *A fragmented disk has many files stored in noncontiguous sectors, which slows the retrieval of data. Defragmenting reorganizes the files so they are located in contiguous sectors and speeds access time.*

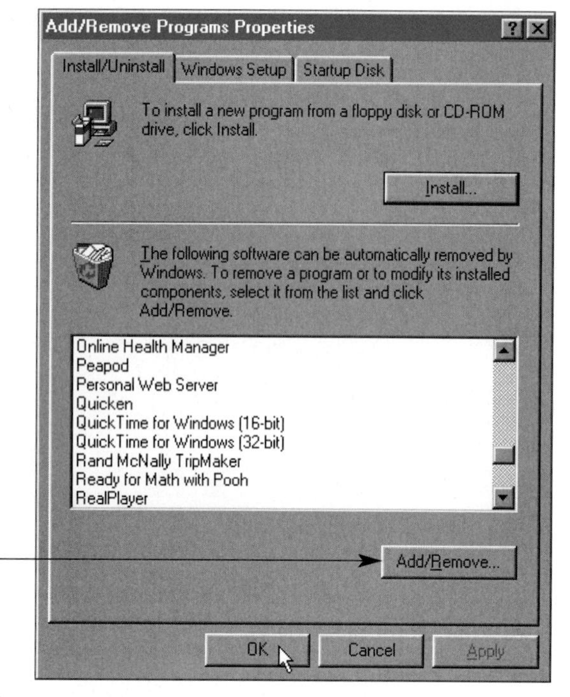

Figure 8-28 With Windows 98, you can uninstall programs by clicking the Add/Remove Programs icon in the Control Panel window.

WEB INFO

For more information on backup utilities, visit the Discovering Computers 2001 Chapter 8 WEB INFO page (**www.scsite.com/dc2001/ch8/webinfo.htm**) and click Backup Utilities.

- An **uninstaller** is a utility that removes an application, as well as any associated entries in the system files (Figure 8-28). When you install an application, the operating system records the information that it uses to run the software in the system files. If you attempt to remove the application from your computer by deleting the files and folders associated with that program without running the uninstaller, the system file entries remain. Most operating systems include an uninstaller; you also can purchase a stand-alone program, such as CyberMedia's UnInstaller.
- A **backup utility** allows you to copy, or backup, selected files or your entire hard disk onto another disk or tape. During the backup process, the backup utility monitors progress and alerts you if additional disks or tapes are needed. Many backup programs will compress files during this process, so the backup files require less storage space than the original files.

  For this reason, backup files usually are not usable in their backed up form. In the event you need to use one of these files, a **restore program**, which is included with the backup utility, reverses the process and returns backed up files to their original form. You should back up files and disks regularly in the event your originals are lost, damaged, or destroyed. Windows 98 includes a backup utility, which also includes a restore program (Figure 8-29).

Figure 8-29 A backup utility allows you to copy files or your entire hard disk to another disk or tape.

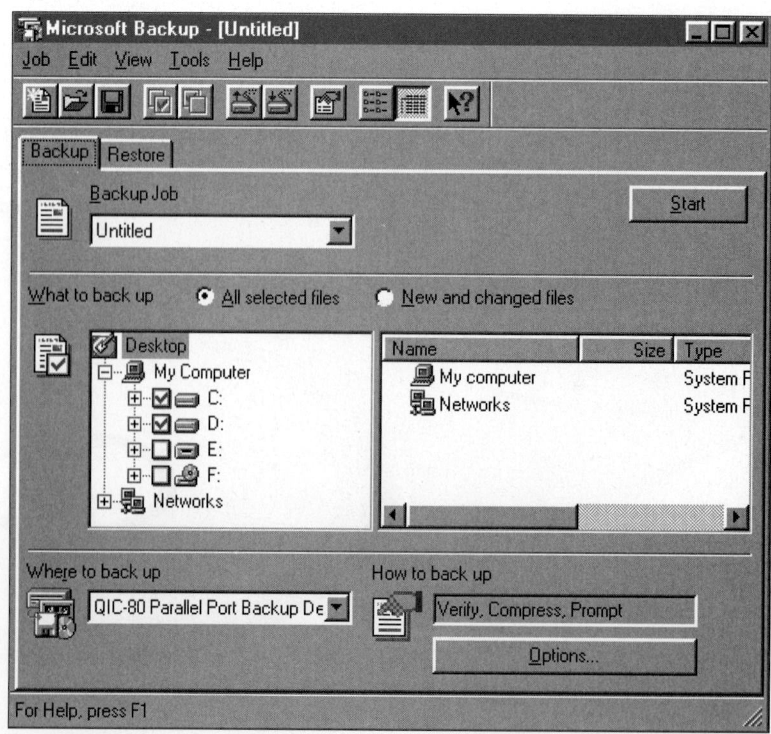

- An **antivirus** program is a utility that prevents, detects, and removes viruses from a computer's memory or storage devices. A **virus** is a program that copies itself into other programs and spreads through multiple computers. Viruses often are designed to damage a computer intentionally by destroying or corrupting its data. Antivirus programs and viruses are discussed in more depth in Chapter 14.
- A **screen saver** is a utility that causes the monitor's screen to display a moving image or blank screen if no keyboard or mouse activity occurs for a specified time period (Figure 8-30). When you press a key on the keyboard or move the mouse, the screen returns to the previously displayed image.

Screen savers originally were developed to prevent a problem called **ghosting**, in which images could be permanently etched on a monitor's screen. Although ghosting is not a problem with today's monitors, screen savers still are used for reasons of security, business, or entertainment. To secure a computer, for example, you can configure your screen saver so that you must enter a password to stop the screen saver and redisplay the previous image. Some screen savers use push technology so you receive updated and new information each time the screen saver displays.

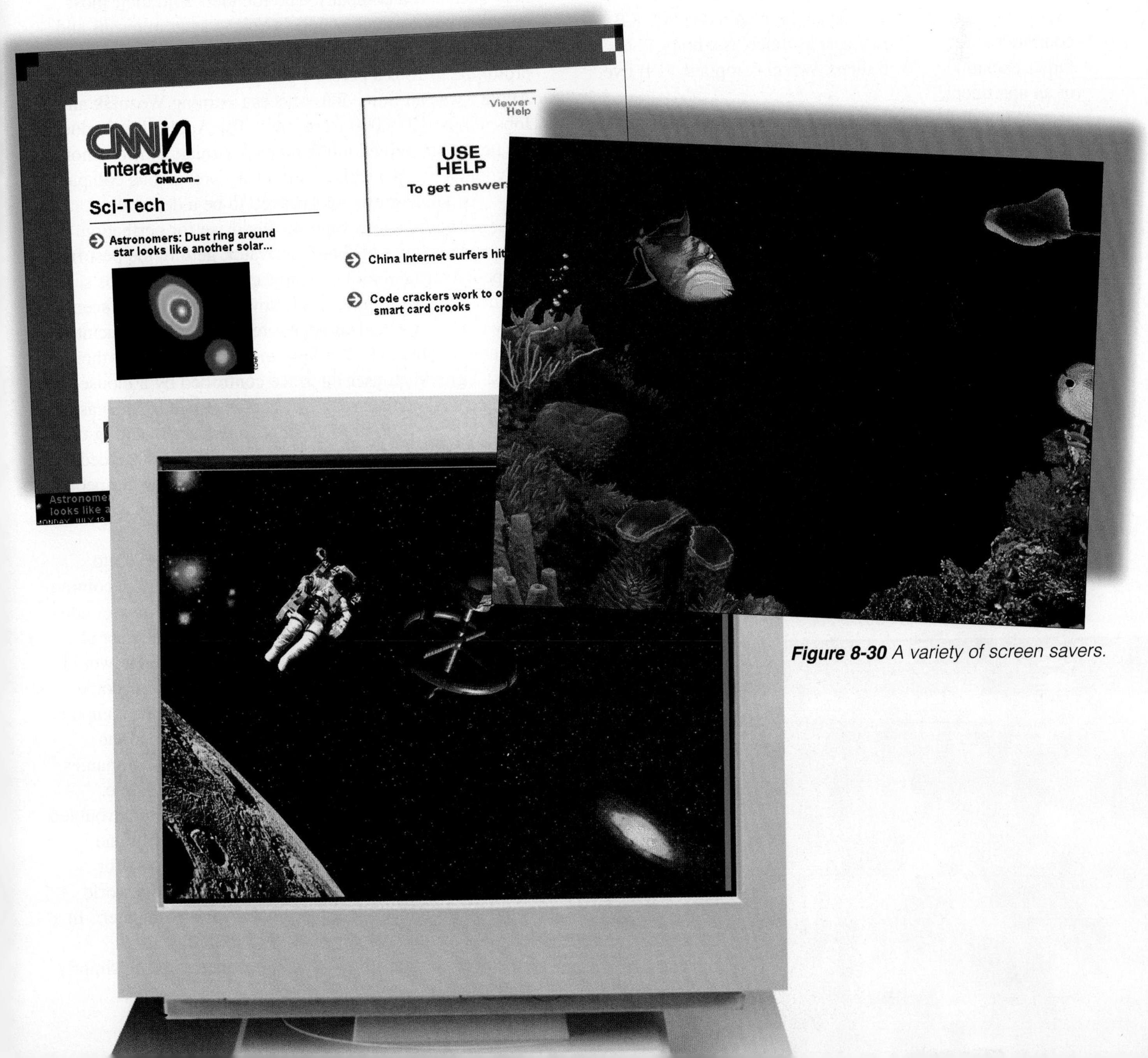

Figure 8-30 *A variety of screen savers.*

TECHNOLOGY TRAILBLAZERS

STEVE WOZNIAK AND STEVE JOBS

Before they were 35-years-old, Steve Wozniak and Steve Jobs built the first desktop personal computer, co-founded Apple Computer, marketed a revolutionary operating system, and became multimillionaires. Not bad.

Steve Wozniak always was interested in computers. Growing up, he devoted much of his time to designing and building sophisticated machines. "It was all self-done; I didn't ever take a course, didn't ever buy a book on how to do it. Just pieced it all together." Wozniak told his father, an engineer, that someday he would own a 4 KB computer. "Yeah, but they cost as much as a house!" his father exclaimed. "Well then," Wozniak replied, "I'll live in an apartment."

After dropping out of the University of California at Berkeley, Wozniak went to work at Hewlett-Packard where he met Steve Jobs, a summer employee. "Woz[niak] was the first person I met who knew more about electronics than I did," Jobs recalls. On his own time, Wozniak had built a *blue box* that, with a toy whistle from Cap'n Crunch cereal boxes, could be used to make free long-distance telephone calls. Jobs helped Wozniak sell the device.

In 1972, Jobs graduated from high school and attended Reed College for a semester. He took classes in philosophy, embraced the contemporary counterculture, and eventually journeyed to India in search of enlightenment. When he returned, he joined Wozniak's Homebrew Computer Club. The club's focus was on small computers that could empower individuals. "Never trust a computer you can't throw out a window," Wozniak said. Unfortunately, small computers were built from kits and programmed with a confusing array of switches. Wozniak envisioned a small computer "laid out like a typewriter with a video screen." The concept of an accessible computer would be a Wozniak benchmark.

Steve Wozniak & Steve Jobs

When Wozniak designed a simple small computer, Jobs suggested they start a company and make the circuit boards. Even if the company failed, at least they could say they once *had* a company. The founders sold their most valuable possessions — Jobs's Volkswagen mini-bus and Wozniak's scientific calculator — for capital and built a prototype in Jobs's parent's garage. Marketed in 1976, the Apple I was an immediate success, earning Wozniak and Jobs almost $775,000 from sales. The Apple II, introduced the next year, earned much more. Wozniak and Jobs not only had a company, they had a very prosperous company.

As Apple grew, Jobs proved to be a demanding manager with an uncompromising drive for perfection. Yet, he also was a brilliant motivator, getting the best from the people around him. Employees joked about Jobs's "reality-distortion field" that allowed him to make seemingly unreasonable ideas appear reasonable. The Macintosh computer, announced in 1984, incorporated one of these ideas: a graphical user interface controlled by a mouse. To Jobs, just as the telegraph had been usurped by the easier-to-use telephone, interfaces requiring typed commands surely would be displaced by graphical user interfaces allowing people simply to point and click. Jobs was right, and the Macintosh was a hit.

Wozniak left Apple in 1985 to spend more time with his family, community projects, and work in computer education. He still serves as an adviser at Apple. Jobs also left in 1985. He would co-found NeXT Software and become Chairman and CEO of Pixar, the computer animation studio that developed the feature film, *Toy Story*. Both companies eventually were acquired by Apple.

In 1997, Jobs returned to a troubled Apple as interim CEO. A wealthy man with only one share of Apple stock, Jobs did not accept the post for financial reasons. Instead, he maintains that "the world would be a slightly better place with Apple Computer." In a short time, a quarterly profit of $45 million was announced. Perhaps the world *is* destined to be a "slightly better place."

COMPANIES ON THE CUTTING EDGE

COMPANIES ON THE CUTTING EDGE

APPLE VS. IBM COMPATIBLES – THE RIVALRY

The Montagues and the Capulets. The Yankees and the Red Sox. Coke and Pepsi. The list of famous feuds is lengthy, but few are more passionate than one that grips the computer industry: Apple computers and IBM compatibles.

Apple iMac

Steve Jobs and Steve Wozniak founded Apple Computer in 1977. Jobs suggested the name Apple, but Wozniak says, "He doesn't always let on where ideas come from, or how they come into his head." The name may have reminded Jobs of time spent in an orchard, or of the Beatles's record label, or simply of his favorite fruit. Whatever the origin, Wozniak maintains that other names "all sounded boring compared to Apple." The company's logo is easier to explain — an apple with a bite (a play on byte) taken out of it.

The company's first computer, the Apple I, had a video interface, built-in ROM, and a cost of $666. The Apple II followed with a similarly simple design, 4 KB of memory, and an interface that allowed it to work with a color television. The Apple II was the first mass-marketed personal computer. With Jobs's encouragement, independent programmers wrote about 16,000 applications for the Apple II. In three years, the Apple II earned almost $140,000,000. Several versions of the Apple II were marketed, and the machine became a school standard.

In 1981, IBM, which up to this time had ignored small computers, introduced the IBM PC. The IBM PC reflected two crucial decisions. First, Intel, an outside company, made the processor. Second, Microsoft created the operating system, DOS, but maintained the right to market DOS to other manufacturers. These decisions led to a whole family of IBM-compatible computers —machines with the same processor and operating system as IBM PCs — from companies such as Compaq, Dell, and Gateway. PC prices fell and IBM and IBM-compatible PCs soon dominated the market, particularly in the business world. The number of applications written to work with DOS exploded. Because neither DOS nor its related applications were compatible with Apple computers, Apple's sales sagged.

IBM PC

Two new Apple efforts, the Apple III and the Lisa, failed. The Lisa's revolutionary graphical user interface, obtained from Xerox, made it easy to use, but the $10,000 price made it hard to buy. Apple's next computer, the Macintosh, also had a graphical user interface. The Macintosh had a larger processor and more memory than IBM PCs, but its greatest strengths were its flexibility, creative capabilities, and user-friendliness. Advertised as, *the computer for the rest of us*, the Macintosh was embraced enthusiastically, especially by customers in the humanities.

Microsoft responded by introducing Windows, a graphical face-lift for the DOS operating system. Apple sued unsuccessfully for copyright infringement, a move Wozniak feels was a mistake. "Had we gone to Microsoft and said, 'do anything the way we've already found is good, for 25 cents' the result might have been a commonality as beneficial to Apple as to Microsoft." Microsoft would improve on its initial Windows offering with *true* graphical operating systems, Windows 95 and Windows 98. *Wintel* (a word describing IBM compatibles that combine Windows operating systems and Intel processors) sales climbed while Apple's trade slumped and company morale tumbled. Nevertheless, Mac OS users remain extremely loyal. They insist their operating system is more stable, error-free, and user friendly than any version of Windows. Windows users counter that more applications are available for use with their operating system, and some insist even programs designed to work with both operating systems often work better with the Windows operating system.

Apple vs. IBM compatibles — the rivalry continues and may not end soon. Until it does, as the carnival barker says, "you pays your money and you takes your choice."

IN BRIEF

www.scsite.com/dc2001/ch8/brief.htm

WEB INSTRUCTIONS: *To display this page from the Web, launch your browser and enter the URL, www.scsite.com/dc2001/ch8/brief.htm. Click the links for current and additional information. To listen to an audio version of this IN BRIEF, click the Audio button to the right of the title, IN BRIEF, at the top of the page. To play the audio, RealPlayer must be installed on your computer (download by clicking here).*

1. What Are Various Types of System Software?

System software consists of the programs that control the operations of the computer and its devices. System software performs a variety of functions, such as running applications and storing files, and serves as the interface between a user, the application software, and the computer's hardware. The two types of system software are operating systems and utility programs.

2. How Is an Operating System Different from a Utility Program?

An **operating system (OS)** is a set of programs containing instructions to coordinate all of the activities among computer hardware resources. The part of the software with which you interact is the **user interface**. Two types of user interfaces are command-line and graphical. With a **command-line interface**, you type keywords or press special keys on the keyboard to enter data and instructions. A **graphical user interface (GUI)** allows you to use **menus** and visual images such as **icons** and buttons to issue commands. A **utility program** is a type of system software that performs a specific task, usually related to managing a computer, its devices, or its programs. Most operating systems include several utility programs. Popular utility programs perform functions such as viewing files, compressing files, diagnosing problems, scanning disks, defragmenting disks, uninstalling software, backing up files and disks, checking for viruses, and displaying screen savers.

3. What Are the Features of Operating Systems?

Various capabilities of operating systems are described as single user, multiuser, multiprocessing, and multitasking. A **single user** operating system allows only one user to run one program at a time. A **multiuser** operating system enables two or more users to run a program simultaneously. A **multitasking** operating system allows a single user to work on two or more applications that reside in memory at the same time. A **multiprocessing** operating system can support two or more CPUs running programs at the same time.

4. What Are the Functions of an Operating System?

An operating system performs a number of basic functions that enable the user and the application software to interact with the computer. Operating systems manage memory, spool print jobs, configure devices, monitor system performance, administer security, and manage storage media and files. **Memory management** optimizes use of random access memory (RAM). **Spooling** increases efficiency by placing **print jobs** in a **buffer** until the printer is ready, freeing the CPU for other tasks. A **device driver** is a small program that configures devices by accepting commands and converting them into commands the device understands. **Plug and Play** is the computer's capability to recognize any new device and assist in the installation of the device. Having Plug and Play support means a user can plug in a device, turn on the computer, and then use, or play, the device without having to configure the system manually. A **performance monitor** assesses and reports information about various system resources and devices. Most multiuser operating systems administer security by allowing each user to **log on**, which is the process of entering a **user name** and **password**. A type of program called a **file manager** performs functions related to storage and file management.

IN BRIEF

www.scsite.com/dc2001/ch8/brief.htm

Student Exercises
WEB INFO
IN BRIEF
KEY TERMS
AT THE MOVIES
CHECKPOINT
AT ISSUE
MY WEB PAGE
HANDS ON
NET STUFF
Special Features
TIMELINE 2001
GUIDE TO WWW SITES
MAKING A CHIP
E-COMMERCE 2001
BUYER'S GUIDE 2001
CAREERS 2001
TRENDS 2001
LEARNING GAMES
CHAT
INTERACTIVE LABS
NEWS
HOME

5. What Are Popular Operating Systems Used Today?

DOS (Disk Operating System) refers to several single user, command-line and menu-driven operating systems developed in the early 1980s for PCs. **Windows 3.x** refers to three early **operating environments** that provided a graphical user interface to work in combination with DOS and simplify its use. **Windows 95** is a true multitasking operating system – not an operating environment – with an improved graphical interface. The **Windows 98** operating system is easier to use than Windows 95 and is more integrated with the Internet. **Windows 2000** is an upgrade to Windows 98 and **Windows NT**, an operating system designed for client-server networks. **Windows Millennium** is an updated version of Windows 98 designed for home computer users to surf the Internet or for entertainment. **Windows CE** is a scaled-down Windows operating system designed for use on wireless communications devices and smaller computers. **Palm OS®** is a popular operating system used with mobile computing devices. The **Mac OS**, a descendant of the first commercially successful graphical user interface, is available only on Apple computers. **OS/2** is IBM's multitasking graphical user interface operating system designed to work with 32-bit processors. **UNIX** is a multiuser, multitasking, command-line operating system developed by scientists at Bell Laboratories. **Linux** is a popular, free, UNIX-like operating system. Novell's **NetWare** is a widely used operating system designed for client-server networks.

6. What Is the Startup Process for a Personal Computer?

Starting a computer involves loading an operating system into memory – a process called **booting**. When the computer is turned on, the power supply sends an electrical signal to devices located in the system unit. The CPU chip resets itself and looks for the ROM chip that contains the **BIOS (basic input/output system)**, which is firmware that holds the startup instructions. The BIOS executes the **power-on self test (POST)** to make sure hardware is connected properly and operating correctly. Results of the POST are compared with data in a CMOS chip on the motherboard. If the POST is completed successfully, the BIOS looks for the boot program that loads the operating system. Once located, the boot program is loaded into memory and executed. The boot program loads the **kernel** of the operating system into RAM. The operating system loads system configuration information from the **registry** for each hardware device. The remainder of the operating system is loaded into RAM, the desktop and icons display on the screen, and programs in the StartUp folder are executed.

7. Why Are Some Common Utility Programs Used?

A **file viewer** displays the contents of a file. A **file compression utility** reduces the size of a file. A **diagnostic utility** compiles technical information about a computer's hardware and certain system software programs and then prepares a report outlining any identified problems. A **disk scanner** detects and corrects problems on a disk and searches for and removes unwanted files. A **disk defragmenter** reorganizes files and unused space on a computer's hard disk so data can be accessed more quickly and programs can run faster. An **uninstaller** removes an application, as well as any associated entries in the system files. A **backup utility** copies or backups selected files or the entire hard drive onto another disk or tape. An **antivirus** program prevents, detects, and removes **viruses** (programs often designed to damage a computer) from a computer's memory or storage devices. A **screen saver** causes the monitor's screen to display a moving image on a blank screen if no keyboard or mouse activity occurs for a specific time period.

DISCOVERING COMPUTERS 2001

KEY TERMS

www.scsite.com/dc2001/ch8/terms.htm

WEB INSTRUCTIONS: *To display this page from the Web, launch your browser and enter the URL, www.scsite.com/dc2001/ch8/terms.htm. Scroll through the list of terms. Click a term to display its definition and a picture. Click KEY TERMS on the left to redisplay the KEY TERMS page. Click the TO WEB button for current and additional information about the term from the Web. To see animations, Shockwave and Flash Player must be installed on your computer (download by clicking here).*

antivirus (8.25)
Auto PC (8.14)
background (8.5)
backup utility (8.24)
basic input/output system (BIOS) (8.18)
boot disk (8.20)
boot drive (8.20)
booting (8.17)
buffer (8.6)
cold boot (8.17)
command language (8.3)
command-line interface (8.3)
cross-platform (8.3)
defragmentation (8.23)
device dependent (8.10)
device driver (8.7)
device-independent (8.10)
diagnostic utility (8.22)
Disk Cleanup (8.23)
disk defragmenter (8.23)
Disk Defragmenter (8.23)
Disk Operating System (DOS) (8.11)
disk scanner (8.23)
downward-compatible (8.10)
driver (8.8)
fault-tolerant computer (8.5)
file compression utility (8.22)
file manager (8.10)
file viewer (8.21)
foreground (8.5)
fragmented (8.23)
ghosting (8.25)
graphical user interface (GUI) (8.4)
icon (8.4)
Imaging Preview (8.21)
interrupt request (IRQ) (8.9)
kernel (8.3)
Linux (8.16)
log on (8.10)
Mac OS (8.14)

TO WEB

PERFORMANCE MONITOR: Operating system program that assesses and reports information about various system resources and devices. (8.9)

Macintosh operating system (8.14)
memory management (8.6)
memory-resident (8.3)
menu (8.4)
Microsoft Internet Explorer (8.12)
multiprocessing (8.5)
multitasking (8.5)
multiuser (8.5)
NetWare (8.17)
NT (8.12)
open-source software (8.16)
operating environment (8.11)
operating system (OS) (8.2)
OS/2 (8.14)
OS/2 Warp (8.14)
page (8.7)
paging (8.7)
Palm OS® (8.14)
password (8.10)
performance monitor (8.9)
Plug and Play (8.9)
power-on self test (POST) (8.19)
print job (8.7)
proprietary software (8.10)
queued (8.7)
Quick View (8.21)
registry (8.19)
restore program (8.24)
ScanDisk (8.23)
screen saver (8.25)
single tasking (8.4)
single user (8.4)
software platform (8.2)
spooling (8.7)
swap file (8.6)
system software (8.2)
thrashing (8.7)
uncompress (8.22)
uninstaller (8.24)
UNIX (8.16)
unzip (8.22)
upward-compatible (8.10)
user ID (8.10)
user interface (8.3)
user name (8.10)
user-friendly (8.4)
utility (8.21)
utility program (8.21)
virtual memory (VM) (8.6)
virus (8.25)
warm boot (8.17)
warm start (8.17)
Win95 (8.11)
Win98 (8.12)
Windows (8.11)
Windows 2000 Professional (8.12)
Windows 2000 Server (8.12)
Windows 3.x (8.11)
Windows 95 (8.11)
Windows 98 (8.12)
Windows CE (8.14)
Windows Explorer (8.12)
Windows Millennium (8.14)
Windows NT (8.12)
zipped files (8.22)

Student Exercises
WEB INFO
IN BRIEF
KEY TERMS
AT THE MOVIES
CHECKPOINT
AT ISSUE
MY WEB PAGE
HANDS ON
NET STUFF
Special Features
TIMELINE 2001
GUIDE TO WWW SITES
MAKING A CHIP
E-COMMERCE 2001
BUYER'S GUIDE 2001
CAREERS 2001
TRENDS 2001
LEARNING GAMES
CHAT
INTERACTIVE LABS
NEWS
HOME

AT THE MOVIES

www.scsite.com/dc2001/ch8/movies.htm

WELCOME to VIDEO CLIPS from CNN

WEB INSTRUCTIONS: *To display this page from the Web, launch your browser and enter the URL, www.scsite.com/dc2001/ch8/movies.htm. Click a picture to view a video. After watching the video, close the video window and then complete the exercise by answering the questions about the video. To view the videos, RealPlayer must be installed on your computer (download by clicking here).*

1 Screen Savers

Today's office environment is filled with computer workstations. Because of the many interruptions that take employees away from their computers during the day, computer screens often remain open and idle for long periods. During this idle time, some pretty fascinating screen savers are popping up in the workplace. What is a screen saver? Why were screen savers initially created? Why are they used today? Do you have a screen saver on the computer you use? If you were an employer or instructor would you care whether your employees or students ran their own screen savers? Why or why not?

2 E-mail Virus

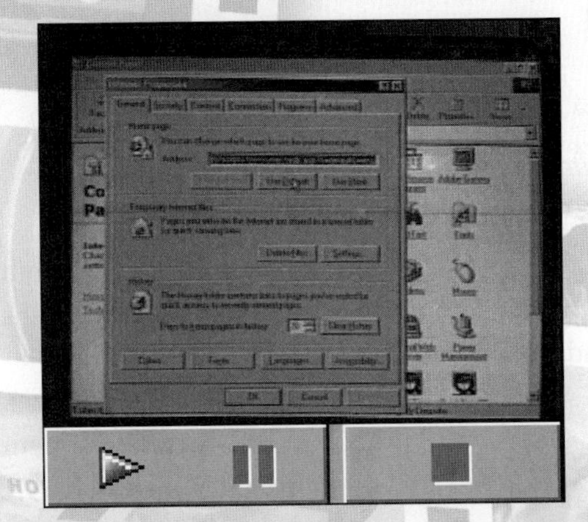

Think of a virus as a program designed intentionally to damage a computer by destroying or corrupting its data. Do you think companies should be held responsible for virus-infected software they sell when the software comes out of a box? Should the software company be held responsible for a virus that is planted by a hacker after the software is running on the computer? Should companies be required to notify customers of known viruses in their software? Could software companies alleviate the virus problem by including antivirus program software in their products that automatically checks for viruses?

3 Biz Hyper Speed Product

Because of the rapid change in online technology, early release of functional, but not perfect, browsers may take place. Advancement in Internet technology is the driving force behind these premature releases. Trying as hard as they can, software development firms nevertheless have difficulty detecting all bugs in new software. Yet, according to the video, consumers do benefit from the rapid release of new software. Do you agree or disagree? Is it acceptable for companies to release software with bugs? If you were a programmer, would you release a program with bugs?

CHECKPOINT

www.scsite.com/dc2001/ch8/check.htm

WEB INSTRUCTIONS: *To display this page from the Web, launch your browser and enter the URL, www.scsite.com/dc2001/ch8/check.htm. Click the links for current and additional information. To experience the animation and interactivity, Shockwave and Flash Player must be installed on your computer (download by clicking here).*

Label the Figure

Instructions: *Identify the information displayed when you boot a computer.*

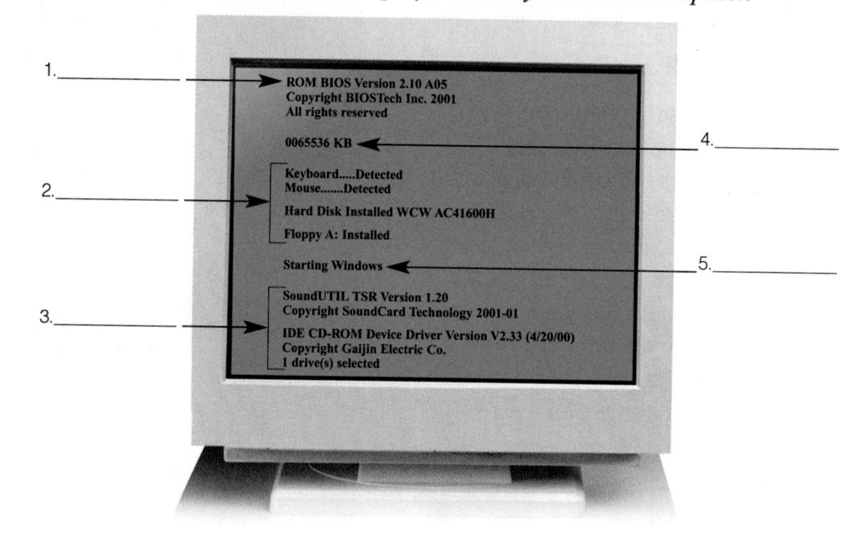

Matching

Instructions: *Match each popular operating system from the column on the left with the best description from the column on the right.*

_______ 1. Linux

_______ 2. Windows 98

_______ 3. Mac OS

_______ 4. OS/2

_______ 5. UNIX

a. Popular, free, GUI operating system whose code is made available to the public.
b. Operating environment that works in combination with an operating system to simplify its use.
c. Widely used network operating system designed for client-server networks.
d. Powerful operating system capable of handling a high volume of transactions in a multiuser environment.
e. IBM's multitasking graphical user interface operating system designed to work with 32-bit processors.
f. Apple computer operating system that set the standard for ease of use.
g. Internet integration allows for optional Web-page like user interface.

Short Answer

Instructions: *Write a brief answer to each of the following questions.*

1. How is a command-line interface different from a graphical user interface? __________ Why is a graphical user interface described as user-friendly? __________
2. What is a fault-tolerant computer? __________ For what type of systems are fault-tolerant computers used? __________
3. How does virtual memory (VM) optimize RAM? __________ What is a swap file? __________
4. How is a device-dependent operating system different from a device-independent operating system? __________ What is a downward-compatible operating system? __________
5. What is a boot disk? __________ Why is it important to have a boot disk available? __________

AT ISSUE

www.scsite.com/dc2001/ch8/issue.htm

SHELLY CASHMAN SERIES®
DISCOVERING COMPUTERS 2001

Student Exercises
WEB INFO
IN BRIEF
KEY TERMS
AT THE MOVIES
CHECKPOINT
AT ISSUE
MY WEB PAGE
HANDS ON
NET STUFF

Special Features
TIMELINE 2001
GUIDE TO WWW SITES
MAKING A CHIP
E-COMMERCE 2001
BUYER'S GUIDE 2001
CAREERS 2001
TRENDS 2001
LEARNING GAMES
CHAT
INTERACTIVE LABS
NEWS
HOME

WEB INSTRUCTIONS: *To display this page from the Web, launch your browser and enter the URL, www.scsite.com/dc2001/ch8/issue.htm. Click the links for current and additional information.*

ONE

At a recent technical conference, a speaker from a noted software company told an audience of information-technology professionals that upgrading to a new operating system would be "seamless." His listeners responded with uncontrollable laughter. Adopting a new operating system seldom is easy. As a result, no matter what the benefits, people often are reluctant to give up their old operating systems. Although reviewers agree that each new version of Windows offers several advantages over previous versions, one of the earliest editions of the operating system, Windows 3.1, remains popular. Why might people be unwilling to embrace new versions of an operating system? How could developers hasten acceptance of a new operating system? If you generally were satisfied with your current operating system, would you upgrade? Why or why not?

TWO

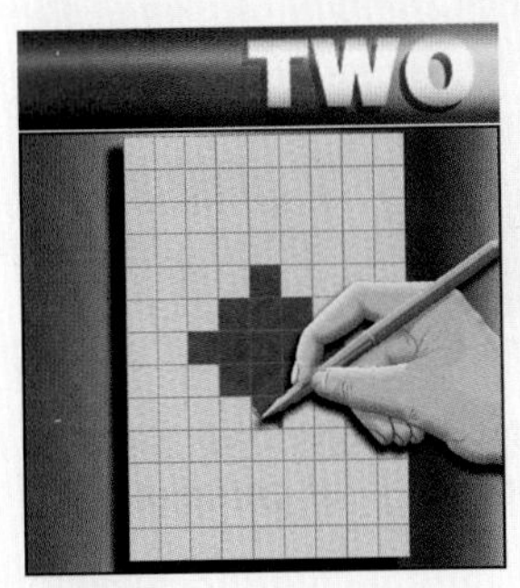

Developing icons for a graphical user interface is not an easy task. Although a good icon need not be a work of art, it must be a memorable symbol of the task it represents. According to Susan Kare, creator of the icons used with many popular GUIs, "The best icons are more like traffic signs than graphic illustrations." Choose *three* of the utilities described in this chapter and, using three sheets of graph paper, create an icon to represent each utility. Let each square on the graph paper stand for a pixel. Color the appropriate squares on the graph paper to create the image for each icon. On the back of the graph paper, explain why the icon is suitable for the utility you have chosen.

THREE

When Microsoft released Windows 98, it claimed the new operating system was *evolutionary*, not *revolutionary*. In other words, instead of being radically different, Windows 98 built on the model of its predecessor, Windows 95. Of the operating systems described in this chapter – DOS, Windows 3.x, Windows 95, Windows 98, Windows 2000, Windows CE, the Mac OS, OS/2, UNIX, Linux, and NetWare – which would you consider revolutionary? Why? Which would you call evolutionary? Why? Based on your experiences with each operating system, is a user better served using a revolutionary or evolutionary operating system? Why?

FOUR

New utility programs are being developed constantly to meet user needs. One new utility guards against computer theft by once a week making a silent call to a control center. If the call emanates from an appropriate number, the call is logged. If the computer has been reported stolen however, the center traces the call to locate the missing computer. What other needs could be addressed by a utility program? Identify three specific tasks (not described in this chapter) computer users would like to have performed that relate to managing or working with hardware, software, or files. Why would these tasks be important to a computer user? What would a utility program do to perform each task? If you were to market the utility program, what would you call it?

FIVE

Futurists claim tomorrow's operating systems may be very different from those we use today. Innovations such as touch screens, speech-recognition capabilities, automatic adaptability to individuals, and even recognition of user emotional states, have been suggested. Some innovators claim operating systems will be simpler, others think they will be more complex. What kind of operating system would you like to see? Write a description of the perfect operating system. Would it be a single-tasking, multitasking, or multiprocessing system? How would it handle such tasks as memory management, configuring devices, monitoring system performance, administering security, and managing storage media? What type of interface would it have? Of the operating systems with which you are familiar, which is most like, or most dislike, the perfect operating system? Why?

DISCOVERING COMPUTERS 2001

Student Exercises
WEB INFO
IN BRIEF
KEY TERMS
AT THE MOVIES
CHECKPOINT
AT ISSUE
MY WEB PAGE
HANDS ON
NET STUFF

Special Features
TIMELINE 2001
GUIDE TO WWW SITES
MAKING A CHIP
E-COMMERCE 2001
BUYER'S GUIDE 2001
CAREERS 2001
TRENDS 2001
LEARNING GAMES
CHAT
INTERACTIVE LABS
NEWS
HOME

CHAPTER 1 2 3 4 5 6 7 8 9 10 11 12 13 14 INDEX

MY WEB PAGE

www.scsite.com/dc2001/ch8/myweb.htm

WEB INSTRUCTIONS: *The icons to the left of the numbered activities on this page indicate the learning system availability. If you are using WebCT, follow the instructions in the exercises below. If you are using the textbook Web site or CyberClass, launch your browser, enter the URL www.scsite.com/dc2001/ch8/myweb.htm, and then click the corresponding icon to the left of the exercise.*

1. Practice Test: Click My Web Page and then click Chapter 8. Click Practice Test. Answer each question and then click the Save Answer button. When completed, click the Finish button. Click the OK button to submit the quiz for grading. Click the View Results button, and then make a note of any missed answers.

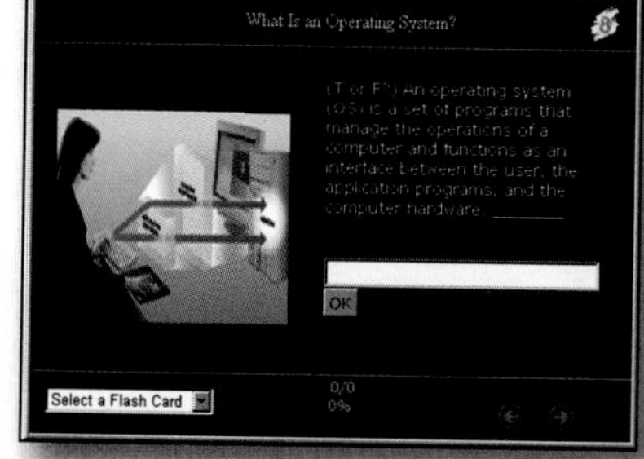

2. Web Guide: Click My Web Page and then click Chapter 8. Click Web Guide to display the Guide to WWW Sites page. Click Reference and then click Webopedia. Search for Operating Systems. Click one of the Operating Systems links. Use your word processing program to prepare a brief report on your findings and submit your assignment to your instructor.

3. Scavenger Hunt: Click My Web Page and then click Chapter 8. Click Scavenger Hunt. Print a copy of the Scavenger Hunt page; use this page to write down your answers as you search the Web. Submit your completed page to your instructor.

4. Who Wants to Be a Computer Genius?: Click My Web Page and then click Chapter 8. Click Computer Genius to find out if you are a computer genius. For directions, click the How to Play button. When you are ready to play, click the Game button. Submit your score to your instructor.

5. Wheel of Terms: Click My Web Page and then click Chapter 8. Click Wheel of Terms to reinforce important terms you learned in this chapter by playing the Shelly Cashman Series version of this popular game. For directions, click the How to Play button. When you are ready to play, click the Game button. Submit your score to your instructor.

6. Career Corner: Click My Web Page and then click Chapter 8. Click Career Corner to display the QuintEssential Careers page. Click one of the tutorials and then complete it. Write a brief report describing what you learned. Submit the report to your instructor.

7. Search Sleuth: Click My Web Page and then click Chapter 8. Click Search Sleuth to learn search techniques that will help make you a research expert. Submit the completed assignment to your instructor.

8. Crossword Puzzle Challenge: Click My Web Page and then click Chapter 8. Click Crossword Puzzle Challenge. Complete the puzzle to reinforce skills you learned in this chapter. For directions, click the How to Play button. When you are ready to play, click the Game button. Submit the completed puzzle to your instructor.

9. Flash Cards: This exercise uses CyberClass. Follow the instructions at the top of this page and then click the CyberClass icon to the left; or ask your instructor for logon instructions. Click Flash Cards on the Main Menu. Click the plus sign. Answer all the questions in any two subjects of your choice. Click the Close button in the upper-right corner to close all windows. Notice the many other exercises at this site. Complete all other exercises as assigned by your instructor.

Student Exercises
WEB INFO
IN BRIEF
KEY TERMS
AT THE MOVIES
CHECKPOINT
AT ISSUE
MY WEB PAGE
HANDS ON
NET STUFF
Special Features
TIMELINE 2001
GUIDE TO WWW SITES
MAKING A CHIP
E-COMMERCE 2001
BUYER'S GUIDE 2001
CAREERS 2001
TRENDS 2001
LEARNING GAMES
CHAT
INTERACTIVE LABS
NEWS
HOME

CHAPTER 1 2 3 4 5 6 7 **8** 9 10 11 12 13 14 INDEX

HANDS ON

www.scsite.com/dc2001/ch8/hands.htm

WEB INSTRUCTIONS: *To display this page from the Web, launch your browser and enter the URL, www.scsite.com/dc2001/ch8/hands.htm. Click the links for current and additional information.*

One

About Windows

This exercise uses Windows 95 or Windows 98 procedures. Double-click the My Computer icon on the desktop. When the My Computer window displays, click Help on the menu bar and then click About Windows 95 or About Windows 98. Answer the following questions:

- To whom is Windows licensed?
- How much physical memory is available to Windows?
- What percent of the system resources are free?

Click the OK button. Close the My Computer window.

Two

Using a Screen Saver

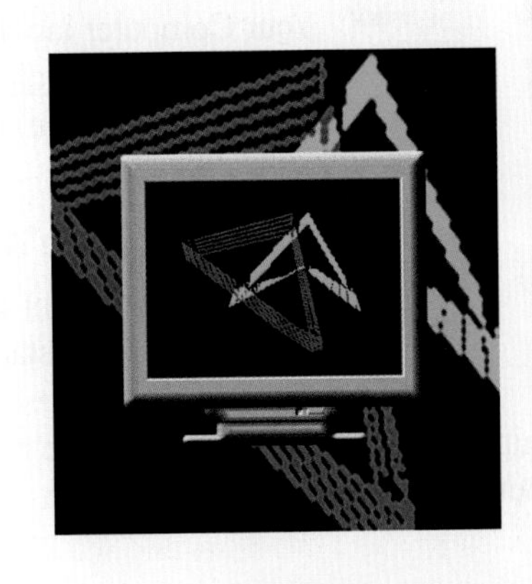

Right-click an empty area on the desktop and then click Properties on the shortcut menu. When the Display Properties dialog box displays, click the Screen Saver tab. Click the Screen Saver box arrow and then click Mystify Your Mind or any other selection. Click the Preview button to display the actual screen saver. Move the mouse to make the screen saver disappear. Answer the following questions:

- How many screen savers are available in your Screen Saver list?
- How many minutes does your system wait before activating a screen saver?

Click the Cancel button in the Display Properties dialog box.

Three

Changing Desktop Colors

Right-click an empty area on the desktop and then click Properties on the shortcut menu. When the Display Properties dialog box displays, click the Appearance tab. Perform the following tasks: (1) Click the question mark button on the title bar and then click the Scheme box. When the pop-up window displays, right-click it. Click Print Topic on the shortcut menu and then click the OK button in the Print dialog box. Click anywhere to remove the pop-up window. (2) Click the Scheme box arrow and then click Rose (large) to display the color scheme in Figure 8-31. Find a color scheme you like. Click the Cancel button.

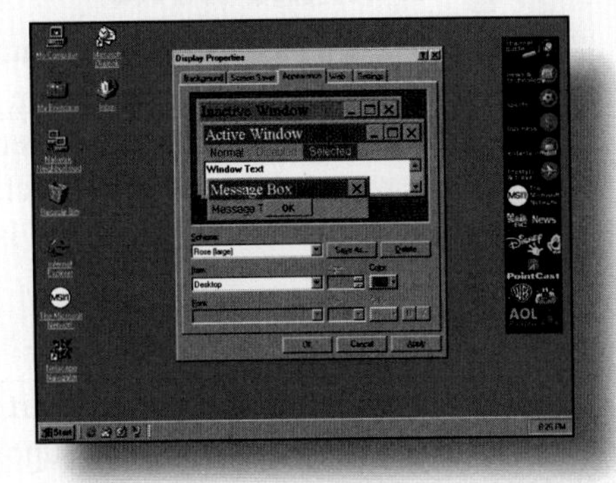

Figure 8-31

Four

Customizing the Desktop for Multiple Users

This exercise uses Windows 98 procedures. If more than one person uses a computer, how can you customize the desktop for each user? Click the Start button on the taskbar and then click Help on the Start menu. Click the Contents tab. Click the Exploring Your Computer book. Click the The Windows Desktop book. Click the Customizing for Multiple Users book. Click an appropriate Help topic and read the Help information to answer each of the following questions:

- How can you display a list of users at startup?
- How can you add personalized settings for a new user?
- How can you change desktop settings for multiple users?

Click the Close button to close the Windows Help window.

DISCOVERING COMPUTERS 2001

Student Exercises
WEB INFO
IN BRIEF
KEY TERMS
AT THE MOVIES
CHECKPOINT
AT ISSUE
MY WEB PAGE
HANDS ON
NET STUFF

Special Features
TIMELINE 2001
GUIDE TO WWW SITES
MAKING A CHIP
E-COMMERCE 2001
BUYER'S GUIDE 2001
CAREERS 2001
TRENDS 2001
LEARNING GAMES
CHAT
INTERACTIVE LABS
NEWS
HOME

CHAPTER 1 2 3 4 5 6 7 8 9 10 11 12 13 14 INDEX

NET STUFF

www.scsite.com/dc2001/ch8/net.htm

WEB INSTRUCTIONS: *To display this page from the Web, launch your browser and enter the URL, www.scsite.com/dc2001/ch8/net.htm. To use the Evaluating Operating Systems lab or the Working at Your Computer lab from the Web, Shockwave and Flash Player must be installed on your computer (download by clicking here).*

EVALUATING OPERATING SYSTEMS LAB

1. Shelly Cashman Series Evaluating Operating Systems Lab

Follow the instructions in NET STUFF 1 on page 1.48 to start and use the Evaluating Operating Systems lab. If you are running from the Web, enter the URL, www.scsite.com/sclabs/menu.htm; or display the NET STUFF page (see instructions at the top of this page) and then click the EVALUATING OPERATING SYSTEMS LAB button.

WORKING AT YOUR COMPUTER LAB

2. Shelly Cashman Series Working at Your Computer Lab

Follow the instructions in NET STUFF 1 on page 1.48 to start and use the Working at Your Computer lab. If you are running from the Web, enter the URL, www.scsite.com/sclabs/menu.htm; or display the NET STUFF page (see instructions at the top of this page) and then click the WORKING AT YOUR COMPUTER LAB button.

PICTURES

3. A Picture's Worth a Thousand Words

Although she is not a programmer, Susan Kare's impact on the modern graphical user interface has been substantial. Kare is the person responsible for many of the icons used in modern graphical interfaces. According to *Forbes* magazine, "When it comes to giving personality to what otherwise might be cold and uncaring office machines, Kare is the queen of look and feel." Click the PICTURES button to learn more about Susan Kare and her approach to developing icons.

IN THE NEWS

4. In the News

When Windows 2000 was launched in March, 2000, hundreds queued up at computer outlets. It is unclear, however, whether the anticipation was caused by the new operating system or by the promotions many dealers offered – one vendor gave system buyers the chance to also purchase a computer for $20. Click the IN THE NEWS button and read a news article about the impact, quality, or promotion of an operating system. What operating system was it? What was done to sell the operating system? Is the operating system recommended? Why or why not?

WEB CHAT

5. Web Chat

In 1999, a U.S. District Court ruled that Microsoft used its software monopoly (the company supplies almost 90 percent of the world's operating systems) to hurt competitors and consumers. The Justice Department had alleged that, by making Internet Explorer an integral part of the Windows operating system, Microsoft engaged in unfair competition against rival browser suppliers, such as Netscape. Microsoft countered that its innovations were for the good of consumers. "Fortunately, this is a case where we've got customer benefit on our side," Bill Gates said. The impact of, and Microsoft's response to, the District Court's ruling is unclear. Many believe, however, that by opening the door to future lawsuits, the decision impedes innovation by all software developers. Should there be restrictions on what can, and cannot, be included with an operating system? Why? How will the District Court's ruling affect future software development? Click the WEB CHAT button to enter a Web Chat discussion related to this topic.

WEB INSTRUCTIONS: *To gain World Wide Web access to additional and up-to-date information regarding this special feature, launch your browser and enter the URL shown at the top of this page.*

Buyer's Guide 2001
How to Purchase, Install, and Maintain a Personal Computer

The decision to buy a personal computer system is an important one – and finding and purchasing a personal computer system suited to your needs will require an investment of both time and money. In general, personal computer buyers fall into three categories: first-time buyers, replacement buyers, and upgrade buyers. A recent survey of North American consumers found that the largest of the three categories, first-time buyers, make up 40 percent of the PC market. Surprisingly, the survey also discovered that most first-time buyers have little computer experience. In fact, more than 70 percent of first-time home computer buyers do not even use a computer at work.

As with many buyers, you may have little computer experience and find yourself unsure of how to proceed. The following guidelines are presented to help you purchase, install, and maintain a computer system. These guidelines also apply to the purchase of a laptop computer. Purchasing a laptop also involves some additional considerations, which are addressed later in this special feature.

How to Purchase a Personal Computer

1. Determine what application software products you will use on your computer. Knowing what software applications you plan to use will help you decide on the type of computer to buy, as well as to define the memory, storage, and other requirements. Certain software products, for example, can run only on Macintosh computers, while others run only on a PC with the Windows operating system. Further, some software products require more memory and disk space than others, as well as additional input/output devices.

When you purchase a computer system, it may come bundled with several software products (although not all will). At the very least, you probably will want software for word processing and a browser to access the World Wide Web. If you need additional applications, such as a spreadsheet, database, or presentation graphics, consider purchasing a software suite that offers reduced pricing on several applications.

Before selecting a specific package, be sure the software contains the features necessary for the tasks you want to perform. Many Web sites and magazines, such as those listed in Figure 1, provide reviews of software products. These sites also frequently have articles that rate computer systems and software on cost, performance, and support.

Type of System	Web Site	URL
PC	Computer Shopper	www.zdnet.com/computershopper/edit/howtobuy
	PC World Magazine	www.pcworld.com
	Tech Web Buyer's Guides	www.techweb.com
	Byte Magazine	www.byte.com
	PC Computing Magazine	www.zdnet.com/pccomp
	PC Magazine	www.zdnet.com/pcmag
	Yahoo! Computers	shopping.yahoo.com/computers
	Family PC Magazine	familypc.zdnet.com
	Microsoft Network	www.eshop.msn.com
	Tips on Buying a PC	www.css.msu.edu/pc-guide.html
Macintosh	Byte Magazine	www.techweb.com/wire/apple
	Ziff-Davis	www.zdnet.com/zdnn/mac
	Macworld Magazine	www.macworld.com
	Apple	www.apple.com

For an updated list of hardware and software reviews and their Web sites, visit www.scsite.com/dc2001/ch8/buyers.htm.

Figure 1 Hardware and Software Reviews

2. Before buying a computer system, do some research. Talk to friends, coworkers, and instructors about prospective computer systems. What type of computer system did they buy? Why? Would they recommend their system and the company from which they bought it? You also should visit the Web sites or read reviews in the magazines listed in Figure 1. As you conduct your research, consider the following important criteria:

- Speed of the processor
- Size and types of memory (RAM) and storage (hard disk, floppy disk, CD-ROM, DVD-ROM, Zip® drive)
- Input/output devices included with the system (e.g., mouse, keyboard, monitor, printer, sound card, video card)
- Communications devices included with the system (modem, network interface card)
- Any software included with the system
- Overall system cost

3. Look for free software.
Many system vendors include free software with their systems. Some sellers even let you choose which software you want. Remember, however, that free software has value only if you would have purchased the software even if it had not come with the computer.

Figure 2 Some mail-order companies, such as Dell Computer, sell computers online.

4. If you are buying a new computer system, you have several purchasing options: buying from your school bookstore, a local computer dealer, a local large retail store; or ordering by mail via telephone or the World Wide Web. Each purchasing option has certain advantages. Many college bookstores, for example, sign exclusive pricing agreements with computer manufacturers and, thus, can offer student discounts. Local dealers and local large retail stores, however, more easily can provide hands-on support. Mail-order companies that sell computer systems by telephone or online via the Web (Figure 2) often provide the lowest prices but extend less personal service. Some major mail-order companies, however, have started to provide next-business-day, onsite services. A credit card usually is required to buy from a mail-order company. Figure 3 lists some of the more popular mail-order companies and their Web site addresses.

Type of System	Company	URL	Telephone Number
PC	Computer Shopper	www.computershopper.com	
	Compaq	www.compaq.com	1-800-888-0220
	CompUSA	www.compusa.com	1-800-266-7872
	dartek.com	www.dartek.com	1-800-531-4622
	Dell	www.dell.com	1-800-678-1626
	Gateway	www.gateway.com	1-800-846-4208
	Micron	www.micron.com	1-800-964-2766
	Packard Bell	www.packardbell.com	1-888-474-6772
	Quantex	www.quantex.com	1-800-346-6685
Macintosh	Apple Computer	store.apple.com	1-800-795-1000
	Club Mac	www.clubmac.com	1-800-258-2622
	MacBase	www.macbase.com	1-800-951-1230
	Mac Connection	www.macconnection.com	1-888-213-0260
	Mac Exchange	www.macx.com	1-888-650-4488

For an updated list of new computer mail-order companies and their Web sites, visit www.scsite.com/dc2001/ch8/buyers.htm.

Figure 3 New computer mail-order companies

Figure 4
Used computer mail-order companies

Company	URL	Telephone Number
American Computer Exchange	www.amcoex.com	1-800-786-0717
Boston Computer Exchange	www.bocoex.com	1-617-625-7722
United Computer Exchange	www.uce.com	1-800-755-3033
eBay	www.ebay.com	

For an updated list, visit www.scsite.com/dc2001/ch8/buyers.htm.

5. If you are buying a used computer system, stick with name brands. Although brand-name equipment can cost more, most brand-name systems have longer, more comprehensive warranties, are better supported, and have more authorized centers for repair services. As with new computer systems, you can purchase a used computer from local computer dealers, local large retail stores, or mail order via the telephone or the Web. Classified ads and used computer brokers offer additional outlets for purchasing used computer systems. Figure 4 lists several major used computer brokers and their Web site addresses.

6. Use a worksheet to compare computer systems, services, and other considerations. You can use a separate sheet of paper to take notes on each vendor's computer system and then summarize the information on a spreadsheet, such as the one shown in Figure 5. Most companies advertise a price for a base system that includes components housed in the system unit (processor, RAM, sound card, video card), disk drives (floppy disk, hard disk, CD-ROM, and DVD-ROM), a keyboard, mouse, monitor, printer, speakers, and modem. Be aware, however, that some advertisements list prices for systems with only some of these components. Monitors, printers, and modems, for example, often are not included in a base system price. Depending on how you plan to use the system, you may want to invest in additional or more powerful components. When you are comparing the prices of computer systems, make sure you are comparing identical or similar configurations.

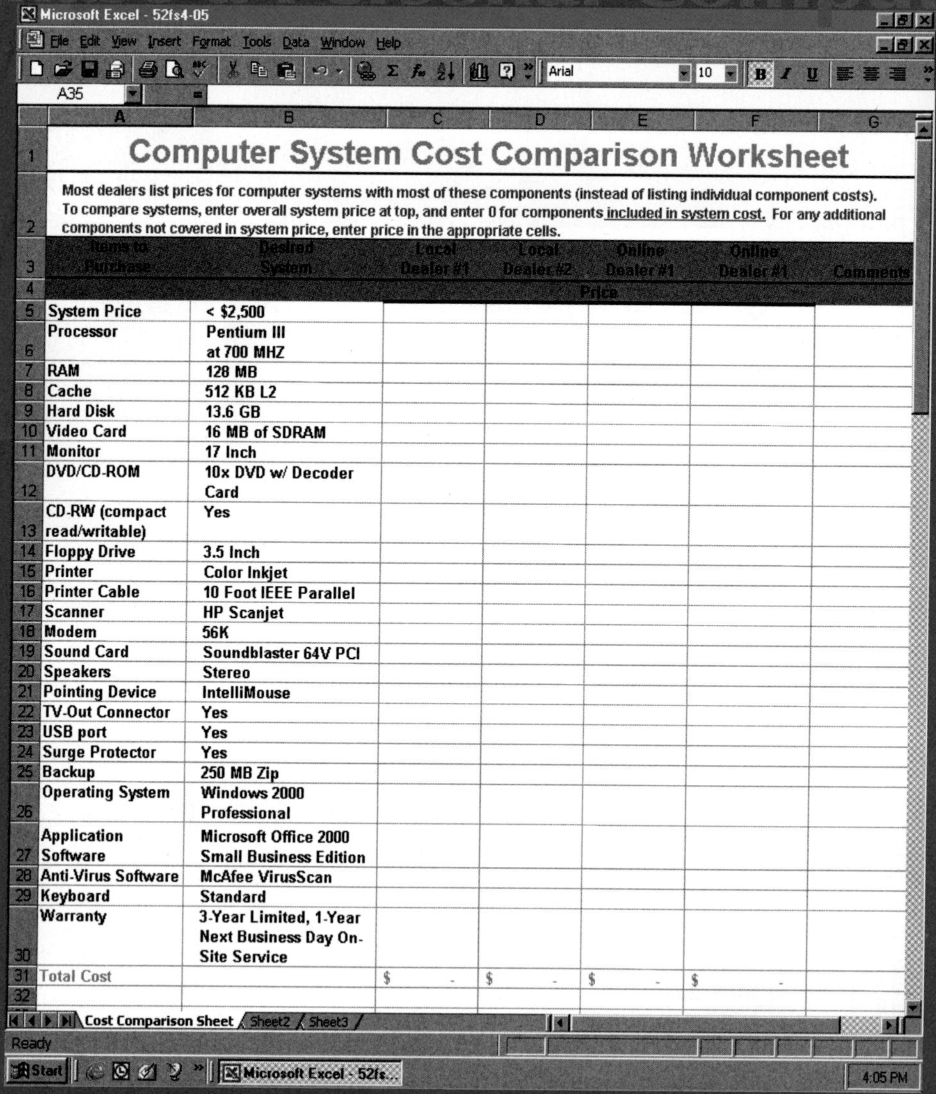
Microsoft Excel - 52fs4-05

Computer System Cost Comparison Worksheet

Most dealers list prices for computer systems with most of these components (instead of listing individual component costs). To compare systems, enter overall system price at top, and enter 0 for components included in system cost. For any additional components not covered in system price, enter price in the appropriate cells.

Items to Purchase	Desired System	Local Dealer #1	Local Dealer #2	Online Dealer #1	Online Dealer #1	Comments
		Price				
System Price	< $2,500					
Processor	Pentium III at 700 MHZ					
RAM	128 MB					
Cache	512 KB L2					
Hard Disk	13.6 GB					
Video Card	16 MB of SDRAM					
Monitor	17 Inch					
DVD/CD-ROM	10x DVD w/ Decoder Card					
CD-RW (compact read/writable)	Yes					
Floppy Drive	3.5 Inch					
Printer	Color Inkjet					
Printer Cable	10 Foot IEEE Parallel					
Scanner	HP Scanjet					
Modem	56K					
Sound Card	Soundblaster 64V PCI					
Speakers	Stereo					
Pointing Device	IntelliMouse					
TV-Out Connector	Yes					
USB port	Yes					
Surge Protector	Yes					
Backup	250 MB Zip					
Operating System	Windows 2000 Professional					
Application Software	Microsoft Office 2000 Small Business Edition					
Anti-Virus Software	McAfee VirusScan					
Keyboard	Standard					
Warranty	3-Year Limited, 1-Year Next Business Day On-Site Service					
Total Cost		$ -	$ -	$ -	$ -	

Cost Comparison Sheet / Sheet2 / Sheet3

Figure 5 A spreadsheet is an effective tool for summarizing and comparing the prices and components of different computer vendors. A copy of the Computer System Cost Comparison Worksheet is on the Discover Data Disk. To obtain a copy of the Discover Data Disk, see the inside back cover of this book for instructions.

7. Be aware of hidden costs. Before purchasing, be sure to consider any additional costs associated with buying a computer, such as an additional telephone line, an uninterruptible power supply (UPS), computer furniture, floppy disks and paper, or computer training classes you may want to take. Depending on where you buy your computer, the seller may be willing to include some or all of these in the system purchase price.

8. Consider more than just price. The lowest cost system may not be the best buy. Consider such intangibles as the vendor's time in business, the vendor's regard for quality, and the vendor's reputation for support. If you need to upgrade your computer often, you may want to consider a leasing arrangement, in which you pay monthly lease fees but upgrade or add on to your computer system as your equipment needs change. If you are a replacement buyer, ask if the vendor will buy your old system; an increasing number of companies are taking trade-ins. No matter what type of buyer you are, insist on a 30-day, no questions-asked return policy on your computer system.

9. Select an Internet service provider (ISP) or online service. You can access the Internet in one of two ways: via an ISP or an online service. Both provide Internet access for a monthly fee that ranges from $5 to $20. Local ISPs offer Internet access through local telephone numbers to users in a limited geographic region. National ISPs provide access for users nationwide (including mobile users), through local and toll-free telephone numbers. Because of their size, national ISPs offer more services and generally have a larger technical support staff than local ISPs. Online services furnish Internet access as well as members-only features for users nationwide. Figure 6 lists several national ISPs and online services. Before you choose an Internet access provider, compare such features as the number of access hours, monthly fees, available services (e-mail, Web page hosting, chat), and reliability.

Figure 6 National ISPs and online services

Company	Service	URL	Telephone Number
America Online	ONLINE	www.aol.com	1-800-827-6364
AT&T IP Services	ISP	www.ipservices.att.com/wss/	1-800-288-3199
CompuServe	ONLINE	www.compuserve.com	1-800-394-1481
Earthlink Network	ISP	www.earthlink.com	1-800-395-8425
GTE Internet	ISP	www.gte.net	1-888-GTE-SURF
MCI	ISP	www.mciworldcom.com	1-800-888-0800
Microsoft Network	ONLINE	www.msn.com	1-800-386-5550
Prodigy	ISP/ONLINE	www.prodigy.com	1-800-PRODIGY
UUNet Technologies	ISP	www.uu.net	1-800-4UUNET4

For information on local ISPs or to learn more on any ISPs and online services listed here, visit The List™ at thelist.internet.com. The List™ — the most comprehensive and accurate directory of ISPs and online services on the Web — compares dial-up services, access hours, and fees for over 5,000 access providers.

For an updated list of ISPs and online service providers, visit www.scsite.com/dc2001/ch8/buyers.htm.

10. Buy a system compatible with the ones you use elsewhere. If you use a personal computer at work or in some other capacity, make sure the computer you buy is compatible. For example, if you use a PC at work, you may not want to purchase a Macintosh for home use. Having a computer compatible with the ones at work or school will allow you to transfer files and spend time at home on work- or school-related projects.

How to Purchase a Personal Computer

11. Consider purchasing an onsite service agreement. If you use your computer system for business or are unable to be without your computer, consider purchasing an onsite service agreement through a local dealer or third-party company. Most onsite service agreements state that a technician will come to your home, work, or school within 24 hours. If your system includes onsite service only for the first year, think about extending the service for two or three years when you buy the computer.

13. Avoid buying the smallest system available. Computer technology changes rapidly, meaning a computer that seems powerful enough today may not serve your computing needs in a few years. In fact, studies show that many users regret they did not buy a more powerful system. Plan to buy a system that will last you for two to three years. You can help delay obsolescence by purchasing the fastest processor, most memory, and largest hard drive you can afford. If you must buy a smaller system, be sure you can upgrade it with additional memory and auxiliary devices as your system requirements grow. Figure 7 includes minimum recommendations for each category of user discussed in this book: Home User, Small Business User, Mobile User, Large Business User, and Power User. The Home User category is divided into two groups, Application Home User and Game Home User.

12. Use a credit card to purchase your system. Many credit cards now offer purchase protection and extended warranty benefits that cover you in case of loss of or damage to purchased goods. Paying by credit card also gives you time to install and use the system before you have to pay for it. Finally, if you are dissatisfied with the system and are unable to reach an agreement with the seller, paying by credit card gives you certain rights regarding withholding payment until the dispute is resolved. Check your credit card agreement for specific details.

BASE SYSTEM COMPONENTS

	Application Home User	Game Home User	Small Business User	Mobile User	Large Business User	Power User
HARDWARE						
Processor	Celeron at 500 MHz	Pentium III at 733 MHz	Pentium III at 650 MHZ	Pentium III at 500 MHz	Pentium III at 800 MHz	Pentium III Xeon at 550 MHz
RAM	96 MB	128 MB	128 MB	128 MB	384 MB	512 MB
Cache	Not applicable	512 KB L2	512 KB L2	512 KB L2	512 KB L2	2 MB L2
Hard Drive	6.4 GB	13.6 GB	13.6 GB	6.4 GB	36 GB	36 GB
Video Graphics Card	16 MB	32 MB	16 MB	16 MB	16 MB	32 MB
Monitor	17"	19"	17"	15" active matrix	19"	21"
DVD/CD-ROM Drive	48X CD-ROM	10X DVD with Decoder Card	48X CD-ROM	6X DVD	48X CD-ROM	10X DVD with Decoder Card
Floppy Drive	3.5"	3.5"	3.5"	3.5"	3.5"	3.5"
Printer	Color inkjet	Color inkjet	8 ppm laser	Portable inkjet	24 ppm laser	8 ppm laser
Fax/modem	56 K	56 K	56 K	56 K	ISDN	ISDN
Sound Card	16-bit	16-bit	16-bit	Built-In	16-bit	16-bit
Speakers	Stereo	Full-Dolby surround	Stereo	Stereo	Stereo	Full-Dolby surround
TV-Out Connector	Yes	Yes	Yes	Yes	Yes	Yes
USB Port	Yes	Yes	Yes	Yes	Yes	Yes
1394 Port	No	Yes	No	No	Yes	Yes
Pointing Device	IntelliMouse	Optical mouse and Joystick	IntelliMouse	Touchpad or Pointing Stick and IntelliMouse	IntelliMouse	IntelliMouse and Joystick
Keyboard	Yes	Yes	Yes	Built-In	Yes	Yes
Backup Disk/Tape Drive	250 MB Zip	1 GB Jaz	1 GB Jaz and Tape	250 MB Zip	2 GB Jaz and Tape	2 GB Jaz and Tape
SOFTWARE						
Operating System	Windows 2000 Professional	Windows 2000 Professional	Windows 2000 Professional	Windows 2000 Professional	Windows 2000 Professional	Windows 2000 Professional
Application Software Suite	Office 2000 Standard	Office 2000 Standard	Office 2000 Small Business Edition	Office 2000 Small Business Edition	Office 2000 Premium	Office 2000 Premium
Internet Access	Cable, Online Service, or ISP	Cable, Online Service, or ISP	Cable	Online Service or ISP	LAN/WAN (T1/T3)	LAN
OTHER						
Surge Protector	Yes	Yes	Yes	Portable	Yes	Yes
Warranty	3-Year Limited, 1-Year Next Business Day On-Site Service	3-Year Limited, 1-Year Next Business Day On-Site Service	3-year On-Site Service	3-Year Limited, 1-Year Next Business Day On-Site Service	3-year On-Site Service	3-year On-Site Service
Other		Headset		Docking Station		
				Carrying case		
Optional Components for all Categories						
digital camera						
multifunction device (MFD)						
scanner						
uninterruptable power supply						
ergonomic keyboard						
network interface card						
TV/FM tuner						
compact disc rewritable (CD-RW)						
video camera						
IrDa port						
mouse pad/wrist rest						

Figure 7 Base system components and optional components

How to Purchase a Laptop Computer

If you need computing capability when you travel, you may find a laptop computer to be an appropriate choice. The guidelines mentioned in the previous section also apply to the purchase of a laptop computer (Figure 8). The following are additional considerations unique to laptops.

Figure 8 Laptop computer

1. **Purchase a laptop with a sufficiently large active-matrix screen. Active-matrix screens** display high-quality color that is viewable from all angles. Less expensive, passive matrix screens sometimes are hard to see in low-light conditions and cannot be viewed from an angle. Laptop computers typically come with a 13.3-inch, 14.1-inch, or 15.4-inch display. For most users, a 14.1-inch display is satisfactory. If you intend to use your laptop as a desktop replacement, however, you may opt for a 15.4-inch display. If you travel a lot and portability is essential, consider that most of the lightest machines are equipped with a 13.3-inch display. Regardless of size, the resolution of the display should be at least 800 x 600 pixels.

2. **Experiment with different pointing devices and keyboards.** Laptop computer keyboards are far less standardized than those for desktop systems. Some laptops, for example, have wide wrist rests, while others have none. Laptops also use a range of pointing devices, including pointing sticks, touchpads, and trackballs. Before you purchase a laptop, try various types of keyboard and pointing devices to determine which is easiest for you to use. Regardless of the pointing device you select, you also may want to purchase a regular mouse unit to use when you are working at a desk or other large surface.

3. **Make sure the laptop you purchase has a CD-ROM or DVD-ROM drive.** Loading software, especially large software suites, is much faster if done from a CD-ROM or DVD-ROM. Today, most laptops come with either an internal or external CD-ROM drive; others have an internal and external unit that allows you to interchange the 3.5-inch floppy drive and the CD-ROM drive. An advantage of a separate CD-ROM drive is that you can leave it behind to save weight. Some users prefer a DVD-ROM drive to a CD-ROM drive. Although DVD-ROM drives are more expensive, they allow you to read CD-ROMs and to play movies using your laptop.

4. **If you plan to use your laptop both on the road and at home or in the office, consider a docking station.** A docking station usually includes a floppy disk drive, a CD-ROM or DVD-ROM drive, and a connector for a full-sized monitor. When you work both at home and in the office, a docking station is an attractive alternative to buying a full-sized system. A docking station essentially turns your laptop into a desktop, while eliminating the need to transfer files from one computer to another.

How to Purchase a Laptop Computer

5. If necessary, upgrade memory and disk storage at the time of purchase. As with a desktop computer system, upgrading your laptop's memory and disk storage usually is less expensive at the time of initial purchase. Some disk storage systems are custom designed for laptop manufacturers, meaning an upgrade might not be available a year or two after you purchase your laptop.

6. If you are going to use your laptop on an airplane, purchase a second battery. Two batteries should provide enough power to last through most airplane flights. If you anticipate running your laptop on batteries frequently, choose a system that uses **lithium-ion batteries** (they last longer than nickel cadmium or nickel hydride batteries).

7. Purchase a well-padded and well-designed carrying case. An amply padded carrying case will protect your laptop from the bumps it will receive while traveling. A well-designed carrying case will have room for accessories such as spare floppy disks; an external floppy disk, CD-ROM, or DVD-ROM drive; a user manual; pens; and paperwork (Figure 9).

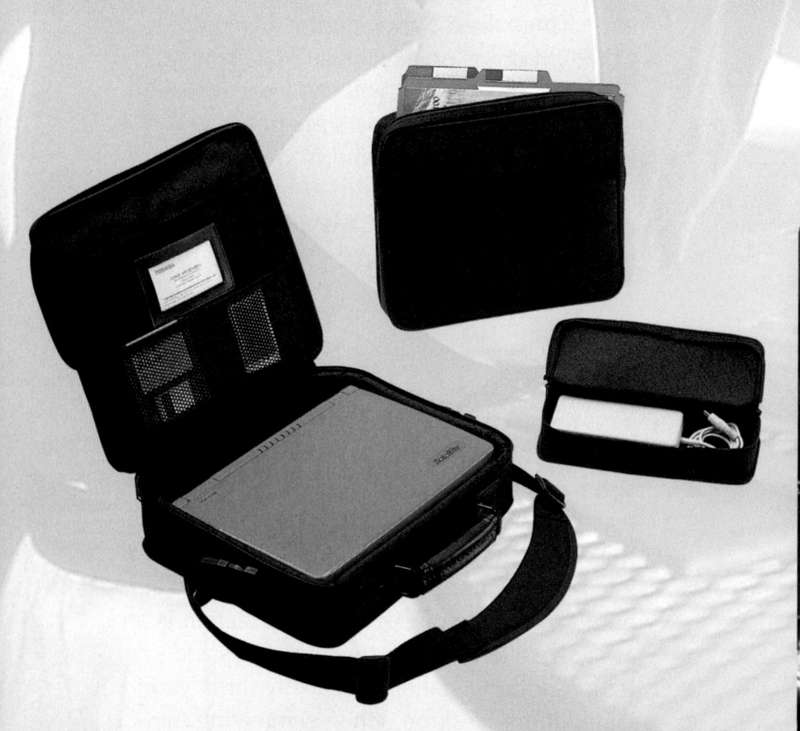

Figure 9 Well-designed carrying case

8. If you travel overseas, obtain a set of electrical and telephone adapters. Different countries use different outlets for electrical and telephone connections. Several manufacturers sell sets of adapters that will work in most countries (Figure 10).

Figure 10 Set of electrical and telephone adapters

9. If you plan to connect your laptop to a video projector, make sure the laptop is compatible with the video projector. Some laptops will not work with certain video projectors; others will not allow you to display an image on the laptop and projection device at the same time (Figure 11). Either of these factors can affect your presentation negatively.

Figure 11 Video projector

How to Install a Personal Computer

1. Read the installation manuals before you start to install your equipment. Many manufacturers include separate installation instructions with their equipment that contain important information. You can save a great deal of time and frustration if you make an effort to read the manuals.

2. Do some research. To locate additional instructions on installing your computer, review the computer magazines or Web sites listed in Figure 12 to search for articles on installing a computer system.

WEB SITE	URL
Getting Started/Installation	
Once You've Bought It	www.newsday.com/plugin/c101main.htm
HelpTalk Online	www.helptalk.com
Ergonomics	
Ergonomic Computing	cobweb.creighton.edu/training/ergo.htm
Healthy Choices for Computer Users	www-ehs.ucsd.edu/vdttoc.htm
Video Display Health Guidelines	www.uhs.berkeley.edu/facstaff/ergonomics/ergguide.html

For an updated list of reference materials, visit www.scsite.com/dc2001/ch8/buyers.htm.

Figure 12 Web references on setting up and using your computer

3. Set up your computer in a well-designed work area, with adequate workspace around the computer. **Ergonomics** is an applied science devoted to making the equipment and its surrounding work area safer and more efficient. Ergonomic studies have shown that using the correct type and configuration of chair, keyboard, monitor, and work surface will help you work comfortably and efficiently, and help protect your health. For your computer workspace, experts recommend an area of at least two feet by four feet. Figure 13 illustrates additional guidelines for setting up your work area.

4. Install bookshelves. Bookshelves above and/or to the side of your computer area are useful for keeping manuals and other reference materials handy.

5. Have a telephone outlet and telephone near your workspace so you can connect your modem and/or place calls while using your computer. To plug in your modem to dial up and access the World Wide Web, you will need a telephone outlet close to your computer. Having a telephone nearby also helps if you need to place business or technical support calls while you are working on your computer. Often, if you call a vendor about a hardware or software problem, the support person can talk you through a correction while you are on the telephone. To avoid data loss, however, do not place floppy disks on the telephone or near any other electrical or electronic equipment.

Figure 13 A well-designed work area should be flexible to allow adjustments to the height and build of different individuals. Good lighting and air quality also are important considerations.

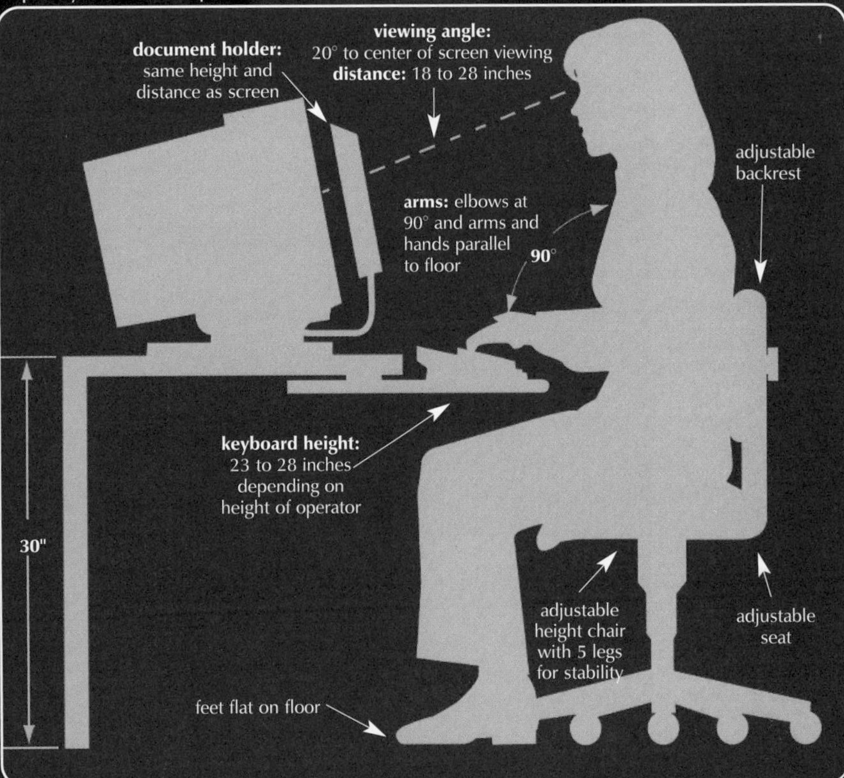

How to Install a Personal Computer

6. While working at your computer, be aware of health issues. Working safely at your computer requires that you consider several health issues. To minimize neck and eye discomfort, for instance, obtain a document holder that keeps documents at the same height and distance as your computer screen. To provide adequate lighting that reduces eye strain, use non-glare light bulbs that illuminate your entire work area. Figure 14 lists additional computer user health guidelines.

Computer User Health Guidlines

1. Work in a well-designed work area. See Figure 13 on the previous page.
2. Alternate work activities to prevent physical and mental fatigue. If possible, change the order of your work to provide some variety.
3. Take frequent breaks. Every fifteen minutes, look away from the screen to give your eyes a break. At least once per hour, get out of your chair and move around. Every two hours, take at least a fifteen-minute break.
4. Incorporate hand, arm, and body stretching exercises into your breaks. At lunch, try to get outside and walk.
5. Make sure your computer monitor is designed to minimize electromagnetic radiation (EMR). If it is an older model, consider adding EMR reducing accessories.
6. Try to eliminate or minimize surrounding noise. Noisy environments contribute to stress and tension.
7. If you frequently use the telephone and the computer at the same time, consider using a telephone headset. Cradling the telephone between your head and shoulder can cause muscle strain.
8. Be aware of symptoms of repetitive strain injuries: soreness, pain, numbness, or weakness in neck, shoulders, arms, wrists, and hands. Do not ignore early signs; seek medical advice.

Figure 14 Following these health guidelines will help computer users maintain their health.

7. Obtain a computer tool set. Computer tool sets include any screwdrivers and other tools you might need to work on your computer. Computer dealers, office supply stores, and mail-order companies sell these tool sets. To keep all the tools together, get a tool set that comes in a zippered carrying case.

8. Save all the paperwork that comes with your system. Keep the documents that come with your system in an accessible place, along with the paperwork from your other computer-related purchases. To keep different-sized documents together, consider putting them in a manila file folder, large envelope, or sealable plastic bag.

9. Record the serial numbers of all your equipment and software. Write the serial numbers of your equipment and software on the outside of the manuals packaged with these items. As noted in Figure 16 on the next page, you also should create a single, comprehensive list that contains the serial numbers of all your equipment and software.

10. Complete and send in your equipment and software registration cards. When you register your equipment and software, the vendor usually enters you in its user database. Being a registered user not only can save you time when you call with a support question, it also makes you eligible for special pricing on software upgrades.

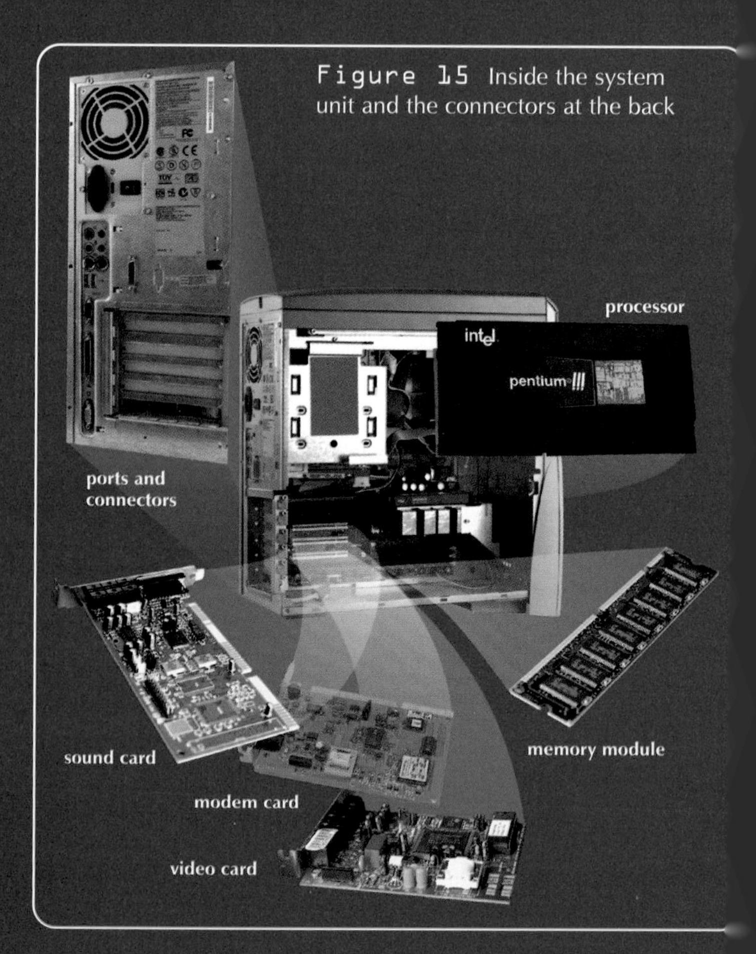

Figure 15 Inside the system unit and the connectors at the back

11. Keep the shipping containers and packing materials for all your equipment. Shipping containers and packing materials will come in handy if you have to return your equipment for servicing or must move it to another location.

12. Identify device connectors. At the back of your system, you will find a number of connectors for your printer, monitor, mouse, telephone line, and so forth (Figure 15). If the manufacturer has not identified them for you, use a marking pen to write the purpose of each connector on the back of the computer case.

13. Install your system in an area where you can maintain the temperature and humidity. You should keep the system in an area with a constant temperature between 60°F and 80°F. High temperatures and humidity can damage electronic components. Be careful when using space heaters, for example, as the hot, dry air they generate can cause disk problems.

14. Keep your computer area clean. Avoid eating and drinking around your computer. Also, avoid smoking. Cigarette smoke can cause damage to the floppy disk drives and floppy disk surfaces

15. Check your home or renter's insurance policy. Some renter's insurance policies have limits on the amount of computer equipment they cover. Other policies do not cover computer equipment at all if it is used for business. In this instance, you may want to obtain a separate insurance policy.

How to Maintain Your Personal Computer

1. Start a notebook that includes information on your system. Keep a notebook that provides a single source of information about your entire system, both hardware and software. Each time you make a change to your system, such as adding or removing hardware or software or altering system parameters, record the change in your notebook. Include the following items in your notebook.

- Vendor support numbers from your user manuals
- Serial numbers of all equipment and software
- User IDs, passwords, and nicknames for your ISP or online service, network access, Web sites, and so on
- Vendor and date of purchase for all software and equipment
- Trouble log that provides a chronological history of equipment or software problems
- Notes on any discussions with vendor support personnel

Figure 16 provides a suggested outline for the contents of your notebook.

PC OWNER'S NOTEBOOK OUTLINE

1. Vendors
Vendor
City/State
Product
Telephone #
URL

2. Internet and online services information
Service provider name
Logon telephone number
Alternate logon telephone number
Technical support telephone number
User ID
Password

3. Web site information
Web site name
URL
User ID
Password
Nickname

4. Serial numbers
Product
Manufacturer
Serial #

5. Purchase history
Date
Product
Manufacturer
Vendor
Cost

6. Software log
Date installed/uninstalled

7. Trouble log
Date
Time
Problem
Resolution

8. Support calls
Date
Time
Company
Contact
Problem
Comments

9. Vendor paperwork

Figure 16 To keep important information about your computer on hand and organized, use an outline such as this sample outline.

How to Maintain Your Personal Computer

2. Before you work inside your computer, turn off the power and disconnect the equipment from the power source. Working inside your computer with the power on can affect both you and the computer adversely. Thus, you should turn off the power and disconnect the equipment from the power source before you open a computer to work inside. In addition, before you touch anything inside the computer, you should touch an unpainted metal surface such as the power supply. Doing so will help discharge any static electricity that could damage internal components.

3. Keep the area surrounding your computer dirt and dust free. Reducing the dirt and dust around your computer will reduce the need to clean the inside of your system. If dust builds up inside the computer, remove it carefully with compressed air and a small vacuum. Do not touch the components with the vacuum.

4. Back up important files and data. Use the operating system or utility program to create an emergency or rescue disk to help you restart your computer if it crashes. You also regularly should copy important data files to disks, tape, or another computer.

5. Protect your system from computer viruses. A computer virus is a potentially damaging computer program designed to infect other software or files by attaching itself to the software or files with which it comes in contact. Virus programs are dangerous because often they destroy or corrupt data stored on the infected computer. You can protect your computer from viruses by installing an antivirus program.

6. Keep your system tuned. Most operating systems include several system tools that provide basic system maintenance functions. One important system tool is the disk defragmenter. Defragmenting your hard disk reorganizes files so they are in contiguous (adjacent) clusters, making disk operations faster. Some programs allow you to schedule system maintenance tasks for times when you are not using your computer. If necessary, leave your computer on at night so the system can run the required maintenance programs. If your operating system does not provide the system tools, you can purchase a stand-alone utility program to perform basic system maintenance functions.

7. Learn to use system diagnostic tools. Diagnostic tools help you identify and resolve system problems, thereby helping to reduce your need for technical assistance. Diagnostic tools help you test system components, monitor system resources such as memory and processing power, undo changes made to system files, and more. As with basic maintenance tools, most operating systems include system diagnostic tools; you also can purchase or download many stand-alone diagnostic tools.

APPENDIX

CODING SCHEMES AND NUMBER SYSTEMS

CODING SCHEMES

As discussed in Chapter 3, a computer uses a coding scheme to represent characters. This section presents the ASCII, EBCDIC, and Unicode coding schemes and discusses parity.

ASCII and EBCDIC

Two widely used codes that represent characters in a computer are the ASCII and EBCDIC codes. **The American Standard Code for Information Interchange**, called ASCII (pronounced ASK-ee), is the most widely used coding system to represent data. Many personal computers and minicomputers use ASCII. The **Extended Binary Coded Decimal Interchange Code**, or **EBCDIC** (pronounced EB-see-dic) is used primarily on mainframe computers. Figure A-1 summarizes these codes. Notice how the combination of bits (0s and 1s) is unique for each character.

When the ASCII or EBCDIC code is used, each character that is represented is stored in one byte of memory. Other binary formats exist, however, that the computer sometimes uses to represent numeric data. For example, a computer may store, or pack, two numeric characters in one byte of memory. The computer uses these binary formats to increase storage and processing efficiency.

Unicode

The 256 characters and symbols that are represented by ASCII and EBCDIC codes are sufficient for English and western European languages but are not large enough for Asian and other languages that use different alphabets. Further compounding the problem is that many of these languages used symbols, called **ideograms**, to represent multiple words and ideas. One solution to this situation is Unicode. **Unicode** is a 16-bit code that has the capacity of representing more than 65,000 characters and symbols. Unicode represents

ASCII	SYMBOL	EBCDIC
00110000	0	11110000
00110001	1	11110001
00110010	2	11110010
00110011	3	11110011
00110100	4	11110100
00110101	5	11110101
00110110	6	11110110
00110111	7	11110111
00111000	8	11111000
00111001	9	11111001
01000001	A	11000001
01000010	B	11000010
01000011	C	11000011
01000100	D	11000100
01000101	E	11000101
01000110	F	11000110
01000111	G	11000111
01001000	H	11001000
01001001	I	11001001
01001010	J	11010001
01001011	K	11010010
01001100	L	11010011
01001101	M	11010100
01001110	N	11010101
01001111	O	11010110
01010000	P	11010111
01010001	Q	11011000
01010010	R	11011001
01010011	S	11100010
01010100	T	11100011
01010101	U	11100100
01010110	V	11100101
01010111	W	11100110
01011000	X	11100111
01011001	Y	11101000
01011010	Z	11101001
00100001	!	01011010
00100010	"	01111111
00100011	#	01111011
00100100	$	01011011
00100101	%	01101100
00100110	&	01010000
00101000	(	01001101
00101001	)	01011101
00101010	*	01011100
00101011	+	01001110

Figure A-1

all the world's current languages using more than 34,000 characters and symbols (Figure A-2). In Unicode, 30,000 codes are reserved for future use, such as ancient languages, and 6,000 codes are reserved for private use. Existing ASCII coded data is fully compatible with Unicode because the first 256 codes are the same. Unicode currently is implemented in several operating systems, including Windows NT and OS/2, and major system developers have announced plans eventually to implement Unicode.

	041	042	043	044	045	046	047
0	А	Р	а	р		Ѡ	Ѱ
1	Б	С	б	с	ё	ѡ	ѱ
2	В	Т	в	т	ђ	Ѣ	Ѳ
3	Г	У	г	у	ѓ	ѣ	ѳ
4	Д	Ф	д	ф	є	Ѥ	Ѵ
5	Е	Х	е	х	ѕ	ѥ	ѵ
6	Ж	Ц	ж	ц	і	Ѧ	Ѷ
7	З	Ч	з	ч	ї	ѧ	ѷ
8	И	Ш	и	ш	ј	Ѩ	Ѹ
9	Й	Щ	й	щ	љ	ѩ	ѹ
A	К	Ъ	к	ъ	њ	Ѫ	Ѻ
B	Л	Ы	л	ы	ћ	ѫ	ѻ
C	М	Ь	м	ь	ќ	Ѭ	Ѽ
D	Н	Э	н	э		ѭ	ѽ
E	О	Ю	о	ю	ў	Ѯ	Ѿ
F	П	Я	п	я	џ	ѯ	ѿ

Figure A-2

Parity

Regardless of whether ASCII, EBCDIC, or other binary methods are used to represent characters in memory, it is important that the characters be stored accurately. For each byte of memory, most computers have at least one extra bit, called a **parity bit**, that is used by the computer for error checking. A parity bit can detect if one of the bits in a byte has been changed inadvertently. While such errors are extremely rare (most computers never have a parity error during their lifetime), they can occur because of voltage fluctuations, static electricity, or a memory failure.

Computers are either odd- or even-parity machines. In computers with odd parity, the total number of on bits in the byte (including the parity bit) must be an odd number. In computers with even parity, the total number of on bits must be an even number (Figure A-3). The computer checks parity each time it uses a memory location. When the computer moves data from one location to another in memory, it compares the parity bits of both the sending and receiving locations to see if they are the same. If the system detects a difference or if the wrong number of bits is on (e.g., an odd number in a system with even parity), an error message displays. Many computers use multiple parity bits that enable them to detect and correct a single-bit error and detect multiple-bit errors.

NUMBER SYSTEMS

This section describes the number systems that are used with computers. Whereas thorough knowledge of this subject is required for technical computer personnel, a general understanding of number systems and how they relate to computers is all most users need.

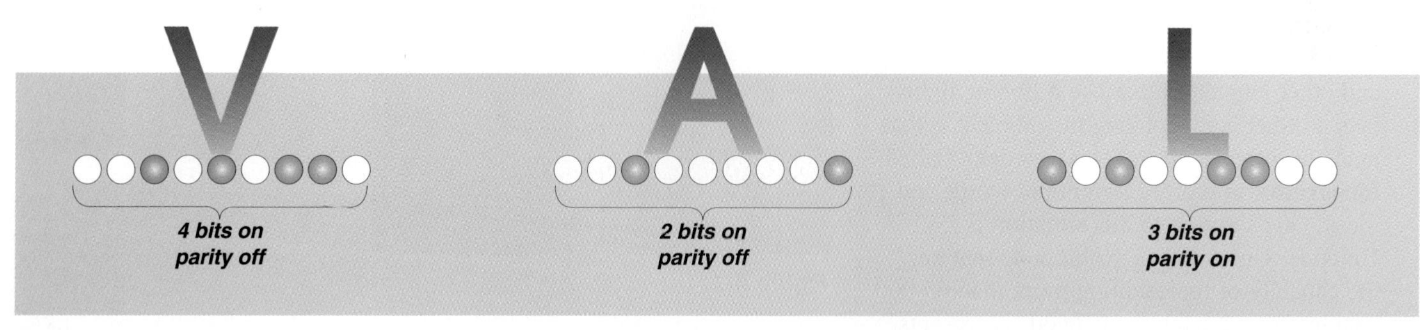

Figure A-3

As you have learned, the binary (base 2) number system is used to represent the electronic status of the bits in memory. It also is used for other purposes such as addressing the memory locations. Another number system that commonly is used with computers is **hexadecimal** (base 16). The computer uses the hexadecimal system to communicate with a programmer when a problem with a program exists, because it would be difficult for the programmer to understand the 0s and 1s of binary code. Figure A-4 shows how the decimal values 0 through 15 are represented in binary and hexadecimal.

The mathematical principles that apply to the binary and hexadecimal number systems are the same as those that apply to the decimal number system. To help you better understand these principles, this section starts with the familiar decimal system, then progresses to the binary and hexadecimal number systems.

DECIMAL	BINARY	HEXADECIMAL
0	0000	0
1	0001	1
2	0010	2
3	0011	3
4	0100	4
5	0101	5
6	0110	6
7	0111	7
8	1000	8
9	1001	9
10	1010	A
11	1011	B
12	1100	C
13	1101	D
14	1110	E
15	1111	F

Figure A-4

The Decimal Number System

The decimal number system is a base 10 number system (deci means ten). The base of a number system indicates how many symbols are used in it. The decimal number system uses 10 symbols: 0 through 9. Each of the symbols in the number system has a value associated with it. For example, 3 represents a quantity of three and 5 represents a quantity of five.

The decimal number system also is a positional number system. This means that in a number such as 143, each position in the number has a value associated with it. When you look at the decimal number 143, the 3 is in the ones, or units, position and represents three ones or (3 x 1); the 4 is in the tens position and represents four tens or (4 x 10); and the 1 is in the hundreds position and represents one hundred or (1 x 100). The number 143 is the sum of the values in each position of the number (100 + 40 + 3 = 143). The chart in Figure A-5 shows how you can calculate the positional values (hundreds, tens, and units) for a number system. Starting on the right and working to the left, the base of the number system, in this case 10, is raised to consecutive powers (10^0, 10^1, 10^2). These calculations are a mathematical way of determining the place values in a number system.

When you use number systems other than decimal, the same principles apply. The base of the number system indicates the number of symbols that are used, and each position in a number system has a value associated with it. By raising the base of the number system to consecutive powers beginning with zero, you can calculate the positional value.

power of 10	10^2	10^1	10^0
positional value	100	10	1
number	1	4	3

1 4 3 =

$(1 \times 10^2) + (4 \times 10^1) + (3 \times 10^0) =$

$(1 \times 100) + (4 \times 10) + (3 \times 1) =$

$100 + 40 + 3 = 143$

Figure A-5

The Binary Number System

As previously discussed, binary is a base 2 number system (bi means two), and the symbols it uses are 0 and 1. Just as each position in a decimal number has a place value associated with it, so does each position in a binary number. In binary, the place values, moving from right to left, are successive powers of two ($2^0, 2^1, 2^2, 2^3$) or (1, 2, 4, 8). To construct a binary number, you place ones in the positions where the corresponding values add up to the quantity you want to represent; you place zeros in the other positions. For example, in a four-digit binary number, the binary place values are (from right to left) 1, 2, 4, and 8. The binary number 1001 has ones in the positions for the values 1 and 8 and zeros in the positions for 2 and 4. Therefore, the quantity represented by binary 1001 is 9 (8 + 0 + 0 + 1) (Figure A-6).

The Hexadecimal Number System

The hexadecimal number system uses 16 symbols to represent values (hex means six, deci means ten). These include the symbols 0 through 9 and A through F (Figure A-4 on page A.3). The mathematical principles previously discussed also apply to hexadecimal (Figure A-7).

The primary reason why the hexadecimal number system is used with computers is because it can represent binary values in a more compact and readable form and because the conversion between the binary and the hexadecimal number systems is very efficient.

An eight-digit binary number (a byte) can be represented by a two-digit hexadecimal number. For example, in the ASCII code, the character M is represented as 01001101. This value can be represented in hexadecimal as 4D. One way to convert this binary number (4D) to a hexadecimal number is to divide the binary number (from right to left) into groups of four digits; calculate the value of each group; and then change any two-digit values (10 through 15) into the symbols A through F that are used in hexadecimal (Figure A-8).

Figure A-6

power of 2	2^3	2^2	2^1	2^0
positional value	8	4	2	1
binary	1	0	0	1

1 0 0 1 =

$(1 \times 2^3) + (0 \times 2^2) + (0 \times 2^1) + (1 \times 2^0) =$

$(1 \times 8) + (0 \times 4) + (0 \times 2) + (1 \times 1) =$

$8 + 0 + 0 + 1 = 9$

power of 16	16^1	16^0
positional value	16	1
hexadecimal	A	5

A 5 =

$(10 \times 16^1) + (5 \times 16^0) =$

$(10 \times 16) + (5 \times 1) =$

$160 + 5 = 165$

Figure A-7

positional value	8421	8421
binary	0100	1101
decimal	4	13
hexadecimal	4	D

Figure A-8

INDEX

AC adapter: External power supply that converts AC power into DC power. **3.28**
Accelerated Graphics Port (AGP): Bus designed by Intel to improve the transmission speed of 3-D graphics and video. **3.27**
Accelerator expansion card, 3.20
Access time: The speed at which the processor can access data from memory. **3.19**-20, **6.3**
CD-ROM, 6.21
hard disks, 6.13
memory devices, 6.3
Accessibility Properties dialog box: Dialog box used in Windows that allows physically challenged users to set options to make Windows easier to use. **5.24**
Accounting software: Used by companies to record and report financial transactions. 2.2, **2.22**-23
packages, 2.7
Accuracy, computer, 1.8
Active Desktop™, 8.13
Active-matrix display/screens: LCD screens that produce color by the use of a separate transistor for each pixel, allowing the display of high-quality color viewable from all angles. **5.5, 8.43**
ActiveX control: Small program that can be downloaded and run in a browser, adding multimedia capabilities to Web pages. **7.12,** 7.14
Adapter card, *see* **Expansion card**
Add-in, *see* **Expansion card**
Add-on, *see* **Expansion card**
Address: A unique number identifying the location of a byte in memory. **3.14**
Address(es)
e-mail, 2.33-34, 7.23-24, 7.28
mailing lists, 7.28-29
Web, 2.35
Address book (e-mail programs): Created by user and contains a list of names and e-mail addresses. **2.34**
Address book (personal information manager software): Software used to enter and maintain names, addresses, and telephone numbers. **2.21**
groupware and, 2.33
Address bus, 3.24
Adobe GoLive, 2.23
Adobe Illustrator, 2.23
Adobe InDesign, 2.23
Adobe PageMaker, 2.23
Adobe PageMill, 2.23, 7.22
Adobe PhotoDeluxe, 2.27
Adobe Photoshop, 2.23
Adobe Premiere, 2.23
Advanced Research Projects Agency, *see* **ARPA**
Advertising
banners, 7.59
e-commerce and, 7.52, 7.59
VR models used to create, 7.18
on Web, 2.35
Allen, Paul, 1.36, 1.37
All-in-one computer: Less-expensive desktop computer that combines the monitor and system unit into a single device; ideal for the casual home user. **1.20**
Alpha microprocessor: Microprocessor developed by Digital Equipment Corporation with clock speeds from 300 to 700 MHz; used primarily in workstations and high-end servers. **3.9,** 3.25
Altair computer, 1.51
AltaVista, 7.30
AMD processors, 3.5, 3.9
American Standard Code for Information Interchange, *see* **ASCII**
Americans with Disabilities Act (ADA): Federal law that requires any company with 15 or more employees to make reasonable attempts to accommodate the needs of physically challenged workers. **4.26**
America Online (AOL), 1.17, 2.36, 2.45, 7.7, 7.52
America Online, Inc., 2.45
Analog: The use of continuous signals to represent data and information. **3.11**
Analog sound, 5.2
Analog video signals, 5.3
Andreessen, Marc, 1.54, 2.44
Animated GIF: Type of animation which combines several images into a single GIF file. **7.14**
Animation: Still images displayed in rapid sequence, giving them the illusion of motion. 1.35, **5.2, 7.14**
clip galleries containing, 2.31
Web page, 7.12, 7.14, 7.22
Anonymous FTP: FTP site that allows anyone to transfer some, if not all, available files. **7.25**
Antivirus program: Program that protects a computer against viruses by identifying and removing any computer viruses found in memory, on storage media, or on incoming files. **8.25**
Apple Computer, 2.3, 2.4
history of, 1.51, 1.53, 1.56, 8.26-27
iMac, 1.57
operating systems, 8.14, 8.27
purchasing computer system and, 8.38
SuperDrive, 6.7
Apple Macintosh, 1.20, 1.37, 1.53, 1.57, 3.9
Applet: Short program executed inside of another program that usually runs on the client (user's computer), as opposed to running on a Web server, but an applet is compiled, which means it usually runs faster than a script. **7.12,** 7.14
Application, *see* **Application software**
Application service provider (ASP): Third-party organization that manages and distributes software and services on the Web. **2.43**
Application software: Programs designed to perform specific tasks for users, such as creating documents, analyzing finances, sending messages, and creating Web pages. **1.12, 2.1**-43
accessed on Web, 2.43
booting and, 8.19
categories of, 2.2
communications, 2.32-41
graphics, 2.23-25
home, personal, and educational use, 2.27-32
integrated, 2.28
multimedia, 2.26
operating system and, 2.3, 8.2-5
productivity, 2.7-23
purchasing computer and, 8.38
RAM needed for, 3.16-17
suite, 2.21
system, *see* System software
types of, 2.2-3
uninstalling, 8.24
upward-compatible, 8.10
Appointment calendar: Feature of personal information management (PIM) software used to schedule activities for a particular day and time. **2.21**
groupware and, 2.33
Architectural applications, 1.35
VR models used in, 7.18
Arithmetic/logic unit (ALU): Component of CPU that performs the execution process of the machine cycle. 3.5, **3.7**
Arithmetic operations: Operations performed by the arithmetic/logic unit including addition, subtraction, multiplication, and division. **3.7**
ARPA (Advanced Research Projects Agency): Agency of the U.S. Department of Defense that built ARPANET, the original networking project that was the beginning of the Internet. **7.4**
ARPANET: Started by ARPA in 1969, was the beginning of the Internet. ARPANET effectively linked together scientific and academic researchers in the U.S. 1.51, **7.4**
Arrow keys: Keys on the keyboard that allow movement of the insertion point left, right, up, or down. **4.5**
Article: Message in a newsgroup. **7.27**
Art Web sites, 2.57
arts domain abbreviation, 7.10
ASCII (American Standard Code for Information Interchange): The most widely used coding system to represent data. **3.12, A.1**
AT Attachment, *see* **ATA (AT Attachment)**
At symbol (@), in e-mail address, 2.34
ATA (AT Attachment): Attachment that integrates an IDE controller into the disk drive. **6.14**
Atanasoff, John V., 1.48
Athlon™ processor, 3.5, 3.9
AU: Format for audio files on the Internet. **7.16**
Audience handouts, presentations and, 2.20
Audio: Any music, speech, or other sound that is stored and produced by the computer. **2.26, 5.2, 5.20**
compressed files, 4.24, 7.16, 7.17
MP3 files, 7.16
streaming, 7.16
Web page, 7.12, 7.16-17
Audio CD, 4.22, 6.18, 6.19, 6.20, 6.22
Audio CD player, audio input using, 4.22
Audio clips, clip galleries containing, 2.31
Audioconferencing, *see* **Internet telephone service**
Audio editing software: Software used to modify audio clips, and usually includes filters, which are designed to enhance audio quality. 2.2, 2.23, **2.26**
Audio files, 4.22
Audio input: Process of entering (recording) music, speech, or sound effects, via a device that plugs into a port on a sound card. **4.22**
Audio output devices: Computer components that produce music, speech, or other sounds, such as beeps. **5.20**-21
Authorware: Multimedia authoring software package that provides the tools needed to build interactive multimedia training and educational programs. 2.23
AutoCorrect, 2.11
Autodesk AutoCAD, 2.23
Autodesk Planix Complete Home Suite, 2.27
AutoFormat, 2.11
Automatic teller machine (ATM): Self-service banking machine attached to a host computer through a telephone network. **5.24**
AutoPC: A device mounted onto a vehicle's dashboard that is powered by Windows CE and is directed by voice commands. **8.14**
AutoSave: Feature that automatically saves open documents at specified time periods. **2.9**
Auxiliary storage, *see* **Storage**

Backbone: High-speed network that connects regional and local networks to the Internet; other computers then connect to these regional and local networks to access the Internet. **7.5, 7.8**
Backbone provider, *see* **National ISP**
Background: Application that is running but not being used, when running multiple applications. **8.5**
Backup: Duplicate of a file, program, or disk that can be used if the original is lost, damaged, or destroyed. **6.9**
floppy disk, 6.9
RAID, 6.15
removable disks, 6.14
utility for, 8.24, 8.48
Backup utility: Utility program that copies or backs up selected files or an entire hard disk onto another disk or tape. **8.24,** 8.48
Bad sector, 6.5, 6.11

Band printer: Impact line printer that prints fully-formed characters when hammers strike a horizontal, rotating band that contains shapes of numbers, letters, and other characters. **5.13**
Banking
automatic teller machines (ATMs) used in, 5.24
magnetic ink character recognition (MICR) reader used in, 4.19
microfiche/microfilm storage used by, 6.27
online, 2.28, 7.3
Banner ads: Online advertisements at another Web site, often hyperlinked to the advertiser's Web site. **7.59**
Bar charts: Charts that display bars of various lengths to show the relationship of data. **2.14**
Bar code: Identification code that consists of a set of vertical lines and spaces of different widths, which represents data that identifies a product. **4.17**
Bar code reader, 5.24
Bar code scanner: Scanner that uses laser beams to read bar codes. **4.17**
Bardeen, John, 1.49
BASIC (Beginner's All-purpose Symbolic Instruction Code): Programming language developed in the mid-1960s for use as a simple, interactive problem-solving language. 1.50
Basic input/output system, *see* **BIOS**
Batteries, laptop computer, 8.44
Bay: Open area inside the system unit used to install additional equipment. **3.27**
Beans, *see* **JavaBeans**
Bechtolsheim, Andreas, 6.28
Bell Laboratories, 8.16
Berners-Lee, Tim, 1.53, 7.36
Berry, Clifford, 1.48
Bina, Eric, 2.44
Binary: Numbering system that uses just two unique digits, 0 and 1. **3.12,** A.4
BIOS (basic input/output system): Firmware contained in ROM chips that contains the computer's startup instructions. **8.18**
Bit (binary digit): Smallest unit of data the computer can handle. **3.12**
transferred along buses, 3.24
Bitmap: Rows and columns of dots that represent the results of a scanned document. **4.14**
Blind users, 4.26
Bookstores, online, 7.64
Boot disk: Floppy disk containing operating system commands that will start the computer; used if unable to boot from the hard disk. **8.20**
Boot drive: Drive from which computer boots; normally the C drive. **8.20**
Booting: Process of starting or resetting a computer that involves loading an operating system into memory. **8.17**-20
Border: Decorative line or pattern along one or more edges of a page or around an image. **2.10**
Bot, *see* **Intelligent agent**
Braille printer: Printer for blind users that outputs information in Braille onto paper. **5.25**
Brattain, Walter, 1.49
Brick-and-mortar businesses: Organizations that do not have a Web site. **7.48,** 7.49
Bricklin, Dan, 1.52
Brochureware: Web sites used as electronic billboards to advertise and provide information about business's product; the sites are not interactive and can be displayed equally well on paper. **7.49**
Broderbund Print Shop Pro Publisher, 2.27
Broderbund 3D Home Design Suite, 2.27
Browser, *see* **Web browser**
B-to-B infomediaries, *see* **Business-to-business infomediaries**
Buffer: Area of memory or storage where data and information is placed while waiting to be transferred. **8.6**
Bulletin board system (BBS), e-commerce and, 7.48
Bunny suits: Special protective clothing worn by people who work in chip manufacturing laboratories. **3.42**
Bus: Electrical channel that allows the various devices inside and attached to the system unit to communicate with each other. **3.24**-27
expansion, 3.26-27
Business computers, networked, 1.15
Business/productivity software, 2.2
Business services, e-commerce and, 7.52
Business users, 1.55
connecting to Internet through, 7.7
Internet access and, 2.35
Internet linking, 7.2
large, *see* Large business users
RAM needs of, 3.16
small, *see* Small business users
Business-to-business (B-to-B or B2B) e-commerce: E-commerce that consists of businesses providing goods and services to other businesses, such as online advertising, recruiting, credit, sales, market research, technical support, and training. **7.19,** 7.50-51
Business-to-business infomediaries: Company that establishes an electronic hub to match and interface among buyers and sellers in a particular industry segment. Also called information infomediaries. **7.51**
Business-to-consumer (B-to-C or B2C) e-commerce: E-commerce that consists of the sale of goods to the general public. Transactions occur instantaneously and globally, saving money for both the business and the consumer. **7.19,** 7.50
Business Web sites, 2.57
Bus width: Size of a bus; determines the number of bits that can be transmitted at one time. **3.25**
Button: Graphical element on desktop that user can click to cause a specific action to take place. **2.4,** 2.7, 8.4
Byte(s): A grouping of eight bits, providing enough different combinations of 0s and 1s to represent 256 individual characters. **3.12,** 3.13, 3.14-15

Cable companies, Internet structure and, 7.5
Cable modem: Modem that sends and receives data over the cable television (CATV) network. 7.7
Cache: High-speed memory between the processor and main RAM that speeds computer processes by storing frequently used instructions and data. **3.17**-18
configuring, 3.18
disk, 6.13
Cache store, 3.17
CAD software, *see* **Computer-aided design (CAD) software**
Calculations, worksheet, 2.13
Calculators, LCD displays used in, 5.4
Camera
digital, *see* Digital camera
video, 4.24, 5.3
Capacity: The size of a storage device, measured by the number of bytes it can hold. **6.3**
CD-ROM, 6.20
DVD-ROM, 6.23
floppy disk, 6.3
hard disk, 6.3, 6.10
PC Card, 6.25
Car, Auto PC for, 8.14
Card, *see* **Expansion card**
Careers
e-commerce, 7.64
Web sites, 2.57
Case, Stephen M., 2.45
Cathode ray tube (CRT): Glass tube that is the core of a CRT monitor. 1.6, **5.3**-4
CD, *see* **Compact disc**
CD player, audio input using, 4.22
CD-R (compact disc-recordable): Technology that allows a user to write on (record items onto) a compact disc using his or her own computer. The disc can be written on only once, and it cannot be erased, but can be read from an unlimited number of times. **6.22**
CD-R drive: Drive that uses laser technology to read items from and write on both audio CDs and standard CD-ROMs. **6.22**
CD-ROM (compact disc read-only memory): Storage media that is a silver-colored compact disk that uses the same laser technology as audio CDs for recording and reading items, and can contain text, graphics, and video, as well as sound. 1.8, **6.19**-22
CD-ROM drive: Drive that uses laser technology to read items on a CD-ROM, and transmit the data, instructions, and information to another device. 1.8, 6.18, **6.20**-21
bays for, 3.27
purchasing laptop with, 8.43
CD-ROM player: Device that uses laser technology to read CD-ROMs. **6.20**
CD-RW (compact disc-rewritable): Erasable compact disc that can be written on (items recorded onto) multiple times. 1.8, **6.22**
Celeron™ processor: Intel processor with a clock speed of 400-500 MHz; designed for less expensive PCs. 3.5, **3.9,** 3.10, 3.25
Cell: In a worksheet, the intersection of a column and row. **2.12**
Cellular telephone
flash memory in, 3.18
smart, 5.5, 7.34
Central processing unit (CPU): A chip or chips on the motherboard containing the control unit and the arithmetic/logic unit, that interpret and carry out the basic instructions that operate a computer. **1.6,** 1.19, **3.5**
booting and, 8.18
disk cache and, 6.13
input and, 4.2
installation and upgrades, 3.10
memory and, 6.2
speed, 3.7, 3.8, 3.9, 3.10, 3.17
superscalar, 3.7, 3.8
system bus and, 3.26
Central processing unit (CPU) chip, connectors for, 3.4
Centronics interface, 3.23
Certificate, *see* **Digital certificate**
Certificate authority (CA): Authorized company or person that issues and verifies digital certificates. **7.20**
Character: Any symbol that requires one byte of computer storage space. **5.2**
coding scheme for, 3.12-13
Characters per second (cps): Measure of speed of the speed of dot-matrix printers. **5.13**
Chart, 5.2
Charting: Feature of spreadsheet software allowing display of data in a chart that shows the relationship of the data in graphical, rather than numerical, form. **2.14**-15
Chat: Real-time typed conversation that takes place on a computer. 7.3, **7.29**
Chat client: Program on user's computer allowing connection to a chat server. **7.29**
Chat room: Communications medium, or channel, that permits users to chat with each other. 1.27, 2.2, **7.29**
Chat tools, 7.17
Checkout: Process of completing e-commerce transaction. **7.54**
Chip: Small piece of semiconducting material, usually no bigger than one-half-inch square, on which one or more integrated circuits are etched. 1.6, 1.19, **3.4,** 3.41-44
future of, 3.30
history of development, 1.50, 1.51
manufacture of, 3.8, 3.18, 3.31, 3.41-44

Chip for chip upgrade: Upgrade in which the existing processor chip is replaced with a new processor chip. **3.10**
Circuit board
chip connection, 3.4
RAM, 3.16
Cisco Systems, 7.52
Clark, James, 1.54, 2.44, 6.29
Cleaning floppy disk drive, 6.8
Clean rooms: Laboratories for chip manufacturing that are 1,000 times cleaner than a hospital operating room, with less than one dust particle per cubic foot of air. **3.42**
Click: Process of pressing the primary mouse button. 2.4, **4.8**
click2learn.com Digital Video Producer, 2.23
click2learn.com Flash, *see* **Flash**
click2learn.com ToolBook: One of the more widely used multimedia authoring software packages, developed by click2learn.com, that uses a graphical user interface and an object-oriented approach. 2.23
Click-and-mortar businesses: Businesses that have both a physical and an online presence. **7.48**
Click-through: Process that occurs when a visitor to a Web site clicks an ad to move to the advertiser's Web page. **7.59**
Client: Computer in a client/server network that can request services from the server. 1.22, **7.7**
Client/server network: Network in which one or more computers are designated as a server(s) and the other computers on the network, called clients, can request services from the server. 8.12, 8.17
Clip, video, 4.24
Clip art: A collection of drawings, diagrams, or photographs that can be inserted in documents. **2.10**
collection of, 2.31
presentation graphics, 2.18-19
in Web pages, 7.20
Clip art/image gallery: Collection of clip art and photographs; many clip art/image galleries also provide fonts, animations, sounds, video clips, and audio clips. 2.27, **2.31**
Clip gallery: Includes clip art images, pictures, video clips, and audio clips, typically organized by categories. **2.19**
Clipboard: Temporary storage location for text during cut, copy, and paste operations. **2.8**
Clock cycle: Each tick, or electronic pulse of the system clock. **3.8**
Clock rate, *see* **Clock speed**
Clock speed: Speed at which processor executes instructions. Also called clock rate. **3.8**
bus, 3.26
processor, 3.8, 3.9, 3.10
Cluster: Grouping of two to eight sectors on a disk, a cluster is the smallest unit of space used to store data. **6.5**
contiguous, 8.48
noncontiguous, 8.23
CMOS, *see* **Complementary metal-oxide semiconductor**
COBOL (COmmon Business-Oriented Language), 1.50
Coding scheme, for representing data, 3.12-13
Cold boot: Turning on a computer after it has been powered off completely. **8.17-**19
College Web sites, 2.35
Color
DTP and, 2.24, 2.25
ink-jet printers, 5.14
laser printers, 5.16
video cards and, 5.8
Color library: Included with desktop publishing (DTP) software, a standard set of colors used by designers and printers to ensure that colors will print exactly as specified. **2.25**
Color monitor: Monitor that displays text, graphics, and video information in color. **5.3**
Color scanner: Scanner that converts color images into a digitized format for use in multimedia applications. 4.14, 4.15
Column(s)
database, 2.15
word processing document, 2.11
worksheet, 2.12
Column charts: Charts that display bars of various lengths to show the relationship of data. Also called bar charts. **2.14**
com domain abbreviation, 7.10
Command: Instruction given to a computer program to cause the program to perform a specific action. **2.4, 4.2**
command-line interface, 8.3
executing, 3.6, 3.7
GUI, 8.4
keyboard, 4.3-5
memory-resident, 8.3
Command language: When working with a command-line interface, the set of commands used to interact with the computer. **8.3**
Command-line interface: User interface that requires the typing of keywords or pressing of special keys on the keyboard to enter data and instructions. **8.3,** 8.11, 8.16
Comments, tracking in word processing document, 2.11
Commercial off-the-shelf software, *see* **Packaged software**
Communications, 1.8-9
Communications devices: Hardware that enables computer users to communicate and exchange data, with another computer. **1.8**
Internet connections and, 7.7
Communications lines
Internet and, 7.4, 7.6, 7.8
interrupt request (IRQ), 8.9
Communications network, *see* Network(s)
Communications protocol, *see* **Protocol**
Communications software: Programs that manage the transmission of data, instructions, and information between computers. 2.2, 2.32-41
Compact disc (CD): Flat, round, portable, metal storage medium that stores items using microscopic pits (indentations) and land (flat areas) in the middle layer of the disc. Land causes light to reflect, which is read as binary digit 1, and pits absorb light, read as binary digit 0. 1.8, **6.16-**22
audio, *see* Audio CD
PhotoCD, 6.21
Compact disc read-only memory, *see* **CD-ROM**
Compact disc-recordable, *see* **CD-R (compact disc-recordable)**
Compact disc-rewritable, *see* **CD-RW (compact disc-rewritable)**
Compact flash card: PC Card used by digital cameras to store pictures, which are then transferred to a computer by inserting the card into the computer's PC Card slot. **6.24**
Compaq Computer Corporation, 1.20, 1.52, 1.57, 8.27
Comparison operations: Operations performed by the arithmetic/logic unit that compares one data item to another to determine if the first item is greater than, equal to, or less than the other item. **3.7**
Complementary metal-oxide semiconductor (CMOS): Battery-powered memory chip used to store configuration information about the computer, the current date and time, and other startup information needed when the computer is turned on. **3.19**
booting and, 8.19
Compression
file, 4.22, 4.24, 7.14, 7.26, 8.22, 8.24
video, 7.17
CompuServe, 1.52
Computer(s): Electronic machine, operating under the control of instructions stored in its memory, that can accept data, manipulate the data, produce results, and store the results for future use. **1.3**
categories of, 1.19-24
everyday life and, 1.2
examples of use, 1.25-35
fault-tolerant, 8.5
history of, 1.48-57
installing system, 8.45-47
introduction to using, 1.1-37
maintaining system, 8.47-48
network, 7.33
purchasing system, 8.38-42
speed of, 3.7, 3.8, 3.9, 3.10, 3.17
starting, 1.10, 8.17-20
tool set, 8.46
Web sites about, 2.57
Computer-aided design (CAD) software: Application software used to create engineering, architectural, and scientific designs. 2.2, 2.23, **2.24**
Computer literacy: A knowledge and understanding of computers and their uses. **1.3**
Computer output microfilm (COM) recorder: Device that records images onto microfilm or microfiche. **6.26**
Computer program: A series of instructions that tells the computer what to do. 1.4, **1.9-14**
Computer programmers, *see* **Programmers**
Computer Science Corporation, 1.50
Computer virus, *see* **Virus**
Computing Web sites, 2.57
Configuration
cache, 3.18
CMOS storing information about, 3.19
RAM, 3.16-17
system, 8.7-8, 8.18-19
Connector: Used to attach a cable to a device. 3.4, **3.22**
installing computer and, 8.47
Consumer-to-consumer (C-to-C or C2C) e-commerce: E-commerce activity that allows consumers to sell to and purchase directly from one another. **7.19,** 7.50
Content portal, 7.55, 7.56
Context-sensitive: Help information related to the current task being attempted. **2.42**
Context-sensitive menus: Menus that display a list of commands commonly used to complete a task related to the current activity or selected item. Also called shortcut menus. **2.6**
Continuous-form paper: Paper in which each sheet is connected together; used by most dot-matrix printers. **5.13**
Continuous speech: High-quality speech recognition software that allows a user to speak in a flowing conversational tone. **4.23**
Controller, hard disk, 6.14
Controller expansion card, 3.20
Control unit: Component of the CPU that directs and coordinates most of the computer's operations. **3.6-**7
Cookie: Small file stored by a Web server on a user's computer that contains data about the user, used to track information about Web site viewers. **7.31-**32, **7.59**
CoolTalk, 7.17
Coprocessor: Special processor chip or circuit board designed to assist the processor in performing specific tasks. **3.11**
Copy machine, multifunction device as, 5.23
Cordless mouse: Mouse that uses infrared or radio waves to communicate with a receiver. Also called wireless mouse. **4.10**
CorelCENTRAL, 2.7
Corel Corporation, Linux operating system and, 8.17
CorelDRAW, 2.23
Corel Gallery, 2.27
Corel Paradox, 2.7
Corel PHOTO-PAINT, 2.27
Corel Presentations, 2.7
Corel Quattro Pro, 2.7
Corel WEB SiteBuilder, 2.23
Corel WordPerfect, 2.7
Corel WordPerfect Suite, 2.7
Countries, domain name abbreviations, 7.10

Creating (document): Process of developing a document by entering text, inserting images, and performing other tasks using an input device. **2.8,** 2.10
Credit card
e-commerce use, 7.20, 7.57
purchasing computer using, 8.42
encrypted information, 7.20
Criteria: In a query operation, restrictions that data must meet while being retrieved. **2.17**
Cross-platform: Applications that run identically on more than one operating system. **8.3**
CRT monitor: Display device that consists of a screen housed in a plastic or metal case. **5.3**-4.
Currency (data type): In a database, fields containing dollar and cents amounts or numbers containing decimal values. **2.16**
Custom software: Application software developed by the user or at the user's request. **1.13**
Customer surveys, e-commerce and, 7.59
Cybersquatting: Practice of domain name speculation. **7.59**
Cylinder: On a hard disk, the location of a single track through all platters. **6.12**
Cyrix, 3.9

Daisy chained devices, 3.24, 6.14
Data: Words, numbers, images, and sounds that are not organized and have little meaning individually; processed into information. **1.3, 4.2**
charting spreadsheet, 2.14-15
electronic data interchange, *see* Electronic data interchange
entering database, 2.17
input, 4.1-27
machine cycle and, 3.6
output of, 5.2-25
representation, 3.11-13
source document, 4.13
storage of, 6.2-27
traveling on Internet, 7.7-8
worksheet, 2.12
Database: Data organized in a manner that allows access, retrieval, and use of that data. **2.15**
organization of, 2.16
Database management system (DBMS): Software used to create a computerized database; add, delete and manipulate data and create forms and reports. **2.15**
Database software: Software used to create a computerized database; add, delete and manipulate data and create forms and reports. 1.12, 2.2, **2.15**-17
packages, 2.7
Data bus, 3.24
Data collection device: Devices designed and used to obtain data directly at the location where the transaction or event takes place. **4.20**
Data communications, *see* Communications
Data projector: Output device that takes an image displayed on a computer screen and projects it onto a larger screen. **5.21**
Data transfer rate: Measure of the speed of a CD-ROM drive, which is the time it takes the drive to transmit data, instructions, and information to another device. **6.21**
Data type: In a database, the type of data that a field can contain (text, numeric, currency, date, a memo, and hyperlink or link data). **2.16**
Data warehouse: Data storage strategy that centralizes the computing environment, in which large megaservers store data, information, and programs, and less powerful client devices connect to the megaservers to access these items. **6.27**
Date (data type): In a database, fields containing month, day, and year information. **2.16**
DAT tape drive, 6.24
Daughterboard: Small circuit board that plugs into the motherboard. **3.10**
Daughterboard upgrade: Upgrade in which the new processor chip is located on a daughterboard. **3.10**
Decimal number system: Base 10 number system. **A.3**
Decoding: The process of translating the instruction into commands the computer understands. **3.6**
Defragmentation: Reorganizing the disk so files are stored in contiguous sectors. **8.23,** 8.48
Dell Computer Corporation, 1.20, 8.27
Density (disk): Storage capacity of a disk. **6.5**-6
Desktop: Onscreen work area that uses graphical elements such as icons, buttons, windows, menus, links, and dialog boxes. **2.4**
Desktop computer: Computer designed to fit entirely on or under a desk or table. **1.20**-22
all-in-one, 1.20
CD-ROM drive in, 6.18
DVD-ROM drive in, 6.18
hard disk in, 6.10
keyboards for, 4.3-6
large businesses, 1.32
monitors for, 5.4, 5.5
system unit in, 3.2
Desktop publishing, 1.35
monitor size and, 5.4
Desktop publishing software, large business use, 1.32
Desktop publishing (DTP) software (personal): Software for home and small business users designed for small-scale desktop publishing projects. 2.2, 2.27, **2.30**
Desktop publishing (DTP) software (professional): Software for professional designers and graphic artists to design and produce sophisticated documents containing text, graphics, and brilliant colors. 2.2, 2.23, **2.24**-25
Desktop video (DTV), *see* **Videoconferencing**
Device(s), *see* Hardware
Device dependent: Operating systems developed by manufacturers specifically for the computers in their product line. **8.10**
Device driver: Used by the operating system, a small program that accepts commands from another program, then converts the commands into commands that the device understands. Also called driver. **8.7**-8, 8.19
Device-independent: Operating systems that will run on many computers. **8.10**
Diagnostic utility: Utility program that compiles technical information about a computer's hardware and certain system software, then prepares a report outlining any identified problems. **8.22,** 8.48
Dialog box: Special window displayed by a program to provide information, present available options, or request a response. **2.6**
Dial-up access: Use of a computer and a modem to dial into an ISP or online service over a regular telephone line. **7.7**
Dicing: During chip manufacturing, using a diamond saw to cut wafers into chips, called die. **3.44**
Dictionaries
reference software, 2.32
spell checker, 2.10
Die: Individual chips that wafers are cut into, during chip manufacturing. **3.44**
Digital: Electronic devices that understand only two discrete states: on and off; describes most computers. **3.11**-12
Digital camera: Camera used to take pictures and store the photographed images digitally instead of on traditional film. **4.20**-21
flash memory in, 3.18
PC Cards for, 6.24
printing photographs from, 5.20
Digital cash (e-cash), *see* **Electronic money**
Digital certificate: Electronic credential that ensures a user is legitimate and the transfer of confidential materials is secure. **7.20**
Digital Equipment Corporation (DEC), 3.9
history of, 1.50, 1.57
Digital light processing (DLP) projector: Data projector that uses tiny mirrors to reflect light, producing crisp, bright, colorful images that remain in focus and can be seen clearly even in a well-lit room. **5.21**
Digital sound, 5.2
Digital subscriber line (DSL), 7.7
Digital television signals, 5.11
Digital video disc-ROM, *see* **DVD-ROM (digital video disc-ROM)**
Digital video signals, 5.3
Digital watches, LCD displays used in, 5.4
Digitizer, *see* **Graphics tablet**
Digitizing tablet, *see* **Graphics tablet**
Dijsktra, Edsger, 1.50
Direct access: Access method in which a particular data item or file can be located immediately, without having to move consecutively through items stored in front of the desired data item or file. **6.23**
Discrete speech: The process of speaking slowly and separating each word with a short pause. **4.23**
Discussion board, *see* **Message board**
Disintermediation: Process of allowing producers to communicate directly with both suppliers and consumers, eliminating the need for intermediaries. The Internet allows such a process to occur. **7.50**
Disk: Storage medium that is a round, flat piece of plastic or metal with a magnetic coating on which items can be written. 1.7, **6.2**
boot, 8.20
floppy, *see* **Floppy disk**
hard, *see* **Hard disk**
Disk cache: Portion of memory that the CPU uses to store frequently accessed items. **6.13**
Disk cartridge, *see* **Removable hard disk**
Disk Cleanup: Windows disk scanner utility that searches for and removes unnecessary files such as temporary files. **8.23**
Disk controller: Special-purpose chip and its associated electronic circuits that manages (controls) the transfer of items from the disk to the rest of the computer. **6.14**
Disk defragmenter: Utility program that reorganizes the files and unused space on a computer's hard disk so data can be accessed more quickly and programs can run faster. **8.23,** 8.48
Disk Defragmenter: Windows disk defragmenter. **8.23**
Disk drive, 1.7
bays for, 3.27
Disk pack: Collection of removable hard disks mounted in the same cabinet as the computer or enclosed in a large stand-alone cabinet. **6.14**
Disk scanner: Utility program that detects and corrects both physical and logical problems on a hard disk or floppy disk, and searches for and removes unwanted files. **8.23**
Diskette, *see* **Floppy disk**
Display devices: Output devices that visually convey text, graphics, and video information. **5.3**-11
ergonomics and, 5.10-11
flat-panel, 5.4-6
high-definition television, 5.11
monitors, 5.3-7
video cards and, 5.8-10
Distributional channels: The paths of goods from manufacturer to consumer. **7.50**
DLT tape drive, 6.24

Docking station: A platform into which a laptop is placed, containing connections to peripherals such as a keyboard, monitor, printer, and other devices. **1.31,** 3.29
Document
DTP, 2.24-25
including audio file in, 4.22
image processing of, 4.16
scanned, 4.14-16
source, 4.13
turnaround, 4.16-17
Web publishing, 7.20
word processing, 2.8-10
Domain
abbreviations, 7.10
top-level, 7.9-10
Domain name: Text version of an IP address. Also called Web address. **7.8, 7.59**
e-mail and, 2.33-34
speculation in (cybersquatting), 7.59
URL, 7.11
Domain name system (DNS): System that registers domain names. **7.10**
Domain name system servers (DNS servers): Internet computer that that stores domain names. **7.10**
Dopants: The materials added to treat, or dope silicon with impurities, altering the way that chips conduct electricity. **3.44**
Dope: Treating silicon with impurities to alter the way chips conduct electricity. **3.44**
DOS (Disk Operating System): Single user operating systems developed in the early 1980s for PCs. **8.11,** 8.12, 8.27
Dot-matrix printer: Impact printer that produces printed images when tiny wire pins on a print head mechanism strike an inked ribbon. **5.12**-13
Dot pitch: Measure of image clarity, which is the vertical distance between each pixel on a monitor. **5.7**
Dots per inch, digital camera and, 4.21
Dots per inch (dpi), printer: Measure of printer resolution. **5.14**
Dots per inch (dpi), scanner: Measurement of scanner resolution. **4.15**
Double-data rate (DDR) SDRAM (SDRAM II): Type of DRAM that is faster than SDRAM. **3.16**
Download (digital camera): Process of transferring a copy of stored pictures from a digital camera to a computer by connecting a cable between the two and using special software included with the camera. **4.20**
Downloading: Using a Web browser, the process of receiving information onto a computer from a server on the Internet. **2.37**
Downloading files, 7.3
compressed, 7.14, 7.26, 8.22
FTP and, 7.25
sound, 7.16
video, 7.17
Downward compatible (floppy disk drive): The ability of a floppy disk drive to recognize and use earlier media. **6.7**
Downward-compatible (operating system): Operating system that recognizes and works with application software that was written for an earlier version of the operating system. **8.10**
Drafting, printers used in, 5.19
DRAM, *see* **Dynamic RAM**
Drawing
graphics tablet used for, 4.13
using paint/image editing software, 2.25
Drawings
digital representation of, 5.2
DTP and, 2.25
printing, 5.19
Drive bay: Open area inside the system unit used to install an additional device such as a disk drive. **3.27**
Driver, *see* **Device driver**
Drum scanners, 4.15
.drv extension, 8.8
Dual inline memory module (DIMM): Type of circuit board containing SDRAM chips and two sets of contacts; inserted into the motherboard. 3.4, **3.16**
Dumb terminal: Terminal with no processing power, that cannot function as an independent device. **1.24, 5.23**
DVD (Digital Video Disc)
history of, 1.56
rewritable, 6.23
DVD-R (DVD-recordable): Compact disc that can be written to once, and read (played) many times. **6.23**
DVD-R drives, 6.23
DVD-RAM: Rewritable version of a DVD that can be erased and recorded on multiple times. **6.23**
DVD-ROM (digital video disc-ROM): Compact disc with an extremely high storage capacity, able to store from 4.7 GB to 17 GB. 1.8, **6.22**-23
DVD-ROM drive: Drive that uses laser technology to read items on a DVD-ROM or CD-ROM, and transmit the data, instructions, and information to another device. 1.8, 6.18, **6.22**
bays for, 3.27
purchasing laptop with, 8.43
Dvorak keyboard: Keyboard that places the most frequently typed letters in the middle of the typing area. **4.6**
Dye-sublimation printer: Nonimpact printer that uses heat to transfer colored dye to specially coated paper. **5.18**
Dynamic RAM (DRAM): Type of RAM that must be re-energized constantly or it loses its contents. **3.16**
access times, 3.19
video cards using, 5.9

Earthlink, 7.52
EBCDIC, *see* **Extended Binary Coded Decimal Interchange Code**
E-cash, *see* **Electronic money**
Eckert, J. Presper, 1.48
E-commerce (electronic commerce): The conducting of business activities online, including shopping, banking, investing, and any other venture that represents a business transaction or uses electronic data interchange; also called Internet commerce, or iCommerce. Four types of e-commerce are: business-to-consumer, consumer-to-consumer, business-to-business, and within-business. 1.56, **7.19**-20, 7.47-64
creating online store, 7.55-59
growth of, 7.47, 7.48-49
interactions, 7.50-51
learning more about, 7.64
market sectors, 7.52-53
processes in, 7.54-59
ECP (Extended Capabilities Port), 3.23
Editing
audio, 2.26
images, 2.25, 2.30, 5.2
sound, 4.22
video, 2.26
Editing (document): Process of making changes to a document's existing content. **2.8,** 2.10
Education
e-commerce and, 7.64
Web sites about, 2.57
Educational information, accessing using Internet, 7.2
Educational institutions
connecting to Internet through, 7.7
Internet linking, 7.2
Educational software: Software designed to teach a particular skill. 1.26, 2.2, **2.32**
Educational use, software for, 2.27-32
edu domain abbreviation, 7.10
EDVAC computer, 1.48
E-finance: Business conducted by financial institutions on the Internet. **7.52**
Electromagnetic radiation (EMR): Magnetic field that travels at the speed of light, produced by monitors because they use electricity. **5.10**
Electronic commerce, *see* **E-commerce**
Electronic credit: Variation on e-money that is a credit-card payment scheme for use on the Web. Also called electronic wallet. **7.20,** 7.57
Electronic data interchange (EDI): Transmission of business documents or data over communication lines. **7.19,** 7.48
Electronic magazine (e-zine), 7.64
Electronic money (e-money): Means of paying for goods and services over the Internet using a digital certificate. Also called digital cash, or e-cash. **6.26, 7.20,** 7.57
Electronic storefront: Online business selling products or services with site that contains descriptions, graphics, and a shopping cart. When ready to complete a sale, the customer enters personal and credit card data through a secure Web connection. **7.19, 7.54**
Electronic wallet, *see* **Electronic credit**
Electrostatic plotter: Plotter that produces high-quality drawings, using a row of charged wires (called styli) to draw an electrostatic pattern on specially coated paper and then fuse toner to the pattern. **5.19**
E-mail (electronic mail): Transmission or exchange of messages and files via a computer network or communications network. 1.16, **2.33**-34, **7.23**-25
attaching audio file to, 4.22
attachments, 4.22, 7.24, 8.22
e-commerce and, 7.48, 7.59
home users, 1.27
Internet used for, 7.3, 7.23-25
large business users, 1.33
mailing lists, 7.28-29
mobile users and, 2.34
rules, 2.34
spam, 7.30, 7.59
Web communities and, 7.29
E-mail address: Combination of a user name and a domain name, identifying a user so he or she can receive Internet e-mail. **2.33**-34, **7.24**
E-mail program/software: Program used to create, send, receive, forward, store, print, and delete e-mail messages. 2.2, **2.33, 7.24**-25
Emergency disk, *see* **Rescue disk**
E-money, *see* **Electronic money**
Emoticons: Combination of symbols and letters used in e-mail to express emotion, such as :) representing a smile, and :(representing a frown. **2.34, 7.30**
Employment Web sites, 2.57
Encoding schemes: Included with most e-mail software, the conversion of Internet e-mail attachments to binary format so the attachments can arrive at the recipient's computer in the same format in which they left the sender's computer. **7.25**
Encryption: Process of converting readable data into unreadable characters to prevent unauthorized access. 7.20, 7.33, **7.57**
Encyclopedias, reference software and, 2.32
ENERGY STAR program: Program developed by the U.S. Department of Energy and the U.S. Environmental Protection Agency that encourages manufacturers to create energy-efficient devices that require little power when they are not in use. **5.11**
Engineering applications, 1.35
monitor size and, 5.4
printers used in, 5.19
Enhanced keyboards: Describes most desktop computer keyboards, meaning they have twelve function keys along the top, two CTRL keys, two ALT keys, and a set of arrow and additional keys between the typing area and the numeric keypad. **4.6**
Enhanced resolution, 4.15
ENIAC computer, 1.48
Enterprise ASP, 2.43

Gas plasma monitors: Flat-panel monitors that use gas plasma technology, providing a layer of gas deposited between two sheets of material. When voltage is applied, the gas glows and produces the pixels that form an image. **5.6**
Gates, Bill, 1.36, 1.37, 1.52
Gateway, 1.20, 8.27
Gender changer: Device used to join two connectors that are either both female or both male. **3.22**
Gerstner, Louis, 4.28, 4.29
Ghosting: Problem with older monitors in which images could be permanently etched onto a monitor's screen. **8.25**
GIF (Graphics Interchange Format): File that is a graphical image saved using compression techniques to reduce the file size for faster downloading from the Web. **7.14**
animated, 7.14
GoTo.com, 2.41
gov domain abbreviation, 7.10
Government, Internet structure and, 7.5
Government offices, Internet linking, 7.2
Government Web sites, 2.58
e-commerce, 7.64
Grammar checker, 2.11
Graphical data, graphics tablet used for, 4.13
Graphical user interface (GUI): User interface that combines text, graphics, and other visual cues to make software easier to use. **1.11, 2.3, 4.3, 8.4,** 8.12
history of, 1.53
pointer and, 4.7
Graphic arts applications, printers used in, 5.19
Graphic design/multimedia software, 2.2
Graphics: Digital representations of nontext information such as drawings, charts, and photographs that contain no movement or animation. **5.2, 7.12**-14
animated, 5.2, 7.14
clip galleries containing, 2.31
downloading Web page and, 2.37
editing, 5.2
e-mail and, 7.24
files, 7.14
monitor resolution and, 5.7
word processing software and, 2.10
Web page, 7.12-14
Graphics card, *see* **Video card**
Graphics Interchange Format, *see* **GIF**
Graphics software, 2.23-25
Graphics tablet: Flat, rectangular, electronic plastic board used to input drawings, sketches, or other graphical data. **4.13**
Gray scaling: On a monochrome monitor, the use of many shades of gray from white to black to form images. **5.3**
Groupware: Software that helps groups of people on a network collaborate on projects and share information, in addition to providing PIM functions. 2.2, **2.32**-33
Grove, Andy, 3.30, 3.31
GUI, *see* **Graphical user interface**
Handheld computer: Portable computer designed to fit in user's hand. **1.23,** 2.21
history of, 1.55
home use, 1.27
keyboards for, 4.6
LCD displays for, 5.4, 5.5
pen input, 4.13
small business users, 1.28
Web-enabled, 1.22, 1.31, 7.34-35
Windows CE for, 8.14
Handwriting, optical character recognition and, 4.16
Handwriting recognition software: Software that translates the letters and symbols used in handwriting into character data the computer can use; used with pen computers. **4.13**
Hard copy: Name given to printed information because the information exists physically and is a more permanent form of output than that presented on a display device. **5.11**
Hard disk: Primary storage media for software programs and files consisting of several inflexible, circular disks, called platters, on which items are stored electronically. 1.7, **6.10-16**
access times, 3.19-20
capacity, 6.3
characteristics of, 6.11
defragmentation of, 8.23, 8.48
formatting, 6.11
as memory, 3.14
operating system on, 8.3
operation of, 6.12-13
RAID, 6.15
removable, 1.7-8, 6.14
software on, 1.10
virtual memory, 8.6
Hard disk controller (HDC): Special-purpose chip and its associated electronic circuits that manages (controls) the transfer of items from the hard disk to the rest of the computer. **6.14**
Hard disk drive
bays for, 3.27
history of, 1.52
Hardware: The electric, electronic, and mechanical equipment that makes up a computer. **1.4,** 1.5-8
booting, 8.18-19
communications devices, 1.8
components, 1.5-8
configuring, 8.7-8, 8.19
e-commerce and, 7.56
input devices, 1.5-6
installing, 8.8, 8.45-47
Internet infrastructure and, 7.52
maintaining, 8.47-48
manufacturers, 1.20
operating system and, 8.2-10
output devices, 1.6
purchasing, 8.38-44
registration cards, 8.46
registry, 8.13, 8.19
serial numbers, 8.46
storage devices, 1.7-8
system software and, 8.2
system unit, 1.6-7
Windows support, 8.13
Head crash: On a hard disk, the process of the read/write head touching the surface of a platter, usually resulting in a loss of data or sometimes loss of the entire drive; caused by contamination of the hard disk. **6.13**
Header: Text at the top of each page. **2.11**
Head-mounted pointer: Pointer mounted on a user's head used to control the pointer or insertion point on a screen-displayed keyboard, which is a graphical image of the standard keyboard. **4.26**
Headset: Small speakers placed over the ears and plugged into a port on the sound card, so only the user can hear the sound from the computer. **5.21**
Health
installing computer system and, 8.45, 8.46
keyboards and, 4.6
monitors and, 5.10
Web sites about, 2.58
Health guides, 2.32
Hearing-impaired users, output devices for, 5.24
Heat pipe: Small device used to cool laptop computers. **3.11**
Heat sink: Small ceramic or metal component with fins on its surface designed to absorb and ventilate heat produced by electrical components, such as the processor. **3.10**-11
Help, Online, 2.42-43
Helper application: Program that runs multimedia elements in a window separate from a browser; extends the multimedia capabilities of the browser. **7.12,** 7.13
Hertz (Hz): Measure of monitor refresh rate, the number of times per second the screen is redrawn. **5.7**
Hewlett, William, 5.26-27
Hewlett Packard Company
history of, 1.53, 5.26-27
PCL (Printer Control Language) and, 5.16
Hexadecimal: Base 16 number system used by computer to communicate with a program when a problem with a program exists. **A.3,** A.4
Hidden drive bay, *see* **Internal drive bay**
HiFD (High FD): Floppy disk drive technology that can read from and write on standard 3.5-inch floppy disks as well as its own high-capacity disks. **6.10**
High-definition television (HDTV): Type of television that works with digital broadcasting signals and supports a wider screen and higher resolution display than a standard television. HDTV is capable of receiving text, graphics, audio, and video, and can be used as a monitor while browsing the Internet. **5.11**
High-density floppy disk, 6.6
High FD, *see* **HiFD (High FD)**
High-level format: Formatting step for a hard disk, during which the operating system issues a command to define the file allocation table (FAT) for each partition. **6.11**
History Web sites, 2.59
Hit: Any Web page listed as the result of a search. **2.40**
Hoff, Ted, 1.51
Home, working from, 1.56
Home design/landscaping software: Software used by homeowners or potential homeowners to assist with the design or remodeling of a home, deck, or landscape. 2.2, 2.27, **2.31**
Home page: Starting page for a Web site that provides information about the site's purpose and content, often able to be customized to display areas of interest to the user. **2.37, 7.10**
Home use, software for, 2.27-32
Home users: Users having a desktop computer at home that is used for different purposes by various family members. 1.25, **1.26**-27, 1.55
input devices for, 4.27
Internet access and, 2.35
legal software, 2.29
output devices for, 5.25
processor speed, 3.9
RAM needs, 3.17, 3.29
storage devices and, 6.27
Hopper, Grace, 1.49, 1.50
Host, *see* **Host node**
Host computer: Computer that a dumb terminal is connected to; information is sent to the host computer which performs the processing, then sends the output back to the dumb terminal. **5.23**
Host node: Any computer directly connected to a network. **7.4**
Hot plugging: The ability to add and remove devices while a computer is running. **3.21**
Hot swapping, *see* **Hot plugging**
.htm: File extension for HTML document. **7.20**
HTML, *see* **Hypertext markup language**
.html: File extension for HTML document. **7.20**
http:// (hypertext transfer protocol): Start of most Web page URLs, the communications standard used to transfer pages on the Web. **2.40, 7.11**
https://, 7.57
Humidity, controlling, 8.47
Humor Web sites, 2.59
Hyperlink: Contained in most Web pages, a built-in connection to another related Web page. Also called link. 1.18, **2.37**-38, **7.10**
Hyperlink (data type): In a database, fields containing a Web address that links to a Web page; also called link. **2.16**
Hypertext markup language (HTML): Language used to create and format Web pages, using a set of special codes, called tags. 2.44, **7.20-23**
Hypertext transfer protocol (http://): Start of most Web page URLs, the communications standard used to transfer pages on the Web. **2.40, 7.11**

IBM
DOS and, 8.27
floppy disk introduction by, 6.4
history of, 1.49, 1.50-53, 4.28-29
Intel chips used by, 3.31
OS/2 and OS/2 Warp, 8.14-16
PC-DOS and, 8.11
PC introduction, 8.27
processor speed and, 3.8
IBM-compatible PCs, 1.52, 4.29, 8.27
IBM Personal Computer, 4.29
IBM ThinkPad, 4.29
iCommerce, *see* **E-commerce**
Icon: Small image that displays on the screen to represent a program, an instruction, or some other object. **1.11, 2.4,** 4.3, **8.4**
Ideograms: Symbols used by Asian and other foreign languages to represent multiple words and ideas. **A.1**
Illustration software: Software used to draw pictures, shapes, and other graphical images using various tools on the screen such as a pen, brush, eyedropper, and paint bucket. **2.25**
iMac, 1.57
Image(s)
digital camera, 4.20-21
scanning, 4.14-16
3-D, 7.18
Image editing software: Software used to alter graphics by including enhancements such as blended colors, animation, and other special effects. 2.2, 2.23, **2.25, 5.2**
personal, 2.27, 2.30
Image gallery, *see* **Clip art/image gallery**
Image processing: Use of scanners for capturing, storing, analyzing, displaying, printing, and manipulating images (bitmaps). **4.16**
Image processing system: Electronic filing cabinet that provides access to exact reproductions of original documents. **4.16**
Imaging, *see* **Image processing**
Imaging Preview: Windows file viewer for graphics files. **8.21**
Imation, 6.10
Impact printer: Printer that forms characters and graphics on a piece of paper by striking a mechanism against an ink ribbon that physically contacts the paper. **5.12**-13
Import: Process of bringing clip art into a word processing document. **2.10**
data into database, 2.17
DTP document and, 2.25
Individuals, Internet linking, 7.2
Industry Standard Architecture bus, *see* **ISA (Industry Standard Architecture) bus**
info domain abbreviation, 7.10
Information: Data that is organized, has meaning, and is useful. **1.4,** 4.2
input of, 4.1-27
output of, 5.1-25
processing of, 3.1-20
storage of, 6.2-27
Informational resources: Web sites that are used for product support and other informational activities. **7.49**
Information intermediaries, *see* **Business-to-business infomediaries**
Information processing cycle: Four basic operations of a computer: input, process, output, and storage. **1.4**
input and, 4.1-27
output and, 5.1-25
processing and, 3.1-20
storage and, 6.1-27
InfoSeek, 7.30
Infrared (IR)
IrDA port and, 3.24
mouse using, 4.10
wireless keyboards using, 4.6
Infrared Data Association, *see* **IrDA**
Ink-jet printer: Nonimpact printer that forms characters and graphics by spraying tiny drops of liquid ink on a piece of paper. **5.14**-15
portable, 5.18
Input: Any data or instructions entered into the memory of a computer. **1.4, 4.2**-27
accuracy of, 1.8
audio, 4.22-23
digital cameras, 4.20-21
GUI and, 8.4
joystick, 4.12
mouse, 4.7-10
pen, 1.23, 4.12-13
physically challenged users, 4.26
pointing devices, 4.7-11
pointing stick, 4.11
processed into output, 5.2
reading devices, 4.13-20
scanners, 4.13-18
storage devices used for, 1.7, 6.3
terminal and, 5.23
touchpad, 4.11
touch screen, 4.12
trackball, 4.10
types of, 4.2-3
video, 4.24-25
Input device: Any hardware component that allows user to enter data, programs, commands, and user responses into a computer. **1.5**-6, **4.3**-27
operating system and, 2.3, 8.2
type of user and, 4.27
Insertion point: Symbol that indicates where on the screen the next character that is typed will display. The symbol may be a vertical bar, rectangle, or an underline. **4.5**
Install (the program): Process of running a small setup program that comes with software. **1.9**
Installation, of computer system, 8.8, 8.45-47
Instructions
input of, 4.1-27
output of, 5.1-25
processing of, 3.1-20
storage of, 6.2-27
Instant messaging (IM): Service that notifies user when one or more people are online and then allows user to exchange messages or files with them or join a private chat room with them. 7.3, 7.24, **7.29,** 7.30
Instruction cycle, *see* **Machine cycle**
Instruction time (I-time): The time it takes to fetch an instruction; part of the machine cycle. **3.6**
Insurance policy, covering computers, 8.47
Integrated circuit (IC): Microscopic pathway capable of carrying electrical current; can contain millions of elements such as transistors, diodes, capacitors, and resistors, which act as electronic switches, or gates, that open or close the circuit for electronic signals. **3.4, 3.41**
Integrated CPU: New type of microprocessor that combines functions of a CPU, memory, and a graphics card on a single chip. These chips are designed for less-expensive personal computers and smaller-sized computers such as a set-top box. **3.9**
Integrated Drive Electronics (IDE): One of the most widely used hard disk controllers; support up to four hard disks and can transfer data, instructions, and information to the disk at rates up to 66 MB per second. **6.14**
Integrated Services Digital Network (ISDN): Set of standards for digital transmission of data over analog telephone lines; an ISDN line is a dedicated telephone line that provides faster transmission rates than regular telephone lines. 7.7
Integrated software: Software that combines applications such as word processing, spreadsheet, and database into a single, easy-to-use package. 2.2, 2.27, **2.28**
Intel Corporation
history of, 1.51, 1.53, 1.54, 1.56, 1.57, 3.30-31
IBM and, 4.29, 8.27
processors, 3.9
Intel-compatible microprocessors: Microprocessors manufactured by companies such as Cyrix and AMD that have the same internal design or architecture as Intel processors, and perform the same functions, but often are less expensive. **3.9**
Intelligent agent: Software that independently asks questions, pays attention to work patterns, and carries out tasks on behalf of a user. In e-commerce, intelligent agents (sometimes called bots) analyze customer preferences and tailor the storefront to match. **7.59**
Intelligent smart card: Smart card that contains a CPU and has input, process, output, and storage capabilities. **6.25**
Intelligent terminal: Terminal that has not only a monitor and keyboard, but memory and a processor that has the capability of performing some functions independent of the host computer. **5.23**
Interactive games, 2.32
Interactive learning, 2.26
Interface, disk controller as, 6.14
Interface card, *see* **Expansion card**
Interlacing: Technique used by some older monitors to refresh images using an electron beam to draw only half the horizontal lines with each pass. **5.7**
Internal cache, *see* **Level 1 (L1) cache**
Internal drive bay: Drive bay concealed entirely within the system unit. **3.27**
Internal modem: Expansion card that is inserted into an expansion slot on a computer's motherboard. The modem then attaches to a telephone outlet with a standard telephone cord. **3.20**
Internet: The world's largest network; a worldwide collection of networks that links together millions of computers by means of modems, telephone lines, wireless technology, and other communications devices and media; provides connections to businesses, the government, educational institutions, and individuals. Also called the Net. 1.9, **1.16, 7.2**-33
access, 2.35, 2.36, 7.7, 7.52, 8.41
addresses, 7.8-10
advisory groups and organizations, 7.5
connecting to, 1.17
data traveling on, 7.7-8
e-mail, 2.33, 2.34
future of, 1.58
HDTV used while browsing, 5.11
history of, 1.51, 1.53, 1.56, 7.4-5
infrastructure, 7.52
market sectors and, 7.52-53
number of users, 1.58, 7.2-3
operation of, 7.6-10
power users, 1.35
protocol, 7.7
retailing on, *see* E-commerce
security and privacy issues, 7.32, 7.52, 7.56
television used to connect to, 1.22, 1.55
videoconferencing using, 4.25
Web sites about, 2.59
wireless access, 7.7
World Wide Web and, 1.18
Internet2 (I2): The result of collaboration among more than 170 universities in the United States, along with several industry and government partners, I2 is an extremely high-speed network whose purpose is to develop and test advanced Internet technologies for research, teaching, and learning. **7.5**
Internet appliance, *see* **Web appliance**
Internet commerce, *see* **E-commerce**
Internet Explorer, 1.56, 1.57
Internet files, searching for, 2.40
Internet mall, 7.55, 7.56
Internet relay chat (IRC), 7.29
Internet service provider (ISP): Organization that has a permanent connection to the Internet and provides temporary connections to individuals and companies for a fee. 1.17, **2.36, 7.6**

e-mail and, 7.24
local, 7.6
national, 7.6, 7.52
purchasing computer and, 8.41
server, 2.33
Web site storage, 7.11, 7.23, 7.55, 7.56
wireless communication and, 7.34-35
Internet telephone service: Use of the Internet (instead of a public telephone network) to connect a calling party and one or more called parties, allowing user to talk to people for just the cost of the Internet connection. **7.16**
Internet telephone software: Software used in conjunction with a sound card to digitize and compress Internet telephone conversation and then transmit the digitized audio over the Internet to called parties. **7.16**
Internet telephony, *see* **Internet telephone service**
Interpolated resolution, 4.15
Interrupt request (IRQ): Communications line between a device and the CPU. **8.9**
Intrabusiness e-commerce: E-commerce exchanges within an organization. **7.51**
Intuit QuickBooks, 2.7, 2.22
Intuit Quicken, 2.27
Intuit TurboTax, 2.27, 2.42
Intuit Webware, 2.7
In-vehicle devices, Windows CE for, 8.14
Inventory, POS terminal and, 5.24
Investing, online, 7.3
I/O expansion card, 3.20
Iomega Corporation, 6.10, 6.14
Ion implementation: Process used to add dopants, which are materials used to treat silicon with impurities, in order to create areas on a chip that will conduct electricity. **3.44**
IP (Internet protocol) address: Numeric address for each computer location on the Internet, consisting of four groups of numbers, each separated by a period. **7.8**
IrDA (Infrared Data Association): Organization that developed standards for the IrDA port. **3.24**
IrDA port: Port that transmits data from wireless devices using infrared light waves, instead of cables. **3.24,** 4.10
ISA (Industry Standard Architecture) bus: Most common and slowest expansion bus, used to connect a mouse, modem card, sound card, or low-speed network card directly to the ISA bus or through an ISA bus expansion slot. **3.26**
ISP, *see* **Internet service provider**
Itanium™ processor: Intel processor used in workstations, and servers on a network. 3.5, **3.9,** 3.10, 3.25
I-time (instruction time): The time it takes to fetch an instruction; part of the machine cycle. **3.6**
Java: Compiled object-oriented scripting language used to write stand-alone applications, as well as applets and servlets. 1.55, 6.28, **7.12**
Jaz® disk: Popular, reasonably priced, removable hard disk from Iomega that can store up to 2 GB. **6.14**
Jewel box: Protective case for compact disc. **6.18**
Jobs, Steve, 1.51, 8.26-27
Joint Photographic Experts Group, *see* **JPEG**
Joy, Bill, 6.28
Joystick: Pointing device that is a vertical lever mounted on a base; used for game software such as driving or flight simulator. **4.12**
JPEG (Joint Photographic Experts Group): File that is a graphical image saved using compression techniques to reduce the file size for faster downloading from the Web. **7.14**
Jumping, using links, 2.38, 2.44
Juno Online, 7.52

K, *see* **Kilobyte**
KB, *see* **Kilobyte**
Kemeny, John, 1.50
Kernel: Core of the operating system; responsible for managing memory, files, and devices; maintaining the computer's clock; starting applications; and assigning the computer's resources. **8.3**
boot process and, 8.18, 8.19
Keyboard: Primary input device on a computer. 1.5, **4.3**-6
keys found on, 4.4
laptop computers, 3.2, 4.3, 4.6, 8.43
physically challenged users and, 4.26
port, 3.21
terminal and, 5.23
types, 4.6
Keyguard: Device placed over the keyboard, allowing the hand to rest on the keyboard without accidentally pressing any keys; also guides a finger or pointing device so only one key is pressed at a time. Used by persons with limited hand mobility. **4.26**
Keyword: Specific word, phrase, or code that a program understands as an instruction. **4.2**
command-line interface and, 8.3
Keywords (search): Word or phrase entered in a search engine text box to find a Web page or pages. **2.40**-41
Khosla, Vinod, 6.28
Kilobyte (KB, or K): Equal to 1,024 bytes. **3.14**-15
Kiosk: Computerized information or reference center that allows users to select various options to browse through or find specific information; used to provide information to the public. **1.33,** 4.12
Kiplinger's Home Legal Advisor, 2.27
Kiplinger TaxCut, 2.27
Label: Text entered in a cell that identifies data and helps organize a worksheet. **2.12**
Label printer: Small printer that prints on an adhesive-type material to make labels. **5.20**
LAN, *see* **Local area network**
LAN adapter, *see* Network interface card (NIC)
Landscaping software: Software used by homeowners or potential homeowners to assist with the design or remodeling of a landscape. 2.2, 2.27, **2.31**
Landscape orientation: Printed page that is wider than it is tall, with information printed across the widest part of the paper. **5.11,** 5.13
Landscaping software, *see* Home design/landscaping software
Laptop computer: Portable computer designed for mobility and small enough to fit on a lap, that can be just as powerful as a desktop computer, but is generally more expensive; also called notebook computer. **1.23**
batteries, 8.44
keyboard, 3.2, 4.3, 4.6, 8.43
large businesses, 1.32
LCD displays for, 5.4, 5.5
mobile users, 1.31
PC Card, 3.21, 6.24
pointing device, 4.10, 4.11, 4.29
purchasing, 8.43-44
system unit, 3.2, 3.28-29
touchpad, 4.11
Large business users: Businesses having hundreds or thousands of employees, in offices across a region, the country, or the world. The company may have an equally large number of computers, connected in a local area network or a wide area network, enabling communications between employees at all locations. Most large businesses have their own Web sites to showcase products, services, and selected company information, allowing customers to access information without having to speak to a company employee. Some large businesses also use kiosks to provide information to the public. 1.25, **1.32**-33
input devices for, 4.27
output devices for, 5.25
storage needs of, 6.3, 6.27
system needs, 3.29
Large-format printer: Nonimpact printer that creates photo-realistic quality color prints; operates like an ink-jet printer, but on a much larger scale. **5.19**
LaserJet printer, 5.27
Laser printer: High-speed, high-quality non-impact printer that prints using a laser beam and powdered ink, called toner. The laser beam produces an image on a drum inside the printer by creating an electrical charge on the drum wherever it hits, causing toner to stick to the drum. The image is then transferred to the paper through a combination of pressure and heat. **5.16**-17, 5.27
history of, 1.53
LCD (liquid crystal display): Method for LCD display that uses special molecules (called liquid crystals) deposited between two sheets of material. When an electric current passes through them, the molecules twist, causing some light waves to be blocked and allowing others to pass through, which then creates the desired images on the screen. 5.4-**5**
LCD display: Flat-panel display that creates images using a liquid crystal display. A liquid crystal display (LCD) has special molecules (called liquid crystals) deposited between two sheets of material. When an electric current passes through them, the molecules twist, causing some light waves to be blocked and allowing others to pass through, which then creates the desired images on the screen. LCD displays are commonly used in laptop computers, handheld computers, digital watches, and calculators. **5.5**
LCD projector: Data projector that uses liquid crystal display technology, attaches directly to a computer, and uses its own light source to display the information shown on the computer screen. **5.21**
Learning
educational software for, 2.32
hyperlinks and, 2.37-38
interactive, 2.26
linear, 2.37
multimedia applications used for, 2.26
multimedia presentations and, 2.26
nonlinear, 2.37-38
online, 2.35
software, 2.42-43
Legal software: Software that assists in the preparation of legal documents and provides legal advice to individuals, families, and small businesses. 2.2, 2.27, **2.29**
Letter quality (LQ): Print output acceptable for business letters; generally produced by nonimpact printers. **5.12,** 5.14-19
Level 1: The simplest RAID storage design called mirroring, which has one backup disk for each disk, offering data duplication if a drive should fail. **6.15**
Level 1 (L1) cache: Cache built directly into the processor chip. **3.17**
Level 2 (L2) cache: Cache that is not part of processor chip. **3.17**
Light pen: Handheld input device that contains a light source or can detect light. **4.12**
Linear learning, 2.37
Line charts: Charts that show a trend over a period of time, as indicated by a rising or falling line. **2.14**
Line printer: High-speed impact printer that prints an entire line at a time. **5.13**
Lines per minute (lpm), 5.13

Link: Contained in most Web pages, a built-in connection to another related Web page. Also called hyperlink. **2.37, 7.10**
URL and, 2.39
Link (data type): In a database, fields containing Web address that links to Web page; also called hyperlink. **2.16**
Linux: Popular, free, UNIX-like GUI operating system. 1.54, **8.16**
Liquid crystal display, *see* **LCD**
LISTSERVs: Mailing lists named after a popular mailing list software product. **7.29**
Lithium-ion batteries: Batteries that last longer than nickel cadmium or nickel hydride batteries; used for laptop computers. **8.44**
Loading operating system, 2.3, 3.18
Local area network (LAN): Network that connects computers in a limited geographical area. 1.15
connecting to Internet through, 7.7
e-mail, 2.33
history of, 1.51
large businesses, 1.32
small businesses, 1.28
Local bus: High-speed expansion bus used to connect higher speed devices such as hard disks. **3.27**
Local ISP: Internet service provider that provides one or more local telephone numbers to provide access to the Internet. **7.6,** 8.41
Local networks, Internet and, 7.5, 7.8
Local/Regional ASP, 2.43
Location conversion, *see* **Phased conversion**
Logical operations: Operations performed by the arithmetic/logic unit working with conditions and logical operators such as AND, OR, and NOT. **3.7**
Log on: The process of entering a user name and password into the computer; used with multiuser operating systems. **8.10**
Lotus Approach, 2.7
Lotus Development Corporation, 1.52, 4.29
Lotus FastSite, 2.23
Lotus Freelance Graphics, 2.7
Lotus Organizer, 2.7
Lotus SmartSuite, 2.7, 2.21
Lotus Word Pro, 2.7
Lotus 1-2-3, 1.52, 2.7
Low-level format: First formatting step for a hard disk; organizes both sides of each platter into tracks and sectors to define where items will be stored on the disk. **6.11**
Lycos, 7.30
LZW compression, 7.14

Machine cycle: Set of four basic operations repeated by the control unit: (1) fetching an instruction, (2) decoding the instruction, (3) executing the instruction, and if necessary (4) storing the result. The time it takes to fetch is called instruction time, or I-time, and the time it takes to decode and execute is called execution time or E-time. Adding together I-time and E-time gives the total time required for a machine cycle. The machine cycle is also called the instruction cycle. **3.6**
Macintosh operating system: Developed by Apple Computer, the first commercially successful graphical user interface; released with Macintosh computers in 1984. 2.3, 2.4, **8.14,** 8.27
Mac OS: Name given in recent years to the Macintosh operating system; the latest version includes multitasking, built-in networking support, electronic mail, and enhanced multimedia capabilities. **8.14,** 8.27
Macro (application program): Series of statements that instructs an application how to complete a task, recorded and saved as a sequence of keystrokes and instructions; used to automate routine, repetitive, or difficult tasks in an application. **2.13**
Macromedia, Inc., 2.23
Macromedia Authorware, 2.23
Macromedia Director, 2.23
Macromedia Dreamweaver, 2.23
Macromedia Flash, 2.23
Macromedia Freehand, 2.23
Magazines, electronic, 7.64
Magnetic disk
floppy disk, 6.4-10
hard disk, 6.10-16
Magnetic tape: Storage media that is a magnetically coated ribbon of plastic, capable of storing large amounts of data and information at a low cost. Since magnetic tape uses sequential access, which is much slower than other types of access, it is no longer used as a primary method of storage, but for long-term storage and backup. A tape drive is needed to read from and write data and information onto a tape. **6.23-**24
Magnetic-ink character recognition (MICR): Reader used to read text printed with magnetized ink; used almost exclusively by the banking industry for check processing. **4.19**
Mailbox (e-mail): E-mail storage location usually residing on the computer that connects user to a local area network or the Internet, such as the server operated by an Internet service provider. **2.33, 7.24**
Mailing list: Group of e-mail names and addresses given a single name. **7.28-**29
Mail-order, purchasing computer using, 8.39
Mail server: Server operated by an Internet service provider or online service that contains mailboxes for e-mail. **2.33, 7.24**
Mainframe: A large, expensive, very powerful computer that can handle hundreds or thousands of connected users simultaneously; can also act as a server in a network environment. **1.24**
CPU functions in, 3.5
Maintenance, computer system, 8.47-48
Male connectors: Connectors that have one or more exposed pins. **3.22**
Manufacturing, chip, 3.8, 3.18, 3.31, 3.41-44
Margins: In a document, the portion of a page outside the main body of text, on the top, bottom, and sides of the paper. **2.10**
Markups, *see* **Tags**
Marquee: Text that is animated to scroll across the screen. **7.14**
Mask: During chip manufacturing, image that is produced during a photographic process that reduces a large master design for an integrated circuit to a microscopic size. The mask is then printed onto the photoresist. **3.43**
Mass storage, *see* **Storage**
Mauchly, John W., 1.48
MB, *see* **Megabyte**
Mechanical mouse: Mouse that has a rubber or metal ball on its underside. When the ball rolls in a certain direction, electronic circuits in the mouse translate the movement of the mouse into signals that are sent to the computer. 4.7, **4.9**
Media, e-commerce and, 7.52
Medical guides, 2.32
Medicine, Web sites about, 2.58
Meetings, videoconferencing and, 4.25
Megabyte (MB): Equal to approximately one million bytes. **3.15**
Megahertz (MHz): Measure of clock speed, equal to one million ticks of the system clock. **3.8,** 3.17
Memo (data type): In a database, fields containing freeform text of any type or length. **2.16**
Memory: Temporary holding place in the system unit for instructions to be executed and the data to be used with those instructions, while they are being processed by the CPU. A byte is the basic storage unit in memory. **1.6, 3.14-**20
access times, 3.19-20
adding, 3.20
address, 3.14
buffers, 8.6
CMOS, 3.19
CPU and, 3.6
expansion card, 3.20
input and, 4.2
laser printers, 5.16
loading operating system into, 8.17-20
nonvolatile, 3.15, 3.18, 6.2
operating system and, 2.3, 8.3, 8.6-7
purchasing computer and, 8.38
RAM, *see* RAM
ROM, *see* ROM
storage versus, 6.2
transferring items into, 6.3
video cards, 5.9
virtual, 8.6
volatile, 3.15, 3.16, 6.2
Memory cache: Cache that speeds computer processes by storing frequently used instructions and data. Also called cache store, or RAM cache. **3.17**
Memory card: Smart card that has only storage capabilities. **6.25**
Memory chips, 3.14
Memory management: Operating system function that optimizes the use of RAM. **8.6-7**
Memory module, *see* RAM chips
Memory-resident: Any program that remains in memory while the computer is running, including the operating system kernel and programs such as calendars and calculators that are accessed frequently or need to be accessed quickly. **8.3**
Memory slots, 3.4
Menu: Set of available commands or options, from which a user can select one or more. **2.4,** 2.7, **8.4**
context-sensitive, 2.6
shortcut, 2.6
sub-, 2.4
Menu-driven: Programs that provide menus as a means of entering commands. **4.3**
Merchant account: Bank account used by e-commerce merchant to accept and hold money until an online transaction is complete. **7.57**
Message board: Popular Web-based type of discussion group that does not require a newsreader and is provided free of charge; that is, a user can post messages, reply to messages search for messages by keyword, and browse messages by category or topic. Also called discussion board. **7.28,** 7.29, 7.30
MetaCreations Painter 3D, 3.23
Metropolitan Area Exchanges, 7.8
Microcode: Instructions used to program a PROM chip. **3.18**
Microfiche: Media used to store microscopic images of documents on a four inch by six inch sheet of film. The images are recorded on the film using a computer output microfilm (COM) recorder. The images can be read only with a microfiche reader. **6.26**
Microfilm: Media used to store microscopic images of documents on a 100- to 215-foot roll of film. The images are recorded on the film using a computer output microfilm (COM) recorder. The images can be read only with a microfilm reader. **6.26**
Microphone, 1.6
audio input using, 4.22-23
port, 3.21

Microprocessor: Name sometimes given to the CPU for a personal computer, which usually is contained on a single chip. In addition to the control unit and arithmetic/logic unit, a microprocessor usually contains the registers and system clock. 1.19, **3.5,** 3.8-11
comparison of, 3.9-11
Microsoft Access 2000, 2.7
Microsoft Corporation, 1.36
ActiveX controls and, 7.12
antitrust lawsuit, 1.39
DOS and, 8.11
history of, 1.37, 1.52, 1.53, 1.54, 1.55, 1.56, 1.57
IBM and, 4.29
Windows development, 8.12-14
Microsoft Chat, 7.29
Microsoft Encarta, 2.27
Microsoft Excel 2000, 2.7
Microsoft FrontPage 2000, 2.23, 2.32, 7.22
Microsoft Internet Explorer: Windows 2000 Web browser. 1.18, 1.37, 2.34, 2.37, 2.39, 7.11, **8.12**
Microsoft Money, 2.27
Microsoft NetMeeting, 7.17
Microsoft Network, The (MSN), 1.17, 2.36, 7.7
Microsoft Office 2000, 1.57, 2.7, 2.21, 7.23
Microsoft Outlook 2000, 2.7
Microsoft PhotoDraw 2000, 2.27
Microsoft PowerPoint 2000, 2.7
Microsoft Publisher 2000, 2.27
Microsoft Visual Basic, 1.14
Microsoft Visual FoxPro, 2.7
Microsoft WebTV, 2.45
Microsoft Webware, 2.7
Microsoft Windows, *see* Windows
Microsoft Word 2000, 2.7
Microsoft Works, 2.27
MIDI devices, 4.22
MIDI (musical instrument digital interface) port: Serial port designed to connect the system unit to a musical instrument, such as an electronic keyboard. **3.24**
Mid-tower, 1.20
mil domain abbreviation, 7.10
Millennium bug, 1.57
Millisecond (ms): One millionth of a second; measure of hard disk access time. **3.20,** 6.3
Minicomputer: A more powerful and larger computer than a workstation, often supporting up to 4,000 connected users at the same time. **1.24**
introduction of, 1.50
Mini-tower, 1.20
MIPS: Million instructions per second; measure of computer's speed. **3.7**
Mirroring: The simplest RAID storage design, which has one backup disk for each disk, if a drive should fail. **6.15**
MMX™ technology: Technology contained in most of today's processors; a set of instructions built into the processor so it can manipulate and process multimedia data more efficiently. 1.56, **3.9**
Mobile users: Users who travel to conduct business and use a computer while on the road. Often use a laptop computer equipped with a modem. 1.25, **1.30**-31
e-mail and, 2.34
input devices for, 4.27
laptop computers used by, 3.28
large businesses, 1.32
modems and, 1.23
offline browsing by, 7.18
output devices for, 5.18, 5.25
storage devices and, 6.27
system needs, 3.29
Modem, 1.8
cable, 7.7
connecting to Internet using, 2.36, 7.7
expansion card, 3.20
history of, 1.52
installing computer system and, 8.45
mobile users, 1.23
port, 3.23
Moderated newsgroup: Newsgroup that has its contents reviewed by a moderator. **7.27**
Moderator: Person who reviews contents of articles before they are posted to a newsgroup. **7.27**
Money, electronic, *see* Electronic money (e-money)
Monitor
CRT, *see* CRT monitor
ergonomics, 5.10-11
gas plasma, 5.6
LCD, 5.4-5
port, 3.21
quality, 5.6-7
terminal, 5.23
Monochrome monitor: Monitor that displays text, graphics, and video information in one color (usually white, amber, or green) on a black background; less expensive than color monitors. **5.3**
Moore, Gordon, 3.30, 3.31
Mosaic, 1.54, 2.44
Mosby's Medical Encyclopedia, 2.27
Motherboard: Circuit board that contains electronic components in the system unit. 1.6, **3.4**, 3.10, 8.18
Motion picture movies, storing on DVD-ROM, 6.22
Motorola microprocessor: Microprocessor manufactured by Motorola, found in Apple Macintosh and Power Macintosh systems. **3.9**
Mouse: Input device used to control the movement of the pointer on the screen and make selections from the screen. 1.6, **4.7**-10, 8.4
port, 3.21, 4.10
types, 4.7, 4.9
using, 4.8
Mouse pad: Rectangular rubber or foam pad that provides traction for a mouse. **4.7**
Mouse pointer, *see* **Pointer**
Movie(s), Webcams and, 1.6
Moving Pictures Experts Group, *see* **MPEG**
MP3: Popular technology used to compress audio; files compressed using this format have an extension of .mp3. **7.16**
MPEG (Moving Picture Experts Group): Popular video and audio compression standard. Files in the MPEG format typically have an .mpg extension. **4.24, 7.17**
mpg file extension, 7.17
MPR II standards, 5.10
Ms, *see* **Millisecond**
MS-DOS, 1.37, 1.52, 8.11
MSN.com, 1.17, 2.36, 7.7
Multifunction device (MFD): Single device that provides the functionality of a printer, scanner, copy machine, and perhaps a fax machine. **5.23**
Multimedia: Any computer-based presentation or application that integrates text, graphics, animation, audio, and video. **1.34,** 2.2
processor speed and, 3.9-10
RAM needs, 3.17
on Web, 7.12-18
Windows and, 8.12
Multimedia authoring software: Used to create electronic interactive presentations with text, images, video, audio, and animation. 1.35, 2.2, 2.23, **2.26**
Multimedia developer, 1.35
Multiprocessing: Operating system that can support two or more CPUs running programs at the same time. **8.5,** 8.16
Multisession: CD-ROM on which additional data, instructions, and information can be written on the disc at a later time; includes PhotoCD, CD-R, and CD-RW. **6.21**-22
Multitasking: Operating system that allows a single user to work on two or more applications that reside in memory at the same time. **8.5,** 8.12
Multiuser: Operating system that enables two or more users to run a program simultaneously. **8.5,** 8.16
Museums, Web sites about, 2.60
Music
audio CD, 6.18, 6.19, 6.20, 6.22
audio input and, 4.22
output, 5.2, 5.20
speakers and, 1.6
stored on DVD, 6.23
Musical instrument digital interface (MIDI) port, *see* **MIDI (musical instrument digital interface) port**
Nanosecond (ns): One billionth of a second; measurement for memory access times. **3.19**
National ISP: Internet service provider that provides local telephone numbers for Internet access in most major cities and towns nationwide; some also provide a toll-free telephone number. **7.6,** 8.41
National Science Foundation, 7.4-5
Near letter quality (NLQ): Print that is slightly less clear than letter quality. **5.12**
Net, *see* **Internet**
Netiquette: Short for Internet etiquette, the code of acceptable behaviors on the Internet. **7.30**
NetPC, *see* **Network personal computer**
Netscape Communications Corporation, 1.54, 2.44, 2.35, 2.45
Netscape Communicator, 1.18, 1.54, 1.57, 2.34, 2.35, 7.11
Netscape Netcenter, 7.30
NetWare: A widely used network operating system for client-server networks. **8.17**
Network: A collection of computers and devices connected together via communications devices and media such as cables, telephone lines, modems, and satellites, allowing users to share resources. Sometimes a network is wireless. 1.9, **1.14, 7.2**
backbone, 7.5
business computers, 1.15
client-server, 8.12, 8.17
expansion card, 3.20
groupware and, 2.32-33
large business users, 1.32
mainframe and, 1.24
operating systems for, 8.12, 8.14, 8.16, 8.17
RAID used by, 6.15
sending document on, 2.10
videoconferencing using, 4.25
Windows and, 8.12
Network Access Points, 7.8
Network architecture, *see* **Network topology**
Network computer (NC): A scaled-down, less expensive version of a personal computer designed specifically to connect to a network. Sometimes called a thin-client computer. **1.22, 7.33**
Network interface card (NIC), 7.7
Network personal computer (NetPC): Used in business applications that rely on the server for software and storage. **7.33**
News, accessing using Internet, 7.2
Newsgroup: Online area in which users conduct written discussions about a subject. 2.2, **7.27**
Newsreader: Program used to access a newsgroup to read or add a message; also keeps track of articles read or not read by a user. Included with most browsers. **7.27**
News server: Computer that stores and distributes newsgroup messages. **7.27**
Next Generation Internet (NGI): An Internet-related research and development project funded and led by the United States federal government, NGI is an extremely high-speed network whose purpose is to develop and test advanced Internet technologies for research, teaching, and learning. **7.5**
NIC, *see* Network interface card
Niche-based product, 2.45
Nonimpact printers: Printers that form characters and graphics on paper without actually striking the paper. Some spray ink, while others use heat and pressure to create images. **5.14**-20
Noninterlaced monitors, 5.7
Nonlinear learning, 2.37-38

Personal Digital Assistant (PDA): Handheld computer, often supporting personal information management (PIM) applications. **1.23**
flash memory in, 3.18
Web-enabled, 1.23, 1.28, 1.31, 7.34-35
Personal DTP software: Software designed for small-scale desktop publishing projects. **2.30**
Personal finance software: Simplified accounting program, used to pay bills; balance checkbook; track personal income and expenses; track and evaluate financial plans. 2.2, 2.27, **2.28**-29
Personal identification number (PIN), 5.24
Personal information manager (PIM): Includes an appointment calendar, address book, and notepad to organize personal information. .2.2, **2.21**
groupware and, 2.32-33
packages, 2.7
Personal services
e-commerce and, 7.52
Personal use, software for, 2.2, 2.27-32
Personal Web page: An individual's Web page. 7.11, **7.20**
Phillips Magnavox, 1.55
Photo albums, Web communities and, 7.29
PhotoCD: Compact disc that only contains digital photographic images saved in PhotoCD format. **6.21**
Photo-editing software: Popular type of image editing software used to edit digital photographs. **2.30**
Photographs
clip galleries containing, 2.31
digital cameras and, 4.20-21
digital representation of, 5.2
DTP, 2.25
PhotoCD and, 6.21
printing, 5.20
in Web pages, 7.20
See also Webcams
Photolithography: During chip manufacturing, the process that patterns almost every chip layer into the shape of electronic components. **3.43**
Photo printer: Color printer that produces photo lab quality pictures directly from a digital camera, or taken from a scanned image. **5.20**
Photoresist: During chip manufacturing, the soft, light-sensitive, gelatin-like emulsion that wafers are coated with after being placed in a diffusion oven. **3.43**
Physically challenged users
input devices for, 4.26
output devices for, 5.24-25
speech recognition systems for, 4.23
Picture cards: PC Cards used by digital cameras to store pictures. **6.24**
Pie charts: Charts that have pie shapes cut into slices, used to show the relationship of parts to a whole. **2.15**
Piggyback upgrade: Upgrade in which the new processor chip is stacked on top of the old processor chip. **3.10**
Pin grid array (PGA) chip package, 3.4
Pipelining: Process of CPU executing a second instruction before the first instruction is completed, resulting in faster processing. **3.7**
Pixel: Short for picture element, each dot that is a single point in an electronic image. **5.4**, 5.6-7
PKZIP™, 8.22
Platter: Several inflexible, circular disks stacked on top of one another that make up a hard disk, made of aluminum, glass, or ceramic, and coated with a material that allows items to be magnetically recorded on its surface. **6.10**, 6.12
Player: Program contained in most current operating systems that can play the audio in MP3 files. **7.16**
Plug and Play: Computer's capability to configure expansion cards and other devices as they are installed. **3.20**, 3.24, **8.9**
Plug-in: Program that runs multimedia elements within a browser window; extends the multimedia capabilities of the browser. **7.12**, 7.13
PNG (portable network graphics): File that is a graphical image saved using compression techniques to reduce the file size for faster downloading from the Web, and supports multiple colors and resolutions. **7.14**
Point: Measurement of font size equal to about 1/72 of an inch in height. **2.9**
Point-and-shoot camera: Digital camera, providing acceptable quality photographic images for the home or small business user. **4.21**
Pointer: In a graphical user interface, a small symbol on the screen that can move or select items using the pointing device. Also called mouse pointer. 1.6, **4.7**, 4.26
Pointing device: Input device that allows user to control a pointer on the screen. **4.7**-11, 4.29
Pointing stick: Pressure-sensitive pointing device positioned between keys on the keyboard. **4.11**
Point of presence (POP): Local telephone number for Internet access. **7.6**
Point-of-sale (POS) terminal: Records purchases at the point where the consumer purchases the product or service; can serve as input to other computers. **5.24**
Politics, Web sites about, 2.58
POP (Post Office Protocol): Communications protocol used to retrieve e-mail from a mail server. **7.24**
POP3 (Post Office Protocol 3): Newest version of POP. **7.24**
Port: Interface, or point of attachment to the system unit for external devices such as a keyboard, monitor, printer, mouse, and microphone. 3.3, **3.21**-24
Portable computer: Personal computer small enough to carry. **1.22**-23
Portable network graphics, *see* **PNG**
Portable printer: Small, lightweight, nonimpact printer. **5.18**
Portable storage media, 6.4-10, 6.14
Portal: Web site designed to offer a variety of Internet services from a single, convenient location. 7.3, 7.24, **7.29**-30
content, 7.55, 7.56
wireless, 7.34-35
Portrait orientation: Printed page that is taller than it is wide. **5.11**, 5.13
Port replicator: Device used with laptop computers that allows the connection of many peripheral devices. **3.29**
Posting: Process of adding an article to a newsgroup. **7.27**
Post Office Protocol 3, *see* **POP3**
Post Office Protocol, *see* **POP**
PostScript: Page description language for laser printers. **5.16**
Power Macintosh, 3.9
PowerPC processor, 3.9, 3.25
Power-on self test (POST): During booting process, a series of tests executed by BIOS to make sure the computer hardware is connected properly and operating correctly. **8.19**
Power supply: Converts wall outlet AC power into DC power, which is the type of power required by a computer. **3.28**, 8.18, 8.48
Power users: Users that require the capabilities of a workstation or other powerful computer. 1.25, **1.34**
input devices for, 4.27
output devices for, 5.25
RAM needs, 3.17, 3.29
storage devices and, 6.27
system needs, 3.29
Presentation, multimedia, 2.26
Presentation graphics software: Software that creates documents called presentations. The presentations can be viewed as slides. 1.12, 2.2, **2.18**-20
mobile users, 1.31
packages, 2.7
Primary cache, *see* **Level 1 (L1) cache**
Primary mouse button, 4.8
Print(s): During chip manufacturing, the process of producing each wafer layer's circuit pattern on photoresist by projecting ultraviolet light through a glass mask. **3.43**
Printer: Output device that produces text and graphics on a physical medium. 1.6, **5.11**-20
Braille, 5.25
flash memory, 3.18
impact, 5.12-13
large-format, 5.19
multifunction device as, 5.23
nonimpact, 5.14-20
port, 3.21, 3.22
portable, 5.18
special-purpose, 5.20
spooling and, 8.7
Printing: Process of sending a file to a printer to generate output. 2.6, **2.10**
Print job: Document that is being printed. **8.7**
Printout, *see* **Hard copy**
Print quality, 5.12
Print spooler: Program that manages and intercepts print jobs and places them in queue. **8.7**
Privacy, e-commerce and, 7.59
Processing
parallel, 3.11
word size and, 3.25
Processor, *see* **Central processing unit (CPU)**
Productivity software: Software designed to make people more effective and efficient while performing daily activities. 1.27, 2.2, **2.7**-23
mobile users, 1.31
small business users, 1.28
Program: Series of instructions that tells a computer how to perform the tasks necessary to process data into information. 1.4, **1.9**-14, **4.2**
backup of, 6.9
hard disk storage, 6.10
menu-driven, 4.3
Program files, *see* **File**
Programmable read-only memory (PROM): Blank ROM chip on which user can permanently place items. PROM chips can be programmed only once. **3.18**
Programmable terminal, *see* **Intelligent terminal**
Programmer(s): Person that uses a programming language to write the instructions necessary to direct the computer to process data into information. **1.13**
Programming languages, 1.14
history of, 1.49-50
Project management software: Software used to plan, schedule, track, and analyze the events, resources, and costs of a project. **2.22**
packages, 2.7
Projectors, data, 5.21
PROM, *see* **Programmable read-only memory**
Protocol: Set of rules and procedures for exchanging information among computers. **7.7**, 7.25
file transfer, 7.25
Post Office, 7.24
Secure Electronics Transaction, 7.57
Secure Sockets Layer, *see* Secure Sockets Layer
TCP/IP, 7.7
URL, 7.11
Proprietary software: Software that is privately owned and limited to a specific vendor or computer model. **8.10**
Public-domain software: Free application software donated for public use. **1.13**, 2.2
Puck: Device used to draw on a graphics tablet; looks similar to a mouse. **4.13**

Pull technology, 7.18
Purchasing
computer system, 8.38-42
laptop computer, 8.43-44
printers, 5.13
Push technology: Process in which Web-based content is downloaded automatically to user's computer at regular intervals or whenever the site is updated. **7.18**

Q-DOS, 1.37
QIC tape drive, 6.24
QuarkXPress, 2.23
Query: Specific set of instructions for retrieving data from a database and displaying, printing, or storing it. **2.17**
Queued: Lining up of multiple print jobs in the buffer. **8.7**
Quick View: Windows file viewer for text files. **8.21**
QWERTY keyboard: Computer keyboard layout named after the first six leftmost letters on the top alphabetic line of the keyboard. **4.6**

Radio waves, mouse using, 4.10
RAID (redundant array of independent disks): Group of two or more integrated hard disks into a single unit that acts like a single large hard disk. **6.15**
RAM (random access memory): Describes memory chips in the system unit. When computer is powered on, certain operating system files are loaded from a storage device such as a hard disk into RAM, and remain in RAM as long as the computer is running. **3.15**-16
boot process and, 8.18, 8.19
configuring, 3.16-17
dynamic (DRAM), 3.16
guidelines, 3.17
operating system managing, 8.6-7
RDRAM (Rambus® DRAM), 3.16
static (SRAM), 3.16
thrashing and, 8.7
video cards using, 5.9
RAM cache, 3.17
RAM chips, 3.16
Rambus® DRAM, *see* RDRAM
Rambus® inline memory module (RIMM): Circuit board that contains RDRAM chips. **3.16**
Rand McNally StreetFinder, 2.27
Rand McNally TripMaker, 2.27
Random access: Access method that immediately locates data items and files without having to move consecutively through items stored in front of the desired data item or file. **6.23**
Random access memory, *see* **RAM**
RDRAM (Rambus® DRAM): Newer type of DRAM that is even faster than SDRAM, used by most computers today. **3.16**
Reading: Process of a storage device transferring data, instructions, and information from a storage medium into memory. **6.3**
CD-R, 6.22
CD-ROM, 6.20
CD-RW, 6.22
DVD, 6.23
floppy disk, 6.5, 6.7-8
hard disk, 6.11
tape, 6.23
Read-only memory, *see* **ROM**
Read/write: Storage media, such as magnetic disk, that a user can access (read) data from and place (write) data on a disk any number of times. **6.5**
Read/write head: Floppy disk drive mechanism that reads items from or writes items on the floppy disk. **6.8**
hard disks, 6.12, 6.13
head crash and, 6.13
RealAudio: Standard for transmitting audio data on the Internet; supported by many Web browsers. **7.16**
RealPlayer, 7.16, 7.26
Real-time: Time when the people involved in chat are online simultaneously. **7.29**
RealVideo: Standard for transmitting video data on the Internet; supported by many Web browsers. **7.17**
Recalculation, worksheet, 2.14
rec domain abbreviation, 7.10
Record: Group of related fields. **2.15**
Recordable compact disc, 6.22
Recordable DVD, 6.23
Recorded: Process of writing standard CD-ROM contents by the manufacturer, which can be read and used, not erased or modified. **6.19**
Recorded software, 1.9
Red Hat Software, Linux operating system and, 8.17
Redundant array of independent disks, *see* **RAID**
Reference software: Provides valuable and thorough information for users; includes encyclopedias, dictionaries, health/medical guides, and travel directories. 1.26, 2.2, 2.27, **2.32**
Reference Web sites, 2.60
Refresh rate: The speed with which the monitor redraws images on the screen; the faster the refresh rate. **5.7**
Regional networks, Internet and, 7.5, 7.8
Registers: Temporary storage locations used by the CPU to hold data and instructions. 3.5, **3.8**
Registration cards, hardware and software, 8.46
Registry: In Windows, files that contain system configuration information. 8.13, **8.19**
Reintermediation: Process of creating new distribution channels to facilitate business interactions. **7.50**
Reliability, computer, 1.8
Removable hard disk: Disk drive in which hard disks are enclosed in plastic or metal cases so they can be removed from and inserted into the drive. Also called disk cartridge. 1.7-8, **6.14**
Repetitive strain injury, 4.6, 8.46
Replace: Feature included with word processing software that is used in combination with the find feature to substitute existing characters or words with new ones. **2.10**
Research
accessing using Internet, 7.2
installing computer system and, 8.44
purchasing computer system and, 8.38
Resolution, digital camera, 4.21
Resolution (monitor): Measure of a monitor's sharpness and clarity, related directly to the number of pixels it can display. **5.6**-7
Resolution (printer)
ink-jet printers, 5.14
laser printers, 5.16
Resolution (scanner): The sharpness and clearness of a scanned image, determined by the density of the dots. **4.15**
Resources: Hardware devices, software programs, data, and information that are shared by users on a computer network. **1.15**
Restore program: Program included with a backup utility that reverses the backup process. **8.24**
Retailing, *see* **E-commerce**
Revenue stream: Method used by a business to generate income. **7.49**
Rewritable CD-ROM, 6.22
Rewritable DVD, 6.23
ROM (read-only memory): Memory chips that can be only read, but not modified. ROM is nonvolatile memory, meaning its contents are not lost when power to the computer is turned off. **3.18,** 6.2
access times, 3.19
booting and, 8.18
Router: Device that connects multiple networks, providing the fastest available path for communications traffic. **7.7**
Row
database, 2.15
worksheet, 2.12

Sales tax, e-commerce and, 7.58
Satellite companies, Internet structure and, 7.5
Saved/saving: Process of copying items needed for future use from memory to a storage device before power is turned off. **2.9, 3.16**
ScanDisk: Windows disk scanner utility that detects and corrects problems. **8.23**
Scanner: Light-sensing input device that reads printed text and graphics, then translates the results into a form the computer can use. 2.25, 4.13, **4.14**-16
bar code, 4.17-18
color, 4.14, 4.15
multifunction device as, 5.23
optical character recognition and, 4.16
Science, virtual reality and, 7.18
Science Web sites, 2.60
Screen: Front of the cathode ray tube that is coated with tiny dots of phosphor material that glow when electrically charged. Inside the CRT, an electron beam moves back and forth across the back of the screen, causing the dots to glow, which produces an image on the screen. **5.3**-4
insertion point on, 4.5
pointer on, 4.7, 4.8
touch, 4.12
Screen-displayed keyboard, 4.26
Screen saver: Utility program that causes the monitor's screen to display a moving image or blank screen if no keyboard or mouse activity occurs for a specified time period. **8.25**
Scrolling: Process of moving different portions of a document on the screen into view. **2.10**
SCSI (small computer system interface): Hard disk controllers that can support multiple disk drives, as well as other peripherals such as scanners, high-capacity disk drives, CD-ROM/DVD-ROM drives, tape drives, and printers. **6.14**-15
SCSI port: High-speed parallel port used to attach peripheral devices. **3.24**
SDRAM, *see* **Synchronous DRAM**
SDRAM II, *see* **Double-data rate (DDR) SDRAM**
Search: Feature included with word processing software that locates all occurrences of a character, word, or phrase. **2.10**
Search engine: Software program that finds Web sites, Web pages, and Internet files. **2.40, 7.12**
Search text: Word or phrase entered in a search engine text box to find a Web page or pages; the search engine then displays a list of all Web pages that contain the word or phrase. **2.40**-41
Secondary mouse button, 4.8
Secondary storage, *see* **Storage**
Sector: Small arcs that a track on a disk is broken into; capable of holding 512 bytes of data. **6.5**
bad, 6.5, 6.11
compact disc, 6.17
contiguous, 8.23, 8.48
fragmented hard disk and, 8.23
noncontiguous, 8.23
Secure Electronics Transaction (SET): Protocol used to encrypt the data in an e-commerce transaction. **7.57**
Secure Sockets Layer (SSL): Popular Internet encryption method that provides two-way encryption along the entire route data travels to and from a user's computer, using a private key to encrypt the data. **7.57**
Security
e-commerce, 7.52, 7.56, 7.57, 7.58
Internet, 7.32, 7.52, 7.56, 7.57, 7.58
operating systems administering, 8.10
removable disks and, 6.14

Semiconductor: Material that is a conductor or an insulator; describes silicon. **3.44**
Sequential access: Access method for magnetic tape in which data is read or written consecutively. The tape must be forwarded or rewound to a specific point to access a specific piece of data. **6.23**
Serial numbers, saving, 8.46
Serial port: Type of interface used to connect a device to the system unit; able to transmit only one bit of data at a time. **3.23**
Server: A computer that manages the resources on a network, controlling access to the software, printers, and other devices as well as providing a centralized storage area for programs and data. **1.22, 7.7**
Alpha microprocessor in, 3.9
domain name system, 7.10
e-commerce and, 7.55, 7.56
e-mail, 2.33
FTP, 7.25
Internet service provider, 2.33
mail, 7.24
mainframe as, 1.24
minicomputer as, 1.24
operating system for, 8.16
RAID used by, 6.15
Web, 7.11, 7.23
Service agreement, purchasing computer and, 8.42
Set-top box: Web appliance for the home that incorporates Internet access into a television set. The television set serves as the monitor. 1.22, 1.55, **7.33**
Setup program, 1.10
Shareware: Application software that is distributed free for a trial period. **1.13,** 2.2
Sheet-fed scanners, 4.15
Shockley, William, 1.49
Shopping
bar code scanners and, 4.17
online, 1.27, 2.35, 7.3, 7.19-20. *See also* E-commerce
POS terminal used in, 5.24
Web sites for, 2.61
Shopping cart: Component of online business display that allows consumer to collect purchases. **7.19, 7.54**
Shortcut menus: Menus that display a list of commands commonly used to complete a task related to the current activity. **2.6**
Shugart, Alan, 1.50, 1.52
Shutter: Piece of metal on a 3.5-inch floppy disk that covers an opening in the rigid plastic shell. When a floppy disk is inserted into a floppy disk drive, the drive slides the shutter to the side to expose a portion of both sides of the floppy disk's recording surface. **6.8**
Shuttle-matrix printer: Impact line printer that prints by moving a series of print hammers back and forth horizontally at incredibly high speeds; used for high-volume output. **5.13**
Signature: Additional information appended to e-mail messages. **2.34**
Silicon: Raw material used to make computer chips. **3.42**
Silicon Graphics, Inc., 6.29
Single edge contact (SEC) cartridge, 3.4
Single inline memory module (SIMM): Type of circuit board containing DRAM chips and one set of contacts; inserted into the motherboard. **3.16**
Single-session: Description of most standard CD-ROMs, because all items must be written on the disc at the time it is manufactured. **6.21**
Single-switch scanning display, 4.26
Single tasking, *see* **Single user**
Single user: Operating system that allows only one user to run one program at a time. **8.4**
Sketches, graphics tablet used for, 4.13
Slides, presentation, 2.18, 2.19
Small business users: Businesses having less than fifty employees that use computers to manage their resources effectively, often providing a desktop personal computer. 1.25, **1.28**-29
DTP, 2.30
input devices for, 4.27
legal software, 2.29
output devices for, 5.25
storage needs of, 6.3, 6.27
system needs, 3.29
tax preparation software, 2.29
Small computer system interface, *see* **SCSI**
Small computer system interface (SCSI) port, *see* **SCSI port**
Small office/home office (SOHO): Term sometimes used for a small business. **1.28**
Smart card: Storage media about the size of a credit card that stores data on a thin microprocessor embedded in the card. **6.25**-26
Smart pager: Two-way pager a little larger than a standard pager that can be used to send and receive e-mail and receive news alerts from the Web. **7.34**
Smart phone: Cellular phone that, in addition to tracking phone calls and their costs, allows user to send and receive messages on the Internet and browse Web sites specifically configured for display on a phone. 1.22, 1.31, **7.34**
SMTP (simple mail transfer protocol): Communications protocol for e-mail, used by the mail server to determine how to route the e-mail message through the Internet. **7.24**
Socket, motherboard, 3.10
Soft copy: Information shown on a display device; called soft copy because the information exists electronically and is displayed for a temporary period of time. **5.3**
Software: A series of instructions that tells the hardware how to perform tasks, processing data into information. **1.4,** 1.9-14
business uses, 1.28-29
development, 1.13-14
e-commerce, 7.55, 7.56
free, 1.13, 2.2
Internet infrastructure and, 7.52
memory-resident, 8.3
power users, 1.35
purchasing computer and, 8.38
recorded, 1.9
registration cards, 8.46
setup program. 1.10
Webware, 2.2, 1.13, 2.42, 2.7
Software application: Programs designed to perform specific tasks for users. **2.2**
Software package: Packaged application software that can be purchased from software vendors in retail stores or on the Web. **2.2.**
Software platform: The operating system used on a computer. **8.2,** 8.38
Software suite, 2.2, 2.7
Sony Electronics Inc., 1.55, 6.10
Sort: Process of organizing data or a set of records in a particular order. **2.17**
Sound
audio input and, 4.22
clip galleries containing, 2.31
editing, 4.22
expansion card, 3.20
MIDI and, 3.24
output, 1.6, 5.2, 5.20
speakers and, 1.6
stored on DVD, 6.23
Web page and, 7.12, 7.16-17, 7.22
Sound card: Expansion card used to enhance the sound-generating capabilities of a personal computer by allowing sounds to be input through a microphone and output through speakers. **3.20,** 4.22
Sound files, Web and, 7.16
Sound waves, 5.2
Source document: Document that contains the data to be processed. **4.13**
Spam: Unsolicited e-mail message or newsgroup posting sent to many recipients or newsgroups at once; the Internet's version of junk mail. **7.30, 7.59**
Speaker(s), 1.6, 5.20-21
Speaker-dependent software: Speech recognition program that makes a profile of the user's voice; the user has to train the computer to recognize his or her voice. **4.23**
Speaker-independent software: Speech recognition program that has a built-in set of word patterns, so the user does not have to train the computer to recognize his or her voice. **4.23**
Special-purpose printers, 5.20
Specialist ASP, 2.43
Speech
audio input and, 4.22
output, 5.2, 5.20
Speech recognition: Describes the computer's capability of distinguishing spoken words. Also called voice recognition. 2.11, **4.22**-23, 4.26
Speed
CD-ROM drive, 6.21
compact disc, 6.17
computer, 1.8
dot-matrix printer, 5.13
EIDE controllers, 6.14
hard disks, 6.13
ink-jet printer, 5.15
performance monitor and, 8.9
processor, 3.7, 3.8, 3.9, 3.10, 3.17
SCSI controllers, 6.14-15
shuttle-matrix printer, 5.13
storage device, 6.3
thrashing and, 8.7
Spelling checker: Feature that compares words in a document to an electronic dictionary that is part of the word processing software. **2.10**
Spoiler: Internet message revealing a solution to a game or ending to a movie or program. **7.30**
Spooling: Process of print jobs being placed in a buffer instead of being sent immediately to the printer. **8.7**
Sports Web sites, 2.61
Spreadsheet software: Organizes data in rows and columns. The rows and columns collectively are called a worksheet. 1.12, 2.2, **2.12**-15
packages, 2.7
SRAM, *see* **Static RAM**
SSE instructions: Included in Intel's latest processors to improve the performance of multimedia, the Web, and 3D graphics. **3.10**
Stand-alone: Computer that is not connected to a network and has the capability of performing the information processing cycle operations (input, process, output, and storage). **1.21**
Starting computer, 1.10
Static RAM (SRAM): Type of RAM that does not need energizing as often as DRAM, and is faster and more reliable. Used for special purposes because it is much more expensive than DRAM. **3.16,** 3.17, 3.19
Storage: Disks, tapes, or other material capable of storing items such as data, instructions, and information for future use; storage holds these items when they are not being processed. **1.4,** 1.8, **6.2**-27
audio files, 4.22
buffers, 8.6
measuring, 3.14-15
memory versus, 6.2
operating system managing, 8.6, 8.10
programs in, 4.2
purchasing computer and, 8.38
registers, 3.8
unit of, 3.12, 3.14
video files, 4.24
Storage area network (SAN): High-speed network that connects storage devices. **6.27**
Storage device: Hardware device used to record and retrieve data, instructions, and information to and from a storage medium. **1.7**-8, **6.2**
Storage medium: The physical material on which data, instructions, and information are stored. **1.7, 6.2**
compact discs, 6.16-22
floppy disks, 6.4-10
hard disks, 6.10-16
microfiche, 6.26-27
microfilm 6.26-27

PC Cards, 6.24-25
smart cards, 6.25-26
tapes, 6.23-24
store domain abbreviation, 7.10
Stored program concept: The role of memory to store both data and programs. **3.14**
Storefront, e-commerce and, 7.55, 7.59
Storing: Fourth operation in the machine cycle, performed by the control unit; the process of writing the results to memory. **3.6**
Streaming: Process of transferring data in a continuous and even flow, which allows users to access and use a file before it has been transmitted completely. **7.16**
Streaming audio: Transfer of audio data in a continuous and even flow, which allows a user to listen to the sound as it downloads to a computer. **7.16**
Streaming video: Transfer of video data in a continuous and even flow, which allows a user to view longer or live video images as they are downloaded to a computer. **7.17**
Striping: RAID storage design that splits items across multiple disks in the array, improving disk access times. **6.15**
Structure, table: In a database, a general description of the records and fields in a table, including the number of fields, field names, field lengths, and data types. **2.16**
Structured design, 1.50
Studio camera: Stationary digital camera used for professional studio work. **4.21**
Stylus, *see* **Pen**
Submenu: Commands that displays when a user points to a command on a previous menu. **2.4**
Submission service: Web-based business that usually offers free registration of a Web site with several search engines, or registration to hundreds of search engines for a fee. **7.23**
Subscribe (mailing list): Add your e-mail name/address to a mailing list. **7.28**
Subscribe (newsgroup): Process of saving a newsgroup in a newsreader so it can be accessed easily in the future. **7.27**
Subwoofer, 5.21
Suite, software: Collection of application software sold as a single package. 1.12, **2.21**
Sun Microsystems, 1.55, 6.28, 7.12
Webware, 2.7, 2.43
Supercomputer: The fastest, most powerful category of computer, and the most expensive. **1.24**
CPU functions in, 3.5
SuperDisk™ drive: High-capacity disk drive developed by Imation that uses 120 MB SuperDisk™ floppy disks. **6.10**
Superscalar: Processors that can execute more than one instruction per clock cycle. 3.7, **3.8**
Super video graphics array (SVGA): Standard for most of today's monitors, using a suggested resolution of 800 x 600, 1024 x 768, 1280 x 1024, or 1600 x 1200. **5.9**
Supply chains: Interrelated network of suppliers that work together to create a product. **7.51**
Support tools, 2.42-43
Surfing the Web: Process of displaying pages from one Web site after another. **2.38**
Swap file: Area of the hard disk used for virtual memory; used to swap (exchange) data and program instructions between memory and storage. **8.6**-7
Synchronous DRAM (SDRAM): Type of fast dynamic RAM that is synchronized to the system clock. **3.16**
Synthesizer: Peripheral or chip that creates sound from digital instructions. **3.24**
System administrator, system security and, 8.10
System board: Circuit board that contains most of the electronic components in the system unit. Also called motherboard. **3.4**
System bus: Bus that connects the CPU to main memory. **3.26**
System clock: Small chip used by the control unit to synchronize, or control the timing of, all computer operations; generates regular electronic pulses, or ticks, that set the operating pace of components in the system unit. The faster the system clock, the more instructions the CPU can execute per second. 3.5, **3.8**
SDRAM and, 3.16
System maintenance functions, 8.48
System performance, operating system monitoring, 8.9
System software: Programs that control the operations of the computer and its devices, serving as the interface between the user, application software, and the computer's hardware. The two types of system software are operating systems and utility programs. **1.10**-12, **2.2**-3, **8.2**-25
operating systems, 8.2-20
utility programs, 8.21-25, 8.48
System unit: Box-like case that houses the electronic components of the computer that are used to process data; includes the processor, memory module, expansion cards, ports, and connectors. **1.6**, **3.2**-29
bays, 3.27
booting and, 8.18
buses, 3.24-27
connecting device to, 3.21-24
expansion cards, 3.20-21
expansion slots, 3.20-21
hard disk in, 6.10
laptop computers, 3.2, 3.28-29
memory, 3.14-20
motherboard, 3.4
ports, 3.21-24
power supply, 3.28
processor, 3.2, 3.5-11
Systems analyst: Person responsible for designing and developing a program or an information system. **1.13**

Table
file allocation, 6.6, 6.11
word processing document, 2.11
Table, database: Data in a database that is organized in rows and columns. Rows in a table are called records, and columns are called fields. 2.15
Tags: Codes used by HTML that specify links to other documents, as well as how the Web page displays. Also called markups. **7.20**
Tape, *see* **Magnetic tape**
Tape cartridge: Small, rectangular, plastic housing for tape that is an inexpensive, reliable form of storage often used for personal computer backup. **6.23**
Tape drive: Drive used to read from and write data and information onto a tape. Common types of tape drives are QIC, DAT, and DLT. **6.23**
Tape player, audio input using, 4.22
Taxes, e-commerce and, 7.58
Tax preparation software: Software that guides individuals, families, or small businesses through the process of filing federal taxes. 2.2, 2.27, **2.29**
TCP/IP (transmission control protocol/Internet protocol): Communications protocol used to define packet switching on the Internet. **7.7**
Telecommunications, *see* **Communications**
Telecommunications provider, Web hosting by, 7.55, 7.56
Telecommuting, 1.56
Telephone, smart phone, 5.5, 7.34
Telephone calling card, 6.26
Telephone companies, Internet structure and, 7.5
Telephone lines
fax machines and, 5.22
installing computer system and, 8.45
Internet connection, 2.36, 7.7
Telephone number, Internet connection and, 7.6
Telephone service, Internet, 7.16
Telephone system, automated, 1.32
Television
connecting to Internet through, 1.22, 1.55
high-definition, 5.11
Internet access using, 7.35
viewing on monitor, 3.20
Telnet: Program or command that allows a user to connect to a remote computer on the Internet, by entering a user name and password. Once connected, the user's computer acts like a terminal directly linked to the remote computer. **7.26**-27
Temperature, controlling, 8.47
Templates, 2.11, 7.23
Terman, Fred, 5.26
Terminal: Device that performs both input and output, consisting of a monitor (output), a keyboard (input), and a video card. The three categories of terminals are dumb terminals, intelligent terminals, and special-purpose terminals. **1.24, 5.23**-24
dumb, 1.24
Text: Output consisting of characters used to create words, sentences, and paragraphs. **5.2**
downloading Web page and, 2.37
IP address, 7.8
search, 2.40-41
Web page, 7.12
wordwrap, 2.10
Text (data type): In a database, fields containing letters, numbers, or special characters. **2.16**
TFT displays: Name sometimes given to active-matrix displays because of the thin-film transistor (TFT) technology used. **5.5**
The Microsoft Network, *see* Microsoft Network, The
Thermal dye transfer printer, *see* **Dye-sublimation printer**
Thermal printer: Nonimpact printer that generates images by pushing electrically heated pins against heat-sensitive paper. The print quality is low and the images tend to fade over time, but thermal printers are ideal for use in small devices such as adding machines. **5.17**
Thermal transfer printer, *see* **Thermal wax-transfer printer**
Thermal wax-transfer printer: Nonimpact printer that generates rich, nonsmearing images by using heat to melt colored wax onto heat-sensitive paper. Also called thermal transfer printer. **5.18**
Thesaurus, 2.11
Thin-client computer, *see* **Network computer (NC)**
Thin-film transistor, 5.5
1394 bus: Bus that eliminates the need to install expansion cards into expansion slots. **3.27**
1394 port: Also called FireWire, port that can connect multiple types of devices such as digital video camcorders, digital VCRs, color printers, scanners, digital cameras, and DVD drives to a single connector. Also supports Plug and Play. **3.24**
32-bit processors, 8.12
Thrashing: Time spent by an operating system in paging, instead of executing application software; may indicate need for more RAM. **8.7**
Thread: In a newsgroup, the original article and all subsequent related replies; also called threaded discussion. **7.27**
Threaded discussion, *see* **Thread**
3Com Palm Desktop, 2.7
3DNow!™ technology: Included in AMD's latest processors to improve the performance of multimedia, the Web, and 3D graphics. **3.10**
Thumbnail: Small version of a larger graphical image; can be clicked to display the full-sized image. **7.14**
Title bar: At the top of a window, the horizontal space that contains the window's name. **2.4**

Toggle keys: On keyboard, keys that can be switched between two different states. **4.5**
Toner: Powdered ink used by laser printers; packaged in a cartridge. **5.17**
ToolBook, *see* **click2learn.com ToolBook**
Tool set, installing computer and, 8.46
Top-level domain (TLD): Contained in a domain name, an abbreviation that identifies the type of organization that operates the site. **7.9**
Torvalds, Linus, 1.54
Toshiba, 1.20
Touchpad: Small, flat, rectangular pointing device sensitive to pressure and motion. A touchpad senses the movement of a finger on its surface to simulate mouse operations; can be attached to a personal computer, but more often found on laptop computers. Also called trackpad. **4.11**
Touch screen: Monitor that has a touch-sensitive panel on the screen; interaction with the computer occurs when user touches areas of the screen with a finger. **4.12**
Tower model: Desktop computer that has a tall and narrow system unit that is designed to be placed on the floor vertically. **1.20,** 3.2
Track: Narrow recording band that forms a full circle on the surface of a disk. The disk's storage locations are then divided into pie-shaped sections, which break the tracks into small arcs called sectors. A sector can store 512 bytes of data. **6.5**
compact disc, 6.17
floppy disk, 6.5
hard disk, 6.12
Trackball: Stationary pointing device with a ball mechanism on its top. **4.10,** 4.26
Trackpad, *see* **Touchpad**
Traffic: Bulk of the communications activity on the Internet. **7.5**
Transaction server: Safe system in e-commerce where the customer enters personal, e-cash, and/or credit card data. **7.54**
Transistors: Element of integrated circuit that act as electronic switches, or gates, that open or close the circuit for electronic signals. **3.4,** 3.31, 3.41
history of, 1.49, 1.50
LCD displays using, 5.5
Transmission media, 1.8
Travel directories, 2.32
Travel Web sites, 1.29, 2.61
Turing, Alan, 1.48
Turnaround documents: Documents designed to be returned (turned around) to the organization that created and sent them. **4.16**-17
Tutorials: Step-by-step instructions using real examples that show how to use an application. Some tutorials are available in printed manuals, while others are software-based or Internet-based. **2.43**
TV tuner expansion card, 3.20
Type I cards: PC Cards used to add memory capability to a laptop computer. **3.21,** 6.25
Type II cards: PC Cards containing communications devices such as modems. **3.21,** 6.25
Type III cards: PC Cards used to house devices such as hard disks; currently have storage capacities of more than 520 MB. **3.21, 6.25**
Typing area, on keyboard, 4.4

Unclassified Web sites, 2.61
Uncompress: Process of restoring a compressed file to its original form; also called unzip. **8.22**
Unformat disk, 6.6
Unicode: Coding scheme capable of representing all the world's current languages. **3.12, A.1-2**
Uniform Resource Locator (URL): Unique address for Web page. A URL consists of a protocol, domain name, and sometimes the path to a specific Web page or location in a Web page. A browser retrieves a Web page using its URL, which tells the browser where the Web page is located, making it possible to navigate using links, since links are associated with a URL. **2.39**-40, **7.11**
Uninstaller: Utility program that removes an application, as well as any associated entries in the system files. 1.11, **8.24**
UNIVAC I computer, 1.49
Universal serial bus (USB): Bus that eliminates the need to install expansion cards into expansion slots. USB devices are connected to each other outside the system unit and then a single cable attaches to the USB port. **3.27**
Windows 98 and, 8.13
Universal serial bus (USB) port: Port that can connect up to 127 different peripheral devices with a single connector. **3.24**
UNIX: Multiuser, multitasking operating system developed in the early 1970s. 6.28, **8.16**
Unsubscribe: Process of user removing e-mail name and address from a mailing list. **7.28**
Unzip, *see* **Uncompress**
Upgrade
memory, 3.20
processor, 3.10
purchasing computer system and, 8.42, 8.44, 8.46
Upload: Process of copying Web page from user's computer to the Web server. **7.23**
files to FTP site, 7.25
Upward compatible (floppy disk drive): The ability of a floppy disk drive to recognize and use newer media. **6.7**-8
Upward-compatible (software): Application software that was written for an earlier version of the operating system, but runs under the new version. **8.10**
URL, *see* **Uniform Resource Locator**
Usenet: The entire collection of Internet newsgroups. **7.27**
User(s): A person that communicates with a computer or uses the information it generates. **1.4**
categories of, 1.25-35
input devices for types of, 4.27
number on Internet, 7.2-3
registered with vendor, 8.46
User-ID, *see* **User name**
User interface: Combination of hardware and software that allows a user to communicate with a computer system; controls how the user enters data or instructions and how information and processing options are presented on the screen. **1.11, 2.3, 8.3**-4
command-line, 8.3, 8.11, 8.16
graphical, *see* Graphical user interface
User name: Unique combination of characters, such as letters of the alphabet or numbers, that identifies one specific user; sometimes used for e-mail messages. Also called user-ID. **2.34, 8.10**
User response: Instruction issued by user to the computer by replying to a question posed by a computer program. Based on user's response, the program performs certain actions. **4.3**
User-friendly: Characteristics of a graphical user interface that makes it easy to learn and work with. **8.4**
Utility program: Type of system software that performs a specific task, usually related to managing a computer, its devices, or its programs; also called utility. Most operating systems contain several utility programs. 1.11, 2.3, 6.16, **8.21**-25, 8.48

Validation: Process of comparing data to a set of rules or values to determine if the data is accurate. **2.17**
Value: Numerical data contained in the cells of a worksheet. **2.13**
VCR, 5.3
Vendors
purchasing computer and, 8.38-41
software, 1.13
VeriSign, 7.52
Vertical Market ASP, 2.43
VESA local bus: First standard local bus, used primarily for video cards; has been replaced by the PCI bus, which is more versatile and faster. **3.27**
Video: Photographic images that are played back at speeds of 15 to 30 frames per second and provide the appearance of full motion. **2.26, 5.3**
e-mail and, 7.24
entertainment software and, 2.32
files, 4.24, 7.17
home computers and, 4.25
standards, 5.9
storing on DVD-ROM, 6.22
streaming, 7.17
Webcams, 1.6, 4.25
Web page, 7.12, 7.17, 7.22
Video adapter, *see* **Video card**
Video camera, 5.3
home computers and, 4.25
video input using, 4.24
Video capture card: Expansion card that converts an analog video signal into a digital signal that a computer can understand, enabling connection of a video camera or VCR to a computer and manipulation of the video input. 3.20, **4.24,** 5.3
Video card: Expansion card that converts digital output into an analog video signal that is sent through a cable to the monitor, which displays an image on the screen. Also called graphics card, or video adapter. **3.20, 5.8**-10
AGP, 3.27
terminal and, 5.23
Video chats: Supported by some chat rooms, where users can see each other as they chat. **7.29**
Video clips, 4.24
clip galleries containing, 2.31
Videoconferencing: Meeting between two or more geographically separated individuals who use a network or the Internet to transmit audio and video data. **4.25,** 7.17
Video digitizer: Device used to capture an individual frame from a video and then save the still picture in a file. **4.24**
Video editing software: Software used to modify a segment of a video, called a clip, and typically includes audio editing capabilities. 2.2, 2.23, **2.26**
Video input: Process of entering a full-motion recording into a computer and storing the video on a hard disk or some other medium. The video is captured by plugging a video camera, VCR, or other video device into a video capture card, then played back or edited using video-editing software. **4.24**
Video input device, 4.25, 5.3
Video projector, connecting laptop to, 8.44
Video RAM (VRAM): Special RAM used by higher-quality video cards to improve the quality of graphics. **5.9**-10
Viewable size: Reference of monitor size, viewable size is the diagonal measurement of the cathode ray tube inside the monitor and is larger than the actual viewing area provided by the monitor. **5.4**
Virtual mall, 7.55, 7.56
Virtual memory (VM): Used by some operating systems to optimize RAM, the allocation of a portion of the storage medium, usually the hard disk, to function as additional RAM. **8.6**
Virtual reality (VR): The use of a computer to create an artificial environment that appears and feels like a real environment and allows the user to explore a space and manipulate the environment. Advanced forms of VR require the user to wear specialized headgear, body suits, and gloves to enhance the experience of the artificial environment. **7.17**-18

Virtual Reality Modeling Language (VRML): Language that defines how 3-D images display on the Web. **7.18**
Virus: Program that copies itself into other programs and spreads through multiple computers; often designed to damage a computer by destroying or corrupting its data. **8.25,** 8.48
VisiCalc, 1.52
Visio Technical, 2.23
Visually impaired users
input devices for, 4.26
output devices for, 5.25
Voice, speakers and, 1.6
Voice chats: Supported by some chat rooms, where users can hear each other as they chat. **7.29**
Voice recognition, *see* **Speech recognition**
Volatile memory: Type of memory in which the contents are lost when the power is turned off; example is RAM. **3.15,** 3.16, **6.2**
Volume Business ASP, 2.43
von Neumann, John, 1.48
VRAM, *see* **Video RAM**
VRML, *see* **Virtual reality modeling language**
VR World: 3-D site that contains infinite space and depth. **7.18**

Wallet: A small program that stores user's address and credit card information on the user's hard disk. When a purchase is made on the Web, the user chooses from credit cards in the wallet, and the information contained in the wallet is transferred from the user's computer to the vendor's computer. **7.20**
WAN, *see* **Wide area network**
Warm boot: Process of booting by restarting, or resetting, a computer that is already on. Also called warm start. **8.17**
Warm start, *see* **Warm boot**
WAV: Format for audio files on the Internet. **7.16**
WAV files, *see* **Waveforms**
Waveforms: Audio files stored by Windows, called WAV files. Once a sound is saved in a file, it can be played or edited, attached to an e-mail message, or included in a document. **4.22**
Weather, Web sites on, 2.61
Web, *see* **World Wide Web (WWW)**
Web address, *see* **Domain name**
Web appliance: Device designed specifically to connect to the Internet; the most popular are set-top boxes, smart phones, smart pagers, and Web-enabled PDAs. Also called Internet appliance. **1.22,** 1.27, **7.33**-34
Web application, see Webware
Web audio applications: Individual sound files included in Web pages. **7.16**
Web browser: Software application used to access and view Web pages. Also called browser. 1.18, 2.2, **2.34**-41, **7.11**
history of, 1.54, 1.56, 2.44
large business use, 1.33
multimedia capabilities, 7.12
purchasing computer system with, 8.38
toolbars, 2.39
Windows 98 and, 8.13
Web browser-like features, GUI having, 8.4, 8.13
Webcam: Video cameras on home computers that allow the user to see people at the same time they communicate on the Internet, as well as allowing user to edit videos, create a movie, and take digital photographs. Also called PC Cameras. 1.5, 1.6, 1.7, **4.25**
Web publishing using, 7.20
Webcasting: Concept of using push and pull technologies to broadcast audio, video, and text information, enabling user to receive customized Web content that is updated regularly and automatically. **7.18**-19
Web clipping: Information user requests from a Web site that is targeted specifically for display on a Web-enabled device. **7.35**
Web community: Web site geared toward a specific group of people with similar interests or relationships; usually offer a message board, chat room, e-mail, and photo albusm to facilitate communications among members. **7.29**
web domain abbreviation, 7.10
Web-enabled device, communicating with, 1.22, 1.23, 1.28, 1.31, 7.34
Web folder: Office 2000 shortcut to a Web server. **7.23**
Web hosting services: Companies that provide storage for Web pages for a reasonable monthly fee. **7.23, 7.55**-56
Webmaster: Individual responsible for developing Web pages and maintaining a Web site. **7.11**
Web page: Electronic document on the Web that can contain text, graphical images, sound, and video, as well as connections (links) to other Web pages. **2.35,** 2.37, 1.18, **7.10**
developing using word processing package, 2.11
downloading, 2.37
multimedia, 2.27, 7.12-18
navigating using links, 2.37-38
personal, 7.11, 7.20
publishing, 7.20-23
searching for, 2.40
secure, 7.33
uploading, 7.23
URL, 2.39
Web page authoring software: Software used to create Web pages, in addition to organizing, managing, and maintaining Web sites. 2.2, 2.23, **2.27,** 7.22-23
Web publishing: The development and maintenance of Web pages. 7.11, **7.20**-23
Web server: Computer that delivers (serves) requested Web pages. **2.40, 7.11**
Web pages stored on, 7.11, 7.23
Web site: Collection of related Web pages that can be accessed electronically. 1.18, **2.35, 7.10**
application software Help information links on, 2.44
application software stored on, 2.42
computer industry careers, 7.64
managing, 7.23, 7.59
registering, 7.23
small businesses, 1.29
stored, 7.11, 7.23
Web site management software: Software that provides advanced Web site management features, such as managing users, passwords, chat rooms, and e-mail. **7.23**
WebTV™, 1.22, 1.55, 2.45, 7.34
Web video applications: Individual video files, such as movie or television clips, that must be downloaded completely before being played on a computer. **7.17**
Websters Gold, 2.27
Webware: Packaged software stored by companies on a Web site, accessed on the Web by users visiting the Web site. The company's sometimes also stores a user's data and information at their site. Access to the software sometimes is provided for free, but some companies rent use of the software on a monthly basis, some charge a per user access fee, and others charge a one-time fee. Also call Web application. 2.2, 1.13, **2.42**
packages, 2.7
What-if analysis: Process in which certain values in a spreadsheet are changed in order to reveal the effects of those changes; used to assist in decision making. **2.14**
Whiteboard: During videoconferencing, window on computer screen that displays notes and drawings on all participants' screens, providing multiple users with an area on which they can write or draw. **4.25,** 7.17
Wide area network (WAN): Network that covers a large geographical area (such as a city or country) using a communications channel that combines telephone lines, microwave, satellites, or other transmission media. The Internet is the world's largest WAN. 1.15, 1.32
WillMaker, 2.27
Win95, *see* **Windows 95**
Win98, *see* **Windows 98**
Window: Rectangular area of the screen that is used to display a program, data, and/or information. **2.4**
Windows: Most widely used personal computer operating system and graphical user interface; developed by Microsoft Corporation. Also called Microsoft Windows. 1.10, **2.3, 8.12,** 8.27
desktop, 2.4
formatting disks using, 6.7
history of, 1.37, 1.54, 1.55, 1.56
monitor characteristics and, 5.7, 5.8
physically challenged users modifying settings, 5.24-25
purchasing computer and, 8.38
Sound Recorder, 4.22
utilities, 6.16
Windows 2000: Operating system for network servers that is an upgrade to the Windows 98 and Windows NT operating systems; includes all the features of Windows 98 plus: Wizards to guide user through administrative activities; programs to monitor network traffic and applications; ability to work with multiple CPUs using multiprocessing; tools for Web site creation and management; and support for user and account security. 1.8, **8.12**
Windows 2000 Professional: Operating system computers connected to a network that is an upgrade to the Windows 98 and Windows NT operating systems; includes all the features of Windows 98 plus: Wizards to guide user through administrative activities; programs to monitor network traffic and applications; ability to work with multiple CPUs using multiprocessing; tools for Web site creation and management; and support for user and account security. **8.12**
Windows 3.x: Three early versions of Microsoft Windows: Windows 3.0, Windows 3.1, and Windows 3.11 that were not operating systems, but rather were operating environments, designed to be graphical user interfaces that worked with DOS. **8.11**
Windows 95: Developed by Microsoft, a true multitasking operating system with an improved graphical user interface; also called Win95. Windows 95 was written to take advantage of 32-bit processors, making programs run faster. Windows 95 also supported more efficient multitasking; included support for networking; Plug and Play technology; longer file names; and e-mail. 1.37, 1.55, **8.11**-12, 8.27
Windows 98: Developed by Microsoft, an upgrade to the Windows 95 operating system that was easier to use and more integrated with the Internet; provided faster system startup and shutdown; better file management; and support for new multimedia technologies. Also called Win98. 1.11, 1.37, 1.56, 1.57, **8.12,** 8.27
creating boot disk, 8.20
installing new hardware, 8.8
uninstall programs, 8.24
utility programs, 8.21, 8.22, 8.23-24

Windows 2000: Upgrade to the Windows 98 and Windows NT operating systems. Like Windows 98 and Windows NT, Windows 2000 is a complete multitasking operating system that has a graphical user interface. Two basic versions of Windows 2000 are: the Windows 2000 Server family, and Windows 2000 Professional. 1.58, **8.12**-14
Windows 2000 Professional: Version of Windows 2000 for stand-alone business desktop or laptop computers, as well as for computers connected to the network, called the clients. 1.58, **8.12**
Windows 2000 Server family: Version of Windows 2000 for various levels of network servers that includes Windows 2000 Server, Windows 2000 Advanced Server, and Windows 2000 Datacenter Server. **8.12**
Windows CE: Scaled-down Windows operating system designed for use on wireless communications devices and smaller computers such as handheld computers, in-vehicle devices, and network computers. **8.14**
Windows Explorer: Windows file manager; has a Web browser look and feel. **8.12,** 8.21
Windows Millennium: Updated version of Windows 98 designed for home computer users to surf the Internet or for entertainment. **8.14**
Windows NT: Developed by Microsoft, an operating system designed for client-server networks; also called NT. 1.37, 1.55, **8.12**
Wintel, 8.27
WinZip®, 7.26, 8.22
Wireless communications devices, Windows CE for, 8.14
Wireless Internet access, 7.7, 7.34-35
Wireless keyboards: Keyboards that do not use cables, but instead transmit data using infrared light waves. **4.6**
Wireless mouse, *see* **Cordless mouse**
Wireless network, 1.14
Wireless portal: Single Web site that attempts to provide all information a wireless user (using a Web-enabled device) might require. **7.35**
Wireless technology, future of, 1.58
Wireless transmission media: Transmission media that send communications signals through the air or space using radio, microwave, and infrared signals; include broadcast radio, cellular radio, microwaves, communications satellites, and infrared. 3.24, 4.6
Wireless Web communication, 7.34-35
Wizard: Automated assistant included with many software applications that helps a user complete a task by asking questions and then automatically performing actions based on the answers. **2.43,** 7.23
Woofer, 5.21
Word processing software: A widely used application software package; used to create, edit, and format textual documents. 1.12, 2.2, **2.8**-10
packages, 2.7
purchasing computer system and, 8.38
Web page authoring features in, 7.22
Word size: Number of bits the CPU can process at a given time. **3.25**-26
Wordwrap: In word processing software, the automatic positioning of typed text that extends beyond the right page margin at the beginning of the next line; allows the continuous typing of words in a paragraph without having to press the ENTER key at the end of each line. **2.10**
Work area, well-designed, 8.45
Worksheet: Collection of rows and columns in spreadsheet software. **2.12**
Workstation: Expensive and powerful desktop computer designed for work that requires intense calculations and graphics capabilities. **1.21**
Alpha microprocessor for, 3.9
power users, 1.34-35
World Wide Web (WWW): One of many services available on the Internet, the World Wide Web is a worldwide collection of electronic documents (Web pages) that have built-in links to other Web pages and is the part of the Internet that supports multimedia; also called the Web. Connected users browse pages on the Web to access a wealth of information, news, live radio and video, research, and educational material, as well as sources of entertainment and leisure. 1.18, **2.35, 7.10**-23
connecting to, 2.36-37
DTP software and, 2.24
growth of, 7.49
history of, 1.53, 1.56
home users on, 1.27
large business use, 1.33
multimedia on, 7.12-18
purchasing computer on, 8.39
retailing on, *see* E-commerce
searching for information on, 2.40-41
small business use, 1.29
surfing, 2.38
wireless communication, 7.34-35
World Wide Web Consortium (W3C), 1.54, 7.5
WorldGate, 7.52
Wozniak, Steve, 1.51, 8.26-27
Wrist injury, 4.6
Write-protect notch: Small opening in the corner of a floppy disk with a tab that user slides to cover or expose the notch; used to protect a floppy disk from being accidentally erased. If the notch is exposed, or open, the drive cannot write on the floppy disk. If the notch is covered, or closed, the drive can write on the floppy disk. **6.6,** 6.8
Writing: Process of a storage device transferring data, instructions, and information from memory to a storage medium. When used for writing, the storage device functions as an output source. **6.3**
CD-R, 6.22
CD-ROM, 6.19
CD-RW, 1.8, 6.22
DVD, 6.23
floppy disks, 6.5, 7-8
hard disk, 6.11
tape, 6.23

Xeon™ processor: Intel processor with a clock speed of 500-1000 MHz, used in workstations, and servers on a network. 3.5, **3.9,** 3.10, 3.25
Xerox PARC, 1.51

Y2k Bug, *see* **Millennium Bug**
Yahoo!: A leading global Internet media that is visited by millions of people worldwide every day, and is one of the more popular destinations on the Internet. Yahoo!'s navigational guide and directory is a tailor-made database that runs on the UNIX platform, providing links to other Web sites. 7.30, 7.37
Yahoo! shopping, 7.60-63
Yang, Jerry, 7.37
Year 2000 Bug, *see* **Millennium Bug**

Zero-insertion force (ZIF) socket: Socket that has a small lever or screw designed to facilitate the installation and removal of chips, such as a processor chip. **3.10**
Zip® disk: 3.5-inch floppy disk that can store 100 MB of data; used in Zip® drives. 1.7, **6.10**
Zip® drive: Special high-capacity disk drive developed by Iomega Corporation that uses a 3.5-inch Zip® disk, which can store 100 MB of data. **6.10**
bays for, 3.27
.zip extension, 8.22
Zipped files: Name given to zipped files because they usually have a .zip extension. **8.22**

PHOTO CREDITS

Chapter 1: *Chapter opener* Bob Daemmrich/Stock Boston; *Figure 1-1a* Anthony Wood/Stock Boston; *Figure 1-1b* Andy Sacks/Tony Stone Images; *Figure 1-1c* Mark Richards/PhotoEdit; *Figure 1-1d* David Young-Wolff/PhotoEdit; *Figure 1-1e* Charles Gupton/Tony Stone Images; *Figure 1-2* Scott Goodwin Photography; *Figure 1-4a* ©1999 PhotoDisc, Inc.; *Figure 1-4b* Courtesy of Intel Corporation; *Figure 1-4c* Courtesy of Intel Corporation; *Figure 1-4d* Courtesy of Intel Corporation; *Figure 1-4e* Phil A. Harrington/Peter Arnold, Inc.; *Figure 1-5* Scott Goodwin Photography; *Figure 1-6a* Courtesy of Seagate Technology; *Figure 1-6b* Courtesy of Iomega Corporation; *Figure 1-7* Courtesy of Toshiba America Information Systems, Inc.; *Figure 1-12* Bonnie Kamin; *Figure 1-17* PhotoEssentials, Inc.; *Figure 1-19* Courtesy of International Business Machines; *Figure 1-20* Courtesy of Apple Computer, Inc.; *Figure 1-21a* Courtesy of Hewlett Packard Company; *Figure 1-21b* Courtesy of International Business Machines; *Figure 1-22* Courtesy of Gateway, Inc.; *Figure 1-24* Courtesy of Philips Consumer Electronics Company; *Figure 1-25* Courtesy of International Business Machines; *Figure 1-26* Courtesy of Hewlett Packard Company; *Figure 1-27* Courtesy of International Business Machines; *Figure 1-28* Courtesy of Hewlett Packard Company; *Figure 1-29* Courtesy of International Business Machines; *Figure 1-30* Courtesy of Los Alamos National Laboratory; *Figure 1-33* Courtesy of R-F Link, Inc.; *Figure 1-35* Churchill & Klehr/The Liaison Agency; *Figure 1-40a* Frank Wing/The Liaison Agency; *Figure 1-40b* Courtesy of Copyright © Nokia, 2000; *Figure 1-40c* Bonnie Kamin; *Figure 1-40d* Michael A. Keller Studios/ The Stock Market; *Figure 1-40e* Ed Lallo/Tony Stone Images; *Figure 1-41* Courtesy of Toshiba America Information Systems, Inc.; *Figure 1-42* Courtesy of Toshiba American Information Systems, Inc.; *Figure 1-43a* Bob Schatz/The Liaison Agency; *Figure 1-44b* Courtesy of Sony Electronics, Inc.; *Figure 1-46* Courtesy of Kiosk Information Systems, Inc.; *Figure 1-47a* Spencer Grant/PhotoEdit; *Figure 1-47b* Kevin Horan/Tony Stone Images; *Figure 1-47c* Richard Pasley/Stock Boston; *Figure 1-48* Courtesy of Environmental Systems Research Institute, Inc. Copyright 1996, 1997. All rights reserved; *Figure 1-53* Courtesy of Microsoft Corporation; *Figure 1-54* Courtesy of Microsoft Corporation; ***Timeline:*** *1937* Courtesy of Iowa State University; *1943* The Computer Museum; *1945* Courtesy of the Institute for Advanced Studies; *1946* Courtesy of the University of Pennsylvania Archives; *1947* Courtesy of International Business Machines; *1951* Courtesy of Unisys Corporation; *1952* Courtesy of the Hagley Museum and Library; *1953* Courtesy of M.I.T. Archives; *1957* Courtesy of International Business Machines; *1957* Courtesy of the Department of the Navy; *1958* Courtesy of M.I.T. Archives; *1958* Courtesy of International Business Machines; *1959* Courtesy of International Business Machines; *1960* Courtesy of The Hagley Museum and Library; *1964* Courtesy of International Business Machines; *1965* Courtesy of Dartmouth College News Services; *1965* Courtesy of Digital Equipment Corporation; *1968* Courtesy of International Business Machines; *1970* Courtesy of International Business Machines; *1971* Courtesy of Intel Corporation; *1971* The Computer Museum; *1975* Courtesy of InfoWorld; *1976* Courtesy of Apple Computer, Inc.; *1976* Courtesy of Apple Computer, Inc.; *1979* The Computer Museum; *1980* Courtesy of International Business Machines; *1980* Courtesy of Microsoft Corporation; *1981* Courtesy of International Business Machines; *1982* Courtesy of Hayes; *1983* Courtesy of Lotus Development Corporation; *1983* ©1982, Time, Inc.; *1984* Courtesy of International Business Machines; *1984* Courtesy of Hewlett Packard Company Archives; *1984* Courtesy of Apple Computer Inc.; *1987* Courtesy of Compaq Computer Corporation; *1989* Copyright ©1997 - 1998 W3C (MIT, INRIA, Keio). All Rights Reserved; *1989* Courtesy of Intel Corporation; *1992* Courtesy of Microsoft Corporation; *1993* Courtesy of Intel Corporation; *1993* Jim Clark/The Liaison Agency; *1993* Courtesy of Netscape Communications; *1994* Courtesy of Netscape Communications Corporation; *1995* Courtesy of Microsoft Corporation; *1995* Courtesy of Sun Microsystems, Inc.; *1996* Courtesy of 3Com Corporation; *1996* Reuters/Rick T. Wilking/Archive Photos; *1996* Courtesy of Microsoft Corporation; *1996* Courtesy of Web TV Networks Inc.; *1997* Courtesy of Intel Corporation; *1997* I. Uimonen/Sygma; *1997* Courtesy of International Business Machines; *1997* Motion Picture and Television Archives; *1998* Courtesy of Microsoft Corporation; *1998* Courtesy of Dell Computer; *1998* ©Corel Corporation; *1998* Courtesy of Apple Computer Inc.; *1998* ©1999 PhotoDisc, Inc.; *1999* Courtesy of Microsoft Computer Corporation; *2000* Courtesy of Microsoft Corporation; **Chapter 2:** *Figure 2-22* Courtesy of 3Com Corporation; *Figure 2-23* Courtesy of Microsoft Corporation; *Figure 2-23* Courtesy of Lotus Development Company; *Figure 2-25* Courtesy of Intuit; *Figure 2-27* Courtesy of AutoDesk, Inc.; *Figure 2-28* Courtesy of Adobe Systems; *Figure 2-30* Courtesy of Adobe Systems; *Figure 2-31* Courtesy of Asymetrix Learning Systems, Inc.; *Figure 2-34* Courtesy of Nolo Press, Inc.; *Figure 2-35* Courtesy of Block Financial Group Corp., makers of Kiplinger TaxCut; *Figure 2-39* Courtesy of Broderbund Software; *Figure 2-58* Bonnie Kamin; *Figure 2-59* Jim Clark/The Liaison Agency; *Figure 2-60* Courtesy of America Online, Inc.; **Chapter 3:** *Figure 3-1* Courtesy of International Business Machines; *Figure 3-2b* Courtesy of Intel Corporation; *Figure 3-3* Courtesy of Intel Corporation; *Figure 3-4c* Phil A. Harrington/Peter Arnold, Inc.; *Figure 3-4b* Courtesy of Intel Corporation; *Figure 3-4b* ©1999 PhotoDisc, Inc.; *Figure 3-4d* Courtesy of Intel Corporation; *Figure 3-6a* Courtesy of Intel Corporation; *Figure 3-6b* Courtesy of Intel Corporation; *Figure 3-6c* Courtesy of Intel Corporation; *Figure 3-6d* Courtesy of Advanced Micro Devices, Inc.; *Figure 3-7* David Young Wolff/Tony Stone Images; *Figure 3-17* AP/Wide World Photos; *Figure 3-20* Scott Goodwin Photography; *Figure 3-21* Courtesy of Microsoft Corporation; *Figure 3-24* Courtesy of Intel Corporation; *Figure 3-26* ©1999 PhotoDisc, Inc.; *Figure 3-28* Scott Goodwin Photography; *Figure 3-30* Bonnie Kamin; *Figure 3-39* Scott Goodwin Photography; *Figure 3-40* Courtesy of PC Notebook; *Figure 3-41* Scott Goodwin Photography; *Figure 3-42* Shelly Harrison; *Figure 3-44* Courtesy of Intel Corporation; *Figure 3-45a* Courtesy of Intel Corporation; *Figure 3-45b* Courtesy of Andrew S. Grove/Intel Corporation; ***Computer Chip Feature:*** *Figure 1* ©1999 PhotoDisc, Inc.; *Figure 2* ©1999 PhotoDisc, Inc.; *Figure 3* ©1999 PhotoDisc, Inc.; *Figure 4* Courtesy of International Business Machines; *Figure 5* Hank Morgan/Science Source/Photo Researchers, Inc.; *Figure 6* Courtesy of Intel Corporation; *Figure 7* Courtesy of International Business Machines; *Figure 8* Courtesy of International Business Machines; *Figure 9* Courtesy of International Business Machines; *Figure 10* Courtesy of International Business Machines; *Figure 11* Rosenfeld Images LTD/Science Photo Library/Photo Researchers, Inc; **Chapter 4:** *Figure 4-1* Courtesy of Intel Corporation; *Figure 4-2* Courtesy of Logitech; *Figure 4-7a* Courtesy of International Business Machines; *Figure 4-7b* Courtesy of Hewlett Packard Company; *Figure 4-8* Courtesy of Microsoft Corporation; *Figure 4-13* Courtesy of Truedox Technology Corporation; *Figure 4-14* David Young Wolf/PhotoEdit; *Figure 4-16* Scott Goodwin Photography; *Figure 4-17* Courtesy of International Business Machines; *Figure 4-18* Courtesy of Gravis Gaming Devices/Kensington Technology Group; *Figure 4-19* Michael Newman/PhotoEdit; *Figure 4-20* Courtesy of MicroTouch Systems, Inc.; *Figure 4-21a* Courtesy of 3Com Corporation; *Figure 4-21b* ©1999 PhotoDisc, Inc.; *Figure 4-22a* Michael Newman/PhotoEdit; *Figure 4-22b* Zigy Kaluzny/Tony Stone Images; *Figure 4-26a* Courtesy of Microtek Lab, Inc.; *Figure 4-26b* Courtesy of Visioneer, Inc.; *Figure 4-26c* Courtesy of Howtek, Inc.; *Figure 4-28* Scott Goodwin Photography; *Figure 4-29* BMS Data Handling, Inc.; *Figure 4-29b* Courtesy of Scantron Corporation; *Figure 4-30a* Cassey Cohen/PhotoEdit; *Figure 4-30b* Courtesy of Symbol Technologies, Inc.; *Figure 4-30c* Novastock/PhotoEdit; *Figure 4-32* Courtesy of PenStock ECR; *Figure 4-33* Courtesy of NCR; *Figure 4-35* Courtesy of Trimble Corporation; *Figure 4-36a* Courtesy of Casio, Inc.; *Figure 4-36b* Courtesy of Eastman Kodak Company; *Figure 4-36c* Courtesy of Sony Electronics, Inc.; *Figure 4-38* Brian Smith/Stock Boston; *Figure 4-40* Courtesy of International Business Machines; *Figure 4-42* Courtesy of Play, Inc.; *Figure 4-43* Courtesy of Intel Corporation; *Figure 4-44* Courtesy of Orcca Technologies, Inc.; *Figure 4-45* Courtesy of Prentke Romich Company; *Figure 4-47* The Liaison Agency; *Figure 4-48* Peter Aaron/Esto Photographics /Courtesy of International Business Machines; **Chapter 5:** *Figure 5-1* Courtesy of Intel Corporation; *Figure 5-2* Courtesy of International Business Machines; *Figure 5-5* Courtesy of Dell Computer Corporation; *Figure 5-6* Courtesy of Casio, Inc.; *Figure 5-6b* Courtesy of Sprint PCS; *Figure 5-7* Courtesy of NEC Technologies, Inc.; *Figure 5-8* Courtesy of Fujitsu; *Figure 5-19* Courtesy of Epson America, Inc.; *Figure 5-20* Courtesy of Genicom Corporation; *Figure 5-21* Courtesy of Hewlett Packard Company; *Figure 5-24a* Courtesy of Hewlett Packard Company; *Figure 5-24b* Courtesy of International Business Machines; *Figure 5-26* Scott Goodwin Photography; *Figure 5-27* Courtesy of Primera Technology, Inc.; *Figure 5-28* Courtesy of Citizen America; *Figure 5-29a* Courtesy of CalComp Technology; *Figure 5-29b* Courtesy of CalComp Technology; *Figure 5-29c* Courtesy of CalComp Technology; *Figure 5-30a* Courtesy of Hewlett-Packard Company; *Figure 5-30b* Courtesy of DYMO; *Figure 5-31* Courtesy of Dell Computer Corporation; *Figure 5-32* Courtesy of Telex Communications, Inc.; *Figure 5-33a* Courtesy of In Focus Systems, Inc.; *Figure 5-33b* Jerome Hart Photography/Courtesy of InFocus, Inc.; *Figure 5-34* Bob Daemmrich/The Image Works; *Figure 5-36* Courtesy of Hewlett Packard Company; *Figure 5-37* Michael Rosenfeld/Tony Stone Images; *Figure 5-38a* Bob Daemmrich/The Image Works; *Figure 5-38b* Myrleen Ferguson/PhotoEdit; *Figure 5-40* Courtesy of Index; *Figure 5-41* Courtesy of Hewlett Packard Company; *Figure 5-42* Courtesy of Hewlett Packard Company; **Chapter 6:** *Figure 6-5* Scott Goodwin Photography; *Figure 6-5c* Courtesy of Toshiba America Information Systems, Inc.; *Figure 6-12* Bonnie Kamin; *Figure 6-12b* Courtesy of Iomega Corporation; *Figure 6-13* Courtesy of Seagate Technology; *Figure 6-18* Scott Goodwin Photography; *Figure 6-19* Courtesy of Adaptec Inc.; *Figure 6-20* Courtesy of AC & NC Corporation; *Figure 6-29* Courtesy of Eastman Kodak Company; *Figure 6-30* Courtesy of Pioneer Electronics; *Figure 6-32* Courtesy of Imation Corporation, Data Storage Products; *Figure 6-33* Scott Goodwin Photography; *Figure 6-34* Courtesy of m4Data Inc.; *Figure 6-36* Courtesy of International Business Machines; *Figure 6-37* Courtesy of Kingston Technology Company, Inc.; *Figure 6-38* Courtesy of Gemplus; *Figure 6-39* Courtesy of Tritheim Technologies; *Figure 6-40* Courtesy of Eastman Kodak Company; *Figure 6-42* Courtesy of Sun Microsystems, Inc.; *Figure 6-43* Courtesy of Silicon Graphics, Inc.; **Chapter 7:** *Figure 7-1* PhotoEssentials, Inc.; *Figure 7-5a* Bob Schatz/The Liaison Agency; *Figure 7-5b* Churchill & Klehr/The Liaison Agency; *Figure 7-5c* ©1999 PhotoDisc, Inc.; *Figure 7-5d* Andy Sacks/Tony Stone Images; *Figure 7-7* PhotoEssentials, Inc.; *Figure 7-19* Courtesy of Microsoft Corporation; *Figure 7-42* Courtesy of Wyse Technology, Inc.; *Figure 7-43* Courtesy of Philips Consumer Electronics Company; *Figure 7-44* Copyright © Nokia, 2000; *Figure 7-45* Courtesy of Sprint PCS; *Figure 7-46* Courtesy of 3Com Corporation; *Figure 7-47* Copyright ©1997-1998 W3C (MIT, INRIA, Keio). All rights reserved; *Figure 7-48* Courtesy of Yahoo!; **Chapter 8:** *Figure 8-1* Churchill & Klehr/The Liaison Agency; *Figure 8-2* Courtesy of Filemaker, Inc.; *Figure 8-15* Courtesy of Clarion; *Figure 8-16* Courtesy of Apple Computer, Inc.; *Figure 8-17* Courtesy of International Business Machines; *Figure 8-31a* Jeffrey Braverman; *Figure 8-31b* Courtesy of Apple Computer, Inc.; *Figure 8-32a* Terry Heferman/Courtesy of Apple Computer, Inc.; *Figure 8-32b* Courtesy of International Business Machines; ***Buyer's Guide 2000:*** *Pages 8-37-8-48* ©1999 PhotoDisc, Inc.; *Figure 8* Courtesy of International Business Machines; *Figure 9* Courtesy of Toshiba America, Inc.; *Figure 10* Courtesy of Xircom Corporation; *Figure 11* Courtesy of In Focus Corporation.